Condensed MATTER THEORIES

VOLUME 8

A Continuation Order Plan is available for this series. A continuation order will bring
delivery of each new volume immediately upon publication. Volumes are billed only
upon actual shipment. For further information please contact the publisher.

Condensed
MATTER
THEORIES
VOLUME 8

Edited by

Lesser Blum
University of Puerto Rico
Rio Piedras, Puerto Rico

and

F. Bary Malik
Southern Illinois University
Carbondale, Illinois

Springer Science+Business Media, LLC

Proceedings of the Sixteenth International Workshop on Condensed Matter Theories, held June 1-5, 1992, in San Juan, Puerto Rico

Library of Congress Catalog Card Number 87-656591 (ISSN 0893-861X)

ISBN 978-1-4613-6274-6   ISBN 978-1-4615-2934-7 (eBook)
DOI 10.1007/978-1-4615-2934-7
© 1993 Springer Science+Business Media New York
Originally published by Plenum Press in 1993
Softcover reprint of the hardcover 1st edition 1993

PREFACE

The XVI International Workshop on Condensed Matter Theories (CMT) was held in San Juan, Puerto Rico between June 1 and 5, 1992. It was attended by about 80 scientists from all over the world.

The Workshop was started in 1977 by V.C. Aguilera-Navarro, in Sao Paolo, Brazil, as the Panamerican Workshop on Condensed Matter Theories, to promote the exchange of ideas and techniques of groups that normally do not interact, such as people working in the areas of Nuclear Physics and Solid state Physics, Many Body Theory, or Quantum Fluids, and Classical Statistical Mechanics, and so on. It had also the purpose of bringing together people from different regions of the globe. The next CMT Workshop was held in 1978 in Trieste, Italy, outside of America. But the next four met in the American continent: Buenos Aires, Argentina (1979), Caracas, Venezuela (1980), Mexico City, Mexico (1981), and St. Louis, Missouri (1982). At this time the scope and the participation had increased, and the name was changed to the "International" Workshop in CMT. The 1983 edition took place in Altenberg, Germany. The following CMT workshops took place in Granada, Spain (1984), San Francisco, California (1985), Argonne, Illinois (1986), Oulu, Finland (1987), Taxco, Mexico (1988), Campos do Jordao, Brazil (1989), Elba Island, Italy (1990), and Mar del Plata, Argentina (1991).

There were 48 invited talks in this Workshop. In spite of the temptations of beautiful beaches, the attendance was always large, and the discussions lively up to the last minute. We heard a number of outstanding lectures on solid state physics, density functional theory, diffusion limited aggregation, cellular authormata, superconductivity, quantum fluids (Bose and Fermi liquids), theory of inhomogeneous classical fluids, non-equilibrium and equilibrium fluids, phase transitions, nuclear structure, lattices, etc. and had a number of excellent posters on these topics. We thank all the participants for their contributions.

The enthusiam and support of the committee members has made this Workshop possible: We should mention in particular, A. Plastino, and M. de Llano, who provided the necessary leadership. Locally, invaluable support was provided by the Centro de Recursos para la Ciencia e Ingenieria. We sincerely thank Dr. Manuel Gomez, its director, for financial and logistical supports. Mrs. A.R. Catarich of the centro did an excellent job in the organization. Mr. Osvaldo Rosario helped enormously with the financial organization. Mr. Marc Legault, a graduate student of one of us (LB) was an invaluable help in the organization of the program, the meeting and the proceedings. Ms. I. Howald and Ms. Julie Porter have helped us with the preparation of the final version of the manuscript. We thank them all, very sincerely.

Financial supports from the U.S. Army Research Office, the Office of Naval Research, the EPSCoR program funded by the National Science Foundation, University of Puerto Rico Piedras Campus and Baxter Laboratories are thankfully acknowledged.

The Editors

CONTENTS

SOLID STATE PHYSICS, SUPERCONDUCTIVITY

Lattice Theories, Phase Transitions

Cellular Automata

Fundamental Quantum Mechanics

ELECTRONIC STRUCTURE OF HIGHLY CORRELATED SYSTEMS

L. M. Falicov and J. K. Freericks[*]

Department of Physics, University of California at Berkeley,
and Materials Sciences Division, Lawrence Berkeley Laboratory,
Berkeley, California, 94720, USA

ABSTRACT

Three different but related problems are discussed in this contribution, all related to the so-called Anderson, Hubbard , and $t-J$ Hamiltonians — the prototype Hamiltonians for systems with highly correlated electrons: (1) The relationship — based on the well known canonical (Schrieffer-Wolff) transformation — between the Anderson model (in the small-hybridization and large-Coulomb-interaction regime) and the local moment, the Kondo, the Hubbard and the $t-J$ models, in particular the phenomena of rare-earth magnetism, intermediate valence, and heavy fermions; (2) The exact solution of these Hamiltonians in the periodic small-cluster approximation and the conditions for the existence of the heavy-fermion phenomenon; (3) The metamagnetic transition in heavy-fermion systems.

1. INTRODUCTION

Crystalline compounds that involve lanthanide or actinide ions display a rich variety of physical phenomena: long-range-ordered local magnetic moments [1] (ferromagnets, antiferromagnets, spiral or canted spin arrangements, *etc.*); the Kondo effect (a strong interaction between conduction electrons and local moments that manifests itself in an anomalous resistivity [2] and the quenching of magnetic moments everywhere [3,4]); heavy fermions [5,6] (materials characterized by a huge "density of states at the Fermi level"); intermediate valence [7] (strong charge fluctuations that produce, on average, only a fraction of an electron per ion); and band theory [8] (electrons that are approximated well by noninteracting particles).

Lanthanide and actinide compounds possess ions with localized (atomic) f-orbitals that do not overlap with the corresponding f-orbitals on neighboring ions, but do hybridize with the extended states of the conduction-band electrons. The f-electrons interact very strongly with each other via a screened (on-site) Coulomb interaction U that acts only between two f-electrons that is localized about the same lattice site. The Coulomb energy is larger than any other in the problem ($U > 10$ eV) so that at each site only two possible occupations of the f-shell exist: $(4f)^n$, and $(4f)^{n+1}$. For the sake of definiteness and simplicity a single f orbital site is considered here (n is taken to be 0); doubly occupied f-orbitals, because of the large U, are effectively forbidden. The physics of such an electronic system is described by the lattice (or periodic) Anderson impurity model [9]

$$H_A = \sum_{k\sigma} \varepsilon_k a_{k\sigma}^\dagger a_{k\sigma} + \varepsilon \sum_{i\sigma} f_{i\sigma}^\dagger f_{i\sigma} + U \sum_i f_{i\uparrow}^\dagger f_{i\uparrow} f_{i\downarrow}^\dagger f_{i\downarrow}$$
$$+ \sum_{ik\sigma} [V_{ik} f_{i\sigma}^\dagger a_{k\sigma} + V_{ik}^* a_{k\sigma}^\dagger f_{i\sigma}] \quad , \tag{1}$$

in the large-U limit. (This limit, as understood here, implies both $U \to \infty$ and $\varepsilon + U \to \infty$, so that there is never more than one electron per f-orbital.) The parameters and operators in Eq. (1) include the conduction-band creation (annihilation) operators $a_{k\sigma}^\dagger$ ($a_{k\sigma}$) for a conduction electron in an extended state with wavevector k, spin σ, and energy ε_k; the localized electron creation (annihilation) operators $f_{i\sigma}^\dagger$ ($f_{i\sigma}$) for localized electrons in an atomic orbital centered at lattice site i with energy ε; the on-site Coulomb interaction U; and the hybridization integral V_{ik} that mixes together the localized and extended states. (As already mentioned, the degeneracy of the f-electrons is neglected here; additional f-electron orbitals may easily be added without changing the qualitative nature of the model.) The hybridization integral V_{ik} is assumed to be of the form

$$V_{ik} = \exp(i\,\mathbf{R}_i \cdot \mathbf{k})\, V\, g(k)/\sqrt{N} \quad , \tag{2}$$

with $g(k)$, the form factor, a dimensionless function of order one, and N the number of lattice sites. The Fermi level E_F is defined to be the maximum energy of the filled conduction band states, in the limit $V \to 0$, and the origin of the energy scale is chosen so that $E_F = 0$. The conduction-band density of states per site at the Fermi level is then defined to be ρ.

2. THE SCHRIEFFER-WOLFF TRANSFORMATION OF THE LATTICE ANDERSON IMPURITY MODEL: LOCAL MOMENTS, KONDO LATTICE, HEAVY FERMIONS, INTERMEDIATE VALENCE, AND BAND THEORY

The lattice Anderson impurity model can be exactly diagonalized in two limits: in the noninteracting case ($U\rho \to 0$) the Hamiltonian is a quadratic form in the fermionic operators, $i.e.$ an independent particle problem, and can be diagonalized by a change in one-particle basis; in the zero hybridization limit ($V\rho \to 0$) the Hamiltonian decouples into two independent systems (extended electrons and localized electrons), with explicitly diagonal sub-Hamiltonians for each subsystem. The large-interaction ($U\rho \gg 1$), small-hybridization ($V\rho \ll 1$) limit of the Anderson model is the physically relevant regime for studying lanthanide and actinide compounds. Exact diagonalization studies [10] (on small systems) show that the lattice Anderson impurity model can describe all of the physical phenomena (local moments, Kondo effect, heavy fermions, intermediate valence, and band theory) of lanthanide and actinide compounds simply by varying the parameter $\varepsilon\rho$ from large negative values to large positive values. In fact, the lattice Anderson impurity model, in the infinite-U and $V\rho \ll 1$ limit, is characterized by five regimes depending upon the localized-electron energy ε (see Figure 1):

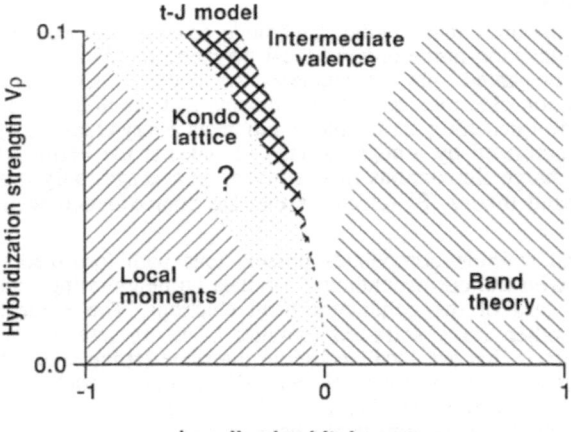

Figure 1. A schematic diagram showing the regions of parameter space where local magnetic moments, Kondo lattice [11,12] (quenched moments everywhere), $t-J$ model, intermediate valence, and one-electron band theory are expected to occur in the small hybridization, large interaction strength limit of the lattice Anderson impurity model. The horizontal axis records the localized f-electron energy (relative to the Fermi level) and the vertical axis records the hybridization strength. Heavy-fermion behavior may occur in the transition region between the Kondo lattice and intermediate valence ($t-J$ model regime). A fine-tuning of the parameters and the crystal structure of the (effective) $t-J$ model is required to produce a heavy-fermion system.

(A) $\varepsilon\rho\ll-V\rho<0$ — isolated local moments [1,8]. In this limit the f-orbitals are occupied by a single electron ($<f_{i\uparrow}^{\dagger}f_{i\uparrow}+f_{i\downarrow}^{\dagger}f_{i\downarrow}>=1$), and (effectively) decoupled from the conduction electrons. There are everywhere local magnetic moments. The local moments interact with each other via exchange, superexchange, and/or Ruderman-Kittel-Kasuya-Yosida (RKKY) interactions [1] with ferromagnetic or antiferromagnetic exchange integrals. The rich magnetic structure of the rare-earth metals appears in this regime.

(B) $-V\rho\ll\varepsilon\rho\ll-V^{2}\rho^{2}<0$ — the regime where the Kondo effect holds for impurities and where a Kondo lattice [11,12] may exist. In this regime there are no charge excitations in the localized orbitals ($<f_{i\uparrow}^{\dagger}f_{i\uparrow}+f_{i\downarrow}^{\dagger}f_{i\downarrow}>=1$), but the local moments interact strongly with conduction-electron spins at the Fermi level to quench the magnetic moments everywhere.

(C) $\varepsilon\rho\approx-V^{2}\rho^{2}$ — the $t-J$ model [6]. In this regime the localized orbitals are almost singly occupied ($<f_{i\uparrow}^{\dagger}f_{i\uparrow}+f_{i\downarrow}^{\dagger}f_{i\downarrow}>=1-\nu$, $\nu\ll1$) and have broadened into a strongly correlated narrow band in which all electronic transport takes place; the conduction band is (effectively) decoupled and acts only as a buffer that determines the concentration of electrons in the narrow band.

(D) $-V^{2}\rho^{2}\ll\varepsilon\rho<V^{2}\rho^{2}$ — intermediate valence [7]. As ε is increased to the Fermi level and beyond the occupation of the localized orbitals becomes nonintegral ($0<<f_{i\uparrow}^{\dagger}f_{i\uparrow}+f_{i\downarrow}^{\dagger}f_{i\downarrow}><1$) and the system is in the intermediate valence regime.

(E) $0<V^{2}\rho^{2}\ll\varepsilon\rho$ — band theory [8]. As ε is increased well beyond the Fermi level all localized states are essentially unoccupied ($<f_{i\uparrow}^{\dagger}f_{i\uparrow}+f_{i\downarrow}^{\dagger}f_{i\downarrow}>\approx0$) and (effectively) decoupled from the conduction electrons. The ground state is determined by one-electron band theory for the extended electronic states, which exhibit a very small hybridization with the f-states.

Heavy-fermionic behavior [5,6] occurs, *under certain circumstances*, in regime (C), the transition region from the Kondo-lattice [11] regime (B) to the intermediate-valence regime (D). The many-body ground state is characterized by a huge number of low-lying excited states that have many different spin configurations (a partial decoupling of spatial and spin degrees of freedom) and is close to a disproportionation instability. This produces a very large coefficient to the term linear in temperature in the specific heat, a large magnetic susceptibility, and poor metallic conductivity.

The five regimes outlined above may be established by employing a Schrieffer-Wolff [13] canonical transformation to the lattice Anderson impurity model. Since the details of this transformation are well known [13,14], only an outline will be presented here.

The Anderson Hamiltonian H_{A} is divided into two terms $H_{A}=H_{0}+H_{hyb}$ where H_{hyb} is the last term in Eq. (1). A canonical transformation $H'=\exp(S)H_{A}\exp(-S)$ is performed with S chosen to satisfy $[H_{0},S]=H_{hyb}$. One finds (to lowest order in V) that $H'=H_{0}+\frac{1}{2}[S,H_{hyb}]$ or

$$H'-H_{0}=-\sum_{ikk'}J_{iikk'}\,\psi_{fi}^{\dagger}\sigma\psi_{fi}\cdot\psi_{k}^{\dagger}\sigma\psi_{k'} \tag{3a}$$

$$-\sum_{ii'k\sigma}[W_{ii'kk}+\frac{1}{4}J_{ii'kk}(f_{i-\sigma}^{\dagger}f_{i-\sigma}+f_{i'-\sigma}^{\dagger}f_{i'-\sigma})]f_{i\sigma}^{\dagger}f_{i'\sigma} \tag{3b}$$

$$+\sum_{ikk'}[W_{iikk'}+\frac{1}{4}J_{iikk'}\psi_{fi}^{\dagger}\psi_{fi}]\psi_{k}^{\dagger}\psi_{k'} \tag{3c}$$

$$+\frac{1}{4}\sum_{ikk'\sigma}[K_{iikk'}f_{i\sigma}^{\dagger}f_{i-\sigma}^{\dagger}a_{k\sigma}a_{k'-\sigma}+h.c.] \tag{3d}$$

where the spinors are defined to be

$$\psi_{k}\equiv\begin{bmatrix}a_{k\uparrow}\\a_{k\downarrow}\end{bmatrix}\quad,\quad\psi_{fi}\equiv\begin{bmatrix}f_{i\uparrow}\\f_{i\downarrow}\end{bmatrix}\quad, \tag{4}$$

σ denotes the Pauli spin matrices, and the coefficients are

$$J_{ii'kk'}\equiv V_{ik}V_{i'k'}^{*}\left[\frac{1}{\varepsilon_{k}-\varepsilon-U}+\frac{1}{\varepsilon_{k'}-\varepsilon-U}-\frac{1}{\varepsilon_{k}-\varepsilon}-\frac{1}{\varepsilon_{k'}-\varepsilon}\right]\quad, \tag{5a}$$

3

$$K_{ii'kk'} \equiv V_{ik} V_{i'k'} \left[\frac{1}{\varepsilon_k - \varepsilon - U} + \frac{1}{\varepsilon_{k'} - \varepsilon - U} - \frac{1}{\varepsilon_k - \varepsilon} - \frac{1}{\varepsilon_{k'} - \varepsilon} \right] \quad , \tag{5b}$$

$$W_{ii'kk'} \equiv \tfrac{1}{2} V_{ik} V_{i'k'}^* \left[\frac{1}{\varepsilon_k - \varepsilon} + \frac{1}{\varepsilon_{k'} - \varepsilon} \right] \quad . \tag{5c}$$

The last term (3d) can always be neglected (in the large-U limit) because it only connects configurations that have zero electrons at site i to configurations with two electrons at site i, which are explicitly forbidden.

The canonical transformation of the lattice Anderson impurity model is approximated well by the lowest order term (in V) when $|V^2 \rho / \varepsilon| < 1$. In this region of parameter space the localized orbitals have an occupancy per site close to one ($\varepsilon < 0$) or close to zero ($\varepsilon > 0$). When the occupancy is close to one, the operator $\psi_{fi}^\dagger \psi_{fi}$ may be replaced by unity and both the term (3c) and the diagonal ($i = i'$) terms in (3b) may be absorbed into a renormalized H_0. The remaining terms in Eq. (3) describe spin scattering of the conduction electrons at the Fermi level by the localized moments (3a), and direct hopping terms within the (narrow) f-band (3b). When the occupancy is close to zero — the band-theory regime — the operators $\psi_{fi}^\dagger \psi_{fi}$, $\psi_{fi}^\dagger \sigma \psi_{fi}$, and $f_i^\dagger f_{i'}$ may all be replaced by zero. The only important terms remaining in Eq. (3) are the changes in the one-particle band-structure arising from (3c).

The local-moment regime corresponds to the case where $\varepsilon \rho \ll -V \rho < 0$ so that $|J_{ii'kk'}| \ll 1$ for k and k' at the Fermi surface. To lowest order, the ground state consists of one f-electron per site and a conduction band filled up to energy E_F. The spin flipping of the local moments by the conduction-band electrons at the Fermi surface (3a) is weak and all other terms in Eq. (3) can be neglected. The interaction (3a) between localized spins and conduction electrons then leads, through H_0, to a variety of localized-spin exchange interactions [15]. These, in turn, determine, at low enough temperatures, the long-range magnetic order.

The Kondo-lattice model [11,12] corresponds to the case where $-V \rho \ll \varepsilon \rho \ll -V^2 \rho^2 < 0$, the f-orbitals are singly occupied {so that the hopping term (3b) may be neglected}, and the density of states of the conduction electrons is not negligible at the Fermi level, so that the spin scattering term (3a) is the most important correction term. The local-moment spins strongly interact with the conduction electrons (at the Fermi surface) which quench the magnetic moments everywhere.

The $t-J$ model occurs in the region where $\varepsilon \rho \approx -V^2 \rho^2 < 0$. The localized states broaden into a narrow band with an occupancy of *nearly* one electron per site; the density of states of the conduction electrons at the Fermi level is negligible, so that the spin-scattering term (3a) may be neglected. In this case, the hopping term (3b) is the most important correction term. The conduction electrons are decoupled from the f-band and act only as a buffer that determines the filling of the f-band. The hole density ν required for the hopping term (3b) to be more important than the spin-scattering term (3a) is

$$\left| \frac{\sum_{FS} g^2(k) / \varepsilon}{\sum_{BZ} g^2(k) \exp(i\, k \cdot \tau) / (\varepsilon_k - \varepsilon)} \right| \ll \nu \ll 1 \quad , \tag{6}$$

where τ is a nearest-neighbor translation vector, FS denotes a summation over wavevectors that lie on the Fermi surface only, and BZ denotes a summation over all wavevectors in the Brillouin zone.

This narrow f-band is described by the large-U limit of the Hubbard model [16], which in turn becomes the $t-J$ model [17]

$$H_{t-J} = -\sum_{ij\sigma} t_{ij} (1 - f_{i-\sigma}^\dagger f_{i-\sigma}) f_{i\sigma}^\dagger f_{j\sigma} (1 - f_{j-\sigma}^\dagger f_{j-\sigma}) + \sum_{ij} J_{ij}\, S_i \cdot S_j \quad , \tag{7}$$

with

$$t_{ij} = \sum_k W_{ijkk} = \sum_k \frac{V_{ik}^* V_{jk}}{\varepsilon_k - \varepsilon} \quad , \tag{8}$$

and $J_{ij} = 4|t_{ij}|^2/U$. Note that the hopping matrix elements t_{ij} in Eq. (8) are short-ranged,

4

i.e., strongly peaked functions of the separation $|R_i - R_j|$ between lattice sites i and j, as expected.

A *necessary* condition for heavy-fermion behavior is that the parameters of the Anderson model fall into the range where the mapping onto the $t-J$ model is valid, but this condition is *not sufficient*. The solutions of the $t-J$ model must also possess a very large number of low-lying excitations with many different spin configurations. This latter condition requires a *fine-tuning* of the parameters in the $t-J$ model and depends strongly upon the geometry and connectivity of the lattice. For example, exact-diagonalization calculations on small clusters [18,19] — see below — indicate that strongly frustrated lattices, for occupancy close to one electron per site, are the best candidates for heavy-fermionic behavior (*e.g.*, the system with 7 electrons in an eight-site face-centered-cubic cluster possesses solutions with strongly enhanced low-temperature specific heat, quasielastic spin excitations, and poor metallic conductivity; the ground state may be magnetic or nonmagnetic).

The intermediate-valence regime occurs when the Schrieffer-Wolff transformation may not be truncated to lowest order, $|\varepsilon\rho| \ll V^2\rho^2$, or its expansion may not be valid at all. In this case, all parameters are equally important and the full many-body problem must be solved. The average occupation (per site) of an f-orbital decreases from 1 to 0 as $\varepsilon\rho$ is increased. This regime has been studied by mapping onto a Fermi liquid [7] and by exact-diagonalization on small clusters [10].

Finally in the region where $\varepsilon\rho \gg V^2\rho^2$, the Schrieffer-Wolff transformation is once again valid to lowest order and one finds that the f-orbitals are completely empty and (effectively) decoupled from the conduction band. The system is described by one-electron theory [8] for the conduction-band electrons alone. This is the regime where ordinary density-functional theory is an excellent tool for calculating the electronic properties.

3. PERIODIC SMALL-CLUSTER APPROACH TO MANY-BODY PROBLEMS IN GENERAL, AND HEAVY-FERMION SYSTEMS IN PARTICULAR

It is known from the basic duality between real and reciprocal spaces [8] that a microcrystal of N sites with periodic boundary conditions has eigenstates that can be classified by N k-vectors, distributed periodically throughout the Brillouin zone. Conversely, the sampling of N points, periodically distributed in the Brillouin zone is equivalent to solving a problem in real space, in a microcrystal of N sites with periodic boundary conditions. This method has been extensively used by the authors, their collaborators, and others [10, 18-21] to solve a variety of problems, both model Hamiltonians and realistic situations [22-24].

The cluster is chosen to be small enough that the many-body Hamiltonian may be exactly diagonalized but (hopefully) large enough that the physics of the infinite lattice is captured. For the heavy-fermion case discussed below, the mapping of the Anderson model (1) onto a $t-J$ model reduces the size of the Hilbert space by a factor of $(3/16)^N$ which allows larger clusters to be studied.

An understanding of exactly how to extrapolate the results for a small-cluster calculation to the thermodynamic limit $(N \rightarrow \infty)$ has not yet been achieved. It is nonetheless obvious that some very interesting effects are observed in these small clusters (small k-space sampling), and one such effect is the appearance of heavy-fermion behavior.

3(a). A Heavy-Fermion Case in a Small-Cluster t-J Model

The small-cluster approach has been applied to the $t-J$ model [18-19]. A very good example of a heavy-fermion system lies in an eight-site face-centered cubic-lattice cluster with seven electrons [18-19]. When the hopping parameters and antiferromagnetic superexchange parameters are chosen to be

$$t_{ij} = \begin{bmatrix} t > 0, & i,j = \text{first-nearest neighbors}, \\ t' = 0.1t, & i,j = \text{second-nearest neighbors}, \\ 0, & \text{otherwise}, \end{bmatrix}$$

$$J_{ij} = \begin{bmatrix} J, & i,j = \text{first-nearest neighbors}, \\ 0, & \text{otherwise}, \end{bmatrix} \quad (9)$$

(out of a total of 1024 states) that is split-off from the higher-energy excitations and which

5

include many different spin configurations (see Table 1). These many-body states are degenerate at $J = 0$ but the degeneracy is partially lifted for finite J, with low-spin configurations favored (energetically) over high-spin configurations. It is worth remarking that even with the antiferromagnetic interaction included, the spin 1/2 ground state remains accidentally degenerate, with a degeneracy of 14, *i.e.*, 14% of the available states remain in the ground-state manifold.

Table 1. Low-energy manifold of many-body eigenstates, at zero magnetic field, for the model heavy-fermion system discussed in the text. The notation is that of References [18,19].

Energy	Total Spin	Degeneracy	Spatial Symmetry Label
$-6t + 6t' - 3J$	½	14	$\Gamma_2 \oplus X_1 \oplus X_2$
$-6t + 6t' - 2J$	½	16	L_3
$-6t + 6t' - 1.5J$	1 ½	32	$\Gamma_{12} \oplus X_1 \oplus X_2$
$-6t + 6t' - 0.5J$	1 ½	16	L_2
$-6t + 6t' + J$	2 ½	18	X_2

4. HEAVY-FERMIONS IN MAGNETIC FIELDS: THE METAMAGNETIC TRANSITION

Heavy-fermion systems have been an active area of research for both experimentalists [5] and theorists [6] since their discovery in the mid-1970's. Heavy-fermion systems are characterized by huge coefficients (γ) to the term linear in T in the specific heat, quasi-elastic spin excitations (large magnetic susceptibility), and poor metallic conductivity. These features may be qualitatively described by a Fermi liquid with a very large density of states at the Fermi level [6]. Heavy-fermion systems may become superconductors (UPt_3, UBe_{13}, $CeCu_2Si_2$, URu_2Si_2, etc.), possess long-range magnetic order (UPt_3, URu_2Si_2, $NpBe_{13}$, U_2Zn_{17}, etc.), or remain paramagnetic metals ($CeRu_2Si_2$, $CeAl_3$, $CeCu_6$, etc.) at low temperatures.

Recent experimental work has concentrated on the properties of heavy-fermion systems in high magnetic fields [25-28]. A "transition" is observed (the so-called metamagnetic transition) at a characteristic magnetic field B_c in $CeRu_2Si_2$ ($B_c = 7.8T$), UPt_3 ($B_c = 21T$), and URu_2Si_2 ($B_c = 36T$). The transition is characterized by a magnetic-field dependence of the coefficient γ, the elastic coefficients, and the magnetic properties. At the critical field B_c, the coefficient γ has a single peak, the elastic coefficients are softened, and the magnetic fluctuations change character. The magnetization shows a steplike structure as a function of magnetic field strength. This contribution presents a many-body theory (*without* the assumptions of Fermi-liquid theory) that describes all of the above electronic properties of heavy-fermion systems (except superconductivity) and their field dependence.

Every heavy-fermion system is composed of ions with localized f-orbitals (lanthanides and actinides) that do not overlap with the corresponding f-orbitals on neighboring ions, but do hybridize with the extended states of the conduction-band electrons. The f-electrons interact very strongly with each other via a screened (on-site) Coulomb interaction U that acts only between two f-electrons that are localized about the same lattice site. Doubly occupied f-orbitals are effectively forbidden, since the Coulomb energy is larger than any other energy in the problem ($U > 10$ eV). The physics of such an electronic system is therefore described by the periodic Anderson impurity model [9] in the large-U ($U \rightarrow \infty$) limit, Eq. (1).

Heavy-fermionic behavior may occur in the restricted region [13,14] of parameter space where $-V\rho \ll \varepsilon\rho \ll -V^2\rho^2 < 0$. The localized orbitals are *almost* singly occupied ($<f_{i\uparrow}^\dagger f_{i\uparrow} + f_{i\downarrow}^\dagger f_{i\downarrow}> = 1-\nu$; $\nu \ll 1$) and the conduction electron density of states at the Fermi level is small. In this case, the kinetic energy of the holes that hop within a narrow "effective" band dominates over the magnetic spin-spin interactions and the spin-flipping terms of the Kondo effect.

The renormalized Schrieffer-Wolff transformation of the lattice Anderson model is then well described by a $t-J$ model in the limit $|V^2\rho/\varepsilon| \ll 1$. When $\rho\varepsilon \ll -V\rho < 0$, the renormalized magnetic interactions J between the local moments of the f-electrons dominate. The local moments interact with each other via all forms of exchange interactions, which determine, at low enough temperatures, the long-range magnetic order. As $\rho\varepsilon$ increases, two effects occur:

(i)   the kinetic energy of the holes in the narrow f-band become important; and

(ii)   a residual Kondo effect begins to quench the local magnetic moments.

In this regime,

$$-V\rho \ll \varepsilon\rho \ll -V^2\rho^2 \quad ,$$

the Anderson model is approximated well by the full $t–J$ model. The conduction electrons are decoupled from the f-band and act only as a buffer that determines the filling of the f-band. This picture is supported by numerical evidence found in exact solutions [19-21] of the lattice Anderson model on four-site clusters (see the next section).

4(a). Heavy-Fermionic Behavior in the t–J Model

A heavy-fermion system is characterized by a many-body ground state with very large number of low-lying excited states that have many different spin configurations (a partial decoupling of spatial and spin degrees of freedom). The localized states broaden into a strongly correlated narrow band in which all electronic transport takes place; the conduction band is (effectively) decoupled and acts only as a buffer that determines the concentration of electrons in the narrow band. The formation of a heavy-fermion ground state (and its low-lying excitations) require a fine-tuning of the parameters in the (effective) $t–J$ model and depends strongly upon the geometry and connectivity of the lattice.

4(b). The Metamagnetic Transition

As seen in the example of the previous section (Table I), the ground-state manifold of a heavy-fermion system contains a very large number of almost degenerate state. In that example there are 96 many-body states — out of a total of 1024 — which are degenerate for $J=0$. A finite J value partially lifts that degeneracy y, with low-spin configurations energetically favored over high-spin configurations.

A magnetic field (in the z-direction) partially lifts the degeneracy even more, since the many-body eigenstates with z-component of spin m_z have an energy

$$E(B) = E(0) - m_z g \mu_B B \equiv E(0) - m_z bJ \quad , \tag{10}$$

in a magnetic field B. The symbols g, μ_B, and b denote the Landé g-factor, Bohr magneton, and dimensionless magnetic field, respectively. The high-spin eigenstates are energetically favored in a strong magnetic field and level crossings occur as a function of b.

The phenomena described above are all of the necessary ingredients for a metamagnetic transition. The heavy-fermion system is described by a ground state with nearly degenerate low-lying excitations of many different spin configurations. The antiferromagnetic superexchange pushes high-spin states up in energy with splittings on the order of J. The magnetic field pulls down these high-spin states (with maximal m_z) and generates level crossings in the ground state. In the region near the level crossings, there is an increase in the density of low-lying excitations that produces a peak in the specific heat as a function of b. The magnetization and spin-spin correlation functions both change abruptly at the level crossings.

To illustrate the metamagnetic transition for the simple model above, the specific heat and magnetization are calculated as a function of the magnetic field (at a fixed low temperature). The specific heat satisfies

$$\frac{C_V(b)}{k_B} = \beta^2 \left[\frac{\sum_n E_n^2 \exp(-\beta E_n)}{\sum_n \exp(-\beta E_n)} - \left\{ \frac{\sum_n E_n \exp(-\beta E_n)}{\sum_n \exp(-\beta E_n)} \right\}^2 \right] \quad , \tag{11}$$

where k_B is Boltzmann's constant, β is the inverse temperature ($\beta \equiv 1/k_B T$) and E_n is the energy of the nth many-body eigenstate in a magnetic field b (the summations are restricted to the 96 eigenstates in Table 1). Similarly the magnetization is expressed by

$$M(b) = \frac{\sum_n m_z \exp(-\beta E_n)}{\sum_n \exp(-\beta E_n)} \quad , \tag{12}$$

where m_z is the z-component of spin for the nth many-body eigenstate. The results for the specific heat and magnetization are given in Figures 2 and 3, respectively, at the tempera-

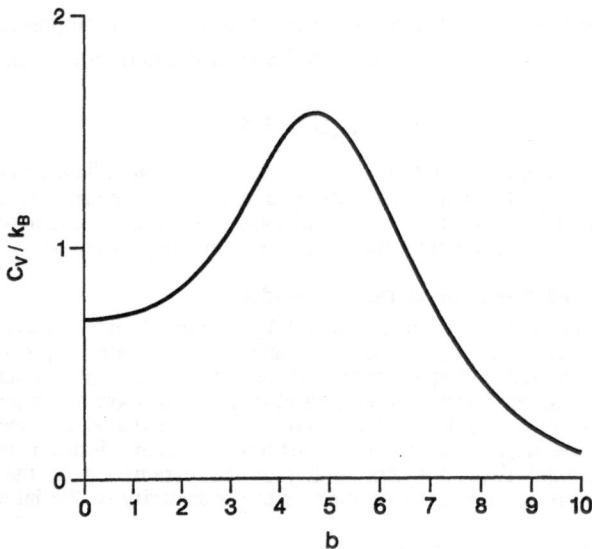

Figure 2. Calculated specific heat as a function of magnetic field for the heavy-fermion model. The temperature is $T = J/k_B$. The horizontal axis contains the dimensionless magnetic field and the vertical axis contains the dimensionless specific heat C_V/k_B. Note the single peak in the specific heat, characteristic of the high-temperature regime (temperature larger than the energy-level spacings).

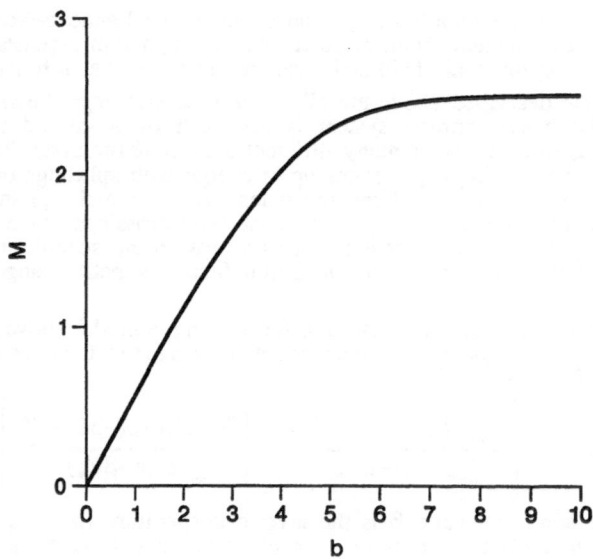

Figure 3. Calculated magnetization as a function of magnetic field at a temperature $T = J/k_B$. Note the smooth transition in the magnetization, characteristic of the high-temperature regime.

ture where $\beta J = 1$. Results for the magnetization at a lower temperature, $\beta J = 5$ are given in Figure 4.

The specific heat at the higher temperature has a single broad peak as a function of magnetic field with the center of the peak moving to larger values of b and the zero-field intercept becoming smaller as the temperature increases. The magnetization smoothly changes from a value of zero to a value of 5/2 as a function of magnetic field, showing little structure.

At lower temperatures the magnetization shows steps at the various values of the field where there are ground-state level crossings.

The results fit the experimental data [25-28] extremely well. The specific-heat measurements resemble the result of Figure 2 with a single-peak structure and the magnetization measurements resemble the "low-temperature" result (Figure 4) with noticeable steps. This is to be expected since magnetization measurements take place at a *constant* low temperature while specific-heat measurements require measurements over a temperature range.

Note that the low-field region ($b < 1$) is not faithfully represented by a small-cluster calculation, since the discreteness of the energy levels will always produce a linear magnetization.

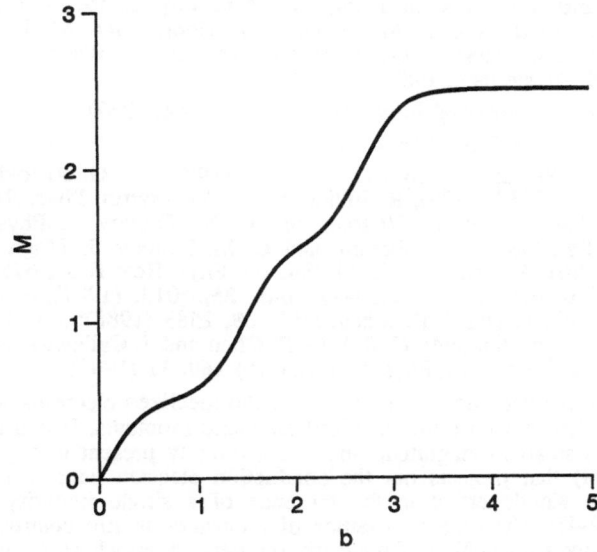

Figure 4. Calculated magnetization as a function of magnetic field at a temperature $T = J/5k_B$. Note the steplike transitions in the magnetization at each level crossing, characteristic of the low-temperature regime.

4(c). Discussion

The physics of the metamagnetic transition can be described as follows: a heavy-fermion system is composed of a ground-state with nearly degenerate low-lying excitations of many different spin configurations; the weak antiferromagnetic superexchange interaction slightly favors low-spin arrangements over high-spins (at zero magnetic field); a magnetic field pulls down the high-spin configurations causing (multiple) level crossing(s) in the ground state and producing a peak in the many-body density of states. The result is a peak in the specific heat (and possibly a richer structure at lower temperatures), steplike transitions in the magnetization, and abrupt changes in ground-state correlation functions.

5. ACKNOWLEDGMENTS

The authors acknowledge stimulating discussions with R. Fye and A. R. Mackintosh. This research was supported at the Lawrence Berkeley Laboratory, by the Director, Office of Energy Research, Office of Basic Energy Sciences, Materials Sciences Division, U.S. Department of Energy, under contract No. DE-AC03-76SF00098.

6. REFERENCES AND FOOTNOTES

[*] Present address: Department of Physics and Institute for Theoretical Physics, University of California, Santa Barbara, California, 93106-4030, USA

[1] J. Jensen and A. R. Mackintosh, *Rare Earth Magnetism: Structures and Excitations*, (Clarendon Press, Oxford, 1991), International Series of Monographs on Physics, Vol. 81.

[2] J. Kondo, in *Solid State Physics*, ed. F. Seitz, H. Ehrenreich and D. Turnbull, (Academic, New York, 1969), Vol. 23, p. 184.

[3] K. G. Wilson, Rev. Mod. Phys. **47**, 773 (1975) .

[4] P. B. Wiegmann im *Quantum Theory of Solids*, (Mir, Moscow, 1982), p. 238; N. Andrei, K. Furuya, and J. H. Lowenstein, Rev. Mod. Phys. **55**, 331, (1983); N. Andrei, K. Furuya, and J. H. Lowenstein, Rev. Mod. Phys. **55**, 331 (1983).

[5] G. R. Stewart, Rev. Mod. Phys. **56**, 755 (1984).

[6] C. M. Varma, Comments Sol. St. Phys. **11**, 221 (1985); P. A. Lee, T. M. Rice, J. W. Serene, L. J. Sham, and J. W. Wilkins, Comments Cond. Matt. Phys. **12**, 99, (1986); P. Fulde, J. Keller, and G. Zwicknagl, in *Solid State Physics*, ed. H. Ehrenreich and D. Turnbull, (Academic, New York, 1988), Vol. 41, p. 1.

[7] D. M. Newns and A. C. Hewson, J. Phys. F: Metal Physics **10**, 2429, (1980); *Valence Fluctuations in Solids*, ed. L. M. Falicov, W. Hanke, and M. B. Maple (North-Holland, Amsterdam, 1981); *Valence Instabilities*, ed. P. Wachter and H. Boppart (North-Holland, Amsterdam, 1982).

[8] C. Kittel, *Quantum Theory of Solids* (Wiley, New York, 1987).

[9] P. W. Anderson, Phys. Rev. **124**, 41, (1961).

[10] S. Asada and S. Sugano, J. Phys. C **11**, 3911, (1978); R. G. Arnold and K. W. H. Stevens, *ibid.* **12**, 5037 (1979); R. Jullien and R. M. Martin, Phys. Rev. B **26**, 6173, (1982); J. C. Parlebas, R. H. Victora, and L. M. Falicov, J. Physique **47**, 1029, (1986); J. C. Parlebas, R. H. Victora, and L. M. Falicov, J. Magn. Magn. Mater. **54/57**, 405, (1986); A. Reich and L. M. Falicov, Phys. Rev. B **34**, 6752, (1986); P. K. Misra, D. G. Kanhere, and J. Callaway, *ibid.* **35**, 5013, (1987); J. Callaway, D. P. Chen, D. G. Kanhere, and P. K. Misra, *ibid.* **38**, 2583 (1988); J. Callaway and D. P. Chen, J. Appl. Phys. **64**, 5944 (1988); D. P. Chen and J. Callaway, Phys. Rev. B **38**, 11869 (1988); J. C. Parlebas, Phys. Stat. Sol. (b) **160**, 11 (1990).

[11] The Kondo effect arises from a resonance of the localized electrons with the conduction electrons that quenches the localized magnetic moments. It is a non-perturbative effect which is small in magnitude and is not directly present in any model (such as the $t-J$ model) that projects out the conduction electron degrees of freedom. The physics of the Kondo effect in the presence of a single impurity is well known (References [2-4]). The case of a lattice of impurities is still controversial. Large-N expansions show a quenching of magnetic moments everywhere (Reference [12]) but are always in the atomic limit (Reference [6]) and do not illustrate what happens in the dense "impurity" limit.

[12] S. Doniach, in *Theoretical and Experimental Aspects of Valence Fluctuations and Heavy Fermions*, edited by C. Gupta and S. K. Malik (Plenum, New York, 1987) p. 179.

[13] J. R. Schrieffer and P. A. Wolff, Phys. Rev. **149**, 491, (1966).

[14] L. C. Lopes, R. Jullien, A. K. Bhattacharjee, and B. Coqblin, Phys. Rev. B **26**, 2640 (1982); J. M. Wills and B. R. Cooper, *ibid.* **36**, 3809, (1987).

[15] C. E. T. Gonçalves da Silva and L. M. Falicov, J. Phys. C: Solid State Phys. **5**, 63, (1972).

[16] J. Hubbard, Proc. R. Soc. London, Ser. A **276**, 238 (1963); **277**, 237 (1964); **281**, 401 (1964); **285**, 542 (1965); **296**, 82, (1967); **296**, 100 (1967).

[17] A. B. Harris and R. V. Lange, Phys. Rev. **157**, 295 (1967); A. H. MacDonald, S. M. Girvin, and D. Yoshioka, Phys. Rev. B **37**, 9753 (1988); A. M. Oleś, *ibid.* **41**, 2562 (1990); A. H. MacDonald, S. M. Girvin, and D. Yoshioka, *ibid.* **41**, 2565, (1990).

[18] A. Reich and L. M. Falicov, Phys. Rev. B **37**, 5560 (1980); **38**, 11199, (1988).

[19] J. K. Freericks and L. M. Falicov, Phys. Rev. B **42**, 4960 (1990).

[20] L. M. Falicov, in *Recent Progress in Many-body Theories*, edited by A. J. Kallio, E. Pajanne, and R. F. Bishop (Plenum, New York, 1988), Vol. 1, p. 275.

[21] J. Callaway, Physica B **149**, 17 (1988).

[22] R. H. Victora and L. M. Falicov, Phys. Rev. Lett. **55**, 1140 (1985).

[23] E. C Sowa and L. M. Falicov, Phys. Rev. B **35**, 3765 (1987); *ibid.* **37**, 8707 (1988).

[24] C. Chen and L. M. Falicov, Phys. Rev. B **40**, 3560 (1989); C. Chen, Phys. Rev. Lett. **41**, 1320 (1990).

[25] C. Paulsen, A. Lacerda, L. Puech, P. Haen, P. Lejay, J. L. Tholence, and A. de Visser, J. Low Temp. Phys. **81**, 317 (1990); P. Haen, J. Voiron, F. Lapierre, J. Flouquet, and P. Lejay, Physica B **163**, 519 (1990); G. Bruls, D. Weber, B. Lüthi, J. Flouquet, and P. Lejay, Phys. Rev. B **42**, 4329 (1990).

[26] T. Müller, W. Joss, and L. Taillefer, Phys. Rev. B **40**, 2614 (1989); H. P. van der Meulen, Z. Tarnawski, A. de Visser, J. J. M. Franse, J. A. A. J. Perenboom, D. Althof, and H. van Kempen, Phys. Rev. B **41**, 9352 (1990); J. J. M. Franse, M. van Sprang, A. de Visser, and A. A. Menovsky, Physica B **163**, 511 (1990).

[27] F. R. de Boer, J. J. M. Franse, E. Louis, A. A. Menovsky, J. A. Mydosh, T. T. M. Palstra, U. Rauchschwalbe, W. Shlabitz, F. Steglich, and A. de Visser, Physica B **138**, 1 (1986); A. de Visser, F. R. de Boer, A. A. Menovsky, and J. J. M. Franse, Solid State Commun. **64**, 527 (1987); K. Sugiyama, H. Fuke, K. Kindo, K. Shimohata, A. A. Menovsky, J. A. Mydosh, and M. Date, J. Phys. Soc. Japan **59**, 3331 (1990).

[28] H. Fujii, H. Kawanaka, T. Takabatake, E. Sugiura, K. Sugiyama, and M. Date, J. Magn. and Magn. Mat. **87**, 235 (1990).

DENSITY FUNCTIONAL THEORY AND
GENERALIZED WANNIER FUNCTIONS

Walter Kohn

Department of Physics
University of California
Santa Barbara, CA 93106

INTRODUCTION

In this lecture I would like to sketch a current effort of reformulating density functional theory (DFT) so as to make it practically applicable to very large molecules or atomic clusters. At the present time the computing time, T, for systems of N_a atoms behaves approximately as N_a^3, limiting calculations to N in the range of $10^{2\pm1}$. Our hope is that, in the new scheme, T will behave *linearly with* N_a, allowing calculations for systems of $10^{5\pm2}$ atoms.

1. Quick Warm-Up on Basics

The starting point of DFT is the *Hohenberg-Kohn theorem:* [1] Consider the class of all electronic groundstates for all possible external potentials, $v(r)$, and all possible numbers of electrons, N. A knowledge of the ground state density distribution, $n(r)$, uniquely determines the underlying potential, $v(r)$, as well as (trivially) N.* Thus $n(r)$ uniquely determines the entire Hamiltonian and hence all quantities derivable from the Hamiltonian, such as ground state energy E and many-body wave-function, Ψ; excited states; correlation functions etc.

Combining this theorem with the Rayleigh Ritz variational principle for the ground state leads to the *Hohenberg-Kohn variational principle*: For every possible ground state density, $n(r)$, there exists a functional $F[n(r)]$ such that, for a given $v(r)$ and N, the expression

$$E_{v(r)}[n(r)] \equiv \int n(r)v(r)dr + F[n(r)] \tag{1}$$

has the following properties

1. $E_v(r)[n(r)]$ assumes its minimum value if $n(r)$ is the correct ground state density.**

---

* If the ground state is degenerate the theorem applies to any one of the many possible densities $n(r)$.

** In the case of a degenerate ground state the minimum is obtained for any of the possible densities, $n(r)$.

Condensed Matter Theories, Vol. 8, Edited by
L. Blum and F.B. Malik, Plenum Press, New York, 1993

2. This minimum is the ground state energy.

This variational principle can be reformulated in the form of the *Kohn-Sham (KS) self-consistent equations:*[2] We first define the exchange-correlation functional $E_{xc}[n(r)]$ as follows:

$$F[n(r)] = T_s[n(r)] + \frac{1}{2}\int \frac{n(r)\ n(r')}{|r - r'|}\ dr dr' + E_{xc}[n(r)],$$

where $T_s[n(r)]$ is the ground state kinetic energy of *non-interacting* electrons with density distribution $n(r)$. Then one must solve the following self-consistent KS equations:

$$(-\frac{1}{2}\nabla^2 + v_{\text{eff}}(r) - \epsilon_i)\phi_i(r) = 0 \qquad (2)$$

$$n(r) = \sum_1^N |\phi_i(r)|^2 \qquad (3)$$

$$v_{\text{eff}}(r) = v(r) + \int \frac{n(r')}{|r - r'|}\ dr' + \frac{\delta E_{xc}[n(r)]}{\delta n(r)}, \qquad (4)$$

where the summation over i runs over the N states with the lowest ϵ_i. The self-consistent $n(r)$ is the correct ground state density and the ground state energy is given by

$$E = \sum_1^N \epsilon_i - \frac{1}{2}\int \frac{n(r)n(r')}{|r - r'|}\ dr dr' + \left(E_{xc}[n(r)] - \int \frac{\delta E_{xc}[n(r)]}{\delta n(r)}\ n(r)dr \right) \qquad (5)$$

To make practical use of these equations we require sufficiently accurate and simple approximations to $E_{xc}[n(r)]$. Fortunately these exist. The simplest is the local density approximation (LDA):

$$E_{xc}^{LDA}[n(r)] = \int \epsilon_{xc}(n(r))n(r)dr \qquad (6)$$

where $\epsilon_{xc}(n)$ is the exchange-correlation energy per particle of a uniform electron gas of density n. (It is known over a wide range of n with an accuracy of the order of 10^{-3})[3]. This approximation regards the electrons as uniform over a characteristic microscopic distance. Despite its apparent simplicity and crudeness it has been found to be very useful. Various improvements have been proposed, many of them replacing $\epsilon_{xc}(n(r))$ in Eq. (6) by a function of the form $\epsilon_{xc}(n(r), |\nabla n(r)|)$[4], which approximately allows for local *density gradients*.

2. The Use of Generalized Wannier Functions Instead of Eigenfunctions

Whatever approximation for $E_{xc}[n(r)]$ is chosen, the most time consuming step in solving the KS equations (2) is the solution of the single particle Schroedinger equations (2).

$$(-\frac{1}{2}\nabla^2 + v_{\text{eff}}(r) - \epsilon)\phi(r) = 0, \qquad (2)$$

where we have temporarily suppressed the subscript i. If these are calculated by a variational Ansetz

$$\phi(r) = \sum_{m=1}^{M} A_m \chi_m(r), \tag{7}$$

one is led to M simultaneous equations in the M unknowns A_m; the coefficients of these equations depend linearly on the energy eigenvalue ϵ. To achieve a given accuracy requires a time $T \propto M^3$[5]. For a given accuracy one needs $M \propto N_a$; hence $T \propto N_a^3$.

We note, however, two important facts. The quanties needed to calculate the energy E, Eq. (5) are only $n(r)$ and $\Sigma_1^N \epsilon_i$, not the eigenfunction $\phi_i(r)$ and eigenvalues ϵ_i themselves. Now let U_{ij} be an arbitrary unitary matrix and define

$$\psi_i(r) \equiv \sum U_{ij}\phi_j(r). \tag{8}$$

Then

$$n(r) = \sum_1^N |\phi_i(r)|^2 = \sum_1^N |\psi_i(r)|^2 \tag{9}$$

and

$$\sum \epsilon_i = \sum |\phi_i, H\phi_i) = \sum (\psi_i, H\psi_i), \tag{10}$$

where H is the Hamiltonian in Eq. (2).

For periodic crystals it is well known that the so-called Wannier functions (WF), w_ℓ[6], are well localized, orthonormal functions which are unitary transforms (or equivalents) of the eigenfunctions ϕ_k belonging to a given band:

$$w_\ell(r) = w(r - R_\ell) = \sum_k \phi_k(r) \, e^{-ikR_\ell}, \tag{11}$$

where R_ℓ is a lattice point. Thus

$$n(r) = \sum_\ell |w(r - R_\ell)|^2, \tag{12}$$

and

$$\sum \epsilon_i = N(w(r), Hw(r)). \tag{13}$$

It has been shown[7] that well-localized unitary functions, $w_\ell(r)$, which we call generalized Wannier functions (GWF), exist also for a large class of *non-periodic* bounded or unbounded systems. We call such systems Wannier representable (WR). The condition for their existence is, loosely speaking, the separation of the set of N ϵ_i's from other eigenvalues by a sufficient *gap*. These GWF's are not unique, different sets of GWF's, w_ℓ and $\tilde{w}_\ell$ being unitarily equivalent.

A very important additional property of the GWF's is what I call their *"short-sightedness"*:[8] the GWF $w_\ell(r)$ depends only on the shape of $v_{\text{eff}}(r)$ for r *near* the center of mass of $|w_\ell(r)|^2$. In this respect they contrast with the eigenfunctions $\phi_i(r)$ which are "normal-sighted": $\phi_i(r_0)$ depends on $v_{eff}(r)$ for r both near to *and far* from r_0.

Using the localization and shortsightedness of the GWF's we are developing a direct iterative algorithm, starting from an initial crude set $w_\ell^{(0)}$ and leading after a number of computational steps of order N_a to the final set w_ℓ with specified accuracy. (The eigenfunctions ϕ_i are *not* required.)

Finally, we remark that the property of shortsightedness implies the property of transferability. Suppose for example that we know the WF's $w_\ell^A, w_{\ell'}^B$ for the periodic solids... AAAA...and...BBBB. Consider now the sandwich structure AAAA BBBB AAAAA. Then, except near the ends and near the AB interfaces, the GWF's will be very similar to the w_ℓ^A and w_ℓ^B. (Again, no such transferability exists for the eigenfunctions ϕ_i.)

ACKNOWLEDGEMENTS

This work was supported by National Science Foundation Grant NSF-DMR90-01502.

REFERENCES

1. P. Hohenberg and W. Kohn, Phys. Rev. **136**, B 864 (1964).

2. W. Kohn and L. J. Sham, Phys. Rev. **140A** 1133 (1965).

3. S. H. Vosko et al., Can. J. Phys. **58**, 1200 (1980).

4. D. C. Langreth and M. J. Mehl, Phys. Rev. B **28**, 1809 (1983); Erratum: Phys. Rev. B **29**, 2310 (1984).

5. W. H. Press, et al., <u>Numerical Recipes</u>, Cambridge University Press (1986).

6. G. Wannier, Phys. Rev. **82**, 191 (1937); W. Kohn, Phys. Rev. B7, 4388 (1973).

7. W. Kohn and J. Orffroy, Phys. Rev. B8, 2485 (1973); J. J. Reur and W. Kohn, Phys. Rev. B **10**, 448 (1974).

8. W. Kohn and J. Orffroy, Phys. Rev. B8, 2485 (1973); J. J. Rehr and W. Kohn, Phys. Rev. B **10**, 448 (1974).

SHORT-RANGE STRUCTURE AND DYNAMICS OF METAL FILMS

T. L. Ainsworth, E. Krotscheck, and W. M. Saslow

Center for Theoretical Physics and Department of Physics
Texas A&M University
College Station, TX 77843   USA

The surface energies, correlations, and collective excitations of metal films and interfaces are calculated using an optimized variational theory that self-consistently accounts for both short-range and long-range correlations. Application of the recent Shore-Rose [Phys. Rev. Lett. **66**, 2619 (1991)] modification of the jellium model for metal surfaces yields excellent agreement between calculated and experimental surface energies of simple metals.

The present theory, which includes pair correlation effects, goes beyond local-density-functional theory. In the surface region the electronic states reflect the properties of the parent material, not merely the local electronic density. This is depicted clearly by the local "correlation hole". We have studied the correlation hole in some detail, and find it to be anisotropic in the surface region. This anisotropy causes notable changes in the surface energies.

In addition, we calculate the dynamic structure factor $S(\mathbf{q}_{\parallel}, \omega, z, z')$, and thereby determine the properties of the collective excitations. Both the dispersion relations and the transition densities $\delta\rho(z)$ are obtained. From this we identify both surface and volume modes and their relative weights.

1. INTRODUCTION

Electron structure theories are classified roughly into three categories: single-particle models, random-phase approximation, and fully-interacting theories. Single-particle models are popular in band-structure calculations where the structure of the many-particle system is believed to be dominated by the background of ion cores and the effects of electron-electron interactions are assumed to be weak. Any many-particle effects arising from the interaction between individual electrons are neglected.

The simplest way to include some many-body effects is to treat the long-range screening in the random-phase approximation (RPA). This approach is both conceptually and computationally simple. The RPA leads to the correct high-density limit of the correlation energy of jellium and the correct long-wavelength limit of the plasmon dispersion relation. It is therefore no surprise that single-particle models and the RPA were for a long time the favorite tools of solid state theorists.

Many-particle effects have previously been included in electron structure calculations in various ways, several popular methods are: variational and Green's function Monte Carlo calculations, coupled-cluster methods, and variational theories. There is general consensus on the equation of state of the uniform electron gas neutralized by a uniform positive charge background — the jellium model.

Non-uniform geometries provide further challenges: The computational effort is increased dramatically due to the broken symmetry. The theoretical basis for the theory of non-uniform electronic systems (and, for that matter, any quantum many-particle system) is provided by the Kohn-Hohenberg density functional theory[1], which states that the ground-state energy of a quantum many-particle system can be considered a functional of the external forces acting on the system, and the one-body density $\rho_1(\mathbf{r})$,

$$E = E[\rho_1(\mathbf{r})].\tag{1.1}$$

Although the Kohn-Hohenberg theorem is in principle exact, it requires specification of the energy functional at least in the vicinity of the physical density $\rho_1(\mathbf{r})$. Most popular is the *local density approximation*[2-4] (LDA) which assumes that the energy functional (or parts of it) can be approximated by the energy of the *bulk jellium* at the local density. The approximation is valid when the variation of the density is small compared with the correlation length. Another, less frequently stated but no less important, prerequisite of any local density approximation is that the system under consideration can actually exist in a uniform phase at any density. While this is true for the electron gas, the same statement is *not* true for self-bound many-body systems like the helium liquids and nuclear matter. It is therefore a challenge to many-body theorists to develop theories which do not suffer from the intrinsic problems of the local density approximation.

Some time ago, in order to overcome the limitations of the LDA, we[5,6] constructed a variational theory based on the Jastrow-Feenberg wave function[7]

$$\Psi_0(1,\ldots,N) = \exp\left\{\frac{1}{2}\sum_i u_1(\mathbf{r}_i) + \frac{1}{2}\sum_{i<j} u_2(\mathbf{r}_i,\mathbf{r}_j)\right\}\Phi_0(1,\ldots,N),\tag{1.2}$$

where $\Phi_0(1,\ldots,N)$ is a Slater determinant of single-particle wave functions $\phi_i(j) \equiv \phi_i(\mathbf{r}_j)\chi(i)$, $(i,j = 1,\ldots,N)$. The $\chi(i)$ are the spin-eigenfunctions.

The spatial single-particle orbitals $\phi_i(\mathbf{r}_j)$ and the *pair correlation functions* $u_2(\mathbf{r}_i,\mathbf{r}_j)$ are determined by minimization of the ground-state energy,

$$\frac{\delta}{\delta u_2(\mathbf{r}_1,\mathbf{r}_2)}\frac{\langle\Psi_0|H|\Psi_0\rangle}{\langle\Psi_0|\Psi_0\rangle} = 0 \text{ and } \frac{\delta}{\delta\phi_n(\mathbf{r})}\frac{\langle\Psi_0|H|\Psi_0\rangle}{\langle\Psi_0|\Psi_0\rangle} = 0.\tag{1.3}$$

The one-body factor $u_1(\mathbf{r})$ is redundant, but can be used to improve the convergence of cluster expansions of the variational energy expectation value. Such cluster expansions of the variational energy expectation value and the resummation of cluster integrals to infinite order are necessary to implement the unconstrained variational prescriptions (Eq. (1.3)) which determine the correlations and the single-particle basis. In the most sophisticated version of the Fermi hypernetted-chain (FHNC) theory the correlation energy of the bulk electron gas is predicted with an accuracy of a few percent [8-10]. In non-uniform geometries, additional approximations have been made[6]. These approximations, which have been discussed at length in Ref. 6, yield an overall accuracy of better than 10% compared to exact results for the bulk electron gas[11] and experimental correlation energies for ten-electron atoms[12].

2. SURFACE ENERGIES

The simplest jellium model of a metal assumes that the ion lattice may be represented by an average charge distribution $\rho_+(\mathbf{r})$, which is a step function. The Hamiltonian of the system is then

$$
H = -\sum_i \frac{\hbar^2}{2m}\nabla_i^2 + \frac{1}{2}\int d^3r_1 d^3r_2 \rho_+(\mathbf{r}_1)\rho_+(\mathbf{r}_2)v_c(|\mathbf{r}_1 - \mathbf{r}_2|)
$$
$$
-\sum_i \int d^3r\, v_c(|\mathbf{r}_i - \mathbf{r}|)\rho_+(\mathbf{r}) + \frac{1}{2}\sum_{i<j} v_c(|\mathbf{r}_i - \mathbf{r}_j|)\,. \tag{2.1}
$$

Here, $v_c(r) \equiv e^2/r$ is the Coulomb interaction. In Refs. 6 and 13, we calculated the surface energy of the jellium model by first computing the ground-state energy of symmetric slabs of varying widths: $d = 14\,a_0\,r_s$, $12\,a_0\,r_s$, $10\,a_0\,r_s$ and $8\,a_0\,r_s$ (a_0 is the Bohr radius). From the ground-state energy as a function of the slab width d, which is proportional to the particle number N, we extract both a surface energy per unit area, σ_u, and a bulk energy per particle, E_∞,

$$
E[N] = 2\,\sigma_u\,A + E_\infty\,N\,. \tag{2.2}
$$

The asymptotic energy per particle E_∞ should, and does, agree with that obtained by an independent calculation of the homogeneous electron gas.

In these calculations, the FHNC theory predicted surface energies that were remarkably different from the predictions of the LDA. In order to reconcile the source of the discrepancy between the local density approximation and the Fermi HNC theory, we carried out an array of test calculations[13]. These results are summarized in Fig. 1:

(a) LDA calculations of the surface energy employing bulk jellium equations of state derived from the RPA, the FHNC theory, and Monte Carlo calculations. We found no significant differences in the surface energies among these choices for the bulk equation of state.

(b) A simple RPA calculation of the surface energy. By simple RPA we mean the summation of all Feynman ring-diagrams, using the bare Coulomb potential as the driving particle-hole interaction. Single-particle orbitals were taken from LDA calculations or determined self-consistently by minimizing the RPA energy with respect to the single-particle basis.

(c) A full RPA calculation using a screened Coulomb interaction in the summation of the ring diagrams. The particle-hole interaction is written in the form

$$
V_{ph}(\mathbf{r}_1, \mathbf{r}_2) = V_c(|\mathbf{r}_1 - \mathbf{r}_2|)C(\mathbf{r}_1, \mathbf{r}_2)\,, \tag{2.3}
$$

where $C(\mathbf{r}_1, \mathbf{r}_2)$ is a *local screening function*. Here the local screening function is determined from the bulk electron gas in the local density approximation, *i.e.*

$$
C(\mathbf{r}_1, \mathbf{r}_2) = C_{bulk}(|\mathbf{r}_1 - \mathbf{r}_2|; \rho_1((\mathbf{r}_1 + \mathbf{r}_2)/2))\,. \tag{2.4}
$$

(d) A full RPA calculation of the surface energy employing a screened Coulomb interaction. However, the local screening function is computed from the full FHNC theory for the non-uniform electron system.

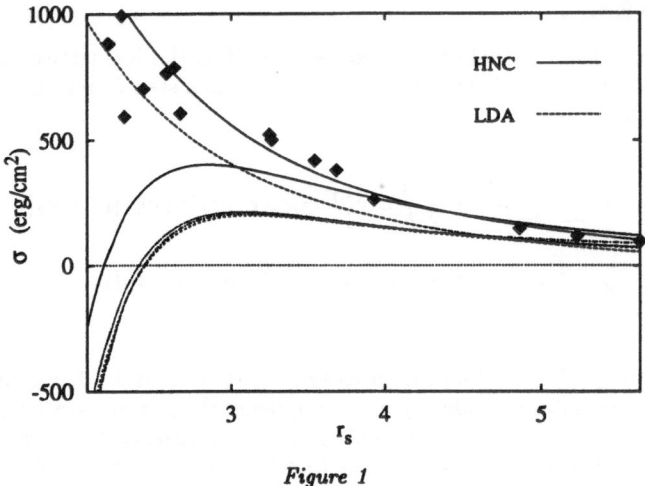

Figure 1

Surface energy of simple metals as a function of the electron density r_s. The four lower curves are results from calculation for pure jellium slabs in FHNC (solid line), simple RPA, LDA, and locally screened RPA (dashed lines). The upper two curves are surface energies obtained in the "ideal-metal" model in FHNC (solid line) and LDA (dashed line).[14] The diamonds represent experimental data from Tyson and Miller (Ref. 15) for Al $(r_s = 2.060)$, Ga $(r_s = 2.192)$, Zn $(r_s = 2.280)$, Pb $(r_s = 2.300)$, In $(r_s = 2.425)$, Cd $(r_s = 2.574)$, Mg $(r_s = 2.624)$, Hg $(r_s = 2.664)$, Li $(r_s = 3.249)$, Ca $(r_s = 3.263)$, Sr $(r_s = 3.546)$, Ba $(r_s = 3.690)$, Na $(r_s = 3.931)$, K $(r_s = 4.862)$, Rb $(r_s = 5.229)$, and Cs $(r_s = 5.630)$

The three cases (a-c) and their variations due to different choices for the single particle orbitals produced virtually identical values for the surface energies. In this connection, it is important to note that the RPA is diagrammatically a proper subset of the FHNC theory. The fourth calculation essentially exhausted the discrepancy between the (locally screened) RPA and the FHNC results, thus suggesting that *the deviation of the short-range structure of the electron gas from that for the uniform system at the same density is the essentially the cause for the discrepancy between the surface energies.* We will later show examples for the correlation hole in the surface after the remaining aspects of theory have been discussed.

Before we proceed with our discussion of a more realistic model of the metal surface, we comment briefly on a recent Green's functions Monte Carlo (GFMC) calculation of aluminum by Li *et al.* (Ref. 16). Li *et al.* conclude that the Langreth-Mehl[17] modification of density functional theory agrees more closely with the GFMC surface energy than the FHNC calculation. Their finding is consistent with our "ideal metal" result in that the FHNC surface energy for aluminum is somewhat too high.

Unfortunately, the Monte Carlo calculation has been carried out in a density regime where the surface energy is a very rapidly varying function of r_s. The simplest version of FHNC for non-uniform geometries — the one which we employ — provides an accuracy of the order of 10% for the electron correlation energy. For $r_s \sim 2$, *e.g.* aluminum, the total energy is approximately zero. Thus the good accuracy achieved by FHNC for the correlation energy could be overshadowed by the large-scale cancellations between the kinetic and Coulomb energies and the correlation energy. Only for aluminum is the total energy approximately 10% of the correlation energy; at all lower densities the correlation energy is not so strongly cancelled by the kinetic and Coulomb terms.

If a discrepancy between FHNC and GFMC exists at larger values of r_s then that would be the first case where our version of FHNC is *not* accurate to within a few percent for an electronic system. This would be extremely interesting since it would imply new many-body physics for electronic systems.

The jellium model of the metal surface is not overly realistic; in particular it predicts an unstable surface for high-density metals. A simple remedy of the situation was recently suggested by Shore and Rose[18,19] who pointed out that the predictions of the jellium model for the surface energy of simple metals can be significantly improved by requiring that the electron gas be in mechanical equilibrium. The theory of "ideal metals" is based on the observation that the electron gas is, at all densities except its equilibrium density around $r_s \approx 4$, under an external pressure which must be compensated by a mechanical stress in the positive background. Shore and Rose take these stresses into account by introducing a dipole barrier of strength $V_0 = e_{jell} - \mu_{jell}$ at the surface of the metal, where e_{jell} is the energy per particle for jellium at the r_s value under consideration and μ_{jell} the corresponding chemical potential. In other words, the "ideal metal" model merely supplements this step function by a dipole barrier of strength V_0, which can be obtained from calculations of the correlation energy of bulk electrons.

The strength of the dipole barrier is fixed by the requirement that the jellium be in mechanical equilibrium — the first-order variation of the energy is zero when two pieces of jellium, initially touching, are slightly displaced. For clarity, a short derivation is presented: Consider jellium with average electron density n_0 and energy $e_{jell}(n_0)$. Cut it, and partially overlap the pieces. To lowest-order the change in energy per particle between the overlapped and juxtaposed configurations is

$$e_{jell}(\overline{n}) - e_{jell}(n_0) + V_0 \frac{\overline{n} - n_0}{n_0} = 0 \,, \tag{2.5}$$

where $\overline{n}$ is the average density in the overlapping regions and V_0 is the strength of the new surface dipole. Expanding $e_{jell}(\overline{n})$ about n_0 and solving for V_0 yields

$$V_0 = -n_0 \frac{\partial e_{jell}}{\partial n}\bigg|_{n_0} = e_{jell}(n_0) - \mu_{jell}(n_0) \,, \tag{2.6}$$

where $\mu_{jell}(n_0)$ is the jellium chemical potential. Additional details and discussions of possible limitations of the "ideal metal" model are given in Refs. 18, 19 and 20. It is important to note that the strength of the surface dipole potential depends solely upon r_s; therefore *no* parameters are introduced by the "ideal metal" model. Theoretical estimates for chemical potentials, cohesive energies, work functions, and surface energies of metals have been calculated using local-density-functional theory. The theoretical predictions of the "ideal metal" model are substantially improved over the predictions of density functional theory without the dipole barrier. However, there is room for improvement. Encouraged by the success of the "ideal metal" model of Refs. 17 and 18, we have applied the FHNC theory to the same model.

A collection of our results for the four slab widths mentioned above is given in Fig. 1. The calculated surface energies are compared with the density-functional calculation of Shore and Rose (Refs. 18 and 19) and the experimental estimates of Tyson and Miller (Ref. 15). The agreement between theoretical predictions and experiments is obviously excellent; for all but very high densities there is no room for further improvement. This provides strong support for the "ideal metal" model and encourages the use of FHNC theory to study dynamic effects at metal surfaces.

3. LOCAL SCREENING

We conclude that our significant corrections to the surface energy of simple metals arise from the difference between the local screening function in a uniform system at fixed density, and the local screening function in the surface. To demonstrate the effect of non-uniformity, we show here two extreme cases: Contour plots of the local screening function $C(\mathbf{r}, \mathbf{r}')$ for thin jellium slabs of about one atomic layer ($d = 2(a_0 r_s)$), and for thick jellium slabs ($d = 12(a_0 r_s)$) when one of the electrons is located on the jellium edge.

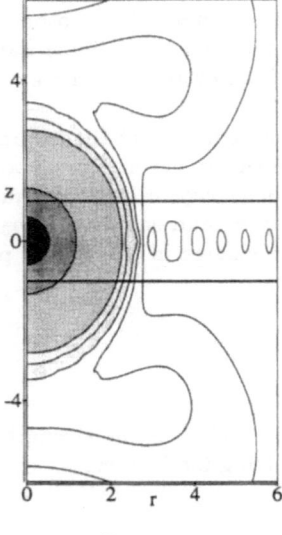
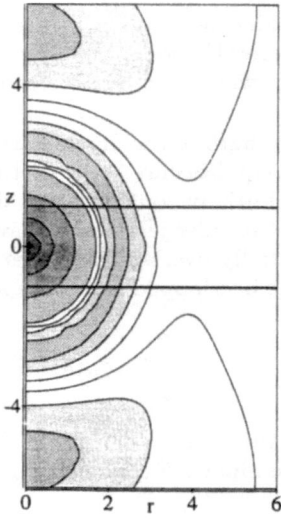

Figure 2a   *Figure 2b*

Contour plots of the local screening function $C(\mathbf{r}, \mathbf{r}')$ in an aluminum slab (Fig. 2a) and cesium slab (Fig. 2b). The center of mass of the two electrons is located at $z = 0$, and the two heavy horizontal lines indicate the jellium edge. Darker areas correspond to lower values of $C(\mathbf{r}, \mathbf{r}')$, except the annular rings with $r \sim 2$ in Fig. 2b which represent slight positive values of $C(\mathbf{r}, \mathbf{r}')$.

Fig. 2 shows contour plots for the thin slabs. Inspection of the data on which the figures are based reveals that the correlation hole is, in the case of the aluminum slab, about 20% wider in the direction perpendicular to the slab than in the direction parallel to the slab. As expected, the anisotropy is less (only about 10%) for the lower-electron-density cesium slab.

Further illustration of the anisotropy of the local screening function and a comparison with the local screening function in the bulk metal is presented in Fig. 3. We

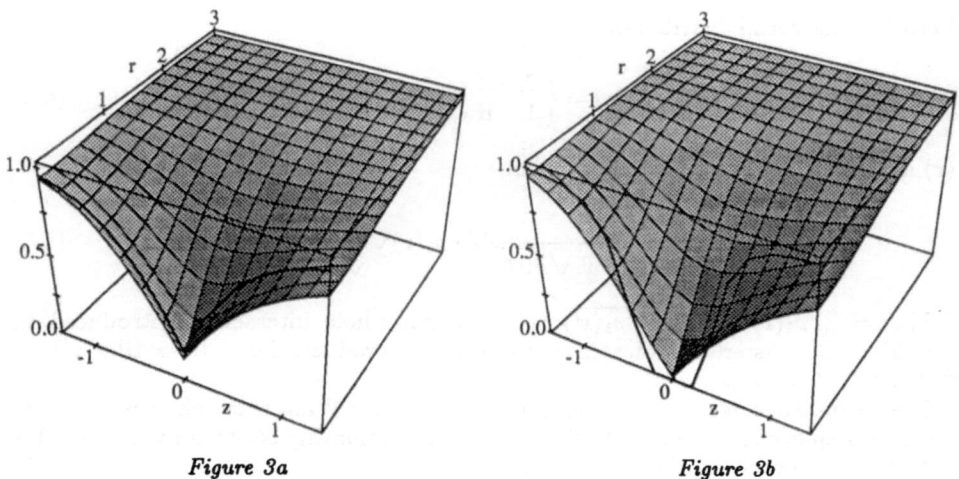

Figure 3a   *Figure 3b*

The local screening function is shown in Fig. 3a for an aluminum slab of width $d = 12r_s a_0$. One particle is located on the jellium edge $z = 0$. Areas with negative z are in the metal, those with positive z outside the jellium. The heavy line on the face of the figure is the trace of the local screening function in the corresponding bulk material. Fig. 3b is similar to Fig. 3a but for a cesium slab of width $12r_s a_0$.

consider an electron located *on the jellium edge* of a reasonably wide jellium slab of width $d = 12$ $(r_s a_0)$. Two features are apparent: The local screening hole is *larger* in the direction out of the slab, and the local screening is qualitatively different from the correlation hole in the bulk material. These features are more pronounced in the cesium slab than in the aluminum one.

The main conclusion that we draw from this part of our work is: *The local density approximation for the particle-hole interaction is inadequate to calculate the surface properties of simple metals.* We are confident of the qualitative validity of our observations, however more quantitative comparisons need to be made before we can estimate the error bars on our results. The level of FHNC approximation used here is just the *minimum* level at which these effects can be studied. However, one may systematically improve the calculation by using more sophisticated versions of the FHNC theory. It is also not clear whether the assumption of *local* and *static* screening is sufficient.

4. COLLECTIVE EXCITATIONS

The central equation of the Fermi HNC theory is derived from Fermi paired-phonon analysis (PPA)[6]

$$H_1 \left[S_F^{-1} * H_1 * S_F^{-1} + 2\hat{V}_{ph} \right] \Psi_\omega = (\hbar\omega)^2 \Psi_\omega. \tag{4.1}$$

In Eq. (4.1), we have introduced the static structure function $S_F(\mathbf{r}, \mathbf{r}')$ of the non-interacting model system,

$$\sqrt{\rho_1(\mathbf{r})} S_F(\mathbf{r}, \mathbf{r}') \sqrt{\rho_1(\mathbf{r}')} \equiv \rho_1(\mathbf{r})\delta(\mathbf{r} - \mathbf{r}') - |\rho_1(\mathbf{r}, \mathbf{r}')|^2, \tag{4.2}$$

where $\rho_1(\mathbf{r}, \mathbf{r}')$ is the one-body density matrix

$$\rho_1(\mathbf{r}, \mathbf{r}') = \sum_m \theta(m)\psi_m(\mathbf{r})\psi_m^*(\mathbf{r}') \tag{4.3}$$

and $\theta(m)$ is the Fermi distribution

$$\theta(m) = \begin{cases} 0, & \text{if } \epsilon_m > \epsilon_F \\ 1 & \text{if } \epsilon_m < \epsilon_F \end{cases} \qquad (4.4)$$

$H_1(\mathbf{r})$ is an effective one-body Hamiltonian,

$$H_1(\mathbf{r}) = -\frac{\hbar^2}{2m} \frac{1}{\sqrt{\rho_1(\mathbf{r})}} \nabla \cdot \rho_1(\mathbf{r}) \nabla \frac{1}{\sqrt{\rho_1(\mathbf{r})}}, \qquad (4.5)$$

and $\hat{V}_{ph} = \sqrt{\rho_1(\mathbf{r})} V_{ph}(\mathbf{r}, \mathbf{r}') \sqrt{\rho_1(\mathbf{r}')}$ is the particle-hole interaction introduced in Eq. (2.3). The asterisk indicates a convolution product, i.e. $[A * B](\mathbf{r}, \mathbf{r}') \equiv \int d^3 r'' A(\mathbf{r}, \mathbf{r}'') B(\mathbf{r}'', \mathbf{r}')$.

The purpose of this section is to clarify two questions: What is the physical meaning of the eigenfrequencies $\hbar\omega$ of Eq. (4.1)? What is the relationship between the Fermi-PPA Eq. (4.1) and the random phase approximation (RPA)?

To be specific, we assume that the single-particle states $\psi_m(\mathbf{r})$ have been generated from a local one-body Schrödinger equation, and that all orbitals are filled up to the Fermi energy ϵ_F. The response function of the non-interacting system is

$$\chi_0(\mathbf{r}, \mathbf{r}'; \omega) = \sum_{mn} \frac{\theta(m) - \theta(n)}{\omega - \epsilon_n + \epsilon_m} \psi_m(\mathbf{r}) \psi_m^*(\mathbf{r}') \psi_n^*(\mathbf{r}) \psi_n(\mathbf{r}'). \qquad (4.6)$$

We obtain the full response function by the RPA equation

$$\chi(\mathbf{r}, \mathbf{r}'; \omega) = \chi_0(\mathbf{r}, \mathbf{r}'; \omega) + \int d^3 r_1 d^3 r_2 \chi_0(\mathbf{r}, \mathbf{r}_1; \omega) V_{ph}(|\mathbf{r}_1 - \mathbf{r}_2|) \chi(\mathbf{r}_2, \mathbf{r}'; \omega). \qquad (4.7)$$

The collective density fluctuations $\delta\rho_1(\mathbf{r})$ are obtained from

$$\int d^3 r' \epsilon(\mathbf{r}, \mathbf{r}'; \omega) \delta\rho(\mathbf{r}') = 0 \qquad (4.8)$$

where $\epsilon(\mathbf{r}, \mathbf{r}'; \omega)$ is the dielectric function

$$\epsilon(\mathbf{r}, \mathbf{r}'; \omega) = \delta(\mathbf{r} - \mathbf{r}') - \int d^3 r_1 \chi_0(\mathbf{r}, \mathbf{r}_1; \omega) V_{ph}(\mathbf{r}_1, \mathbf{r}'). \qquad (4.9)$$

The first problem is to show in which approximation Eqs. (4.6) and (4.7) are equivalent to the PPA Eq. (4.1) of the FHNC theory. In the homogeneous electron gas, it is well known that the PPA can be obtained from the RPA in the so-called "collective" or "mean spherical" approximation which means that one replaces $\chi_0(\mathbf{r}, \mathbf{r}'; \omega)$ by the approximation $\chi_0^{MSA}(\mathbf{r}, \mathbf{r}'; \omega)$, which has only one pole whose strength is chosen such that

$$\int \mathrm{Im}\, \chi_0(\mathbf{r}, \mathbf{r}'; \omega) d\omega = \int \mathrm{Im}\, \chi_0^{MSA}(\mathbf{r}, \mathbf{r}'; \omega) d\omega$$

$$\text{and} \quad \int \omega \,\mathrm{Im}\, \chi_0(\mathbf{r}, \mathbf{r}'; \omega) d\omega = \int \omega \,\mathrm{Im}\, \chi_0^{MSA}(\mathbf{r}, \mathbf{r}'; \omega) d\omega. \qquad (4.10)$$

We show here that the same is true in the non-uniform case. To shorten the expressions, we introduce $\bar{\theta}(n) \equiv 1 - \theta(n)$ and write $\chi_0(\mathbf{r}, \mathbf{r}'; \omega)$ in the form

$$\chi_0(\mathbf{r}, \mathbf{r}'; \omega) = 2 \sum_{mn} \theta(m) \bar{\theta}(n) \frac{(\epsilon_n - \epsilon_m)}{\omega^2 - (\epsilon_n - \epsilon_m)^2} \psi_m(\mathbf{r}) \psi_m^*(\mathbf{r}') \psi_n^*(\mathbf{r}) \psi_n(\mathbf{r}'). \qquad (4.11)$$

For the first integral of Eq. (4.10) we get

$$\int \mathcal{I}m\,\chi_0(\mathbf{r},\mathbf{r}';\omega)\frac{d\omega}{2\pi} = \sum_{mn}\theta(m)\bar{\theta}(n)\psi_m(\mathbf{r})\psi_m^*(\mathbf{r}')\psi_n^*(\mathbf{r})\psi_n(\mathbf{r}')$$

$$= \sqrt{\rho_1(\mathbf{r})}S_F(\mathbf{r},\mathbf{r}')\sqrt{\rho_1(\mathbf{r}')}. \tag{4.12}$$

The second sum rule is more complicated to evaluate. We assume that the single-particle states have been obtained from *local* one-body equation

$$\left[-\frac{\hbar^2}{2m}\nabla^2 + V_H(\mathbf{r})\right]\psi_m(\mathbf{r}) = \epsilon_m(\mathbf{r}). \tag{4.13}$$

One finds, after some algebra,

$$\int \omega\,\mathcal{I}m\,\chi_0(\mathbf{r},\mathbf{r}';\omega)\frac{d\omega}{2\pi} = -\frac{\hbar^2}{2m}\nabla\cdot[\rho_1(\mathbf{r})\nabla] \tag{4.14}$$

which is the expected generalization of the ω^1 sum rule for the non-uniform system. The next problem is to identify the approximation for $\chi_0(\mathbf{r},\mathbf{r}';\omega)$ which leads from the RPA Eq. (4.7) to the PPA Eq. (4.1). In the bulk, one thinks of the particle-hole projection operator $\theta(m)\bar{\theta}(n)$ as a weight function, and approximates

$$\frac{\sum_{mn}\theta(m)\bar{\theta}(n)\left[\omega - \epsilon_m + \epsilon_n\right]^{-1}}{\sum_{mn}\theta(m)\bar{\theta}(n)} \approx \left[\frac{\sum_{mn}\theta(m)\bar{\theta}(n)\left[\omega - \epsilon_m + \epsilon_n\right]}{\sum_{mn}\theta(m)\bar{\theta}(n)}\right]^{-1} \tag{4.15}$$

This and similar approximations have been applied abundantly in solid state physics, see for example Ref. 21 for a review. Applying the same approximation for the response function of the non-uniform system and employing Eq. (4.12), we find

$$\sum_{mn}\theta(m)\bar{\theta}(n)\frac{1}{\omega \pm (\epsilon_n - \epsilon_m)}\psi_m(\mathbf{r})\psi_m^*(\mathbf{r}')\psi_n^*(\mathbf{r})\psi_n(\mathbf{r}')$$

$$\approx \sqrt{\rho_1(\mathbf{r})}S_F(\mathbf{r},\mathbf{r}')\sqrt{\rho_1(\mathbf{r}')}$$

$$* \left\{\sum_{mn}\theta(m)\bar{\theta}(n)\left[\omega \pm (\epsilon_n - \epsilon_m)\right]\psi_m(\mathbf{r})\psi_m^*(\mathbf{r}')\psi_n^*(\mathbf{r})\psi_n(\mathbf{r}')\right\}^{-1} \tag{4.16}$$

$$* \sqrt{\rho_1(\mathbf{r})}S_F(\mathbf{r},\mathbf{r}')\sqrt{\rho_1(\mathbf{r}')} \equiv \chi_0^\pm(\mathbf{r},\mathbf{r}';\omega)$$

and

$$\sqrt{\rho_1(\mathbf{r})}\left[\chi_0^{MSA}\right]^{-1}(\mathbf{r},\mathbf{r}';\omega)\sqrt{\rho_1(\mathbf{r}')} = \frac{1}{2}\left[(\hbar\omega)^2 H_1^{-1} - S_F^{-1} * H_1 * S_F^{-1}\right](\mathbf{r},\mathbf{r}'), \tag{4.17}$$

where $\chi_0^{MSA} \equiv \chi_0^+ - \chi_0^-$. The approximation Eq. (4.16) for $\chi_0^{MSA}(\mathbf{r},\mathbf{r}';\omega)$ satisfies the sum rules Eq. (4.10). Using the "mean spherical approximation" χ_0^{MSA} for the particle-hole propagator, we find

$$\left[\chi_0^{MSA}\right]^{-1}(\mathbf{r},\mathbf{r}_1;\omega) * \epsilon(\mathbf{r}_1,\mathbf{r}';\omega) = \left[\chi_0^{MSA}\right]^{-1}(\mathbf{r},\mathbf{r}_1;\omega) - V_{ph}(|\mathbf{r}_1 - \mathbf{r}'|)$$

$$= \frac{1}{2}\frac{1}{\sqrt{\rho_1(\mathbf{r})}}H_1^{-1}\left\{(\hbar\omega)^2 - H_1\left[S_F^{-1}H_1 S_F^{-1} + 2\hat{V}_{ph}\right]\right\}\frac{1}{\sqrt{\rho_1(\mathbf{r}')}}. \tag{4.18}$$

Thus, with $\Psi_\omega(\mathbf{r}) \equiv \delta\rho_1(\mathbf{r})/\sqrt{\rho_1(\mathbf{r})} = 2\delta\sqrt{\rho_1(\mathbf{r})}$ the RPA Eq. (4.7) becomes the eigenvalue problem of Eq. (4.1). In this sense, the eigenvalues of the PPA equation can also be interpreted as the collective modes of the system, and the eigenfunctions identified with the density fluctuations. In both cases, PPA and RPA, the static structure function is obtained by frequency integration of the response function,

$$
\begin{aligned}
\int \chi(\mathbf{r}, \mathbf{r}'; \omega) d\omega &= \sum_{mn} \theta(m) \bar{\theta}(n) \psi_m(\mathbf{r}) \psi_m^*(\mathbf{r}') \psi_n^*(\mathbf{r}) \psi_n(\mathbf{r}') \\
&= \sqrt{\rho_1(\mathbf{r})} S(\mathbf{r}, \mathbf{r}') \sqrt{\rho_1(\mathbf{r}')} \\
&= \sqrt{\rho_1(\mathbf{r})} \sum_{\epsilon} [H_1 \psi_\epsilon(\mathbf{r})] [H_1 \psi_\epsilon(\mathbf{r}')] \sqrt{\rho_1(\mathbf{r}')} .
\end{aligned}
\tag{4.19}
$$

A related, but conceptually different derivation of the same excitation spectrum can be based on the Feynman theory of collective excitations. Feynman[22] starts with a local *ansatz*

$$
\Psi_X(1, \ldots, N) = F(\mathbf{r}_1, \ldots, \mathbf{r}_N) \Psi_0(1, \ldots, N),
$$
$$
F(\mathbf{r}_1, \ldots, \mathbf{r}_N) = \sum_{i=1}^{N} f(\mathbf{r}_i; t)
\tag{4.20}
$$

for the wave function $\Psi_X(1, \ldots, N)$ of the excited state. Here, $f(\mathbf{r}_i)$ is the *excitation function*, and $\Psi_0(1, \ldots, N)$ the exact ground state wave function, or a variational wave function where the pair correlations have been optimized. With these assumptions, the energy difference

$$
\hbar\omega = \frac{\langle \Psi_X | H | \Psi_X \rangle}{\langle \Psi_X | \Psi_X \rangle} - \frac{\langle \Psi_0 | H | \Psi_0 \rangle}{\langle \Psi_0 | \Psi_0 \rangle} = \frac{\langle \Psi_0 | [F, [T, F]] | \Psi_0 \rangle}{\langle \Psi_0 | F^2 | \Psi_0 \rangle}
\tag{4.21}
$$

is an upper bound for the correlation energy. Minimizing the excitation energy with respect to the excitation function $f(\mathbf{r})$ leads to the eigenvalue problem

$$
H_1(\mathbf{r}) \Psi_\omega(\mathbf{r}) = \hbar\omega \int d^3 r' S(\mathbf{r}, \mathbf{r}') \Psi_\omega(\mathbf{r}')
\tag{4.22}
$$

where $\Psi_\omega(\mathbf{r}) = \sqrt{\rho_1(\mathbf{r})} f(\mathbf{r})$. It is readily verified that the eigenstates of the PPA Eq. (4.1) and the generalized Feynman relation Eq. (4.22) are identical.

Eq. (4.1) provides a simple way to reduce the problem to the RPA or to study the local density approximations. The basic definition of the particle-hole interaction is

$$
V_{ph}(\mathbf{r}, \mathbf{r}') = \frac{\delta^2 E_c}{\delta \rho(\mathbf{r}) \delta \rho(\mathbf{r}')}
\tag{4.23a}
$$

In the HNC approximation, the particle-hole interaction is calculated by self-consistently summing ring and ladder diagrams. While the definition Eq. (4.21) is identical with the HNC approximation only in the case that all diagrams are summed, the inconsistency is generally small. In simple RPA, the particle-hole interaction is approximated by the bare Coulomb potential,

$$
V_{ph}^{RPA}(\mathbf{r}, \mathbf{r}') = V_c(|\mathbf{r} - \mathbf{r}'|),
\tag{4.23b}
$$

whereas the LDA suppliments the bare Coulomb potential with a δ-function interaction,

$$
V_{ph}^{LDA}(\mathbf{r}, \mathbf{r}') = V_c(|\mathbf{r} - \mathbf{r}'|) + \left. \frac{d^2 \epsilon_c(\rho)}{d\rho^2} \right|_{\rho=\rho_1(\mathbf{r})} \delta(\mathbf{r} - \mathbf{r}') .
\tag{4.23c}
$$

Therefore all three varieties of particle-hole interactions (Eqs. (23)) are easily contrasted and compared.

5. PLASMON DISPERSION

We now turn to the discussion of the collective excitations of metal films. The calculation of the dynamic structure function has been carried out as follows: We first calculate the structure of the film by solving self-consistently the PPA equation. This produces the one-body Hamiltonian $H_1(\mathbf{r})$ and the static structure function $S(\mathbf{r}, \mathbf{r}')$, and also provides a first estimate of the collective modes $\Psi_\omega(\mathbf{r})$ and their corresponding

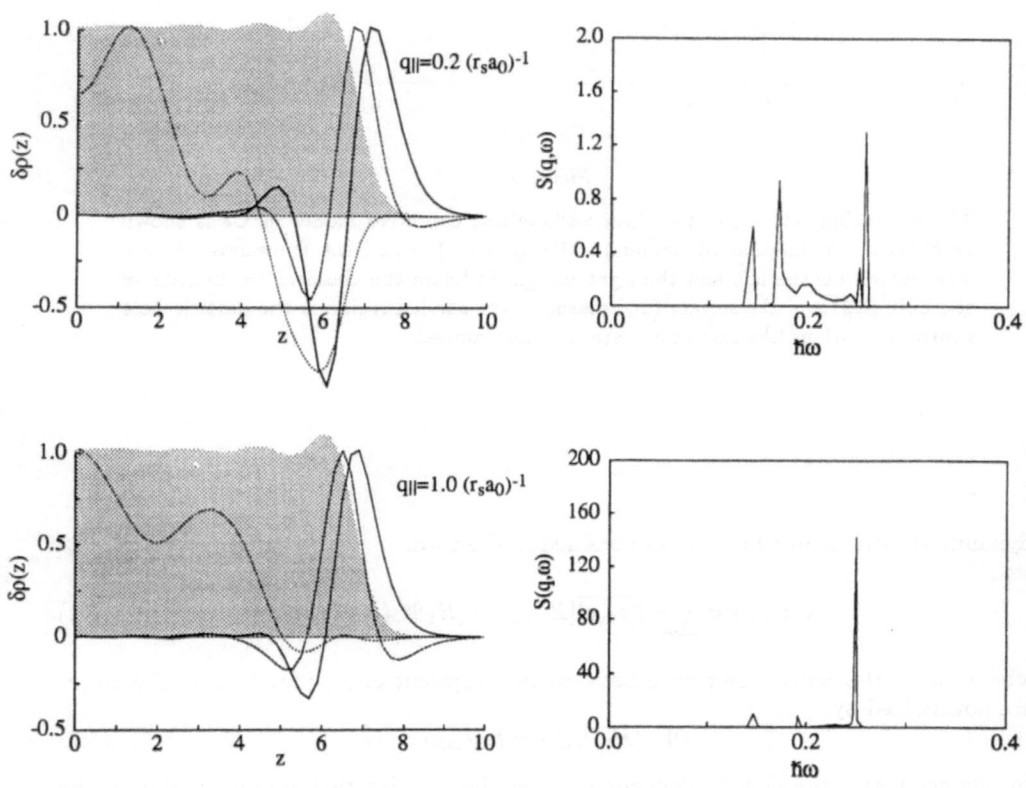

Figure 4

The figures show two typical spectra and the corresponding transition densities for cesium. The upper pair of figures corresponds to a very long wavelength, whereas the lower pair shows the spectrum at a wavenumber of $1\ (r_s a_0)$. The shaded region in the left pair of figures is the electron density of the background, the solid, dashed, and dotted lines are the transition densities of the lowest, next lowest, and highest energy peak shown in the spectra on the right. The normalization of the transition densities is arbitrary.

frequencies. However, discretizing in a small box restricts the collective modes to those whose energy is such that the corresponding wave function vanishes identically at the boundary of the box. Since the asymptotic behavior of $H_1(\mathbf{r})$ and $S(\mathbf{r}, \mathbf{r}')$ is known, we can now increase the sampling box size and solve in a larger box the generalized Feynman relation Eq. (4.22). Thus we obtain a discrete, but arbitrarily dense, spectrum. The

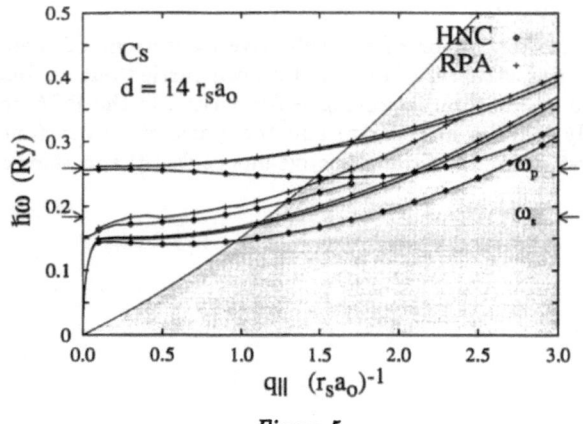

Figure 5

The dispersion relation of the three well-defined collective modes in Cs is shown in HNC approximation (diamonds), RPA (crosses), and LDA (unmarked lines). The arrows on the left and the right margin indicate the classical frequencies of the bulk (ω_p) and the surface (ω_s) plasmon, the shaded region is the particle-hole continuum, where the excitations are Landau damped.

dynamic structure function is then obtained in the form

$$S(\mathbf{r},\mathbf{r}';\omega) = \frac{1}{\Delta\omega}\sqrt{\rho_1(\mathbf{r})}[H_1\psi_\omega(\mathbf{r})][H_1\Psi_\omega(\mathbf{r}')]\sqrt{\rho_1(\mathbf{r}')} \tag{5.1}$$

where $\Delta\omega$ is the central difference between two adjacent energy levels, and the $\psi_\omega(\mathbf{r})$ are normalized by

$$(\Psi_\omega|H_1|\Psi_{\omega'}) = \hbar\omega\delta_{\omega,\omega'} . \tag{5.2}$$

In our geometry, the density depends only on the coordinate z perpendicular to the slab, *i.e.* $\rho_1(\mathbf{r}) = \rho(z)$. In this situation, the dynamic structure function for grazing angles

$$S(q,\omega) = |\delta\rho_\omega(q_\parallel)|^2 \tag{5.3}$$

is of special interest. Here, we have used

$$\delta\rho_\omega(q_\parallel) = \frac{\hbar^2 q^2}{2m\Delta\omega(q_\parallel)}\int d^3r\sqrt{\rho(z)}\Psi_\omega(\mathbf{r})e^{i\mathbf{q}_\parallel\cdot\mathbf{r}_\parallel} . \tag{5.4}$$

We have studied three typical cases: symmetric slabs of Al, Na and Cs, each of width $d = 14 a_0 r_s$. It should be noted that all energies considered here are well above the work function, in other words all spectra are continuous. However, we will see that many of the modes are characterized by very sharp peaks in the dynamic structure function and can be identified quite clearly.

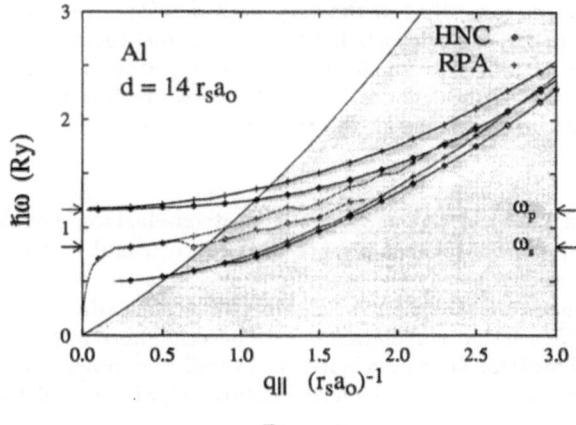

Figure 6

Same as Fig. 5 but for aluminum. The predictions of the LDA are almost indistinguishable from the RPA and are not shown.

Of course, many-particle correlations have the most pronounced effect on the low-density system Cs. Therefore we discuss this case first. Fig. 4 depicts two typical spectra and corresponding transition densities; the spectra at other wavelengths are similar but with different relative weights. Even though all excitations are in the continuum, the sharp maxima are remarkable.

Common to all of our spectra is that we can identify *two* surface modes and one volume mode. It is, of course, tempting to identify the volume mode with the bulk plasmon, $\omega \approx \omega_{pl}$ and one of the surface modes with the surface plasmon, $\omega_s \approx \omega_{pl}/\sqrt{2}$. Of course, at very long wavelengths the finite extension of our film in the z-direction should become visible — the dispersion relation becoming $\omega \sim \sqrt{q_{\parallel}}$. Our identification becomes clearer by examining the dispersion relations of these three modes, which we obtain by connecting the peaks in the spectrum as a function of momentum. Fig. 5 shows the these dispersion relations for cesium; we also include in this figure the predictions of the RPA and the LDA dispersion relations.

We see that the volume mode and one of the surface modes approach the predicted long-wavelength limits, it is therefore legitimate to identify these modes with the bulk and the surface plasmon. It is interesting to note that, at intermediate wavelengths, the surface plasmon merges with the bulk plasmon, thus a clean distinction between the two is not possible at shorter wavelengths. On the other hand, the second surface mode is quite visible throughout the spectrum.

Fig. 5 shows also a comparison with the RPA and, as far as possible, with the LDA. As expected, the RPA starts to fail at intermediate wavelengths. It is interesting to observe that the LDA provides hardly any improvement over the RPA; apparently a δ-function interaction is *not* a good approximation for the correlation effects shown in Figs. 2 and 3.

The picture changes somewhat in the very high density Al slabs, see Fig. 6. The RPA becomes exact in the high-density limit, $r_s \to 0$, and therefore is a much better approximation. The Al films are much narrower on an atomic length scale than the Cs films. Therefore the surface modes in the Al slabs are less well defined. This is indicated by the larger uncertainties in the surface plasmon.

6. SUMMARY

We have studied the local screening of the electron-electron correlations in "ideal metals". Apparently, the Shore-Rose picture provides a useful and simple method to stabilize the jellium, and to justify the study of excitations. At the level of the present work, we have confined our attention to the very basic ingredients, *i.e.* the dynamic structure function. We see that correlation effects become more apparent the closer one looks at the physical system. Or equivalently, the predictive power of local density approximations becomes less impressive — notwithstanding wide use of LDA in electronic structure calculations.

The dynamic structure function is, of course, only the first step when making contact to experiments. We are presently developing a theory for inelastic scattering of electrons from a metallic or thin-film surface, in the limit where the surface is translation-invariant. Employing the distorted-wave first-Born approximation applied to the screened Coulomb interaction, the transition rate for inelastic scattering of electrons may be obtained. This calculation involves the dynamic structure factor as well as the surface-distorted incident and scattered wave-functions, and the dynamically screened Coulomb interaction. Calculations are underway of the electron-energy-loss spectrum, which have been measured recently by Plummer and co-workers.[23]

ACKNOWLEDGEMENT

This work was supported, in part, by the National Science Foundation grants PHY-8806265 and PHY-9108066 and the Texas Advanced Research Program under Grant 010366-012. Stimulating discussions and correspondence with R. M. Martin, J. P. Perdew, H. B. Shore, and W. Plummer are gratefully acknowledged.

REFERENCES

1. P. Hohenberg and W. Kohn, Phys. Rev. **136B**, 864 (1964).
2. W. Kohn and L. J. Sham, Phys. Rev. **140A**, 1133 (1965).
3. N. D. Lang and W. Kohn, Phys. Rev. **B1**, 4555 (1970).
4. W. Kohn and P. Vashishta, in *Theory of the Inhomogeneous Electron Gas*, ed. by S. Lundqvist and N. H. March (Plenum, New York, 1983).
5. E. Krotscheck, Phys. Rev. **B31**, 4267 (1985).
6. E. Krotscheck, W. Kohn, and G.-X. Qian, Phys. Rev. **B32**, 5693 (1985).
7. E. Feenberg, *Theory of Quantum Fluids* (Academic, New York, 1969).
8. L. J. Lantto, Phys. Rev. **B 22**, 1380 (1980).
9. J. G. Zabolitzky, Phys. Rev. **B 22**, 2353 (1980).
10. E. Krotscheck, Annals of Physics (New York) **155**, 1 (1984).
11. D. M. Ceperley and B. J. Alder, *Phys. Rev. Lett.* **45**, 566 (1980).
12. Tao Pang, C. E. Campbell, and E. Krotscheck Chem. Phys. Lett. **163**, 537 (1989).
13. E. Krotscheck and W. Kohn, Phys. Rev. Lett. **57**, 862 (1986).
14. T. L. Ainsworth and E. Krotscheck, Phys. Rev. **B45**, 8779 (1992).
15. W. R. Tyson and W. A. Miller, Surf. Sci. **62**, 267 (1977).
16. X.-P. Li, R. J. Needs, Richard M. Martin and D. M. Ceperley, Phys. Rev. **B45**, 6124 (1992).
17. D. C. Langreth and M. J. Mehl, Phys. Rev. **B28**, 1809 (1983).
18. H. B. Shore and J. H. Rose, Phys. Rev. Lett. **66**, 2519 (1991).

19. H. B. Shore and J. H. Rose, Phys. Rev. **B43**, 11605 (1991).
20. J. C. Soler, Phys. Rev. Lett. **67**, 3044 (1991); H. B. Shore and J. H. Rose, Phys. Rev. Lett. **67**, 3045 (1991).
21. P. Fulde, *Electron Correlation in Molecules and Solids* Springer, Berlin 1991.
22. R. P. Feynman, Phys. Rev. **95**, 262 (1954).
23. J. A. Gaspar, A. G. Eguiluz, K.-D. Tsuei and E. W. Plummer, Phys. Rev. Lett. **67**, 2854 (1991).

DYNAMIC MECHANISMS OF DISORDERLY GROWTH

Dedicated to the people of Puerto Rico and their music

H.E. Stanley,* S. Havlin,*+ J. Lee,* and S. Schwarzer*

*Center for Polymer Studies and Department of Physics
Boston University, Boston, MA 02215 USA

+Department of Physics
Bar-Ilan University, Ramat-Gan Israel

The purpose of this talk is to present a brief overview of our group's recent research into dynamic mechanisms of disorderly growth, an exciting new branch of condensed matter physics in which the methods and concepts of modern statistical mechanics are proving to be useful. Our strategy has been to focus on attempting to understand a single model system—diffusion limited aggregation (DLA). This philosophy was the guiding principle for years of research in phase transitions and critical phenomena. For example, by focusing on the Ising model, slow progress was made; this progress eventually led to understanding a wide range of critical point phenomena, since even systems for which the Ising model was not appropriate turned out to be described by variants of the Ising model (such as the XY and Heisenberg models). So also, we are optimistic that whatever we may learn in trying to "understand" DLA will lead to generic information helpful in understanding general aspects of dynamic mechanisms underlying disorderly growth.

At the outset, I would like to acknowledge my indebtedness to my collaborators, who have kindly consented to join me as co-authors on this progress report, and to the organizers for having invited me to return to a land whose people and especially thier music are so wonderful.

1. GROWTH PROBABILITIES: SIMULATIONS OF DLA CLUSTERS

Like the Ising model, the rule defining DLA is simple.[1] At time 1, we place in the center of a computer screen a white pixel, and release a random walk from a large circle surrounding the white pixel. The four perimeter sites have an equal *a priori* probability p_i to be stepped on by the random walk. Accordingly, we write

$$p_i = \frac{1}{4} \qquad (i = 1, \ldots, 4). \tag{1a}$$

The rule is that when the random walker steps on a perimeter site, it sticks irreversibly. This forms a cluster of mass $M = 2$, with $N_p = 6$ perimeter sites, henceforth called *growth sites*. Now the probabilities are *not* all identical: each of the growth sites of the two tips has growth probability $p_{max} \cong 0.22$, while each of the four growth sites on the sides has growth probability $p_{min} \cong 0.14$.

Just because the third particle is *more likely* to stick at the tip does not mean that the next particle *will* stick on the tip. Indeed, the most that one can say about the cluster is to specify the *growth site probability distribution* (GSPD)—i.e., the set of numbers,

$$\{p_i\} \qquad i = 1 \ldots N_p, \tag{1b}$$

where p_i is the probability that perimeter site ("growth site") i is the next to grow, and N_p is the total number of perimeter sites ($N_p = 4, 6$ for the cases $M = 1, 2$). The recognition that the set of $\{p_i\}$ gives us essentially the *maximum* amount of information we can have about the system is connected to the fact that tremendous attention has been paid to these p_i.

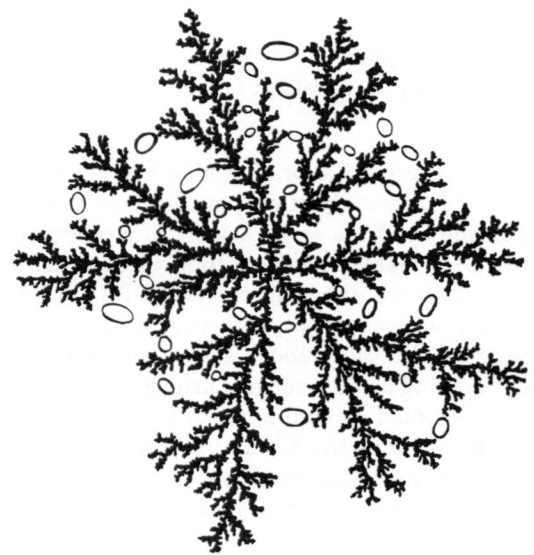

Fig. 1. Off-lattice DLA cluster of 10^5 sites, indicating some of the "necks" that serve to delineate "voids." After Ref. 17.

If the DLA growth rule is iterated 10^5 times, then we obtain a large cluster (Fig. 1) characterized by a range of growth probabilities that spans several orders of magnitude—from the tips to the fjords—so that histograms of these growth probabilities typically use a logarithmic scale for the abscissa (Fig. 2).

Diffusion limited aggregation (DLA) has become important for describing a wealth of diverse physical and chemical phenomena.[2] Recently, several phenomena of *biological* interest have also attracted the attention of DLA *aficionados*. These include the growth of bacterial colonies,[3] the retinal vasculature,[4] and neuronal outgrowth.[5] The last example is particularly intriguing since if evolution chose DLA as the morphology for the nerve cell, then perhaps we can understand "why" this choice was made. What evolutionary advantage does a DLA morphology convey? Can we use the answer to this question to better design the next generation of computers? Already we appreciate that a fractal object is the most efficient way to obtain a great deal of

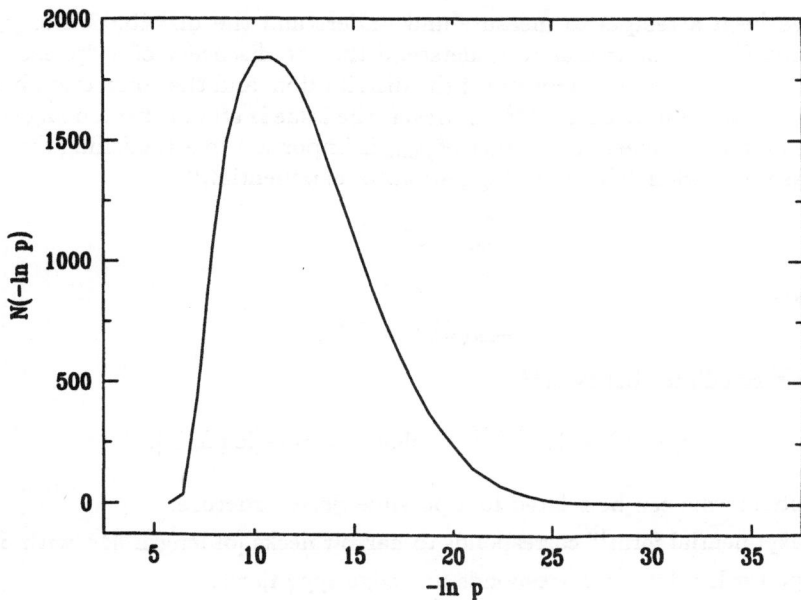

Fig. 2. $N(-\ln p)$ for 3D DLA. $N(-\ln p)d\log p$ is the average number of growth sites with growth probabilities p' such that $-\ln p < -\ln p' < -\ln p + d\ln p$. Here, we display data for 50 off-lattice clusters of mass 15 015.

intercell "connectivity" with a minimum of "cell volume," so one immediate question is "which" fractal did evolution select, and why?

We will save time and space by resisting the temptation at this point to "pull out the family photo album" to show lots of realizations. Instead, we may refer the interested reader (and their non-specialist colleagues) to the album *Les Formes Fractales* (and its English translation *Fractal Forms*) prepared in connection with a hands-on exhibition of the same name that is currently on view at the *Palais de la Découverte* in Paris.[6]

As with many models in statistical mechanics, the theoretical challenge is as important as the experimental realizations in "hooking" theorists. And as with many statistical mechanical models, the "defining rule" in DLA is simple even though the "consequences of that rule" are extremely rich. Understanding how such a rich consequence can follow from such a simple rule is indeed an irresistible challenge.

In the case of DLA, this challenge is enhanced by the fact that—unlike other models with simple rules (such as the Ising model)—in DLA there is no Boltzmann factor so we can more easily explain and understand since one does not have to know any physics beforehand. Indeed, it initially surprises almost everyone who sees DLA develop in real time on a computer screen that a complex outcome (at the *global* level of a "form") seems to bear no obvious relation to the details of the simple rule that produced this form—at least, I have known no one who predicted exactly the "form" of DLA from knowledge only of the rule.

Despite some progress,[7-14] no genuine understanding of DLA has emerged.

There have been attempts to measure and understand the distribution of p_i, and its lower cutoff p_{min}, as well as to understand the the discovery of a "phase transition" in the behavior of the moments of the distribution, and the connection between multifractality and multiscaling.[8−14] However, the issue is still far from being settled.

For example, the mass dependence of p_{min} is important in establishing the nature of the phase transition. There are suggestions of exponential,[10]

$$p_{min} \sim e^{-AM^x}, \tag{2a}$$

power law[11]

$$p_{min}(L) \sim M^{-\alpha}, \tag{2b}$$

and an "intermediate" behavior[12]

$$p_{min}(L) \sim M^{-\log M} \qquad [\log p_{min} \sim -(\log M)^2]. \tag{2c}$$

Each of these forms can be related to a possible fjord structure:

(a) The exponential form[10] corresponds to narrow necks [of length M^β with $\beta > 0$],

(b) The power law form[11] corresponds to wedge type fjords.

(c) The "intermediate" behavior[12] can be explained in terms of a structural model of DLA, which has self-similar *voids* connected by *necks*, as explained in the next section.[12,13]

The scaling form of the complete growth site probability distribution $\{p_i\}$ is also of interest. Trunfio and Alstrøm,[10] Mandelbrot and Evertsz[10] and Schwarzer *et al*[12] proposed different types of possible behavior.

2. THE "VOID-NECK" MODEL OF DLA STRUCTURE

Our own group's numerical results (e.g., Fig. 3) support possibility (c), and may provide a clue for the underlying puzzle of understanding DLA structure.[12] Specifically, we have proposed a "void-neck" model of DLA[12] in order to explain the result (2c). The void-neck model states that each fjord is characterized by a hierarchy of voids separated from each other by narrow "necks" or "gateways." The key feature of the model are: (i) The distribution of voids must be *self-similar*, (ii) The voids are separated by necks ("channels," or "gateways"): a random walker can pass from one void to the next only by passing through a gateway.

What is the evidence supporting the void-neck model of DLA growth dynamics?

(1) First, we note that if necks "dominate", then (2a) would have to be satisfied. The numerics rule this out.

(2) Second, we note that if self-similar voids dominate, then (2b) would have to be satisfied. Again, the numerics rule this out.

(3) Photos of large DLA clusters reveal the presence of such voids and necks [Fig. 1]. Moreover, when the DLA mass is doubled, we find that outer branches "grow together" to form new necks (enclosing larger and larger voids).

(4) The void-neck model can be *solved* under the approximation that the voids are strictly self-similar and the gates are narrow. The solution demonstrates that $\log p_{min} \propto (\log M)^2$.

(5) The void-neck model is consistent with a recent calculation[14] suggesting that DLA structures can be partitioned into two zones:

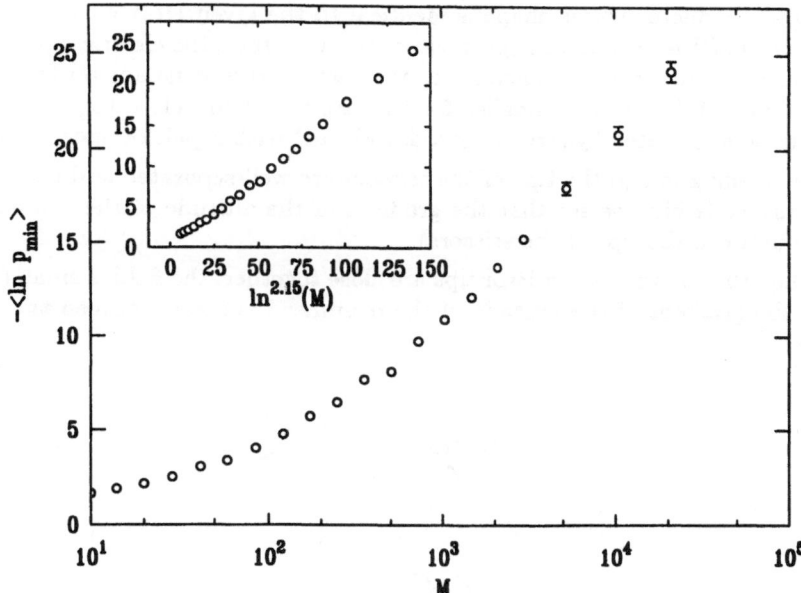

Fig. 3. Quenched average $-\langle \ln p_{\min} \rangle$ of the minimum growth probability $p_{\min}$ in 19 2D off-lattice DLA clusters as a function of the cluster mass M, where $10 < M < 21\,000$. Note the upward curvature in the log-log representation, which indicates that $p_{\min}$ decays faster than any power-law as function of M. The inset shows the same quantity, but plotted vs. $(\ln M)^{2.15}$; $p_{\min}$ displays straight-line behavior from $M \simeq 90$ up to the maximum $M = 21\,000$, so the functional dependence of $p_{\min}$ on M can be written as $\ln p_{\min} \sim \ln^{2.15} M$. The error on the exponent 2.15 is about ± 0.2.

(a) An inner *finished zone*, typically with $r \leq R_g$ (where R_g is the radius of gyration), for which the growth is essentially "finished" in the sense that it is overwhelmingly improbable that future growth will take place.

(b) An outer *unfinished zone* (typically $r \geq R_g$) in which the growth is unfinished.

Thus future growth will almost certainly take place in the region $r > R_g$. Now $2R_g \approx \frac{1}{2} L$, where L is the spanning diameter. Hence only about 1/4 the total "projected area" of DLA is finished, the rest of the DLA being *unfinished*. We suggest that the finished region will be created from the unfinished region by tips in the unfinished region growing into juxtaposition (thereby forming voids). The growth of DLA is fixed by the growth probabilities, which are of course largest on the tips.

Indeed, two tips will grow closer and closer until their growth probabilities become so small that no further narrowing will occur. This observed phenomenon can be perhaps better understood if one notes that the growth probabilities $\{p_i\}$ of a given DLA cluster are identical to normalized values of the electric field $\{E_i\}$ on the surface of a charged conductor whose shape is identical to the given DLA cluster. Thus as two arms of the DLA "conductor" grow closer to each other, the electric field at their surface must become smaller (since $E_i \propto \nabla \phi_i$, where $\phi \equiv$ constant on the surface of the conductor). That E_i is smaller for two arms that are close together can be graphically demonstrated by stretching a drumhead* with a pair of open scissors.

(1) If the opening is big, the tips of the scissors are well-separated and the field on the surface is big (we see that the gradient of the altitude of the drumhead is large between the tips of the scissors).

(2) On the other hand, if the scissor tips are close together, the field is small (we see that the gradient of the altitude of the drumhead is small between the scissor tips).

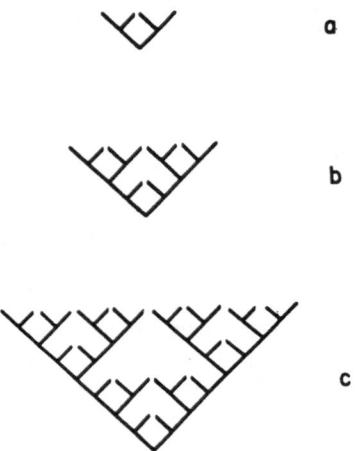

Fig. 4. Construction of a deterministic hierarchical model for DLA. (a) The generator and the first generation of the model; (b) the second generation; and (c) the third generation (after Ref. 13).

3. EXACTLY-SOLVABLE DETERMINISTIC MODEL OF DLA

Lee et al.[13] developed a hierarchy of deterministic fractal models designed to capture some of the essentials in the structure of DLA. These models, whose key ingredients are narrow necks and self-similar voids, are generalizations of the model presented in Refs. 12 and 13. *We find an analytic solution of the growth site probability distributions for the entire family of models.* This distribution is found to have the same form as that of DLA clusters recently measured by Schwarzer et al.[12], $n(\alpha, M) \sim \exp[-A\alpha^\gamma (\ln M)^{-\delta}]$, where $n(\alpha, M)d\alpha$ is the number of growth sites with $\alpha < -\ln p_i / \ln M < \alpha + d\alpha$. The two exponents ($\gamma = 2 \pm 0.3$, $\delta = 1.3 \pm 0.3$) found numerically by Schwarzer et al.[12] to characterize the distribution, are found analytically by Lee et al.[13] to be 2 and 1 respectively. It is possible that the form of the distribution and the exponents are determined *only* by the presence of the necks and

---

* A convenient drumhead is obtained by stretching panty hose across a circular sewing hoop (R. Selinger, private communication)

self-similar voids, independent of further details of models. The agreement between the distribution (and its exponents) of DLA and the models provides further support for the void-neck description of the structure of DLA.

The model is defined as follows. The first generation [Fig. 4(a)], consists of three wedges, is the generator of the model. In order to get the next generation, we replace every wedge in the first generation with the generator (Fig. 4(b)). The third generation is obtained by replacing every wedge in the second generation with the generator [Fig. 4(c)]. In general, one can obtain generation n by replacing all the wedges generation $n - 1$ with the generator.

In order to obtain the growth site probability distribution $n(\alpha, M)$, Lee *et al.*[13] expand the distribution, using the Cauchy identity, in terms of Gauss polynomials. Each distribution for one Gauss polynomial "marginally" overlaps with all the other terms. This fact permits the re-summation of the expansion to get a closed form for $n(\alpha, M)$. Due to the "marginal" overlap, one cannot obtain the exact amplitudes for the distribution, but we find approximate values, which are in good agreement with exact numerical data.

One interesting point to emerge is that the distribution is the same (except amplitudes) for the entire family of models studied in Ref. 13. Since the common ingredient for the entire family is a hierarchy of self-similar "voids" separated by narrow necks, it is possible that the form of $n(\alpha, M)$ obtained here is just a consequence of the void-neck feature, and is independent of further details of a model.

4. SCALING PROPERTIES OF THE PERIMETER OF DLA: THE "GLOVE" ALGORITHM

Almost all the sites of a fractal lie on its surface. This simple observation explains why fractals are of great importance in a wide range of disciplines. In biology, matter exchange takes place across membranes and often requires large contact areas of the participating systems: oxygen diffuses into the blood in lung tissue and trees absorb nutrients through their widely branched root network. In chemistry, reaction rates depend on the surface that the reacting species expose to each other; the surface of a catalyst plays a central role in catalytically controlled reactions; important for applications is also the use of porous media as electrodes for batteries.

Very recently, Schwarzer *et al.*[15] introduced a "glove" algorithm and used it to carry out a systematic study of various properties of the DLA and percolation perimeters. They developed an algorithm—the "glove" method—which can be applied to study topological properties of any fractal (or self-affine) surface. In particular, the glove method can be used to determine:

(i) the total perimeter of a fractal, the set of all nearest neighbor sites of the fractal, and a generalization thereof to neighboring sites of higher order — here we find a scaling relation which also suggests a novel method for the determination of the fractal dimension of an object;

(ii) the accessible perimeter of a fractal, which is the set of the perimeter sites that can be reached from the exterior of the object, and a generalization thereof to neighbor sites of higher order — this quantity has been studied experimentally on porous media and fresh fractures, and theoretically on percolation clusters.

(iii) the "lagoon"-size distribution, where "lagoons" are generalizations of the notion of voids to the case of *loopless* fractals and describe the regions of a fractal inaccessible to probes with a given particle size. The glove algorithm also enables us to identify unambiguously "necks" in a fractal structure.

Determination of the ℓ-Perimeter

We begin by representing the investigated object in discretized form on a lattice and label its sites with the index 0 [black sites in Fig. 5]. In the first step, we find all the nearest-neighbor sites of the object and label them $\ell = 1$, as shown in Fig. 5a. Those sites that are nearest neighbors of sites with $\ell = 1$ and not already labeled are identified as $\ell = 2$ sites. We repeat the procedure and obtain ℓ values for all sites surrounding the object [see Fig.5a]. The number ℓ associated with every lattice site is called the topological distance of the site to the object. We will use the term "ℓ-perimeter" to refer to the set of sites with label ℓ. Schwarzer et al.[15] argued that $\Pi(M, \ell)$, the number of sites of the ℓ-perimeter, should obey a scaling law of the form $\Pi(M, \ell)/\ell \sim f(\ell/M^{1/d_f})$, where $f(u) \sim u^{-d_f}$ for $u \to 0$ and $f(u) \to const$ for $u \to \infty$. Simulations of 21 2D off-lattice DLA clusters (and also 50 2D percolation hulls) support this relation.

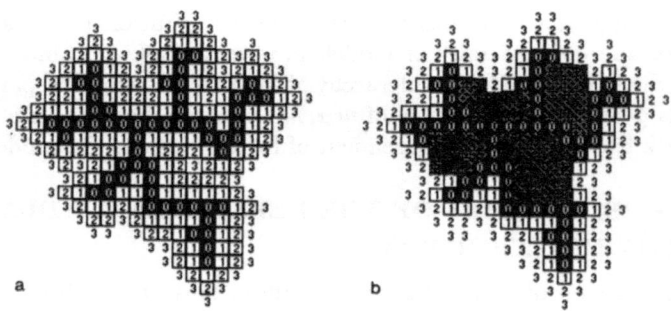

Fig. 5. (a) To illustrate the construction of the ℓ-perimeter we show the object and the ℓ-perimeters with $1 \le \ell \le 3$. Sites belonging to the investigated object are displayed in black and bear the label 0. Other numbers denote the order of the ℓ-perimeter. (b) Lagoons are constructed by placing flexible gloves, one lattice unit thick, on the object. The gloves cannot penetrate through narrow openings. In our example, the 2-glove has "sealed" all openings, and we identify the enclosed space as lagoons (grey sites). After Ref. 15.

The ℓ-Gloves

Next we describe the procedure to determine the "ℓ-gloves" of the object. In the first step, instead of labeling all the neighbor sites, we place a flexible "glove," one lattice unit thick, on the object. In general, since gloves cannot penetrate through the narrowest openings (less or equal to $\sqrt{2}$ lattice constants wide), they cannot cover the object completely. The second glove is placed on the union of object and 1-glove. We iterate the covering process to obtain ℓ-gloves up to any desired order. Fig. 5(b) illustrates the glove algorithm by showing the gloves of order $\ell = 1, 2, 3$ for a small DLA cluster. Note that the 1-glove penetrates the "fjords" of the object but the 2-glove cannot. Unlike the ℓ-perimeter, the ℓ-glove comprises only those sites that can be reached from the exterior without stepping on any other site of the ℓ-perimeter, and so forms a connected subset of the ℓ-perimeter. Like the ℓ-perimeter, the subsequent gloves explore fewer and fewer details of the surface of the object. The

ℓ-perimeter and the ℓ-glove are identical for large ℓ (greater than the largest "neck" width of the object; see below).

Necks and Lagoons

Significantly, the glove algorithm can be used to extend to the case of loopless fractals the notion of voids as empty spaces in multiply connected fractals. Imagine, e.g., a circle with a small opening of width $w = 2\ell_o$, a simple example of a loopless object. Cover the surface with gloves, one after the other. When the number of gloves reaches ℓ_o, the glove cannot penetrate into the opening. We denote by "lagoon" the set of points left in the interior — now inaccessible from the exterior. The number of enclosed sites is the lagoon size s. The sites, where glove ℓ_0 touches itself, identify a "neck." Note that there exists a one-to-one correspondence of lagoons and necks, so that each lagoon has a unique neck width w given by $w = 2\ell_o$. For the object in Fig. 5b, all the necks of lagoons have width $w = 4$. Schwarzer et al.[15] found that the lagoon size distribution in DLA is consistent with a self-similar structure of the aggregate, but that even for large lagoons the most probable width of the necks that separate the lagoons from the exterior is very small.

5. MULTIFRACTAL SCALING OF 3D DLA

In a recent work,[16] the multifractal spectrum of the growth probability of 3D off-lattice DLA was studied. The results indicate that, in contrast to 2D DLA, there appears to be *no* phase transition in the multifractal spectrum. Why? In both 2D and 3D, "necks" are created by side branches in DLA that grow closer and closer until their growth probabilities become so small that no further narrowing occurs. However, in 3D, even if there are points where tips from different branches of the aggregate come close or meet, there is no significant screening of growth due to this configuration, because no volume is cut off from the exterior and particles can enter the cluster from a direction perpendicular to the loop. Simply stated, one cannot cut off a volume with branches in the same way one can cut off an area. Thus we interpret the apparent absence of a phase transition for 3D as the effect of the topological differences between 2 and 3 dimensions. We further note that as d increases, d_f becomes closer to $d-1$; the higher d is, the less dense the clusters are, since $\rho(R) \sim R^{d_f - d}$. Thus it is tempting to conjecture that $d = 2$ is a "lower critical dimension" in the sense that there is a phase transition for $d = 2$ but power-law scaling for all $d > 2$.

In sum, for 3D, even when tips from different branches are close, there is no significant screening of growth, since particles can enter from directions perpendicular to the loop, suggesting a power-law dependence of $p_{\min}$ on the mass M of the cluster. Thus the apparent absence of a phase transition in 3D DLA can be interpreted as due to the topological difference between 2 and 3 dimensions. We have calculated the $\{p_i\}$ for 50 off-lattice 3D DLA clusters, and compared our analysis to the 2D case which is believed to undergo a phase transition. We find the 3D case is quite different. Specifically, we find

(i) the local slopes $\tau(q, M) \equiv \partial \ln Z / \partial \ln M$ do not diverge for $q < 0$ as they do in 2D (here Z denotes the qth moment of the distribution $\{p_i\}$),

(ii) The Legendre transform function $f_L(\alpha) \equiv q\alpha - \tau$ (where $\alpha \equiv \partial \tau / \partial q$) has no systematic mass dependence, as it has in 2D, and

(iii) $p_{\min}$ has a power-law singularity in M, following Eq. (2b), in contrast to the 2D case, where $p_{\min}$ vanishes much faster, according to (2c).

6. SUMMARY

In summary, we have (i) one "firm" numerical result, $\log p_{min} \sim (\log M)^2$, given by Eq. (2c). We have also (ii) an analytic argument that this behavior follows from a void-neck model of DLA structure in which there exist self-similar voids separated by necks whose width does not scale. We also have (iii) a plausibility argument that the tips of DLA grow together until they are separated by a distance which is typically a few pixels, as well as visual evidence supporting this picture, and (iv) some understanding of why 3D is different than 2D in terms of the inability of the necks in 3D to cut off a volume of space.

Acknowledgements

We are grateful to A. Bunde and H. E. Roman for collaboration on the key formative stages of this project in 2D (which are reviewed in Ref. 18), to M. Wolf and P. Meakin for collaboration on the 3D extensions, and to A. Aharony, P. Alstrøm, A.-L. Barabási, R. Blumenfeld, A. Coniglio, T. C. Halsey, A. B. Harris, G. Huber, D. Stauffer, P. Trunfio, and T. Vicsek for helpful discussions and to ONR and NSF for financial support.

REFERENCES

1. T. A. Witten and L. Sander, Phys. Rev. Lett. **47**, 1400 (1981).

2. P. Meakin, in *Phase Transitions and Critical Phenomena*, eds. C. Domb and J. L. Lebowitz (Academic, Orlando, 1988), Vol. 12; J. Feder, *Fractals* (Pergamon, New York, 1988); H. E. Stanley and N. Ostrowsky, eds. *Random Fluctuations and Pattern Growth: Experiments and Models* (Kluwer Academic Publishers, Dordrecht, 1988); T. Vicsek, *Fractal Growth Phenomena* (World, Singapore, 1989); H. E. Stanley and N. Ostrowsky, eds. *Correlations and Connectivity: Geometric Aspects of Physics, Chemistry and Biology* [Proceedings 1990 Cargèse NATO ASI, Series E: Applied Sciences, Vol. 188] (Kluwer, Dordrecht 1990); *Fractals and Disordered Systems*, eds. A. Bunde and S. Havlin (Springer, Heidelberg 1991).

3. H. Fujikawa and M. Matsushita, J. Phys. Soc. Japan **58**, 3875 (1989).

4. F. Family, B. R. Masters, and D. E. Platt, Physica D **38**, 98 (1989).

5. F. Caserta, H. E. Stanley, W. Eldred, G. Daccord, R. Hausman, and J. Nittmann, Phys. Rev. Lett. **64**. 95 (1990); see also the brief report H. E. Stanley, F. Caserta, W. Eldred, G. Daccord, R. Hausman, and J. Nittmann, Bull. Am. Phys. Soc. **34**, 716 (1989).

6. For photographs of natural systems described by DLA, see Plates 9–12, 15, 20–24, 41–45 and 50 of E. Guyon and H. E. Stanley: *Les Formes Fractales* (Palais de la Découverte, Paris, 1991) [English Translation: *Fractal Forms* (Elsevier, Amsterdam 1991)]. See also Plates 3, 7, 8 and 11 of D. Stauffer and H. E. Stanley, *From Newton to Mandelbrot: A Primer in Modern Theoretical Physics* (Springer Verlag, Heidelberg, 1990). For a color-coded map of the $\{p_i\}$ for large DLA clusters in both circular and strip geometry, see B. B. Mandelbrot and C. J. G. Evertsz, Nature **348**, 143 (1990).

7. See, e.g., L. A. Turkevich and H. Scher, Phys. Rev. Lett. **55**, 1026 (1985); A. Coniglio, in *On Growth and Form: Fractal and Non-Fractal Patterns in Physics*, eds H. E. Stanley and N. Ostrowsky (Nijhoff, Dordrecht, 1985), p. 101; G. Parisi and Y. C. Zhang, J. Stat. Phys. **41**, 1 (1985); Y. Hayakawa, S. Sato and M.

Matsushita, Phys. Rev. A **36**, 1963 (1987); T. C. Halsey, Phys. Rev. Lett. **59**, 2067 (1987); L. Pietronero, A. Erzan, and C. J. G. Evertsz Phys. Rev. Lett. **61**, 861 (1988).

8. P. Meakin, H. E. Stanley, A. Coniglio and T. A. Witten, Phys. Rev. A **32**, 2364 (1985); T. C. Halsey, P. Meakin and I. Procaccia, Phys. Rev. Lett. **56**, 854 (1986); C. Amitrano, A. Coniglio and F. di Liberto, Phys. Rev. Lett. **57**, 1016 (1986); P. Meakin, A. Coniglio, H. E. Stanley, and T. A. Witten, Phys. Rev. A **34**, 3325 (1986).

9. J. Lee and H. E. Stanley, Phys. Rev. Lett. **61**, 2945 (1988); J. Lee, P. Alstrøm, and H. E. Stanley, Phys. Rev. A **39**, 6545 (1989); B. Fourcade and A. M. S. Tremblay, Phys. Rev. Lett. **64**, 1842 (1990).

10. R. Blumenfeld and A. Aharony, Phys. Rev. Lett. **62**, 2977 (1989); P. Trunfio and P. Alstrøm, Phys. Rev. B **41**, 896 (1990); B. Mandelbrot and C. J. G. Evertsz, Physica A **177**, 386 (1991); C. J. G. Evertsz, P. W. Jones and B. B. Mandelbrot, J. Phys. A **24**, 1889 (1991); B. B. Mandelbrot, Physica A **168**, 95 (1990); B. B. Mandelbrot, C. J. G. Evertsz and Y. Hayakawa, Phys. Rev. A **42**, 4528 (1990); C. J. G. Evertsz and B. B. Mandelbrot, Physica A **185**, 77 (1992); C. J. G. Evertsz and B. B. Mandelbrot, J. Phys. A **25**, 1981 (1992); C. J. G. Evertsz, B. B. Mandelbrot and L. Woog, Phys. Rev. A **45**, 5798 (1992).

11. A. B. Harris and M. Cohen, Phys. Rev. A **41**, 971 (1990); A. L. Barabási and T. Vicsek, J. Phys. A **23**, L729 (1990); R. Ball and R. Blumenfeld, Phys. Rev. A **44**, 828 (1991).

12. (a) S. Schwarzer, J. Lee, A. Bunde, S. Havlin, H. E. Roman, and H. E. Stanley, Phys. Rev. Lett. **65**, 603 (1990); (b) S. Schwarzer, J. Lee, S. Havlin, H. E. Stanley, P. Meakin, Phys. Rev. A **43**, 1134 (1991). (c) See also the recent work of M. Wolf [Phys. Rev. A **43**, 5504 (1991); Phys. Rev. A **46**, xxx (1992)] which confirms many features of our 2D results.

13. J. Lee, S. Havlin, H. E. Stanley and J. E. Kiefer, Phys. Rev. A **42**, 4832 (1990); J. Lee, S. Havlin and H. E. Stanley, Phys. Rev. A **45**, 1035 (1992).

14. A. Coniglio and M. Zannetti Physica A **163**, 325 (1990); C. Amitrano, A. Coniglio, P. Meakin and M. Zannetti, Phys. Rev. B **44**, 4974 (1991); for a discussion of multifractality and multiscaling in terms of the localization of growth sites in DLA clusters, see J. Lee, A. Coniglio, S. Schwarzer and H. E. Stanley, "Localization of Growth Sites in DLA Clusters: Multifractality and Multiscaling" Nature (submitted).

15. S. Schwarzer, S. Havlin, and H. E. Stanley, "Scaling Properties of the Perimeter of Diffusion Limited Aggregation," Phys. Rev. A (submitted)

16. S. Schwarzer, M. Wolf, S. Havlin, P. Meakin and H. E. Stanley, Phys. Rev. A **46** R-3016 (1992); S. Schwarzer, S. Havlin and H. E. Stanley, "Multifractal scaling of 3D diffusion-limited aggregation" in *International Conference on Fractals and Disordered Systems* (Hamburg, Germany, 1992); Physica A **191**, xxx (1992).

17. C. Amitrano, P. Meakin and H. E. Stanley, Phys. Rev. A **40**, 1713 (1989).

18. H. E. Stanley, A. Bunde, S. Havlin, J. Lee, E. Roman, and S. Schwarzer, Physica A **168**, 23 (1990); S. Havlin, A. Bunde, E. Eisenberg, J. Lee, H. E. Roman, S. Schwarzer and H. E. Stanley, "Multifractal Fluctuations in the Dynamics of Disordered Systems," in *Proc. STATPHYS-18*, Berlin, August 1992; Physica A (in press).

PAIRING AND BCS THEORY IN AN EXACTLY-SOLUBLE MANY FERMION MODEL

C. Esebbag[1], M. de Llano[2a] and R.M. Carter (nee Quick)[2]

[1] Departamento de Física Teórica
Universidad Autónoma de Madrid
28049 Madrid
Spain
[2] Department of Physics
University of Pretoria
Pretoria 0002
South Africa

INTRODUCTION

We consider the exactly soluble one-dimensional (1D) fermion fluid [1] with pairwise attractive delta interactions for two main reasons. Firstly, in the case of two distinct fermion species (for example the two spin states of the electron) dynamical similarities exist between the present 1D model and 3D electron fluid jellium model. In particular for weak coupling the model reproduces the essential singularity familiar from standard 3D low-temperature superconductivity [2].

Secondly, the model displays a crossover transition from the strong-coupling extreme of tightly-bound, non-interacting local Bose pairs to the weak-coupling limit of large, overlapping Cooper pairs. Such crossover transitions are of interest in their own right and have also been seen in Fermi systems with attractive interactions, both in two [3] and three [4] dimensions and in the extended negative-U Hubbard model [5]. Notably for the Hubbard model, the weak-coupling limit of overlapping Cooper pairs is associated with conventional low-temperature superconductivity. On the other hand, the strong-coupling limit of an ideal Bose condensate of tightly bound electron pairs describes the formation of bipolarons on a lattice in high temperature superconductivity due to local on-site pairing. Furthermore, the observation that a correct theory of high-temperature superconductivity should interpolate between the Bose gas and Cooper pair regimes is suggested by recent experimental studies [6] on *all* "exotic" superconductors, whether 2D- or 3D-like.

This work falls naturally into two sections, namely the model at zero and at finite temperature. At zero temperature, it is known that Bardeen-Cooper-Schrieffer

(BCS)-like[7] theories reproduce both the Bose gas and the Cooper pair extremes correctly, not only in the present 1D model but also for the Fermi gas in 2 and 3D, and the extended negative-U Hubbard model [2, 3, 4, 5]. However one is now able to test BCS theory in the intermediate region since for the present model the exact results are also known [8].

In contrast, the prediction of the critical temperature for these systems is still an open question [3], particularly in the intermediate region. Certainly, for the 3D Fermi gas one can interpolate the finite temperature behaviour between the two limits if one goes beyond mean field theory [4]. Then in the Cooper pair or weak-coupling limit the critical temperature is the BCS critical temperature for pair breaking. In contrast, in the Bose gas or strong-coupling limit the critical temperature is the Bose-Einstein condensation temperature associated with the motion of the center of mass of the pairs (the ionisation of the bound pairs occurs at a far higher temperature). More relevant for the present work, however, is the 2D Fermi gas for which Bose condensation does not occur. Here the picture is considerably more complex [9] as will be outlined in the section on finite temperature behavior. Also given in the same section are the results of finite temperature BCS theory in the present 1D model. This can only yield the correct critical temperature in the weak coupling limit where the transition is due to Cooper-pair breaking. To clarify these issues future work will focus on the determination of the exact critical temperature.

In the next section we summarize the properties of the exactly- soluble many-fermion model; in the third section we present the zero temperature results including a comparison between the exact and BCS results; the subsequent section discusses the model at finite temperature and lastly the final section gives our conclusions.

THE MODEL HAMILTONIAN

Consider a system of $N(\gg 1)$ spin 1/2 fermions of mass m and degeneracy $\nu = 2$ in a box of length L interacting via a pairwise attractive delta interaction. The Hamiltonian for the system is then

$$H = -\frac{\hbar^2}{2m} \sum_{i=1}^{N} \frac{d^2}{dx_i^2} - v_0 \sum_{i<j} \delta(x_i - x_j). \tag{1}$$

If one defines dimensionless coordinates

$$x_i' = \rho x_i, \tag{2}$$

where $\rho = \frac{N}{L}$ is the density, one can write a dimensionless Hamiltonian as

$$H' \equiv mH/\hbar^2\rho^2 = -\frac{1}{2} \sum_{i=1}^{N} \frac{d^2}{dx_i'^2} - \lambda \sum_{i<j} \delta(x_i' - x_j'), \tag{3}$$

where

$$\lambda \equiv \frac{mv_0}{\hbar^2 \rho} \tag{4}$$

is a dimensionless coupling constant ($0 \leq \lambda < \infty$). This parameter then completely specifies the properties of the system. From the form for λ, one can see that high (low) particle density is associated with weak (strong) coupling, precisely as is the case for 3D jellium. The zero temperature properties of the system can then be calculated exactly as detailed below.

The exact ground state energy $E(N)$ can be numerically determined [2] as a function of λ using the Gaudin equations [8] and the results for the dimensionless ratio

$$\epsilon(\lambda) \equiv E(N)/|E_0(N)|, \tag{5}$$

where $E_0(N)$ is the energy evaluated at $\rho = 0$, are given in Figure 1. The variable λ^{-1} has a straightforward physical interpretation. Since the wave function for an isolated pair is $exp[-mv_0|x_1 - x_2|/2\hbar^2]$, $\lambda^{-1} \equiv \hbar^2\rho/mv_0$ is simply the ratio of the pair radius to the average interparticle spacing, $L/N \equiv 1/\rho$.

Given the pressure $P \equiv -\partial E/\partial L$, one can also determine the chemical potential (Gibbs free energy per particle at zero temperature), by

$$\mu = \frac{E(N) + PL}{N} = \frac{E(N)}{N} + \rho\frac{\partial[E(N)/N]}{\partial\rho}. \tag{6}$$

One already knows that the BCS approximation gives the exact results in the two extremes of λ [2]. Firstly, in the weak-coupling limit $\lambda \to 0$ ($\rho \to \infty$) the system comprises many weakly-interacting overlapping Cooper pairs and the energy is essentially that for an ideal Fermi gas,

$$E(N) = N\frac{1}{3}\frac{\hbar^2 k_F^2}{2m} = N\frac{\hbar^2\pi^2}{24m}\rho^2, \tag{7}$$

where $\rho = 2k_F/\pi$ was used. Furthermore in this limit the chemical potential (6) becomes identical to the Fermi energy

$$\mu = E_F \equiv \frac{\hbar^2 k_F^2}{2m} = \frac{\hbar^2\pi^2}{8m}\rho^2 > 0. \tag{8}$$

Secondly, in the strong-coupling limit $\lambda \to \infty$ ($\rho \to 0$) the system is an ideal Bose gas of $[N/2]$ tightly-bound local Bose pairs and the energy is given as

$$E(N) = E_0(N) = \lfloor N/2 \rfloor E_0(2). \tag{9}$$

Here $\lfloor N/2 \rfloor$ is the number of pairs, defined as the nearest integer less than or equal to $N/2$, and $E_0(2)$ is the binding energy of a single pair

$$E_o(2) \equiv -mv_0^2/4\hbar^2. \tag{10}$$

The chemical potential can then be calculated using (6), (9) and (10) and is equal to half the pair binding energy, namely

$$\mu = \frac{1}{2}E_0(2) < 0. \tag{11}$$

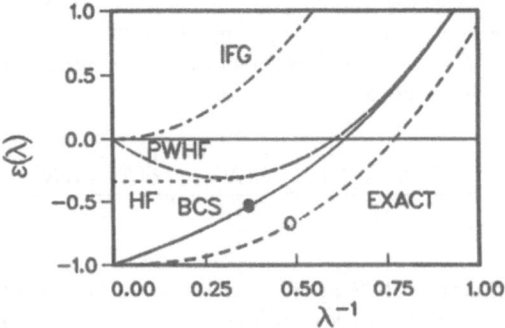

Figure 1. The dimensionless ground state energy per particle $\epsilon(\lambda)$ given exactly, given by BCS theory, given by the plane-wave Hartree-Fock (PWHF) approximation, given by the full Hartree-Fock (HF) approximation and the ideal Fermi gas (IFG) value, all as a function of the dimensionless parameter, $\lambda^{-1} = \frac{\hbar^2\rho}{mv_0}$. The open (closed) circle marks the crossover point at which the chemical potential changes sign for the exact (BCS) calculation, namely at $\lambda^{-1} \simeq 0.47(0.36)$.

Clearly the value of λ at which μ becomes negative can be identified as a *crossover point* in this model between the BCS regime and the Bose gas regime.

Experimentally, both for low- and high- temperature superconductors only *pairs* are important. Since the 1D model described above only permits "clusters" of at most *two* fermions, it is appropriate for testing 2D and 3D theories where such pairing is assumed.

ZERO TEMPERATURE RESULTS

Zero Temperature BCS Theory

We now present the BCS theory for $0 < \lambda < \infty$ in the 1D model at zero temperature. We follow the general derivation and notation of Fetter and Walecka [10] throughout. For (1) the relevant matrix elements are

$$V_{k_1 k_2 k_3 k_4} = \frac{v_0}{L} \delta_{k_1 + k_2, k_3 + k_4},$$ (12)

from which one obtains the Hartree-Fock single-particle energies as

$$\epsilon_k = \epsilon_k^0 - \frac{1}{2} \rho v_0,$$ (13)

where $\epsilon_k^0 = \frac{\hbar^2 k^2}{2m}$ is the kinetic energy. These energies can be measured relative to the chemical potential μ if one defines

$$\xi_k \equiv \epsilon_k - \mu.$$ (14)

The BCS gap parameter for (1) then becomes

$$\Delta = \frac{v_0}{L} \sum_k u_k v_k,$$ (15)

where the BCS transformation coefficients u_k and v_k are given by

$$v_k^2 = 1 - u_k^2 = \frac{1}{2} \left[1 - \frac{\xi_k}{E_k} \right],$$ (16)

and

$$E_k = \sqrt{\xi_k^2 + \Delta^2}$$ (17)

are the quasiparticle energies. Substituting these expressions into the definition (15) for the gap, and replacing the sum over k by an integral over the density of states, gives the BCS gap equation

$$1 = \frac{v_0}{2L} \sum_k \frac{1}{E_k} \to \frac{v_0}{4\pi} \int_{-\infty}^{+\infty} dk \, [\xi_k^2 + \Delta^2]^{-\frac{1}{2}}.$$ (18)

The chemical potential can then be determined self-consistently using the number conservation equation

$$N = 2 \sum_k v_k^2 \rightarrow \frac{L}{2\pi} \int_{-\infty}^{+\infty} dk \left[1 - \frac{\xi_k}{\sqrt{\xi_k^2 + \Delta^2}} \right]. \tag{19}$$

Note that only in the limit of weak coupling is the chemical potential given [3] by the Fermi energy, i.e., $\mu = E_F$. In general, one must determine the chemical potential and the gap self-consistently as a function of the dimensionless constant λ.

Comparison of BCS and Exact Results

We first rewrite equations (18) and (19) in terms of a dimensionless gap parameter $\tilde{\Delta}$ and chemical potential $\tilde{\mu}$ defined by

$$\tilde{\Delta} \equiv \frac{\hbar^2}{mv_0^2} \Delta \tag{20}$$

and

$$\tilde{\mu} \equiv \frac{\hbar^2}{mv_0^2} \mu, \tag{21}$$

and then solve for these quantities from the two equations as a function of λ. Since there are no singularities in the solutions of the gap and number equations for any value of λ, this gives a smooth transition from the Cooper pair (BCS) to the Bose gas regimes. The exact solution also displays smooth behaviour as seen in Fig. 1. Table 1 compares the dimensionless chemical potential in the BCS approximation with the exact values. Despite perfect agreement in both extremes, agreement is only moderately good for intermediate λ, $0 < \lambda^{-1} < 1$.

Instead of the gap parameter Δ we consider a physically more relevant quantity, namely, the energy difference between the first excited and the ground state of the quasiparticle spectrum [3] which is simply equal to twice the minimal quasiparticle energy. The "dimensionless" energy gap,

$$\tilde{E}_{gap} = \frac{\hbar^2}{mv_0^2} E_{gap} = \begin{cases} 2\tilde{\Delta} & \tilde{\mu} > 0 \ \text{(BCS regime)} \\ \\ 2\sqrt{\tilde{\Delta}^2 + \tilde{\mu}^2} & \tilde{\mu} < 0 \ \text{(Bose gas regime)} \end{cases} \tag{22}$$

is graphed in Figure 2. As in the case of the 2D Fermi gas [3] a *bifurcation* in $\tilde{E}_{gap}$ as a function of the coupling occurs when $\mu = 0$.

The total ground-state energy [10] for (1) in the BCS approximation is given by

$$E(N) = 2 \sum_k \epsilon_k v_k^2 + \frac{v_0}{L} \left(\sum_k v_k^2 \right)^2 - \Delta \sum_k u_k v_k, \tag{23}$$

Table 1. The dimensionless chemical potential $\tilde{\mu} = \frac{\hbar^2}{mv_0^2}\mu$ derived exactly from (6), derived from BCS theory from (18) and (19), and derived from plane-wave Hartree-Fock theory using (6), all as a function of the dimensionless parameter $\lambda^{-1} = \frac{\hbar^2 \rho}{mv_0}$.

λ^{-1}	Exact $\tilde{\mu}(\lambda)$	BCS $\tilde{\mu}(\lambda)$	PWHF $\tilde{\mu}(\lambda)$
0.05	-0.1242	-0.1120	-0.0219
0.10	-0.1215	-0.0979	-0.0377
0.15	-0.1165	-0.0825	-0.0472
0.20	-0.1088	-0.0657	-0.0507
0.25	-0.0978	-0.0470	-0.0479
0.30	-0.0832	-0.0262	-0.0390
0.35	-0.0643	-0.0030	-0.0239
0.40	-0.0407	0.0232	-0.0026
0.50	0.0220	0.0864	-0.0584
0.60	0.1075	0.1679	0.1441
0.70	0.2169	0.2716	0.2545
0.80	0.3510	0.4004	0.3900
0.90	0.5100	0.5556	0.5493
1.00	0.6938	0.7372	0.7337
1.50	1.9848	2.0259	2.0258
2.00	3.8934	3.9348	3.9348
2.50	6.4191	6.4606	6.4606
5.00	28.3010	28.3425	28.3425

giving for the dimensionless energy per particle (5) the expression

$$\epsilon(\lambda) = \frac{2\lambda}{\pi} \int_{-\infty}^{+\infty} d\tilde{k}\tilde{k}^2 \left[1 - \frac{\tilde{\xi}_k}{\sqrt{\tilde{\xi}_k^2 + \tilde{\Delta}^2}} \right] - 2\lambda^{-1} - 8\lambda\tilde{\Delta}^2 \qquad (24)$$

where

$$\tilde{\xi}_k \equiv \frac{\hbar^2}{mv_0^2}\xi_k = \frac{1}{2}\tilde{k}^2 - \frac{1}{2\lambda} - \tilde{\mu}, \quad \tilde{k} = \frac{\hbar^2 k}{mv_0}. \qquad (25)$$

This was calculated numerically, the results being listed in Table 2 and graphed in Figure 1. Again, although the BCS results are exact in the two extremes $\lambda \to 0$ and $\lambda \to \infty$, and interpolate smoothly between the two limits, there is a visible discrepancy with the exact ground-state energy for intermediate λ. This may be significant for calculations in 2- and 3-D Fermi liquids and in the extended Hubbard model.

To highlight this discrepancy for strong coupling (or low density) one can construct the power series expansions in λ^{-1} for $\epsilon(\lambda)$ and $\tilde{\mu}(\lambda)$. Using the Gaudin equations one can show [2] that

$$\epsilon(\lambda) = -1 + \frac{\pi^2}{12}\lambda^{-2} + O(\lambda^{-3}) \qquad \text{(exact).} \qquad (26)$$

Given (4), (5), (10) and the fact that $\rho = 2k_F/\pi$, the first correction term on the rhs of (26) corresponds to an energy per particle of $\frac{1}{2}[\frac{1}{3}\frac{\hbar^2 k_F^2}{4m}]$, i.e. to one half the energy per particle for non-interacting *fermions* of mass 2m. This is in spite of the *boson* nature of pair clusters and is consistent with a result of Girardeau [11]. From (6) and (21) we have

$$\tilde{\mu}(\lambda) = \frac{1}{8}[\epsilon(\lambda) - \lambda\,\epsilon'(\lambda)] = -\frac{1}{8} + \frac{\pi^2}{32}\lambda^{-2} + O(\lambda^{-3}) \qquad \text{(exact).} \qquad (27)$$

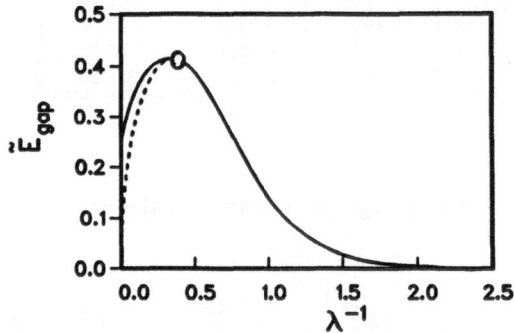

Figure 2. The dimensionless energy splitting between the first excited and ground-state of the quasiparticle spectrum (gap) in the BCS approximation as a function of the dimensionless parameter λ^{-1}. The open circle marks the bifurcation point at which the analytic form for the gap changes. The full curve gives the true gap (22), and the dashed branch denotes the expression $2\tilde{\Delta}$.

We next examine the BCS theory. In the limit $\lambda^{-1} \to 0$ one can obtain [12] the following limiting forms

$$\tilde{\Delta}^2(\lambda) = \frac{1}{4}\lambda^{-1} - \frac{3}{8}\lambda^{-2} + O(\lambda^{-3}) \quad \text{(as } \lambda \to \infty) \qquad \text{(BCS)} \qquad (28)$$

Table 2. The dimensionless ground-state energy $\epsilon(\lambda)$ for the exact solution, BCS theory and the plane-wave Hartree-Fock approximation (PWHF), all as a function of the dimensionless parameter, $\lambda^{-1} = \frac{\hbar^2 \rho}{m v_0}$.

λ^{-1}	Exact $\epsilon(\lambda)$	BCS $\epsilon(\lambda)$	PWHF $\epsilon(\lambda)$
0.05	-0.9978	-0.9488	-0.0918
0.10	-0.9909	-0.8947	-0.1671
0.15	-0.9784	-0.8374	-0.2260
0.20	-0.9595	-0.7766	-0.2684
0.25	-0.9333	-0.7117	-0.2944
0.30	-0.8989	-0.6421	-0.3039
0.35	-0.8551	-0.5674	-0.2970
0.40	-0.8011	-0.4866	-0.2736
0.50	-0.6587	-0.3037	-0.1775
0.60	-0.4652	-0.0858	-0.0156
0.70	-0.2158	0.1753	0.2120
0.80	0.0931	0.4872	0.5055
0.90	0.4636	0.8560	0.8648
1.00	0.8971	1.2858	1.2899
1.50	4.0320	4.4021	4.4022
2.00	8.7994	9.1594	9.1594
2.50	15.2073	15.5617	15.5617
5.00	71.9030	72.2467	72.2467

and

$$\tilde{\mu}(\lambda) = -\frac{1}{8} + \frac{1}{4}\lambda^{-1} + O(\lambda^{-2}) \quad \text{(as } \lambda \to \infty\text{)} \qquad \text{(BCS)} \qquad (29)$$

One already notes a term in λ^{-1} in the BCS expression for $\tilde{\mu}$ not present in the exact expansion (27). Using these expressions one can similarly obtain the following power series for the dimensionless energy,

$$\epsilon(\lambda) = -1 + \lambda^{-1} + \frac{1}{2}\lambda^{-2} + O(\lambda^{-3}) \quad \text{(as } \lambda \to \infty \text{)} \qquad \text{(BCS)} \qquad (30)$$

Again, one notes a term in λ^{-1} not present in the expansion for the exact solution (26) confirming our observation that BCS gives only moderately good agreement for $0 < \lambda^{-1} < 0.3$. In addition, the exact coefficient of the λ^{-2} term in (26) is $\pi^2/12 \simeq 0.8$, and not $1/2$ as given by BCS. Despite this, Fig. 1 shows that the BCS ground-state energy agrees fairly well with the exact result even in this strong-coupling region and far better than either the plane-wave Hartree-Fock (PWHF) result to be discussed below or even the full Hartree Fock (HF) result given in ([13]).

In contrast, in the weak-coupling or Cooper pairing regime ($\lambda^{-1} \gtrsim 3$) we obtain very good agreement between the BCS and the exact results (See Table 2). In this regime it can be shown that [12] the gap displays the characteristic BCS-type essential singularity familiar from low-temperature superconductivity in 3D, namely

$$\Delta \simeq 8 \; E_F \exp[-\frac{\pi^2}{2\lambda}] \quad (\text{as } \lambda \to 0) \tag{31}$$

As in 2D [3] and 3D [7] there is an extra factor of 2 in the exponent compared to the Cooper pairing result for the binding energy [2] of a Cooper pair

$$\Delta_C \simeq 8 \; E_F \exp[-\frac{\pi^2}{\lambda}] \ll \Delta \quad (\text{as } \lambda \to 0) \tag{32}$$

This singular behavior of the Cooper pair binding energy Δ_C also occurs [14] in the BCS model interaction problem in 1-, 2- and 3-D. For $\lambda^{-1} \gtrsim 1.5$, the chemical potential and ground-state energy are given to better than 1% accuracy by setting $\Delta = 0$ in the number equation (19) and the energy equation (24). Making this approximation readily reduces the BCS transformation parameters v_k^2 (16) to the step function

$$v_k^2 = \theta(\mu - \frac{\hbar^2 k^2}{2m}), \tag{33}$$

so that the BCS results become identical to those of plane-wave Hartree-Fock, which are

$$\epsilon(\lambda) = \frac{\pi^2}{3}\lambda^{-2} - 2 \; \lambda^{-1} \qquad \text{(PWHF)} \tag{34}$$

$$\tilde{\mu}(\lambda) = \frac{\pi^2}{8}\lambda^{-2} - \frac{1}{2}\lambda^{-1} \qquad \text{(PWHF)} \tag{35}$$

where (35) follows from (34) through (6). The PWHF energy (34) is just the expectation value of (1) in a Slater determinant of plane waves. These two quantities are listed in Tables 2 and 1 respectively, and (34) is graphed in Fig. 1. Also graphed in Figure 1 is the ideal Fermi gas (IFG) result which is just the first term on the rhs of (34), namely $\pi^2\lambda^{-2}/3$. One notes that PWHF and BCS are essentially identical for $\lambda^{-1} \gtrsim 1.5$, and agreement with the exact results to within 0.5% was achieved for $\lambda^{-1} \gtrsim 5$.

FINITE TEMPERATURE RESULTS

Strong Coupling Transitions and Bose Condensation

In the strong coupling limit the system behaves as an ideal Bose gas of fermion pairs and for this reason we outline the possible transitions for Bose gases. As is well known, Bose-Einstein condensation does not occur for an ideal Bose gas in one or two dimensions [15]. For this reason Bose-Einstein condensation is not expected in the present 1D model. This absence also holds for the 2D Fermi gas [9]. We note in passing that Bose-Einstein condensation into the lowest state can occur for an interacting Bose gas in one dimension [18] despite the absence of true long-range order in one or two dimensions [17].

It should also be noted that the absence of long range order [17] does not preclude the existence of Kosterlitz-Thouless-type transitions which are related to the excitation spectrum of the Bose gas. An example is the superfluid transition present for an interacting 2D Bose gas [16]. Such transitions are not yet fully understood for the 2D Fermi gas and should of course be considered in the present 1D model.

Finite Temperature BCS Theory

The finite temperature BCS theory gives the critical temperature for the breaking of Cooper pairs and is a straightforward extension of zero temperature theory. Again we use the notation of Fetter and Walecka [10]. Introducing the quasiparticle creation operators

$$\alpha_k = u_k a_{k\uparrow} - v_k a^\dagger_{-k\downarrow}, \qquad \beta_{-k} = u_k a_{-k\downarrow} + v_k a^\dagger_{k\uparrow} \qquad (36)$$

where u_k and v_k are the BCS transformation parameters given by (16), the quasiparticle vacuum $|O>$ is defined by

$$\alpha_k|O>= 0, \qquad \beta_k|O>= 0 \qquad \forall k. \qquad (37)$$

The transition from the zero to the finite temperature theory is achieved by simply making the replacements for the vacuum expectation values shown below

$$< O|\alpha^\dagger_k \alpha_k|O >=< O|\beta^\dagger_k \beta_k|O >= 0 \rightarrow f_k$$
$$< O|\alpha_k \alpha^\dagger_k|O >=< O|\beta_k \beta^\dagger_k|O >= 1 \rightarrow (1 - f_k), \qquad (38)$$

where f_k is the quasi-particle occupation probability given by

$$f_k = \frac{1}{\exp[\beta E_k] + 1}, \qquad (39)$$

E_k is the quasiparticle energy (17) and $\beta \equiv (k_B T)^{-1}$.

The finite-temperature gap is defined by

$$\Delta(T) = \frac{v_0}{L} \sum_k u_k v_k (1 - 2f_k), \qquad (40)$$

which replaces (15). This can be rewritten, substituting for the BCS coefficents u_k and v_k from equation (16) and the quasiparticle energies E_k from equation (17), to give the finite temperature gap equation

$$1 = \frac{v_0}{2L} \sum_k \frac{1}{E_k}(1 - 2f_k) \rightarrow \frac{v_0}{4\pi} \int_{-\infty}^{+\infty} dk \, [\xi_k^2 + \Delta^2]^{-\frac{1}{2}} \tanh(\frac{1}{2}\beta\sqrt{\xi_k^2 + \Delta^2}). \quad (41)$$

The temperature at which the gap disappears is the critical temperature T_c and is given implicitly by setting $\Delta(T_c)$ equal to zero in the gap equation to yield

$$1 = \frac{v_0}{4\pi} \int_{-\infty}^{+\infty} dk \, [\xi_k^2]^{-\frac{1}{2}} \tanh(\frac{1}{2}\beta_c\sqrt{\xi_k^2}). \quad (42)$$

Again, the number conservation equation

$$N = 2 \sum_k [u_k^2 f_k + v_k^2(1 - f_k)] \rightarrow \frac{L}{2\pi} \int_{-\infty}^{+\infty} dk \left[1 - \frac{\xi_k}{\sqrt{\xi_k^2 + \Delta^2}} \tanh(\frac{1}{2}\beta\sqrt{\xi_k^2 + \Delta^2})\right],$$
$$(43)$$

determines the chemical potential but now also as a function of the temperature T.

Results of Finite Temperature BCS Theory

Equations (41) and (43) can be solved numerically to give the gap and chemical potential as a function of temperature for any value of λ. Figure 3 shows the ratio of the gap to its zero temperature value as a function of the ratio of the temperature to its critical value T_c for two values of λ representing weak and strong coupling. As in conventional 3D low-temperature superconductivity, the form of the curve is practically independent of the coupling strength.

Equation (42) was solved numerically for the critical temperature and Table 3 shows the dimensionless critical temperature

$$\tilde{T}_c(\lambda) \equiv \frac{\hbar^2}{mv_0^2}T_c(\lambda). \quad (44)$$

as a function of λ. Also given are the dimensionless chemical potential at the critical temperature $\tilde{mu}_c(\lambda) \equiv \tilde{mu}(T_c)$ and the ratio of the energy gap at zero temperature to the critical temperature.

Of particular interest are the two extremes. Firstly in the limit of weak coupling (high density), $\lambda^{-1} \rightarrow \infty$, one can analytically derive the relation

$$\frac{\tilde{E}_{gap}}{\tilde{T}_c} = \pi \exp[-\gamma] \approx 3.53 \quad (45)$$

where $\gamma \approx 0.577$ is the Euler constant, using the same method as Friedberg and Lee [19]. This relation is of course familiar from standard 3D low-temperature superconductivity. Secondly in the limit of strong coupling (low density), $\lambda^{-1} \rightarrow 0$, one can prove that $T_c \rightarrow 0$. This then implies that the ratio of the energy gap to the critical temperature tends to infinity in this limit since $E_{gap} \rightarrow \frac{1}{4}\frac{mv_0^2}{\hbar^2} \neq 0$.

CONCLUSIONS

We have considered the soluble 1D many-fermion fluid with pairwise attractive δ-function interactions. This model is particularly pertinent for two reasons. Not only is it dynamically similar to 3D "jellium", but it also only admits pair clusterings which must be assumed in 2 and 3D models for superconductivity to explain the magnetic flux quantisation measurements both in conventional and "exotic" superconductors.

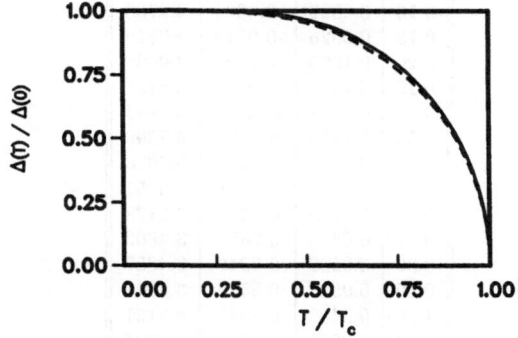

Figure 3. The dimensionless ratio of the BCS gap to its value at zero temperature, $\frac{\Delta(T)}{\Delta(0)}$, as a function of the scaled temperature $\frac{T}{T_c}$ where T_c is the critical temperature for two different values of the dimensionless variable λ^{-1}. The solid curve refers to $\lambda^{-1} = 0.05$ and the dashed curve to $\lambda^{-1} = 2$.

At zero temperature we have compared the results for the chemical potential and ground state energy for the exact solution and the BCS theory. The BCS results are exact in the two extremes of coupling/density and moderately good for intermediate values. It is exponentially close to the plane-wave Hartree-Fock (PWHF) approximation for weak coupling and/or high density, and notably superior to PWHF for strong coupling and/or low density. This latter regime consists of an almost ideal gas of tightly-bound bosonic pairs of fermions, and its appearance is signalled by the otherwise positive chemical potential becoming negative. The crossover to this regime is smooth, but the BCS energy gap "bifurcates" into an analytically distinct form when the boson (pair) radius is about one third of the inter-fermion spacing.

Given the existence of an ideal Bose gas of pairs for strong coupling (low density) we believe that Bose condensation does not occur, in agreement with the results for the 2D Fermi gas [9]. We also present the results for finite temperature BCS theory and show that the BCS critical temperature for the breaking of Cooper pairs approaches zero in the strong coupling or Bose gas limit and the familiar BCS value in the weak coupling or Cooper pairing limit.

Table 3. The dimensionless critical temperature, $\tilde{T}_c$, the dimensionless chemical potential at the critical temperature, $\tilde{\mu}_c$, and the ratio of the energy gap to the critical temperature, $\frac{E_{gap}}{T_c}$, all calculated as a function of the dimensionless parameter, $\lambda^{-1} = \frac{\hbar^2 \rho}{m v_0}$ using finite temperature BCS theory.

λ^{-1}	$\tilde{T}_c$	$\tilde{\mu}_c$	$\frac{E_{gap}}{T_c}$
0.01	0.0422	-0.1210	6.2658
0.05	0.0665	-0.1069	4.6708
0.10	0.0822	-0.0895	4.2733
0.15	0.0926	-0.0719	4.0919
0.20	0.1000	-0.0536	3.9696
0.25	0.1053	-0.0343	3.8727
0.30	0.1087	-0.0137	3.7943
0.35	0.1106	0.0085	3.7399
0.40	0.1111	0.0328	3.7018
0.50	0.1077	0.0901	3.6051
0.60	0.0985	0.1648	3.5178
0.70	0.0844	0.2651	3.4803
0.80	0.0683	0.3946	3.4857
0.90	0.0531	0.5516	3.5011
1.00	0.0402	0.7348	3.5131
1.50	0.0077	2.0258	3.5274
2.00	0.0012	3.9348	3.5277
2.50	0.0002	6.4606	3.5278
5.00	0.0000	9.6033	3.5278

ACKNOWLEDGEMENTS

MdeLl gratefully acknowledges discussions with Professors S. A. Moszkowski and S. Pittel, as well as a NATO Research Grant. MdeLl and RMQ also gratefully acknowledge the financial support of the Foundation for Research Development, Pretoria.

NOTES

[a] On leave from Physics Department, North Dakota State University, Fargo ND 58105, USA.

REFERENCES

[1] J. B. McGuire, "State function and spectral properties of a two-component interacting Fermi gas," *J. Math. Phys.* 31:164 (1990).

[2] M. Casas, C. Esebbag, A. Extremera, J. M. Getino, M. de Llano, A. Plastino and H. Rubio, "Cooper pairing in a soluble one-dimensional many-fermion model," *Phys. Rev. A* 44:4415 (1991).

3 M. Randeria, J.-M. Duan and L.-Y. Shieh, "Bound States, Cooper Pairing and Bose Condensation in Two Dimensions," *Phys. Rev. Lett.* 62:981 (1989); 62:2887(E) (1989); "Superconductivity in a two-dimensional Fermi gas: Evolution from Cooper pairing to Bose condensation," *Phys. Rev. B* 41:327 (1990); K Miyake, "Fermi Liquid Theory of Dilute Submonolayer ^{3}He on Thin ^{4}He II Film," *Prog. Theor. Phys.* 69:1794 (1983).

4 P. Noziéres and S. Schmitt-Rink, "Bose Condensation in an Attractive Fermion Gas : From Weak to Strong Coupling Superconductivity," *J. Low Temp. Phys.* 59:195 (1985).

5 R. Micnas, J. Ranninger and S. Robaskiewicz, "Superconductivity in narrow-band systems with local nonretarded attractive interactions," *Rev. Mod. Phys.* 62:113 (1990).

6 Y. J. Uemura et al, "Basic Similarites among Cuprate, Bismuthate, Organic, Chevrel-Phase, and Heavy-Fermion Superconductors Shown by Penetration-Depth Measurements," *Phys. Rev. Lett.* 66:2665 (1991); "Magnetic-field penetration depth in K_3C_{60} mesaured by muon spin relaxation," *Nature* 352:605 (1991).

7 J. Bardeen, L. N. Cooper and J. Schrieffer, "Theory of Superconductivity," *Phys. Rev.* 108:1175 (1957).

8 M. Gaudin, "Un Systeme a une dimension de fermions en interaction," *Phys. Lett. A* 24:55 (1967).

9 S. Schmitt-Rink, C. M. Varma and A. E. Ruckenstein, "Pairing in Two Dimensions," *Phys. Rev. Lett.* 44:445 (1989); J. W. Serene, "Stability of two-dimensional Fermi liquids against pair fluctuations with large total momentum," *Phys. Rev. B* 40:10873 (1989); A. Tokumitu, K. Miyake and K. Yamada, " Crossover between Cooper-Pair Condenstaion and Bose-Einstein Condensation of 'Di-Electronic Molecules' in Two-Dimensional Superconductors," *Prog. Theor. Phys.* 106:63 (1991).

10 A. L. Fetter and J. D. Walecka, "Quantum Theory of Many Particle Systems," McGraw-Hill, New York (1971), *p.330ff*.

11 M. Girardeau, " Relationship between Systems of Impenetrable Bosons and Fermions in One Dimension," *J. Math. Phys.* 1:516 (1960).

12 C. Esebbag, M de Llano and R. M. Quick, "Test of Bardeen-Cooper-Schrieffer theory in an exactly soluble many-fermion model," University of Pretoria preprint (1992).

13 G. Gutierrez and A. Plastino "A Systematic Approach to the Hartree-Fock Problem in the Thermodynamic Limit," *Ann. Phys.* 133:332 (1981).

14 C. Esebbag, J. M. Getino, M. de Llano, S. A. Moszowski, U. Oseguera, A. Plastino and H. Rubio, "Cooper pairing in one, two, and three dimensions," *J. Math Phys.* 33:1221 (1992).

15 R. K. Pathria, "Statistical Mechanics," Pergamon Press, Oxford (1984).

16 D. S. Fisher and P. C. Hohenberg, "Dilute Bose gas in two dimensions," *Phys. Rev. B.* 37:4936 (1988).

17 P.C. Hohenberg, "Existence of Long-Range Order in One and Two Dimensions," *Phys. Rev.* 158:33 (1967).

18 J.M. Luttinger and H. K. Sy, "Bose-Einstein Condensation in a One-Dimensional Model with Random Impurities," *Phys. Rev. A.* 7:712 (1973).

19 R. Friedberg and T. D. Lee, "Gap energy and long-range order in the boson-fermion model of superconductivity," *Phys. Rev. B.* 40:6745 (1989).

CORRELATIONS IN COUPLED ELECTRON LAYERS

D. Neilson[a], L. Świerkowski[a]* and J. Szymański[b]

[a] School of Physics
University of New South Wales
Kensington 2033
Australia

[b] Telecom Research Laboratories
770 Blackburn Road
Clayton Vic. 3168
Australia

I. INTRODUCTION

In strongly interacting electron systems the buildup of correlations can fundamentally alter the nature of the states of the system compared with the metallic systems. Since the strength of the correlations is determined from the competition between the Fermi energy and the Coulomb potential they become increasingly significant for low densities. It is possible to assist the buildup of highly correlated states by introducing an external field, for example a magnetic field.

The Wigner solid is a particularly dramatic example of the correlation buildup: the homogeneous electron liquid at zero temperature undergoes a phase transition to a crystalline lattice which is made up purely of the electrons. A major difficulty in experimentally fabricating the pure electron solid is that the density of the electrons has to be extremely small before potential energy effects start to compete with the kinetic energy. This is due to the tiny electron mass.

We have proposed[1] that the strongly correlated region for the pure electron liquid could be accessed experimentally using existing technology without needing to resort to external fields by designing systems with two parallel electron layers replacing the single isolated layer. With two layers the electrons in one layer couple to the electrons in the other layer so that each layer acts as a polarizable background for the other. For a given density the correlations are much stronger in two layer systems than for a single layer and in particular the critical density for the phase transition to the Wigner crystal state can be higher, making it more easily accessible experimentally.

The two layer system has further interesting aspects. Because each layer acts as polarizable backgrounds for the other layer the ground state can take the form of charge density waves (CDW). These inhomogeneous density distributions occur at densities much higher than the critical density for the Wigner transition[1]. The transconductance associated with a current in one layer inducing a current in the other layer can also be studied[2].

We introduce an interlayer pair correlation function $g_{12}(\mathbf{r})$ which is defined as the probability, given an electron at the origin in the first layer, that there is an electron in the second layer at a distance $|\mathbf{r}|$ parallel to the layers. The vector $\mathbf{r}$ will always be taken to be the component of the separation in the direction parallel to the layers. The intralayer pair correlation function $g_{11}(\mathbf{r})$ is the probability that two electrons in the same layer are separated by the distance $|\mathbf{r}|$.

The function $g_{12}(\mathbf{r})$ is fundamentally different from the $g_{11}(\mathbf{r})$ for two reasons. Firstly, the usual sum rule $\int d^2r[1 - g_{11}(\mathbf{r}] = 1/n_1$ (n_l is the electron density in the lth layer) applies to the intralayer correlation function due to the assumed presence of an electron at the origin. However there is no "self" electron at the origin of the other layer and so the corresponding sum rule for the interlayer pair correlation function is $\int d^2r[1 - g_{12}(\mathbf{r})] = 0$. Secondly, the Coulomb interaction between electrons from different layers is not singular at the origin. Consequently, we expect the interlayer correlations to be weaker than the single layer and intralayer correlations.

In our previous work[1] we used the Monte Carlo pair correlation function[3] for the single layer to determine the effective interaction between electrons within the same layer while treating the interlayer interaction within RPA. In the present calculation we continue to use the Monte Carlo data to determine the intralayer effective interaction but now introduce the STLS self-consistent procedure to estimate the interlayer effective interaction. If the interlayer correlations do remain weak then STLS may continue to give a fairly good estimate for $g_{12}(\mathbf{r})$ provided $g_{11}(\mathbf{r})$ is independently known to sufficient accuracy.

II. THEORY

We consider a system of electrons whose motion in the z direction is confined by the presence of two quantum wells adjacent to each other, producing two parallel electron layers. Within linear-response mean-field theory the total potential acting on electrons in a particular layer $l = 1$ or $l = 2$ consists of the external potential $V_l^{\text{ext}}(\mathbf{q},\omega)$ plus the interaction induced by changes in electron density in the other layer, $l' \neq l$. The induced electron density in the lth layer $\delta n_l(\mathbf{q},\omega)$ is thus

$$\delta n_l(\mathbf{q},\omega) = -\chi_l(\mathbf{q},\omega)\left[V_l^{\text{ext}}(\mathbf{q},\omega) + \sum_{l \neq l'}[1 - G_{ll'}(\mathbf{q})]V_{ll'}(\mathbf{q})\delta n_{l'}(\mathbf{q},\omega)\right], \tag{1}$$

where $\chi_l(\mathbf{q},\omega)$ is the response function for a single isolated layer l and $G_{ll'}(\mathbf{q})$ for $l \neq l'$ is a static local field which modifies the interaction between two electrons from different layers.

The response function of the two layer system is defined by

$$\delta n_l(\mathbf{q}, \omega) = - \sum_{l'=1}^{2} \chi_{ll'}(\mathbf{q}, \omega) V_{l'}^{ext}(\mathbf{q}, \omega). \tag{2}$$

Comparing Eqs. (1) and (2) we see that $\chi_{ll'}(\mathbf{q}, \omega)$ can be obtained by inverting the matrix

$$\left[\chi^{-1}(\mathbf{q}, \omega)\right]_{ll'} = \frac{1}{\chi_l(\mathbf{q}, \omega)} \delta_{ll'} + [1 - G_{ll'}(\mathbf{q})] V_{ll'}(\mathbf{q})(1 - \delta_{ll'}). \tag{3}$$

The interlayer local field $G_{ll'}(\mathbf{q})$, with $l' \neq l$, is calculated using the method of Singwi, Tosi, Land and Sjölander (STLS) [4] which was later extended to the case of a two component plasma [5]. In the absence of tunneling between the layers our two-layer system can be formally viewed as a two component plasma with bare interactions given by $V_{ll'}(\mathbf{q})$.

Central to the STLS scheme is the *Ansatz* that the density-density correlation function can be approximated by a product of two densities and the static pair correlation function (see, for example, Ref. 6). We write

$$< \delta\hat{n}_l(\mathbf{r}, t)\delta\hat{n}_{l'}(\mathbf{r}', t) > \approx \delta n_l(\mathbf{r}, t) g_{ll'}(\mathbf{r} - \mathbf{r}')\delta n_{l'}(\mathbf{r}', t), \tag{4}$$

where the symbol $\hat{}$ distinguishes the operators $\delta\hat{n}_l(\mathbf{r}, t)$ from their expectation values $\delta n_l(\mathbf{r}, t)$. If $g_{ll'}(\mathbf{r} - \mathbf{r}')$ is replaced everywhere by unity then Eq. (4) leads to the Hartree approximation.

Using Eq. (4) together with the relation between the pair correlation function and the static structure factor,

$$g_{ll'}(\mathbf{r}) = 1 + \frac{1}{\sqrt{n_l n_{l'}}} \int \frac{d^2 q}{(2\pi)^2} e^{i\mathbf{q}\cdot\mathbf{r}}[S_{ll'}(\mathbf{q}) - \delta_{ll'}], \tag{5}$$

one obtains the expression for the local field [4,5,6]:

$$G_{ll'}(\mathbf{q}) = -\frac{1}{\sqrt{n_l n_{l'}}} \int \frac{d^2 k}{(2\pi)^2} \frac{\mathbf{q} \cdot \mathbf{k}}{q^2} \frac{V_{ll'}(\mathbf{k})}{V_{ll'}(\mathbf{q})}[S_{ll'}(|\mathbf{q} - \mathbf{k}|) - \delta_{ll'}]. \tag{6}$$

The fluctuation-dissipation theorem can be written as

$$S_{ll'}(\mathbf{q}) = \frac{1}{\sqrt{n_l n_{l'}}} \frac{\hbar}{\pi} \int_0^\infty \mathrm{Im}\chi_{ll'}(\mathbf{q}, \omega) d\omega. \tag{7}$$

Equations (3), (6) and (7) form a closed set of equations for the local fields $G_{ll'}(\mathbf{q})$ which can be solved self-consistently. The above approach can be straightforwardly extended to the case of arbitrary number of layers, as well as to the case of finite numbers of occupied subbands.

We use this self-consistent procedure to calculate the inter-layer local field $G_{ll'}(\mathbf{q})$ for $l \neq l'$. In order to to specify the response function for a single isolated layer, $\chi_l(\mathbf{q}, \omega)$, we follow the procedure introduced in Refs. 1,7. First $\chi_l(\mathbf{q}, \omega)$ is written in the form [4]:

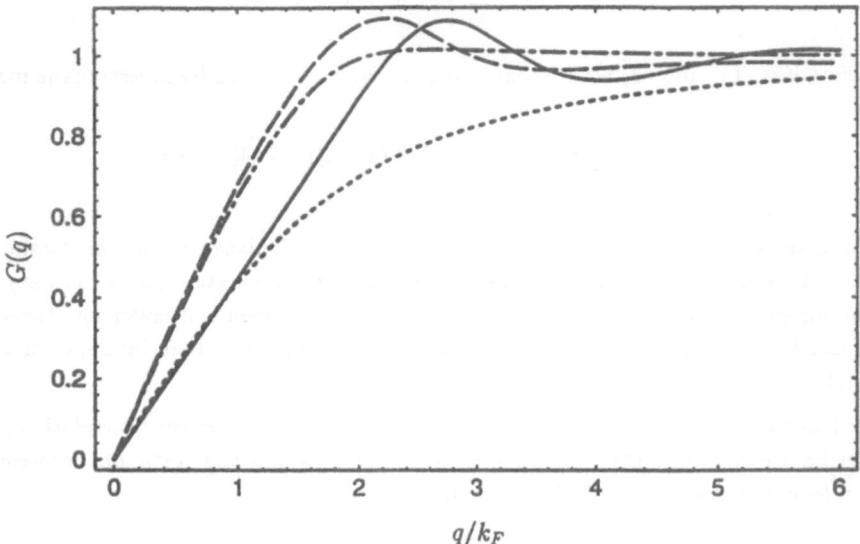

Figure 1. Solid line: static local field $G(q)$ for single layer at $r_s = 40$ (solid line). Dashed line: local field $\bar{G}(q)$ calculated from Eq. 6. Chain line: STLS local field $G_{STLS}(q)$. Dotted line: Hubbard-like local field $G_H(q)$.

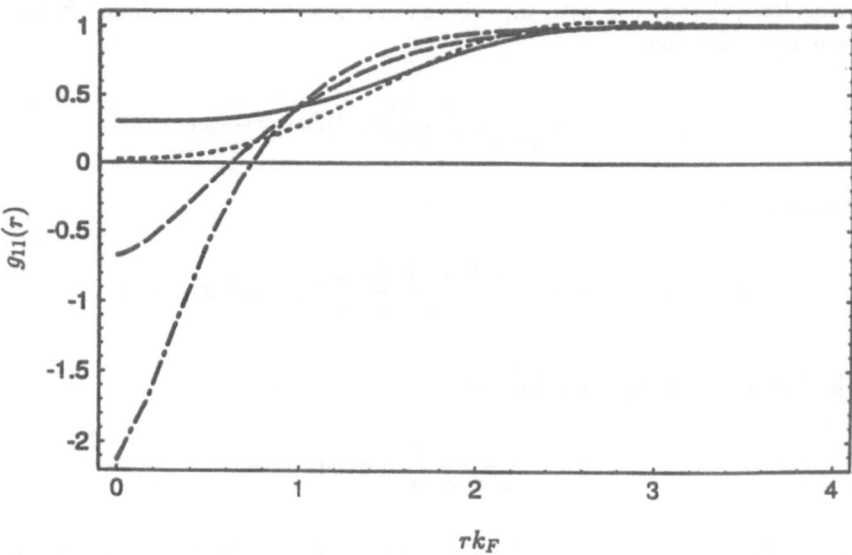

Figure 2. Comparison of different calculations for the intralayer pair correlation function $g_{11}(r)$ at $r_s = 5$. Solid line: present result. Dashed line: STLS. Chain line: RPA. Dotted line: $g(r)$ for the single layer calculated by Ref. 3. $g_{11}(r)$ goes negative for small r for both the RPA and STLS.

$$\chi_l(\mathbf{q}, \omega) = \frac{\chi_l^{(0)}(\mathbf{q}, \omega)}{1 + V_{ll}(\mathbf{q}) \left[1 - G_{ll}(\mathbf{q}) \right] \chi_l^{(0)}(\mathbf{q}, \omega)} , \tag{8}$$

where $\chi_l^{(0)}(\mathbf{q}, \omega)$ is the Lindhard function for the two-dimensional system [8]. Ref. 3 gives the pair correlation function $g_{ll}(r)$ for the ground state of the two-dimensional electron liquid for a range of densities down to the Wigner crystallization point. Using Eqs. (5), (7) and (8) and the $g(r)$ data in Ref. 3 we can uniquely specify the intra-layer local field factor $G_{ll}(\mathbf{q})$.

Another method of finding local fields for the two-layer system, would consist in substituting Eq. (8) into Eq. (3) and then solving the Eqs. (3), (6) and (7) self-consistently for *both* the intra- and inter-layer local fields $G_{ll'}(\mathbf{q})$. This is a natural extension of the STLS approach but as we shall see in the next section it does not appear to be appropriate for low density systems close to the Wigner transition. This generalized STLS approach could however be used for coupled layer systems at higher densities.

III. RESULTS AND DISCUSSION

Once we know the pair correlation function $g(r)$ for a single layer using Ref. 3 we can use Eqs. 7 and 8 to uniquely determine the local field $G(\mathbf{q})$. Fig. 1 shows $G(\mathbf{q})$ for a single layer at $r_s = 40$. Shown for comparison is $\bar{G}(\mathbf{q})$ calculated from the approximate expression, given by Eq. (6) with $l = l'$ and $S_{ll}(\mathbf{q})$ determined from the data in Ref. 3. The disagreement between the two curves indicates that STLS is not a good approximation for the intralayer local field in this very strongly correlated region since the STLS self-consistency condition is that the two functions $G(\mathbf{q})$ and $\bar{G}(\mathbf{q})$ should be the same. The local field $G_{STLS}(\mathbf{q})$ confirms this. $G_{STLS}(\mathbf{q})$ is calculated using the complete STLS self-consistent iterative procedure. It can be seen that it develops no peak. Finally a local field derived from a Hubbard-like expression, $G_H(\mathbf{q}) = |\mathbf{q}|/\sqrt{\mathbf{q}^2 + \alpha^2 k_F^2}$ is shown. The parameter α has been adjusted so that the compressibility calculated in Ref. 3 is recovered.

In Fig. 2 we compare the results of different calculations of the intralayer pair correlation function $g_{11}(r)$ at the density $r_s = 5$. The $g_{11}(r)$ calculated using the STLS self-consistent scheme goes negative at small r. This unphysical result is indicative of a breakdown in the approach for all $r_s \geq 5$. In contrast, the $g(r)$ for the single layer as calculated in Ref. 3 goes to zero for small r and goes smoothly to unity for increasing r. Finally the $g_{11}(r)$ calculated using our combination method of Monte Carlo results and STLS is shown. This $g_{11}(r)$ is quite similar to the corresponding function $g(r)$ for the single isolated layer and like that function it remains positive as r goes to zero.

Figure 3 shows our intralayer $g_{11}(r)$ and also the interlayer $g_{12}(r)$ for densities $r_s = 10$ and 30 for three values of the interlayer spacing, $(a - a_c)/a_c = 0.01$, 0.1 and 1 where a_c is the critical spacing for the transition to the inhomogeneous ground state. For spacings very close to the transition the oscillatory peaks in $g_{11}(r)$ become pronounced.

In the case of two layers the function $\chi_{ll'}(\mathbf{q}, \omega)$ is a 2×2 matrix which can be diagonalized to determine the eigenstates. Figure 4 shows the static diagonal element $\chi_-(\mathbf{q})$ in which the density modulations δn_l in the adjacent layers are π out of phase. This configuration has a lower energy than the other configuration in which the density

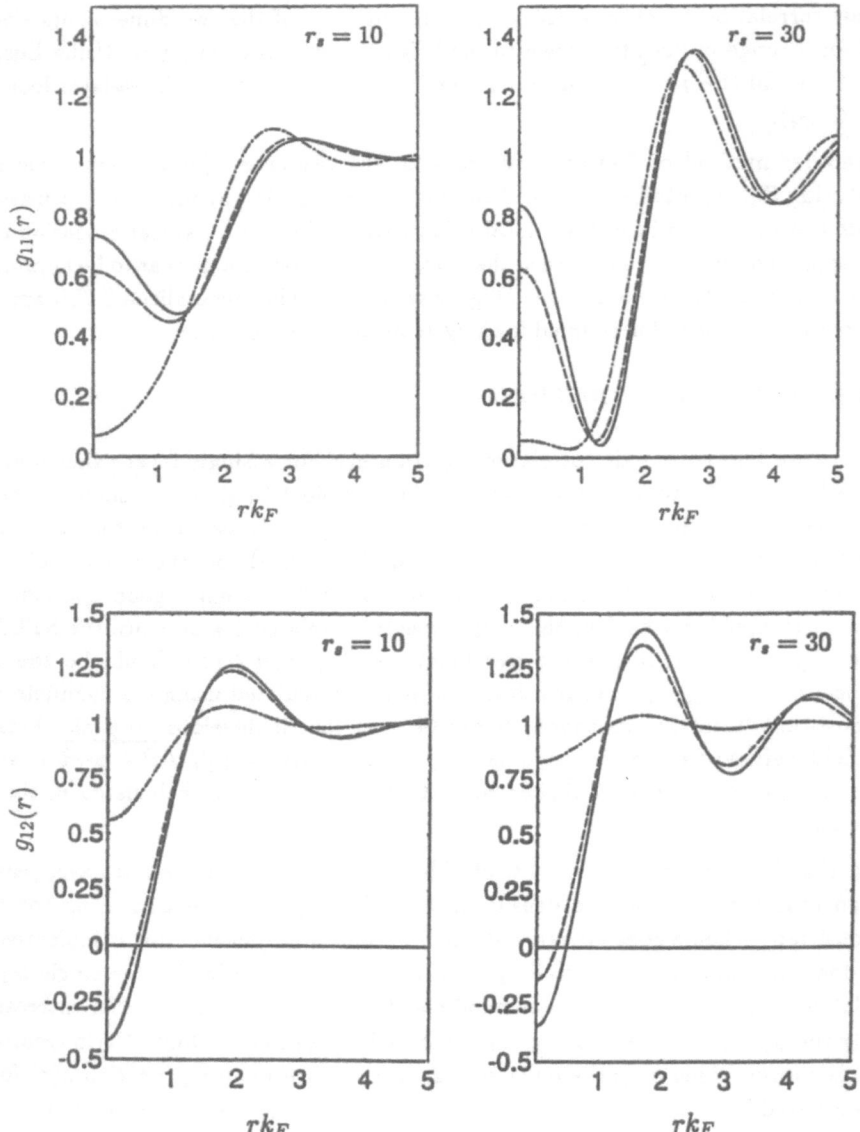

Figure 3. The intralayer $g_{11}(r)$ and the interlayer $g_{12}(r)$ for densities $r_s = 10$ and 30 for three values of the interlayer spacing, $(a - a_c)/a_c = 0.01, 0.1, 1$ (solid, dashed, and chain line, respectively). a_c is the critical spacing for the transition to the inhomogeneous ground state.

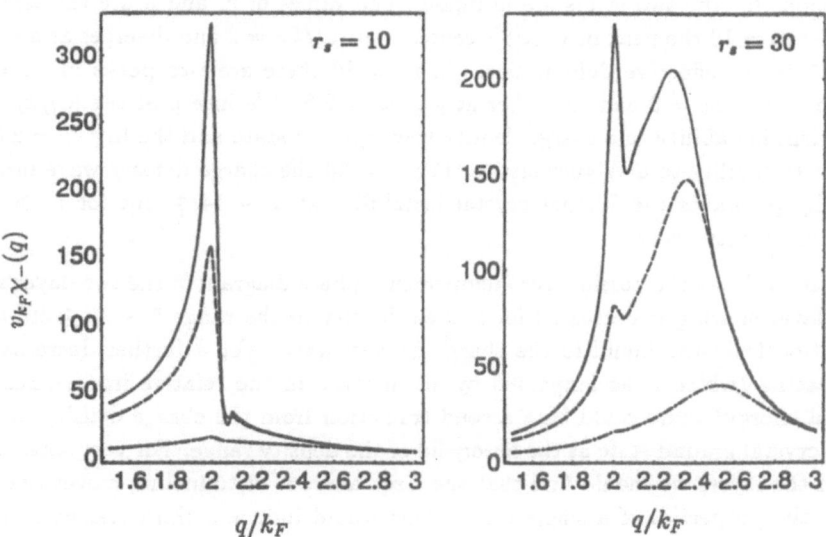

Figure 4. The static response function $\chi_-(q)$ for densities $r_s = 10$ and 30 and $(a - a_c)/a_c = 0.01, 0.1, 1$. The factor $v_{k_F} = 2\pi e^2/k_F$. For $r_s = 10$ the peak in $\chi(q)$ is centered at $|q|/k_F = 2.0$. For $r_s = 30$ there are two peaks in $\chi(q)$, one at $|q|/k_F = 2.0$ and one at $|q|/k_F \sim 2.5$.

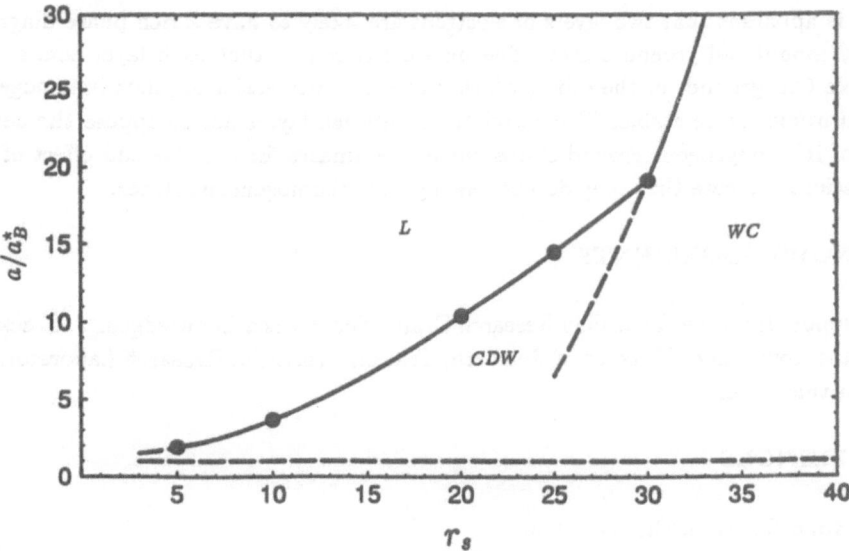

Figure 5. Phase diagram for two parallel electron layers in GaAs. a is the spacing between the layers. The three phase regions shown are L: the liquid state, CDW: the charge density wave ground state and WC: the Wigner crystal. For fixed r_s in the range $5 \lesssim r_s \lesssim 30$, as the layer spacing is decreased there is a transition from the liquid to a charge density wave. For $r_s > 30$ the transition from the liquid is into a Wigner crystal state.

modulations for the two layers are in phase. The values of r_s and a are the same as in Fig. 3. At $r_s = 10$ the peak in $\chi_-(q)$ is centered at $|q|/k_F = 2$ and diverges at $a \sim 3.7a_B^*$, where a_B^* is the effective Bohr radius. At $r_s = 30$ there are two peaks in $\chi_-(q)$, one centered at $|q|/k_F = 2$ and the other at $|q|/k_F = 2.5$. We interpret the $|q|/k_F = 2$ as an incipient instability to a charge density wave ground state and the $|q|/k_F = 2.5$ as an incipient instability to a Wigner lattice. For $r_s = 30$ the charge density wave instability eventually preempts the Wigner crystal instability at $a \sim 14a_B^*$ but for $r_s > 30$ the Wigner instability prevails.

Figure 5 shows the possible zero temperature phase diagram of the two–layer system. As the layer spacing is decreased for a fixed density in the range $5 \lesssim r_s \leq 30$, there is a transition from the liquid to the charge density wave. For a further decrease in the layer spacing (which is accompanied by an increase in the relative importance of the potential energy) there could be a second transition from the charge density wave to a Wigner crystal ground state at the lower end of the density range. For very small spacing between the layers we should find that the large amount of tunneling moves the system towards the properties of a single layer. This would induce a third transition back to the liquid state. For $r_s > 30$, as the layer spacing is decreased the transition from the liquid is directly into a Wigner crystal state. For very small layer spacings there should again be a transition back to the liquid. In Fig. 5 the solid line is the boundary we have determined between the uniform liquid phase and the inhomogeneous phase. Broken lines represent the other boundaries we expect to be present, a boundary between the charge density wave phase and the Wigner phase, and, for small a, the transition from the inhomogeneous phases back to the liquid phase due to the increase in tunneling.

It is apparent that two layers of electrons are likely to have a rich phase diagram of density modulated ground states. The physical reason is that each layer acts as a polarizable background for the other, which makes it much easier for such inhomogeneous configurations to be stable. The correlations between layers act to oppose the development of inhomogeneous ground states but our estimates for the size and effect of these correlations indicate that they do not destroy the inhomogeneous states.

ACKNOWLEDGEMENTS

Support from the Australian Research Grants Scheme is acknowledged. J. S. acknowledges the approval of Director of Research, Telecom Australia Research Laboratories, to publish this paper.

REFERENCES

*Australian Research Grants Fellow·

[1] L. Świerkowski, D. Neilson and J. Szymański , Phys. Rev. Lett. 67, 240–243 (1991).

[2] H.C.A. Oji, A.H. Mac Donald and S.M. Girvin, Phys. Rev. Lett. 58, 824 (1987).

[3] B. Tanatar and D.M. Ceperley, Phys. Rev. B 39, 5005 (1989).

[4] K.S. Singwi, M.P. Tosi, R.H. Land, and A. Sjölander, Phys. Rev. **176**, 589 (1968).

[5] A. Sjölander and J. Stott, Phys. Rev. B **5**, 2109 (1972); P. Vashishta, P. Bhattacharyya, and K.S. Singwi, Il Nuovo Cimento, **23 B**, 172 (1974).

[6] K.S. Singwi and M.P. Tosi, in *Solid State Physics*, edited by H. Ehrenreich, F. Seitz and D. Turnbull (Academic, New York, 1981), Vol. **36**, p. 177.

[7] D. Neilson, L. Świerkowski, A. Sjölander and J. Szymański , Phys. Rev. B **44**, 6291 (1991).

[8] F. Stern, Phys. Rev. Lett. **18**, 546 (1967).

MEAN FIELD PREDICTIONS OF THE DILUTE tJ MODEL FOR HIGH TEMPERATURE SUPERCONDUCTIVITY

J. W. Halley*, S. Davis*,# and X.-F. Wang *

*School of Physics and Astronomy
University of Minnesota
Minneapolis, MN 55455

#Department of Mathematics
Cornell University
Ithaca, NY

We describe mean field calculations of the properties of the dilute tJ model for high T_c superconductors. We first discuss a physical motivation for the model in terms of the effects of oxygen vacancies and the appropriate parametrization for a qualitative description of high T_c superconductors. We present results on the local carrier density, the transition temperature as a function of carrier concentration, oxygen vacancy concentration and elastic scattering impurity concentration. These are compared with experiment on high T_c materials. We conclude that the model is a promising framework for understanding high T_c materials microscopically.

INTRODUCTION

A conspicuous feature of the new superconductors [1-2] which is usually assumed to be an inessential complication by theorists is that virtually all of the high temperature materials have a very high degree of spatial disorder leading to mean free paths which are at best a few times the superconducting coherence length. This means at least that one should be developing theoretical methods and models for taking theoretical account of the microscopic disorder. There is also a possibility [4] that some kinds of point defects or twin boundaries might act like "pairing centers" and enhance the pairing interactions leading to the superconductivity itself. For both these reasons we have been developing both mean field and correlated calculational methods and models for taking account of the effects of point defects in models of high T_c superconductors.

In this paper we describe recent results on a dilute tJ model for high temperature superconductivity with particular emphasis on new mean field results which can be compared directly with experiment. In the next section we describe the model and briefly review our mean field computational techniques. The following section describes the parametrization of the model. Results are then described on the transition temperature as a function of the concentration of special bonds, on the local charge density as a function of temperature, on the transition temperature as a function of randomly distributed elastically scattering impurities and on the transition temperature as a function of carrier concentration. In each case we compare results with experimental quantities. We conclude with a discussion.

DILUTE t-J MODEL

The model[5] arises if one assumes that the current carrying holes in the copper oxygen planes of high temperature superconductors are attracted to each other through an exchange interaction at the randomly placed sites of oxygen vacancies. Physical arguments that suggest these assumptions are discussed in more detail elsewhere[5]. Then the model consists of carriers moving with nearest neighbor hopping on a square lattice with an exchange interaction J between a randomly selected fraction p of the bonds. The onsite energy , called ϵ at these special bonds is allowed to be different from that at the other sites. The Hamiltonian is

$$H - \mu N = \sum_{i,\sigma}(\epsilon(i) - \mu)d_{i,\sigma}^{\dagger}d_{i,\sigma} + \sum_{i,j,\sigma}(b(i,j)d_{i,\sigma}^{\dagger}d_{j,\sigma} + h.c.) + . \tag{1}$$

$$- \sum_{i,j}(J(i,j)/2)(d_{i\uparrow}^{\dagger}d_{j\downarrow}^{\dagger} - d_{i\downarrow}^{\dagger}d_{j\uparrow}^{\dagger})(d_{j\downarrow}d_{i\uparrow} - d_{j\uparrow}d_{i\downarrow})$$

We suppose that J(ij) is zero except at a randomly selected fraction p of bonds where its value $J(ij) = J$ is a parameter of the model. b_{ij} is made the same at all nearest neighbor pairs and set equal to t (usually we take t=1 to establish the energy scale). $\epsilon(i)$ is zero at sites away from special bonds but finite and $= \epsilon$ on sites next to special bonds. μ is the chemical potential. We will suppose here, on the basis of the physical arguments mentioned ,that μ lies near the top of the band. To specify the mean field theory, we define the retarded functions[6] :

$$G_{\sigma}^{ij}(t) = -i\Theta(t) < \{d_{i,\sigma}(t), d_{j,\sigma}^{\dagger}(0)\} >$$

$$F_{\sigma}^{ij}(t) = -i\Theta(t) < \{d_{i,\sigma}^{\dagger}(t), d_{j,-\sigma}^{\dagger}(0)\} >$$

We obtain equations of motion for the functions $G_{ij} = G_{\uparrow}^{ij} + G_{\downarrow}^{ij}$ and $F_{ij}(t) = F_{\uparrow}^{ij} - F_{\downarrow}^{ij}$:

$$i\hbar dG_{ij}/dt = 2\hbar\delta(t)\delta_{i,j} + \epsilon(i)G_{ij} + \sum_{\delta}b(i+\delta,i)G_{i+\delta j} + \sum_{\delta}(J(i+\delta,i)/2)\Delta_{i,\delta}F_{i+\delta\ j}. \tag{2}$$

$$i\hbar dF_{ij}/dt = -\epsilon(i)F_{ij} - \sum_{\delta}b(i+\delta,i)F_{i+\delta\ j} + \sum_{\delta}(J(i+\delta,i)/2)\Delta_{i,\delta}^{*}G_{i+\delta\ j}. \tag{2}$$

with the initial conditions $G_{ij}(0) = -i2\delta_{ij}$, $F_{ij}(0) = 0$. The gap equation is

$$\Delta_{ij}^{*} = -i\int_{-\infty}^{\infty}\frac{F_{ij}(\omega + i\epsilon) - F_{ij}(\omega - i\epsilon)}{e^{\beta\omega} + 1}d\omega. \tag{3}$$

where

$$F_{ij}(\omega + i\epsilon) = 1/2\pi \int_{-\infty}^{\infty} F_{ij}(t)e^{i\omega t} dt$$

In order to reduce the number of equations, we define the sums $F_i = \sum_j c_j F_{ij}$, $G_i = \sum_j c_j G_{ij}$. Then $F_i(0) = 0$, $G_i(0) = -i2c_i$.. To calculate the average Δ [6] at the special sites we choose the c_i as follows. All c_i associated with sites not next to a special site are zero. Associate a random number ϕ_i evenly distributed between 0 and 2π with each site next to a vacancy. Now consider a pair of sites next to a special bond. Label the sites 1 and 2. Set $c_1 = e^{i\phi_2}$ and $c_2 = e^{i\phi_1}$. The phase factors $e^{i\phi}$ are a calculational device only and are not to be confused with the phase factors associated with the gap function itself. Choosing c_i equal to zero away from vacancies is done only because we are choosing to evaluate Δ self-consistently at those sites and does not mean that the gap is zero or neglected elsewhere on the lattice in our calculation.

With choices of the coefficients c_i described above, cancellation of random phases [6] gives the gap equation: We define $F_r(t) = \sum_i \pm e^{-i\phi_i} F_1$ where the plus sign is used if i is adjacent to an x bond and the minus sign is used if i is adjacent to a y bond. Then gap equation takes the form

$$\Delta^* = \frac{-1}{2N_s\beta\hbar} \int_0^\infty \frac{F_r(t)}{sinh(t\pi/\beta\hbar)} dt. \qquad (4)$$

where N_s is the number of oxygen vacancies. By use of the preceding expressions we obtain the equations of motion for F_1 and G_1 which are essentially identical to the equation of motion written above.

We solve these equations by simple numerical integration in the time domain. The model favors equal phases of the gap for vacancies on parallel bonds but phases of the gap differing by π for vacancies on perpendicular bonds. After a self-consistent solution for the average gap Δ has been found using the methods outlined above, then the density of states can be found by essentially the same techniques[5].

PARAMETRIZATION

We set the parameter b (more commonly called t) $= 1$ throughout, thus measuring energies in units of b. Thus the energy unit is of the order of 1 eV. The parameter ϵ is chosen so that the one electron states on the sites next to the special bonds associated with finite J (and with oxygen vacancies in the physical interpretation) have diagonal energies near the top of the band. In our interpretation the band is predominantly oxygen like, but the sites next to the special bonds are copper like. Thus we may compare ϵ with ϵ_{pd} recalling that we are working in an electron representation, but must bear in mind that, because the copper sites in question are next to oxygen vacancies the value of ϵ would be expected to be lower than the value of ϵ_{pd}. The physical picture requires that $\epsilon < \mu$, otherwise the oxygen vacancies would not be screened as discussed in the introduction. These arguments by no means fix the parameter ϵ uniquely. In most of the results presented in the next section, we take $\epsilon/b = 3.5, \mu/b = 3.8$. The value of ϵ used is not very different from reported estimates of ϵ_{pd} though these estimates vary over a wide range.

The value of the concentration p of oxygen vacancies in the plane is not known with accuracy better than $\pm1\%$ for any high temperature superconductor. Experimental reports on samples claiming p is zero are not uncommon but this limitation on the experimental resolution must be born in mind. If it is definitively shown that high T_c superconductors exist with $p = 0 \pm 10^{-4}$ then this model will have been shown to be irrelevant to the basic physics of the superconductivity. This does not appear to have occurred at present. In the calculations reported below, we mainly report results for the value $p = .04$ which is less than the value of 5% , which is the highest reported value [7] of which we are aware. We have made exploratory calculations on a range of p values up to 5%. The qualitative results are similar to those reported below as long as p is of the order of a few percent.

The value of J is completely unknown. We find that the mean field model produces unphysical results when J is greater than about 2.0 (the exact upper limit depends on the other parameters) and we usually choose J =1.5 so that it is as large as possible while remaining in the physical region. This choice is made for computational convenience, because calculations with small J require very long time integrations which are numerically very expensive. On the other hand, a J of order unity is not completely inconceivable physically: The J for the ordinary copper sites is known to be an extraordinarily large number of the order of .5 eV. Though the physics of the J in this model is quite different, this renders a value of order unity somewhat plausible. In fact, J could be substantially smaller and still yield a satisfactory T_c from an experimental point of view. The transition temperatures, however, cannot be taken seriously in comparison with experiments at this stage of model building in any case.

Finally, the physical picture in the introduction requires that the fermi level μ be near the top of the (oxygen like) band, so that the number of holes should be in the range to the characteristically small carrier hole densities observed in high T_c materials. We have studied the model for values of μ in a range satisfying this criterion (usually $\mu/b = 3.8$).

RESULTS

In Figure 1 we show $T_c(p)$. The nonmonotonicity of T_c as a function of p can be understood as a density of states effect[8] . The plateau which appears this Figure increases the physical plausibility of the model siginificantly, because it means that the approximate experimental reproducibility of T_c from sample to sample (in which the oxygen vacancy concentration certainly varies by a few percent) might not invalidate the present approach.

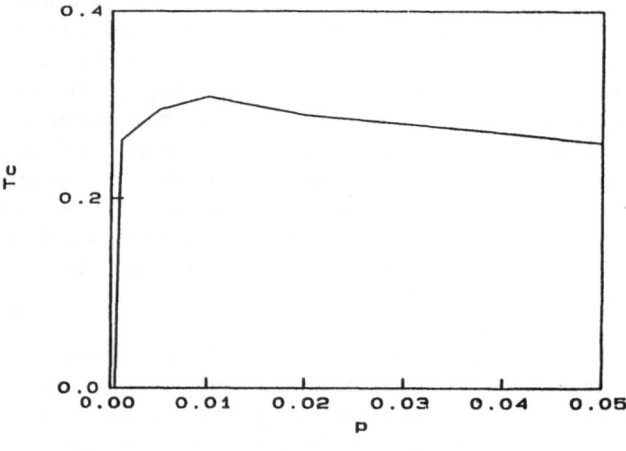

Figure 1. $T_c(p)$.

The local charge density at the special sites is of interest, because it is believed to be related to the positron annihilation rate for positrons trapped at vacancy sites [9-15] For some time, experiments have existed[9-15] which showed that a kink exists at T_c in the temperature dependence of this annihilation rate. The annihilation rate is in turn proportional to the electron density at the trapping site. In one paper[15], the measured rates were converted to a local electron density at the vacancy as shown in Figure 2a. In Figure 2b, we show a calculation of the temperature dependence of the local charge density as calculated in the model with the same parameters used for most of the other calculations reported here ($\mu/b = 3.8, p = .04, \epsilon/b = 3.5, J = 1.5$). A kink occurs at T_c, as qualitatively expected in this model, in which the phase transition produces increased phase coherence of the order parameter at the special "oxygen vacancy" sites. The scale of T_c is unrealistic but one sees that in other respects the temperature dependence of the calculated local charge density is in semiquantitative agreement with this experimental data on YBCO. Other investigators see similar effects in positron annihilation in $YBa_2Cu_3O_{7-\epsilon}$ but in $La_{2-\epsilon}Sr_\epsilon CuO_4$ and the thallium compounds, the reported effect of superconductivity on the positron lifetime has the opposite sign. These variations in the sign of the positron annihilation lifetime effect in various high T_c materials may be associated with differences in the trapping behavior of the positrons in these materials.

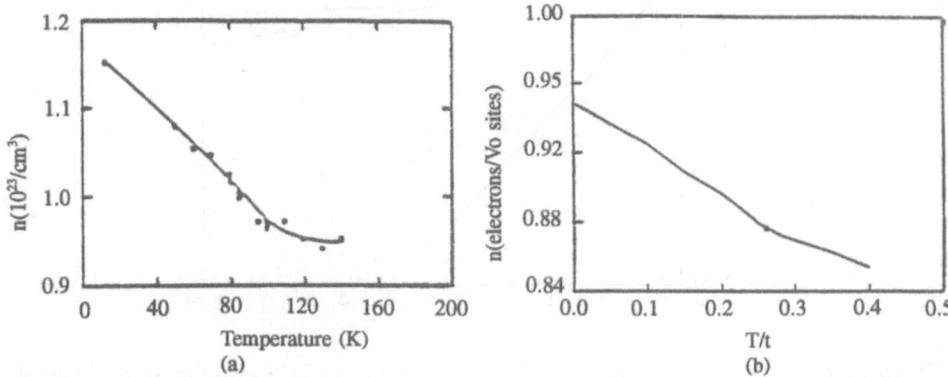

Figure 2. a) Experimentally deduced electron density at oxygen vacancies from reference 15. b) Calculation of the same quantity in the model.

A model for the effects of additional, nonmagnetic impurities on the superconductivity in this model has been described earlier[16]. Qualitatively, the expectation is that such nonmagnetic impurities should be more effective in destroying superconductivity than they would be in a model with a spatially uniform pairing force, because the nonmagnetic impurities can effectively act at the "weak links" between the regions of large pair correlations. To explore this effect with the present parametrization, we have found self-consistent solutions to the same BCS -like equations described in the previous section, in the case that the model is modified so that ϵ_i takes three values: 0 (at "oxygen" sites, with probability 1-2p-x), ϵ (at "copper" sites next to oxgen vacancies; as in the preceding section, these occur at the neighbors of "oxygen vacancies" occurring on the bonds of the lattice with probability p) and ϵ' (at "zinc" sites, distributed at random with site probability x.) The model thus differs from the one described in the preceding section only in the addition of randomly positioned sites with a value ϵ' different from 0 or the value ϵ assigned to "copper" sites next to oxygen vacancies. (In contrast, in reference 16, we used an Anderson model of continuously distributed values of ϵ_i to describe the effects of other impurites.) Because x and p are both much less than one in the calculations we do, we neglect any possible effect of improbable sites which are both at the ends of "special" bonds and which are selected for the value $\epsilon_i = \epsilon'$. To establish that the conjectured effects of random heterogeneous distribution of the pairing force on the sensitivity of T_c to the new "zinc" impurities, we also solved the BCS equation for a uniform pairing force in the same model, with J adjusted to give the same T_c at x=0. For ϵ' we used the value 2.5/b.

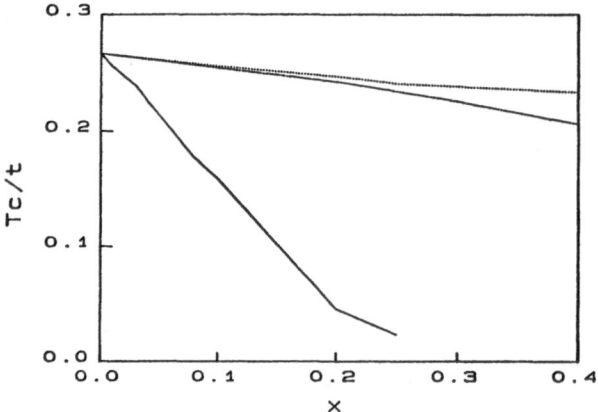

Figure 3. $T_c(x)$ for s (broken line) and d (dotted line) wave BCS ordering in a tJ model with uniform J and in the present model (solid line).

In Figure 3 we show the effects of the added impurities on T_c in the case of s and d wave pairing in the case of homogeneous pairing and in the case of heterogeneous pairing with the parameters described in the previous section ($\mu/b = 3.8$). One sees in Figure 3 that the result of the heterogeneous pairing model is to very sharply increase the sensitivity of T_c to the presence of on site disorder of this type, as we conjectured and partly demonstrated earlier.

In Figure 4a we show experimental data [17] on T_c as a function of substitutional concentration for a number of cationic substituents for copper in YBCO and in Figure 4b we show the corresponding calculation in the model as discussed in the preceding section.

The trends are quite strikingly similar, and (Figure 3) T_c is much more sensitive to spin independent elastic scatterers than it is in a model with uniform pairing forces. (Because we have used an artificially large value of J for numerical reasons, the magnitudes of T_c in Figure 4b are unrealistically large.) No kind of fine tuning of the parametrization of the model was made to produce the result in Figure 4b.

We note that the model qualitatively predicts that, at fixed p, ("oxygen vacancy" concentration) varying the carrier concentration (controlled by μ in this model) can be expected to lead, as μ increases, to a rise in T_c as μ comes into resonance with the available level on the "Cu^{1+}" sites next to the special bonds ("oxygen vacancies") as controlled by ϵ in the model and then to a fall as μ moves far above this level. Such a rise and fall of T_c with carrier concentration does occur , of course, in all high T_c materials, though it could have several origins and it is by no means clear that the oxygen vacancy concentration in the

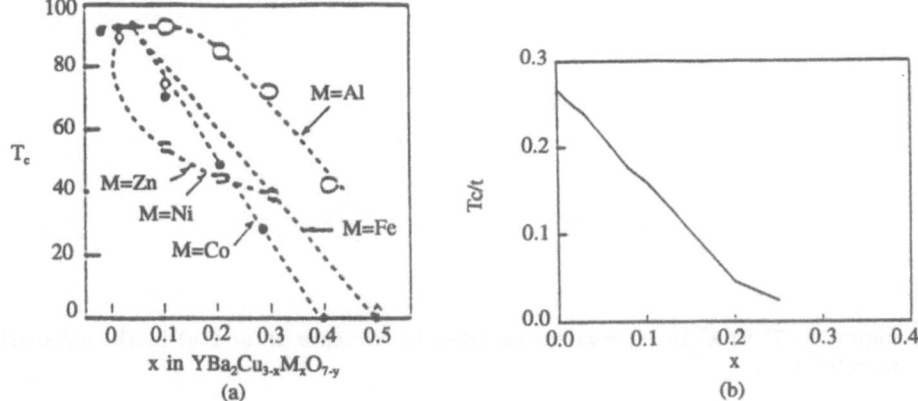

Figure 4. a) Experimental data on $T_c(x)$ compared with b) calculation in the present model.

plane remains fixed as the carrier concentration is varied. Nevertheless, we made such a calculation within the model, varying μ around the value 3.8b used elsewhere in these calculations and fixing the other parameters (p=.04, J/b=1.5, $\epsilon/b = 3.5$) with results shown in Figure 5b for comparison with experiments on T_c as a function of carrier concentration (as determined from Hall effect measurements[18] on thin films of $DyBa_2Cu_3O_{7-\delta}$) as shown in Figure 5a. Here again, the results are qualitatively quite similar to the experimental ones and we stress that no special adjustments were made to fit the data.

DISCUSSION AND CONCLUSIONS

A first question about models of this sort is whether the required oxygen vacancies exist in the vast array of cuprate superconductors known experimentally. In this regard, we point out that only a few percent of oxygen vacancies are required, and that existing experimental methods do not determine oxygen vacancy concentration to better than a few percent. It should be noted that the model refers to in-plane oxygen vacancies. Apical oxygen vacancies[19] might also play such a role. (In the case of compounds of the La_2CuO_4 type, it is reported[20] that vacancies in the copper oxygen planes, as postulated here, could account as well for the experimental results reported in reference 19 as the apical oxygen vacancies postulated in reference 19 .) Vacancies in the chain planes of 123 type compounds almost certainly would not have the effects required for the present model. Experimentally,

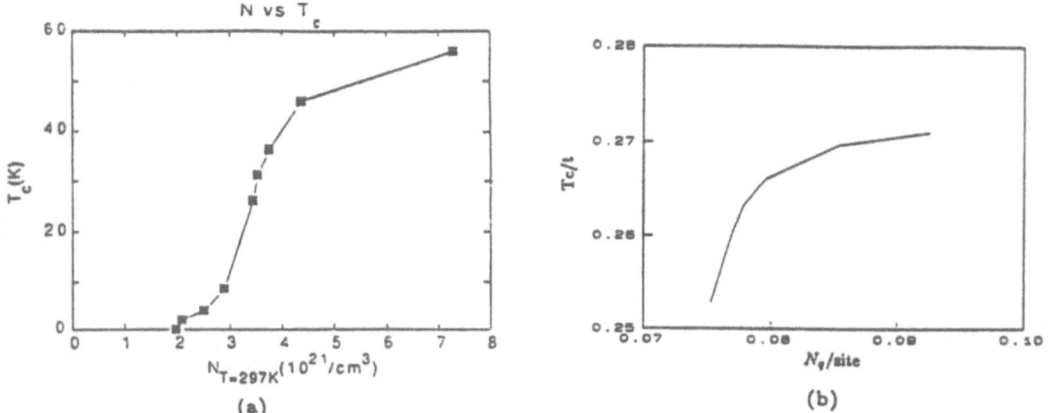

Figure 5. $T_c(n)$ a) in experiments reported in reference 18 and b) in the calculation reported here.

of course, one produces superconductivity in 123 compounds by *adding* oxygen to the chain planes. This would not contradict the assumptions of the present model if we imagine that adding oxygen in the chain planes may drive some oxygen off the superconducting copper oxygen sheets, producing the in plane oxygen vacancies required in the model. Such a hypothesis could be tested by Madelung energy calculations which we have not undertaken. Point defect energy calculations have been undertaken by Baetzold[21]. In $YBa_2Cu_3O_7$, he reports vacancy formation energies for oxygen sites 1 (chain plane), 4 (apical), 2 and 3 (sheets) of 17.23, 17.34 21.90 and 21.76 eV respectively. More importantly, however, he reports the considerably lower energy of 1.19 eV for formation of a Frenkel defect in which the oxygen at site 1 in the chain plane is moved to the oxygen at site 5, also in the chain plane. The hypothesis discussed in this paragraph postulates the (probably nonequilibrium) formation of Frenkel defects in which an oxygen at site 2 or 3 is moved to site 5. No calculation of the energy cost of such a Frenkel defect is reported in reference 21.

ACKNOWLEDGEMENTS

We wish to thank the Minnesota Supercomputer Institute and the Electric Power Research Institute for support.

REFERENCES

1. J. G. Bednorz and K. Muller, Z. Phys. B64, 189 (1986)
2. M. K. Wu et al, Phys. Rev. Lett. 58, 908 (1987)
3. J. W. Halley and H. B. Shore, Phys. Rev. B37, 525 (1988)
4. J. C. Phillips, Phys. Rev. Lett. 58, 1028 (1987)
5. J. W. Halley, S. Davis and X-F Wang, (unpublished)
6. J. W. Halley and H. B. Shore, Phys. Rev. B37, 525 (1987)
7. M. A. Beno, L. Soderheim, D. W. Capone II. D. G. Hinks, J. D. Jorgensen, J. D. Grace, I. K. Schuller, C. U. Segre and K. Zhang, Appl. Phys. Lett 51, 57 (1987)
8. J. W. Halley, X. F. Wang and S. Davis (unpublished)
9. Y. C. Jean, et al ,Phys, Rev. B36, 3994 (1987)
10. Y. C. Jean et al, Phys. Rev. Lett. 60, 1069 (1988)
11. Y. C. Jean et al, J. Phys. Condens. Mater. 1, 2696 (1989)
12. Y. C. Jean, et al Phys. Rev. Lett. 64, 1593 (1990)
13. D. R. Harshman, L. F. Schneemeyer, Y. V. Waszzak, Y. C. Jean, M. J. Fluss, R. H. Howell and A. L. Wachs, Phys. Rev. B38, 848 (1988)
14. E. C. von Stetten, S. Berko, S. S. Li, R.R. Lee, J. Brynstead, D. Singh, H. Krakauer, W. E. Pickett, and R. E. Cohen, Phys, Rev. Lett. 60, 2198 (1988)
15. S. G. Usmar, P. Sferlazzo, K. G. Lynn and A. R. Moodenbaugh, Phys. Rev. B 36, 8854 (1987)
16. J. W. Halley, S. Davis and S. Sen, Physica B165&166, 999 (1990)
17. J. M. Tarascon and B. G. Bagley in "Chemistry of Superconductor Materials",T. A. Vanderah, editor, Noyes Publications, Park Ridge, New Jersey (1992), Chapter 8, p.310
18. T. Wang, K. M. Beauchamp, D. D. Berkley, B. R. Johnson, J. X. Liu, T. Zhang and A. M. Goldman, Phys. Rev. B 43, 8623 (1991)
19. Z. Tan, M. E. Filipkowski, J. I. Budnick, E. K. Heller, D. L. Brewe,B. L. Chamberland, C. E. Bouldin, J. C. Woicik and D. Shi, Phys. Rev. Lett, 64 , 2715(1990)
20. J. I. Budnick, private communication
21. R. Baetzold, Physica C 181, 252 (1991)

COULOMB COUPLING AND VIRTUAL PHONON EXCHANGE

IN SPATIALLY SEPARATED 2D ELECTRON SYSTEMS

P. Vasilopoulos and H. C. Tso

Concordia University, Department of Physics
1455 de Maisonneuve O., Montréal, Québec, Canada

ABSTRACT

Momentum transfer between two parallel quasi-two-dimensional wells, separated by a barrier of thickness d, is studied at low temperatures when an weak electric field is applied to one of them. Screening is treated self-consistently in the random-phase approximation and a drifted-polarizability model is used for the *nonequilibrium* electron polarizability that characterizes the resulting interlayer coupling. The transfer is mostly due to the direct (screened) interaction and the exchange of virtual phonons between electrons of the two wells. The first interaction has a nearly quadratic temperature (T) dependence, for low T, and varies approximately as d^{-4} at large d. The latter shows a peak at low temperatures and depends very weakly on the separation d. The two interactions together account very well for the recently observed temperature and separation dependences of the interwell momentum transfer rate. It is shown that exchange of real phonons between electrons of the two wells is orders of magnitude weaker a process than that of virtual phonons and cannot account for the observations. The evaluated dependence of the transfer rate on the electron densities is in agreement with the observed one. The induced current in the drag well is also evaluated as function of temperature and separation.

INTRODUCTION

The Coulomb coupling between two *spatially* separated electron-gas layers has been predicted long ago[1] to influence the transport properties of these layers. When an electric field is applied to one of the layers it induces, through momentum transfer, a *contactless* current in the other. This was confirmed recently in a system consisting of a semi-infinite three-dimensional (3D) layer and a quasi-two-dimensional (2D) electron gas 300Å apart with the electric field applied to either of them[2].

The observation of Ref. [2] motivated further research. On the theoretical side Ref. [3] treated succesfully the 3D-2D coupling pertinent to the experiments of Ref.

[2], Ref. [4] the coupling between two quantum wires in the presence of a magnetic field, and Ref. [5] the coupling between a one-dimensional (1D) and an i-dimensional (iD) ($i = 1, 2, 3$) electron gas. Experimentally, Ref. [6] reported the observation of a *contactless* current in a 2D-2D electron-hole system. On the other hand, in a 2D-2D electron-electron system using high impedance circuits no current was allowed to flow in one of the wells; as a result a voltage developed in one of them, called the drag well, due to the accumulation of electrons swept along the direction of the electric field which was applied to the drive well[7]. This voltage is *opposite* to the resistive voltage in the drive well and balances the drag due to the interlayer interactions that are responsible for momentum transfer.

The first theoretical calculations[7, 8] pertinent to the experiments of Ref. [7] considered only the direct Coulomb interaction between the two wells and found that the interwell transfer rate should have a nearly T^2 dependence and a d^{-4} dependence on the separation d between the two wells. However, the observed[7] deviations of this rate from the predicted ones and especially the d dependence made it clear that an additional mechanism was involved in the momentum transfer. Exchange of real phonons between electrons in the two wells was estimated to be too small to account for the observations[7]. It was later shown that virtual phonon exchange between the wells could account for the observations[9, 10].

In this paper we present a complete treatment of the observations of Ref [7] and give the details missing from the brief Ref. [9]. In addition, we evaluate the current induced in the drag well, if allowed to flow, as function of temperature and separation. In the next section we present the momentum balance equations. In Sec. III we first give the relaxation frequencies for impurity scattering, for the direct Coulomb and phonon-mediated interactions, and for the usual electron-phonon interaction; this is followed by the induced current in terms of these frequencies. Numerical results follow in Sec. VI. Finally, a summary and a brief discussion are given in the last section.

MOMENTUM BALANCE EQUATIONS

We consider a system of two quasi 2D quantum wells as shown in Fig. 1. For

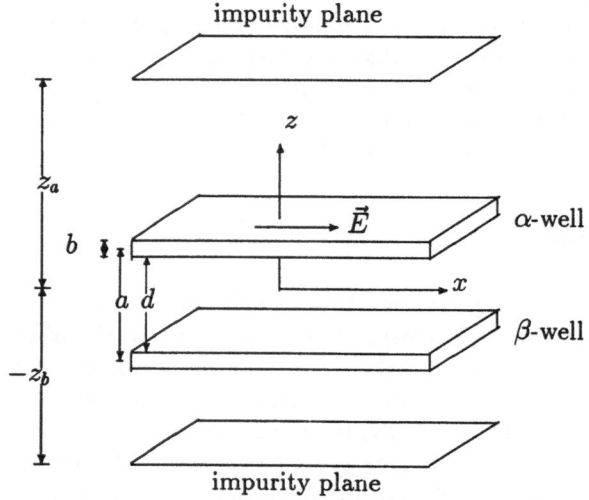

Fig.1. The geometry of double-quantum well structure.

simplicity we take the wells as square and infinitely deep since we do not expect the coupling to depend sensitively on the shape of the wells. We assume that only the lowest one-electron levels are occupied in the wells. Moreover, since the reported experiments involve rather large separations d we neglect tunneling. This makes the role of the other interlayer interactions easier to assess.

The many-body Hamiltonian describing this system is

$$H(t) = H_\alpha(t) + H_\beta(t) + H_{\alpha\beta}(t) + H_p(t) + H_i(t) , \tag{1}$$

where H_α is the Hamiltonian of the electrons in the drive (α) well and contains electron-electron, electron-impurity, and electron-phonon interactions; H_β is the counterpart for the electrons in the drag (β) well; $H_{\alpha\beta}$ is the Coulomb interaction between electrons in the α and β well; H_p and H_i are the phonon and impurity Hamiltonians, respectively. The electron self-energy in the drive well is illustrated[11] as follows:

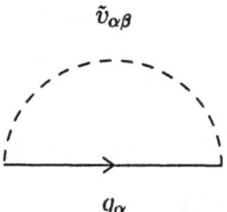

Fig. 2. The electron self-energy in the drive well.

Here g_α is the thermodynamic Green's function of the electrons in the drive well and $\tilde{v}_{\alpha\beta}$ is the *nonequilibrium* screened interaction between electrons in the α well and the medium which includes electrons in the β well, phonons, and impurities. The electric field $\vec{E} = E\hat{x}$ is included in the one-electron Hamiltonians, contained in the many-body counterparts of Eq. (1), by means of a time-dependent vector potential $\vec{A}_\gamma(t) = -(t_1 - t_0)E_\gamma\hat{x}, \gamma = \alpha, \beta$, with $E_\alpha = E$ and $E_\beta = 0$. Applying the operators $\lim_{2\to1+}\{-(\bar{e}/2m^*_\gamma)[\vec{\nabla}_1 - \vec{\nabla}_2 2i\bar{e}\vec{A}_\gamma(t_1)]\}$ to the equations of motion for g_γ we obtain the momentum balance equations for electrons in the drive and the drag well, respectively:

$$-\frac{m^*_\alpha}{\bar{e}}\frac{\partial\vec{j}_\alpha}{\partial t} + en_\alpha\vec{E} = \vec{F}^{\alpha\beta} + \vec{F}^{\alpha\beta}_p \tag{2}$$

$$-en_\beta\frac{V_D\vec{j}_\beta}{L\,j_\beta} - \frac{m^*_\beta}{\bar{e}}\frac{\partial\vec{j}_\beta}{\partial t} = \vec{F}^{\beta\alpha} + \vec{F}^{\beta\alpha}_p . \tag{3}$$

Here, V_D is the measured drag voltage, L the length of the specimen, n_α (n_β) the 2D density of the drive (drag) well, $\vec{j}_\alpha$ ($\vec{j}_\beta$) the corresponding 2D current density, $\vec{E}$ the electric field applied only in the drive well, $\bar{e} = e/\sqrt{\epsilon_\infty}$, and t_0 the time when $\vec{E}$ is switched on. $F^{\alpha\beta}_p$ is the coupling force due to the (virtual) phonon-mediated interaction. The force due to direct Coulomb, phonon (without the virtual part), and impurity scatterings is

$$\vec{F}^{\alpha\beta}(t_1) = \int dz_1 \int_{t_0}^{t_1} d(3)\, [\Pi^\alpha_> \vec{\nabla}_1 \tilde{v}_{\alpha\beta} > - \Pi^\alpha_< \vec{\nabla}_1 \tilde{v}_{\alpha\beta} <] \tag{4}$$

expressed in terms of $\Pi^\alpha \equiv \Pi^\alpha(1;3)$, the *nonequilibrium* polarizability, and $\tilde{v}_{\alpha\beta} \equiv \tilde{v}_{\alpha\beta}(1;3)$, the *nonequilibrium* screened interaction between electrons in the drive well and the medium. The arguments in $\Pi^\alpha(1;2)$ and $\tilde{v}_{\alpha\beta}(1;2)$ represent the correlation between the space-time coordinates $(\vec{x}_1, t_1)$ and $(\vec{x}_2, t_2)$ and the subscripts $>$ or $<$ refer to the corresponding functions when t_1 is larger or smaller than t_2, respectively. The integration $\int_{t_0}^{t_1} d(3)$ is a shorthand for $\int d\vec{x}_3 \int_{t_0}^{t_1} dt_3$. The appearance of the drag voltage V_D in Eq. (3) is due to the accumulation of electrons at one end of the drag well. If the system is uniform, i.e., if there is no electron accumulation at any end, which happens when the current is allowed to flow in the drag well, this term vanishes identically. The nonequilibrium screened interaction[12],[4] in real-time domain $\tilde{v}_{\alpha\beta} \gtrless$ is given by

$$\tilde{v}_{\alpha\beta} \gtrless (1;2)$$
$$= \int d(3) \int d(4) \bar{v}'_{\alpha i +}(1;3) n_i^{2D} \delta(x_3 - x_4) \bar{v}'_{\alpha i -}(4;2) \delta(z_3 - z_i)$$
$$+ \int d(3) \int d(4) \bar{v}_+(1;3)[\Pi^\alpha_\lessgtr(3;4) + \Pi^\beta_\lessgtr(3;4)] \bar{v}_-(4;2)$$
$$+ \sum_{\vec{k},\lambda} \frac{1}{4\pi^2} \int d\omega_1 \int d\omega_2 \bar{M}'_{\alpha +}(\vec{k};\omega_1) e^{i\vec{k}\cdot(\vec{x}_1 - \vec{x}_2) - i\omega_1(t_1 - t_3)}$$
$$\times 2\omega_\lambda D_\lambda \gtrless (\vec{k}; t_3 - t_4) \bar{M}'_{\alpha -}(\vec{k};\omega_2) e^{-i\omega_2(t_4 - t_2)} , \qquad (5)$$

In Eq. (5), n_i^{2D} is the 2D density of impurities located at $z = z_i$. Further, the retarded (advanced) screened electron-impurity interaction $\bar{v}'_{\alpha i +}$ ($\bar{v}'_{\alpha i -}$), the screened Coulomb interaction $\bar{v}_+$ ($\bar{v}_-$) and the screened electron-phonon interaction $\bar{M}'_{\alpha +}$ ($\bar{M}'_{\alpha -}$) are related to the retarded (advanced) equilibrium inverse dielectric function K_+ (K_-) by the following equations:

$$\bar{v}'_{\alpha i \pm}(1;2) = \int d(3) K_\pm(1;3) v_{\alpha i \pm}(\vec{x}_3 - \vec{x}_2)\delta(t_3 - t_2) \qquad (6)$$

$$\bar{v}_\pm(1;2) = \int d(3) K_\pm(1;3)\delta(t_3 - t_2) \frac{\bar{e}^2}{|\vec{x}_3 - \vec{x}_2|} \qquad (7)$$

$$\bar{M}'_{\alpha \pm}(\vec{k};\omega) = \int d(\vec{x}_1 - \vec{x}_2) \int d(t_1 - t_2) K_\pm(1;3)$$
$$\times \sum_{\vec{k}} \frac{1}{2\pi} \int d\omega \, e^{i\vec{k}\cdot(\vec{x}_3 - \vec{x}_2) - i\omega(t_3 - t_2)} M'_{\alpha \pm}(\vec{k};\omega) \qquad (8)$$

with $(\zeta = q_\| b, \ v(q_\|) = 2\pi \bar{e}^2/q_\|)$

$$M_\alpha(\vec{q};\omega) = M^\alpha_\lambda(\vec{q};\omega_\lambda)[1 - v(q_\|)F(q_\|)(1 + e^{-q_\| a})\Pi^{\beta (0)}(\vec{q}_\|;\omega)] , \qquad (9)$$

$$F(q_\|) = [3\zeta + \frac{8\pi^2}{\zeta} - \frac{32\pi^4}{\zeta^2} \frac{(1 - e^{-\zeta})}{(\zeta^2 + 4\pi^2)}]/(\zeta^2 + 4\pi^2) , \qquad (10)$$

$M^\alpha_\lambda(\vec{q};\omega_\lambda)$ the matrix element of the electron-phonon interaction, and $\Pi^{\beta (0)}(\vec{q}_\|;\omega)$ the *equilibrium* polarizability of electrons in the drag well.

For weak electric fields, we may write $\vec{j}_\gamma = n_\gamma \bar{e} \vec{v}_d^\gamma$ with $\vec{v}_d^\gamma$ the drift velocity in the γ well, and employ the drifted-polarizability model[4] for the *nonequilibrium* polarizabilities,

$$\Pi_>^\gamma(1;2) = -2i \sum_{\vec{q}_\parallel} |\xi_\gamma(z_1)|^2 \xi_\gamma(z_2)|^2 e^{i\vec{q}_\parallel \cdot (\vec{x}_{1\parallel} - \vec{x}_{2\parallel})} e^{-i(E_{\vec{k}_\parallel}^\gamma - E_{\vec{k}_\parallel - \vec{q}_\parallel}^\gamma - \vec{q}_\parallel \cdot \vec{v}_d^\gamma)(t_1 - t_2)}$$

$$\times [1 - f_\gamma(E_{\vec{k}_\parallel}^\gamma)] f_\gamma(E_{\vec{k}_\parallel - \vec{q}_\parallel}^\gamma) \tag{11}$$

where E_γ is the electron energy and ξ_γ the one-electron wave function in the γ well. For $\Pi_<^\gamma(\cdots)$, we obtain Eq. (11) with the Fermi-Dirac factors interchanged. The linearized total force is then given by

$$\vec{F}^{\alpha\beta} = (\Omega_{\alpha\beta}^c + \Omega_{\alpha\beta}^p) m_\alpha^*(\vec{v}_d^\alpha - \vec{v}_d^\beta) + (\Omega_{\alpha i} + \Omega_{\alpha p}) m_\alpha^* \vec{v}_d^\alpha , \tag{12}$$

with $\Omega_{\alpha\beta}^c, \Omega_{\alpha\beta}^p$, $\Omega_{\alpha p}$, and $\Omega_{\alpha i}$, the relaxation frequencies per unit area due to direct Coulomb and phonon-mediated coupling, phonon and impurity scattering, respectively. The force $\vec{F}^{\beta\alpha}$ is obtained from Eq. (12) by interchanging the indices α and β. The first term in Eq. (12) induces a current: electrons in the drive well acquire monmentum from the electric field ($E_\alpha \neq 0, v_d^\alpha \neq 0$) and drag the electrons in the β well ($E_\beta = 0$) by transfering to them part of their momentum ($v_d^\alpha > v_d^\beta$) via the direct and phonon-mediated interactions. In the steady state and zero drag current limit, Eq. (3) becomes

$$- e n_\beta V_D / L = (\Omega_{\alpha\beta}^c + \Omega_{\alpha\beta}^p) v_d^\alpha \tag{13}$$

and V_D / v_d^α depends only on $\Omega_{\alpha\beta}^c$ and $\Omega_{\alpha\beta}^p$.

RELAXATION FREQUENCIES

Impurity scattering

For the system we consider, we have two layers of remote impurities located to the left of the left well and to the right of the right well as illustrated in Fig. 1. The geometry was chosen such as to correspond to the experiments of Ref. [7]. We assume that the impurity potential is Coulombic and weak enough that it can be treated within the self-consistent Born approximation. We obtain the pertinent relaxation frequency $\Omega_{\gamma i}$ as

$$\Omega_{\gamma i} = \frac{8\pi^2 \bar{e}^2 n_i^{2D}}{m_\alpha^*} \sum_{\vec{q}_\parallel} \frac{|J(q_\parallel)|^2}{\epsilon_+(\vec{q}_\parallel; 0)} \frac{\partial Im[\Pi_+^{\gamma(0)}(\vec{q}_\parallel; \omega)]}{\partial \omega}\bigg|_{\omega=0} , \tag{14}$$

and the form factor $J(q_\parallel)$ in the form

$$J(q_\parallel) = \int dz_1 \int dz_2 \, e^{-q_\parallel |z_1 - z_2|} [\delta(z_2 - z_a) + \delta(z_2 - z_b)] , \tag{15}$$

where z_a and z_b are the z-coordinates of the remote-impurity planes.

Coulomb scattering and equilibrium inverse dielectric function

To treat Coulomb scattering self-consistently, we have to determine the inverse dielectric function of the system under consideration. We employ the inversion condition,

$$\int dz_3 \, \epsilon_\pm(\vec{q}_\parallel, z_1, z_3; \omega) K_\pm(\vec{q}_\parallel, z_3, z_2; \omega) = \delta(z_1 - z_2) \tag{16}$$

with $\epsilon_\pm$ the retarded/advanced dielectric function and $K_\pm$ the retarded/advanced inverse $\epsilon_\pm$. In the random-phase approximation (RPA) the dielectric function can be written as

$$\epsilon_\pm(\vec{q}_\parallel, z_1, z_2; \omega) = \quad \delta(z_1 - z_2) + v(q_\parallel) \int dz_3 \, |\xi_\alpha(z_3)|^2 |\xi_\alpha(z_2)|^2 e^{-q_\parallel |z_1 - z_3|} \Pi_\pm^{\alpha \, (0)}$$
$$+ v(q_\parallel) \int dz_3 \, |\xi_\beta(z_3)|^2 |\xi_\beta(z_2)|^2 e^{-q_\parallel |z_1 - z_3|} \Pi_\pm^{\beta \, (0)} , \tag{17}$$

where

$$\Pi_\pm^{\gamma \, (0)} = 2 \sum_{\vec{k}_\parallel} \frac{f_\gamma(E_{\vec{k}_\parallel}^\gamma) - f_\gamma(E_{\vec{k}_\parallel - \vec{q}_\parallel}^\gamma)}{\omega + E_{\vec{k}_\parallel}^\gamma - E_{\vec{k}_\parallel - \vec{q}_\parallel}^\gamma \pm i0^+} ; \tag{18}$$

$E_{\vec{k}_\parallel}^\gamma = \hbar^2 k_\parallel^2 / 2m_\gamma^*$ and f_γ is the Fermi function of electrons in the γ well. Inversion of Eq. (16) gives

$$K_\pm(\vec{q}_\parallel, z_1, z_2; \omega) = \quad \delta(z_1 - z_2) + v(q_\parallel) \int dz_3 \, |\xi_\alpha(z_3)|^2 \Pi_\pm^{\alpha \, (0)} H_\pm^\alpha(\vec{q}_\parallel, z_2; \omega)$$
$$+ v(q_\parallel) \int dz_3 \, |\xi_\beta(z_3)|^2 \Pi_\pm^{\beta \, (0)} H_\pm^\beta(\vec{q}_\parallel, z_2; \omega) , \tag{19}$$

where

$$H_\pm^{\alpha,\beta}(\vec{q}_\parallel, z_2; \omega) = \quad \frac{[1 - v_{\beta\beta} \Pi_\pm^{\beta,\alpha \, (0)}] |\xi_{\alpha,\beta}(z_2)|^2 + v_{\alpha\beta} \Pi_\pm^{\beta,\alpha \, (0)} |\xi_{\beta,\alpha}(z_2)|^2}{\epsilon_\pm(\vec{q}_\parallel; \omega)} , \tag{20}$$

$$\epsilon_\pm(\vec{q}_\parallel; \omega) = \quad [1 - v_{\alpha\alpha} \Pi^{\alpha \, (0)}][1 - v_{\beta\beta} \Pi^{\beta \, (0)}] - v_{\alpha\beta} \Pi^{\beta \, (0)} v_{\beta\alpha} \Pi^{\alpha \, (0)} , \tag{21}$$

and

$$v_{\alpha\beta} = v(q_\parallel) \int dz_1 \int dz_2 \, |\xi_\alpha(z_1)|^2 e^{-q_\parallel |z_1 - z_2|} |\xi_\beta(z_2)|^2 . \tag{22}$$

Employing the inverse dielectric function and the drifted-polarizability model[8] in Eq. (4), we obtain

$$\Omega_{\alpha\beta}^c = -\frac{\hbar}{\pi m_\alpha^*} \int_{-\infty}^{\infty} d\omega \sum_{q_\parallel} \frac{|F(q_\parallel) v(q_\parallel) e^{-q_\parallel a}|^2}{|\epsilon_+(\vec{q}_\parallel; \omega)|^2} q_\parallel^2 \frac{\partial n(\omega')}{\partial \omega'} \Big|_{\omega' = -\omega} Im[\Pi_+^{\alpha \, (0)}] Im[\Pi_+^{\beta \, (0)}] , \tag{23}$$

where $n(\omega)$ is the Bose-Einstein distribution function and $\Pi_+^{\alpha \, (0)} \equiv \Pi_+^{\alpha \, (0)}(\vec{q}_\parallel)$. At low temperatures, the energy transfer ω is close to zero due to the presence of the Fermi factors in both Π_+^α and Π_+^β; this leads to [4]

$$\Omega_{\alpha\beta}^c \sim \frac{4m_\alpha^*}{\hbar^3} \sum_{\vec{q}_\parallel, k_\alpha, k_\beta} \frac{|v(q_\parallel) e^{-q_\parallel a}|^2}{|\epsilon_+(\vec{q}_\parallel; 0)|^2} (k_B T^2) q_\parallel \frac{\partial f_\alpha(E_{k_\alpha}^\alpha)}{\partial E_{k_\alpha}^\alpha} \frac{\partial f_\beta(E_{k_\beta}^\beta)}{\partial E_{k_\beta}^\beta} . \tag{24}$$

We see that $\Omega_{\alpha\beta}^c$ has approximately a T^2 dependence. As for the separation dependence we obtain, in the limit $d \to \infty$, that $\Omega_{\alpha\beta}^c$ behaves like d^{-4}. As pointed out in Ref. [8] this d dependence reflects the corresponding one of the dielectric function given by Eq. (17). These T and d dependences were also reported in Ref. [6].

Phonon-mediated electron-electron coupling

Apart from interacting directly the electrons in the two wells can exchange real or virtual phonons. This leads to two additional terms in the interwell interaction. The total phonon-mediated screened interaction $v_{\alpha\beta}^p(q_\parallel; \omega; t)$ reads

$$v_{\alpha\beta}^{p}(\vec{q}_{\parallel};\omega) \equiv \sum_{q_z,\lambda} |I(-iq_z)|^2 \frac{|M_\alpha'(\vec{q};\omega)|^2}{|\epsilon(\vec{q}_{\parallel};\omega;t)|^2} D_\lambda(\vec{q};\omega)[1 + \frac{\delta\tilde{\omega}_\lambda^\beta}{v_d^\beta}] \tag{25}$$

where

$$\delta\tilde{\omega}_\lambda^\beta \simeq \frac{2}{\hbar} \sum_{q_z} \frac{|M_\alpha'(\vec{q};\tilde{\omega}_\lambda)|^2}{|\epsilon_+(\vec{q}_{\parallel};\tilde{\omega}_\lambda)|^2} \frac{|I(-iq_z)|^2}{\epsilon_+(\vec{q}_{\parallel};\tilde{\omega}_\lambda)} \frac{\partial\Pi_+^\beta(\vec{q}_{\parallel};\tilde{\omega}_\lambda)}{\partial\tilde{\omega}_\lambda} \vec{q}_{\parallel} \cdot \vec{v}_d^\beta . \tag{26}$$

The first and second terms in the square brackets of Eq. (25) correspond to exchange of virtual and real phonons, respectively. Here $D_\lambda(\vec{q};\omega) = 2\omega_\lambda/(\omega^2 - \tilde{\omega}_\lambda^2)$ is the dressed phonon propagator and the *nonequilibrium* dielectric function is given by

$$\epsilon(\vec{q}_{\parallel};\omega;t) = \quad [1 - v_{\alpha\alpha}\Pi^\alpha(\vec{q}_{\parallel};\omega;t)][1 - v_{\beta\beta}\Pi^\beta(\vec{q}_{\parallel};\omega;t)]$$

$$-v_{\alpha\beta}\Pi^\beta(\vec{q}_{\parallel};\omega;t)v_{\beta\alpha}\Pi^\alpha(\vec{q}_{\parallel};\omega;t) , \tag{27}$$

further, for the modified phonon frequency $\tilde{\omega}_\lambda$ we have

$$\tilde{\omega}_\lambda^2 = \omega_\lambda^2 - 2\omega_\lambda \sum_{q_z} \frac{|M_\lambda^\alpha(\vec{q})|^2 |I(-iq_z)|^2}{\epsilon_+(\vec{q}_{\parallel};0)} [\Pi^{\alpha\,(0)}(\vec{q}_{\parallel};0) + \Pi^{\beta\,(0)}(\vec{q}_{\parallel};0)] \tag{28}$$

and for the form factor $I(-q_z)$ $(\eta = q_z b)$

$$I(-iq_z) = 8\pi^2 \sin(\eta/2)/[\eta(\eta^2 - 4\pi^2)] . \tag{29}$$

$\Pi^{\alpha,\beta}(\vec{q}_{\parallel};\omega;t)$ is the *nonequilibrium* polarizability of electrons in the drive (drag) well and $z_{\alpha i}$ the distance between the impurity plane and the center of the drive well. As can be seen from Eq. (25), the phonon term can be very large at certain range of energy transfer, i.e., when $\omega \sim \tilde{\omega}_\lambda$. In this case, the self-energy illustrated in Fig. (2), based on the assumption that $\tilde{v}_{\alpha\beta}$ is weak, is not valid. Higher-order terms due to phonon scattering must be taken into account.

Virtual phonon exchange. To simplify the computation, we employ the thin-well approximation and the jellium model for the term in Eq. (25) proportional to $|M(...)|^2$ denoted by v_0. To understand the validity of the model, we compute M_α from first principles using the monopole term of the electron-ion interaction instead of the usual parametrized interaction $|M_\alpha(\vec{q})|^2 = D^2q/2\rho s$ since D, the deformation potential, is the Fourier transform of the electron-ion interaction times the number density N of the lattice. In other words, $|M_\alpha(\vec{q})|^2 = 8(\pi\bar{e}^2N)^2/q^3\rho s$. Here ρ is the average lattice density and s the speed of sound. With this, the modified phonon modes become

$$\tilde{\omega}_\lambda^2 = \omega_\lambda^2 - v(q_{\parallel})\omega_0^2[\Pi^{\alpha\,(0)}(\vec{q}_{\parallel};0) + \Pi^{\beta\,(0)}(\vec{q}_{\parallel};0)]/\epsilon_+(\vec{q}_{\parallel};0) , \tag{30}$$

where $\omega_0^2 = 4\pi\bar{e}^2N^2/\rho$ is the plasma frequency. With $\rho = 5.36$ g/cm^3 and $N \sim 4.5 \times 10^{22} cm^{-3}$ for GaAs, $\omega_0 \sim 10^{12} sec^{-1}$. The dominant contribution to the drag comes from $\omega \sim k_B T \sim \omega_\lambda$ and $k_B T \sim 7 \times 10^{11} sec^{-1} \sim \omega_0$ at $T \sim 5$ K which justifies the use of the jellium model. When $\omega_\lambda = \omega_0$ (jellium model), $\tilde{\omega}_\lambda^2$ will further reduce to $\tilde{\omega}_\lambda^2 = \omega_\lambda^2/\epsilon_+(q_{\parallel};0)$, and the first term in the square bracket of Eq. (25)

$$v_0(q_{\parallel};\omega) \simeq \frac{v(q_{\parallel})}{q_{\parallel}\epsilon_+(\vec{q}_{\parallel};0)} \frac{\tilde{\omega}_\lambda^2}{\omega^2 - \tilde{\omega}_\lambda^2} . \tag{31}$$

Eq. (31) is only an approximation obtained within the jellium model for very thin-wells. The deviations Δv_0 from this value of v_0 can be represented by $\Delta v_0 = f_c(q,\omega)v_0$.

A full evaluation of $f_c(q, \omega)$ is heavily involved numerically speaking. To reduce the already involved numerical work we assume that it is approximately equal to the constant f_c; then the interaction including these deviations is written as

$$v_0(q_\parallel; \omega) \simeq \frac{v(q_\parallel)}{q_\parallel \epsilon_+(\vec{q}_\parallel; 0)}(1 + f_c)\frac{\tilde{\omega}_\lambda^2}{\omega^2 - \tilde{\omega}_\lambda^2} . \tag{32}$$

Because of the divergence of v_0 for $\omega \simeq k_B T \simeq \tilde{\omega}_\lambda$, four-point vertex corrections [see Eq. (149) of Ref. [11]] must be taken into account. Upon considering these corrections, the self-energy of the electrons in the drive well becomes

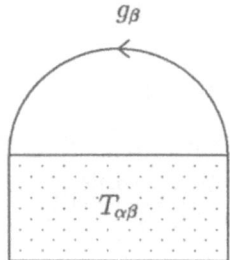

Fig. 3.The electron self-energy in the drive well beyond the jellium model and with four-point vertex corrections.

and the nonlocal interaction between electrons in different wells, $T_{\alpha\beta}$, obeys the self-consistent[13],[11] equatio n

$$T_{\alpha\beta}(1, 2; 3, 4) = v_0(1; 2)\delta(1; 3)\delta(2; 4)$$
$$+ \int d(5) \int d(6)\, v_0(1; 2)g_\alpha(1; 5)g_\beta(2; 6)\, T_{\alpha\beta}(5, 6; 3, 4) , \tag{33}$$

with

$$v_0(1; 2) = \sum_{\vec{q}_\parallel} \frac{1}{2\pi} \int d\omega\, e^{i\vec{q}_\parallel \cdot (\vec{x}_1 - \vec{x}_2) - i\omega(t_1 - t_2)} v_0(\vec{q}_\parallel; \omega) . \tag{34}$$

Diagramatically, $T_{\alpha\beta}$ is equal to

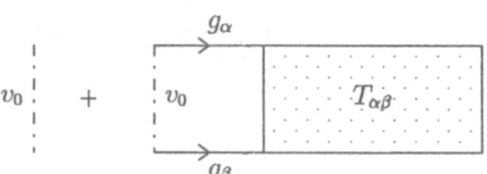

Fig. 4.The interaction $T_{\alpha\beta}$ between electrons in the α and β wells.

Solving Eq. (33) is very difficult. We may simplify it using the contact[13],[14] approximation

$$v_0(1; 2) = v_0\, \delta(|z_1 - z_2| - a)\delta(t_1 - t_2) \quad \text{for } \omega^2 < \tilde{\omega}_\lambda^2$$
$$= 0 \quad \text{for } \omega^2 > \tilde{\omega}_\lambda^2$$

with $v_0 = \lim_{\vec{q}_\parallel \to 0} v_0(\vec{q}_\parallel; 0)$. In this case, $T_{\alpha\beta}$ reduces to

$$T_{\alpha\beta}(1,2;3,4) = \sum_{\vec{k}_\parallel} \int \frac{d\omega}{2\pi} e^{i\vec{k}_\parallel \cdot (\vec{x}_{1\parallel} - \vec{x}_{3\parallel}) - i\omega(t_1 - t_3)} \tilde{v}_0(\vec{k}_\parallel; \omega) \delta(\vec{x}_{1\parallel} - \vec{x}_{2\parallel}) \delta(t_1 - t_2)$$

$$\times \delta(|z_1 - z_2| - a)\delta(\vec{x}_{3\parallel} - \vec{x}_{4\parallel})\delta(|z_3 - z_4| - a)\delta(t_3 - t_4) \quad (35)$$

where

$$\tilde{v}_0(\vec{k}_\parallel; \omega) = \frac{2\pi\hbar^2}{m^*} \frac{\xi_0}{\omega - \xi_0 - \hbar k_\parallel^2/4m^*}, \quad (36)$$

where $\xi_0 = \tilde{\omega}_\lambda \, exp[-\hbar^2/v_0(0;0)m^*\bar{e}^2]$. Diagramatically, we have the self-energy reduced to

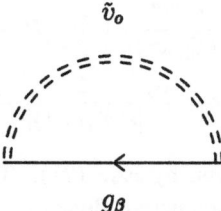

Fig. 5. The electron self-energy in the drag well using the contact approximation.

Consequently, the contribution to the relaxation frequency, in addition to $\Omega_{\alpha\beta}^c$, is

$$\Omega_{\alpha\beta}^p \simeq -\frac{\hbar}{2\pi m^*} \int_{-\infty}^{\infty} d\omega \sum_{\vec{q}_\parallel} |\tilde{v}_{0+}(P_0; \omega)|^2 q_\parallel^2 \frac{\partial n(\omega')}{\partial\omega'} \times Im[\Pi_+^{\alpha(0)}] Im[\Pi_+^{\beta(0)}], \quad (37)$$

with $P_0 = (2m^*|E_F^\alpha - E_F^\beta|)^{1/2}/\hbar$ and $\tilde{v}_{0+}(P_\parallel; \omega) = \tilde{v}_0(P_\parallel; \omega + i0^+)$; $E_F^{\alpha,\beta}$ are the corresponding Fermi energies of the electrons in the two wells.

Real phonon exchange. The second term of Eq. (25) can be evaluated as in previous part. As can be seen. however, this term is proportional to the the fourth power of the electron-phonon interaction and therefore much smaller than the virtual phonon term proportional to the second power. We have evaluated numerically and found that it is about three orders of magnitude smaller than the first term thus confirming the estimates of Ref. [7]. Since it is really so small the relevant expressions for the frequency will not be given here.

Electron-phonon interaction

Electrons in one well can interact with the phonons of the system without exchanging them with electrons in the other well. This corresponds to the usual electron-phonon interaction which can be labelled 'incoherent' in the present context. The relaxation frequency $\Omega_{\alpha p}$ corresponding to this 'incoherent' scattering is obtained as

$$\Omega_{\alpha p} = \frac{2}{m_\gamma^*} \sum_{\vec{q}} |I(-iq_z)|^2 \frac{|M_\gamma'(\vec{q}; \tilde{\omega}_\lambda)|^2}{|\epsilon_+(\vec{q}_\parallel; \tilde{\omega}_\lambda)|^2} q_\parallel^2 \frac{\partial n(\tilde{\omega}_\lambda)}{\partial\tilde{\omega}_\lambda} Im[\Pi_+^{\gamma(0)}(\vec{q}_\parallel; \tilde{\omega}_\lambda)]. \quad (38)$$

It turns out that for the temperature region of interest this contribution is not significant compared to the impurity scattering. Besides, this part contributes only to the induced current and does not affect the coupling at all.

Total interaction and induced current

When a current is allowed to flow in the drag well the resistive to its flow force, as given by the right-hand side of Eq. (4), will have contributions from all interactions and especially from impurities at low temperatures. As we argued above the contributions from real phonon exchange and the usual 'incoherent' electron-phonon interaction can be neglected. Then the total nonequilibrium interaction $\tilde{v}_{\alpha\beta}(\vec{q}_{\parallel}; \omega; t)$ is given approximately by

$$
\tilde{v}_{\alpha\beta}(\vec{q}_{\parallel}; \omega; t) \simeq \quad F(q_{\parallel}) v(q_{\parallel}) \frac{e^{-q_{\parallel} a}}{\epsilon(\vec{q}_{\parallel}; \omega; t)} + \sum_{q_z, \lambda} |I(-iq_z)|^2 \frac{|M'(\vec{q}; \omega)|^2}{|\epsilon(\vec{q}_{\parallel}; \omega; t)|^2} D_\lambda(\vec{q}; \omega)
$$
$$
+ |J(q_{\parallel})|^2 v(q_{\parallel}) n_i^{2D} \frac{1}{|\epsilon(\vec{q}_{\parallel}; \omega; t)|^2} e^{-2q_{\parallel} z_{\alpha i}} \tag{39}
$$

where the dielectric function is given by Eq. (21). This and Eq. (4) give the total frictional force and consequently the current densities j_α and j_β. The first and second term in Eq. (39) are, respectively, due to the direct Coulomb and (virtual) phonon-mediated interactions; the last term is due to impurity scattering. Using the definitions $I_i = j_\beta\, w$ and $I_a = j_\alpha\, w$, for the *induced* and *applied* current, respectively, we obtain in matrix notation

$$
\begin{pmatrix} I_i \\ I_a \end{pmatrix} = w \begin{pmatrix} n_\beta \Omega_{\beta\alpha} \\ n_\alpha(\Omega_{\alpha i} + \Omega_{\beta\alpha}) \end{pmatrix} \times \frac{\bar{e}^2 E}{m^*[(\Omega_{\alpha i} + \Omega_{\alpha\beta})(\Omega_{\beta i} + \Omega_{\beta\alpha}) - (\Omega_{\alpha\beta}\Omega_{\beta\alpha}]} \tag{40}
$$

with $\Omega_{\alpha\beta} \equiv \Omega_{\alpha\beta}^c + \Omega_{\alpha\beta}^p$, and w the width of the sample.

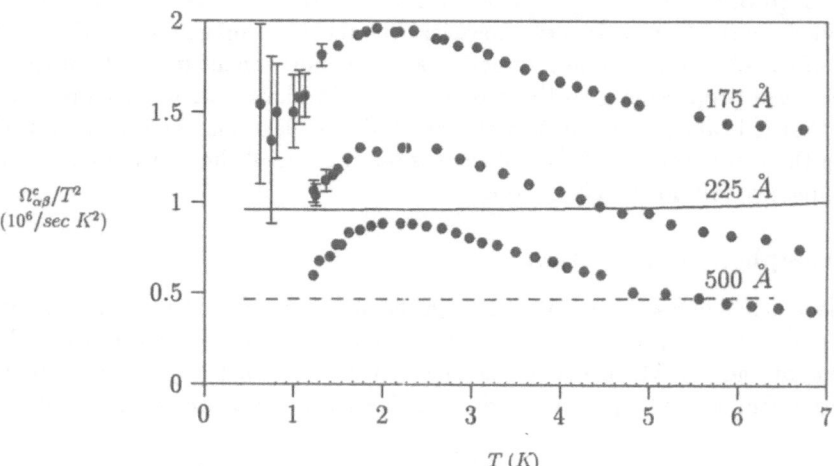

Fig. 6. Relaxation frequency $\Omega_{\alpha\beta}^c$, due to the direct Coulomb interaction, per temperature squared and unit area as function of temperature at separations of $175\mathring{A}$ (solid curve), $225\mathring{A}$ (dashed curve), and $500\mathring{A}$ (dotted curve). The solid circles are the experimental results of Ref. [7].

NUMERICAL RESULTS

Zero drag current

When no current is allowed to flow in the drag well, as in the experiments of Ref. [7], we have $v_d^\beta = 0$ and the coupling force in Eq. (12), with α and β interchanged, measures the drag voltage given by Eq. (13). The coupling frequencies $\Omega_{\alpha\beta}^c$ and $\Omega_{\alpha\beta}^p$ are evaluated numerically from Eqs. (23) and (37), respectively. In Fig. (6), we present $\Omega_{\alpha\beta}^c/T^2$ as function of the temperature for three different separations together with the experimental results of Ref. [7] shown by the solid circles.

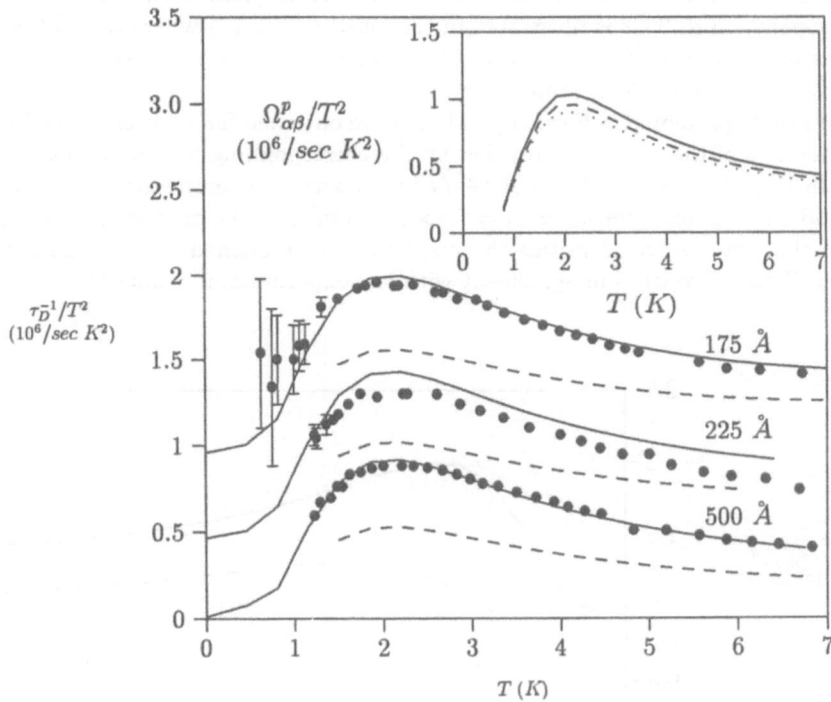

Fig. 7. The total relaxation frequency τ_D^{-1}, due to both the direct and (virtual) phonon-mediated Coulomb couplings, per temperature squared and unit area as function of temperature for the same parameters as in Fig. 6. The solid curves are the theoretical results for $f_c = .053$ and the dashed ones for $f_c = 0$. The solid circles are the experimental results of Ref. [7] . Inset: the contribution to τ_D^{-1}/T^2 due to virtual phonon exchange, $\Omega_{\alpha\beta}^p$, as function of temperature. The solid, dashed, and dotted curves correspond to separations $175\overset{\circ}{A}$, $225\overset{\circ}{A}$, and $500\overset{\circ}{A}$, respectively.

The well width is $100\overset{\circ}{A}$, the effective mass $m^* = .068m_0$, and the electron densities $n_\alpha = n_\beta = 1.5 \times 10^{11}/cm^2$. As can be seen this interaction alone cannot explain two important aspects of the experimental results: the hump of the curves at about $T = 2K$ and the amplitude especially for the $d = 500\overset{\circ}{A}$ curve where the theoretical result is about two orders of magnitude smaller than the experimental one. The latter discrepancy is due to the d^{-4} dependence of $\Omega_{\alpha\beta}^c$. These two discrepancies made it clear[7],[9] that an additional coupling mechanism was present in the experiments. We notice in passing that a similar temperature dependence was predicted in Ref. [16] for an electron-hole system at zero separation.

In Fig. (7), we show again the experimental results (solid circles) and the theoretical ones obtained by evaluating $\Omega_{\alpha\beta}^{p}$ from Eq. (37) and adding it to $\Omega_{\alpha\beta}^{c}$ of Fig. (6). The dashed curves are obtained when the jellium model is used for the interaction v_0 and the solid curves when corrections beyond this model are included in v_0 as given by Eq. (32). The adjustable parameter f_c, which is zero in the jellium model, is equal to 0.053 and is the *same* for all curves. It is seen that the agreement between theory and experiment is very good for all solid curves but rather poor for the dashed ones as far as their maximum is concerned. Another feature of the experimental results explained by the theory is the insensitivity of this virtual-phonon exchange mechanism to the separation d. This is illustrated in the inset of Fig. (7) where $\Omega_{\alpha\beta}^{p}/T^2$ is shown alone as function of temperature; the solid, dashed, and dotted curves correspond to $d = 175\text{Å}$, 225Å, and 500Å, respectively.

The results presented in Figs. (6) and (7) were obtained for *equal* electron densities. A further test of the theory is provided by the corresponding results for *unequal* densities. In Fig. (8), we plot the total τ_D^{-1}/T^2 as function of temperature for $d = 175\text{Å}$. The solid curve is again for $n_\alpha = n_\beta = 1.5 \times 10^{11}/cm^2$ and the dashed one $n_\beta = 2n_\alpha/3$. We see that for *unequal* densities the maximum is broadened and the amplitude is reduced. This behavior is in agreement with the experimental results[17].

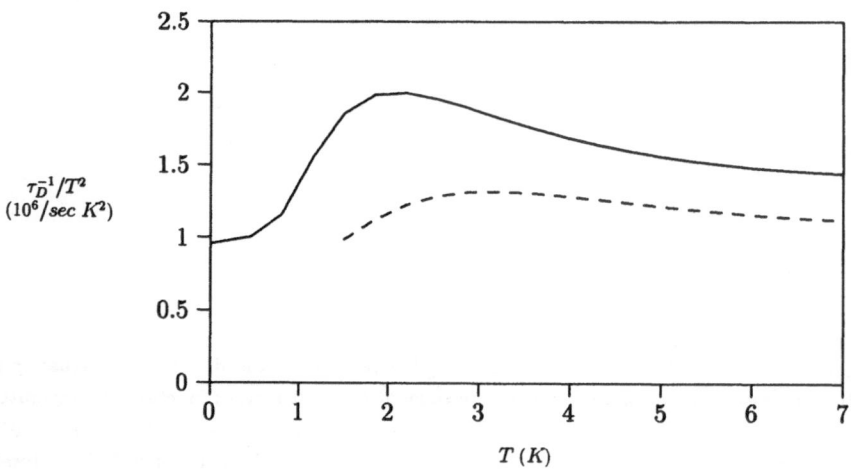

Fig. 8. τ_D^{-1}/T^2 as function of temperature for $d = 175\text{Å}$ and $n_\alpha = 1.5 \times 10^{11}/cm^2$. The solid and dashed curves are for $n_\beta = n_\alpha$ and $n_\beta = 2n_\alpha/3$, respectively.

Induced current

When the current is allowed to flow in the drag well, as in the experiments of Refs. [2, 6] we have $v_d^\beta \neq 0$ and the term proportional to the voltage V_d, in Eq. (3) vanishes. The induced then current I_i is given in terms of the applied one I_a by Eq. (40).

In Fig. (9), we plot the ratio I_i/I_a, divided by T^2, as function of temperature and in Fig. (10) the same ratio as function of the separation d.

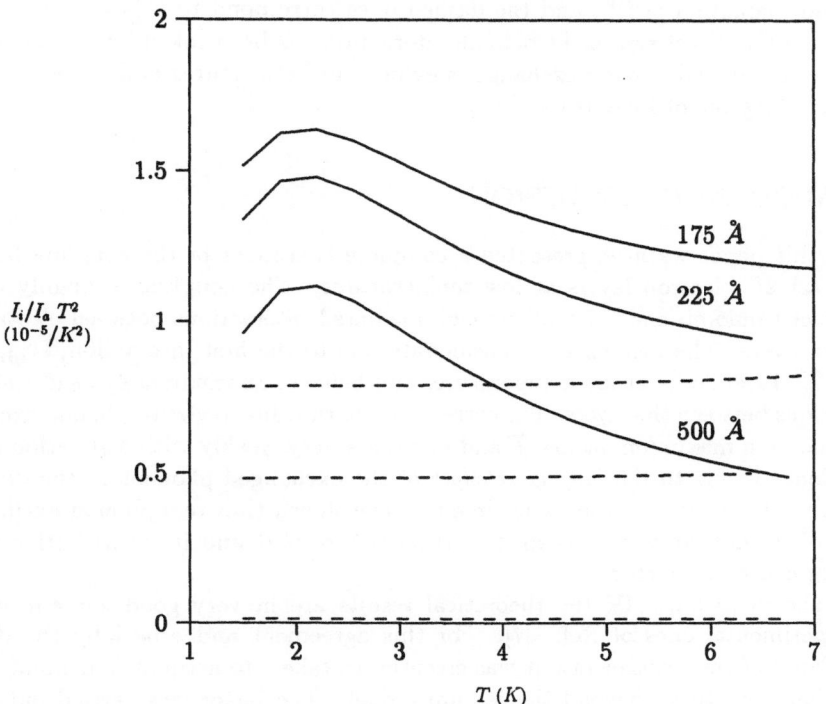

Fig. 9. I_i/I_a, divided by T^2, as function of temperature for $n_\alpha = n_\beta = 1.5 \times 10^{11}/cm^2$ and separations of 175 Å, 225 Å, and 500 Å. The solid curves include the factor $f_c = 0.053$ in the evaluation of $\Omega^p_{\alpha\beta}$ and the dashed ones correspond to $\Omega^p_{\alpha\beta} = 0$.

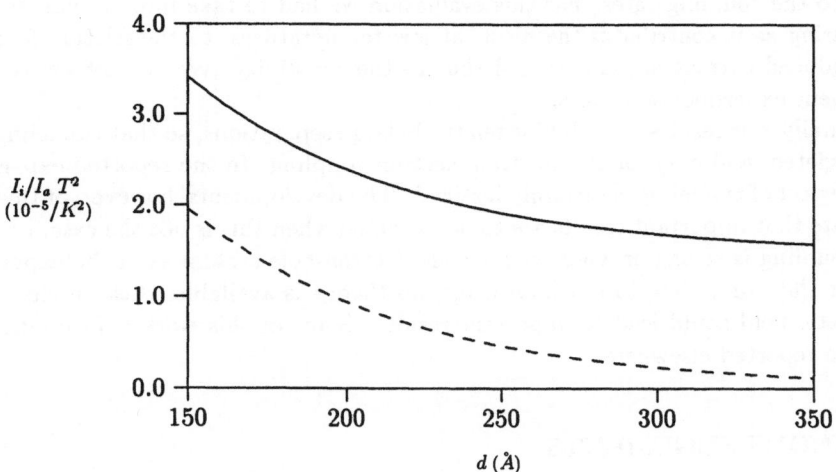

Fig. 10. I_i/I_a divided by T^2 as function of separation at $T = 2\ K$. The solid curve includes the factor $f_c = 0.053$ in the evaluation of $\Omega^p_{\alpha\beta}$ and the dashed one corresponds to $\Omega^p_{\alpha\beta} = 0$.

The parameters used are the same as in Fig. (6). The solid curves include the factor f_c in the evaluation of $\Omega^p_{\alpha\beta}$ and the dashed ones correspond to $\Omega^p_{\alpha\beta} = 0$, i.e., in the latter only the direct screened Coulomb interaction has been taken into account. The effect of the virtual-phonon exchange is evident in both figures and agrees with the corresponding one of Figs. (6) and (7).

SUMMARY AND DISCUSSION

In this paper we have presented a complete treatment of the coupling between two quasi 2D electron layers at low temperatures. The coupling is mainly due to the direct Coulomb and (virtual) phonon-mediated interactions between electrons of the two layers. The momentum transfer rate due to the first interaction, $\Omega^c_{\alpha\beta}$, has a nearly T^2 dependence at low temperatures and behaves approximately as d^{-4} at large separations between the layers. The corresponding rate due to virtual phonon exchange, $\Omega^p_{\alpha\beta}$, shows a maximum at low T and decreases very weakly with separation d; this dependence is due to the *long wavelength* of the exchanged phonons in the direction normal to the layers. At the same time we have shown that real phonon exchange is orders of magnitude weaker than its virtual counterpart and its contribution to the coupling can be neglected.

As shown in Sec. IV the theoretical results are in very good agreement with the experimental ones of Ref. [7]. For this agreement and especially the density dependence of the transfer rate it was essential to take into account four-point vertex corrections and to go beyond the jellium model. The latter was carried out using, for numerical tractability, the adjustable parameter f_c. However, since its value, with which agreement was obtained, is very small, close to its jellium model value ($f_c = 0$), and the *same* for all curves it indicates that the approach is correct. We hope to present elsewhere a physical argyment for its determination.

In addition to the coupling transfer rates discussed above, we have evaluated the induced current and showed that the virtual phonon contribution is as important to it as to the coupling rates. For this evaluation we had to take into account impurity scattering as it contributes the most, at low temperatures, to the friction force that the induced current encounters and thus to the resistivity. We are not aware of any pertinent experimental results.

Finally, our results are valid for relatively large separations, so that tunneling could be neglected, and only for the electron-electron coupling. In the reported experiments the neglect of tunneling is certainly justified. The developments, however, of this paper indicate that important results are to be expected when this is not the case, i.e., when the coupling is strong or when we have an electron-hole system as in the experiments of Ref. [6], for which, to our knowledge, no theory is available. Also, inclusion of a magnetic field could lead to important results. Some of this work is in progress and will be reported elsewhere.

ACKNOWLEDGEMENTS

This work was supported by NSERC Grant No. OPG0121756.

REFERENCES

[1] M. B. Pogrebinskii, Sov. Phys. Semic. **11**, 372 (1977); P. J. Price, Physica **117b**, 750 (1983); I. I. Boiko and Yu. M. Sirenko, Phys. Stat. Sol. (b) **159**, 805 (1990).

[2] P. M. Solomon, P. J. Price, D. J. Frank, and D. C. La Tulipe, Phys. Rev. Lett. **63**, 2508 (1989).

[3] B. Laikhtman and P. Solomon, Phys. Rev. B **41**, 9921 (1990); D. L. Maslov, ibid **45**, 1911 (1992); I. I. Boiko, P. Vasilopoulos, and Yu. M. Sirenko, ibid **45**, (1992).

[4] H. C. Tso and P. Vasilopoulos, Phys, Rev. B **45**, 1333 (1992); P. Vasilopoulos and H. C. Tso, Surf. Sci. 263, 368 (1992).

[5] Yu. M. Sirenko and P. Vasilopoulos, Phys. Rev. B **46**, (1992).

[6] U. Sivan, P. M. Solomon, and H. Shtrikman, Phys. Rev. Lett. **68** 1196 (1992).

[7] T. J. Gramila, J. Eisenstein, A. H. MacDonald, L. N. Pfeiffer, and K. West, Phys. Rev. Lett. **66**, 1216 (1991); Surf. Sci. **263**, (1992).

[8] P. Vasilopoulos and H. C. Tso, Springer-Verlag, (1992).

[9] H. C. Tso, P. Vasilopoulos, and F. M. Peeters, Phys. Rev. Lett. **68**, 2516 (1992).

[10] A. H. MacDonald, private communication; T J. Gramila, Bull. of Am. Phys. Soc. **37**, 526 (1992).

[11] H. C. Tso, and N. J. M. Horing, Phys. Rev. **44**, 8886 (1991).

[12] H. C. Tso, Ph.D. Thesis, Stevens Institute of Technolgy, 1990.

[13] L. P. Kandanoff, and G. Baym, Quantum Statistical Mechanics (Benjamin, New York, 1962), p. 177.

[14] G. D. Mahan, Many-Particle Physics (Plenum Press, New York, 1981), p. 780.

[15] H. C. Tso, and N. J. M. Horing, Phys. Lett. **A152**, 498 (1991).

[16] H. C. Tso, and N. J. M. Horing, Phys. Rev. **44**, 11358 (1991).

[17] T. J. Gramila et al., EP2DS9, Japan (1991).

REFERENCES

EXCHANGE ENERGY: A NON-LOCAL SELFCONSISTENT QUANTAL

AND SEMICLASSICAL APPROACH

A. Puente, M. Casas and H. Krivine[(+)]

Departament de Física, Universitat de les Illes Balears
E-07071 Palma de Mallorca, Spain

[(+)] Division de Physique Théorique,1 Institut de Physique
Nucléaire - 91406 Orsay - Cedex, France

ABSTRACT

Using the Density Matrix Expansion we have built a non local Energy Density Functional without problems of divergence in the large gradient limit. We have applied this functional to the study of ground state properties and the dipole excitation energy of some metallic clusters, in the quantal and in the semiclassical approaches.

1. INTRODUCTION

The Density Functional Theory (DFT) provides an effective one particle description of interacting fermion systems[1,2], and it can be also applied to boson systems[3]. In practice for many systems this theory is used within some approximation. The basic one is the Local Density Approximation (LDA), which assumes that the Exchange Correlation Energy per particle is that of the homogeneous electron gas with density ρ.

The LDA is applied with success to electronic systems running from atoms, molecules and metal clusters; nevertheless some limitations are clear i.e. Atomic exchange energies are underestimated by as much as 10-15% (as compared to the Hartree Fock results) and binding energies are typically too high.

A lot of work exists which goes beyond the LDA by investigating non local corrections via density gradient terms, see for instance[4−6]. These approximations contain generally some coefficients which are determined by least-square fitting to the exact Hartree-Fock exchange energies.

There are also some difficulties to incorporate these corrections into a computational scheme, specially in a fully self-consistent way[7], considering functional derivatives of these corrections into an effective one-particle potential, because in some cases

Keywords: Energy Density Functional, Exchange Energy
1 Unité de Recherche des Universités Paris 11 et Paris 6, Associée au CNRS

the exchange potential becomes divergent in the large gradient limit.

The interest for including corrections to the exchange-energy is renewed because recently[8] many studies over metal clusters have been made. The study of these systems in the jellium model gives that more of the 70% of the total mean field is due to the exchange potential.

In this work we start from the exact Hartree-Fock exchange energy for spin saturated systems and we use, the method originally introduced in Nuclear Physics, the Density Matrix Expansion (DME)[9] in order to improve the LDA approximation. The non diagonal density in the DME approach depends of an arbitrary momentum k because of the truncation of the expansion. The choice of k is usually that of the infinite system, but using this choice it is not possible to include the second order term of DME, because it diverges for the Coulomb potential. Another choice of k have been proposed (i.e. Campi Bouyssy)[10] which implies that the second order of the DME vanishes.

We want to show that if we incorporate the DME second order corrections in a fully selfconsistent way, the exchange energy per particle does not depend crucially on the value of k, and by taking it to be constant it is possible to include the next order of the DME.

We have performed selfconsistent calculations in both the quantal and the semiclassical approaches.

The plan of the article is as follows. In section II we present a short reminder of the density functional formalism and we derive variationally an effective one-particle potential.

In section III we present the same equations in the semiclassical approach using the Weizsäcker expansion for the kinetic energy density. In section IV we compare quantal and semiclassical results for metal clusters.

Section V contains the summary, conclusions and future trends of this work.

2. QUANTAL EFFECTIVE ONE PARTICLE POTENTIAL

In the Density Functional Theory (DFT) the total energy of the ground state is written as:

$$E[\rho(\bar{r})] = T_s[\rho(\bar{r})] + \int v_j(\bar{r})\rho(\bar{r})d\bar{r} + 1/2 \int \frac{\rho(\bar{r})\rho(\bar{r}')}{|\bar{r} - \bar{r}'|} d\bar{r}d\bar{r}' + E_{xc}[\rho(\bar{r})] \qquad (2.1)$$

where $T_s[\rho(\bar{r})]$ is the kinetic energy of a noninteracting electron system of density $\rho(\bar{r})$ in some external potential $v_j(\bar{r})$, the third term is the classical Coulomb repulsion energy, and the last term corresponds to the exchange and correlation energy.

Usually, $E_{xc}[\rho(\bar{r})]$ is split into separate exchange and correlation terms. The correlation energy has been given in various approximate forms, in this work we use for metallic clusters the Wigner formula. The exchange energy in the LDA is taken as the Fock result of a homogeneous infinite system. Indeed the exchange energy for a finite spin saturated system is written as

$$E_x = -\frac{1}{2\nu} \int \int \frac{\rho^2(\bar{r}, \bar{r}')}{|\bar{r} - \bar{r}'|} d\bar{r}d\bar{r}' \qquad (2.2)$$

where $\rho(\bar{r}, \bar{r}') = \sum_\alpha \Phi_\alpha(\bar{r}) \Phi_\alpha^*(\bar{r}')$ is the non-diagonal density matrix and ν is the degeneration factor. The computation of this exchange term is cumbersome because of the angular dependence on $\bar{r}$ and $\bar{r}'$, and a lot of approximation have been proposed for $\rho(\bar{r}, \bar{r}')$, one of the most accurate is the Density Matrix Expansion DME[9]. In this approach, the angular average of $\rho(\bar{r}, \bar{r}')$ is expanded around $\bar{R} = \frac{\bar{r}+\bar{r}'}{2}$ in terms of the relative distance $\bar{s} = \bar{r} - \bar{r}'$, and one obtains

$$\rho(\bar{R}, \bar{s}) = \rho(\bar{R})\, \hat{j}_1(ks) + \frac{1}{6}s^2 \hat{j}_3(ks) \left[\frac{1}{4}\Delta\rho(\bar{R}) - \tau(\bar{R}) + \frac{3}{5}k^2\rho(\bar{R}) \right] + \cdots \quad (2.3)$$

$$\text{where} \quad \hat{j}_1(x) = (2l+1)!! \frac{j_l(x)}{x^l}$$

$$\text{and} \quad \tau(\bar{R}) = \sum_\alpha |\bar{\nabla}\phi_\alpha(\bar{R})|^2$$

The value of $\rho(\bar{R}, \bar{s})$ depends on k because of the truncation of the expansion. For a homogeneous infinite system only the first term of the expansion (2.3) contributes to the exchange energy. Choosing for k that of the infinite system $(3\pi^2\rho(R))^{1/3}$, the first term of this expansion coincides with the Slater approach used in the LDA. Nevertheless, using this value of k it is not possible to include the non-local correction given by the second term, because for the Coulomb potential the term in j_3^2 diverges.

Another choice of k was proposed by the Campi-Bouyssy[10] approach, where k is fixed by the condition that the second term vanishes. The expression for the density matrix is formally identical to the Slater one

$$\rho(\bar{R}, \bar{s}) = \rho(\bar{R})\, \hat{j}_1(\hat{k}s) \quad (2.4)$$

$$\text{with} \quad \hat{k} = \left[\frac{5}{3\rho(\bar{R})} \left(\tau(\bar{R}) - \frac{1}{4}\Delta\rho(\bar{R}) \right) \right]^{1/2}$$

In this work we take k as a constant in (2.3) and the exchange energy for the Coulomb potential can be computed without problems of divergence up to second order in the DME expansion

$$E_x = -\int (E_{x1} + E_{x2} + E_{x3})d\bar{R} \quad (2.5)$$

where

$$E_{x1} = \frac{9\pi}{2\nu k^2}\rho^2$$

$$E_{x2} = \frac{35\pi}{6\nu k^4}\rho \left[\frac{3}{5}k^2\rho + \frac{1}{4}\Delta\rho - \tau \right]$$

$$E_{x3} = \frac{(35)^2\pi}{48\nu k^6} \left[\frac{3}{5}k^2\rho + \frac{1}{4}\Delta\rho - \tau \right]^2$$

In the case of metallic clusters we use for the correlation term the Wigner formula

$$E_c = \int \epsilon_c(\bar{R})\, d\bar{R} = -\int \frac{0.44\rho}{7.8 + (3/4\pi\rho)^{1/3}} d\bar{R} \quad (2.6)$$

and the external potential v_j is created by a spherical jellium

We explore in a perturbation scheme the importance of the non local corrections. For sodium clusters we take values of k around the Fermi momentum k_F computed at the saturation density of the cluster. The contribution of the non local terms $(E_{x2} + E_{x3})$ is about 20% of the exchange energy per particle and, for values of k around k_F one has $E_{x3} < E_{x2}$. Nevertheless, for values such that $k << k_F$, there is a cancellation between the two terms and the contribution of E_{x3} can be dominant. In this work the contribution of E_{x3} is taken into account as a perturbation.

The single particle equations are derived by requiring that (2.1) be stationary with respect to variations of the single particle functions $\phi^*(R)$ subject to the constraint of normalization $\int \rho d\bar{r} = N$. Because (2.1) involves ρ, τ and $\Delta\rho$ one must consider variations in $\phi^*, \bar{\nabla}\phi^*$ and $\Delta\phi^*$. Assuming spherical symmetry $\phi = (u/r)Y$, the resulting Euler equations computed only with the first two terms of (2.5), are:

$$-\frac{\hbar^2}{2m^*}\left(u'' - \frac{l(l+1)}{r^2}u\right) - \left(\frac{\hbar^2}{2m^*}\right)' u' + U_q u = \epsilon u \qquad (2.7)$$

where

$$U_q(r) = \{v_j + \int \frac{\rho(\bar{r}')}{|\bar{r} - \bar{r}'|}d\bar{r}' - \frac{16\pi}{\nu k^2}\rho + \frac{35\pi}{6\nu k^4}[\tau - 1/4\Delta\rho] + \frac{\delta\epsilon_c}{\delta\rho}$$
$$+ \frac{1}{2r}\left(\frac{\hbar^2}{2m^*}\right)' - \frac{1}{4}\left(\frac{\hbar^2}{2m^*}\right)''\}$$

and m^* is an effective mass coming from the inclusion of the non local terms in the DME

$$\frac{\hbar^2}{2m^*} = \frac{\hbar^2}{2m} + \frac{35\pi\rho}{6\nu k^4} \qquad (2.8)$$

Making the standard choice of $k = (\frac{6\pi^2\rho}{\nu})^{1/3}$ this effective mass becomes divergent at $r \to \infty$ ($\frac{\hbar^2}{2m^*} = \frac{\hbar^2}{2m} + \frac{35\pi}{6\nu(3\pi^2)^{4/3}}\rho^{-1/3}$). This divergence is obtained including only the first two terms in equation (2.5).

We now take $k = k_F$ and, as it is shown in fig. 1 the effective mass is ~ 0.4 m at the interior of the cluster going to m at the surface, this coincides with the Hartree-Fock prediction[11].

From eq (2.7) it is possible to define a local equivalent potential, making the usual change in the radial function $u(r) = [\frac{m^*(r)}{m}]^{1/2}u^L(r)$, and $u^L(r)$ becomes a solution of the standard Schrödinger equation with a local equivalent potential defined as:

$$V(r,\epsilon) = \frac{m^*(r)}{m}\left[U_q(r) - \frac{m^*(r)}{2\hbar^2}\left(\frac{\hbar^2}{2m^*(r)}\right)'^2 + \frac{1}{2}\left(\frac{\hbar^2}{2m^*(r)}\right)''\right] + \left(1 - \frac{m^*(r)}{m}\right)\epsilon \quad (2.9)$$

Notice that the state dependence of this potential is related to the non local effect by the factor $\frac{m^*(r)}{m}$ and when $\frac{m^*(r)}{m} \to 1, V(r,\epsilon) \to U_q(r)$

In fig. 2 we show this non local potential for the three last occupied shells of a sodium cluster with $N = 92$ particles.

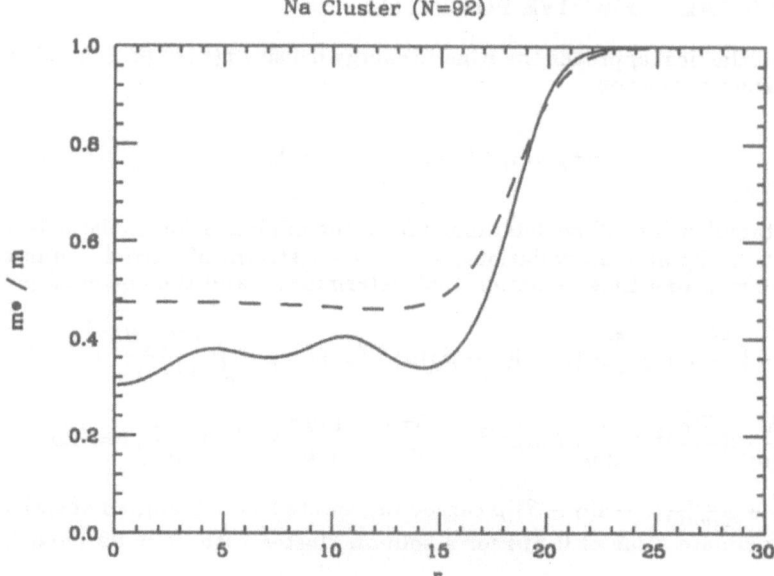

Fig. [1] Comparison between quantal (solid line) and semiclassical (dashed line) effective mass ($\frac{m^*(r)}{m}$) for a sodium cluster. In both approaches the value of k is taken at the saturation density.

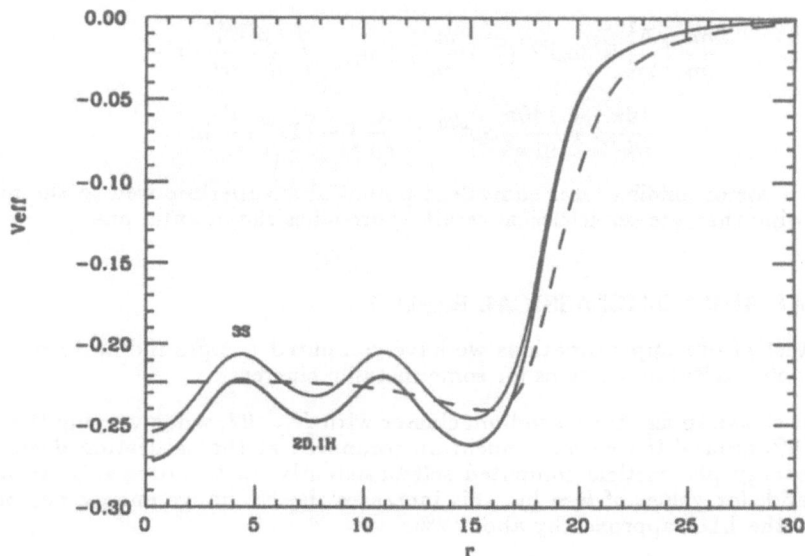

Fig. [2] Local equivalent potential $V(r,\epsilon)$ for the three last occupied shells of a sodium cluster (dashed line). For comparison the semiclassical result (dashed line) is also superimposed.

3. SEMICLASSICAL EFFECTIVE POTENTIAL

In the semiclassical approach the Kinetic energy density $\tau(\bar{r})$ of (2.1) is substituted by the Weizsäcker expansion

$$\tau(\bar{r}) = \alpha\rho^{5/3} + \beta\frac{|\bar{\nabla}\rho|^2}{\rho} + \gamma\Delta\rho \tag{3.1}$$

The ground state density will be determined from the minimization of the total energy with respect to ρ, $\bar{\nabla}\rho$ and $\Delta\rho$ variations, with the constraint of particle number conservation. The resulting Euler equation that determines ρ and the chemical potential μ is:

$$-\{\frac{\hbar^2}{m}\beta + \frac{35\pi}{12\nu k^4}(1 - 4(\gamma - \beta))\rho\}\nabla^2\rho + \{v_j + \int\frac{d\bar{r}'\,\rho(\bar{r}')}{|\bar{r} - \bar{r}'|} +$$

$$+\frac{\hbar^2}{2m}\beta(\frac{\nabla\rho}{\rho})^2 + \frac{\hbar^2}{2m}5/3\alpha\rho^{2/3} - \frac{16\pi\rho}{\nu k^2} + \frac{140\pi}{9\nu k^4}\alpha\rho^{5/3} + \frac{\delta\epsilon_c}{\delta\rho}\}\rho = \mu\rho \tag{3.2}$$

The factor $\frac{\hbar^2}{m} + \frac{35\pi}{12\nu\beta k^4}(1 - 4(\gamma - \beta))\rho$ can be interpreted as defining an effective mass with an approximate value of $0.5m$ for a sodium cluster with $N = 92$ particles (see fig. 1).

By comparison with the result obtained using only the first term of equation (2.5) we can define from (3.2) a local equivalent potential:

$$U(r, \mu) = \frac{\hbar^2}{2m}\beta\left(\frac{\nabla\rho}{\rho}\right)^2(\frac{m^*}{m} - 1)+$$

$$\frac{m^*}{m}\{\frac{\hbar^2}{2m}5/3\alpha\rho^{2/3}(1 - \frac{m^*}{m}) + v_j + \int\frac{\rho(\bar{r}')}{|\bar{r} - \bar{r}'|}d\bar{r}' -$$

$$-\frac{16\pi}{\nu k^2}\rho + \frac{140\pi}{9\nu k^4}\alpha\rho^{5/3} + \frac{\delta\epsilon_c}{\delta\rho}\} + (1 - \frac{m^*}{m})\mu \tag{3.3}$$

In fig. 2 the corresponding local equivalent potential is superimposed to the quantal result, showing that the semiclassical result approaches the quantal one.

4. QUANTAL AND SEMICLASSICAL RESULTS

As a test of our approximations we have computed the ground state properties and the dipole excitation energies for some metallic clusters.

As it is shown in fig. 3 for a sodium cluster with $N = 92$, when varying the values of k in a 10% around the Fermi momentum computed at the saturation density, the exchange energy per particle computed selfconsistently up to order j_1j_3, varies less than 3%, and, for values of $k \sim k_F$, this increases the Exchange energy per particle obtained in the LDA approach by about 2%.

The inclusion of the E_{x3} term as a perturbation increases the exchange energy per particle in a 3% .

The same trends are confirmed in the semiclassical results, nevertheless by taking $\beta = 1/4$ and $\gamma = 1/3$ in the Weiszäcker expansion the semiclassical Exchange energy

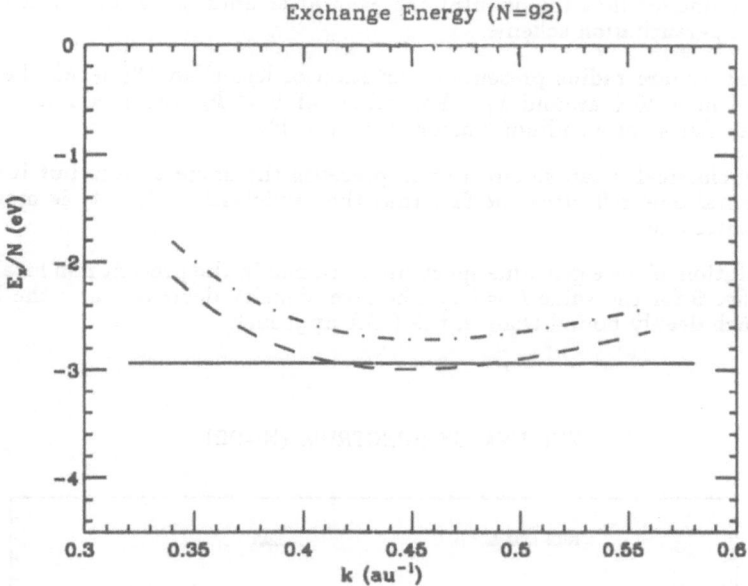

Fig. [3] Quantal (dashed line) and semiclassical (dot-dashed line) exchange energy per particle as a function of k for a sodium cluster. For comparison the LDA prediction (continuous line) is also superimposed.

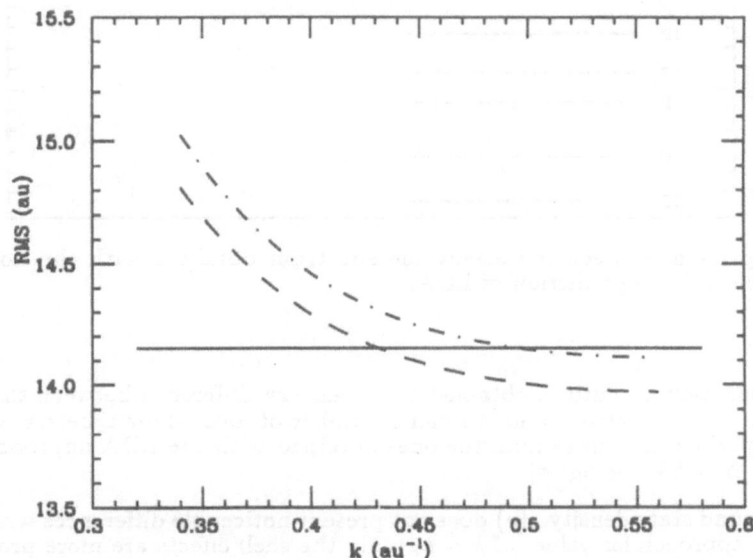

Fig. [4] Quantal (dashed line) and semiclassical (dot-dashed line) mean square radius for a sodium cluster. The continuous line is the LDA prediction.

per particle is smaller that the quantal one even after adding the contribution of the E_{x3} term in a perturbation scheme.

The mean square radius presents a variation of less than 1% when the value of k is increased in a 40% around k_F. For values of $k < k_F$ the mean square radius increases. (see fig. 4 for a sodium cluster with $N = 92$).

The semiclassical mean square radius presents the same trends but it is bigger than the quantal one, reflecting the fact that the semiclassical density is more diffuse than the quantal one.

The evolution of the eigenvalue spectrum obtained in the present non local method is shown in fig. 5 for the value $k = k_F$. The level density decreases and the low-lying levels are much deeply bound than in the LDA approach.

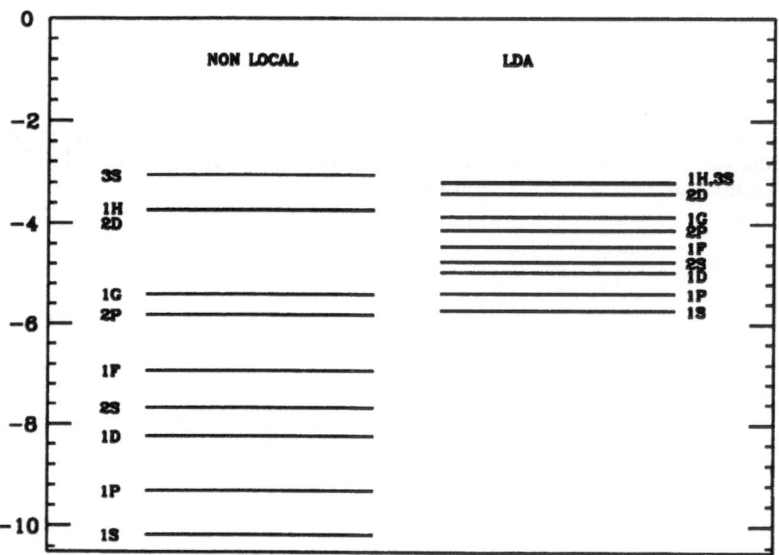

Fig. [5] Comparison between the eigenvalue spectrum obtained with the non local approach and the prediction of LDA.

The ionization potentials, obtained as the energy difference between the clusters with N and $N - 1$ electrons and the same number of ions, show a better agreement with the experimental values than the ones obtained with the LDA approach, except for the case $N = 58$ (see fig. 6).

The ground state density $\rho(r)$ does not present noticeable differences with respect to the LDA approach for values of $k \sim k_F$ only the shell effects are more pronounced. Nevertheless, for $k < k_F, \rho(r)$ becomes more diffuse, but for these values of k the contribution to the exchange energy of the E_{x3} term becomes dominant and it cannot be treated as a perturbation.

Using the sum rules technique and a generalized scaling transformation[12] we have computed an upper bound of the dipole excitation energy as $E_D = \sqrt{\frac{m_k}{m_{k-2}}}$ where m_k is the moment of order k of the strength function S(E).

As it is well known, odd moments of S(E) can be obtained with RPA precision as expectation values of some operator in the ground state[13]. For the case of the dipole we have chosen the operator $Q = rY_{10}$ and we have computed the sum rules m_1 and m_3.
As it is easily shown m_1 is proportional to the electron number N

$$m_1 = \frac{\hbar^2}{2m} \frac{3N}{4\pi} \tag{4.1}$$

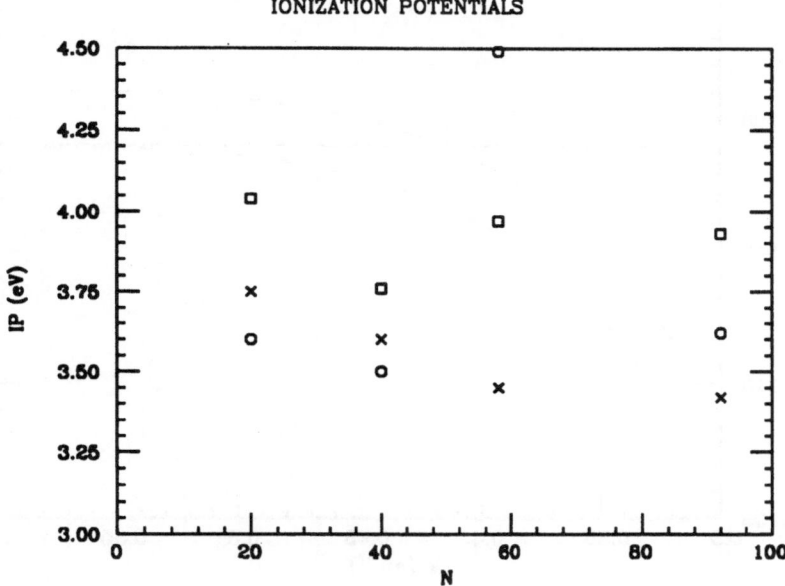

IONIZATION POTENTIALS

Fig. [6]   Ionization potentials obtained with the non local approach (circles) and the prediction of LDA (squares) for Sodium clusters with N=20,40,58 and 92 atoms. For comparison the experimental [15] values are also shown (crosses) .

The only contribution to m_3 [14] is the jellium-electron term because the translational symmetry is only broken by the jellium field. The rest of the electron-electron terms vanish because the operator rY_{10} corresponds to a global translation of electrons

$$m_3 = -2\pi \left(\frac{\hbar^2}{m}\right)^2 \int_0^\infty dr_2 \rho'(r_2) \int_0^{r_2} dr_1 r_1^2 \rho_j(r_1) \tag{4.2}$$

where $\rho_j(r)$ is the jellium density and $\rho'(r)$ is the first derivative of the electronic density.

In fig. 7 we show the evolution of the dipole energy as a function of k for a cluster with $N = 92$.

For values of $k < k_F$ the dipole energy decreases as a consequence of the increasing of the diffusivity of the ground state density, approaching the values of the polarizability to the experimental results. Nevertheless at this values of k the contribution of E_{z3} cannot be taken into account as a perturbation.

We hope that the selfconsistent inclusion of the E_{z3} contribution will allow us to compute in this small range of k values. Nevertheless as it is clear in (4.2) the jellium density plays also a crucial role in the determination of this dipole energy.

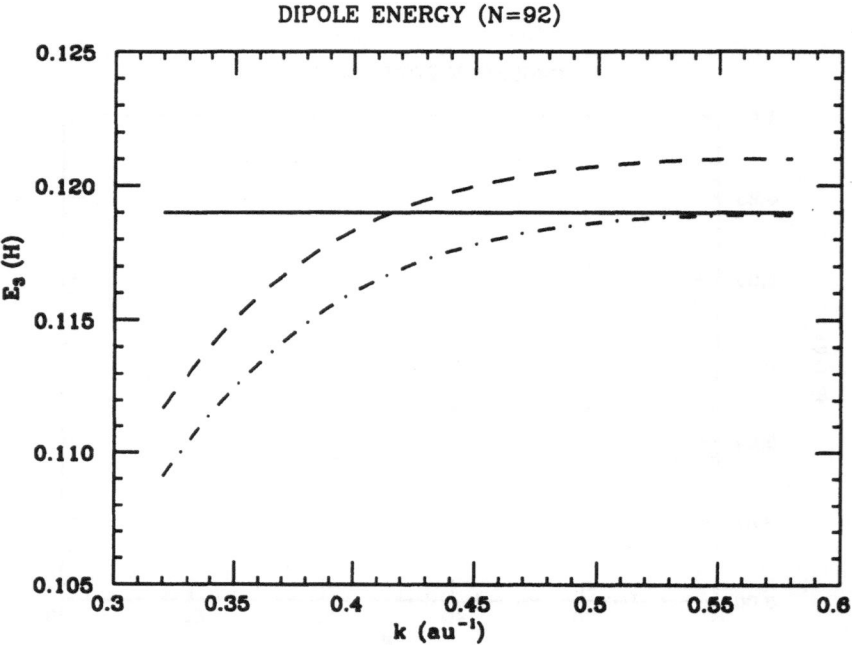

Fig. [7] Evolution of the dipole energy as a function of k for a sodium cluster. Quantal result (dashed line), semiclassical approach (dot-dashed line) and LDA prediction (continuous line).

5. SUMMARY

Starting from the Density Matrix Expansion it is possible to construct an Energy Density Functional with a non local exchange term, even for the case of the Coulomb potential.

The incorporation of the DME corrections in a fully selfconsistent way shows that the exchange energy per particle does not depend crucially of the value of k and taking it to be constant all orders of the DME can be included without problems of divergence.

The use of this non local Density Functional gives very similar results in quantal and in the semiclassical approaches.

The preliminary results obtained with metallic clusters accounting for only the first non local contribution to the exchange energy shows that the exchange energy per particle increases with respect to the LDA result. The density becomes more diffuse for values of $k < k_F$ nevertheless in this interval the contribution of the next term of the DME must be taken into account in a selfconsistent scheme. Work in this direction is in progress.

In the future we are planning to use the Campi Bouyssy approach in the same context and to extend this study to atoms.

ACKNOWLEDGEMENTS

We are indebted to J. Treiner, N. Pavloff, Ll. Serra and M. de Llano for discussions. This work has been supported by DGICYT (Spain) grant (PS 90-0212) and by the IN2P3-CICYT exchange program.

REFERENCES

[1]- "Density Functional Methods in Physics" eds E.M. Dreizler and J. Providencia, Vol. 123 of NATO Advanced Study Institut Series B: Physics (Plenum, NY. 1985).

[2]- "Energy Density Functional Theory of Many-Electron Systems" E. Kryachko and E. Ludeña, (Kluwer, Dordrecht 1990).

[3]- S. Stringari and J. Treiner, J. Chem. Phys. 87 (1987) 5021 and Phys. Rev. B36 (1987) 8369.

[4]- A. D. Becke, Phys. Rev. A38 (1988) 3098.

[5]- J. P. Perdew, Phys. Rev. Lett. 55 (1985) 1665.

[6]- Ch. Lee and Z. Zhou, Phys. Rev. A44 (1991) 1536.

[7]- P. Mlynarski and D. R. Salahub, Phys. Rev.B43 (1991) 1399.

[8]- "Small Particles and Inorganic Clusters", Z. Phys. D19 (1991).

[9]- J. Negele and D. Vautherin, Phys. Rev. C5 (1972) 1472.

[10]- X. Campi and A. Bouyssy, Phys. Lett. 73B (1978) 263.

[11]- M. S. Hansen and H. Nishiska, prepint (1991).

[12]- M. Casas and J. Martorell, Nucl. Phys. A490 (1988) 329.

[13]- O. Bohigas, A. M. Lane and J. Martorell, Phys. Rep. 51 (1979) 267.

[14]- Ll. Serra, F. Garcias, M. Barranco, J. Navarro, C. Balbás and A. Mañanes, Phys. Rev. B39 (1989) 8247.

[15]- M.L. Homer, J.L. Persson, E.C. Honea , R.L. Whetten , Z. Phys. D22 (1991).

BOGOLONS IN A MULTICOMPONENT MANY FERMION

SYSTEM AND THE ENERGY GAP

M. Moreno[a], G. Carmona, R. M. Méndez-Moreno, S. Orozco,
M. A. Ortíz[b] and M. de Llano[c*]

a.*Instituto de Física, Universidad Nacional Autónoma de México,
Apartado Postal 20-364, 01000 México, D. F.*
b. *Depto. de Física, Facultad de Ciencias, Universidad Nacional
Autónoma de México, Apartado Postal 70-543, 04510 México, D. F.*
c. *Departamento de Física de la Materia Condensada, C-III,
Universidad Autónoma de Madrid, Cantoblanco, 28049 Madrid, España*

ABSTRACT

The energy gap equation for a many fermion system with several intrinsic components
is derived in the framework of the Bogoliubov-Valatin transformation. The solution for a
model in which the resulting effective one-particle operator is required to satisfy a Dirac
supersymmetry is obtained. The Dirac supersymmetric system can in principle surmount
the *phonon barrier* on transition temperatures making it a viable candidate for higher-T_c
superconductivity.

INTRODUCTION

Since the discovery of high-T_c superconductivity by Bednorz and Müller [1] there has
been vigorous activity both theoretically and experimentally in search of a (pairing) mechanism which leads to the high transition temperatures observed in copper-oxide superconductors. Experiments on high-T_c superconductor materials reveal important information
on both the structure and the dynamics of such materials[2]. A common characteristic of
the high-T_c superconductor oxides is their quasi-two-dimensional (2D) electronic structure
with a strongly nested Fermi surface[3]. Because superconductivity in these materials occurs mainly in planes, a simple model for these consists of a set of parallel two-dimensional
electron-gas layers interacting with (acoustic) lattice vibrations[4]. The interlayer (presumably Josephson) coupling remains a necessary element because of the Mermin and
Wagner theorem[5, 6] which precludes any form of long-range-order (LRO) in the pure
2D electron gas at finite temperature. The LRO is restored by three-dimensional phonons,
or equivalently by interlayer electron-electron interactions which eliminate fluctuations[7].

The anomalous isotope effect observed in various cuprates has been considered as an
important piece of evidence for non-phononic mechanisms in high-temperature superconductivity. Nevertheless, recent experimental results indicate that some of these effects can
be understood in terms of a band structure density of states, as Tsuei *et al.* have shown[8],
wherein a singularity near the Fermi level along with a conventional Bardeen-Cooper-
Schrieffer (BCS) weakly-coupled phonon-mediated pairing is considered responsible for

Condensed Matter Theories, Vol. 8, Edited by
L. Blum and F.B. Malik, Plenum Press, New York, 1993

high T_c superconductivity. This model agrees with many of the superconducting and normal state characteristics in these materials thus rendering the new cuprates as BCS-like superconductors. Unfortunately, this singularity has not been observed experimentally.

Recently, the new high T_c superconductor materials were studied on the basis of an abnormal occupancy in the three dimensional (3D) fermion gas model [9, 10]. The use of the well-known Cooper pair equation[11] together with generalized Fermi surface topologies can lead to a behavior for T_c which is different from the usual BCS one, for which T_c scales with the Debye temperature, $T_D \approx 300^0 K$. In fact, it can lead to a much larger scale that goes as $\sqrt{T_D T_F}$, where $T_F \approx 10,000^0 K$. Consequently, this model can produce values of T_c higher than those expected from the traditional BCS "phonon barrier" of 25 to $30^0 K$ without invoking stronger electron-phonon coupling nor exotic interaction mechanisms.

Because in the new high T_c materials superconductivity occurs mainly in planes, we have also explored the consequences of an abnormal Fermi sea occupation for a 2D electron gas. This is important because in the 3D·case the modification to T_c depends on $\sqrt{T_D T_F}$ and for the 2D fermion gas a smaller T_F is obtained [12]. When this model is applied to the 2D electron gas high T_c values as in the new cuprate superconductors can be reproduced. A parameterization of a representative set of high-T_c superconductors has been carried out. The most important feature of it is that the energy scale of the departure from a normal Fermi occupancy $(1 - \alpha^2)E_F$ is of the order of the Debye energy. This lends credibility to the main model hypothesis because it requires an energy scale accessible to the crystal. This result has the effect of restoring the Debye energy as the overall scale that determines higher T_c values. Nonetheless, the net effect of the assumed abnormal occupancy is to overcome the so-called "phonon barrier" of BCS theory.

A serious limitation of this type of calculation is imposed by the Cooper-Bethe-Goldstone formalism that takes into account the many body nature of the problem only through the Pauli principle for *two* fermions in a Fermi gas.

In this work we study the consequences of a BCS mechanism for a general multicomponent fermion system in the canonical transformation formalism proposed by Bogoliubov and Valatin [13, 14]. This is known to render a consistent many body calculation and should enlighten the physical origin as well as the limitations of the tighter bound Cooper pair predicted by the anomalous occupation hypothesis. When one restricts the resulting one particle operator with a Dirac supersymmetry a generic form of the diagonalized operator can be obtained. This form allows higher values of the energy gap and suggests that higher transition temperature between the normal and superconductor phases can be achived.

It is worthwhile to point out that the connection between superconductivity and supersymmetry is an idea that has been pioneered by Nambu [15]. This idea has been investigated as a clue to the underlying physical mechanism of the interacting boson model. The result is that a broken supersymmetry is required. The Dirac supersymmetry [16, 17, 18, 19] can considered as either a specific form of supersymmetry breaking (because it breaks the positive definiteness of the energy spectra of standard supersymmetry) or, perhaps more appropiately, as a generalization of the usual supersymmetry to cases when the Hamiltonian is a fermionic operator. At the same time this supersymmetry breaking restores the existence of a gap in the energy spectrum of the system.

DIRAC SUPERSYMMETRY

A general supersymmetric (Susy) Hamiltonian, H_S satisfies

$$[Q, H_S] = 0 = [Q^\dagger, H_S] \tag{1}$$

with Q and $Q^\dagger$ fermionic operators, i.e.such that $Q^2 = 0 = Q^{\dagger 2}$. In particular a supersymmetric Hamiltonian can be of the form

$$H_S = \{Q, Q^\dagger\} \tag{2}$$

and

$$[Q^\dagger Q, QQ^\dagger] = 0 \tag{3}$$

where the $QQ^\dagger$ and $Q^\dagger Q$ commute with H_S. If the ground state is nondegenerate and normalizable, Susy is unbroken. The states $|n\rangle$, $|n_+\rangle = Q^\dagger |n\rangle$ and $|n_-\rangle = Q|n\rangle$ are degenerate, but $|n_+\rangle$ or $|n_-\rangle$ are null.

The connection between the Dirac equation and the supersymmetric quantum mechanics can be provided [17] if the Dirac Hamiltonian can be written as

$$H = Q + Q^\dagger + \Lambda, \tag{4}$$

with λ a hermitean operator. These operators must satisfy the following anticommutation relations.

$$\{Q, \Lambda\} = \{Q^\dagger, \Lambda\} = 0. \tag{5}$$

Then the Hamiltonian commutes with $QQ^\dagger$ and $Q^\dagger Q$ and we get for the squared Hamiltonian

$$H^2 = \{Q, Q^\dagger\} + \Lambda^2 =: h^2 + \Lambda^2, \tag{6}$$

with $h^2 = \{Q, Q^\dagger\}$ and where h is required to be an even root of this operator. If we take the definition $\hat{\Lambda} = \frac{\Lambda}{\sqrt{\Lambda^2}}$, the Foldy-Wouthuysen Hamiltonian is written

$$H_{FW} = \hat{\Lambda}\sqrt{\{Q, Q^\dagger\} + \Lambda^2}, \tag{7}$$

for this form to hold one needs a Λ with non-null eigenvalues. A Foldy-Wouthuysen transformation (FWT) is generated by

$$iS = \hat{\Lambda}(Q + Q^\dagger)\theta, \tag{8}$$

where θ must satisfy

$$\tan(2h\theta_{FW}) = \frac{h}{\sqrt{\Lambda^2}}. \tag{9}$$

In our notation θ has units of h^{-1} and from the last equation it follows that

$$\sqrt{\Lambda^2}\,\theta_{FW} = f(\frac{h^2}{\Lambda^2}), \tag{10}$$

which shows that θ_{FW} exists with the required properties and it follows the usual triangle construction, see Fig. 1.

One can write a matrix representation for Q and $Q^\dagger$ in the form

$$Q = \begin{pmatrix} 0 & 0 \\ A & 0 \end{pmatrix} \quad Q^\dagger = \begin{pmatrix} 0 & A^\dagger \\ 0 & 0 \end{pmatrix}. \tag{11}$$

Because $\hat{\Lambda}$ commutes with H_{FW} and its eigenvalues are ± 1, Eq.(7) implies that, in general, the energy spectra has two branches; between them there exists a gap determined by the lowest eigenvalues of Λ^2 and h^2. If $\hat{\Lambda}$ has negative eigenvalues and H^2 is not bounded from above one gets an unstable ground state for the Susy Hamiltonian. The usual remedy for this situation is to introduce a Dirac sea. The important feature of the Susy interactions is that the definition of this sea is independent of the size of the couplings, we call this property the stability of the Dirac sea.

From the hermiticty of Λ and the conditions in Eq. (5) one can show that the FW Hamiltonian in Eq. (7) anticommutes with Q and $Q^\dagger$. The standard definition of a Susy Hamiltonian is that it must commute with the fermionic operators Q and $Q^\dagger$, as in Eq. (1). The anticommutation condition implies that, in order to construct observables which are constants of motion in the Dirac-Susy Hamiltonian Eq. (4), one must form bosonic operators out of the fermionic ones.

Very important is that the positive and negative energy states do not mix, and the gap size is not decreased. Fig. 2 pictures the Dirac sea in the free and the cases with interaction. After the second quantization is done, one gets a stable Dirac sea.

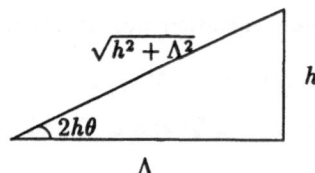

Figure 1. Triangle construction for the Foldy-Wouthuysen transformation.

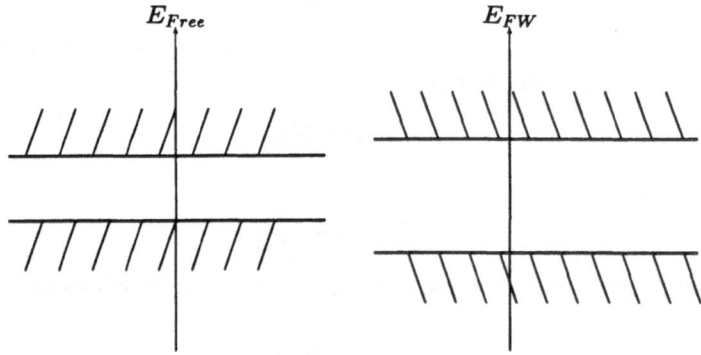

Figure 2. Schematic Dirac seas for the free case and Susy Dirac potentials.

SUPERCONDUCTIVITY

In the BCS theory of superconductivity the instability in the Fermi fluid is caused by an effective attractive electron-electron or hole-hole interaction. This force arises from the electron-phonon interaction, which is the standard mechanism. Cooper electron pairs (with zero total spin and center-of-mass momenta) are predominant, and of vital importance to the theory.

The essential feature of the Bogoliubov-Valatin (BV) theory is to study the Fermi sea instability through a canonical transformation which mixes electron and hole states of opposite quantum numbers (k ↑ and −k ↓). The physical basis of this transformation relies on the picture that the electrons that form the Fermi sea are escaping it and transform into bound Cooper pairs. The new states, the so called bogolons, are related to the electron states by

$$
\begin{pmatrix} \alpha_k \\ \beta_{-k} \end{pmatrix} = \begin{pmatrix} u_k & -v_k \\ v_k & u_k \end{pmatrix} \begin{pmatrix} a_{k\uparrow} \\ a^\dagger_{-k\downarrow} \end{pmatrix},
\tag{12}
$$

the transformation matrix must be orthogonal in order to preserve the anticommutation relations for the bogolon operators, α and β, and one usually further restricts it to be orthogonal, were possible we will follow the notation of Fetter and Walecka Ref. [20]. The BV transformation can be shown to describe the instability of the electron Fermi sea [21]. Formally this is taken into account by a new vacuum state that satisfies

$$
\alpha_k |0> = \beta_k |0> = 0.
\tag{13}
$$

At zero temperature the thermodynamic potential $\Omega(T = 0, V, \mu)$, would be given by the expectation value of the operator

$$
\begin{aligned}
\hat{K} &= \hat{H} - \mu \hat{N} \\
&= \sum_{k\lambda} a^\dagger_{k\lambda} a_{k\lambda} (\epsilon^0_k - \mu) \\
&\quad - \frac{1}{2} \sum_{\substack{\lambda_1, \lambda_2, \lambda_3, \lambda_4 \\ k_1 + k_2 = k_3 + k_4}} <k_1\lambda_1 k_2\lambda_2 |V| k_3\lambda_3 k_4\lambda_4> a^\dagger_{k_1\lambda_1} a^\dagger_{k_2\lambda_2} a_{k_4\lambda_4} a_{k_3\lambda_3}
\end{aligned}
\tag{14}
$$

Substitution of the BV into this equation leads to separation of this operator into a zero, one (H_I), and two body (H_{II}), operators; the usual assumption is that the new two body operator can be treated as a small perturbation. The BV transformation is fixed by the requirement that the one body operator is diagonal in the α, β base

$$
H_I = H_1 + H_2
\tag{15}
$$

$$
H_1 = \sum_k (\alpha^\dagger_k \alpha_k + \beta^\dagger_{-k} \beta_{-k})[(u^2_k - v^2_k)\xi_k + 2u_k v_k \Delta_k]
\tag{16}
$$

$$
H_2 = \sum_k (\alpha^\dagger_k \beta^\dagger_{-k} + \beta_{-k} \alpha_k)[2u_k v_k \xi_k - (u^2_k - v^2_k)\Delta_k],
\tag{17}
$$

where ξ_k measures the Hartree Fock quasi-particle energy with respect to the chemical potential μ and Δ_k is the gap function

$$
\Delta_k = \sum_{k'} <k - k|V|k' - k'> u_k v_k
\tag{18}
$$

This form can be written in a matrix (Nambu) form

$$
H_I = \sum_k \begin{pmatrix} \alpha_k \\ \beta^\dagger_{-k} \end{pmatrix}^\dagger \begin{pmatrix} u_k & v_k \\ -v_k & u_k \end{pmatrix} \begin{pmatrix} \xi_k & \Delta_k \\ \Delta_k & -\xi_k \end{pmatrix} \begin{pmatrix} u_k & -v_k \\ v_k & u_k \end{pmatrix} \begin{pmatrix} \alpha_k \\ \beta^\dagger_{-k} \end{pmatrix},
\tag{19}
$$

the BV transformation is characterized by the requirement that the one-particle Hamiltonian H_I should be diagonal for a non-null value of v_k with respect to the $\alpha\beta$ basis. This implies that

$$
\xi_k \sin 2\chi_k = \Delta_k \cos 2\chi_k
\tag{20}
$$

where the usual identifications $u_k = \cos \chi_k$ and $v_k = \sin \chi_k$ have been made. The single particle or bogolon excitation energy is then given by

$$E_k = \sqrt{\xi_k^2 + \Delta_k^2} \tag{21}$$

and the gap function Δ_k is determined from the self consistent equation

$$
\begin{aligned}
\Delta_k &= \sum_{k'} < k - k|V|k' - k' > u_k v_k \tag{22} \\
&= \frac{1}{2} \sum_{k'} < k - k|V|k' - k' > \frac{\Delta_k}{E_k},
\end{aligned}
$$

these equations always have the normal solution $\Delta_k = 0$ and $E_k = \xi_k$; the superconducting solution is characterized by $\Delta_k \neq 0$. A simple realistic nontrivial solution is obtained for an interaction of the form

$$< k - k|V|l - l > = \frac{g}{\Omega} \theta(\hbar \omega_D - \xi_k) \theta(\hbar \omega_D - \xi_l) \tag{23}$$

where g is the (square of the) electron phonon coupling, Ω is a normalization volume or area and ω_D is a cutoff frequency usually associated to the Debye temperature T_D through $k_B T_D = \hbar \omega_D = E_D$. Introducing $N(0)$, the density of states for one spin projection, one obtains to the famous *gap equation*

$$
\begin{aligned}
1 &= \frac{gN(0)}{2} \int_{-E_D}^{E_D} \frac{d\xi}{\sqrt{\Delta^2 + \xi^2}} \\
&= gN(0) \int_0^{E_D} \frac{d\xi}{\sqrt{\Delta^2 + \xi^2}} \\
&= gN(0) \sinh^{-1}(E_D/\Delta) \\
&= gN(0) \log(\frac{E_D}{\Delta} + \sqrt{1 + (\frac{E_D}{\Delta})^2}), \tag{24}
\end{aligned}
$$

solving for Δ one gets

$$\Delta = \frac{2E_D}{e^{1/\lambda_g} - e^{-1/\lambda_g}} \approx 2E_D e^{-1/\lambda_g} \tag{25}$$

where the small coupling approximation $\lambda_g = gN(0) \ll 1$ was done. In a finite temperature formalism this implies

$$T_c \approx 1.13 \hbar \omega_D e^{-1/N(0)g}. \tag{26}$$

For pure elements λ_g is in the interval $\{0.15, 0.6\}$.

MULTICOMPONENT SYSTEMS

We will now generalize the BVT for a system in which the fermions are characterized by a discrete parameter λ that can take N values. For $N = 2$ this could correspond to the spin degree of freedom. However we don't prejudge at this stage the nature of the discrete index. Assume for example that the electrons are confined to a slab of thickness t; in this case the discreteness in λ would correspond to the different excitations of the transverse degree of freedom and, if necessary, to the spin. Another concrete realization corresponds to a layered electron gas, for which λ represents the specific layer in which the electrons are contained.

The thermodynamic potential at zero temperature (or effective Hamiltonian) is

$$\hat{K} = \hat{H} - \mu\hat{N}$$
$$= \hat{T} - \hat{V} \tag{27}$$

with

$$\hat{T} = \sum_{k\lambda} a^\dagger_{k\lambda} a_{k\lambda}(\epsilon^0_k - \mu)$$

$$\hat{V} = -\frac{1}{2} \sum_{\substack{\lambda_1, \lambda_2, \lambda_3, \lambda_4 \\ k_1 + k_2 = k_3 + k_4}} < k_1\lambda_1 k_2\lambda_2 |V| k_3\lambda_3 k_4\lambda_4 > a^\dagger_{k_1\lambda_1} a^\dagger_{k_2\lambda_2} a_{k_4\lambda_4} a_{k_3\lambda_3} \tag{28}$$

Let us now define

$$A_k := \begin{pmatrix} a_{k1} \\ \vdots \\ a_{kN} \\ a^\dagger_{-k1} \\ \vdots \\ a^\dagger_{-kN} \end{pmatrix} \tag{29}$$

and

$$B_k := \begin{pmatrix} \check{\alpha}_k \\ \check{\beta}^\dagger_{-k} \end{pmatrix} = \begin{pmatrix} \check{u}_k & \check{v}_k \\ -\check{v}_k & \check{u}^*_k \end{pmatrix} A_k = U A_k, \tag{30}$$

where $N \times N$ matrices and N vectors are denoted with a check above; the commutation relations for the operators $a_{k\lambda}$ can be expressed in matrix form by

$$A_k A^\dagger_{k'} \pm A^*_{k'} A^T_k = 1\delta_{kk'} \tag{31}$$

the canonicity condition requires that

$$B_k B^\dagger_{k'} \pm B^*_{k'} B^T_k = 1\delta_{kk'} \tag{32}$$

which is satisfied if U is an orthogonal matrix as can be easily verified by sandwiching Eq.(31) between U and $U^\dagger$.

When applied to the Hamiltonian (27) one obtains an operator of the form

$$\hat{K} = H_O + H_I + H_{II} \tag{33}$$

where the three terms in the right hand side correspond to the zero (c-number), one and two bogolon terms respectively. The BVT is defined by the requirement that H_I, the one bogolon term, be diagonal. One has

$$H_I = \sum_k \begin{pmatrix} \check{\alpha}_k \\ \check{\beta}^\dagger_{-k} \end{pmatrix}^\dagger \begin{pmatrix} \check{u}_k & \check{v}_k \\ -\check{v}_k & \check{u}_k \end{pmatrix} \begin{pmatrix} \check{\xi}_k & \check{\Delta}_k \\ \check{\Delta}_k & -\check{\xi}_k \end{pmatrix} \begin{pmatrix} \check{u}_k & -\check{v}_k \\ \check{v}_k & \check{u}_k \end{pmatrix} \begin{pmatrix} \check{\alpha}_k \\ \check{\beta}^\dagger_{-k} \end{pmatrix}. \tag{34}$$

The relevant structure for our purposes is

$$\begin{pmatrix} \check{\xi}_k & \check{\Delta}_k \\ \check{\Delta}_k & -\check{\xi}_k \end{pmatrix} \tag{35}$$

if one identifies

$$Q = \begin{pmatrix} 0 & 0 \\ A & 0 \end{pmatrix} = \begin{pmatrix} 0 & 0 \\ \check{\Delta}_k & 0 \end{pmatrix} \tag{36}$$

$$\Lambda = \begin{pmatrix} \check{\xi}_k & 0 \\ 0 & -\check{\xi}_k \end{pmatrix} \tag{37}$$

one can easily confirm our central result that the standard Bogoliubov-Valatin operator and its [most direct] extrapolation to a multicomponent system have in fact of a Dirac supersymmetry structure. The BVT is totally equivalent in this case to the Foldy-Wouthuysen transformation and its generator can be read form Eqs. (8) and (9). Quite generally the condition (5) implies that

$$[\check{\Delta}, \check{\xi}] = 0. \tag{38}$$

MODELS WITH TWO COMPONENTS

In this section we will consider a couple of models with two components. For this models the components are not [trivially] equivalent to spin. The physical systems to which these models would correspond are in fact inspired in the electron gas confined to a bidimensional slab of finite thickness or to a two layer bidimensional electron gas for which the spin degree of freedom can be neglected.

Let us consider the case for which the fermion degrees of freedom have been diagonalized before the BVT. They describe therefore quasiparticles à la Landau; it has been emphasized by Schrieffer [21] that the pairing mechanism is more adequately applied to these states. Formally these states are described by diagonal matrices $\check{\xi}_k$. In the models we will consider we will further assume that

$$\check{\xi}_k = 1\xi_k;$$

physically this means that the two components of quasielectrons are equivalent before the BVT. Under this condition Eq.(38) is satisfied for an arbitrary matrix $\check{\Delta}_k$. Performing the BVT [or Foldy-Wouthuysen transformation] we *derive* the bogolon Hamiltonian

$$H_{BV} = \sqrt{\xi_k^2 + \Delta_k^2} \tag{39}$$

this is diagonalized by a transformation that diagonalizes Δ_k the most general form of a diagonal two by two matrix is

$$\check{U}\check{\Delta}_k\check{U}^{-1} = \Delta_k^0 1 + \delta_k^0 \sigma_3 \tag{40}$$

The cases for which $\delta_k^0 = 0$ correspond the traditional one component system. On the other hand, $\delta_k^0 \neq 0$ does contain the new phenomena that can occur in the two component case. Physically the relative size of the different components in Δ_k is controlled by essentially the same pairing mechanism. This implies that the largest δ_k^0 should be of the same order of magnitude than Δ_k^0. This implies that the maximum overall effect of the doubling of components should be doubling the eigenvalues of $\check{\Delta}_k$. Now, the transition temperature T_c is essentially determined by the smallest eigenvalue of $\check{\Delta}_k^2$. We therefore look for a $\check{\Delta}_k$ of the form

$$\check{\Delta}_k = \kappa_k^0(a \cdot \sigma) \tag{41}$$

where the *components* of a should, by the above argument, be of order 1 and κ_k^0 will be set by the usual cutoff [Debye] energy. One easily gets that

$$\delta_k^{0^2} = \kappa_k^{0^2}|a|^2 \tag{42}$$

This is clearly the most favorable situation because any contribution to $\check{\Delta}_k$ that comes from the unit matrix will split the eigenvalues. For this case one expects a gap increase of a factor 3, which in turn implies an increase in T_c of order $\sqrt{3}$.

CONCLUSIONS

In this work we have shown that there exists a general connection between the Dirac supersymmetry and superconductivity. The relevant aspects of Dirac supersymmetry were reviewed in particular the remarkable result that there exists a unitary Foldy-Wouthuysen transformation that decouples the positive and negative energy states and which can be explicitly constructed. The relation to non-Dirac (traditional) supersymmetry was presented. For the case of a one component (usual) superconductivity pairing phenomena we showed that the effective quasi-electron Hamiltonian is precisely of the Dirac supersymmetic form, the formalism of the Bogoliubov-Valatin transformation lends itself for this discussion.

We then generalized the relation between superconductivity and Dirac supersymmetry to a multicomponent system. In the case of a two component system it was shown that a simple Ansatz leads to an increase in the gap energy. This increment was estimated to be of order 70% for the two component system. In the case of an N component system we conjecture that the gap increase could be of order $\sqrt{N^2 - 1} \approx N$.

ACKNOWLEDGEMENTS

This work is partially supported by Dirección General de Asuntos del Personal Académico, Project IN102991, Universidad Nacional Autónoma de México (U.N.A.M.) and Programa Universitario de Superconductividad de Alta Temperatura Crítica, U.N.A.M. México D. F., México.

* On leave from Physics Department, North Dakota State University, Fargo, N. D. 58105, USA.

REFERENCES

[1] J. G. Bednorz and K. A. Müller, *Z. Phys. B* **64**, 189 (1986).

[2] J. I. Gersten, *Phys. Rev. B* **37**, 1616 (1988).

[3] J. D. Jorgensen, H. B. Schulter, D. G. Hinks, D. W. Capone II, K. Zhang, M. B. Brodsky and D. J. Scalapino, *Phys. Rev. Lett.* **58**, 1020 (1987); D. J. Kim, *Physica* **148 B**, 278 (1987).

[4] Y. Takahashi and H. Umezawa, *Collective Phenomena* **2**, 55 (Gordon and Breach Science Publishers Ltd., Great Britain, 1975); Z. Ye, H. Y. Chu and H. Umezawa, *Phys. Lett. A* **154**, 421 (1991).

[5] N. D. Mermin and H. Wagner, *Phys. Rev. Lett.* **17**, 1133 (1966).

[6] P. C. Hohenberg, *Phys. Rev.* **158**, 383 (1967).

[7] D. C. Mattis, *Phys. Rev. B* **36**, 745 (1987).

[8] C. C. Tsuei, D. M. Newns, C. C. Chi, and P. C. Pattnaik, *Phys. Rev. Lett.* **65**, 2724 (1990).

[9] V. C. Aguilera-Navarro, M. de Llano and A. Plastino, *Proceedings of the XIV International Workshop in Condensed-Matter Theories* **6**, 227 (1991).

[10] D. A. Agrello, V. C. Aguilera-Navarro, C. Keller, M. de Llano, A. Plastino and J. P. Vary, *Mod. Phys. Lett.* **B5**, 805 (1991).

[11] L. N. Cooper, *Phys. Rev.* **104**, 189 (1956).

[12] G. Carmona, R. M. Méndez-Moreno, S. Orozco, M. A. Ortíz, M. Moreno, M. de Llano and V. C. Aguilera-Navarro, to be published.

[13] N.N. Bogoliubov, *Sov. Phys.-JETP,* **7**, 41 (1958).

[14] J.G. Valatin, *Nuovo Cimento,* **7**, 843 (1958).

[15] M. Mukerjee and Y. Nambu, *Ann. Phys.* **191**, 143 (1989).

[16] M. Moreno and A. Zentella, *J. Phys.* **A 22** L821 (1989).

[17] M. Moreno, R. Martínez and A. Zentella, *Mod. Phys. Lett.* **A 5**, 949 (1990).

[18] R. Martínez, M. Moreno and A. Zentella, *Phys. Rev.* **D 43** (1991) 2036.

[19] M. Moreno and R. M. Méndez-Moreno, *Generalized Supersymmetric Quantum Mechanics,* in *Proc. of the Workshop on High Energy Phenomenology,* Ed. by M. A. Pérez and R. Huerta, (World Scientific, Singapore) in press, 1992. *ibid. Relativistic Equations in External Fields* in *Proc. in honor of M. Moshinsky,* Ed. by A. Frank, T. H. Seligman and B. Wolf, (Springer-Verlag,) in press, 1992.

[20] A. L. Fetter and J. D. Walecka, *Quantum Theory of Many-Particle Systems,* McGraw-Hill, New-York (1971).

[21] J. R. Schrieffer, *Theory of Superconductivity,* W. A. Benjamin, Inc., New York, 1964.

FIRST PRINCIPLES ELECTROSTATIC POTENTIAL OF IMPURITIES IN QUANTUM WELLS

Fredy R. Zypman

University of Puerto Rico at Humacao
Department of Physics and Electronics
Humacao, PR 00791

INTRODUCTION

The method of images is used to calculate the electrostatic potential of an electric charge inside a double barrier structure. This potential is important for finding impurity energy levels in the quantum structure, and has never been used before in exact form. The media were modeled by asigning different dielectric constants to the barriers and to the well. The highly doped contacts were assumed to have infinite dielectric constants. An analytical method to speed up numerical convergence of the series is also presented. This method is based in calculating the Fourier Transform parallel to the interfaces of the potential. This function can be obtained in closed form. The potential is then obtained by inverse Fourier Transform. This integral is computationally more efficient to evaluate than the direct sum of coulomb image terms.

REVIEW OF QUANTUM WELLS AND IMPURITIES

In the past ten years, there has been interest in the study of electron energy levels due to the presence of shallow impurities in quantum wells. Basically, a quantum well[1] is made by the alternate growth of layers of different semiconductors. This can be achieved with MBE (Molecular Beam Epitaxy)[2] system, or MOCVD (Metal-Organic Chemical Vapor Deposition)[3] both high-quality technologies.

One of the most common quantum wells is that made with gallium arsenide (GaAs) and the alloy gallium aluminum arsenide (GaAlAs) (Figure 1). These materials have very high electron mobility, an important property in transport devices. Also they have a very small lattice mismatch, which provides for mechanically stable layers (a large mismatch creates strained layers, which in time will peel off). By varying the concentration of Al in GaAlAs, one can choose the size of the barrier in the quantum well, which corresponds to the conduction band mismatch between GaAlAs and GaAs (Band Gap Engineering).

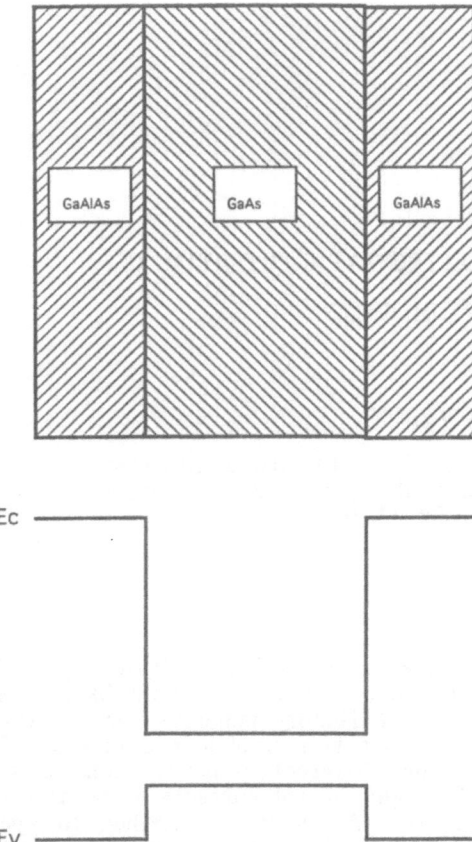

Figure 1. A Quantum Well. The top drawing shows a 3-layer
system of alternated materials. The bottom drawing show
the bottom of the conduction band and the top of the valence band of the system. Electrons in the conduction band
tend to concentrate in the GaAs region.

Impurities in semiconductors serve to change the concentration of
carriers and they can be of two kinds, donors and acceptors. A donor
will release an electron which will go into the conduction band. An ac-
ceptor will take an electron from the valence band, creating a hole
there. From the point of view of electrons, donors are attractive coulomb
centers and acceptors are repulsive ones. Impurities are always present.
Sometimes because we want them, like in modulation-doping[4], and always
because of intrinsic uncertainties of the instrument of growing.

These impurities modify the bound states in quantum wells. If
they are not too strong, they will change the underlying quantum well
band structure just a little and therefore can be treated as a perturbation to the "pure" quantum well states. In order to find these levels, one first has to find the potential in which the electron is
immersed. This potential is due to the impurity, and the quantum well
materials. The difference in dielectric constants (see endnote) between GaAs and GaAlAs creates surface charges at the interfaces. This

total potential has not been used exactly in any of the papers dealing with the energy levels problem[5-14]. It turns out that the electrostatic problem can be solved exactly and be carried out analytically until almost the end of the calculation.

In Section II the electrostatic problem will be explained and solved. Section IV presents the conclusions.

ANALYTICAL SOLUTION

A Coulomb center (attractive or repulsive impurity) is supposed to be located in the well region (Figure 2). A cylindrical coordinate system (r, θ, z) is considered such that the z direction is perpendicular to the interfaces and passes through the impurity. There is no loss of generality with this choice since the system is (without the impurity) translationaly invariant along any direction parallel to the interfaces. The origin is taken at the center of the GaAs layer.

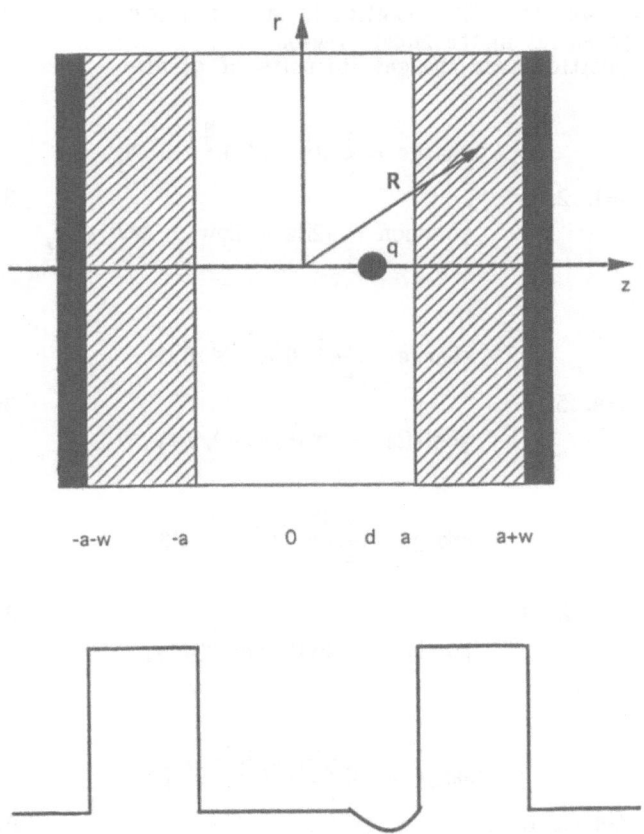

Figure 2. Quantum Well with an attractive impurity. The same QW as in Figure 1 plus an impurity and electric contacts on the sides. The bottom drawing shows a schematics of the conduction band distortion due to the impurity.

We consider the electric properties of the structure as follows: the outermost, highly doped GaAs regions, are taken as having infinite conductivity so that the fields will not penetrate them (these regions are highly doped, to serve as contacts in electronic transport measurements). The remainder three layers will be assigned three different dielectric constants ϵ_1, ϵ_2, ϵ_3. ϵ_1 and ϵ_3 not taken to be equal in the general derivation to account for the situation in which different concentration of Al are used. In most practical situations ϵ_i are equal within 5-10%.

The method of images is used in order to find the desired potential. By systematically simetrizing the original and image charges respect to the interfaces, we construct an infinite sequence of image charges. The electrostatic potential can immediately be formally written down as an infinite series of Coulomb terms:

$$V_j\,(r,\ z) = \frac{1}{\epsilon_j} \sum \frac{\text{charge}}{[r^2+(z-\text{position})^2]^{1/2}} \tag{1}$$

where the sum runs over certain appropriate tags that label the corresponding charge-position pair.

Next we will find explicitly a form for $V_2(r,\ z)$, the electrostatic potential in the middle region.

The positions and charges (in units of q) are:

$$\text{charge} = \tfrac{1}{2} A^{\frac{N}{2}-p}\ B^p\ C^{\frac{N}{2}-r}\ D^r\ \beta_i$$

$i=1,\ 2,\ 3$ $\qquad 0 \leq p \leq \frac{N}{2},\ 0 \leq r \leq \frac{N}{2}$

$$\text{position} = -2Na - 2pw - 2rw + \gamma_i$$

N even

$$\text{charge} = \tfrac{1}{2}C^{\frac{N}{2}-p}D^pA^{\frac{N}{2}-r}B^r\ \beta_i$$

$i=4,\ 5,\ 6$ $\qquad 0 \leq p \leq \frac{N}{2},\ 0 \leq r \leq \frac{N}{2}$

$$\text{position} = 2Na+2pw+2rw+\gamma_i$$

$$\text{charge} = \tfrac{1}{2}A^{\frac{N+1}{2}-p}B^pC^{\frac{N-1}{2}-r}D^r\beta_i$$

$i=1,\ 2,\ 3$ $\qquad 0 \leq p \leq \frac{N+1}{2},\ 0 \leq r \leq \frac{N-1}{2}$

$$\text{position} = 2Na+2pw+2rw-\gamma_i$$

N odd

$$\text{charge} = \tfrac{1}{2}C^{\frac{N+1}{2}-p}D^pA^{\frac{N-1}{2}-r}B^r\ \beta_i$$

$i=4,\ 5,\ 6$ $\qquad 0 \leq p \leq \frac{N+1}{2},\ 0 \leq r \leq \frac{N-1}{2}$

$$\text{position} = -2Na-2pw-2rw-\gamma_i$$

with:

$$\gamma_1 = d \qquad\qquad \beta_1 = 1$$
$$\gamma_2 = -2a-d \qquad\qquad \beta_2 = -C$$
$$\gamma_3 = -2a-2w-d \qquad\qquad \beta_3 = D$$

$$\gamma_4 = d \qquad\qquad \beta_4 = 1$$
$$\gamma_5 = 2a-d \qquad\qquad \beta_5 = -A$$
$$\gamma_6 = 2a+2w-d \qquad\qquad \beta_6 = B$$

$$A = \frac{\epsilon_2-\epsilon_3}{\epsilon_2+\epsilon_3} \;\; ; \;\; B = \frac{\epsilon_2\epsilon_3}{(\epsilon_2+\epsilon_3)^2} \;\; ; \;\; C = \frac{\epsilon_2-\epsilon_1}{\epsilon_2+\epsilon_1} \;\; ; \;\; D = \frac{\epsilon_2\epsilon_1}{(\epsilon_2+\epsilon_1)^2}$$

The sum in Equation [1] converges very slowly due mainly to the presence of the highly doped regions. An alternative way to accomplish the sum, is to first take the "parallel" [to the interfaces] Fourier Transform of the potential:

$$f_2(k,z) \equiv \int d^2\vec{r}\; V_2(r,z)\; e^{-i\vec{k}\cdot\vec{r}}.$$

By noting that:

$$\int d^2\vec{r}\left\{ \frac{1}{[r^2+(z\text{-position})^2]^{1/2}} \right\} e^{-i\vec{k}\cdot\vec{r}} = \frac{2\pi}{k}\, e^{-k|z\text{-position}|} \; ,$$

We can write:

$$f_2(k,z) = \frac{\pi}{\epsilon_2}\frac{1}{k}\sum_{i=1}^{3}\beta_i \cdot \left\{ \sum_{N=2,4,6\ldots}\sum_{p=0}^{N/2}\sum_{r=0}^{N/2} \frac{(N/2)!}{p!(N/2-p)!}\frac{(N/2)!}{r!(N/2-r)!} A^{\frac{N}{2}-p}B^p C^{\frac{N}{2}-r}D^r \cdot \right.$$

$$\cdot\, e^{-k(z-\gamma_i)}\, (e^{-2ka})^N\,(e^{-2kw})^p\,(e^{-2kw})^r +$$

$$+ \sum_{N=1,3,5\ldots}\sum_{p=0}^{\frac{N+1}{2}}\sum_{r=0}^{\frac{N-1}{2}} \frac{\left(\frac{N+1}{2}\right)!}{p!\left(\frac{N+1}{2}-p\right)!}\frac{\left(\frac{N-1}{2}\right)!}{r!\left(\frac{N-1}{2}-r\right)!} A^{\frac{N+1}{2}-p}B^p C^{\frac{N-1}{2}-r}D^r \cdot$$

$$\left. \cdot\, e^{k(z+\gamma_i)}\, (e^{-zka})^N\,(e^{-zkw})^p\,(e^{-zkw})^r \right\} +$$

$$+ \frac{\pi}{\epsilon_2}\frac{1}{k}\sum_{i=4}^{6}\beta_i \cdot \left\{ \sum_{N=2,4,6\ldots}\sum_{p=0}^{N/2}\sum_{r=0}^{N/2} \frac{\left(\frac{N}{2}\right)!}{p!\left(\frac{N}{2}-p\right)!}\frac{\left(\frac{N}{2}\right)!}{r!\left(\frac{N}{2}-r\right)!} C^{\frac{N}{2}-p}D^p A^{\frac{N}{2}-r}B^r \cdot \right.$$

$$\cdot\, e^{k(z-\gamma_i)}\, (e^{-2ka})^N\,(e^{-2kw})^p\,(e^{-2kw})^r +$$

$$+ \sum_{N=1,3,5\ldots}\sum_{p=0}^{\frac{N+1}{2}}\sum_{r=0}^{\frac{N-1}{2}} \frac{\left(\frac{N+1}{2}\right)!}{p!\left(\frac{N+1}{2}-p\right)!}\frac{\left(\frac{N-1}{2}\right)!}{r!\left(\frac{N-1}{2}-r\right)!} C^{\frac{N+1}{2}-p}D^p A^{\frac{N-1}{2}-r}B^r \cdot$$

$$\left. \cdot\, e^{-k(z+\gamma_i)}\, (e^{-2ka})^N\,(e^{-2kw})^p\,(e^{-2kw})^r \right\} + \frac{\pi}{\epsilon_2}\frac{1}{k}\sum_{i=1}^{6} e^{-k|z-\gamma_i|}\,\beta_i \; .$$

123

The terms with factorials represent the number of times a charge in a given position is repeated.

The sums can be performed, and the result is:

$$f_2(k,z) = \frac{2\pi}{k\Delta}\left\{ \begin{array}{l} e^{k(a-d+z)}g(-d,-\epsilon_1)\ \eta(\epsilon_3) \\ + e^{-k(a+z)}g(d,\epsilon_3)\ \eta(-\epsilon_1) \\ + qe^{-k|z-d|} \end{array}\right\}$$

where:

$$g(\delta,\epsilon) \equiv (1-e^{2kw})(e^{2ka}+e^{2k\delta})$$
$$-\frac{\epsilon}{\epsilon_2}\ (1+e^{2kw})(e^{2ka}-e^{2k\delta}).$$

Similarly,

$$f_1(k,z) = \frac{8\pi e^{kw}}{k}\ \frac{g(d,\epsilon_3)}{\Delta}\ \sinh(z+a-w)$$

$$f_3(k,z) = \frac{8\pi e^{k(w-d)}}{k}\ \frac{g(-d,-\epsilon_1)}{\Delta}\ \sinh(z-a-w)$$

The potential can finally be obtained by numerically evaluating:

$$V_j(r,z) = \frac{1}{2\pi}\int_0^{+\infty} kf_j(k,z)\ J_0(kr)\ dk.$$

This integral converges very fast. An example of V(r, z) is shown in Figure 3.

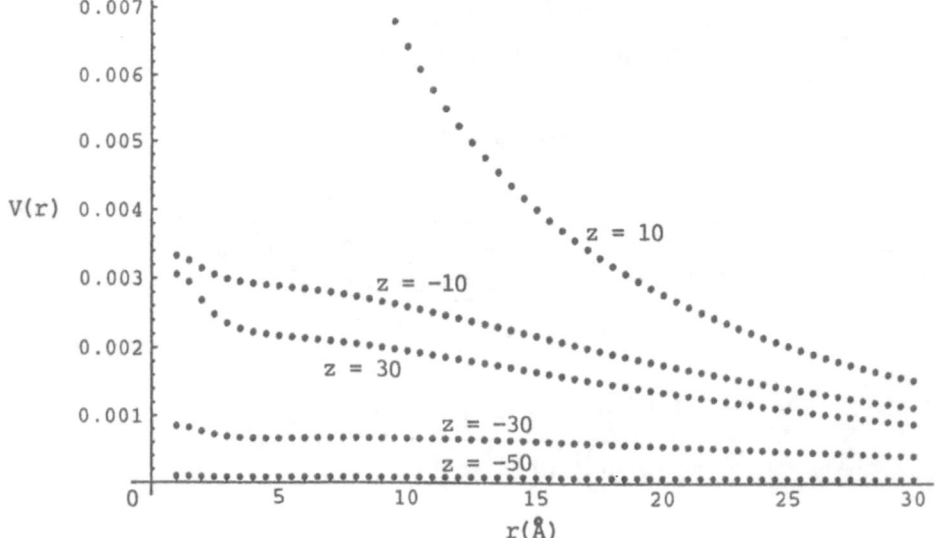

Figure 3. V(r,z) for a=30 Å, d=10 Å, W=20 Å for a GaAlAs/GaAs/GaAlAs QW.

CONCLUSIONS

We have presented here a solution to the electrostatic problem of a shallow impurity located in a double-barrier quantum structure. This solution is useful to be used in Schrodinger Equation to find the energy levels of an electron. Since such a solution has not been reported, we believe it is for lack of V(r,z). Our paper also shows a way to obtain fast numerical convergence.

ACKNOWLEDGMENTS

This work was supported by grant #2703 of the University of Puerto Rico at Humacao.

ENDNOTE

The polarization effects of each material are lumped into its dielectric constant. Since the effective mass of the electron is ten times smaller than the free space mass and the dielectric constant of GaAs is ten times that of vacuum, the radius of the orbit of an electron in the ground state of a hydrogenic atom is 100 times Bohr's radius. Thus the electron sees the average potential of the atomic sites, which can be described by the material's dielectric constant.

REFERENCES

1. C.R. Leavens, R. Taylor, editors. "Interfaces, Quantum Wells, and Superlattices", NATO ASI Series B, Plenum Press, New York-London (1987).
2. E.H.C. Parker, editor. "The Technology and Physics of Molecular Beam Epitaxy", Plenum Press, New York (1985).
3. P.D. Dapkus, *Annu. Rev. mater. Sci.* 12:243 (1982).
4. R. Dingle, H.L. Störmer, A.C. Gossard, W. Wiegmann, *Appl. Phys. Lett.* 33:665 (1978).
5. G. Bastard, *Phys. Rev. B* 24:4714 (1981).
6. C. Mailhiot, Yia-Chung Chang, T.C. McGill, *Phys. Rev. B* 26:4449 (1982).
7. S. Chaudhuri, *Phys. Rev. B* 28:4480 (1983).
8. W.T. Masselink, Yia-Chung Chang, H. Morcoq, *Phys. Rev.* 28:7373 (1983).
9. C. Priester, G. Allan, M. Lannoo, *Phys. Rev. B* 29:3408 (1984).
10. J.A. Brum, G. Bastard, C. Guillemot, *Phys. Rev. B* 30:905 (1985).
11. F. Crowne, T.L. Reinecke, B.V. Shamabrook, *Solid State Comm.* 50:875 (1984).
12. C. Guillemot, *Phys. Rev. B* 31:1428 (1985).
13. N.C. Jarosik, B.D. McCombe, B.V. Shanabrook, J. Comas, John Ralston, G. Wicks, *Phys. Rev. Lett.* 54:1283 (1985).
14. D.S. Chuv, Ying-Chih Lou, *Phys. Rev. B* 34:14504 (1991).

DYNAMICS OF CORRELATED SUPERLATTICES

G. Kalman[1] and K.I. Golden[2]

[1]Department of Physics
Boston College
Chestnut Hill, MA 02167

[2]Dept. of Computer Science and Electrical Engineering
University of Vermont
Burlington, VT 05405

I. INTRODUCTION

Superlattice structures have attracted a great deal of attention over the past decade (For a review see [1]). The semiconductor type I superlattice of interest in this paper is an array of electron layers trapped in potential wells on semiconductor interfaces. For the purpose of most discussions, including the present one, the superlattice can be represented by a model of an infinite array of monolayers separated from each other by a fixed distance, say d; each monolayer is a two-dimensional (2D) one-component plasma (OCP) of N electrons embedded in a neutralizing positive background; the mean areal electron density is given by $n_a = N/A$. Tunneling between neighboring layers will be neglected: this is justifiable as long as $d > \lambda$, the decay length of the electron wave function.

The collective mode structure of the superlattice has been analyzed in a number of works [2,3,4]. All the existing calculations have been performed in the RPA, with the neglect both of intralayer and of interlayer correlations. The remarkable feature of the emerging collective mode structure is that the longitudinal plasmon excitation spectrum which in a 3D system is separated by a finite gap, namely the plasma frequency,

$\omega_p = (4\pi n e^2/m)^{1/2}$ from the 0 frequency, and in a 2D system for k→0 has the characteristic $\omega(k) \sim \sqrt{k}$ dependence, now for k→0 spreads into a band of acoustic modes ($\omega(k) \sim k$), with each mode labeled by a q value [$k = (k_x, k_y)$ is the 2D wave vector parallel to the xy plane of the monolayers and $q = k_z$ is the perpendicular wavenumber]. The only exception is the q = 0 mode which is bulk-like, i.e., its oscillation frequency at k = 0 is the equivalent of the three-dimensional plasma frequency $\omega_p = (4\pi n_a e^2/md)^{1/2}$.

The q ≠ 0 dispersion relation is

$$\omega^2(k\rightarrow 0, q) = k^2 d^2 \omega^2_p \left\{ \frac{1}{2(1-\cos qd)} + \frac{3}{4} \frac{a}{d} \frac{E_{kin}}{e^2/a} \right\} ; \qquad (1)$$

Here $a = (\pi n_a)^{-1/2}$, the two-dimensional Wigner-Seitz radius, and E_{kin} is the kinetic energy of a particle.

Recent works [5,6] have indicated that electron-electron correlations may give rise to important and qualitatively new effects in superlattice structures. In discussing correlations one has to distinguish between intralayer and interlayer correlations. The former is characterized by the conventional r_s parameters ($r_s = a/a_B$, a_B is the effective Bohr radius, $a_B = \dfrac{\hbar^2 \varepsilon}{e^2 \, m_{eff}}$, m_{eff} is the effective electron mass and ε is the dielectric constant of the substrate), while the latter depends both on r_s and on the d/a ratio.

In this paper we analyze the effect of correlations on the collective mode spectrum. In Section II we recapitulate the Fluctuation-Dissipation Theorem for superlattices, and the expression for the long wavelength dielectric response function for a strongly correlated superlattice, formulated recently by Golden and Lu [7,8]. In Section III we show how the ensuing dispersion affected especially by the interlayer correlation invalidates the results of the RPA calculations and provides a finite energy gap, instead of an acoustic behavior, for any q-value [9]. In Section IV we analyze a simple two-layer system and show that it exhibits a similar mode spectrum. Section V re-visits recent results of Swierkowski, Neilson and Szymanski [5] where it was claimed that Wigner crystallization in a 2D electron monolayer is enhanced by the plasma of the adjacent monolayers, and consequently occurs at lower r_s values than for an isolated layer: we show that within the framework of the approximation used in Ref. [5] a dynamical instability accompanies the effect claimed [10]; we speculate as to the effect of the interlayer correlations, discussed in Section III, on the instability and on the more general scenario.

II DIELECTRIC RESPONSE FUNCTION

The longitudinal dielectric function $\varepsilon(kq\omega)$ of the superlattice can be expressed [2,3] in terms of the form factor F(kq), the 2D Coulomb interaction potential (energy) $\varphi(k)$

$$F(kq) = \frac{\sinh kd}{\cosh kd - \cos qd}$$

$$\varphi(k) = \frac{2\pi e^2}{k} \tag{2}$$

and the density response function $\widehat{\chi}(kq\omega)$:

$$\frac{1}{\varepsilon(kq\omega)} = 1 + F(kq)\varphi(k)\widehat{\chi}(kq\omega). \tag{3}$$

In deriving (3), we have ignored the retardation effect; this is a reasonable assumption in view of the fact that the displacement currents are almost always dominated by the electrostatic effects of nearby layers. The high frequency expansion of $\widehat{\chi}\,(kq\omega)$ can be given in terms of the frequency moments $\langle \omega^s \rangle$

$$\mathrm{Re}\widehat{\chi}\,(kq\omega) = -\frac{\langle \omega \rangle \,(kq)}{\omega^2} - \frac{\langle \omega^3 \rangle \,(kq)}{\omega^4}\, \cdots \tag{4}$$

which are defined through

$$\langle \omega^s \rangle \,(kq) = \frac{1}{\pi} \int_{-\infty}^{\infty} d\omega \, \omega^s \, \mathrm{Im}\widehat{\chi}\,(kq\omega) \tag{5}$$

The determination of the frequency moment is made possible by the fluctuation-dissipation theorem [7]

$$\mathrm{Im}\hat{\chi}\,(kq\omega) = \frac{n_a}{2\hbar N\nu} \int_{-\infty}^{\infty} dt\, e^{i\omega t} \langle\, [n^{\dagger}{}_{kq}, n_{kq}(t)] \rangle$$

$$= -\frac{n_a}{\hbar}\,\tanh\!\left[\frac{\hbar\omega}{2k_BT}\right] S(kq\omega), \qquad (6)$$

where n_{kq} - s are Fourier-transforms of the local density operators, and ν is the number of layers in the superlattice; $S(kq\omega)$ is the dynamical structure factor

$$S(kq\omega) = \frac{1}{2N\nu} \int_{-\infty}^{\infty} dt\, e^{i\omega t} \langle\{ n^{\dagger}{}_{kq}, n_{kq}(t) \}\rangle \qquad (7)$$

The $s=1$ and $s=3$ frequency moments can be calculated [7] to be

$$\langle\omega\rangle\,(kq) = -\frac{n_a k^2}{m} \qquad (8)$$

$$\langle\omega^3\rangle\,(kq) = -\frac{n_a k^2}{m}\,\{\, \omega_{p2}^2\,(k)F(kq) + \frac{3k^2}{m}\,E_{kin}$$

$$+ \left(\frac{\hbar k^2}{2m}\right)^2 + \omega^2{}_{p2}\,(k)D(kq)\}. \qquad (9)$$

where $\omega_{p2}(k) = \left(2\pi n_a e^2 k/m\right)^{1/2}$ is the 2D plasma frequency, and E_{kin} is the expectation value of the kinetic energy per particle of the interacting system. $D(kq)$, the dynamical matrix, is the functional of the static structure function

$$S(kq) = \frac{1}{2\pi} \int_{-\infty}^{\infty} d\omega\, S(kq\omega) . \qquad (10)$$

Trading $S(kq)$ for the pair correlation function $g(kq)$

$$1 + ng(kq) = S(kq) \qquad (11)$$

$D(kq)$ can be expressed as

$$D(kq) = \frac{1}{A}\sum_{k'}\frac{1}{\nu}\sum_{|q'|\le\pi/d} F(k'q')\,\frac{(k\cdot k')^2}{k^3 k'}\,\{ g(k-k'\,q-q') - g(k'\,q') \} \qquad (12)$$

A being the surface area of a monolayer.

Following now the mean-field-theory approach of Iwamoto, Krotschek, and Pines [11], the long-wavelength limit of the dielectric response function can be inferred from the moment expansion:

$$\varepsilon(k\to 0, q\omega) = 1 - \frac{F(kq)\omega^2{}_{p2}(k)}{\omega^2 - \omega^2{}_{p2}(k)D(kq) - (3k^2/m)E_{kin}} \qquad (13)$$

Eq. (13) is also in agreement with calculations based on the Quasi-Localized Charge model [12,13,14] for arbitrary k-values. All of the static correlational effects show up in the dynamical matrix $D(kq)$ through the equilibrium pair correlation functions $g(rz)[r =$

(x,y)]. The principal assumption in the derivation of Eq.(13) consists in identifying the correlational part of the static local field correction with the correlational contribution $D(kq)$ of Eq.(12) to the third-frequency-moment sum rule coefficient, thereby guaranteeing satisfaction of the sum rule and correct high-frequency behavior. The importance of the third-frequency-moment sum rule lies in the fact that it is the lowest order one to exhibit particle correlations. In strong coupling regimes characteristic of the crystalline phase of the OCP, it has been shown [13-16] that the correlational contributions to the dispersion of the 2D and 3D plasmon modes are identical to the correlational part of the third-frequency-moment sum rule coefficient. The infinite superlattice, which is intermediate between the 2D and 3D configurations undoubtedly exhibits this same feature. As such, Eq. (13) is expected to provide an especially reliable description of superlattice plasmons in coupling regimes where intralayer particle correlations and layer-layer correlations are strong. The dielectric response function (13), which is valid in both the quantum and classical domains, is taken to be the basis for the calculation of the following Sections.

III MODE STRUCTURE

Ignoring the kinetic energy term in Eq. (12) - justifiable for the strong coupling regime of interest here - the dispersion relation derived from (13) is

$$\omega^2(k\ q) = \omega^2_{p2}(k)\{F(kq) + D(k\ q)\} \tag{14}$$

In the absence of correlations - i.e. in the absence of the $D(kq)$ term in Eq. (14) - the $k\to 0$ limit leads to the acoustic mode described by Eq. (1). To see the significant modification brought about by interlayer correlations, we consider the $k\to 0$ limit of Eq. (12):

$$D(k\to 0, q) = \frac{1}{kd}\overline{D}(q) \tag{15}$$

$$\overline{D}(q) = \frac{1}{A}\sum_{k'}\frac{1}{v}\sum_{|q'|\leq\pi/d}\frac{(k\cdot k')^2}{(kk')^2}F(k'\ q')k'd\ \{g(k'\ q - q') - g(k'q')\}. \tag{16}$$

As indicated, $D(kq)$, unlike its 2D and 3D OCP sum rule counterparts, exhibits a $1/k$-dependent term in this limit, originating from the layer-layer correlations. The origin of this peculiar behavior lies in the fact that, while the superlattice is subject to three-dimensional forces and is characterized by the 3D $S(kq)$ structure function, it obeys 2D dynamics which results in the q-independent $(k\ k'/kk')^2$ factor in $D(kq)$. This $1/k$ term profoundly alters the structure of the $q \neq 0$ plasmon dispersion in the superlattice. The $k = 0$ plasmon oscillation frequency develops now the non-vanishing value

$$\omega(0\ q) = \sqrt{\frac{\overline{D}(q)}{2}}\ \omega_p \tag{17}$$

(which follows from substituting (6) into (2) and setting $\varepsilon(k\to 0, q\ \omega)$ equal to zero). Eq. (7) represents a "correlational" optic plasmon mode [9].

In order to examine the behavior of $\overline{D}(q)$ more closely, we express it in terms of layer-layer correlation functions $g_{m0}(m = 0, \pm 1, \pm 2, \cdots)$.

$$g(k,z) = \sum_m g_{m0}(k)\delta(z - z_m)\ ,$$

$$g(k\ q) = \sum_m g_{m0}(k)e^{iqz_m}\ , \tag{18}$$

$$z_m = md.$$

Introducing (18) in (16) and performing the q-summations, one obtains

$$\overline{D}(q) = -(1 - \cos qd)dI_{10}(d) - (1 - \cos 2qd)dI_{20}(d) - \cdots, \tag{19}$$

$$I_{m0}(d) = \frac{1}{A}\sum_{k'} k' g_{m0}(k') e^{-k'dm} \tag{20}$$

$$= \frac{1}{2\pi} g_{m0}(r) \frac{1}{(r^2 + d^2 m^2)^{3/2}} \left(\frac{3d^2 m^2}{r^2 + d^2 m^2} - 1\right), \quad m = 1,2,\cdots$$

In the weak coupling limit, it can be rigorously shown that the $g_{m0}(k)$'s are negative over the entire k' domain; hence, from Eq. (20) the I'_{m0}-s are negative. As the coupling increases, the layer-layer pair correlation functions are expected to develop an oscillary behavior similar to that observed for the pair correlation functions in two- and three-dimensional Coulomb liquids; simple model calculations of I_{10} (see below) indicate that the I_{m0}-s are negative in the strong coupling regime leading to $\overline{D}(q)$ positive. In the sequel, we need retain only the most important layer 0-layer 0 (intralayer) and layer 1-layer 0 (interlayer) correlational contributions.

The $q \neq 0$ oscillation frequency (17) is now readily calculated from Eqs. (19) and (20). We obtain

$$\omega^2(0\,q) = -\frac{1}{2}\omega^2_p I_{10}(d)d(1 - \cos qd) \geq 0. \tag{21}$$

As to the q = 0 mode, one can show from (13) that, as expected, in the k = 0 limit it is unaffected by the layer-layer correlations and remains the well known 3D bulk plasmon oscillation frequency ω_p.

Eq. (21) is valid both in the quantum and in the classical domains. The physical origin of the disappearance of the acoustic mode can be understood by realizing that the (k=0, q→0) mode can be regarded as a shear wave propagating along the z-direction, made possible by correlations along this direction. In the $q \to 0$ limit, the shear mode should assume an acoustic behavior in q; indeed, by expanding (21) one finds

$$\omega^2 = -\frac{d}{4} I_{10} \omega^2_p q^2 d^2 \tag{22}$$

which is the expected behavior.

At finite wavenumbers, Eq. (14) can also be expressed in terms of $g_{00}(k)$ and $g_{10}(k)$:

$$\omega^2(kq) = \omega^2(0\,q) + \frac{1}{2}\omega^2_p kd \{F(kq) + D_{00}(k) + D_{10}(kq)\} \tag{23}$$

where

$$D_{00}(k) = \frac{1}{A}\sum_{k'} \frac{(\mathbf{k}\cdot\mathbf{k})^2}{k^3 k'} \{g_{00}(\mathbf{k} - \mathbf{k'}) - g_{00}(\mathbf{k'})\} \tag{24}$$

$$= \frac{1}{2k} \int_0^\infty dr \frac{1}{r^2} g_{00}(r) \left\{1 - 4J_0(kr) + 6\frac{J_1(kr)}{kr}\right\}$$

is the isolated 2d layer contribution to D(k q) and

$$D_{10}(k\ q) = \frac{1}{k}\cos qd \int_0^\infty dr \frac{r}{(r^2 + d^2)^{3/2}} g_{10}(r)$$

(25)

$$\times \left\{ [J_0(kr)-1](\frac{3d^2}{r^2+d^2} - 1) + 3J_2(kr)\frac{r^2}{r^2+d^2} \right\}.$$

is the lowest-order layer-layer contribution. Detailed calculations of D(k q) and ω^2(k q) would require the availability of S(k q) or g_{10}(r) data. MC simulations and HNC calculations have provided static structure function and pair correlation function data for the 2D and 3D OCP in the classical [17,18,19,20] and quantum [21,22] domains. However, MC or HNC data for the layered electron liquids of interest here have yet to be generated. The lack of data notwithstanding, one can gain further insight into how layer-layer correlations alter the RPA description of plasmon dispersion by studying simple models for g_{00} (r) and g_{10} (r). Even though the g_{0m} -s are interrelated and constrained by the equivalents of the perfect screening sum rule and the Stillinger-Lovitt condition, a simple model which reflects the most significant effect of strong correlations, namely the creation of a correlational hole, should be able to provide a good qualitative picture of the changed mode structure. Thus, the intralayer correlation function g_{00} (r) and the layer-layer correlation function g_{10} (r) are approximated as

$$g_{00}(r) = \begin{array}{l} -1 \text{ for } r \leq a, \\ \\ 0 \text{ for } r > a, \end{array}$$

(26)

and

$$g_{10}(r) = \begin{array}{l} -1 \text{ for } r \leq \alpha a, \\ \\ 0 \text{ for } r > \alpha a. \end{array}$$

(27)

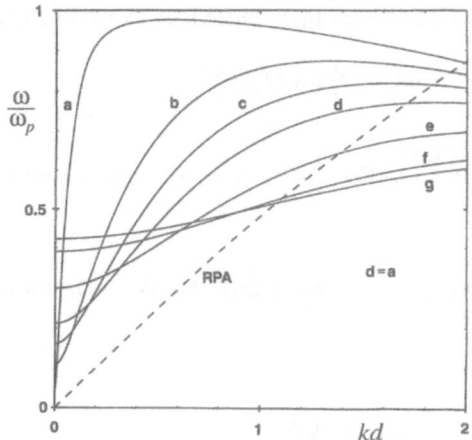

Fig. 1. Typical dispersion curves for a superlattice with interlayer correlation taken into account; a is the interparticle distance in the monolayer, d is the layer-layer separation The curves are labelled by different qd values: from (a) through (g) qd = 5°, 30°, 45°, 60°, 90°, 135°, 180°, respectively. The RPA boundary (corresponding to qd=π) is also shown.

Substituting into (20) and (21), one obtains

$$\omega^2(0\,q) \approx \frac{1}{2}\,\omega_p^2\,\frac{\alpha^2 a^2 d}{(\alpha^2 a^2 + d^2)^{3/2}}\,(1 - \cos qd). \tag{28}$$

We choose $\alpha = 0.5$ (the structure of the dispersion curves is not sensitive to the choice of α) and $(d/a) = 1$ to generate a typical dispersion which is shown in Fig. 1. Viewed in ω,k,q - dispersion space, there evidently exists an upper bound k_{max} to k, dictated by the condition $(\partial\omega/\partial q)_{k=k_{max}} = 0$. The existence of an envelope curve delineating the boundary beyond which no dispersion curve penetrates can be inferred from Fig. 1. This can be compared with the RPA boundary (the $q = \pm\pi/d$ dispersion curve) also displayed here. The exclusion of the region between the RPA and correlational boundaries is a significant signature of the superlattice layer-layer correlations and should be verifiable by experiment.

The dispersion of the collective intraband modes (plasmons) in modulation-doped $GaAs/Al_xGa_{1-x}As$ multiple-quantum-well heterostructures has been determined by inelastic light scattering . The RPA acoustic plasmon as given by Eq. (1) was definitely observed in the Olego-Pinczuk-Gossard-Wiegmann experiments [23]; however, the experiment was not appropriate to distinguish between the correlational plasmons described here and the acoustic behavior predicted by the RPA model.

The correlational plasmon can also be evaluated for a weakly coupled superlattice (although in this case the validity of Eq. (13) is less solidly grounded) by using the classical Debye approximation for the correlation functions:

$$n\,g_{00}\,(k) = -\frac{\kappa}{k}\,\frac{1}{\sqrt{1 + \left(\frac{\kappa}{k}\right)^2 + 2\frac{\kappa}{k}\coth kd}} \tag{29}$$

$$n\,g_{10}\,(k) = \frac{\kappa}{k}\,\sinh kd\,\left\{1 - \frac{\frac{\kappa}{k} + \coth kd}{\sqrt{1 + \left(\frac{\kappa}{k}\right)^2 + 2\frac{\kappa}{k}\coth kd}}\right\} \tag{30}$$

where κ is the 2D Debye wavenumber. Using now (30) to evaluate Eqs. (20), (19) and (17) and assuming $qd = \pi$, one finds

$$\frac{\omega\,(0q)}{\omega_p} = \Gamma^{1/2}\,\left(\frac{d}{a}\right)^{3/2}\,\psi\,\left(\Gamma\frac{d}{a}\right) \tag{31}$$

where the function $\psi\,(x)$ can be calculated from Eqs. (19), (20) with the aid of (30).

IV TWO LAYERS

The main physical effects that distinguish the behavior of the superlattice from that of a simple monolayer, already emerge in the system that consists of two interacting layers. This similarity is emphasized and exploited in the recent work of Swierkowski, Neilson and Szymanski [5]. We now show how the considerably simpler structure of the two-layer system exhibits essentially the same dynamical behavior as the superlattice discussed in the previous Section. The calculation can be conveniently performed in the "Quasilocalized Charge Approximation" which is based on the observation that in the case of strong correlations the particles are trapped in local potential wells and their motion consists mostly of small oscillations around the quasi-equilibrium position [13,14]. Even though the calculation is essentially classical, it is not expected that in the strong coupling regime this substantially affects the general validity of the conclusions.

The analysis can be done in the formalism of a two-component system [13], whose dispersion relation can be written in the form

$$|\Pi| = 0 \tag{32}$$

with

$$\Gamma_{AB} = \omega^2 \delta_{AB} - \sqrt{\frac{n_A n_B}{m_A m_B}}\, \varphi_{AB}(k)\, k^2 - D_{AB}(k) \tag{33}$$

where $D_{AB}(k)$ is the dynamical matrix with a definition slightly different from that of the previous Section.

$$D_{AB}(k) = \frac{1}{A} \sum_C \sum_{k'} \frac{(k \cdot k)^2}{k^2} \left\{ \sqrt{\frac{n_A n_B}{m_A m_B}}\, \varphi_{AB}(k')\, g_{AB}(k-k') - \delta_{AB} \frac{n_C}{m_A} \varphi_{AC}(k')\, g_{AC}(k') \right\} \tag{34}$$

The subscripts A, B ... enumerate the species, $\varphi_{AB}(k)$ is the interaction potential (energy) between species A, B and $g_{AB}(k)$ is the corresponding pair correlation function. The transition to the two-layer system is now effected by identifying

$$\varphi_{11}(k) = \varphi_{22}(k) = \varphi(k)$$

$$\varphi_{12}(k) = \varphi(k)\, e^{-kd}$$

$$n_1 = n_2 = n$$

$$m_1 = m_2 = m \tag{35}$$

$$D_{11}(k) = D_{22}(k) = D(k)$$

$$D_{12}(k) = D_x(k)$$

The dispersion relation (31) now becomes

$$\left\{ \omega^2 - \omega^2_{p2}(k) - D(k) \right\}^2 = \left\{ \omega^2_{p2}(k)\, e^{-kd} + D_x(k) \right\}^2 = 0 \tag{36}$$

with the solutions

$$\omega^2_{\pm} = \omega^2_{p2}(k)\left\{ 1 \pm e^{-kd} \right\} + \left\{ D(k) \pm D_x(k) \right\}^2 \tag{37}$$

It follows from (34) that in the $k \to 0$ limit

$$D(k \to 0) = -D_x(k \to 0)$$

$$= -\frac{1}{A} \sum_{k'} \left(\frac{k \cdot k'}{k\, k'} \right)^2 \omega^2_{p2}(k')\, e^{-kd}\, g_{12}(k') \tag{38}$$

$$= D_0$$

D_0 is a positive constant (since $g_{12}(k)$ is predominantly negative). Thus one can identify the ω_+ solution in (37) as the in-phase oscillations of the two layers with

$$\omega^2_+ (k \to 0) = 2\omega^2_{p2}(k) \tag{39}$$

On the other hand, the ω_- solution is the out-of-phase oscillation of the two layers:

$$\omega^2_- (k \to 0) = \omega^2_{p2}(k)\, dk + 2 D_0 \tag{40}$$

It is clear that in the absence of correlations ($D_0 = 0$) this is the equivalent of the RPA acoustic mode in the superlattice: however, the effect of the correlations is to destroy the acoustic behavior and to create an energy gap.

V DYNAMICAL INSTABILITY

In their recent work Swierkowski, Neilson and Szymanski [5] pointed out that interlayer interaction can enhance the effect of intralayer correlations. They investigated a two-layer system and their analysis indicates that for small enough d/a at large enough r_s values the static density response function $\hat{\chi}(k)$ develops a strong peak in the vicinity of the k value corresponding to the inverse lattice constant of the 2D Wigner lattice. According to the interpretation of Ref. [5] this can be taken as the indication of the onset of crystallization (at an $r_s = 15 \sim 20$ value which would be well below the crystallization limit for a single monolayer).

The major approximation used in Ref. [5] is the complete neglect of interlayer correlations: the mutual interaction of the layers is taken into account only through the average RPA field. Without addressing the question of the appropriateness of this approximation, here we wish to investigate the dynamical consequences of the strong intralayer correlations within this model. Then, in view of the demonstration in Section II of the profound effect of the interlayer correlations on the dynamical mode structure, we may make inferences as to the expected effect of interlayer correlations on the conclusions of Ref. [5].

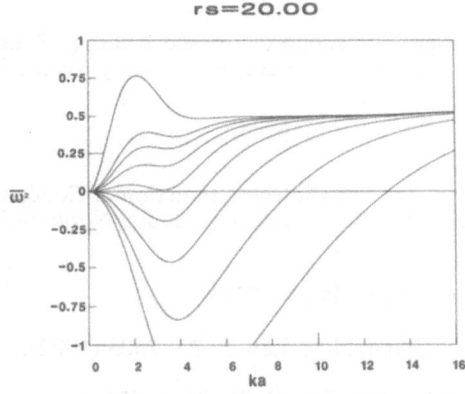

Fig. 2. Dispersion curves for the superlattice calculated from the model where the interlayer correlations are neglected and the intralayer correlations are taken to be identical to those in an isolated monolayer; r_s is the coupling parameter of the monolayer. The curves are labelled by different d/a values: from top to bottom d/a = 2.0, 1.0, 0.9, 0.8, 0.7, 0.6, 0.5, 0.4, 0.3, respectively. Observe the emergence of the dynamical instability ($\omega^2 < 0$) at ka = 3.6

To implement this program we consider the dispersion relation (14). With the assumption of no interlayer correlations ($g_{m0}(k) = \delta_{m0} g(k)$), using (18), Eq. (12) becomes

$$D(kq) = D(k) = \frac{1}{A} \sum_{k'} \frac{(k \cdot k')^2}{k^3 k'} \{g(k-k') - g(k')\}$$ (41)

which is just the 2D dynamical matrix [14]. In contrast to the scenario discussed in Section II where the interlayer correlations cause D(k) to behave as k^{-1} for k→0, here D(k) is negative and $\sim k$ and thus it doesn't qualitatively affect the small-k acoustic behavior of the superlattice plasmon. However, for finite k the situation is quite different: for strong enough coupling D(k) assumes a strongly oscillating behavior causing $\omega^2(k)$ to approach 0 and to eventually become < 0. Fig. 2 shows this behavior; D(k) has been calculated from the Monte Carlo data of Tanatar and Ceperley [22]. One can observe that $\omega^2(k) < 0$ and thus a dynamical instability (softening of the mode) is created around ka = 3.2, when d/a < 0.68. This can be compared with the observation of Ref. [5] where the critical value of d/a at $r_s = 20$ is around d/a = 0.48 and the static instability occurs around ka = 3.38. In spite of the differences in the critical values (which can be partly attributed to the different number of layers in the two models, partly to the different approximations used), it seems that one is faced with the two manifestations of the same physical phenomenon (The relationship between static and dynamical instabilities was discussed by Mermin in [24]).

How does then the switching on of the interlayer correlations affect the phenomenon described? We know from the analysis of Section II that an energy gap develops at k = 0, resulting in the small-k dispersion of Fig. 1. Although the interlayer correlations obviously alter the finite k behavior of the dispersion curves as well, and one cannot simply combine Fig. 1 with Fig. 2, it seems rather plausible that the emergence of the energy gap will create at least more severe requirements for the critical d/a and r_s values associated with the dynamical instability. Whether the dynamical (and the associated static) instability survives at all under the consistent treatment of the intralayer correlations, remains an open question at this time.

ACKNOWLEDGMENT

This work has been partially supported by the National Science Foundation under Grants PHY-9115714 and PHY-9115695 and by the Army Research Office.

REFERENCES

[1]   N. Raj and D.R. Tilley, in "The Dielectric Function of Condensed Systems" ed. by L.V. Keldysh, D.A. Kirzhnitz and A.A. Maradudin, North-Holland, New York, 1989.

[2]   A.L. Fetter, Ann. Phys. (N.Y.) **88**, 1 (1974).

[3]   S. Das Sarma and J.J. Quinn, Phys. Rev. B **25**, 7603 (1982).

[4]   A. Tselis and J.J. Quinn, Phys. Rev. B **29**, 3318 (1984).

[5]   L. Swierkowski, D. Neilson, and J. Szymanski, Phys. Rev. Lett. **67**, 240 (1991).

[6]   C. Zhang and N. Tzoar, Phys. Rev. A **38**, 5786 (1988).

[7]   K.I. Golden and De-xin Lu, Phys. Rev. A **45**, 1084 (1992).

[8]   De-xin Lu and K.I. Golden Phys. Lett. A **160**, 473 (1991).

[9]   K.I. Golden and G. Kalman, to be published.

[10] G. Kalman, K.I. Golden, and Ph. Wyns, to be published.

[11] N. Iwamoto, E. Krotscheck, and D. Pines, Phys. Rev. B **29**, 3936 (1984).

[12] K.I. Golden, De-xin Lu, and G. Kalman, in preparation.

[13] G. Kalman and K.I. Golden, Phys. Rev. A **41**, 5515 (1990).

[14] K.I. Golden, G. Kalman, and Ph. Wyns, Phys. Rev. A **41**, 6940 (1990).

[15] K.I. Golden and De-xin Lu, Phys. Rev. A **31**, 1763 (1985).

[16] K.I. Golden, Phys. Lett **112A**, 397 (1985).

[17] F. Lado, Phys. Rev. B **17**, 2827 (1978).

[18] J.P. Hansen, Phys. Rev. A **8**, 3096 (1973); E.L. Pollock and J.P. Hansen, **8**, 3110 (1973).

[19] H. Totsuji, Phys. Rev. A **17**, 399 (1978).

[20] R.C. Gann, S. Chakravarty, and G.V. Chester, Phys. Rev. B **21**, 326 (1979).

[21] D.M. Ceperley and B. Adler, Phys. Rev. Lett. **45**, 566 (1980).

[22] S. Tanatar and D.M. Ceperley, Phys. Rev. B **39**, 5005 (1989).

[23] D. Olego, A. Pinczuk, A.C. Gossard, and W. Wiegmann, Phys. Rev. B **26**, 7867 (1982).

[24] D. Mermin, Ann. Phys. (N.Y.) **18**, 421 (1962).

CRYSTALLINE STRUCTURES IN

THE DEFORMABLE JELLIUM

S. Orozco, M. A. Ortíz and R. M. Méndez-Moreno

Departamento de Física, Facultad de Ciencias
Universidad Nacional Autónoma de México,
Apartado Postal 20-364, 01000 México, D. F.

ABSTRACT

The electron gas in the deformable jellium is studied with different crystallographic symmetries. The single particle state function can be written in terms of expansions which give charge density centered around different specific three-dimensional lattices as the cubic lattices or the tetragonal ones. Symmetries in less than three dimensions have also been proposed, one can propose an expansion with charge density centered around a square lattice and homogenous density in the orthogonal direction, or expansions with periodic charge density in only one direction and homogenous density in the orthogonal plane. A brief review of the ground state properties with some of these symmetries is presented in this work. A comparison of the results in the cases studied is presented.

INTRODUCTION

The electron gas and its phases has been a matter of great interest for many years since the pionnering work of Wigner[1]. In the recent years the concern with this model has been renewed. In the presence of an external magnetic field it is a useful model in the study of the quantum Hall effect[2, 3]. The electron gas with abnormal occupation is used as a model for some high-T_c superconductor materials[4, 5].

The ground state properties of the electron gas in the jellium have been obtained with several methods. Among then the density functional formalism[6] and the integral-approximant technique[7]. Perhaps the most ambitious calculations with the jellium model are the variational Monte Carlo and the Green's-function Monte Carlo[8]. The essencial asumption of the jellium model is to suppose an inert uniform neutralizing background. A more refined model is the deformable jellium, in which the background is allowed to deform in order to get local charge neutrality. This fact guarantees a lower energy per particle and therefore a more stable system[9, 10, 11]. Using trial functions with different crystallographic symmetries, the self-consistent Hartree-Fock (HF) approximation has been used in the evaluation of ground state properties in the deformable jellium[12]. The HF method gives a general procedure that defines the best orbitals, given a starting point for many approximation schemes[13].

The deformable jellium and the HF method has been exploited by us in the study of the electron gas. Our approach has been to obtain self-consistency with a set of modulating functions that contain the trivial plane wave (PW) as a possible solution. This has turned a very powerful technique that has the capability of describing both the metallic and the low density regions in a unified nonperturbative fashion. Many levels of approximation in the expansion for the state function have been reported[14, 15]. Recently, the convergence of this algebraic HF procedure for the ground state has been obtained at metallic and intermediate densities using an improved expansion in terms of cosine functions[16].

A remarkable achievement of the deformable jellium is the description of the transition from the homogenous phase at high densities into localized states at lower densities. At metallic densities where the PW is the self-consistent ground state the deformable jellium coincides with the uniform jellium. At intermediate densities corrugated (charge density wave type) solutions are obtained. At very low densities the ground state solution approaches the Wigner crystal. In general the corrugated phase of the deformable jellium predicts lower ground state energies per particle than the uniform jellium, and therefore a more stable system[15, 17]. One can understand the improvement brought by the deformable jellium in terms of a simple electrostatic analog, in which the condition of local neutrality is the more favorable energetically. Once in the corrugated region the system resembles the lattice localization.

In this work a brief review of ground state properties of the electron gas in the deformable jellium is done where trial expansions with charge density centered around different crystallographic symmetries are employed. Among these, the simple cubic, the body-centered-cubic[18], the tetragonal and the orthorombic have been studied. Symmetries in less than three dimensions are also studied, as for example the function which gives charge density centered around a square lattice and homogeneos density in the orthogonal direction. We also consider here the expansion which gives periodic charge density along one direction (and homogenous density in the orthogonal plane), which corresponds to the Overhauser's charge density waves(CDW) type solutions. The HF method is used and many levels of approximation in the expansion of the ground state function can be found in the literature. We make a comparison of the results with the various symmetries and discuss some properties as ground state energy per particle, charge density, the transition point to periodic charge density and the coupling parameter.

THEORY

In the deformable jellium model, the divergences which appear in the evaluation of the ground state energy, with the independent particle approximation and Coulomb interaction, cancel automatically. In order to evaluate the matrix elements of the energy when the expansions for the state function are given in terms of plane waves, the deformable jellium is defined in such a way that the direct term of the electron-electron interaction cancels the background energies. In that way the divergencies can be eliminated in the same form as they cancel in the uniform jellium with a PW as solution. This model can be obviously extended to include other type of interactions, as for example the Yukawa-type screened Coulomb interactions[19]. The hypothesis involved in the model and applications of this can be found elsewhere[15, 11].

The deformable jellium together with the Hartree-Fock (HF) approximation provide a systematic method to describe the ground state properties of the electron gas. The single particle state functions are taken to be the usual PW's, multiplyed by modulating functions, where the minimal modulating frequency q_0 is constrained by the orthonormality of the orbitals. The proposed functions are of the form

$$\phi_k(\mathbf{r}) = \frac{e^{i\mathbf{k}\cdot\mathbf{r}}}{\sqrt{V}} \sum_{n_x=-N}^{N} \sum_{n_y=-N}^{N} \sum_{n_z=-N}^{N} C_{n_x n_y n_z} e^{iq_0 \mathbf{n}\cdot\mathbf{r}}. \tag{1}$$

In this general expresion the vectors $\mathbf{k} = i k_x + j k_y + k k_z$ and $\mathbf{r} = i x + j y + k z$. V is the volume in which periodic boundary conditions are imposed. The coefficients $C_{n_x n_y n_z}$ are considered independent of $\mathbf{k}$ and are self-consistently determined by solving the HF equations with the orthonormality condition. In order to satisfy the HF equations the minimal modulating frequency $q_0 = 2k_F$. The upper and lower limits in all the sums are taken equal to N and $-N$ respectively. The vector $\mathbf{n} = i n_x + j n_y + k n_z$ and the term with $n_x = n_y = n_z = 0$ is the PW solution. The proper selection of the contributing terms in each sum, will determine the charge density centered around different lattices.

When the three sums contribute to Eq. (1) the charge density is centered around three dimensional crystalline structures. The full state function with $(2N+1)^3$ terms gives charge density centered around a simple cubic lattice (a, a, a). With other 3D symmetry as the tetragonal $(a, a, 2a)$, the number of terms is smaller than with the simple cubic, in this case we have $(2N + 1)^2(N + 1)$.

If one of the sums is omited, two dimensional crystalline symmetries are obtained and homogenous charge density in the orthogonal direction. In this case the full state function corresponds to a square lattice (a, a) with $(2N + 1)^2$ terms. However with the proper selection of the terms in the two sums it is possible to have other two dimensional symmetries as for example the exagonal. When only one of the components of the vector $\mathbf{n}$ is different from zero, the traditional Overhauser's charge density waves (CDW) are obtained[10].

The orbitals can be rewritten in terms of the equivalent expansion in cosine functions.

$$\phi_k(\mathbf{r}) = \frac{e^{i\mathbf{k}\cdot\mathbf{r}}}{\sqrt{V}} \sum_{n_x=0}^{N} \sum_{n_y=0}^{N} \sum_{n_z=0}^{N} C_{n_x n_y n_z} cos(iq_0 n_x x) cos(iq_0 n_y y) cos(iq_0 n_z z) \tag{2}$$

That equivalence is posible because the self-consistently determined coefficients in Eq. (1) have interesting properties, i.e. $C_{n_x n_y n_z} = C_{-n_x -n_y -n_z}$, with all the posible combinations in the signs of the $n_i's$. This expansion has $(N + 1)^3$ terms with the simple cubic symmetry instead of the $(2N + 1)^3$ terms of the function in Eq.(1) (when the three sums appears in both). Then calculations with the Eq. (2) are more recomendable since they are more economic from a computational point of view.

To obtain the transition point from homogenous to periodic charge density we use several criteria. One is to find the density r_s in Bohr ratios, where all the coefficients in the wave function become zero (in the approximation required), except the PW term. Other criterium which has proved to be quite precise, is the determination of the discontinuity in the coupling parameter at the transition point where the state function changes from PW to periodic one. The coupling parameter is defined as the ratio of the average potential energy to the average kinetic one.

In order to evaluate the ground state energy per particle (in rydbergs) with the expansion given by Eq. (2), the kinetic energy is given by terms of the form

$$\frac{<T>}{N} = A_0 \sum_{n_x=0}^{N} \sum_{n_y=0}^{N} \sum_{n_z=0}^{N} |C_{n_x n_y n_z}|^2 \; [1 + \frac{20}{3}(n_x^2 + n_y^2 + n_z^2)] \tag{3}$$

where A_0 is a numerical constant, and the exchange energy has the form

$$\frac{<V>}{N} = -\frac{A_1}{512} \sum_{n_1} \sum_{n_2} \sum_{n_3} \sum_{n_4} C_{n_1}^* C_{n_2}^* C_{n_3} C_{n_4} \; I(n_1, n_4) \; F(n_1, n_2, n_3, n_4). \tag{4}$$

N is the number of particles and $C_n \equiv C_{n_x n_y n_z}$. We have used $\sum_n$ for $\sum_{n_x} \sum_{n_y} \sum_{n_z}$, $I(n_1, n_4)$ is a function that stems from the integrals of the Coulomb potential in terms of the components of n_1 and n_4. $F(n_1, n_2, n_3, n_4)$ is a sum of terms which are products of Kronecker δ functions in the components n_x, n_y and n_z of the four $n_i's$. As always the exchange term has been greatly simplified taking into account symmetry considerations, and we have to evaluate 8^3 instead of 8^4 terms for each triad of values (n_x, n_y, n_z).

RESULTS AND DISCUSSION

For all the symmetries presented in this work the state function has been self-consistently determined in a wide interval of densities. A relevant general feature is that the self-consistent state function is the PW in the high density region including the metal-like densities. At greater r_s values (in the intermediate and high density regions) periodic solutions are obtained. In order to find convergence for different properties of the system, in a wide range of densities, expansions for the orbitals in Eq. (2) with different number of terms have been considered, changing the upper limit in the sums in the coseno series.

In the determination of the transition point from homogenous to periodic density, for the cubic simple symmetry, the transition r_s value depends in the number of terms in the state function up to the expansion with $\mathcal{N} = 3$ for the upper limit. Begining with this value and for greater expansions the value of the transition point is always the same. For all the other three dimensional symmetries studied here, the r_s value at which the transition to periodic charge density occurs, becomes independent of the number of terms in the expansion begining with functions where the upper limit in the three sums $\mathcal{N} = 3$. This is an interesting result because the number of iterations needed to get self-consistency in the accuracy required, explodes near the transition point. For symmetries in less dimensions the value of the upper limit required to get convergence in the determination of the transition point (to periodic charge density) is greater i.e., $\mathcal{N} = 5$.

In table I the r_s values at which the HF solution changes from PW to periodic one are shown for different symmetries. The simple cubic symmetry (a, a, a) occurs at the lowest value of $r_s = 26$ than the other symmetries. This result. agrees with the value obtained for the 3D system, via the density functional method of Ref. ([20]). Charge density centered around other cubic lattices occur at smaller densities than for the simple cubic, in particular the fcc symmetry is obtained for a value of $r_s > 100$. With the tetragonal symmetry the transition from homogenous to tetragonal density occurs in two steps: first a transition to planar symmetry (square lattice) and homogenous density along the orthogonal direction, and then the transition to the tetragonal one $(a, a, 2a)$ at greater values of the interparticle distance. This is the only symmetry that is achieved in two steps. For the planar and linear symmetries the transitions are at 31.2 and 31.3 respectively[21].

As another criteria to determine the transition point, the coupling parameter has been evaluated. In the region where the HF state function is the PW, the parameter increases linearly with r_s. When the state function becomes a periodic function, the coupling parameter shows a discontinuity. In Fig. 1 the coupling parameter as a function of the interparticle distance, is shown for the simple cubic, square and linear symmetries. In the high density region the parameter is the same for the three cases, since the HF solution is the PW. After the transition three different curves are obtained, and the transition point is well diferenciated in each case.

The ground state energy per particle in rydbergs in terms of r_s, is shown in the Figs.

Table 1: The transition point r_s from PW to periodic solutions.

Symmetry	Transition points, r_s
simple cubic	26
bcc	70
fcc	180
tetragonal	31.2,78.5
square lattice	31.2
one dimensional	31.3

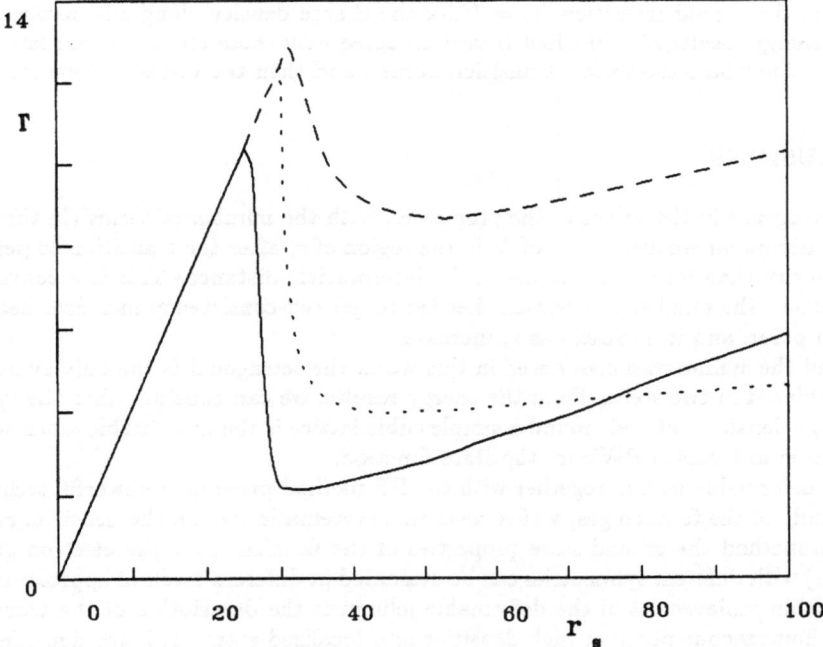

Figure 1. The coupling parameter as a function of r_s. The solid curve is for the simple cubic symmetry, de dotted one for the planar and the dashed curve for the linear symmetry.

2 and 3. In the region where the HF orbitals are not the PW, convergency in the ground state energy is obtained with a smaller value of $\mathcal{N}$ for values of r_s near the transition than at greater values of the interparticle distance parameter. As it was expected the most stable system is that with all the terms in the three sums for a given value of $\mathcal{N}$, that is the simple cubic one. In Fig. 2 the energy for the simple cubic, with $\mathcal{N} = 3$ up to 5 is shown. The convergence in energy for this symmetry is obtained up to $r_s \cong 45$, however to obtain convergence for higher densities it is necessary to increases the value of $\mathcal{N}$.

In Fig. 3 the difference in energy per particle with the plane wave energy $\Delta E = E - E_{PW}$, as a function of r_s is shown for the systems with tetragonal $(a, a, 2a)$, bcc and planar symmetries. In the bcc case it is possible to improve the energy results if the number of terms in the state function with this symmetry is increased. In the planar (square lattice) and linear cases, convergency in the energy was already obtained[21]. It is observed that the energy in the tetragonal and planar symmetries coincides in the region between the first and the second transitions for the tetragonal case. However when the charge density around a tetragonal lattice is obtained, the energy in this case is lower than in the planar symmetry.

In Fig. 4 the charge density is shown for the tetragonal symmetry at $r_s = 80$. It is interesting to observe the size of the *peaks* in the charge density in xy plane $(D_{max} = 36.58)$, as compared to the charge density along the z axis, which is smaller $(D = 22.68)$. However before the second transition $(r_s = 78.5)$, the charge density along z is homogenous. Also the energy results after the first transition agree with those for the square lattice up to the r_s value where the second transition occurs, and then the tetragonal symmetry is obtained.

CONCLUSIONS

Convergence in the values of the properties, with the number of terms (in the state function) occurs for smaller values of $\mathcal{N}$ in the region of r_s after the transition to periodic charge density than for greater values of the interparticle distance. This is a convenient result because the number of iterations needed to get self-consistency increases near the transition point, and it decreases as r_s increases.

Of all the symmetries considered in this work, the tetragonal is the only symmetry that is achieved in two steps. From the energy results, we can conclude that the system with charge density centered around a simple cubic lattice is the more stable when we use an expansion in terms of PW's for the state function.

The deformable jellium together with the HF method provides a powerful technique for the study of the fermion gas, which cover in a systematic way all the densities region. With this method the ground state properties of the fermion gas (the electron gas, in particular) with different symmetries can be evaluated at different levels of approximation. A remarkable achievement of the deformable jellium is the description of the transition from the homogenous phase at high densities into localized states at lower densities. At metallic densities where the PW is the self-consistent ground state the deformable jellium coincides with the uniform jellium.

ACKNOWLEDGEMENTS

This work is supported in part by Dirección General de Asuntos del Personal Académico at Universidad Nacional Autónoma de México (UNAM), project number IN102991, México D. F., México. The authors acknowledge Mr. César A. Zepeda for technical assistance with the figures.

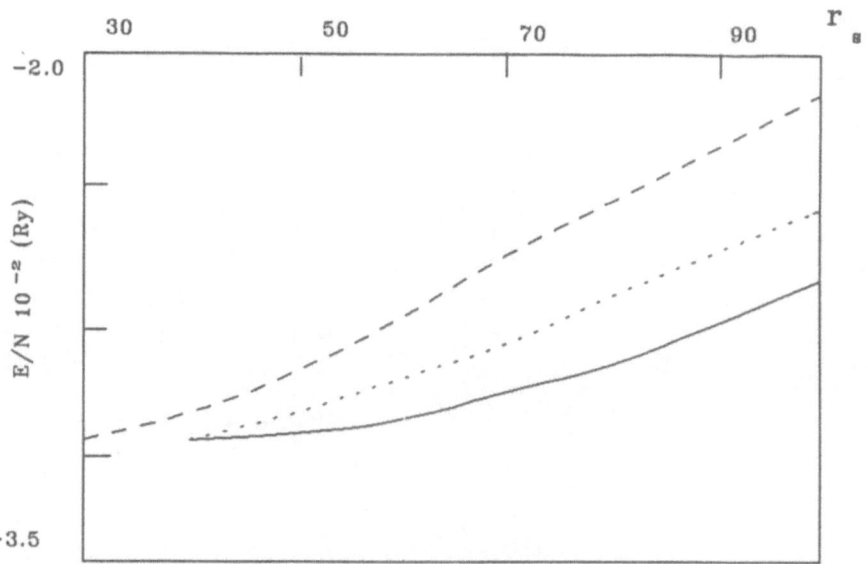

Figure 2. The ground state energy per particle in rydbergs in terms of r_s. The results for $\mathcal{N} = 3, 4, 5$ are shown for the case with charge density centered around a simple cubic lattice.

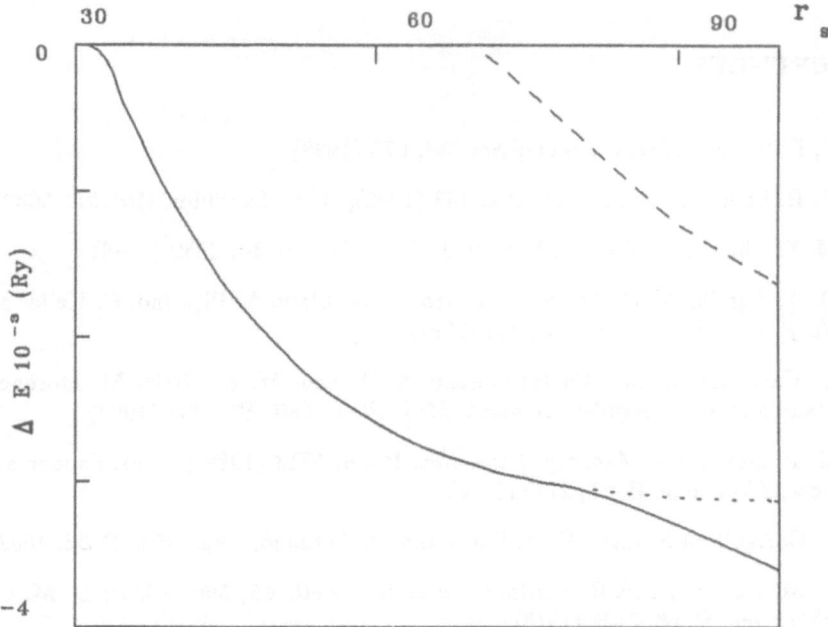

Figure 3. The difference in energy ΔE in rydbergs. The results for the tetragonal $(a, a, 2a)$, the bcc and planar symmetries are shown.

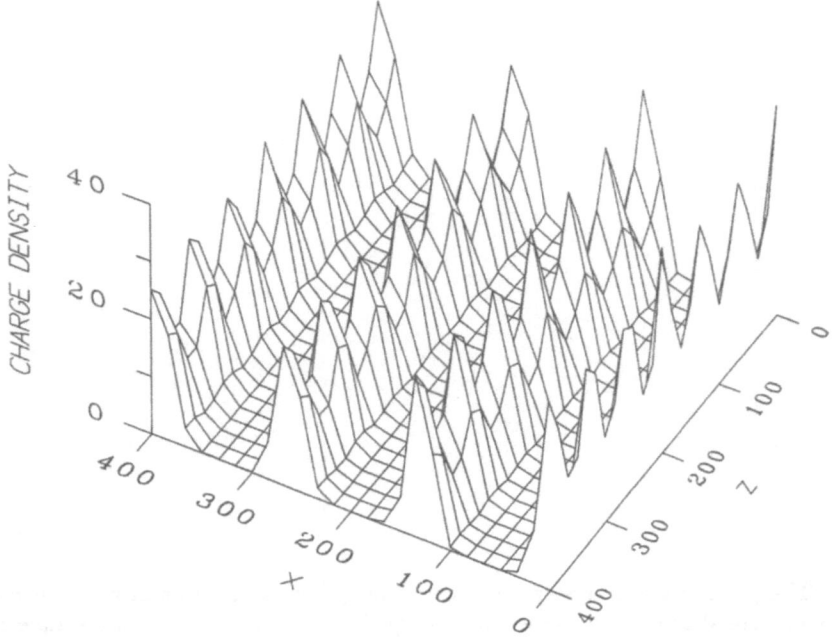

Figure 4. The charge density for the tetragonal symmetry at $r_s = 80$. The size of the *peaks* in the xy plane is different than the charge density along z.

REFERENCES

[1] E. P. Wigner, *Trans. Faraday Soc.* **34**, 678 (1938).

[2] R. B. Laughlin, *Ann. Phys.* **191**, 163 (1989); R. B. Laughlin, *ibid.* **23**, 5632 (1981).

[3] M. P. Chaubey and C. M. Van Vliet, *Phys. Rev.* **B 34**, 3932 (1986).

[4] D. A. Agrello, V. C. Aguilera-Navarro, M. de Llano, A. Plastino, C. Keller and J. P. Vary, *Mod. Phys. Lett.* **B 5**, 805 (1991).

[5] G. Carmona, R. M. Méndez-Moreno, S. Orozco, M. A. Ortíz, M. Moreno, M. de Llano and V. C. Aguilera-Navarro, *Mod. Phys. Lett.* **B6**, 935 (1992).

[6] M. P. Das and J. Mahanty, *Phys. Rev.* **B 38**, 5713 (1988); L. M. Sander and J. H. Rose, *Phys. Rev.* **B 21**, 2739 (1980).

[7] V. C. Aguilera-Navarro, G. A. Baker and M. de Llano, *Phys. Rev.* **B 32**, 4502 (1985).

[8] D. M. Ceperley and B. J. Alder, *Phys. Rev. Lett.* **45**, 566 (1980); D. M. Ceperley, *Phys. Rev.* **B 18**, 3126 (1978).

[9] H. Frohlich, *Proc. R. Soc. London,* Ser. A **223**, 296 (1954).

[10] A. W. Overhauser, *Phys. Rev.* **167**, 691 (1968).

[11] R. M. Méndez-Moreno and M. A. Ortíz, *Rev. Mex. Fis.* **32**, 413 (1986).

[12] V. C. Aguilera-Navarro, M. de Llano and A. Plastino, *Kinam,* **1**, 441 (1979).

[13] D. S. Koltum and J. M. Eisenberg, *Quantum Mechanics of Many Degrees of Freedom* (Wiley, New York, 1988).

[14] R. M. Méndez-Moreno, M. A. Ortíz and M. Moreno, *Phys. Rev.* **A 40**, 2211 (1989).

[15] R. G. Barrera, M. Greter and M. de Llano, *J. Phys.* **C 12**, L715 (1979). *J. of Phys.*

[16] R. M. Méndez-Moreno, M. Moreno and M. A. Ortíz, *Phys. Rev.* **A 44**, 2370 (1991).

[17] R. M. Méndez-Moreno and M. Moreno, *Phys. Rev.* **A 41**, 6662 (1990).

[18] M. A. Ortíz, R. M. Méndez-Moreno y M. Moreno, *Supl. Bol. Soc. Mex. Fis. del XXXI Congreso Nac. de Fis.* (1988).

[19] M. A. Ortíz and R. M. Méndez-Moreno, *Phys. Rev.* **A 36**, 888 (1987).

[20] L. M. Sander, J. H. Rose and H. B. Shore, *Phys. Rev.* **B 18**, 6506 (1978).

[21] S. Orozco, M. A. Ortíz and R. M. Méndez-Moreno, *Phys. Rev.* **A 41**, 5473 (1990).

RECENT PROGRESS IN THE VARIATIONAL STATISTICAL MECHANICS OF BOSE FLUIDS

C. E. Campbell

School of Physics and Astronomy
University of Minnesota
Minneapolis, Minnesota 55455, U. S. A.

INTRODUCTION

In the wake of the successful development of the Feenberg-Jastrow Euler-Lagrange (FJEL) theory of the ground state and low excited states of liquid ^{4}He,[1,2] some years ago we embarked on a finite temperature version of that theory, with the objective being a similarly accurate description of the phase diagram and temperature dependent properties of this strongly correlated boson superfluid.[3,4] In this note we wish to summarize some recent progress in this program, and to introduce a possible new approach.

The near impossibility of dealing with the strong, short-range repulsion between helium atoms within standard many-body perturbation theory based upon the ideal Bose gas states led to the development of the ground and low-excited state theories in coordinate space by Jastrow, Feynman, Feenberg and their coworkers. We now know that the ground state wave function can be represented well by the Jastrow-Feenberg function:

$$\Psi_0(\vec{r}_1, \cdots, \vec{r}_N) = \prod_{i < j}^{N} e^{1/2\, u_2(r_{ij})} \prod_{i < j < k}^{N} e^{1/2\, u_3(\vec{r}_i,\, \vec{r}_j,\, \vec{r}_k)}$$

where u_2 and u_3 are determined by a set of Euler-Lagrange equations:

$$\frac{\delta}{\delta u_2} \frac{\langle \Psi_0 | H | \Psi_0 \rangle}{\langle \Psi_0 | \Psi_0 \rangle} = 0 = \frac{\delta}{\delta u_3} \frac{\langle \Psi_0 | H | \Psi_0 \rangle}{\langle \Psi_0 | \Psi_0 \rangle}.$$

It is this theory which we refer to as the FJEL theory. The excitation spectrum of the fluid, Landau's phonon-roton spectrum, is well described by the approximate excited states[5]

$$|\mathbf{k}\,\rangle = \Psi_\mathbf{k} = \left[A_\mathbf{k}^{(1)}\, \rho_\mathbf{k} + \sum_{\substack{\mathbf{q}_1,\, \mathbf{q}_2 \\ \mathbf{q}_1 + \mathbf{q}_2 = \mathbf{k}}} A_{\mathbf{q}_1,\, \mathbf{q}_2}^{(2)}\, \rho_{\mathbf{q}_1}\, \rho_{\mathbf{q}_2} \right] \Psi_0$$

where $\rho_\mathbf{k}$ is the density fluctuation operator, which in coordinate space is

Condensed Matter Theories, Vol. 8, Edited by
L. Blum and F.B. Malik, Plenum Press, New York, 1993

$$\rho_{\vec{k}} = \sum_{j=1}^{N} e^{i\vec{k}\cdot\vec{r}_j} \quad .$$

When only the first term is retained one obtains the well-known Bijl-Feynman spectrum, while the second term incorporates the Feynman-Cohen backflow (or equivalently, the Jackson-Feenberg anharmonic term).[6-8]

At low temperatures, the thermodynamics of the fluid may be obtained by treating these excitations as non-interacting excitations satisfying Bose statistics. Penrose,[9] and Reatto and Chester,[10] derived the low temperature many-body density matrix of a Bose fluid by observing that, in the absence of elementary excitation interactions, these collective modes produce independent harmonic oscillator density matrices where the collective coordinate is proportional to ρ_k. The results they obtained can be written in the form given in Eq's (1-3) below, where $\gamma(r)$ and the temperature dependence of $u(r)$ are determined entirely by the long wavelength behavior of the elementary excitation spectrum, with the single parameter being the sound velocity.

With these particular choices of $u(r)$ and $\gamma(r)$, this density matrix is limited in its applicability to low temperatures, specifically to temperatures where the number density of these thermally excited excitations is small compared to the number density of the helium atoms. In particular (with one exception) such a naive model cannot describe the phase transitions of the fluid. (The exception is the transition from off-diagonal long range order at T=0 to quasi-off-diagonal long range order in the two-dimensional boson fluid at infinitesimal temperatures.[10]) In order to partially overcome these limitations, it is evident that one needs a more general procedure for determining $u(r)$ and $\gamma(r)$, and indeed for exploring other possible structures of the density matrix which might apply in a larger portion of the phase diagram. A number of years ago I proposed a generalization of the FJEL theory to finite temperatures by invoking the variational principle on the coordinate space representation of the statistical density matrix. In particular I noted that certain features of the low temperature structure of the density matrix are quite general, namely that the density matrix can be written in coordinate space in the form[3]

$$W(\mathbf{r}_1\cdots\mathbf{r}_N ; \mathbf{r'}_1\cdots\mathbf{r'}_N) =$$
$$= \Phi(\mathbf{r}_1\cdots\mathbf{r}_N) \, Q(\mathbf{r}_1\cdots\mathbf{r}_N ; \mathbf{r'}_1\cdots\mathbf{r'}_N) \, \Phi^*(\mathbf{r'}_1\cdots\mathbf{r'}_N) \tag{1}$$

where Q is the incoherence factor which inextricably connects the primed and unprimed coordinates. Moreover, in a Bose system the factors Φ and Q are both real and non-negative in the absence of flow (which is not to be confused with the well-known general feature that the density matrix has real and non-negative eigenvalues, a feature not limited to Bose systems). It is obvious that, at T = 0, $\Phi = \Psi_0$, the exact ground state, while Q = 1 , whereas when T > 1, Q ≠ 1, and both Φ and Q are temperature dependent. As noted above, these features are evident in the low temperature density matrices of Penrose, Reatto and Chester, and should be maintained in a variational theory.

The appropriate finite temperature variational theory for the density matrix is the Gibbs-Delbrück-Moliere variational principal wherein the trial Helmholtz free energy A[W] for a trial density matrix W is minimized with respect to W:

$$A[W] = \left[\text{tr } (HW) + \frac{1}{\beta} \text{tr } (W \ln W) \right] / \text{tr } (W) \quad .$$

When A is varied over all Boson density matrices, the minimum is the exact Helmholtz fee energy $A_0(T,\Omega)$, where Ω is the volume. Thus $A \geq A_0$.

To apply this variational theory to liquid ^{4}He, one must make a judicious choice of trial density matrix. Our first choice was to follow the obvious generalization of the Jastrow function which emerged from the low temperature analysis of Penrose, and Reatto and Chester, so that Φ is taken to have the Jastrow form (the minimum choice for helium):

$$\Phi(r_1 \cdots r_N) = \prod_{i<j}^{N} e^{\frac{1}{2}u(r_{ij})} \tag{2}$$

and the incoherence factor is also of the minimal product form which can link primed to unprimed coordinates:

$$Q(r_1 \cdots r_N ; r'_1 \cdots r'_N) = \prod_{i,j}^{N} e^{\gamma(r'_i - r_j)} . \tag{3}$$

Then the variational functions become $u(r)$ and $\gamma(r)$, with the corresponding Euler-Lagrange equations being:

$$\delta A / \delta u(r) = 0 = \delta A / \delta \gamma(r) .$$

The determination of the internal energy, $U = \text{Tr } W H / \text{Tr } W$, is straightforward, requiring no mathematics beyond that already developed for the ground state theory at the Jastrow level. However the entropy term in the free energy is problematic, and has thus far required a relatively crude but informative approximation, known as the separability approximation, which produces a mean field like expression for the entropy:

$$S = \sum_{k} (1 + n_k) \, ln \, (1 + n_k) - n_k \, ln \, n_k$$

where n_k is a simple function of γ_k, the fourier transform of $\gamma(r)$. One of the Euler-Lagrange equations then takes the form

$$n_k = \frac{1}{e^{\beta \varepsilon(k,\beta)} - 1}$$

where

$$\varepsilon(k,\beta) = \frac{\hbar^2 k^2}{2m \, S(k,\beta)} \coth \frac{\beta \varepsilon(k,\beta)}{2}$$

which is the finite temperature generalization of the Bijl-Feynman spectrum. $S(k,\beta)$ is the finite temperature liquid structure function, which is determined by the second Euler-Lagrange equation, which must be solved simultaneously with this finite T Bijl-Feynman spectrum.

The principal result obtained from solving this set of Euler-Lagrange equations was a phase diagram demarked by a liquid-gas spinodal line.[4] This spinodal line is the locus of points where the compressibility diverges (the isothermal sound velocity vanishes), and thus represents the limit of mechanical stability of the system; at a given density, it gives the lower bound on supercooling before the onset of spinodal decomposition. (While there is some controversy about how one should define such a boundary when discussing spinodal decomposition, the mathematical property of this curve as it emerges from the above analysis can be stated with precision: it is the boundary of the region in which the equilibrium (or metastable) state of the fluid is homogeneous.)

Since the liquid-gas critical point lies on this curve, coinciding with the maximum temperature on the curve, these results produced a microscopic determination of the critical temperature and density. Our results gave a temperature about 1 K below the experimental value, at a density of about 10% below experiment, a level of accuracy which we expected based upon Jastrow Euler-Lagrange results for the ground state. Moreover, the approximation which we invoked for the entropy is equivalent to a mean-field approximation in the elementary excitations, and thus thermodynamic mean-field behavior emerged, although this is a correlated mean field theory (not an oxymoron) since the particle-hole interaction which drives the theory is tamed by the Jastrow correlations. In particular, the $T = 0$ spinodal point, known from previous work on the ground state, does not involve the entropy and thus is not troubled by approximations for the entropy.

This analysis also permitted us to explore the properties of other quantities in the vicinity of the critical point. Particularly interesting is the critical opalescence that is seen in the liquid structure function.

RECENT PROGRESS

The above variational analysis of the Penrose-Reatto-Chester-Jastrow (PRCJ) density matrix is rewarding in that it gives a semi-quantitative, first principles account of the liquid-gas critical point and spinodal line. Within the PRCJ class of density matrices, the part of this approach which is still open for significant improvement is the treatment of the entropy, which is essential to a move away from thermodynamic mean-field results. Toward that end, Clements reformulated the entropy problem in a way which suggests a systematic improvement scheme, although going beyond the present results has not yet proven to be practical.[11]

The above work was carried out with a truncation of the Feenberg decomposition at the lowest (i.e. Jastrow) level. As in the ground state problem, significant quantitative improvements should be expected upon a systematic enlargement of the trial density matrix space. In particular, since Φ and Q are both real and non-negative, they can be written in the Feenberg form and important qualitative improvements upon these earlier results have been obtained recently by including the next higher terms in the Feenberg decomposition of the Φ and Q:

$$\Phi(\vec{r}_1, \cdots, \vec{r}_N) = \prod_{n=1}^{N} \prod_{i_1 < i_n}^{N} e^{1/2 \ u_n(\vec{r}_{i_1} \cdots \vec{r}_{i_n})}$$

$$Q(\mathbf{r}_1 \cdots \mathbf{r}_N \ ; \ \mathbf{r'}_1 \cdots \mathbf{r'}_N) = \prod_{n \neq m}^{N} \prod_{i_1 < \cdots < i_n}^{N} \prod_{j_1 < \cdots < j_m}^{N} e^{\gamma_{n,m}(\mathbf{r'}_{i_1} \cdots \mathbf{r'}_{i_n}, \mathbf{r}_{j_1} \cdots \mathbf{r}_{j_n})}$$

The functions u_n and $\gamma_{n,m}$ can then be determined by Euler-Lagrange equations. As in the case of the ground state, we expect that u_3 is quantitatively important in liquid helium. Moreover, the three-body functions $\gamma_{1,2}$ and $\gamma_{2,1}$ carry the essential information about the back-flow terms in the excitation spectrum which contribute to the free energy through a thermal excitations,[12] and $\gamma_{2,2}$ is the first term which incorporates the effects of interactions between elementary excitations. Recently, Clements et al incorporated these effects into the trial density matrix by replacing the density fluctuation operator by a collective mode operator which includes backflow effects:[13]

$$F_k = \rho_k + \sum_{\substack{q_1, q_2 \\ q_1+q_2=k}} \frac{\gamma_{12}(q_1, q_2)}{\gamma_{11}(k)} \rho_{q_1} \rho_{q_2}$$

The incoherence factor is then taken to be bi-linear in this operator:

$$Q(r_1 \cdots r_N \; ; \; r_1' \cdots r_N') = \prod_k^N e^{1/2N \, \gamma_{11}(k) \left(F_k' F_{-k} + F_k F_{-k}' - F_k' F_{-k}' - F_k F_{-k}\right)}$$

where the primes on the F_k indicate that they are functions of the primed coordinates. Note that Q has been redefined so that it is identically one when the primed coordinates are set equal to the unprimed coordinates, i.e., on the diagonal. In this formulation, $\gamma_{1,2}(k,q)$ is the Fourier transform of $\gamma_{1,2}(|r_1-r_1'|, |r_2-r_1'|)$, while the coordinate space function $\gamma_{2,2}$ is the Fourier transform of $\gamma_{1,2}^2/\gamma_{1,1}^2$. Consequently, $\gamma_{2,2}$ is not an independently variable function in the present formulation.

Because of the use of the Jackson-Feenberg correlated basis function form for the collective coordinate F_k, we call this trial density matrix the CBF density matrix.

As in the earlier work, the entropy for this CBF density matrix is obtained by using the separability approximation, though in this case it is formulated in terms of the collective coordinates F_k instead of ρ_k.[12] Again this approximation leads to a Bose entropy form for the entropy. There are now four Euler-Lagrange equations to be solved, generated by the four independent functions u_2, u_3, $\gamma_{1,1}$, and $\gamma_{1,2}$. The spinodal line and critical point are shifted as expected from the previous ground state calculations. Perhaps most important is the fact that the Bose entropy form once again gives a temperature dependent elementary excitation spectrum through the expression:

$$n_k = \frac{1}{e^{\beta \varepsilon(k,\beta)} - 1}$$

where this excitation spectrum now contains backflow contributions:

$$\varepsilon(k) = \frac{\hbar^2 k^2}{2m \, S(k)} \coth \frac{\beta \varepsilon(k)}{2} +$$

$$+ \frac{1}{N} \sum_q \frac{|\langle \Phi | \rho_q \rho_{k-q} \, \delta H \, \rho_{-k} | \Phi \rangle|^2 \coth \beta \varepsilon(k)/2}{S(k) S(k-q) \left[\varepsilon_{BF}(q) + \varepsilon_{BF}(k-q) - \varepsilon^*(k)\right]}$$

where $\delta H \equiv H - E_0 - \varepsilon_{BF}(k)$ and $\varepsilon_{BF}(k) = \hbar^2 k^2 / [2m \, S(k)]$. It is important to note that, in the limit $T \to 0$, this excitation spectrum goes over to the zero temperature spectrum of the Jackson-Feenberg form including the correlated basis function 3-phonon vertex correction (i.e., backflow) obtained by Chang and Campbell,[5] which was found to be essential (along with the inclusion of u_3) for obtaining the correct density dependence of the roton parameters in liquid ^{4}He. Thus in our earlier work we found that, at the simplest, PRCJ level,[3] the excitation spectrum which emerges from the Bose entropy is the finite temperature generalization of the Bijl Feynman spectrum, while in this most recent CBF level,[11] inclusion of u_3 and $\gamma_{1,2}$ in the density matrix produces a Bose entropy which gives an excitation spectrum which is the finite temperature generalization of the Feynman-Cohen or Jackson-Feenberg spectrum containing backflow corrections. It should be further noted, however, that the results of the CBF density matrix calculation do not produce the temperature dependence of the roton gap which is observed in neutron scattering experiments. Of course one cannot yet formally identify this statistical excitation spectrum with that which is measured by a response function such as neutron scattering. Moreover the spectrum obtained here has a simple form only because the approximation for the entropy produces a Bose entropy form; it is not at all clear how one may extract a statistical excitation spectrum out of a more general and more accurate treatment of the entropy.[11] Similarly, a comparison to neutron scattering results really calls for the calculation of the temperature dependence of the dynamical response function $S(k, \omega)$. Nevertheless, the agreement, obtained here and in the earlier work, of the statistical excitation spectrum in the limit $T \to 0$ with the dynamical excitation spectrum calculated at $T = 0$, strongly suggests that such a formal relationship can be obtained. Moreover, we would expect the temperature dependence of the roton gap to agree more with the neutron results than obtained in our most recent work. One possible explanation for the present disagreement is that, in the absence of an independent variation of $\gamma_{2,2}$, there is no reason to expect that the theory obtains the correct (presumably attractive) interaction between rotons, which produces the decrease of the roton gap with increasing roton occupation (i.e., increasing temperature).

In spite of the successes of this program, it was clear from the beginning that the Jastrow or Jastrow-Feenberg level analysis would not produce a lambda transition.[8,3] Indeed, one can show that such a truncated Jastrow-Feenberg density matrix always exhibits off-diagonal long-range order (ODLRO) in three-dimensions, and quasi-ODLRO in two-dimensions. Moreover, the truncated Feenberg decomposition of the incoherence factor Q fails the important test that it give the correct high temperature limit of the density matrix, which is a permanant (or determinant in the case of fermions) of Gaussians with the scale of the thermal deBroglie wave length.[8,3] More correctly, the ideal Bose gas density matrix is given by

$$Q_{IBG}(\{r\}; \{r'\}) = \text{Perm } \Gamma(|\ r_i - r_j'\ |)$$

where Γ is a Gaussian with thermal deBroglie length scale, and *Perm* represents the sum over coordinate permutations (i.e., taking the permanent of the matrix with matrix elements Γ_{ij}). This result is correct for the ideal Bose gas for all temperatures above the Bose-Einstein temperature, and for temperatures below the Bose-Einstein temperature it is only necessary to include the separate contribution of the zero momentum state which has macroscopic occupation n_0. With interactions present, this continues to be the high temperature limit of the incoherence factor, while the factor Φ has as its high temperature limit the square root of the Boltzmann factor, which is to say that it has the Jastrow form with $u_2 = -\beta V$ (plus a term in the laplacian of V if one is consistent in taking the high temperature limit).[8,3] Feynman noted this structure of the density matrix in his first paper on liquid helium (which we were unaware of when we pointed to these features in our earlier work), and argued that this structure should persist as the temperature is lowered, with suitable modifications in Φ and Q.[14] In particular he noted that Φ will become the ground state wave function at $T = 0$, and Q will be modified by correlation effects, which he modeled by introducing an effective mass into the thermal deBroglie wavelength. Recently, Blendowske and Fliessbach used Feynman's density matrix to evaluate the temperature dependence of the liquid structure factor and radial distribution function in the vicinity of the lambda point which are in qualitative agreement with the anomalous behavior observed experimentally.[15]

In the discussion of our variational work on the PRCJ density matrix I proposed the inclusion of a permanent in the incoherence factor Q in order to have the possibility of a transition from the low temperature off-diagonal long range ordered state to a phase with no ODLRO, i.e., a lambda transition.[8,3] It is now clear that this can be properly called a generalized Feynman density matrix. The proposal was to treat the function Γ in the permanant as a variational function, which in the simplest form of the theory replaces $\gamma_{1,1}$ as the function which drives the statistical mechanics. I noted that the function Γ should behave very much like an effective one-body density matrix, decaying to zero at large r for temperatures above the lambda temperature, but taking on a finite large r value proportional to n_0 below the lambda transition.

Unlike the PRCJ trial density matrix, this generalized Feynman density matrix posed a new mathematical challenge even for calculating the internal energy and radial distribution function. Senger and Ristig solved this problem by using the permutation cycle evaluation of the permanant to produce a diagrammatics with statistical bonds Γ and dynamical bonds [exp u_2] -1, which they were then able to re-sum into a Bose-hypernetted chain form in analogy with the Fermi-hypernetted chain theory developed for Slater-Jastrow functions.[16] Subsequently, Senger et al derived the Γ and u_2 Euler-Lagrange equations under the assumption that Γ has normal fluid behavior, i.e., vanishes in the large r limit.[17] Consequently this analysis is restricted to non-ODLRO states of the system. However, the variational analysis shows that this assumption fails below some temperature where Γ can no longer have normal fluid behavior. These results are in close parallel with the ideal Bose gas behavior, as should be expected since the introduction of the correlation factor exp u_2 would not generally be expected to change the qualitative behavior of the system in-so-far as the existence of ODLRO and the transition between ODLRO and non-ODLRO is concerned. As in the PRCJ case, the most difficult quantity to deal with is the entropy, which again in simplest (and at present only tractable) approximation takes on a Bose entropy form with effective Bose-Einstein function [17]

$$ n_k = \frac{1}{e^{\beta [\varepsilon (k,\beta) - \mu]} - 1} . $$

This constitutes a definition of $\varepsilon(k)$ through n_k, which in turn is defined in terms of the cyclic hypernetted chain functions, which are finally given by the Euler-Lagrange equations for Γ and u_2. An important feature is that this effective Bose-Einstein function is required to satisfy a constraint[17]

$$ \sum_k n_k = N $$

where N is the number of particles in the system, and it is the fact that this sum rule cannot be satisfied for a short-range Γ below a critical temperature which signifies the phase transition. As in the non-ideal Bose gas, this manifests itself in the fact that the sum of the argument of n_k must vanish at k = 0 at the transition temperature:

$$ \varepsilon_k - \mu \Rightarrow 0 \text{ as } k \Rightarrow 0. $$

Although this analysis has not yet been extended to temperatures below where this first happens, I expect that the analysis when carried to lower temperatures will preserve this result and also require that n_k be macroscopic at k =0, corresponding to a finite asymptote for $\Gamma(r)$ as r diverges.

FUTURE PROSPECTS

It would appear from the above analysis that we are very close to having a microscopic variational theory which describes the gross features of the phase diagram and temperature dependent properties of liquid ^{4}He. However there are some clear challenges and questions which emerge from this analysis. One interesting question is that of the ingredients required in a model density matrix to reproduce the correct critical behavior for the Bose fluid. Because of the approximations in the calculation of the entropy, we cannot even say at present what the critical indices are for the three density matrix classes already studied. Clearly the entropy approximation in each case is mean field.

We do not yet know what difficulties will be encountered when the generalized Feynman density matrix is solved in its off-diagonal long-range ordered (i.e., low temperature) phase. There is some uncertainty about whether one should include both the PRCJ type incoherence factor and the generalized Feynman incoherence factor at low temperatures. As in any variational theory, including both of these factors is certain to produce a free energy that is closer to the exact free energy by virtue of the fact that the function space for the variation is enlarged. On the other hand, we recognize that the Feynman density matrix includes single-particle modes and the PRCJ density matrix includes collective (phonon-roton) modes, and these two modes are, according to conventional wisdom, identical (or hybridized) in the ordered phase. When Senger et al include both of these factors in the high temperature phase, the result is two Bose entropies, i.e. a single quasiparticle entropy and a collective mode entropy. On the other hand, the phenomenological evidence is that at low temperatures the thermodynamics (e.g., heat capacity) of liquid ^{4}He is well described by a single, phonon-roton spectrum. We can conjecture that the low temperature analysis, when correctly done, will collapse the two modes into one so that the usual low temperature results are obtained. We can conjecture further that, in the $T \rightarrow 0$ limit, the leading behavior of the PRCJ and generalized Feynman density matrices will be the same, and that the excitation spectrum in the latter case will in fact be the finite temperature Bijl-Feynman result, so that it is not really necessary to include $\gamma_{1,1}$ in the low temperature phase to obtain the correct low T limit.

In all three of the trial density matrices discussed above--PRCJ, CBF, and generalized Feynman--the weak point in the analysis is the determination of the entropy as a functional of the variational functions. Clements has devised a scheme in the PRCJ case which produces correction terms to the separability approximation which are functionally manageable;[11] however the results obtained when the leading corrections are included do not solve the formal problems of the earlier approximations. One such problem is the overabundance of excitations with increasing temperatures. This leads to numerical difficulties on the low density side of the phase diagram in the PRCJ and CBF density matrix analyses, where the spinodal line occurs at temperatures which, though low, are high enough for the density to have many more excitations than particles. An extrapolation of the numerical results indicates that the spinodal line intercepts zero density at a finite temperature. This is probably unphysical, since it would exclude a vapor, but it is unclear whether this result is associated with the choice of density matrix (which is unphysical in that the gas has ODLRO), or the entropy approximation. In this regard, it should be noted that the argument which led to the PRCJ low temperature density matrix, namely the treatment of multiple Feynman phonon states as independent states, is technically invalid since the degree of the polynomial in ρ_k must be no higher than the number of particles, N, an error which becomes significant as the temperature increases. If one can arrange to incorporate this constraint into the PRCJ density matrix one should recover the ODLRO to short-range order transition, including the proper normal vapor-superfluid liquid two-phase region.

To be more specific, the method of correlated bases can be used to create a complete, orthonormal basis in a number representation, with states:[7,8]

$$|n_{k_1}, \ n_{k_2}, \ \cdots, \ n_{k_\alpha}) = \frac{\left[\prod_{i=1}^{\alpha} (\rho_{k_i})^{n_i} - orthog\right]\Phi_0}{\sqrt{N^M\left[\prod_{i=1}^{\alpha} \{n_i! \ S_o(k_i)^{n_i}\}\right]}}$$

where the total number of excitations $M = \sum n_i \leq N$, *orthog* represents a polynomial in the density fluctuation operators of order $M - 1$ with simple which has the effect of orthogonalizing the state to all states of total number of excitations $M' < M$ and the same total wave number, and Φ_0 is of the Feenberg form used above. It is then convenient to define raising and lowering operators $b_k{}^\dagger$ and b_k, respectively:

$$b_{k_\alpha}^\dagger |n_{k_1}\cdots n_{k_\alpha}\cdots) = \sqrt{n_{k_\alpha}+1}\ |n_{k_1}\cdots n_{k_\alpha}+1 \ \cdots)$$

$$b_{k_\alpha}^\dagger |n_{k_1}\cdots n_{k_\alpha}\cdots) = 0 \quad \text{if} \quad \sum_k n_k = N$$

and similarly for the lowering operators. These operators are *almost* boson operators.

One may now define a trial density matrix taking advantage of this method by defining the density matrix in terms of a model Hamiltonian using these operators:

$$\widehat{W}_t = \frac{\exp{-\beta H_t}}{Z_t}$$

where H_t is a functional of these creation and annihilation operators. One then uses the Gibbs-Bogoliubov variational principle, in which H_t is the variational quantity.[18]

A particularly simple application of this approach is to define H_t to have the Bogoliubov form:

$$H_t = \sum_k \left\{ E_k b_k^\dagger b_k + \left[b_k^\dagger b_{-k}^\dagger + b_{-k} b_k\right] v_k \right\}$$

where E_k and v_k are variational functions in addition to u_2 which defines Φ_0, though there is some redundancy between u_2 and v_k. Since this trial Hamiltonian can be brought to diagonal form by a Bogoliubov transformation, one immediately recognizes that the entropy of this density matrix is exactly the usual Bose entropy form, with the additional constraint that the number of excitations is bounded by N. Of course the problem is most easily formulated in the grand canonical ensemble, where a chemical potential naturally takes care of this constraint after one defines appropriate zero momentum creation and annihilation operator. Thus this approach has the great advantage that it is unnecessary to *approximate* the entropy of the trial density matrix; it is precisely an ideal Bose gas entropy with a variationally determined spectrum. The price for this is two-fold: first, it is clearly a thermal mean-field theory, though dynamical correlations are built in; and second, it is necessary to calculate diagonal matrix elements and off-diagonal paired-phonon matrix elements of the exact Hamiltonian in the correlated basis. However there is much experience in the calculation of such matrix elements, and thus there should be no difficulties as serious as the problem of calculating the entropy in the previous approaches. The most significant advantage of this approach is that it appears to be possible to treat the entire phase diagram with this single class of trial density matrices, and the statistical excitation spectrum enters the problem in a natural way. A more ambitious approach would be to include higher terms in H_t, such as $b^\dagger bb$ and $b^\dagger b^\dagger bb$, which would play a role similar to $\gamma_{1,2}$ and $\gamma_{2,2}$ in the variational density matrix. However, there are no results available yet for density matrices of these forms, so our comments about them are purely conjectural at the present time.

ACKNOWLEDGMENTS

I would like to thank Professor L. Blum for organizing this stimulating workshop, and the U. S. Army Research Office for partial travel support in conjunction with this workshop.

REFERENCES

1. C. E. Campbell, in: "Progress in Liquid Physics," C. A. Croxton, ed., Wiley, New York (1978).
2. C. E. Campbell and F. J. Pinski, Nucl. Phys. A 328:21 (1979).
3. C. E. Campbell, K. E. Kürten, M. L. Ristig, and G. Senger, Phys. Rev. B 30:3728 (1984).
4. G. Senger, M. L. Ristig, K. E. Kürten, and C. E. Campbell, Phys Rev. B 33:3728 (1986).
5. C. C. Chang and C. E. Campbell, Phys. Rev. B 13:3779 (1976).
6. References 7 and 8 contain more detailed reviews of the excitation spectrum and experimental probes of that spectrum.
7. C. E. Campbell, in: "Excitations in 2-Dimensional and 3-Dimensional Quantum Fluids," A. F. G. Wyatt and H. J. Lauter, ed., Plenum (1991).
8. C. E. Campbell and B. E. Clements, in: "Elementary Excitations in Quantum Fluids," K. Ohbayashi and M. Watabe, ed., Springer (1989).
9. O. Penrose: in: "Proceedings of the International Conference on Low Temperature Physics," J. R. Dillinger, ed., (University of Wisconsin Press, Madison, Wisconsin, 1958), p. 117.
10. L. Reatto and G. V. Chester, Phys. Rev. 155:88 (1967).
11. B. E. Clements, Ph.D. dissertation, University of Minnesota; and B. E. Clements and C. E. Campbell, to be published.
12. S. Battaini and L. Reatto, Phys. Rev. B 28:1263 (1983).
13. B. E. Clements, E. Krotscheck, J. A. Smith, and C. E. Campbell, to be published.
14. R. P. Feynman, Phys. Rev. 91:1291 (1953).
15. R. Blendowske and T. Fliessbach, J. Phys.: Condens. Matter 4:3661 (1992).
16. G. Senger and M. L. Ristig, in: "Condensed Matter Theories, Vol. 5" V. C. Aguilera-Navarro, ed., Plenum, New York (1990)
17. G. Senger, M. L. Ristig, C. E. Campbell and J. W. Clark, Ann. Phys. (N.Y.), to be published.
18. L. R. Whitney and C. E. Campbell, Physica 108B:1371 (1981).

DENSITY - QUASIPARTICLE INTERPRETATION OF EXCITATIONS IN LIQUID ⁴HE

Henry R. Glyde

Department of Physics and Astronomy
University of Delaware
Newark, Delaware 19716

ABSTRACT

The dispersion curve of the phonon-roton excitations in superfluid ⁴He at low T is well described by the theory of Landau, Feynman and the Correlated Basis Function (CBF) method. However, the temperature dependence of these excitations is quite complicated and not well described by the thermal broadening of a single excitation. We discuss our recent interpretation of this temperature dependence which includes both density and quasiparticle excitations in liquid ⁴He. We also show how the chief relation on which the interpretation is based can be derived in terms of the full quasiparticle-quasiparticle interaction.

INTRODUCTION

Landau[1] first proposed that superfluid ⁴He at low temperature supported excitations having the classical phonon-roton energy dispersion curve. He interpreted these as collective density excitations of the fluid at all wave vectors Q with the phonons and rotons representing different wave vector regions of the same excitation. The P-R energy dispersion curve is shown in Fig. 1.

Feynman[2] provided a microscopic description of this P-R curve in terms of density excitations in the fluid. The excited state was obtained by operating on the ground state by a single density operator, $\rho^+(Q)$. This particularly described the phonon region well. To describe the maxon and roton regions, an excited state consisting of more than one density operator is required.[3] Using a state which is a linear combination of density excitations and suitable variational parameters, a good description of the roton region at higher Q can be obtained using Correlated Basis Function methods.[5,6] Thus, at low T, the phonon region is well described as a density excitation. The maxon/roton region is also well described, but as a significantly modified density excitation.[4-6]

Experimentally, the P-R excitation energy is defined as the position of the sharp peak in the dynamic structure factor, $S(Q,\omega)$, observed in inelastic neutron scattering measurements.[7,8] The scattering intensity observed[9,10] in superfluid ^{4}He at low T at the phonon ($Q = 0.4\,\text{Å}^{-1}$) and at the maxon ($Q = 1.1\,\text{Å}^{-1}$) region is shown in Fig. 2. This shows a well defined sharp peak plus some intensity at higher ω. Both the position of the sharp peak and the broader intensity at higher ω (at low T) is well described by CBF methods.[4]

As first noted by Woods and Svensson,[11] the temperature dependence of $S(Q,\omega)$ is very rich and complicated. It is not well described by the broadening of a single excitation with increasing temperature. In Fig. 3 we show the change of the observed[10] intensity at the maxon with increasing T. This is the extension to higher temperature of the intensity shown on the right side of Fig. 2. Rather than a broadening of the sharp peak with T, the chief effect is that the intensity in the sharp peak decreases with increasing T until it disappears from $S(Q,\omega)$ entirely at $T = T_\lambda$. In normal ^{4}He, only broad scattering remains. This broad scattering is peaked at a higher energy than the position of the sharp peak. Thus, while the sharp peak broadens somewhat with increasing T, the main change between $T = 1.29\,\text{K}$ and $T = T_\lambda$ is that the sharp peak has disappeared from $S(Q,\omega)$ at $T = T_\lambda$. We emphasize that the intensity is broad in normal ^{4}He at $Q = 1.1\,\text{Å}^{-1}$. A similar temperature dependence is observed at higher Q values, particularly at the roton region.

At low Q, in the phonon region, the temperature dependence is different.[9] In Fig. 4 we show the temperature dependence of the intensity observed[9] at $Q = 0.4\,\text{Å}^{-1}$; the left hand side of Fig. 2 At $Q = 0.4\,\text{Å}^{-1}$, the chief effect is a thermal broadening of the dominant sharp peak in $S(Q,\omega)$. In normal ^{4}He, the intensity remains largely confined to a single peak. This peak also remains reasonably sharp — characteristic of scattering from a collective density excitation. Thus, while there may be some change in the intensity at $T = T_\lambda$, the chief temperature dependence in the phonon region is a broadening of the sharp peak.

In normal ^{4}He, the scattered intensity is quite sharp at low Q but becomes very broad as Q increases. At the maxon and higher Q, $S(Q,\omega)$ is broad; characteristic of scattering from weakly interacting particle-hole excitations as seen from Fig. 3. In the next section we discuss a recent interpretation[9,12-14] of the temperature dependence of

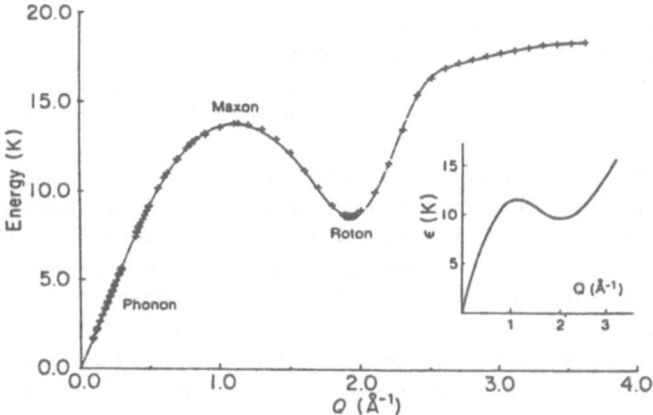

Figure 1 Phonon-maxon-roton dispersion curve in superfluid ^{4}He at low temperature with inset showing Landau's (1947) curve (from Refs. 7 and 26).

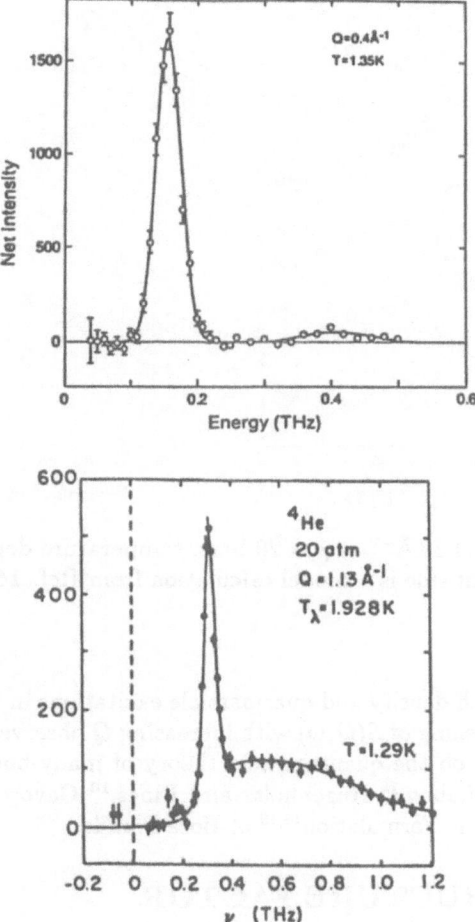

Figure 2 Observed neutron scattering intensity in superfluid ^{4}He at low temperature in the phonon ($Q = 0.4\,\text{Å}^{-1}$) and maxon ($Q = 1.13\,\text{Å}^{-1}$) regions (from Refs. 9 and 10).

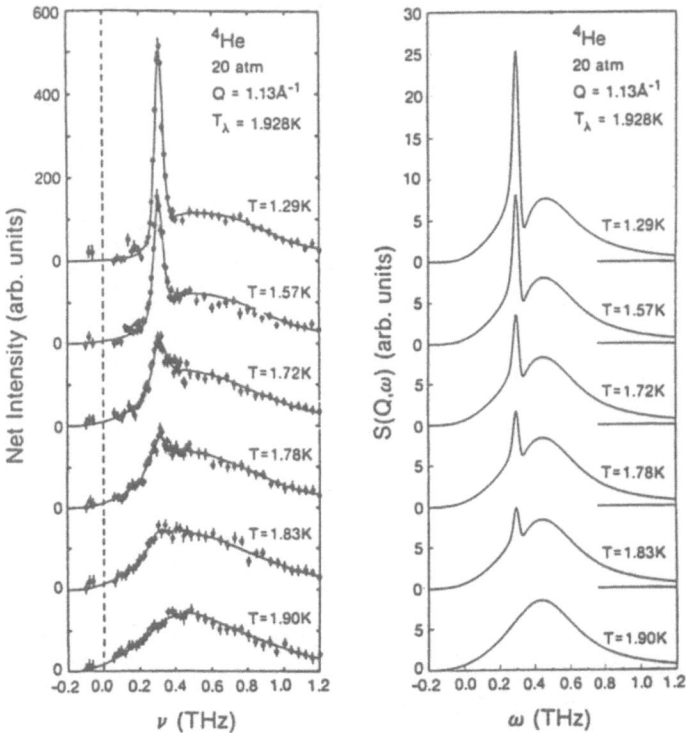

Figure 3   Maxon, $Q = 1.13\,\text{Å}^{-1}$ at $p = 20$ bars, temperature dependence. Left side is data from Ref. 10; right side is a model calculation from Ref. 14.

$S(Q,\omega)$ in terms of both density and quasiparticle excitations in the fluid. It explicitly incorporates the broadening of $S(Q,\omega)$ with increasing Q observed in normal ^{4}He. The interpretation is based on the quantum field theory of many-body physics applied to Bose systems by Bogoliubov,[15] Hugenholtz and Pines,[16] Gavoret and Nozières[17] and the subsequent Dielectric Formulation[18,19] of Bose liquids.

DYNAMIC STRUCTURE FACTOR

The observed $S(Q,\omega)$ is[21]

$$S(Q,\omega) = \frac{1}{2\pi} \int_{-\infty}^{\infty} dt e^{i\omega t} \frac{1}{N} \langle \rho(Q,t)\rho^+(Q,0) \rangle. \tag{1}$$

That is, excitations in the density operator $\rho(Q)$ are observed. The dynamic suscepti-bility corresponding to $S(Q,\omega)$ is

$$\chi(Q,t) = -i\frac{1}{V} \langle T\rho(Q,t)\rho^+(Q,0) \rangle \tag{2}$$

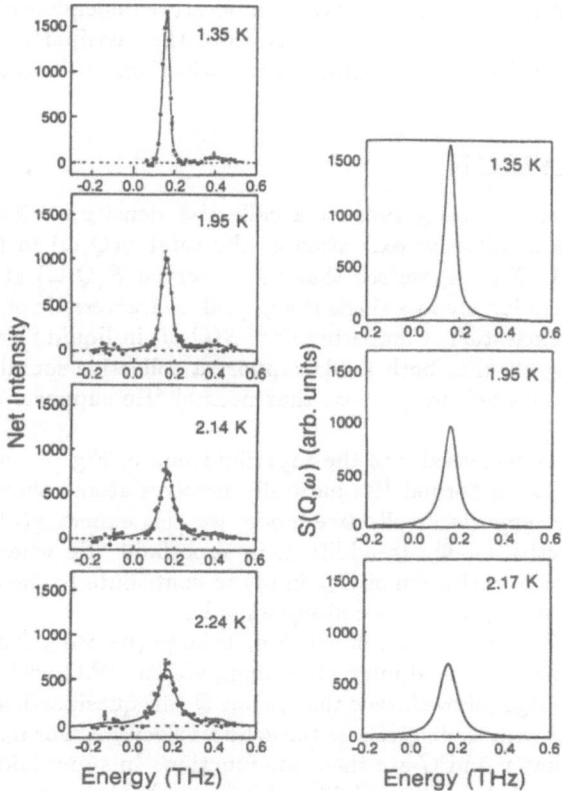

Figure 4 Phonon, $Q = 0.4\,\text{Å}^{-1}$ at SVP, temperature dependence. Left side is data from Ref. 9; right side is a model calculation based on Ref. 14.

and $S(Q,\omega)$ and $\chi(Q,\omega)$ are related by

$$S(Q,\omega) = -\frac{1}{n\pi}[n_B(\omega) + 1]Im\chi(Q,\omega) \tag{3}$$

where $n = N/V$ and $n_B(\omega)$ is the Bose function. That is, $S(Q,\omega)$ is proportional to the imaginary part of $\chi(Q,\omega)$. As shown by Gavoret and Nozières,[17] the $\chi(Q,\omega)$ in superfluid ^{4}He can be expressed as,

$$\chi(Q,\omega) = n_0(T)\Lambda_\alpha(Q,\omega)G_{\alpha\beta}(Q,\omega)_\beta\Lambda(Q,\omega) + \chi'(Q,\omega) \tag{4}$$

where $n_0(T) = N_0(T)/N$ is the condensate fraction. We derive this expression in an appendix. This result shows that in superfluid ^{4}He, where $n_0(T)$ is finite, the observed density $S(Q,\omega)$ contains a term proportional to the single particle Green function, $G_{\alpha\beta}$. Through this term, single quasiparticle excitations, the excitations of $G_{\alpha\beta}$, can be observed in $S(Q,\omega)$. The Vertex functions Λ_α are defined in (A17). $\chi'(Q,\omega)$ is the density response involving atoms having momentum above the condensate only.

Using (4) we now interpret the observed temperature dependence of $S(Q,\omega)$ noted above. In this density-quasiparticle interpretation, the quasiparticles are defined as excitations of $G_{\alpha\beta}$ and the density excitations as excitations of the total $\chi(Q,\omega)$ in (4).

INTERPRETATION

Firstly, liquid ^{4}He could clearly support a collective density excitation as proposed by Landau. This is a collective excitation of the total $\chi(Q,\omega)$ in (4). At low Q it is a sound mode. In Fig. 4, we see that the observed $S(Q,\omega)$ at $Q = 0.4\,\text{Å}^{-1}$ in normal ^{4}He is confined largely to a single sharp peak characteristic of scattering from a collective (density) excitation. Comparing[9,22,23] $S(Q,\omega)$ in liquid ^{3}He and normal ^{4}He at $Q = 0.4\,\text{Å}^{-1}$ suggests that both fluids support a collective sound mode at low Q. Thus, as is widely accepted, we propose that normal ^{4}He supports a sound mode at low Q.

As temperature is decreased into the superfluid phase, Fig. 4 shows that $S(Q,\omega)$ sharpens. The $\chi(Q,\omega)$ in normal ^{4}He naturally involves atoms above the condensate only. If normal ^{4}He supports a collective mode, we also expect $\chi(Q,\omega)$ in superfluid ^{4}He to have a collective mode. In addition, in superfluid ^{4}He where $n_0(T) \neq 0$, we expect the quasiparticle excitation of $G_{\alpha\beta}$ in (4) to contribute to the observed $S(Q,\omega)$ at the same time that $\chi(Q,\omega)$ has a collective mode.

Based on dilute Bose gas models in which n_0 is large ($n_0 \simeq 1$), the $G_{\alpha\beta}$ has a sharp mode resulting from a mean field interaction (e.g., via an RPA) with the atoms in the condensate.[24] Also, Bogoliubov showed that at low Q this quasiparticle energy of $G_{\alpha\beta}$ is linear in Q, $\omega_Q = cQ$ with slope given by the sound velocity c. For $n_0 = 1$, Hugenholtz and Pines showed that χ and G are the same function. In superfluid ^{4}He n_0 is small, $n_0 \lesssim 0.1$. For an interacting Bose Fluid (arbitrary n_0), Gavoret and Nozières[17] and the subsequent Dielectric formulation show that the response of the total $\chi(Q,\omega)$ and of $G_{\alpha\beta}$ are coupled via the condensate. Specifically, at low Q Gavoret and Nozières[17] showed that, due to this coupling, the excitations of G and of χ continue to have the same linear dispersion, $\omega_Q = CQ$ in superfluid ^{4}He; i.e.,

$$\lim_{Q,\omega \to 0} G_{11}(Q,\omega) = \frac{n_0}{n} \frac{mc^2}{\omega^2 - c^2 Q^2 + i\epsilon} \tag{5}$$

$$\lim_{Q,\omega \to 0} \chi(Q,\omega) = \frac{nQ^2/m}{\omega^2 - c^2 Q^2 + i\epsilon} \tag{6}$$

Thus, the quasiparticle response has the same energy dependence as the total observed density response — with width vanishing as Q^2 or greater. Thus, we would expect to observe only a single sharp peak in superfluid ^{4}He of low Q, in spite of the two terms in (4). We may call this single peak in superfluid ^{4}He a joint density-quasiparticle excitation in the low Q phonon region. It arises from a mean field involving interaction with atoms above and in the condensate. We could equally well call it simply a density excitation, since it exists in normal ^{4}He.

We saw that the density excitations observed in normal ^{4}He broaden significantly with increasing Q. At the maxon (see Fig. 3) and the roton Q, $S(Q,\omega)$ is a broad function characteristic of scattering from weakly interacting particles. Comparing[9] with liquid ^{3}He and other fluids, normal ^{4}He does not appear to support a collective density mode in $\chi(Q,\omega)$ for $Q \gtrsim 1\,\text{Å}^{-1}$. If the density response is approximately temperature independent, as it appears[22,23] to be in liquid ^{3}He, then we would expect $\chi'(Q,\omega)$ in (4) to also be a broad function in superfluid ^{4}He for $Q \gtrsim 1,\text{Å}^{-1}$.

Using (4) we interpret the sharp peak in $S(Q,\omega)$ in superfluid ^{4}He in the maxon and roton region as a sharp peak in $G_{\alpha\beta}$ in the first term of (4). This has intensity proportional to $n_0(T)$. This assumes that $G_{\alpha\beta}(Q,\omega)$ remains a sharp function of ω as Q increases. On the right side of Figs. 3 and 4 we present a model calculation[14] of $S(Q,\omega)$ based on this picture. For $G_{\alpha\beta}(Q,\omega)$ we choose a zero order like Green function with zero width at all Q ($\Gamma_{\text{SP}} = 0$). The energy in $G_{\alpha\beta}$ is determined by fitting to the peak region of the low temperature $S(Q,\omega)$. For the "regular" χ' observed in normal ^{4}He we fit a damped RPA like form to the observed $S(Q,\omega)$ having a half width Γ_0. Multi pair contributions to χ' are neglected. We assume[14] these quasiparticle and density excitations are independent of temperature. Using the Dielectric Formulation, a more complicated form of (4) in which G and the "regular" χ' are coupled via the condensate, we obtain[14] the model results shown on the right side of Figs. 3 and 4. In these model results, only $n_0(T)$ varies with T, following the simple Bose Gas dependence. An improved form of $n_0(T)$ could certainly be used.

At the maxon (Fig. 3), the model qualitatively reproduces the observed loss of intensity in the sharp peak with increasing T. There remains intensity at low ω at low T in the model which is not observed. This results from keeping χ' independent of T. However, the basic qualitative feature of the observed temperature dependence is reproduced. At $Q = 0.4\,\text{Å}^{-1}$, the model produces a broadening of a single peak with T. This is because the uncoupled quasiparticle and density energies are chosen to have nearly identical energies and the Γ_0 in normal ^{4}He at $Q = 0.4\,\text{Å}^{-1}$ is small.

In this density-quasiparticle interpretation, the observed single peak in $S(Q,\omega)$ at low Q in the phonon region is interpreted as a joint density/quasiparticle excitation. The quasiparticle response at low Q has the same linear dispersion, $\omega_Q = cQ$, as the sound mode. This was shown by Bogoliubov for a Bose Gas and for an interacting Bose Fluid by Gavoret and Nozières and is certainly not new. In normal ^{4}He, the observed $S(Q,\omega)$ is a density mode only since, when $n_0 = 0$, $G_{\alpha\beta}$ cannot be observed in $S(Q,\omega)$.

As Q increases, we propose that the density response broadens greatly. This is based on the very broad $S(Q,\omega)$ observed in normal ^{4}He for $Q \gtrsim 1\,\text{Å}^{-1}$. We propose that the quasiparticle response of $G_{\alpha\beta}(Q,\omega)$ remains sharp as Q increases. Based on (4) the sharp peak in Fig. 3 is interpreted as a sharp peak in $G_{\alpha\beta}$. Thus, in the maxon-roton region, the sharp peak which defines the maxon-roton energy is a quasiparticle peak, a peak in the single particle response given by G. This is consistent with the view of the P-R excitation as a quasiparticle at all Q, as initially proposed by Bogoliubov[15] for a Bose Gas and discussed by Nozières and Pines[24] for superfluid liquid ^{4}He. In this density-quasiparticle interpretation, we would clearly expect a continuous phonon-roton dispersion curve. In the Dielectric Formulation, the single particle (G) and density (χ) response functions of superfluid ^{4}He have a common denominator. Thus, a sharp peak in one function will appear as a sharp peak in the other. On this basis, a formulation of the density response alone at low T could produce the observed P-R energy whatever the origin of the peak. This would reconcile the CBF and the Density-Quasiparticle interpretation. However, both density and quasiparticle contributions to $S(Q,\omega)$ appear to be needed to explain the temperature dependence of $S(Q,\omega)$.

ACKNOWLEDGEMENTS

It is a pleasure to acknowledge the support of the U.S. Army Research Office at Durham, North Carolina.

APPENDIX

In this appendix, we outline how the Eq. (4) for $\chi(\mathbf{Q},\omega)$ is obtained in terms of the vertex functions $\Lambda(\mathbf{Q},\omega)$ and the general quasiparticle-quasiparticle interaction $\gamma(p,p';Q)$. The $\gamma(p,p';Q)$ is the complete interaction including all possible processes. We consider $T = 0\,\mathrm{K}$ for simplicity and show how the finite temperature result is obtained from it.

We begin with (2) for $\chi(Q,t)$ and substitute $\rho(Q) = \sum_p a_{\mathbf{p}-\mathbf{Q}/2}^+ a_{\mathbf{p}+\mathbf{Q}/2}$ to write $\chi(Q,t)$ in the form

$$\chi(\mathbf{Q},t) = \frac{1}{V^2} \sum_{\mathbf{pp'}} \chi(\mathbf{p},\mathbf{p'};\mathbf{Q},t) \tag{A1}$$

where

$$i\chi(\mathbf{p},\mathbf{p'};\mathbf{Q},t) = V\langle T a_{\mathbf{p}-\mathbf{Q}/2}^+(t) a_{\mathbf{p}+\mathbf{Q}/2}(t) a_{\mathbf{p'}+\mathbf{Q}/2}^+(0) a_{\mathbf{p'}-\mathbf{Q}/2}(0)\rangle \tag{A2}$$

$\chi(\mathbf{p},\mathbf{p'};\mathbf{Q},t)$ describes the creation of a particle-hole pair $(\mathbf{p'}+\mathbf{Q}/2, \mathbf{p'}-\mathbf{Q}/2)$ at $t = 0$, the propagation of the pair from time $t = 0$ to time t when the pair is annihilated from the fluid. Eq. (A2) is a two-body Green function in which one body is a particle and the other is a hole. We can write the general expression for the 2-body Green function, following Gavoret and Nozières,[17] as

$$G_{\alpha\beta\gamma\delta}(1234) = V\langle a_{\mathbf{p}_1}^\alpha(t_1) a_{\mathbf{p}_2}^\beta(t_2) a_{\mathbf{p}_3}^{+\gamma}(t_3) a_{\mathbf{p}_4}^{+\delta}(t_4)\rangle \tag{A3}$$

Here we have introduced the notation $a_p^\alpha = a_p$ if $\alpha = +$ and $a_p^\alpha = a_{-p}^+$ if $\alpha = -$. In its usual form, in which $\alpha = \beta = \gamma = \delta = +$, all the operators in $G_{\alpha\beta\gamma\delta}$ of (A3) remain unchanged. In this case G_{++++} describes the creation of two particles at times $t = t_4$ and $t = t_3$, propagation of these two additional particles in the fluid until they are annihilated at times t_2 and t_1. The $\chi(\mathbf{p},\mathbf{p'},\mathbf{Q},t)$ is clearly a two-body G in which $\alpha = \delta = -$ and $\beta = \gamma = +$, $t_3 = t_4 = 0$ and $t_1 = t_2 = t$. Explicitly,

$$i\chi(\mathbf{p},\mathbf{p'},Qt) = G_{-++-}(\mathbf{p}-\mathbf{Q}/2, \mathbf{p}+\mathbf{Q}/2; \mathbf{p'}+\mathbf{Q}/2, \mathbf{p'}-\mathbf{Q}/2) \tag{A4}$$

with $t_3 = t_4 = 0$ and $t_1 = t_2 = t$.

We may obtain the equation for χ in terms of the single particle $G's$ from the general equation for $G_{\alpha\beta\gamma\delta}(1234)$. This is for Bosons,

$$G_{1234}(1234) = -V[G_{13}(13)G_{24}(24)\delta_{p_1,p_3} + G_{14}(14)G_{23}(23)\delta_{p_1,p_4}]$$
$$-iG_{11'}(11')G_{22'}(22')\gamma_{1'2'3'4'}(1'2'3'4')G_{3'3}(3'3)G_{4'4}(4'4) \tag{A5}$$

In (A5) all the $G_{\alpha\beta}(\mathbf{p},t-t')$ are in the "pt" representation.

$$G_{\alpha\beta}(\mathbf{p},t) = -i\langle T a_{\mathbf{p}}^\alpha(t) a_{\mathbf{p}}^{\beta+}(0)\rangle \tag{A6}$$

$G_{++}(\mathbf{p}, t)$ is the usual Green function, G_{+-} and G_{-+} are anomalous Green functions which are finite when there is a condensate and $G_{--}(\mathbf{p}, t) = G_{++}(-\mathbf{p}, -t)$.

The first term in (A5) is the "zero order" term which represents propagation of the fully renormalized p-h pair without interaction with each other. The second term is the interaction term in which γ is the full interaction vertex. This vertex contains many intermediate states which "mix" the $G_{\alpha\beta}$. That is, the interaction γ_{1234} "flips" the operators and mixes the regular and anomalous $G_{\alpha\beta}$.

We may think of $\gamma_{1234}(1234)$ as analogous to a spin-dependent interaction which can flip the spin and mixes spin states in a Fermi liquid. In this Bose case, because of the sink of particles in the condensate, the "spin" does not have to be conserved in the scattering. Indeed, (A5) is identical to the equation for $G_{\alpha\beta\gamma\delta}(1234)$ in a spin 1/2 Fermi system if all signs are changed (except for the $G(14)G(23)$ term which remains negative) and spin does not need to be conserved in the interaction (see page 92, Reference 25).

We have written (A5) in the pt representation so that we may readily make the correspondence indicated in (A4) and identify the $G_{\alpha\beta}$ in (A6) to obtain the equation for $\chi(\mathbf{p}, \mathbf{p}'; \mathbf{Q}t)$. For example, making the correspondence $(1 = -, 3 = +$ and $p_1 = p - \frac{Q}{2}, p_3 = p' + \frac{Q}{2})$ indicated by comparing (A4) and (A5), the first G in (A5) for $\chi(\mathbf{p}, \mathbf{p}'; \mathbf{Q}t)$ is,

$$
\begin{aligned}
G_{13}(13) &= G_{13}(\mathbf{p}_1, t_1 - t_3)\ \delta_{\mathbf{p},\mathbf{p}_3} \\
&= -i\langle a^1_{\mathbf{p}_1}(t_1) a^{+3}_{\mathbf{p}_3}(0)\rangle\ \delta_{\mathbf{p}_1,\mathbf{p}_3} \\
&= -i\langle a^+_{\mathbf{p}-\mathbf{Q}/2}(t) a^+_{\mathbf{p}'+\frac{\mathbf{Q}}{2}}(0)\rangle\ \delta_{\mathbf{p}',-\mathbf{p}} \\
&= G_{-+}\left(-\mathbf{p} + \frac{\mathbf{Q}}{2}, t\right)
\end{aligned}
\tag{A7}
$$

Making a similar correspondence for all the $G's$ in (A5), we obtain for the Fournier transform $\chi(\mathbf{p}, \mathbf{p}', \mathbf{Q}\omega)$ of (A4) using the notation $p = \mathbf{p}, \omega_p, Q = \mathbf{Q}, \omega$,

$$
i\chi(p, p', Q) = i\chi_0(p, p'; Q) + i\chi_I(p, p', Q)
\tag{A8}
$$

where

$$
\begin{aligned}
i\chi_0(p, p', Q) = &-[G_{-+}(-p + \tfrac{Q}{2})G_{+-}(p + \tfrac{Q}{2})\delta_{p',-p} + G_{--}(-p + \tfrac{Q}{2}) \\
&G_{++}(p + \tfrac{Q}{2})\delta_{p',p}](2\pi)^4
\end{aligned}
\tag{A9}
$$

$$
\begin{aligned}
i\chi_I(p, p'Q) = &-iG_{-1}(-p + \tfrac{Q}{2})G_{+2}(p + \tfrac{Q}{2})\gamma_{1234}(p, p'; Q) \\
&G_{3+}(p' + \tfrac{Q}{2})G_{4-}(-p' + \tfrac{Q}{2})
\end{aligned}
\tag{A10}
$$

Explicitly, here

$$
G_{+-}\left(p + \frac{Q}{2}\right) = G_{+-}(\mathbf{p} + \mathbf{Q}/2, \omega_p + \omega/2)
\tag{A11}
$$

The "zero order" result (A9), may be readily checked by evaluating $\chi(\mathbf{p}, \mathbf{p}', \mathbf{Q}t)$ in (A2) directly in zero order and replacing the zero order $G's$ by full $G's$. The result (A8) is presented graphically in Fig. A1.

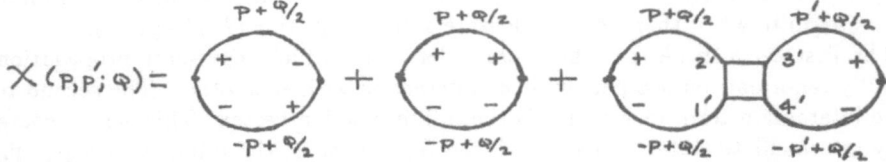

$$\chi(p,p';q) =$$

Figure A1

SINGULAR TERMS

We now identify the "singular" terms in $\chi(Q,\omega)$ due to the existence of a Bose condensate which are proportional to a single $G_{\alpha\beta}$ and lead to the first term in (4). These are the terms in (A8) involving the zero momentum state (the condensate) — i.e., the momentum index p of $G_{\alpha\beta}(p)$ is zero. From (A9) and (A10), this clearly happens when $p = \pm Q/2$ or $p' = \pm Q/2$. Using $G_{\alpha\beta}(0) = -in_0\delta(\omega)$, the singular terms of $\chi_0(p,p';Q)$ are, from (A9),

$$i\chi_0^S(p,p',Q) = in_0[G_{+-}(Q)\delta_{p',-p} + G_{++}(Q)\delta_{p',p}]\delta_{p,\frac{Q}{2}}(2\pi)^4$$
$$+in_0[G_{-+}(Q)\delta_{p',-p} + G_{--}(Q)\delta_{p',p}]\delta_{p,-\frac{Q}{2}}(2\pi)^4 \tag{A12}$$

and

$$\chi_0^S(\mathbf{Q},\omega) = \int \frac{d^4p}{(2\pi)^4} \int \frac{d^4p'}{(2\pi)^4}\chi_0^S(p,p'Q)$$
$$= n_0 \sum_{\alpha\beta} G_{\alpha\beta}(\mathbf{Q},\omega) \tag{A13}$$

We recognize this immediately as the first term in (4) with $\Lambda_\alpha = 1$. The singular terms in the interaction term $\chi_I(p,p',Q)$ of (A10) are,

$$i\chi_I^S(p,p';Q) = -n_0 \sum_\alpha G_{\alpha 2}(Q)\gamma_{234}(Q,p')G_{3+}\left(p'+\frac{Q}{2}\right)G_{4-}\left(-p'+\frac{Q}{2}\right)\delta_{p,\pm\frac{Q}{2}}$$
$$-n_0 G_{-1}(-p+\tfrac{Q}{2})G_{+2}(p+\tfrac{Q}{2})\gamma_{123}(p,Q)\sum_\beta G_{3\beta}(Q)\delta_{p',\pm\frac{Q}{2}} \tag{A14}$$

where

$$\gamma_{234}(Q,p) = \gamma_{234}\left(0,Q;p+\frac{Q}{2},p-\frac{Q}{2}\right) \tag{A15}$$

The $\gamma_{234}(Q,p)$ is a 3-point interaction with the 4th index reaching into the condensate, $p = 0$. The γ in (A15) is written in the particle-hole form and usually denoted as $P(Q,p)$. Following Gavoret and Nozières,[17] we denote the product of the 3-point interaction and two Green functions as

$$L_{\beta+-}(Q,p) = i\gamma_{\beta\gamma\delta}(Q,p)G_{\gamma+}\left(p+\frac{Q}{2}\right)G_{\delta-}\left(-p+\frac{Q}{2}\right)$$
$$L_{-+\delta}(p,Q) = iG_{-\gamma}(-p+\tfrac{Q}{2})G_{+\delta}(p+\tfrac{Q}{2})\gamma_{\gamma\delta\alpha}(p,Q) \tag{A16}$$

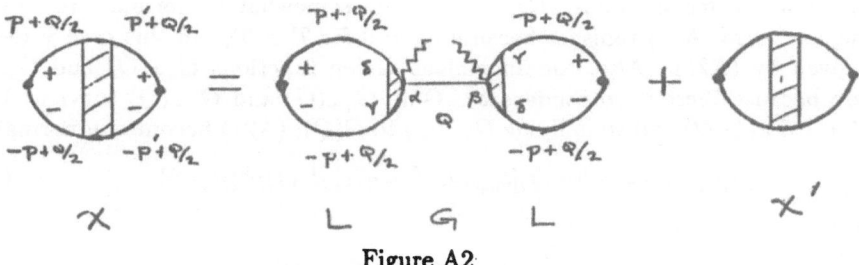

Figure A2

Using this definition,

$$\chi_I^S(Q,\omega) = n_0 \sum_{\alpha\beta} \left[G_{\alpha\beta}(Q) \int d\bar{p} L_{\beta+-}(Q,p) \right.$$

$$\left. + \int d\bar{p} L_{-+\alpha}(p,Q) G_{\alpha\beta}(Q) \right] \qquad (A17)$$

The "zero order" and "interaction" terms may be readily combined as

$$\chi^S(\mathbf{Q},\omega) = n_0 \sum_{\alpha\beta} \Lambda_\alpha(\mathbf{Q},\omega) G_{\alpha\beta}(\mathbf{Q},\omega)_\beta \Lambda(\mathbf{Q},\omega) \qquad (A18)$$

where

$$\Lambda_\alpha(Q,\omega) = \left[1 + \int d\bar{p} L_{-+\alpha}(p,Q\omega) \right] \qquad (A19)$$

This singular term (A17) and (A18) is displayed graphically in Fig. A2. It represents the creation of a *p-h* pair in the fluid (say, by a neutron) given by the two G's in (A16). Through the interaction $\gamma(Q,p)$ this *p-h* pair scatters and emerges as a single quasiparticle (or hole), described by $G_{\alpha\beta}(Q)$, plus a particle going into the condensate. Through the interaction, the single quasiparticle can also interact with a particle in the condensate and emerge as *p-h* pair above the condensate. The "zero order" singular term in (A13) represents free propagation of a quasiparticle — created directly by a neutron.

The remaining terms of χ, denoted by $\chi'(Q,\omega)$ in (4), are obtained from (A8) requiring that all of the momentum indices lie above the condensate. Introducing the function,

$$\chi^0_{-\alpha+\beta}(p,Q) = iG_{-\alpha}\left(-p+\frac{Q}{2}\right) G_{+\beta}\left(p+\frac{Q}{2}\right), \qquad (A20)$$

the equation for $\chi'(p,p',Q)$ from (A8) is, in the presence of a condensate,

$$\chi'(p,p',Q) = [\chi^0_{-++-}(p,Q)\delta_{p,-p'} + \chi^0_{--++}(p,Q)\delta_{p,p'}](2\pi)^4$$
$$+ \sum_{\alpha\beta\gamma\delta} \chi^0_{-\delta+\beta}(p,Q)\gamma_{\alpha\beta\gamma\delta}(p,p',Q)\chi'_{\gamma+\delta-}(p',Q) \qquad (A21)$$

This is clearly a complicated equation. The term $\chi'(Q,\omega)$ in (4) is obtained from (A21) by integrating over p and p' as in (A13).

The equation for the total $\chi(Q,\omega)$ simplifies somewhat in normal ^{4}He. Firstly, the singular term (A18) vanishes because $n_0 = 0$ for $T > T_\lambda$. In this case χ reduces to χ' given by (A21). Also, the anomalous Green functions $G_{-+}(Q)$ and $G_{+-}(Q)$ are zero because there is no condensate. Only $G_{++}(Q)$ and $G_{--}(Q)$ survive. Using $G_{--}(Q) = G_{++}(-Q)$ and simplifying $G_{++}(Q)$ to $G(Q)$, (A21) becomes in normal ^{4}He

$$\chi(p,p',Q) = \chi^0(p,Q)[\delta_{p,p'}(2\pi)^4 + \gamma(p,p',Q)\chi^0(p',Q)] \tag{A22}$$

where

$$\chi^0(p,Q) = iG\left(p - \frac{Q}{2}\right)G\left(p + \frac{Q}{2}\right). \tag{A23}$$

In Fermi and Bose liquid theory it is usual to separate γ as $\gamma = I + I\chi_0\gamma$. I is that part of γ which does not contain χ_0 as an intermediate state. In terms of I, (A8) and (A22) can be written in the form $\chi = \chi_0 + \chi_0 I\chi$. Beginning from these equations the Dielectric Formulation can be developed by making a local approximation to I — say denoted by $v(Q)$. In this way, the DF can be expressed in terms of a realistic and renormalized $v(Q)$. The generalization to Finite temperature can be made following the rules of page 146, Reference 25.

In summary, in superfluid ^{4}He for $T < T_\lambda$ where there is condensate, the full $\chi(\mathbf{Q},\omega)$ is

$$\chi(\mathbf{Q},\omega) = \chi^S(\mathbf{Q},\omega) + \chi'(\mathbf{Q},\omega), \tag{A24}$$

where $\chi^S(\mathbf{Q},\omega)$ is the "singular" part (A18) containing the single particle Green function as a factor and is proportional to $n_0(T)$. The $\chi'(Q,\omega)$ involves atoms lying above the condensate only and is given by (A21). In normal ^{4}He, $\chi^S = 0$ and $\chi(Q,\omega)$ reduces to $\chi'(Q,\omega)$. In normal ^{4}He, the anomalous Green functions vanish and the equation for the full $\chi(Q,\omega) = \chi'(Q,\omega)$ simplifies to (A22). This equation can clearly support a collective density excitation in a strong interaction limit such as $Q \to 0$. These are the results of this appendix. In all cases, the equations involve full renormalized Green functions and the full interaction γ including all possible scattering processes.

REFERENCES

1. L. D. Landau, J. Phys. U.S.S.R. **5**, 71 (1941); 11, 91 (1947).

2. R. P. Feynman, Phys. Rev. **94**, 262 (1954).

3. R. P. Feynman and M. Cohen, Phys. Rev. **102**, 1189 (1956).

4. E. Manousakis and V. R. Pandharipande, Phys. Rev. B **33**, 150 (1986).

5. C. E. Campbell, in *Progress in Liquid Physics* (C.A. Croxton, ed.) p. 213. Wiley, New York, 1978.

6. L. Reatto, Proc. of Quantum Fluids and Solids Conference, June 1992, edited P.E. Sokol (J. Low Temp. Phys. to appear).

7. H. R. Glyde and E. C. Svensson, in *Neutron Scattering*, edited by D. L. Price and K. Sköld, Methods of Exp. Phys., Vol. 23, Part B, (Academic Press, New York 1987) p. 303.

8. A. D. B. Woods and R. A. Cowley, Rep. Prog. Phys. **36**, 1135 (1973).

9. W. G. Stirling and H. R. Glyde, Phys. Rev. B **41**, 4224 (1990).

10. E. F. Talbot, H. R. Glyde, W. G. Stirling and E. C. Svensson, Phys. Rev. B **38**, 11229 (1988).

11. A. D. B. Woods and E. C. Svensson, Phys. Rev. Lett. **41**, 974 (1978).

12. H. R. Glyde and W. G. Stirling, in *Phonons 89*, edited by S. Hunklinger, W. Ludwig and G. Weiss (World Scientific, Singapore, 1990).

13. H. R. Glyde and A. Griffin, Phys. Rev. Lett. **65**, 1454 (1990).

14. H. R. Glyde, Phys. Rev. B **45**, 7321 (1992).

15. N. N. Bogoliubov, J. Phys. U.S.S.R. **11**, 23 (1947).

16. N. Hugenholtz and D. Pines, Phys. Rev. **116**, 489 (1959).

17. J. Gavoret and P. Nozières, Ann. Phys. NY **28**, 349 (1964).

18. A. Griffin and T. H. Cheung, Phys. Rev. A **7**, 2086 (1973).

19. P. Sézpfalusy and I. Kondor, Ann. Phys. NY **82**, 1 (1974).

20. Yu. A. Nepomnyashchii, Sov. Phys. JETP **62**, 289 (1985) and references therein.

21. S. W. Lovesey, *Theory of Neutron Scattering from Condensed Matter*, Vol. 1 (OUP, Oxford 1984).

22. R. Scherm, K. Guckelsberger, B. Fåk, K. Sköld, A. J. Dianoux, H. Godfrin and W. G. Stirling, Phys. Rev. Lett. **59**, 217 (1987).

23. D. W. Hess and D. Pines, J. Low Temp. Phys. **72**, 247 (1988) and references therein.

24. P. Nozières and D. Pines, *Theory of Quantum Liquids*, Vol II (Addison-Wesley, Redwood City, 1990).

25. A. A. Abrikosov, L. P. Gorkov, I. Ye. Dzyaloshinskii, *Quantum Field Theoretical Methods in Statistical Physics* (Pergamon, Oxford, 1965).

26. R. J. Donnelly, J. A. Donnelly, and R. N. Hills, J. Low Temp. Phys. **44**, 471. (1981).

TRAPPING OF VORTICES BY IMPURITIES IN ^{4}He

M. Saarela and F. Kusmartsev

Department of Theoretical Physics,
University of Oulu, SF-90570 Oulu, Finland

L.D. Landau Institute for Theoretical Physics
Moscow,117940, GSP-1, Kosygina 2, V-334, Russia

INTRODUCTION

Vadim Beresinski [1] was the first who recognized that in condensed matter, with a specific two-dimensional (2D) character, there may exist a new type of a phase transition, associating with the creation of vortices. Kosterlitz and Thouless [2] developed the ideas by introducing an intuitive Hamiltonian. They derived scaling equations and described the phase transition as a pairing of vortices. Two-dimensional quantum fluids are good examples of systems with Beresinski-Kosterlitz-Thouless transition. With recent possibilities of making thin films and building up new quantum devices it has becomes very important to understand the physics of these fluids. Topological excitations, which play a key role in many of the phenomena, have become an essential part of these studies.

In the superfluid ^{4}He the Beresinski-Kosterlitz-Thouless phase transition is the mechanism which distroyes the superfluidity [3]. In that phase transition macroscopic amounts of vortex-antivortex pairs are created. The mixture of vortex-antivortex pairs is equivalent with the 2D Coulomb plasma where the Debye screening is responsible of reducing the Coulomb attraction. A similar mechanism screens the vortex-antivortex attraction letting single vortices move indepedently above the critical temperature.

The microscopic structure of the vortex and its interaction with the many-particle background and with antivortices is less understood, but a facinating many-body problem [5]-[7], which we shall study in this work. Our attempt is to develop a many-body formalism for quantum boson fluids, starting from the two-particle interaction, and treat vortices as quasiparticle impurities [4]. We assume that vortices carry *mass* [8], which is set equal to the mass of the expelled superfluid [9]. The amount of the expelled material depends on the vorticity, on the superfluid density, and on the correlations between the superfluid particles. It will be determined by a self-consistent many-body calculation. The quasiparticle assumption also implies that the vortex in a homogeneous superfluid can be put anywhere in the system, and thus the superfluid density is equal to a constant. The quantity of interest then is the *probability of finding* a superfluid particle at a given distance away from the center of vorticity.

The question of trapping an impurity atom into the center of a vortex is studied by comparing chemical potentials in two cases: (1) We calculate how much energy is required to create one vortex and put one impurity atom far away from it, and (2) compare that with the energy required to place an impurity into the fluid with a vortex around it. We ignore here the interaction between the vortex and the impurity particle.

Our formalism is based on the Jastrow-Feenberg type correlated wave functions [10], and we assume that the superfluid ^{4}He is at absolute zero temperature. We take into account only the pair correlations, but the extension to the triplet correlations is staight forward [11]. The wave function of the ground state of the 2D ^{4}He is then a simple real function of N particle coordinates,

$$\Psi_B(\mathbf{r}_1, ..., \mathbf{r}_N) = exp\left[\frac{1}{2}\sum_{\substack{i,j=1 \\ i \neq j}}^{N} u^{BB}(\mathbf{r}_i, \mathbf{r}_j)\right]. \tag{1}$$

Suppose now that stationary rotational flow patterns, vortices, are formed into ^{4}He. In the quantum mechanical description of such excitation one adds a complex phase factor, $\phi(\mathbf{r}_0, \mathbf{r}_j)$ into the ground state wave function $\Psi_B(\mathbf{r}_1, ..., \mathbf{r}_N)$,

$$\Psi_v(\mathbf{r}_0, \mathbf{r}_1, ..., \mathbf{r}_N) = e^{\sum_{j=1}^{N} \phi(\mathbf{r}_0, \mathbf{r}_j)}\Psi_B(\mathbf{r}_1, ..., \mathbf{r}_N). \tag{2}$$

The gradiant of the imaginary part of the phase factor, which is called the vector potential, $\vec{A}(\mathbf{r}_0, \mathbf{r}_j)$, is related to the velocity field $\mathbf{v_s}(\mathbf{r}_0, \mathbf{r}_j)$ of a background ^{4}He particle at the possition $\mathbf{r}_j$ around an arbitrary point $\mathbf{r}_0$,

$$\vec{A}(\mathbf{r}_0, \mathbf{r}_j) \equiv \nabla_j \Im m[\phi(\mathbf{r}_0, \mathbf{r}_j)] = \frac{m_B}{\hbar}\mathbf{v_s}(\mathbf{r}_0, \mathbf{r}_j), \tag{3}$$

where m_B is the ^{4}He mass. The condition that the wave function Ψ_v must be *single valued* requires that the circulation around the point $\mathbf{r}_0$ using any closed path should be an integer multiple of 2π,

$$\oint \vec{A}(\mathbf{r}_0, \mathbf{r}) \cdot d\mathbf{r} = 2\pi q, \tag{4}$$

where the integer q is the vorticity of the circulation.

The real part of the phase desrcibes the response of the superfluid into the vortex. It is the correlation function between the vortex and a backgound particle in the Jastrow-Feenberg theory, and thus we can write the real part of the wave function in the form,

$$\Psi(\mathbf{r}_0, \mathbf{r}_1, ..., \mathbf{r}_N) = e^{\frac{1}{2}\sum_{j=1}^{N} u^{vB}(\mathbf{r}_0, \mathbf{r}_j)}\Psi_B(\mathbf{r}_1, ..., \mathbf{r}_N), \tag{5}$$

where we have used the notation $u^{vB}(\mathbf{r}_0, \mathbf{r}_j) \equiv 2\Re e[\phi(\mathbf{r}_0, \mathbf{r}_j)]$.

The wave function in Eq.(2) requires that each particle has the vorticity q=1. One could insert also more complicated phase factors, which would describe rotation of clusters of two or more particles by using triplet and higher order phase factors, but that would make the basic structure of our approach much less transparent and that is why we ignore them here.

The standard form of the ground state Hamiltonian, H_B, of the superfluid containes the kinetic energy term and the pair interaction between the particles,

$$H_B = -\sum_{i=1}^{N}\frac{\hbar^2}{2m_B}\nabla_i^2 + \sum_{\substack{i,j=1 \\ i<j}}^{N} V^{BB}(|\mathbf{r}_i - \mathbf{r}_j|). \tag{6}$$

The expectation value,

$$E_B = \frac{<\Psi_B|H_B|\Psi_B>}{<\Psi_B|\Psi_B>}, \tag{7}$$

determines the ground state energy of the system.

The imaginary part of the phase factor added into the wave function Ψ_v can be tranformed into an interaction like term in the Hamiltonian,

$$\frac{<\Psi_v|H_B|\Psi_v>}{<\Psi_v|\Psi_v>} = \frac{<\Psi|H_B + \sum_{j=1}^{N}\frac{\hbar^2}{2m_B}|\vec{A}(\mathbf{r}_0, \mathbf{r}_j)|^2|\Psi>}{<\Psi|\Psi>}. \tag{8}$$

Note that the wave function $\Psi(\mathbf{r}_0, \mathbf{r}_1, ..., \mathbf{r}_N)$ without a subindex still includes the real part of the phase.

In getting the above result we have written the kinetic energy term in the following form

$$< \Psi_v | \nabla^2 | \Psi_v > = \frac{1}{2} \left[< \Psi | i \nabla \cdot \vec{A} + 2i \vec{A} \cdot \nabla - \vec{A}^2 + \nabla^2 | \Psi > + herm. \; conj. \right]. \quad (9)$$

The first term on the right hand side disappears in the choice of the Landau gauge. The second term, which is the usual linear term in the vector potential in the case of magnetic field also disappears here, because our wave function Ψ is the wave function for a boson system at zero momentum, and it is a *real* quantity.

By treating the vortex as a quasiparticle, which is not localized to any fixed position in space, requires that we must include into the Hamiltonian a kinetic energy term related to the *zero point motion* of the center of the vortex,

$$H_v = H_B - \frac{\hbar^2}{2m_v} \nabla_0^2 + \sum_{j=1}^{N} \frac{\hbar^2}{2m_B} |\vec{A}(\mathbf{r}_0, \mathbf{r}_j)|^2, \quad (10)$$

with the vortex mass, m_v.

Using the Hamiltonian of Eq.(10) and the wave function of Eq.(5) we can treat the single vortex as an impurity in the superfluid. Its interaction with the background particles is determined by the square of the vector potential, its mass will be the result of the self-consistent calculation of the size of the hole created into the system and its wave function contains the correlations between the vortex and the background particles.

The rest of this paper is devided into three parts. In the next chapter we study the properties of a single vortex and trapping a vortex by an impurity. In the second chapter we derive the vortex-antivortex interaction and give a discussion of its structure, and at the end we have a short summary.

SINGLE VORTEX IN A BOSON FLUID

Vortex as an impurity

Knowing the Hamiltonian and the many-body wave function we can start the many-body treatment of the vortex-background interaction. We will derive the effective direct interaction as well as the interaction induced by the *vitual* phonon creation [12], [13]. The Jastrow-Feenberg theory is a variational theory where the energy required for the creation of a stationary vortex excitation,

$$
\begin{aligned}
\mu^v &= E_v - E_B \\
&= \frac{< \Psi | H_v | \Psi >}{< \Psi | \Psi >} - \frac{< \Psi_B | H_B | \Psi_B >}{< \Psi_B | \Psi_B >},
\end{aligned} \quad (11)
$$

is minimized with respect to the unknown vortex-background correlation function,

$$\frac{\delta \mu^v}{\delta u^{vB}(\mathbf{r}_0, \mathbf{r}_1)} = 0. \quad (12)$$

The basic quantities in the Jastrow-Feenberg theory are the one- and two-particle densities. The one-particle density for the vortex impurity is defined by integrating over all the background coordinates in the wave function,

$$\rho^v(\mathbf{r}_0) = \frac{1}{\mathcal{N}_0} \int d^2 r_1 ... d^2 r_N |\Psi(\mathbf{r}_0, ..., \mathbf{r}_N)|^2 = \frac{1}{\Omega}, \quad (13)$$

and it is equal to the inverse of *the total volume* of the system, Ω, which includes the volume occupied by the vortex. Similarly in the vortex-background two-particle density one integrates over all but one background coordinates.

$$\rho^{vB}(\mathbf{r}_0, \mathbf{r}_1) = \frac{N}{\mathcal{N}_0} \int d^2 r_2 ... d^2 r_N |\Psi(\mathbf{r}_0, \mathbf{r}_1, ..., \mathbf{r}_N)|^2, \quad (14)$$

where $\mathcal{N}_0$ is the normalization integral

$$\mathcal{N}_0 = \int d^2r_0...d^2r_N|\Psi(\mathbf{r}_0, ..., \mathbf{r}_N)|^2. \tag{15}$$

The radial distribution function between the vortex and the background particle, which gives the probability of finding a background particle at the distance $|\mathbf{r}_0 - \mathbf{r}_1|$ away from the vortex center is defined by the relation,

$$\rho^{vB}(\mathbf{r}_0, \mathbf{r}_1) = \rho^v(\mathbf{r}_0)\rho^B(\mathbf{r}_1)g^{vB}(\mathbf{r}_0, \mathbf{r}_1). \tag{16}$$

In the definition we have used the *pure background density*,

$$\rho^B(\mathbf{r}_1) \equiv \frac{N}{\Omega_B}. \tag{17}$$

Definitions (13), (17), (14), and (16) determine the volume integral

$$\int d^2r_1 g^{vB}(\mathbf{r}_0, \mathbf{r}_1) = \Omega_B. \tag{18}$$

When the volume, Ω_B, occupied by ^{4}He particles, the volume for one ^{4}He atom, v_B, and the total volume of the system are known we can calculate the volume required by the vortex, $v_v = \Omega - \Omega_B$. This difference can be calculated from the sequential relation for the vortex radial distribution function of Eq. (16),

$$\int d^2r_1 \rho^B(\mathbf{r}_1)\left[g^{vB}(\mathbf{r}_0, \mathbf{r}_1) - 1\right] = -\frac{v_v}{v_B} \equiv -\beta, \tag{19}$$

where we have introduced the *volume excess factor β*.

The vortex structure function is the Fourier transform of the radial distribution function

$$S^{vB}(k) = \rho^B \int d^2r e^{i\mathbf{k}\cdot\mathbf{r}}\left[g^{vB}(r) - 1\right], \tag{20}$$

and it's value at the origin,

$$S^{vB}(0+) = -\beta, \tag{21}$$

is given by the sequential relation (19).

The volume excess factor tells us how much ^{4}He material is expelled and thus determines the mass of the vortex,

$$m_v = \beta m_B \tag{22}$$

The variational problem in the single impurity limit [14] - [21] is divided into two parts: (1) one optimizes the background wave function given in Eq.(1) by searching for the minimum energy of the background particles, and (2) then minimizes the energy gained or lost by adding one vortex, in other words the *chemical potential* of the vortex. The direct evaluation of the expectation values in Eq.(11) gives us

$$\begin{aligned}
\mu^v = {} & \rho^v \rho^B \int d^2r_0 d^2r_1 g^{vB}(\mathbf{r}_0, \mathbf{r}_1)\Big[\frac{\hbar^2}{2m_B}|\vec{A}(|\mathbf{r}_0 - \mathbf{r}_1|))|^2 \\
& - \frac{\hbar^2}{8}\Big(\frac{1}{m_v}\nabla_0^2 + \frac{1}{m_B}\nabla_1^2\Big)u^{vB}(\mathbf{r}_0, \mathbf{r}_1)\Big] \\
& - \frac{1}{2}(\rho^B)^2 \int d^2r_1 d^2r_2 g^{BB}(\mathbf{r}_1, \mathbf{r}_2)\frac{\hbar^2}{8m_B}(\nabla_1^2 + \nabla_2^2)\Delta u^{BB}(\mathbf{r}_1, \mathbf{r}_2). \tag{23}
\end{aligned}$$

The last term describes the rearrangement of the background correlations due to the vortex. We have assumed that the pair distribution function, $g^{BB}(\mathbf{r}_1, \mathbf{r}_2)$, is optimized and hence we get a contribution only from the change in the correlation function. The correlation function and the pair distribution function are connected through the hypernetted chain (HNC) summation of diagrams,

$$g^{BB}(\mathbf{r}_1, \mathbf{r}_2) = e^{u^{BB}(\mathbf{r}_1, \mathbf{r}_2) + N^{BB}(\mathbf{r}_1, \mathbf{r}_2)}. \tag{24}$$

The vortex appears here in the diagrammatic expansion of the nodal sum, $N^{BB}(\mathbf{r}_1, \mathbf{r}_2)$, as an *internal* point. The change, Δu^{BB}, is then a change in the nodal sum when one internal vortex quasiparticle is added into the system for fixed $g^{BB}(\mathbf{r}_1, \mathbf{r}_2)$,

$$\Delta u^{BB}(\mathbf{r}_1, \mathbf{r}_2) = -\int d^2 r_0 \rho^v(\mathbf{r}_0)\frac{\delta N^{BB}(\mathbf{r}_1, \mathbf{r}_2)}{\delta \rho^v(\mathbf{r}_0)}. \tag{25}$$

In the momentum space the variation has a simple expression,

$$\rho^v \frac{\delta N^{BB}(k)}{\delta \rho^v} = \left(X^{vB}(k)\right)^2. \tag{26}$$

where $X^{vB}(k)$ is the sum of non-nodal diagrams,

$$X^{vB}(k) = \frac{S^{vB}(k)}{S^{BB}(k)}. \tag{27}$$

The HNC summation connects also the correlation and radial distribution functions with one vortex,

$$g^{vB}(\mathbf{r}_0, \mathbf{r}_1) = e^{u^{vB}(\mathbf{r}_0, \mathbf{r}_1) + N^{vB}(\mathbf{r}_0, \mathbf{r}_1)}. \tag{28}$$

This can be used to eliminate the correlation function $u^{vB}(\mathbf{r}_0, \mathbf{r}_1)$ from Eq.(23). Inserting the expression,

$$N^{vB}(k) = \frac{S^{vB}(k)(S^{BB}(k) - 1)}{S^{BB}(k)}, \tag{29}$$

for the nodal sum we can write the vortex chemical potential in the following form,

$$\begin{aligned}
\mu^v &= \int d^2 r_0 d^2 r_1 \Big[\frac{\hbar^2}{2m_B}|\vec{A}(|\mathbf{r}_0 - \mathbf{r}_1|)|^2 g^{vB}(\mathbf{r}_0, \mathbf{r}_1) \\
&+ \frac{\hbar^2}{2m_v}|\vec{\nabla}_0\sqrt{g^{vB}(\mathbf{r}_0, \mathbf{r}_1)}|^2 + \frac{\hbar^2}{2m_B}|\vec{\nabla}_1\sqrt{g^{vB}(\mathbf{r}_0, \mathbf{r}_1)}|^2\Big] \\
&- \int d^2 k\, S^{vB}(k) w_{ind}^{vB}(k).
\end{aligned} \tag{30}$$

where $w_{ind}^{vB}(k)$ is the *phonon induced potential*,

$$w_{ind}^{vB}(k) = -\frac{\hbar^2 k^2}{4}\frac{S^{vB}(k)(S^{BB}(k) - 1)}{S^{BB}(k)}\left(\frac{1}{m_v} + \frac{1}{m_B} + \frac{1}{m_B S^{BB}(k)}\right). \tag{31}$$

The only unknown quantity in the chemical potential is now the vortex-background distribution function. By varying the chemical potential with respect to $\sqrt{g^{vB}(\mathbf{r}_0, \mathbf{r}_1)}$ we obtain the Euler equation. A most convenient form for numerical work is found by introducing a particle-hole potential, which is the effective direct vortex-background interaction,

$$\begin{aligned}
V_{p-h}^{vB}(\mathbf{r}_0, \mathbf{r}_1) &= \frac{\hbar^2}{2m_B}g^{vB}(\mathbf{r}_0, \mathbf{r}_1)|\vec{A}(|\mathbf{r}_0 - \mathbf{r}_1|)|^2 \\
&+ \left(g^{vB}(\mathbf{r}_0, \mathbf{r}_1) - 1\right)w_{ind}^{vB}(\mathbf{r}_0, \mathbf{r}_1) \\
&+ \frac{\hbar^2}{2m_v}|\vec{\nabla}_0\sqrt{g^{vB}(\mathbf{r}_0, \mathbf{r}_1)}|^2 + \frac{\hbar^2}{2m_B}|\vec{\nabla}_1\sqrt{g^{vB}(\mathbf{r}_0, \mathbf{r}_1)}|^2,
\end{aligned} \tag{32}$$

and solving the Euler equation,

$$V_{p-h}^{vB}(k) = -\frac{\hbar^2 k^2}{4}\frac{S^{vB}(k)}{S^{BB}(k)}\left(\frac{1}{m_v} + \frac{1}{m_B S^{BB}(k)}\right), \tag{33}$$

iteratively in momentum space.

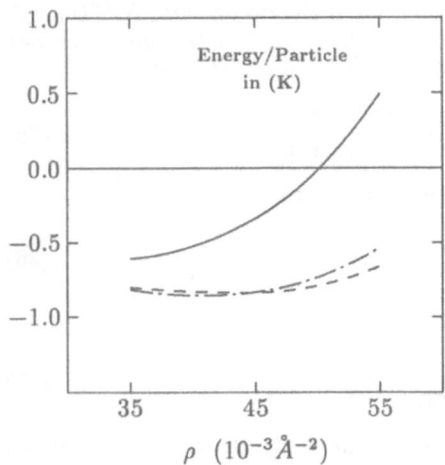

Fig. 1 The energy/particle of 2D ^{4}He as a function of density. The full curve is hypernetted chain result used in this paper. The dashed-dot curve includes the triplet correlations and elementary diagrams and is taken from ref. [20], and the dashed curve gives the Monte Carlo result of ref. [22].

The iteration procedure starts with a given particle-hole potential and the radial distribution function. From Eq.(33) one calculates the vortex structure function $S^{vB}(k)$ and inserts that into Eq.(31). This gives a new induced potential, which is then Fourier transformed and put into the coordinate space definition of the particle-hole potential, Eq.(32), together with the new radial distribution function. That results a new particle hole potential and we are back in the beginning.

Vortex in a 2D boson fluid

The 2D ^{4}He fluid is a model for very thin films. Its properties in the ground state have been studied carefully using the Monte Carlo method [22]. A good agreement with those results is obtained with the extended variational approach where triplet correlations are added into the wave function and the HNC relation of Eq.(24) includes also elementary diagrams [20]. In this work we limit ourselves into the HNC-approximation, which gives reasonable results for the energy as a function of pressure, but fails to saturate at the right density. The comparison of results from different approaches for the energy/particle as a function of density is shown in Fig.1. The calculations are done using the Aziz potential [23].

In 2D fluids the vortex core reduces into a simple dot in the xy-plane,

$$\nabla_j \times \vec{A}(\mathbf{r}_0, \mathbf{r}_j) = 2\pi q \hat{k} \delta(\mathbf{r}_0 - \mathbf{r}_j), \tag{34}$$

where $\hat{k}$ is a unit vector in the z-direction. The choice of the Landau gauge,

$$\nabla_j \cdot \vec{A}(\mathbf{r}_0, \mathbf{r}_j) = 0, \tag{35}$$

determines then the analytic form for the vector potential explicitly,

$$\vec{A}(\mathbf{r}_0, \mathbf{r}_j) = q\hat{k} \times \nabla_j ln|\mathbf{r}_0 - \mathbf{r}_j|. \tag{36}$$

As described in the previous chapter the vortex-background interaction is proportional to the square of the vector potential,

$$V^{vB}(\mathbf{r}_0, \mathbf{r}_j) = \frac{\hbar^2}{2m_B}|\vec{A}(\mathbf{r}_0, \mathbf{r}_j)|^2 = \frac{\hbar^2}{2m_B}\frac{q^2}{|\mathbf{r}_0 - \mathbf{r}_j|^2}. \tag{37}$$

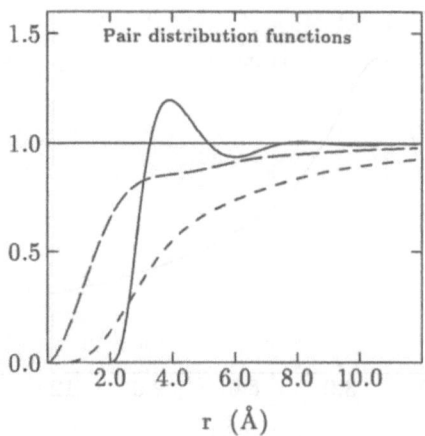

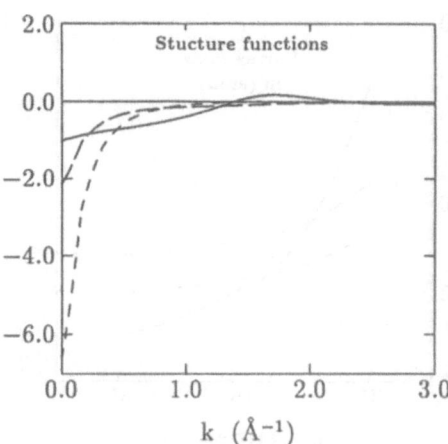

Fig. 2 The pair distribution functions for the pure ^{4}He (solid curve), the single vortex with vorticity q=1 (long dashed curve) and q=2 (short dashed curve) in the 2D ^{4}He.

Fig. 3 The structure functions in the 2D ^{4}He. The notations are the same as in Fig. 2.

The chemical potential calculated using this interaction diverges logaritmically with the size of the system. That is a slow divergence and can be handled explicitly in the numerical work.

$$\mu^v = \mu^v_{reg} + \alpha ln(R/r_c). \tag{38}$$

The first term on the right hand side is the regular part of the chemical potential and the second term shows the logaritmic divergence with the system radius R. The coefficient α depends on the superfluid density and the vorticity,

$$\alpha = \frac{\pi \rho^B q^2 \hbar^2}{m_B}. \tag{39}$$

A fictitious core radius is typically chosen to be $r_c \approx 1$Å. Also the volume occupied by the vortex diverges logaritmically with the size of the system. From Eq.(33) we find that

$$\lim_{k \to 0} V^{vB}_{p-h}(k) = -\frac{\hbar^2 k^2}{4m_B[S^{BB}(k)]^2}S^{vB}(0)$$
$$= m_B c_0^2 \beta = m_v c_0^2 \tag{40}$$

where c_0 is the speed of the zero sound in ^{4}He. The vortex mass can now be calculated from Eq.(32) yielding

$$\frac{m_v}{m_B} = \frac{m_v^{reg}}{m_B} + \frac{\alpha}{m_B c_0^2}ln(R/r_c). \tag{41}$$

Using typical values for ^{4}He we find that the cofficient $\frac{\alpha}{m_B c_0^2} \approx \frac{1}{10}$, which means that for a system with a reasonable size the effective mass is dominated by the first term.

With the explicit form of Eq.(37) for the vortex-background interaction we can solve the Euler equation (33). The iterative solution determines the radial distribution functions shown in Fig. 2 and the structure functions shown in Fig. 3, where we present results for vorticities q=1 and q=2. The corresponding functions for the ^{4}He background are also shown in those figures at the density ρ=0.04Å^{-2}. In all the numerical calculations we use the choice R=36Å, for the radius of the system.

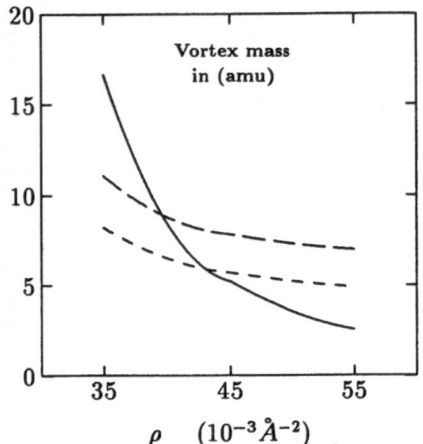

$\rho \quad (10^{-3} \mathring{A}^{-2})$

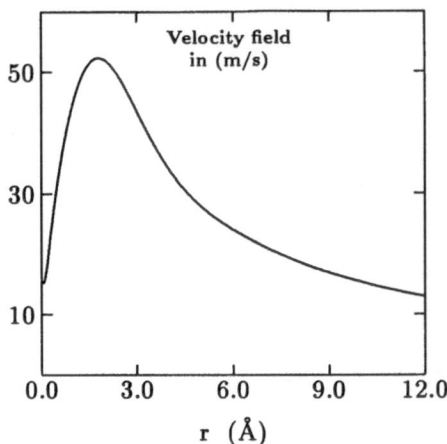

r (Å)

Fig. 4 The vortex mass as function of density. The solid curve is the mass of a single vortex, the long dased curve gives the mass of the vortex, which has a hydrogen isotope impurity at the center, and the short dashed curve gives the mass with the helium isotope at the center.

Fig. 5 The velocity distribution of the superfluid around a single vortex with vorticity q=1 calculated at the density $\rho = 0.04 \mathring{A}^{-2}$.

The radial distribution function starts from zero and approaches the asymptotic value one monotonically. The increase of the vorticity moves the background particles futher away from the center of the vortex as expected intuitively. The value of the structure function at the origin determines the volume excess factor and from that we calculate the vortex mass. That is shown in Fig. 4 as a function of density. It is interesting to notice that there is a strong density dependence in the range of our interest, and at the highest densities the vortex mass is only a fraction of the ^{4}He mass. In Fig. 5. we give results for the superfluid velocity field, the product of the vector potential and the radial distribution function. It has a clear maximum of 52m/s at 1.8Å.

The chemical potential calculated from Eq.(30) for the vorticity q=1 is found both in Fig. 6. and in Fig. 7. together with the results for the ^{3}He and hydrogen isotopes. It shows very little density dependence, which is related to the shrinking of the size of the circulation seen also in the density dependence of the mass.

Trapping of a vortex by an impurity

In the discussion above we assumed that a vortex is created at an arbitrary point in the space. Impurities existing in the fluid may trap them. On the other hand impurities may provide a natural core for the creation of a vortex. The trapping occures if the chemical potential of a single impurity at the center of the vortex is lower than the sum of the chemical potentials of a single impurity and one vortex put far apart into the fluid.

The bare interaction of the trapped impurity-vortex system with the background is the sum of the bare impurity-background interaction and the vortex-background interaction derived from the vector potential.

$$V^{IvB}(\mathbf{r}_0, \mathbf{r}_1) = V^{IB}(\mathbf{r}_0, \mathbf{r}_1) + \frac{\hbar^2}{2m_B}|\vec{A}(\mathbf{r}_0, \mathbf{r}_1)|^2 \qquad (42)$$

This interaction is still a two-particle interaction because both the vortex and the impurity are located at the same position in the space.

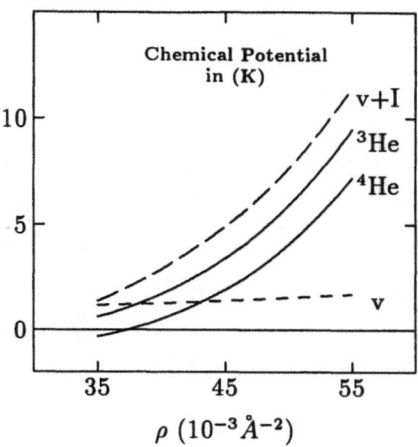

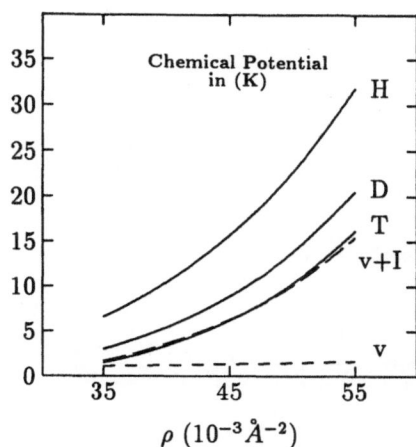

Fig. 6 The chemical potentials for helium isotopes (solid curves), for the single vortex (short dashed curve), and for the vortex with a helium isotope at the center (long dashed curve).

Fig. 7 The chemical potentials for hydrogen isotopes (solid curves), for the single vortex (short dashed curve), and for the vortex with a hydrogen isotope at the center (long dashed curve).

We may now think that the trapped impurity-vortex system behaves like a new kind of quasiparticle. Its mass is determined in the same way as the bare vortex mass, through the expelled volume, and calculated self-consistently from the Euler equation. The solution of the Euler equation for the radial distribution function determines the probability of finding a background particle near this new object. The results for the deuterium impurity using the following Lennard-Jones type interaction [24],

$$V^{IB}(r) = 6.57 \left[\left(\frac{3.19}{r} \right)^{12} - \left(\frac{3.19}{r} \right)^{6} \right], \qquad (43)$$

are shown in Figs. 8 and 9. Similar results hold also for the ^{3}He impurity.

Neighter the deuterium nor any other hydrogen isotope impurity forms a mixture with the ^{4}He. They form, however, bound surface states on films [25], [26] and perhabs that property could be used to study their trapping into vortices experimentally. It is clearly seen that the strong repulsion between the trapped impurity and the background forces the background particles away from the center of the vortex, which is compensated by the peak in the radial distribution function due to the attractive part of the interaction. The density dependence of the trapped vortex mass, shown by the dashed curves in Fig. 4, is therefore less pronounced than for the bare vortex.

In Figs. 6. and 7. we have the results for the chemical potentials of the single impurities and single vortices. Their sum should be compared with the results for the trapped vortices. Clearly all the hydrogen isotopes will be trapped. The ^{3}He impurity is trapped at small densities, but when $\rho^B > 0.045 Å^{-2}$ then ^{3}He prefers to be separated. Of cource, the HNC-approximation used in this calculation is not accurate in relating the density with the pressure where that may occur. For comparison we also show the chemical potential for the ^{4}He particle, which can not be put into the center of the vortex.

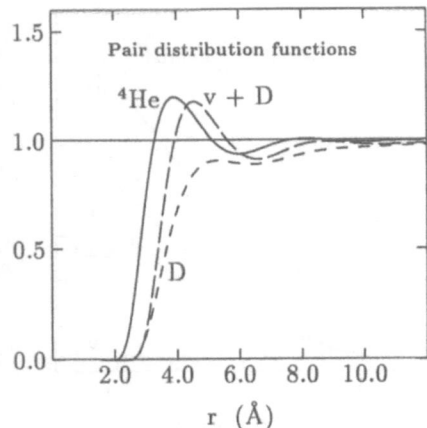

Fig. 8 The pair distribution functions for pure ^{4}He (solid curve), for the deuterium impurity in ^{4}He (short dashed curve) and the vortex trapped by the deuterium impurity (long dashed curve). These curves are calculated at the density 0.04 Å^{-2}.

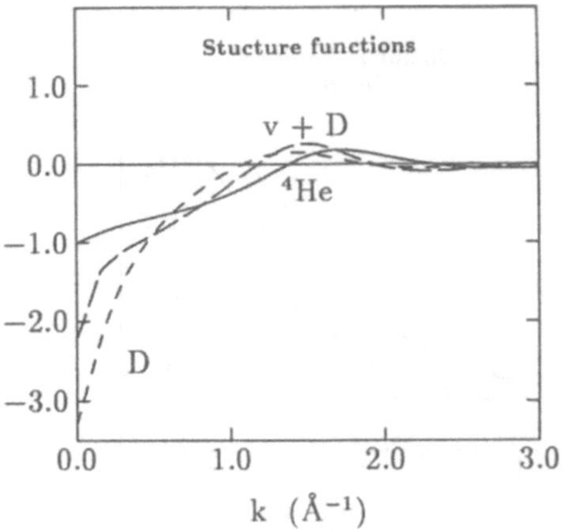

Fig. 9 The structure functions calculated at the density 0.04 Å^{-2}. The notations are the same as in Fig. 8.

VORTEX-ANTIVORTEX PAIR IN A 2D BOSON FLUID

In the 2D fluid it is energetically more favourable to create a vortex-antivortex pair instead of single vortices, because the logaritmic divergence is then canceled. The phase of the many-body wave function containes, in that case, two centers of circulation with opposite vorticities leading to the following vector potential

$$\nabla_j \times \vec{A}(\mathbf{r}_v, \mathbf{r}_a, \mathbf{r}_j) = 2\pi q \hat{k}\Big(\delta(\mathbf{r}_v - \mathbf{r}_j) - \delta(\mathbf{r}_a - \mathbf{r}_j)\Big). \tag{44}$$

We label the centers of the vortex and the antivortex by subindices v and a, respectively.

The square of the vector potential containes, besides the vortex-background interaction, also the vortex-antivortex interaction, which is mediated by a background particle,

$$\frac{\hbar^2}{2m_B}|A(\mathbf{r}_v, \mathbf{r}_a, \mathbf{r}_j)|^2 = V^{vB}(\mathbf{r}_v, \mathbf{r}_j) + V^{vB}(\mathbf{r}_a, \mathbf{r}_j) + V^{vaB}(\mathbf{r}_v, \mathbf{r}_a, \mathbf{r}_j) \tag{45}$$

with

$$V^{vaB}(\mathbf{r}_v, \mathbf{r}_a, \mathbf{r}_j) = -\frac{\hbar^2}{m_B}\frac{(\mathbf{r}_j - \mathbf{r}_v) \cdot (\mathbf{r}_j - \mathbf{r}_a)}{|\mathbf{r}_j - \mathbf{r}_v|^2|\mathbf{r}_j - \mathbf{r}_a|^2}. \tag{46}$$

As in the previous chapter we determine the real part of the phase variationally by minimizing the energy of the vortex-antivortex pair,

$$E^{pair} = E^{B+v+a} - \mu^v - \mu^a - E^B, \tag{47}$$

where E^{B+v+a} is the energy expectation value of the fluid with the vortex-antivortex pair added into it. The chemical potentials μ^v and μ^a are equal, because the positive and negative circulations, associated with the vortex and antivortex, have the same energy, and the last term is the energy of the pure background fluid.

Using the standard techniques within HNC-approximation [27] and the superposition approximation for the triplet distribution function we find the following expression for the effective vortex-antivortex interaction.

$$\begin{aligned} V_{eff}^{va}(\mathbf{r}_v, \mathbf{r}_a) &= \int \rho^B d^2 r_1 g^{vB}(\mathbf{r}_v, \mathbf{r}_1) g^{aB}(\mathbf{r}_a, \mathbf{r}_1) V^{vaB}(\mathbf{r}_v, \mathbf{r}_a, \mathbf{r}_1) \\ &+ w_{ind}^{va}(\mathbf{r}_v, \mathbf{r}_a) \\ &\equiv V_{sing}^{va}(\mathbf{r}_v, \mathbf{r}_a) + V_{reg}^{va}(\mathbf{r}_v, \mathbf{r}_a) \end{aligned} \tag{48}$$

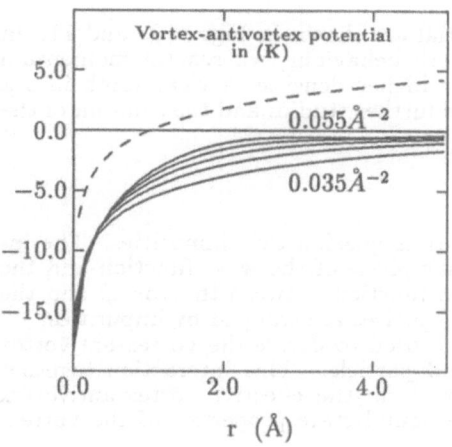

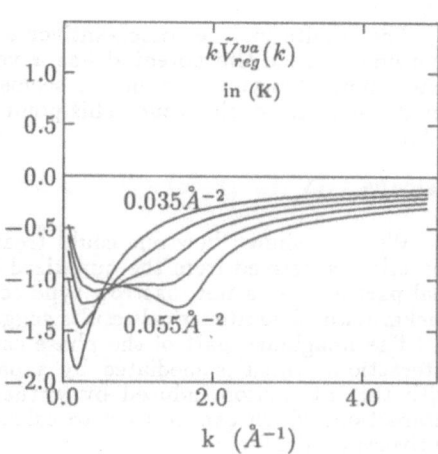

Fig. 10 The vortex-antivortex potential in coordinate space. The dashed curve is the logaritmically divergent part of the interaction calculated at the density 0.04 Å^{-2}. The solid curves give the regular part of the interaction at five different densities with steps of 0.005 Å^{-2}. The lowest and the highest densities are marked in the figure.

Fig. 11 The regular part of the vortex-antivortex interaction in the momentum space at five different densities. The notations are the same as in Fig. 10.

The integral term is a convolution type integral and can be directly integrated in the Fourier space, provided that the Fourier transformations exist. Therefore we write,

$$g^{vB}(\mathbf{r}_v, \mathbf{r}_1)g^{aB}(\mathbf{r}_a, \mathbf{r}_1) = h^{vB}(\mathbf{r}_v, \mathbf{r}_1)h^{aB}(\mathbf{r}_a, \mathbf{r}_1) + h^{vB}(\mathbf{r}_v, \mathbf{r}_1) + h^{aB}(\mathbf{r}_a, \mathbf{r}_1) + 1 \qquad (49)$$

where $h = g - 1$ and calculate the Fourier transform,

$$H^{vB}(k) = F\left\{\frac{h^{vB}(|\mathbf{r}_v - \mathbf{r}_1|)}{|\mathbf{r}_v - \mathbf{r}_1|^2}\right\}. \qquad (50)$$

The last term in Eq.(49) leads to the logaritmic potential used in the Kosterlitz-Thouless theory.

$$\begin{aligned} V_{sing}^{va}(\mathbf{r}_v, \mathbf{r}_a) &= \int \rho^B d^2 r_1 V^{vaB}(\mathbf{r}_v, \mathbf{r}_a, \mathbf{r}_1) \\ &= -4\pi\rho^B \frac{\hbar^2}{m_B}\left[ln\left|\frac{R}{r_c}\right| - ln\left|\frac{\mathbf{r}_a - \mathbf{r}_v}{r_c}\right|\right] \end{aligned} \qquad (51)$$

The full effective interaction must also include the phonon induced interaction, $w_{ind}^{va}(\mathbf{r}_v, \mathbf{r}_a)$. Using the HNC-approximation we can determine the expression for the effective interaction as a function of relative momentum between the vortex-antivortex pair,

$$\begin{aligned} V_{reg}^{va}(k) &= \frac{\hbar^2}{2m_B}\left[2\left(\frac{dH^{vB}(k)}{dk}\right)^2 - 8\pi\rho^B\frac{dH^{vB}(k)}{kdk}\right. \\ &\quad \left. - \frac{k^2}{2}\left[\frac{S^{vB}(k)}{S^{BB}(k)}\right]^2\left(2\frac{m_B}{m_v}S^{BB}(k) + 1\right)\right]. \end{aligned} \qquad (52)$$

This potential has bound states, which can be solved from the Schrödinger equation

$$-\frac{\hbar^2}{2m_v}\nabla^2\Phi(\mathbf{r}) + (V_{eff}^{va}(\mathbf{r}) - E)\Phi(\mathbf{r}) = 0. \qquad (53)$$

The results for the vortex-antivortex potential are shown in Figs. 10. and 11. In coordinate space the potential has a very smooth behaviour, whereas in momentum space more structure is seen. It seems that at higher densities a clear jumb into a smaller size pair could occur. This point requires further studies and the solution of the Schrödinger equation.

SUMMARY

We have shown how one could treat vortices as quasiparticle impurities. The interaction is derived from the quantized imaginary phase of the wave function and the real part acts as a new Jastrow type correlation function between the vortex and the background. The numerical results suggest that vortices are trapped by impurities.

The imaginary part of the phase can also be used to derive the vortex-antivortex interaction, which is mediated by a background particle. This interaction together with the interaction induced by virtual phonon form the effective vortex-antivortex interaction, which can be used to calculate the bound state properties of the vortex-antivortex pair.

ACKNOWLEDGMENTS

We thank A. Kallio and P. Pietiläinen for many valuable discussions. One of us (M.S.) thanks also Finnish Academy of the financial support.

REFERENCES

[1] V.L. Beresinski, Sov. Phys. JETP **32**, 493 (1970).

[2] J.M. Kosterlitz and D. J. Thouless, J. Phys. **C6**, 1181 (1973), and Prog. Low Temp. Phys. **VIIb**, 373 (1978).

[3] For a recent review of experiments see; G. Agnolet, D.F. McQueeney and J.D. Reppy Phys. Rev. **39**, 8934 (1989).

[4] D.P. Arovas, R. Schrieffer, F. Wilczek, and A. Zee, Nucl. Phys. **B251**, 117 (1985).

[5] G.V.Chester, R. Metz, and L. Reatto, Phys. Rev. **175**, 275 (1968).

[6] A.L. Fetter, Ann. Phys. **70**, 67 (1972).

[7] M.M. Salomaa and G.E. Volovik, Rev. Mod. Phys. **59**, 533 (1987).

[8] J-M Duan and A.J. Leggett, Phys. Rev. Lett. **68**, 1216 (1992).

[9] C.M. Muirhead, W.F. Vinen, F.R.S., and R.J. Donnelly, Phil. Trans. R. Soc. Lond. **A 311**, 433 (1984).

[10] E. Feenberg, *Theory of Quantum Liquids* (Academic, New York, 1969).

[11] C. C. Chang and C. E. Campbell, Phys. Rev. **B15**, 4238 (1977).

[12] C. E. Campbell, in *Progress in Liquid Physics* , edited by C. A. Croxton (Wiley, London 1977) Chapter 6.

[13] J. W. Clark, in *Progress in Particle and Nuclear Physics,* edited by D. H. Wilkinson (Pergamon, Oxford 1979), Vol. 2, p. 89.

[14] T.B.Davison and E Feenberg, Phys. Rev. **178**, 106 (1969).

[15] W.E. Masey, C.W. Woo, and H.T. Tan, Phys. Rev. **A1**, 519 (1970).

[16] J.C. Owen, Phys. Rev. **B23**, 5815 (1981).

[17] K. E. Kürten and C. E. Campbell, Phys. Rev. **B26**, 124 (1982).

[18] K. E. Kürten and M.L. Ristig, Phys. Rev. **B27**, 5479 (1983), and Phys. Rev. **B31**, 1346 (1985).

[19] J. Boronat, A. Fabroccini and A. Polls, J. Low Temp. Phys. **74**, 347 (1989).

[20] M. Saarela, in *Recent Progress in Many Body Theories Volume 2*, edited by Y. Avishai, p. 337 (Plenum Press, New York, 1990).

[21] M. Saarela and E. Krotscheck, to be published.

[22] P. A. Whitlock, G. V. Chester M. H. Kalos , Phys. Rev. **B38**, 2418, (1988).

[23] R.A. Aziz, V.P.S. Nain, J.C. Carley, W.L. Taylor, and G.T. McColville, J. Chem. Phys. **70**, 4330 (1979).

[24] P. J. Toennis, W. Weltz, and G. Wolf, Chem. Phys. Lett. **44**, 5 (1986).

[25] I.B. Mantz and D.O. Edwards, Phys. Rev. **B20**, 4518 (1979).

[26] E. Krotscheck, M. Saarela, and J. L. Epstein, Phys. Rev. **B38**, 111 (1988).

[27] J.C. Owen, Phys. Rev. Lett. **47**, 586 (1981).

UNUSUAL ONE–ELECTRON STATES ON THE SURFACE OF LIQUID HELIUM

Eugene P. Bashkin

*Department of Physics and Material Sciences Center
Philipps University, 3550 Marburg, Germany

INTRODUCTION

A 2D system of electrons localized near the surface of liquid and solid insulators like helium[1] and hydrogen[2], has actively been studied for a long time. A consideration the properties of such a 2D electron system is based on the starting idea that the one–electron state corresponds to a bound state for the motion perpendicular to the surface (the z–axis) and to a nearly free motion in the xy–plane. The localization of the electron in the z–direction is due to an electric field of the electron image or to a clamping electric field. Inasmuch as the localization range $\langle z \rangle$ is normally rather large and the electron is on average quite far from the surface, a motion along the surface is just slightly disturbed by the interaction of the electron with surface excitatons (ripplons). At sufficiently high temperatures the scattering of electrons with helium atoms of the vapour can also be rather significant and responsible for the dissipation characteristics of the electron layer. When lowering the temperature the contributions of both scattering mechanisms diminish and vanish at $T = 0$.

As it is well known, liquid ^{4}He cannot stand any impurities. A single exception is the ^{3}He isotope which can dissolve in liquid ^{4}He and thus forms a system of impurity excitations in superfluid helium. When the temperature decreases, ^{3}He quasiparticles are captured by the surface of ^{4}He and a 2D subsystem of ^{3}He excitations is created[3,4]. The interaction between a localized electron and a ^{3}He surface state drastically changes the structure of the one–electron state, which strongly influences all properties of the 2D electron gas at low enough temperatures. The central point of this contention is the statement that the interaction associated with the polarization of a ^{3}He surface quasiparticle by the Coulomb field of an electron causes the latter to localize on the surface impurity. The motion of an electron in the xy–plane becomes finite as well, and a kind of a ^{3}He$^-$ ion with axial symmetry appears. As a result the electron and the ^{3}He atom can propagate along the surface only together, as a whole, although each of them moves in its own "plane" and those two "planes" are parallel to the surface and distant from each other. In this case the translational motion of the electron is strongly affected by the coupled ^{3}He quasiparticle. The scattering of the coupled ^{3}He atom with other surface impurities results in a residual resistance even at $T \longrightarrow 0$. The effect in question may be detected experimentally by measuring the mobility of electrons at low enough temperatures. Photoelectric absorption of radiation connected with the decay of the bound state is considered. At higher areal densities of ^{3}He the electron may again become delocalized in the xy-plane. However, the energy spectrum is different from the bare electron dispersion law.

BINDING ENERGY

The energy of the localization of the electron near the surface and the binding energy of the surface impurity considerably exceed the potential of the charge–impurity interaction. Therefore the starting point of the calculation is that a weak polarization interaction between a ^{3}He surface impurity and electron just slightly disturbs the motion of particles in the z-direction but entirely changes the relative motion of them in the xy-plane and leads to a bound state. In other words, a perturbation theory can be applied in order to describe the perpendicular motion only, but the orbital dynamics of particles in the xy-plane should be found from the exact Schrödinger equation. Quantitatively this means that one has to solve the 2D Schrödinger equation with the electron–impurity potential averaged over the unperturbed wave functions of a bare ^{3}He quasiparticle and electron along the z-coordinate.

The change in the energy of the surface ^{3}He quasiparticle placed in the Coulomb field induced by the electron consists of the polarization term $-\alpha e^2/2r^4$ (where α is the polarizability of an impurity particle) as well as the terms due to the deformation of the surface in the electric field. One may neglect the latter terms because of the following reason. Ignoring the deformation is valid if the radius of the 2D bound state (for motion along the surface) calculated in this approximation is much smaller than the radius characterizing the deformation of the plane surface (the radius of the dimple). Because of the large value of the capillary constant in liquid helium, the necessary requirements can easily be fulfilled.

In view of this, one can represent the electron–impurity hamiltonian in the form:

$$\hat{H}_{e-i} = \hat{H}_e^{(0)}(z_e) + \hat{h}_e^{(0)}(\vec{\rho}_e) + \hat{H}_i^{(0)}(z_i) + \hat{h}_i^{(0)}(\vec{\rho}_i) - \frac{\alpha e^2}{2(\vec{r}_e - \vec{r}_i)^4} \,, \qquad (1)$$

Eq.(1) means that a quantum–mechanical motion of each particle is described in terms of two sets of the separable variables z_k and $\vec{\rho}_k$ (the index $k = i, e$ designates different particles, and $\vec{\rho}$ is the 2D radius–vector in the xy-plane). This, indeed, implies some extra requirements for an external potential acting on the particles. However, if we neglect any deformations of the free surface of liquid ^{4}He and consider it as a flat plane those requirements are obviously satisfied. In accordance with these remarks, we assume that the influence of the electron–impurity interaction on the motion along the z-coordinate can be described in terms of a perturbation theory which, however, does not hold at all in order to calculate the "disturbed" motion in the $\vec{\rho}$-plane.

Let us define the set of the wave functions $\psi_k^{(0)}(z_k)$ as the eigenfunctions of the equation:

$$\hat{H}_k^{(0)}(z_k)\psi_k^{(0)}(z_k) = E_k^{(0)}\psi_k^{(0)}(z_k) \,. \qquad (2)$$

Here $E_k^{(0)}$ are the corresponding eigenvalues of the energy. The solution of the Schrödinger equation $\hat{H}_{i-e}\Psi = E\Psi$ will then be sought in the form:

$$\Psi(z_i \,, \vec{\rho}_i \,; z_e \,, \vec{\rho}_e) = \phi(z_i \,, z_e)\chi(\vec{\rho}_i \,, \vec{\rho}_e) \,,$$
$$\phi(z_i \,, z_e) = \sum_{kn} C_{kn}\psi_{in}^{(0)}(z_i)\psi_{ek}^{(0)}(z_e) \,. \qquad (3)$$

Here the indices k and n designate different quantum–mechanical states. When multiplying the Schrödinger equation by $\psi_{im}^{(0)*}(z_i)\psi_{el}^{(0)*}(z_e)$ and integrating over $dz_e dz_i$ one can easily find:

$$C_{lm}(\hat{h}_i^{(0)} + \hat{h}_e^{(0)})\chi - \sum_{kn} \int \psi_{im}^{(0)*} \psi_{el}^{(0)*} \frac{\alpha e^2}{2(\vec{r}_i - \vec{r}_e)^4} \psi_{in}^{(0)} \psi_{ek}^{(0)} \chi dz_i dz_e = C_{lm}\epsilon_{lm}\chi , \quad (4)$$

where $\epsilon_{lm} = E - E_{im}^{(0)} - E_{el}^{(0)}$. To the first order of the perturbation theory for the motion along the coordinates z_i and z_e, when $C_{lm} = 1$ and $C_{kn} = 0$ if $l \neq k$ and $m \neq n$, we obtain:

$$(\hat{h}_i^{(0)} + \hat{h}_e^{(0)})\chi + <\hat{V}> \chi = \epsilon_{lm}\chi ,$$
$$<\hat{V}> = -\int \psi_{im}^{(0)*} \psi_{el}^{(0)*} \frac{\alpha e^2}{2(\vec{r}_i - \vec{r}_e)^4} \psi_{in}^{(0)} \psi_{ek}^{(0)} dz_i dz_e , \quad (5)$$

When separating as usual the relative motion of particles from the center–of–mass motion one obviously gets:

$$\hat{h}_r^{(0)} \chi_r + \langle \hat{V} \rangle \chi_r = (\epsilon_{lm} - E_c)\chi_r \quad (6)$$

where E_c is the energy corresponding to the center–of–mass motion and the subscript r indicates the quantities characterizing the relative motion of the particles, e.g. $\hat{h}_r^{(0)}$ is related to the kinetic energy of a particle with the reduced mass.

Thus the problem in question reduces to resolving the 2D Schrödinger equation with the effective interaction potential which corresponds to an attraction between the electron and surface ^{3}He impurity. An interesting point is that the effective interaction between an electron and impurity ^{3}He excitation in bulk corresponds to a repulsion[5,6]. This results from the fact that the density of ^{4}He increases in the Coulomb field when the distance from the electron gets smaller. Since the volume per impurity ^{3}He atom in a solution is larger than the volume per ^{4}He atom, the ^{3}He component has to be affected by the Archimedes force in the direction opposite to the gradient of the ^{4}He density. On the contrary, in the case in question the gravitational field and the surface tension play an important role in preventing the free surface of liquid ^{4}He from being disturbed. That is why the effective interaction in the main approximation looks roughly the same as the interaction between an electron and ^{3}He atom in vacuum, i.e. corresponds to an attraction.

Another interesting point is that a repulsive hard core in the electron–impurity interaction does not come into effect. This happens because the electron and ^{3}He particle are on average rather far from each other. The impurity is localized very close to the surface whereas the electron is basically very distant from the surface. At such macroscopic distances the interaction between them is a pure polarization attraction.

Here we will not discuss the structure of the impurity wave function in the z–direction[7,8]. In the main approximation it is quite sufficient to treat the surface states as quasiparticles with some effective mass M, which form a 2D rarefied gas[3,4]. We also neglect all terms proportional to m/M, where m is the electron mass. Within such an approximation we consider the motion of the electron in an external field of the impurity at rest, and the effective interaction potential from the Schrödinger equation (6) obviously reduces to:

$$\langle \hat{V} \rangle = -\frac{\alpha e^2}{2} \int_{-\infty}^{\infty} \frac{\psi_e^{(0)*}(z)\psi_e^{(0)}(z)}{(\rho^2 + z^2)^2} dz . \quad (7)$$

In order to be specific we consider here the electron in the ground state. The equation

189

(6) with the effective potential (7) can easily be solved[6,9], and the energy of the bound s–state is given by the expression[6]:

$$|\epsilon_r| = |\epsilon_{11} - E_c| = \frac{\hbar^2}{m}\langle z^{-2}\rangle exp\left[-\frac{\hbar^2}{m}\frac{4}{\alpha e^2\langle z^{-2}\rangle}\right] .$$

(8)

(Let us point out that in the case when the electron or positive ion is in the interior of the liquid and is localized near the surface by the clamping electric field, the value of M is much smaller than the effective mass of the ion and the effective potential (6) under certain conditions has a number of very deep levels which are classified the same as the spectrum of a 2D oscillator[6]:

$$\epsilon_r = -\frac{\alpha e^2}{2}\langle z^{-4}\rangle + \hbar\omega\left(n + \frac{|l|+1}{2}\right) ,$$

$$\omega^2 = \frac{8\alpha e^2}{M}\langle z^{-6}\rangle .$$

(9)

We, however, will not discuss this case in this paper.)

PHOTOELECTRIC EFFECT

We have just come to a conclusion that a weak attraction due to the polarization of a ^{3}He surface impurity in the Coulomb field of an electron gives rise to the ^{3}He$^-$ bound state. Such a bound state can be considered as an ion with the axial symmetry of electron wave functions. This ion contains one additional electron which, however, can be released by an external electromagnetic radiation. The process, in fact, means a decay of the bound state in the field of an electromagnetic wave and is absolutely similar to the traditional photoelectric effect. The cross–section of ionization of a ^{3}He$^-$ particle is given by the well–known golden rule of quantum mechanics[9]:

$$d\sigma = 2\pi|M_{fi}|^2\delta\left(-|\epsilon_r| + \omega - \frac{p^2}{2m}\right)\frac{d^2p}{(2\pi)^2} .$$

(10)

In Eq. (10) we put $\hbar = 1$, $c = 1$, and M_{fi} is the transition matrix element. It is also assumed that the frequency ω of light is lower than the red boundary for the electron transitions between the discrete levels along the z–direction. In this case the wave functions $\psi_{en}^{(0)}(z_e)$ do not explicitly enter the transition probability because of the normalization condition, and the cross–section of photoelectric absorption (10) depends on the 2D characteristics in the xy–plane only. In the nonrelativistic limit the transition matrix element M_{fi} has the usual form:

$$M_{fi} = i\frac{e}{m}\left(\frac{2\pi}{\omega}\right)\int \chi_{ef}^*(\nabla\vec{e})e^{i\vec{k}\vec{r}}\chi_{ei}d^2r ,$$

(11)

where $\vec{e}$ and $\vec{k}$ are the polarization and momentum of the photon respectively. Since the final electron state possesses the orbital momentum $l = 1$, one can choose the electron wave function χ_{ef} in the simple form $exp\,(i\vec{p}\,\vec{r}\,)$ (exactly as in the case of the photoelectric splitting of deuteron). After this, the differential cross–section (10) reduces to the simple expression:

$$d\sigma = \frac{e^2}{m\omega}(\vec{e}\,\vec{p}\,)^2\chi_{ei}^2(\vec{p} - \vec{k})d\phi ,$$

(12)

where ϕ is the polar scattering angle in the xy–plane, and $\chi_{ei}(\vec{p}\,)$ is the Fourier transform of the initial electron wave function.

The initial wave function χ_{ei} should be found from the Schrödinger equation (6) with the effective interaction (7). In the limiting case in question when the energy of the electron bound state is given by Eq.(8), the initial wave function corresponds to the normalized MacDonald function:

$$\chi_{ei}(\rho) = \frac{\nu}{\pi^{1/2}} K_0(\nu\rho) \,, \; \nu^2 = \frac{2m|\epsilon_r|}{\hbar^2} \,. \tag{13}$$

Substituting $\chi_{ei}(\rho)$ from Eq.(13) into the cross–section (12) yields (in usual units):

$$d\sigma = 8\pi e^2 \frac{(\vec{e}\,\vec{p}\,)}{c\omega} \frac{|\epsilon_r|}{[2m|\epsilon_r| + (\vec{p} - \vec{k})^2]^2} d\phi \,. \tag{14}$$

In the rigorous Eq.(14), the momentum of the photon can be, of course, neglected. Then taking into account the conservation law for the energy one immediately obtains:

$$d\sigma = \frac{4\pi e^2}{m\omega c} \frac{|\epsilon_r|}{\hbar\omega} \frac{\hbar\omega - |\epsilon_r|}{\hbar\omega} (\vec{e}\,\vec{n})^2 d\phi \,, \tag{15}$$

where $\vec{n} = \vec{p}/p$. In the case of unpolarized photons the cross–section should be averaged over all possible polarizations which is equivalent to the replacement $(\vec{n}\vec{e}\,)^2 \longrightarrow (1/2)[\vec{n}_k \times \vec{n}]^2$, where $\vec{n}_k = \vec{k}/k$.

The photoelectric absorption is a good and sensitive instrument to detect the 2D bound states of an electron and a surface ^{3}He impurity. The existence of an abrupt threshold in the absorption of radiation as a function of frequency (the red absorption boundary) provides an extra advantage for carrying out the experiment.

LOW–FREQUENCY FIELDS AND STABILITY OF BOUND STATES

What happens to a ^{3}He$^-$ ion when it is affected by an alternating electric field with the frequency lower than the red boundary $|\epsilon_r|/\hbar$? It is rather obvious that even in the static limit, $\omega \longrightarrow 0$, i.e. in the case of an uniform and stationary electric field, there exists a finite probability of ionization of the 2D bound state. In contrast to the photoelectric effect the phenomena in question is connected with the tunnelling of the electron from the potential well (7) near the impurity to the more distant region where the potential energy of the electron in an external electric field becomes even lower than the bottom of the potential well. Of course, the probability of such a process is exponentially small at low electric fields but it gets much more important when increasing the field.

Measuring the mobility of electrons is the most traditional and common way to study the properties of the 2D layer. The phenomenon of "tunnelling ionization" mentioned above may drastically change the transport characteristics of 2D electrons. At $T << |\epsilon_r|$ all electrons are coupled with the ^{3}He impurities. At low electric field an electron can move along the surface only together with a ^{3}He quasiparticle as a whole. Under these conditions the electric current is actually the motion of the heavy ^{3}He$^-$ ions. Even at $T \longrightarrow 0$ there is a finite resistance (residual resistance) due to the scattering of the ^{3}He$^-$ ions with the other ^{3}He atoms on the surface of ^{4}He. However, when increasing the intensity of an electric field, a number of ^{3}He$^-$ ions get ionized and the light free 2D electrons (with the bare electron mass) are created. These electrons experience a much smaller resistance when $T \longrightarrow 0$. Thus two groups of carriers with mobilities differing strongly from each other, appear when increasing an external electric field.

Further we will neglect the momentum of the photon and consider the influence of the spatially uniform electric field $\mathcal{E} = \mathcal{E}_0 cos\omega t$ on the bound electron in a ^{3}He$^-$ ion. Instead of a scalar potential it is more convenient to describe the electric field in terms of a vector potential in the form:

$$\vec{A} = (-c\mathcal{E}_0\omega^{-1}sin\omega t \, , \, 0 \, , \, 0) \, . \tag{16}$$

Here the external electric field is assumed to be directed along the x–axis. The exact solution of the problem under consideration is rather complicated although it can certainly be found at least in the static limit $\omega = 0$. We will not discuss the rigorous result in this paper and restrict ourselves to calculating the probability of the decay of the bound state taking into account only the exponential term.

In fact, if the intensity $\mathcal{E}_0$ and frequency ω are not very high (the quantitative criterion will be formulated further), the probability of ionization via the tunnelling is exponentially small. If one is not interested in obtaining the preexponential factor, the probability can be estimated by means of the 1D Schrödinger equation in the x–direction only. In accordance with these arguments the corresponding hamiltonian may be presented as:

$$\hat{H} = \frac{1}{2m}\left(-i\hbar\partial_x + \frac{|e|\mathcal{E}_0}{\omega}sin\omega t\right)^2 \, . \tag{17}$$

The boundary condition for the electron wave fuction $\Psi(x,t)$ reduces to the requirement that $\Psi(x \longrightarrow 0 \, , \, t)$ should match the wave function of the bound electron with the energy ϵ_r. The further procedure is entirely similar to the one carried out in Ref.[9] and leads to the result:

$$w \sim exp\left[-\frac{2|\epsilon_r|}{\hbar\omega}f(\gamma)\right] \, , \tag{18}$$

where the functions $f(\gamma)$ and γ are defined as follows:

$$f(\gamma) = \left(1 + \frac{1}{2\gamma^2}\right)arcsinh\gamma - \frac{(1+\gamma^2)^{1/2}}{2\gamma} \, ,$$
$$\gamma = \frac{\hbar\omega\nu}{|e|\mathcal{E}_0} \, . \tag{19}$$

Inasmuch as the probability (18) is assumed to be small, the criterion $\hbar\omega << |\epsilon_r|$ is expected to be fulfilled. In the static limit $\omega = 0$ which is exactly the case of most experiments on the mobility, Eq. (18) yields:

$$w \sim exp\left[-\frac{2\hbar^2\nu^3}{3m|e|\mathcal{E}_0}\right] \, . \tag{20}$$

Eq.(20) for the probability is valid if the magnitude of the external electric field is not too large: $|e|\mathcal{E}_0 << \hbar^2\nu^3/m$. Eq.(20) provides the esimate for the critical electric field in which the regime of heavy ^{3}He$^-$ carriers is expected to get replaced by the regime of light free electrons. The possibility to change the binding energy ϵ_r by applying a clamping electric field perpendicular to the surface provides an experimentalist with the opportunity to choose the most convenient intensity of a field parallel to the surface (and, perhaps, the temperature range) in order to detect the unusual one–electron states in experiments on the photoelectric absorption or on the mobility.

The author greatly appreciates useful and interesting discussions he had with C. Campbell.

*The permanent address: Kapitza Institute for Physical Problems, 117334 Moscow, Russia.

REFERENCES

1. V. S. Edel'man, Soviet Phys. USPEKHI, **23**, 227 (1980).
2. V. V. Zav'yalov and I. I. Smol'yaninov, Soviet Phys. JETP, **65(1)**, 194 (1987).
3. A. F. Andreev, Soviet Phys. JETP, **23**, 938 (1967).
4. D. O. Edwards and W. F. Saam, Prog. Low Temp. Phys., **7A**, 285 (1978).
5. V. B. Schikin, Soviet Phys. USPEKHI, **20**, 226 (1977).
6. E. P. Bashkin. Soviet Phys. JETP, **59(1)**, 94 (1984).
7. E. Krotscheck, M. Saarela and J. L. Epstein, Phys. Rev. Lett., **61**, 1728 (1988); **64**, 427 (1990).
8. N. Pavloff and J. Treiner, J. Low Temp. Phys. **83**, 331 (1991).
9. L. D. Landau and E. M. Lifshitz, Quantum Mechanics, Pergamon, Oxford (1977).

EXCITATIONS OF THE SURFACE OF LIQUID ⁴HE

K. A. Gernoth,[1] J. W. Clark,[1] G. Senger,[2] and M. L. Ristig[2]

[1] McDonnell Center for the Space Sciences
and Department of Physics,
Washington University, St. Louis, MO 63130, USA

[2] Institut für Theoretische Physik,
Universität zu Köln, D−5000 Köln 41, Germany

I. INTRODUCTION

The surface of liquid ⁴He at zero temperature and the ⁴He vapor-liquid interface at temperatures $T > 0$ are currently the objects of substantial experimental and theoretical interest. Part of this activity concentrates on studying ground state properties, such as the thickness of the ⁴He surface and interface, for which some experimental data over a range of temperatures are now available.[1,2] Other experimental and theoretical work explores the elementary excitations supported by these inhomogeneous ⁴He systems. The spectrum of excitations of the surface of liquid ⁴He in the short wavelength regime has been measured in recent experiments on ⁴He films adsorbed on a substrate.[3,4] It is well established by earlier measurements[5−9] that at long wavelengths the dispersion relation of ⁴He surface modes may be described correctly by the laws of hydrodynamics.

In the past, semi-phenomenological pictures based on classical or quantum hydrodynamics had been adopted to describe the spectrum of ⁴He surface excitations at long wavelengths.[10] Ji and Wortis employed a Landau model to analyze the excited states of a ⁴He surface.[11] More recently, the behavior of the spectrum of ⁴He surface excitations in the short wavelength region at energies close to the bulk roton minimum has been investigated within a hydrodynamic description.[12]

A theoretical ab-initio treatment of ⁴He surface modes may start from a generalized Feynman eigenvalue equation for the excited wave functions and the associated excitation energies.[10,13] During the last few years this approach has been mainly applied to ⁴He films with or without a supporting substrate[14−17] and to ⁴He clusters.[18,19] At nonzero temperatures we may employ an appropriate generalization of the Feynman eigenvalue equation to describe a vapor-liquid ⁴He system with a planar interface.[20] At present, backflow effects[21,22] are ignored. Employing the powerful theory of correlated basis functions (CBF theory)[23,24] we may incorporate these backflow effects in future more sophisticated calculations.

The input quantities entering the Feynman description of elementary excitations are the density profile of the ⁴He surface or, at nonzero temperatures, the density profile of the ⁴He vapor-liquid interface, and the spatial two-body distribution function. Density functional theory, employed at zero temperature[25] and at nonvanishing temperatures[26], provides information on the spatial structure of the ^{4}He surface and interface. Density profiles and excited states of ^{4}He clusters at zero temperature[18,19] as well as at nonzero temperatures[27,28] are also extensively studied by applying quantum Monte Carlo techniques.

In this work we focus on an analysis of the elementary excitations of liquid ^{4}He at zero temperature with a planar surface. We do this within the framework of CBF theory. On the variational level it is based on a correlated N-body trial ground state of Hartree-Jastrow type and on Feynman's approximation for the excited states. In the present simple realization of this formalism we either employ parametrized density profiles and Jastrow correlation factors[20] or construct the corresponding optimal quantities by solving associated Euler-Lagrange equations. CBF theory of spatially inhomogeneous quantum liquids[29,30] furnishes the necessary theoretical tools for a systematic optimization of the density profile and the spatial two-body distribution function. The relevant Euler-Lagrange equations may be interpreted as a renormalized Hartree equation for the density profile[29,30] and a paired-phonon (PPA) equation for the two-body quantities.[30] Within this approach, the Feynman equation for the excited states and energies can be equivalently formulated as a renormalized Bogoliubov eigenvalue equation.[31]

In a first step of numerical calculations we employ the parametrized Jastrow factor of short-ranged Schiff-Verlet type[32] used in our earlier work[20] and, by means of an appropriate numerical procedure, solve the renormalized Hartree equation for the density profile of liquid ^{4}He with a planar surface. Exploiting the symmetry of this system, the density profile $\varrho(\mathbf{r})$ is taken to depend only on the component z of the position vector $\mathbf{r}$ in the direction vertical to the (x,y) surface plane, $\varrho(\mathbf{r}) = \varrho(z)$. The bulk ^{4}He liquid density $\varrho_L = \lim_{z \to -\infty} \varrho(z)$ enters the calculation as an externally specified parameter. In the numerical implementation of the formalism this specification is made through the imposition of a boundary condition. A suitably chosen auxiliary potential is applied to the ^{4}He system to balance the pressure inside the bulk liquid. Employing the resulting density profile and the parametrized two-body correlations, the Feynman eigenvalue equation is solved numerically to obtain the spectrum of bulk and surface excitations.

In a second, more sophisticated step within CBF theory, we optimize the ^{4}He surface profile *and* (by means of the PPA condition) the spatial two-body distribution function under the boundary condition $\varrho_L = 0.0218\,\text{Å}^{-3}$ (experimental saturation density). We then proceed to solve the Bogoliubov eigenvalue equation with the optimized density profile and optimized two-body correlations as inputs. The resulting optimal excitation spectrum obeys the familiar hydrodynamic dispersion laws in the long wavelength regime. The excitation energies of surface states form several discrete branches in the energy versus wave number plane, merging smoothly with the continuous part of the spectrum at atomic wavelengths.

Section II gives a brief account of the optimization of the spatial structure of the ^{4}He surface within variational CBF theory and provides the Feynman and renormalized Bogoliubov eigenvalue equations needed to evaluate the excitation spectrum. In Section III we discuss the general features of the excitation spectrum of liquid ^{4}He with a planar surface and display the energy versus wave number diagram. We discuss these results and compare them with other results derived earlier.[12,20] The paper concludes with a description of proposed improvements of the present approach.

II. GROUND AND EXCITED STATES

We adopt a Hartree-Jastrow ansatz Φ_0 for the ground state wave function and minimize the associated energy expectation value $E_0 = \langle \Phi_0 | H | \Phi_0 \rangle / \langle \Phi_0 | \Phi_0 \rangle$ of the Hamiltonian H with respect to this correlated trial wave function. The minimization leads to two Euler-Lagrange equations, a Hartree equation determining the density profile $\varrho(z)$ and a so-called paired-phonon equation determining the optimal spatial distribution function $g(\mathbf{r}_1, \mathbf{r}_2)$. This quantity is generated by a Jastrow pseudo-potential u that, in the presently adopted geometry of a half space, depends only on the coordinates z_1 and z_2 and on the projected relative distance $\eta = \left[(x_1 - x_2)^2 + (y_1 - y_2)^2 \right]^{1/2}$, i.e., $u(\mathbf{r}_1, \mathbf{r}_2) = u(\eta, z_1, z_2)$.

The relation between the quantity u and the spatial distribution function g is given by the hypernetted-chain (HNC) equations.[33] Neglecting the elementary contributions (i.e. in HNC/0 approximation) these equations read explicitly

$$g(\eta, z_1, z_2) = \exp\{u(\eta, z_1, z_2) + N(\eta, z_1, z_2)\}, \tag{1}$$

$$g(\eta, z_1, z_2) = 1 + N(\eta, z_1, z_2) + X(\eta, z_1, z_2), \tag{2}$$

$$N(q, z_1, z_2) - \int_{-\infty}^{\infty} \varrho(z_3) N(q, z_1, z_3) X(q, z_3, z_2)\, dz_3 = \int_{-\infty}^{\infty} \varrho(z_3) X(q, z_1, z_3) X(q, z_3, z_2)\, dz_3. \tag{3}$$

Eq. (2) provides a decomposition of the spatial distribution function g into a nodal part N and a non-nodal part X. The Hankel transforms[20] $N(q, z_1, z_2)$ and $X(q, z_1, z_2)$ associated with the nodal function N and the direct distribution function X, respectively, are related via the Ornstein-Zernike relation (3).[34] As a simple option, we may adopt a parametrized pseudo-potential $u = u_{SV}$ of Schiff-Verlet type[32]

$$u_{SV}(\eta, z_1, z_2) = -\left(\frac{b_0 + b_1 \sqrt{\varrho(z_1)\varrho(z_2)}}{\sqrt{\eta^2 + (z_1 - z_2)^2}}\right)^5 \tag{4}$$

with parameters $b_0 = 2.8\,\text{Å}$ and $b_1 = 9.98\,\text{Å}^4$ taken from Ref. 16. The function (4) inserted into the HNC/0 equations (1)$-$(3) generates the two-body quantities g, N, and X. Alternatively, we may replace the parametrized form (4) by the optimal pseudo-potential, constructed systematically within CBF theory. The explicit construction is based on the paired-phonon equation.

The Euler-Lagrange equation determining the square root of the density profile $\varrho(z)$ takes the form[29,30]

$$\{T_z + V_H(z) + U_{\text{ext}}(z)\}\sqrt{\varrho(z)} = \mu\sqrt{\varrho(z)} \tag{5}$$

of a renormalized Hartree equation with chemical potential μ. The operator T_z is given by $T_z = -(\hbar^2/2m)\,\partial^2/\partial z^2$. The renormalized Hartree potential V_H depends on the bare two-body interaction $v(|\mathbf{r}_1 - \mathbf{r}_2|)$, assumed in this study to be of Lennard-Jones (6,12) type with the standard ^{4}He parameters $\varepsilon_{LJ} = 10.22\,\text{K}$ and $\sigma_{LJ} = 2.556\,\text{Å}$. It further depends on the two-body quantities g, u, N, and X and on the density profile $\varrho(z)$ itself. Consequently eq. (5) constitutes a nonlinear equation for the function $\sqrt{\varrho(z)}$. The operative expression for the renormalized Hartree potential V_H may be taken from Refs. 29, 30, or 16. In this work we solve eq. (5) numerically by means of a Newton-Raphson iteration scheme,[29] imposing the boundary condition $\lim_{z \to -\infty} \varrho(z) = \varrho_L$ for a given bulk density of the liquid. In this case, an external potential U_{ext} is needed to balance the pressure in the ^{4}He liquid and to localize the surface. We use a potential U_{ext} of the form

$$U_{\text{ext}}(z) = \frac{U_0}{1 + \exp[\alpha(z - z_W)]}. \tag{6}$$

The parameter α is chosen as $\left(-2m\left[\mu - V_H(\text{L}_2) - U_{\text{ext}}(\text{L}_2)\right]/\hbar^2\right)^{1/2}$, where L_2 marks one of the boundaries of the finite box $[-\text{L}_1 \le z \le \text{L}_2]$ in which the calculation is performed. The parameter z_W merely positions the Gibbs dividing surface.

In some of our numerical applications we adjust the strength U_0 in such a way that the potential (6) balances the bulk ^{4}He liquid pressure, i.e., we set $U_0 = e_B^{(F)} - \mu_F$, where $\mu_F = \lim_{z \to -\infty} V_H(z)$ and $e_B^{(F)} = \lim_{z \to -\infty} e_F(z)$ are, respectively, the chemical potential and the binding energy per particle of the bulk ^{4}He liquid at density ϱ_L in the absence of an external potential. The quantity $e_F(z)$ is obtained as a functional of the density profile $\varrho(z)$ and of the two-body quantities u, g, N and X from the expression for the total energy of the ^{4}He system per unit surface area, E/A, inside a finite box $[-L_1 \leq z \leq L_2]$, i.e., $E/A = E_F/A + \int_{-L_1}^{+L_2} \varrho(z)U_{\text{ext}}(z)\,dz$ with $E_F/A = \int_{-L_1}^{+L_2} \varrho(z)e_F(z)\,dz$. References 29, 30, or 16 give explicit expressions for E/A as a functional of the Hartree-Jastrow wave function Φ_0.

The paired-phonon condition for the function $\tilde{X}(q,z_1,z_2) = \sqrt{\varrho(z_1)\varrho(z_2)}\,X(q,z_1,z_2)$, reads[30]

$$-\{2\varepsilon_0(q) + H_0(1) + H_0(2)\}\tilde{X}(q,z_1,z_2) + \int_{-\infty}^{\infty} \tilde{X}(q,z_1,z_3)\{\varepsilon_0(q) + H_0(3)\}\tilde{X}(q,z_3,z_2)\,dz_3$$

$$= 2\tilde{V}_{\text{ph}}(q,z_1,z_2). \qquad (7)$$

References 30 and 16 provide concrete expressions for the particle-hole interaction $\tilde{V}_{\text{ph}}$ in terms of the bare two-body potential, the density profile, and the two-body functions u, g, N, and X. The operator H_0 is defined by $H_0(i) = T_{z_i} + w(z_i)$, $i = 1, 2$, with one-body potential $w(z) = -\{T_z\sqrt{\varrho(z)}\}/\sqrt{\varrho(z)}$. The energy $\varepsilon_0(q) = \hbar^2 q^2/2m$ is the kinetic energy of a free ^{4}He atom with momentum $\hbar q$ parallel to the (x,y) surface plane. The sum $\varepsilon_0(q) + H_0$ is the operator for the energy of a single ^{4}He atom moving (with parallel momentum $\hbar q$) in the inhomogeneous background constituted by the density profile $\varrho(z)$.

An appropriate treatment of the excited states of the planar surface of liquid ^{4}He may start from a Feynman ansatz of the form[10, 13]

$$\Phi_{\kappa,q}(\mathbf{r}_1, \mathbf{r}_2, \cdots, \mathbf{r}_N) = \sum_{i=1}^{N} \exp\{i(q_x x_i + q_y y_i)\}\varphi_\kappa(z_i, q) \cdot \Phi_0(\mathbf{r}_1, \mathbf{r}_2, \cdots, \mathbf{r}_N) \qquad (8)$$

for an excited state of N ^{4}He atoms. In accordance with the planar symmetry of the system, the quantum number q characterizes the parallel momentum $\hbar\mathbf{q} = \hbar(q_x, q_y)$ of an excited eigenstate. The additional quantum number κ may assume discrete and/or continuous values. Minimization of the excitation energy $\varepsilon_\kappa(q) = \langle\Phi_{\kappa,q}|H - E_0|\Phi_{\kappa,q}\rangle/\langle\Phi_{\kappa,q}|\Phi_{\kappa,q}\rangle$ with respect to the Feynman ansatz, while keeping the ground state wave function Φ_0 fixed, yields an eigenvalue equation[10, 13, 31] for the excitation energies and the associated wave functions $\psi_\kappa(z, q) = \sqrt{\varrho(z)}\,\varphi_\kappa(z, q)$. This equation may be cast into the form[20]

$$\{\varepsilon_0(q) + H_0(1)\}\psi_\kappa(z_1, q) - \int_{-\infty}^{\infty} \tilde{X}(q, z_1, z_2)\{\varepsilon_0(q) + H_0(2)\}\psi_\kappa(z_2, q)\,dz_2 = \varepsilon_\kappa(q)\psi_\kappa(z_1, q). \qquad (9)$$

We may view eq. (9) as a nonlocal Schrödinger equation for the wave functions $\psi_\kappa(z_1, q)$ and the corresponding excitation energies $\varepsilon_\kappa(q)$. The nonlocal integral operator incorporates – via the correlation function $\tilde{X}$ of the ground state – the collective properties of the excited modes that are induced by the interactions among the ^{4}He atoms. Recall that the operator $\varepsilon_0(q) + H_0$ corresponds to the energy of a single ^{4}He atom moving with parallel momentum $\hbar q$ in the spatially inhomogeneous background produced by the density profile $\varrho(z)$. At a finite temperature $T > 0$ the excitation energy $\varepsilon_\kappa(q)$ appearing in eq. (9) is to be replaced by the generalized expression[20] $\varepsilon_\kappa(q)\tanh[\beta\varepsilon_\kappa(q)/2]$, with $\beta = 1/k_B T$. The density profile $\varrho(z)$ and the non-nodal function $\tilde{X}$ in this case describe the vapor-liquid ^{4}He interface.

If $\tilde{X}$ satisfies the optimization condition (7), the eigenvalue equation (9) may be replaced by the Bogoliubov equation[31]

$$\{\varepsilon_0(q) + H_0(1)\}^2\psi_\kappa(z_1, q) + 2\int_{-\infty}^{\infty} \tilde{V}_{\text{ph}}(q, z_1, z_2)\{\varepsilon_0(q) + H_0(2)\}\psi_\kappa(z_2, q)\,dz_2 = \varepsilon_\kappa(q)^2\psi_\kappa(z_1, q).$$

$$(10)$$

III. RESULTS AND DISCUSSION

The renormalized Hartree equation (5) is solved at density $\varrho_L = 0.0175\,\text{Å}^{-3}$ using the parametrized Jastrow input function (4). The strength U_0 of the external potential (6) is $U_0 = e_B^{(F)} - \mu_F$. The parameter z_W is set to $z_W = 12.0\,\text{Å}$. The numerical calculations are performed within a finite box $[-L_1 \leq z \leq L_2]$ with $L_1 = L_2 = 24.0\,\text{Å}$, employing a numerical procedure that essentially eliminates finite-size effects. In Figure 1 we show the results for the density profile $\varrho(z)$, the renormalized Hartree potential $V_H(z)$, the total one-body potential $V_H(z) + U_{\text{ext}}(z)$, and the energy profile $e_F(z)$. The local density $\varrho(z)$ (solid curve) approaches its asymptotic value ϱ_L as $z \to -\infty$ and vanishes exponentially as $z \to +\infty$. The surface has a $90\% - 10\%$ width of approximately $6.0\,\text{Å}$. The total one-body potential $V_H(z) + U_{\text{ext}}(z)$ (long-dashed curve) approaches the chemical potential $\mu = -5.11\,\text{K}$ deep inside the liquid ($z \to -\infty$). It exhibits a rather shallow well (about $0.5\,\text{K}$ deep) in the center of the surface region before it rises again to vanish asymptotically as $z \to +\infty$. The asymptotic behavior of the energy function $e_F(z)$ (dot-dashed line) follows immediately from its construction. Deep inside the liquid ($z \to -\infty$), $e_F(z)$ approaches the binding energy $e_B^{(F)}$ per particle of the free homogeneous ^{4}He liquid at density ϱ_L. It displays a maximum in the central region of the surface layer, dropping down to its asymptotic value μ as $z \to +\infty$. Note that the chemical potential μ and the (free) binding energy $e_B^{(F)}$ coincide due to our choice of the strength U_0. Using these results for the density profile and the correlation function $\tilde{X}(q, z_1, z_2)$, the Feynman eigenvalue equation (9) is solved numerically, giving the complete spectrum of bulk as well as surface excitations.

Figure 2 displays our numerical results on the energies of the bound surface excitations as a function of wave number q for the ^{4}He system characterized by the bulk density $\varrho_L = 0.0175\,\text{Å}^{-3}$ and correlations generated by the pseudo-potential (4). At a given wave number q the spectrum of excited modes may contain one, two or three bound surface states, with excitation energies that constitute the discrete part of the spectrum at that wave number. The wave function $\psi_\kappa(z, q)$ of a surface mode vanishes in the asymptotic limits $z \to \pm\infty$ and

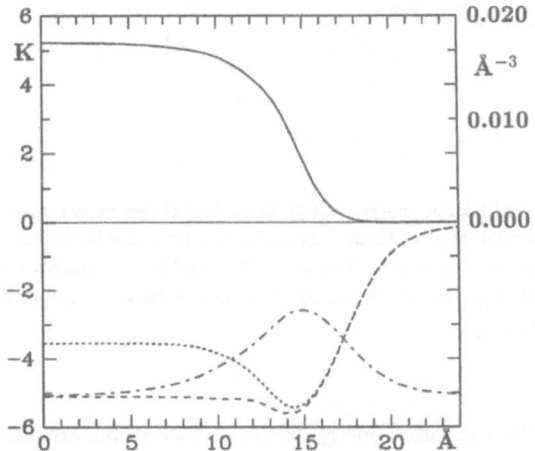

Figure 1. Density profile $\varrho(z)$ (solid curve), renormalized Hartree potential $V_H(z)$ (short-dashed curve), total effective one-body potential $V_H(z) + U_{\text{ext}}(z)$ (long-dashed curve), and energy profile $e_F(z)$ (dot-dashed curve) of liquid ^{4}He obtained by solving the renormalized Hartree equation (5) at given bulk liquid density $\varrho_L = 0.0175\,\text{Å}^{-3}$ using parametrized two-body correlations of form (4).

is essentially confined to the surface layer. The interpolated solid dots in Figure 2 indicate the excitation energies of the surface modes, while the other solid lines represent the separation energy $\varepsilon_0(q) + |\mu|$ and the Feynman dispersion relation $\varepsilon_L(q)$ of liquid ^{4}He at bulk density ϱ_L. The main features of the spectrum of elementary excitations of the surface of liquid ^{4}He are insensitive to the particular shape of the density profile and the particular form of the two-body correlations. To demonstrate this we have performed a similar calculation at a different bulk density, $\varrho_L = 0.0185\,\text{Å}^{-3}$. The calculated spectrum of bound states differs only marginally from the result shown in Figure 2, which corresponds to $\varrho_L = 0.0175\,\text{Å}^{-3}$. To provide a further check we go beyond the ansatz (4) for the pseudo-potential u and evaluate the *optimal* functions $\varrho(z)$, $u(\eta, z_1, z_2)$, $g(\eta, z_1, z_2)$, and the *optimal* excitation energies $\varepsilon_\kappa(q)$.

In this treatment we simultaneously solve the renormalized Hartree equation (5) and the PPA condition (7) in the presence of an external potential of the form (6) at the experimental saturation density $\varrho_L = 0.0218\,\text{Å}^{-3}$. The strength U_0 of potential (6) is determined by the condition $\mu = \mu_F + U_0 = -7.17\,\text{K}$ in order to match the chemical potential μ to the experimental saturation value. The parameter z_W is set to $z_W = 11.4\,\text{Å}$ and the box $[-L_1 \leq z \leq L_2]$ is determined by the lengths $L_1 = 30.8\,\text{Å}$ and $L_2 = 21.0\,\text{Å}$. The $90\% - 10\%$ surface thickness of the resulting density profile is approximately $5.6\,\text{Å}$. With this density profile and the corresponding optimized particle-hole interaction $\tilde{V}_{\text{ph}}$ as input functions, the renormalized Bogoliubov eigenvalue equation (10) is solved to determine the optimal bulk and surface excitation energies. (Recall that in this case the Bogoliubov equation (10) and the Feynman equation (9) are equivalent.[31])

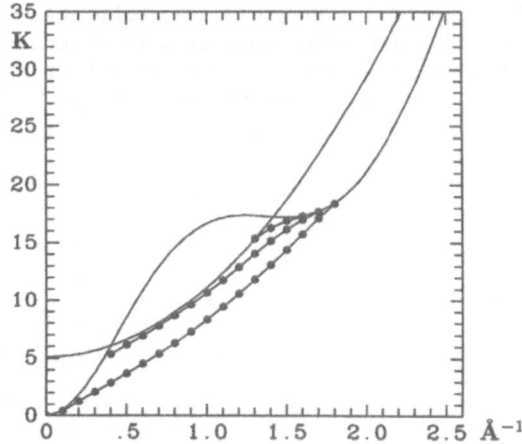

Figure 2. Excitation energy $\varepsilon_\kappa(q)$ of surface states versus wave number q (interpolated solid dots) of liquid ^{4}He calculated from the Feynman equation (9). Input data are the same as for Figure 1. The solid lines represent the separation energy $\varepsilon_0(q) + |\mu|$ and the Feynman dispersion relation $\varepsilon_L(q)$ of the bulk ^{4}He liquid at density ϱ_L.

The results for the excitation energy of the surface states are shown in Figure 3 (solid dots) as a function of wave number q. A comparison of the optimal results with those displayed in Figure 2 confirms our expectation that the general structure of the spectrum is rather insensitive to the detailed features of the surface and the spatial correlations. However, for a quantitative analysis of the dispersion properties of the energetically low-lying excitations it is imperative to employ the CBF optimization procedure. The optimal excitation energy $\varepsilon_L(q)$

of bulk liquid ^{4}He depends linearly on wave number q ($q \leq 0.3\,\text{Å}^{-1}$), obeying the dispersion law $\varepsilon_L(q) = \hbar c q$ of bulk phonons.[24] The lowest-lying branch of surface modes (Fig. 3) follows the hydrodynamic dispersion law

$$\varepsilon_1(q)^2 = \frac{\hbar^2}{m}\left[mc_3^2\,q^2 + \frac{\sigma}{\varrho_L}\,q^3\right] \tag{11}$$

in the range of wave numbers up to about $0.8\,\text{Å}^{-1}$. Thus the lowest-lying branch of surface modes rises linearly at small wave numbers (up to approximately $q = 0.2\,\text{Å}^{-1}$), following the dispersion law $\varepsilon_1(q) = \hbar c_3 q$. The surface waves at such small momenta and energies are driven mainly by the external force. At larger values ($0.2\,\text{Å}^{-1} \leq q \leq 0.8\,\text{Å}^{-1}$) of wave number q or in the absence of an external potential(and thus of surface phonons), one finds the characteristic dispersion properties of capillary waves, i.e., $\varepsilon_1(q) \sim q^{3/2}$. The hydrodynamic dispersion law (11) and its linear part are indicated in Fig. 3 by long-dashed lines. The numerical values for the speed c_3 and quantity σ are $c_3 = 73\,\text{m/sec}$ and $\sigma = 0.12\,\text{K\,Å}^{-2}$.

We may compare the present results on the spectrum of elementary excitations with results reported in Ref. 20 and those of Pitaevskii and Stringari[12]. Within our present analysis, the excitation energies of bound surface states at wave number q (q smaller than the bulk roton wave number $q_R \simeq 1.9\,\text{Å}^{-1}$) remain below the minimum bulk roton energy. This behavior is consistent with the ripplon-roton hybridization mechanism proposed in Ref. 12, which implies that the bulk roton energy is an upper bound for the energies of surface modes below the bulk roton momentum. In case of the optimal excitation spectrum (Fig. 3), the roton energy provides even a universal upper bound for the energies of bound surface states, which – at wave numbers and energies very close to the bulk roton – seem to merge smoothly with the

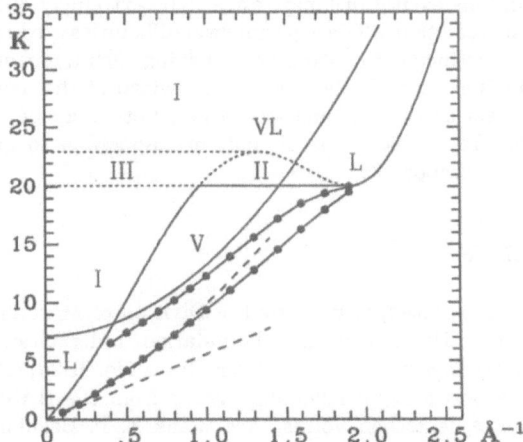

Figure 3. Energy spectrum ε of elementary excitations versus wave number q for liquid ^{4}He with a planar surface at bulk saturation density $\varrho_L = 0.0218\,\text{Å}^{-3}$. The excitation energies of bound surface modes form two discrete branches (solid dots, suitably interpolated) corresponding to wave functions $\psi_1(z,q)$ and $\psi_2(z,q)$ with no or one node in the surface region, respectively. The long-dashed straight line depicts the linear dispersion relation $\varepsilon_1(q) = \hbar c_3 q$; and the other long-dashed line, the hydrodynamic dispersion law (11). The solid line starting at a non-zero energy value at wave number $q = 0$ represents the separation energy $\varepsilon_0(q) + |\mu|$. The continuum modes are separated into excited vacuum (V) states, liquid (L) states, and vacuum-liquid (VL) states. The (plain) solid curves indicate the boundaries of the V, L, and VL domains. The continuum states may be non-degenerate (I), accidentally twofold degenerate (II), or threefold degenerate (III) (see Ref. 20.)

continuum of bulk excitations. The present results on the excitation energies of surface states at large values q, displayed in Figures 2 and 3, are similar to our earlier results reported in Ref. 20 (c.f. Fig. 9 therein). However, the earlier calculations gave a larger number of discrete branches of surface excitations. We ascribe this to the fact that the surface profile used in Ref. 20 has an extremely large 90% − 10% width of 9.4 Å, which is to be compared with the width of 5.6 Å characterizing the surface profile of the present study. The broader the surface, the more branches of surface modes appear in the spectrum of elementary excitations, a finding in concert with the intuitive expectation that a thick surface can accommodate more excited states than a thin one.

We conclude with the following observations. The spectrum of surface excitations at atomic wave lengths ($q \geq 1.0\,\text{Å}^{-1}$) has only recently become accessible to experiment − up to a wave number of approximately $1.5\,\text{Å}^{-1}$ − by neutron scattering from ^{4}He films adsorbed on a substrate.[3,4] The excitation energies of the liquid-vacuum surface of the ^{4}He films investigated in these experiments do not exceed the roton minimum, in agreement with our predictions and those of Ref. 12. Our results are, of course, only valid for a half-space system and cannot be directly compared with these measurements obtained from ^{4}He films. Also, backflow effects,[21,22] important at short wavelengths, are not taken into account within the Feynman picture adopted in the present treatment of elementary excitations. Inclusion of backflow effects is expected to improve considerably upon the results from the Feynman approximation, in the high parallel-momentum regime. In such a refined analysis, a proper choice of the value of the parameter α in the potential (6) could match the 90% − 10% thickness of the surface profile to the experimental value measured in very recent experiments.[1] As we have seen, the number of discrete branches of surface modes depends sensitively on the surface thickness. Future work should explore the detailed manner in which the discrete branches of surface excitation energies merge into the continuum of bulk excitations around the roton minimum. In this context, it may be noted that an experimental investigation of ^{4}He surface modes close to the continuum of bulk excitations may prove to be extremely difficult, enhancing the importance of theoretical predictions. The experimental difficulties are in part associated with the problem of generating an incident neutron beam having such a narrow energy bandwidth that only a surface state is excited. If the energy bandwidth of the neutron beam is larger than the gap between a surface and a bulk excitation, both may be excited, resulting in a superposition of states from which it is difficult or impossible to extract unambiguous information on the surface component.

ACKNOWLEDGMENTS

This work has been supported, in part, by the Division of Materials Research and the Physics Division of the U.S. National Science Foundation under Grant No. PHY90-02863 and by the Deutsche Forschungsgemeinschaft under Grant No. Ri267/20-1. K. A. Gernoth gratefully acknowledges a postdoctoral fellowship award from the BASF Aktiengesellschaft and the Studienstiftung des deutschen Volkes. We thank E. P. Bashkin, K. E. Kürten, E. Krotscheck, P. Leiderer, A. Polls, S. Stringari, L. Szybisz, and J. P. Toennies for stimulating discussions.

REFERENCES

1. L. B. Lurio, T. A. Rabedeau, P. S. Pershan, I. F. Silvera, M. Deutsch,
   S. D. Kosowsky, and B. M. Ocko, *Phys. Rev. Lett.* **68**, 2628 (1992).
2. D. V. Osborne, *J. Phys. Condens. Matter* **1**, 289 (1989).
3. H. J. Lauter, H. Godfrin, V. L. P. Frank, and P. Leiderer,
   Phys. Rev. Lett. **68**, 2484 (1992).
4. H. J. Lauter, H. Godfrin, and P. Leiderer, *J. Low Temp. Phys.* **87**, 425 (1992).
5. K. A. Pickar and K. R. Atkins, *Phys. Rev.* **178**, 399 (1969).

6. P. J. King and A. F. G. Wyatt, *Proc. Roy. Soc.* A **322**, 355 (1971).
7. S. Cunsolo and G. Jacucci, *in:* "Low Temperature Physics LT-13,"Vol. 1, Proc. 13th Int. Conf. on Low Temperature Physics, Boulder, 1972, K. D. Timmerhaus, W. J. O'Sullivan, and E. F. Hammel, eds., Plenum Press, New York (1973).
8. S. T. Boldarev and V. P. Peshkov, *Physica* **69**, 141 (1973).
9. F. Wagner, *J. Low Temp. Phys.* **13**, 317 (1973).
10. D. O. Edwards and W. F. Saam, *in:* "Progress in Low Temperature Physics," D. F. Brewer, ed., North-Holland, Amsterdam (1978).
11. G. Ji and M. Wortis, *Phys. Rev.* B **34**, 7704 (1986).
12. L. Pitaevskii and S. Stringari, *Phys. Rev.* B **45**, 13133 (1992).
13. C. C. Chang and M. Cohen, *Phys. Rev.* B **11**, 1059 (1975).
14. E. Krotscheck, *Phys. Rev.* B **31**, 4258 (1985).
15. J. L. Epstein and E. Krotscheck, *Phys. Rev.* B **37**, 1666 (1988).
16. L. Szybisz and M. L. Ristig, *Phys. Rev.* B **40**, 4391 (1989).
17. E. Krotscheck and C. J. Tymczak, *Phys. Rev.* B **45**, 217 (1992).
18. M. V. Rama Krischna and K. B. Whaley, *J. Chem. Phys.* **93**, 746 (1990).
19. S. A. Chin and E. Krotscheck, *Phys. Rev.* B **45**, 852 (1992).
20. K. A. Gernoth and M. L. Ristig, *Phys. Rev.* B **45**, 2969 (1992).
21. R. P. Feynman, *Phys. Rev.* **94**, 262 (1954).
22. R. P. Feynman and M. Cohen, *Phys. Rev.* **102**, 1189 (1956).
23. J. W. Clark and E. Feenberg, *Phys. Rev.* **113**, 388 (1959).
24. E. Feenberg, "Theory of Quantum Fluids," Academic Press, New York (1969).
25. S. Stringari and J. Treiner, *Phys. Rev.* B **36**, 8369 (1987).
26. A. Guirao, M. Centelles, M. Barranco, M. Pi, A. Polls, and X. Viñas, *J. Phys. Condens. Matter* **4**, 667 (1992).
27. C. L. Cleveland, U. Landman, and R. N. Barnett, *Phys. Rev.* B **39**, 117 (1989).
28. P. Sindzingre, M. L. Klein, and D. M. Ceperley, *Phys. Rev. Lett.* **63**, 1601 (1989).
29. M. Saarela, P. Pietiläinen, and A. Kallio, *Phys. Rev.* B **27**, 231 (1983).
30. E. Krotscheck, G.−X. Quian, and W. Kohn, *Phys. Rev.* B **31**, 4245 (1985).
31. E. Krotscheck, S. Stringari, and J. Treiner, *Phys. Rev.* B **35**, 4754 (1987).
32. D. Schiff and L. Verlet, *Phys. Rev.* **160**, 208 (1967).
33. T. Morita and K. Hiroike, *Prog. Theor. Phys.* **25**, 537 (1961).
34. L. S. Ornstein and F. Zernike, *Proc. Acad. Sci. Amsterdam* **17**, 793 (1914).

MAGNETIC AND PAIRING PROPERTIES OF

LIQUID ^3He: A DENSITY FUNCTIONAL APPROACH

E. S. Hernández[1], M. Barranco[2] and A. Polls[2]

[1] Departamento de Física, Facultad de Ciencias Exactas y
Naturales, Universidad de Buenos Aires, RA-1428, Argentina
[2] Departament d'Estructura i Constituents de la Matèria,
Facultat de Física, Universitat de Barcelona, E-08028 Spain

INTRODUCTION

Inspired by a well established success in nuclear physics, the use of density - dependent effective interactions or density functionals for liquid ^3He makes room to the possibility to obtain, in a selfconsistent fashion, the ground state and thermodynamic properties, as well as the spectrum of collective excitations[1-8]. In this context, we may enumerate excellent fits to the isotherms and coexistence line[1], surface tension and surface width[1-3], zero sound spectrum[4,5], paramagnon spectrum[4], density - dependent Landau parameters in the spin symmetric[4,6,7] and the spin antisymmetric[7] channels, and predictions of collective excitations in ^3He clusters[8].

In contrast to the viewpoint adopted by Weisgerber et al in refs.[3,5], that remain close to the spirit of the pseudopotential theory developed by Pines and coworkers[9], the above mentioned approaches[1,2,4,6-8,10] privilege the traditional Skyrme - like interactions, which have to be conveniently modified to take into account the nonlocality of the interaction in the spin symmetric channel[6,10] as well as the magnetic Landau parameters[7]. In spite of the fact that a zero range interaction is unable to properly reproduce the experimentally determined anomalous zero sound dispersion[11], it can be shown[4] that if one replaces the Skyrme - like energy functional by an appropriate finite - range density functional, one may compute the response function and sum rules in a rather simple manner. This density functional consists of a soft - core Lennard - Jones potential[12] plus the velocity terms of the Skyrme interaction with the appropriate density dependence. In addition, a coarse - graining of the density[13] is necessary to reproduce the observed plateau at transferred momentum $q \approx k_F$, (with k_F the Fermi momentum) in the zero sound dispersion curve[11].

On the other hand, the desirable features of a theoretical description of the atom - atom interaction in liquid ^3He are the ability to acceptably reproduce i) the ground - state, i. e., the saturation point at zero pressure, as well as the equation of state for finite temperatures up to the liquid - gas phase transition; ii) the spectrum of collective excitations in both the spin symmetric and spin - antisymmetric channels, respectively associated to density and to spin fluctuations; iii) the scalar and vector Landau parameters in either spin channel and iv) the scattering properties in the medium, in other

Condensed Matter Theories, Vol. 8, Edited by
L. Blum and F.B. Malik, Plenum Press, New York, 1993

words, the transport coefficients as functions of density or temperature, as well as the the critical curve for the normal - to - superfluid phase transition in the thermodynamic (P, T) plane. The pseudopotential theory[9] has been largely successful to provide reasonable fits to most of these magnitudes.

The good behavior of the Skyrme - plus - Lennard - Jones[4] (herafter, to be referred to as SLJ) interaction in the spin symmetric channel, together with the fact that it has been parameterized so as to fit the Landau coefficients at the saturation point, suggests the investigation of its properties in the spin antisymmetric channel and of its ability to describe particle - particle scattering in the medium. In this respect, we have verified through direct computations of the scattering amplitude between particles of opposite momentum, that one does not approximate the experimental data[14] for the critical temperatures $T_c(P)$ for the Ginzburg - Landau phase transition corresponding to the triplet $l = 1$ channel; furthermore, the effective interaction is strongly attractive in the $^1S_0$ channel, indicating the presence of BCS - like pairing. A careful analysis of the quantities entering the structure of the scattering amplitudes permits one to relate this feature to the behaviour of the SLJ functional in the spin - antisymmetric channel. Indeed, the SLJ functional does not reproduce the density dependence of the magnetic Landau coefficients f_0^a, f_1^a, and does not provide the appropriate q - dependence of the scattering amplitudes $t_0^a(q), t_1^a(q)$.

This drawback can be overcome in a simple manner, that respects the underlying philosophy of density - dependent effective interactions. In ref.[7] a generalization to the functional in[6,10] was proposed that includes, in the particle - particle interaction, a term depending upon the spin density. This is sufficient to give, with a proper choice of the full set of force parameters, the magnetic Landau coefficients in the whole range of densities, from saturation to fusion. In ref.[4] a similar spin- density interaction term was introduced and parameterized in order to fit the f_0^a and f_1^a Landau coefficients only at zero pressure. The purpose of this work is to find a generalization of the density functional SLJ that permits us to adjust the above Landau parameters in the full density range on the one hand, and on the other, to investigate the structure of the density functional that makes possible as well, to adjust the curve $T_c(P)$ of the second - order phase transition.

THE MAGNETIC SKYRME INTERACTION

Let us consider the following modified Skyrme - like interaction

$$V(\vec{r}) = (t_0 + t_0'\rho^\gamma + u_0\rho^\alpha \vec{s}.\vec{s} + v_0\rho^\beta \vec{J}.\vec{J})\delta(\vec{r})$$

$$+\frac{1}{2}(t_1 + t_1'\rho + u_1\rho^\alpha \vec{s}.\vec{s} + v_1\rho^\beta \vec{J}.\vec{J})[\overleftarrow{\nabla}^2 \delta(\vec{r}) + \delta(\vec{r})\overrightarrow{\nabla}^2]$$

$$+(t_2 + t_2'\rho + u_2\rho^\alpha \vec{s}.\vec{s} + v_2\rho^\beta \vec{J}.\vec{J})\overleftarrow{\nabla}.\delta(\vec{r})\overrightarrow{\nabla}, \tag{1}$$

where $\vec{s}(\vec{r})$ and $\vec{J}(\vec{r})$ are, respectively, the local spin and spin current densities. As in refs.[4,7], the spin dependent terms vanish in unpolarized systems; expression (1) generalizes the previous ones in that it contains a tensor field $\vec{J}$, whose inclusion enables us to perform separate fits on the Landau coefficients $f_0^a(\rho)$ and $f_1^a(\rho)$ (cf. Eqs.(4) below) without modifying the set of t- parameters of the SLJ functional.

From the total energy

$$E = \int d\vec{r}\, \mathcal{E}(\vec{r}), \tag{2}$$

where $\mathcal{E}(\vec{r})$ is the energy density functional that contains the particle - particle interaction (1), the effective quasiparticle - quasihole interaction $f_{\vec{k}_1 \vec{k}_2 \vec{k}_3 \vec{k}_4}$, with $\vec{k}_i$ a momentum label, is obtained by functional differentiation with respect to the density. A transformation to particle - hole momentum space and a multipole expansion of the spin symmetric and spin antisymmetric effective interaction

$$f_{\vec{p}\vec{p}'\vec{q}}^{s,a} = \sum_l f_l^{s,a}(q) \, P_l(\cos\theta_L), \tag{3}$$

where $\vec{p}$, $\vec{p}'$ lie on the Fermi surface, q is the transferred momentum, $\theta_L = \vec{p} \cdot \vec{p}'$ is the Landau angle and $P_l(x)$ the l-th Legendre polynomial, yields the interaction amplitudes or q-dependent fields

$$f_0^s(q) = \frac{1}{2}t_0 + \frac{1}{4}t_0'(\gamma + 1)(\gamma + 2)\rho^\gamma + \frac{5(t_1 + 3t_2) + 13(t_1' + 3t_2')\rho}{20}k_F^2$$
$$+ \frac{2(t_1 - 3t_2) + 3(t_1' - 3t_2')\rho}{16}q^2, \tag{4a}$$

$$f_1^s(q) = -\frac{t_1(\rho) + 3t_2(\rho)}{4}k_F^2, \tag{4b}$$

for the spin symmetric channel, and

$$f_0^a(q) = -\frac{1}{2}(t_0 + t_0'\rho^\gamma) - \frac{t_1(\rho) - t_2(\rho)}{4}k_F^2 - \frac{t_1(\rho) + t_2(\rho)}{8}q^2 + \frac{1}{2}u(\rho)k_F^2, \tag{4c}$$

$$f_1^a(q) = \frac{t_1(\rho) - t_2(\rho)}{4}k_F^2 + \frac{1}{2}v(\rho)k_F^2. \tag{4d}$$

In Eqs. (4) we have called

$$t_i(\rho) = t_i + t_i'\rho, \quad i = 1, 2 \tag{5}$$

$$u(\rho) = [u_0 + \frac{3}{10}(u_1 + 3u_2)k_F^2]\rho^{\alpha+2}, \tag{6}$$

and similarly for $v(\rho)$.

The Landau coefficients are the $q = 0$ specifications of Eqs. (4); our best adjustments of the experimental values[15] are displayed in Figs. 1 and 2.

Note that the SLJ functional yields fields $f_l(q)$ of the same appearance as in (4) with the replacements[4]

$$\frac{1}{2}t_0 \longrightarrow \int d\vec{r} \cos(\vec{q}.\vec{r})V(r), \tag{7}$$

where

$$V(r) = V_{LJ}(h_1)(\frac{r}{h_1})^4, \quad r \le h_1, \tag{8a}$$

$$V_{LJ}(r) = 4\epsilon[(\frac{\sigma}{r})^{12} - (\frac{\sigma}{r})^6], \quad r \ge h_1, \tag{8b}$$

and

$$\frac{1}{4}t_0'\rho^\gamma(\gamma + 1)(\gamma + 2) \longrightarrow \frac{1}{4}t_0'\rho^\gamma(\gamma + 1)[\gamma w(q) + 2]w(q), \tag{9a}$$

with the coarse - graining factor[4]

$$w(q) = 3\frac{j_1(qh_2)}{qh_2}. \tag{9b}$$

In these expressions, in addition to the standard Lennard - Jones parameters ϵ, σ one has the matching parameter h_1, to be determined under the condition that the volume integral of the potential in Eqs. (8) yields $t_0/2$, while h_2 is a free parameter that one may select in order to give the best adjustment to the anomalous zero sound dispersion[4].

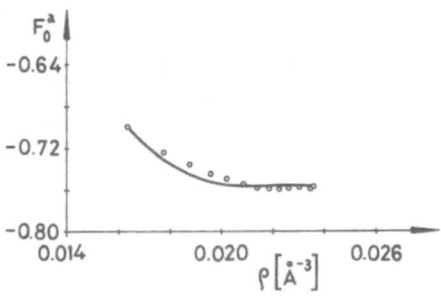

Figure 1. The spin - antisymmetric dimensionless Landau coefficient F_0^a for the best fit with the interaction in Eq. (1) as a function of the density. Experimental data are taken from ref.[15].

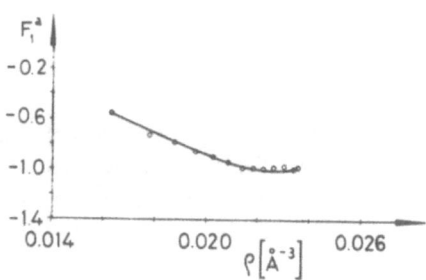

Figure 2. Same as Fig. 1 for the spin - antisymmetric dimensionless Landau coefficient F_1^a.

SCATTERING AMPLITUDES AND CRITICAL TEMPERATURE

In order to calculate the scattering amplitudes in the particle - particle channel, we first compute these quantities in the particle - hole channel according to Landau - Bethe - Salpeter equation[16] in the stationary limit (i. e., for zero frequency perturbations). In the Landau limit, they read

208

$$t_l^{s,a} = \frac{f_l^{s,a}}{1 + F_l^{s,a}/(2l+1)},$$ (10)

for every multipolarity l. As usual, $F_l = N(0)f_l$ is the dimensionless Landau parameter, $N(0)$ being the density of states at the Fermi level.

Secondly, we assume that the q- dependence obtained for the expansion amplitudes $f_l(q)$ in Eq. (3) propagates to the scattering amplitudes through expression (10) for every q in the range $0 \le q \le k_F$. This is a feature of the present model and is compatible with one of the major assumptions of pseudopotential theory[9], where scalar fields $f_0^{s,a}(q)$, responsible of the scattering, are adopted coincident with the corresponding Fourier transforms of conveniently designed spin symmetric and spin antisymmetric potentials $f^{s,a}(\vec{r})$. Our choice of the q- dependent fields consists of extending the above criterium to both the scalar and the vector contributions, with Eq. (1) yielding the effective particle-particle interaction. In addition, one has[17]

$$q = 2k_F sin(\frac{\phi}{2}).$$ (11)

We now evaluate the scattering amplitudes in the particle - particle channel for finite q according to the standard transformation formulae[16]

$$t_s(q) = t^s(q) - 3t^a(q),$$ (12a)

$$t_t(q) = t^s(q) + t^a(q),$$ (12b)

where, for scattering of particles with antiparallel momenta

$$t^{s,a}(q) = t_0^{s,a}(q) - t_1^{s,a}(q).$$ (13)

The pairing matrix elements V_0, V_1 in the $^1S_0$ and $^3P_1$ particle channels are finally obtained as

$$V_0 = \int_{-1}^{1} dx \frac{1+x}{2} t_s(x),$$ (14a)

$$V_1 = \int_{-1}^{1} dx\, x \frac{1+x}{2} t_t(x),$$ (14b)

where the prefactors of the particle - particle scattering amplitudes under the integral carry the proper symmetrization[14,17]. In this context, our scattering amplitudes include finite short wavelength effects beyond the $s - p$ approximation[14,18] due to the presence of q in their arguments.

Finally, the critical temperature for each multipolarity is computed as[14]

$$T_c = 1.13\, \delta_c\, T_F\, exp\left(-\frac{1}{\frac{1}{2}N(0)|V_l|}\right),$$ (15)

with T_F the density -dependent Fermi temperature and δ_c a scaling coefficient.

Our modified SLJ interaction consists of the generalized Skyrme - like force in Eq (1) with incorporated finite range effects as expressed in Eqs. (7) to (9). However, the corresponding density functional, has not proven capable to provide a proper fit to the density - dependent Landau parameters *and* to the critical temperature. A careful exploration of the force parameter space has shown that it is not possible to preserve the qualities of the SLJ density functional regarding the adjustment of the thermody-namical and spectral characteristics in the density channel, the ability of the spin and spin current density interactions to fit the spin antisymmetric Landau coefficients, and to yield as well the pairing phase transition as described by the experimental data[14], with a single parametrization.

The new possibility that we have considered is splitting the particle - particle density dependent interactions in such a way that the quasiparticle effective interaction exhibits different strengths in either particle - hole channel. In other words, we have looked for a modified SLJ functional that yields amplitudes $f_l^{s,a}(q)$ as given by Eqs. (4) with different force parameters t_i, t_i' ($i = 0, 1, 2$) for the symmetric and antisymmetric particle - hole channels. Such a functional may indeed originate in a Skyrme - like interaction of the form (1) with the specific structure

$$V(\vec{r}) = \frac{1}{2}\{[t_0^{(s)}(\rho) + 3t_0^{(a)}(\rho)]\delta(\vec{r}) + \frac{1}{2}[t_1^{(s)}(\rho) + 3t_1^{(a)}(\rho)](\overleftarrow{\nabla}^2\delta(\vec{r}) + \delta(\vec{r})\overrightarrow{\nabla}^2)$$

$$+3[t_2^{(s)}(\rho) - t_2^{(a)}(\rho)]\overleftarrow{\nabla}.\delta(\vec{r})\overrightarrow{\nabla}\}P_s$$

$$+\frac{1}{2}\{[t_0^{(s)}(\rho) - t_0^{(a)}(\rho)]\delta(\vec{r}) + \frac{1}{2}[t_1^{(s)}(\rho) - t_1^{(a)}(\rho)](\overleftarrow{\nabla}^2\delta(\vec{r}) + \delta(\vec{r})\overrightarrow{\nabla}^2)$$

$$+[3t_2^{(s)}(\rho) + t_2^{(a)}(\rho)]\overleftarrow{\nabla}.\delta(\vec{r})\overrightarrow{\nabla}\}P_t$$

$$+V_\sigma, \tag{16}$$

where P_s, P_t are, respectively, the projector operators upon the singlet and triplet par-ticle - particle channels and V_σ is the spin - plus - spin current density -dependent interaction explicitly displayed in (1).

This choice is supported by Bedell and Pines' observation[14] that the existence of triplet pairing in ^3He is related to the nonvanishing relative range $\delta = 1 - \frac{r_{\uparrow\downarrow}}{r_{\uparrow\uparrow}}$, being $r_{\uparrow\downarrow}, r_{\uparrow\uparrow}$ the ranges of the scalar pseudopotentials for antiparallel and parallel spins, respectively. With this in mind, we have freely explored different possibilities to substitute a zero - range force term by a finite - range one in the antisymmetric channel. Our conclusion is that the most convenient choice consists of the replacement,

$$t_1^{(a)}\delta(\vec{r}) \longrightarrow t_1^{(a)}exp(-\frac{r^2\sigma_1^2}{2}), \tag{17}$$

together with coarse graining of the densities accompanying the spin density and the spin current density in the interaction (1). In this case, the best adjustment to the crit-ical temperature data was obtained with Gaussian weigths for the density averaging[13], which amounts to multiplying the density powers $\rho^{\alpha+2}$ and $\rho^{\beta+2}$ in Eqs. (4c) and (4d) by $exp(-q^2/h_u^2)$ and $exp(-q^2/h_v^2)$, respectively.

In Figs. 3 - 5 we display the q - dependent fields $f_0^s(q), f_0^a(q)$ and $f_1^s(q)$, respec-tively, as functions of transferred momentum in units of the Fermi momentum. The vector field $f_1^s(q)$ is just a constant for each density, in the present model (cf. Eq. (4b)), and is not shown. We can appreciate that the general appearance of these fields is compatible with that obtained by Pines and coworkers in the frame of pseudopoten-tial theory[14]. It is also worthwhile recalling that the spin - symmetric fields enter the dispersion relation for the zero sound[4], which thus provides an indirect test of their adequacy to such purpose.

The major test for the density functional here proposed is given by the line of the second order phase transition in the (P, T) thermodynamic plane. In Fig. 6 we show our results, together with the available experimental data. It may be seen that our calculations reproduce both the saturation and the fusion points, and that overall deviations compensate on either side of the apparent inflection point. The rms deviation of the present predictions with respect to the experimental data amounts to 9%, which may be compared with a figure of approximately 15% that can be extracted from the

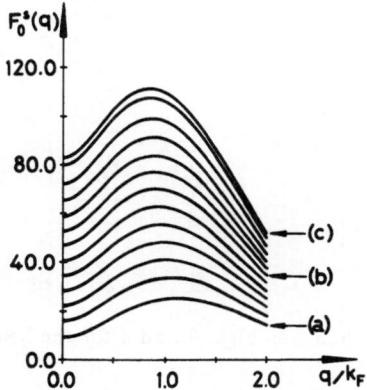

Figure 3. The field $F_0^s(q)$ as a function of transferred momentum (in units of the Fermi momentum) for different densities. The curves labelled (a), (b) and (c) respectively correspond to ρ_0, to 0.0214 $\mathring{A}^{-3}$ and to fusion density.

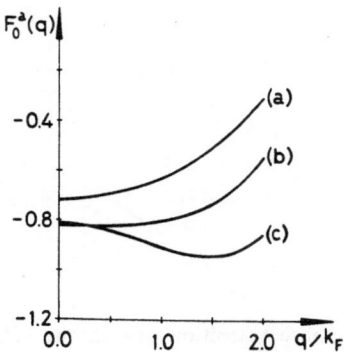

Figure 4. Same as Fig. 3 for the field $F_0^a(q)$.

calculations performed within the frame of pseudopotential theory[14], while for the $s-p$ approximation presented in ref.[14], the deviation is higher than 45%. Similar precisions have been found in the computation of the magnetic Landau coefficients, in spite of the fact that the price to be paid for the fit in Fig. 6 is an average 10% loss of accuracy in the adjustment of these coefficients, with respect to the results in Figs. 1 and 2. It

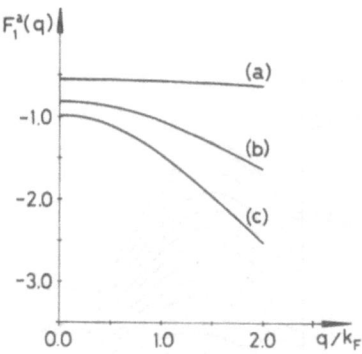

Figure 5. Same as Figs. 3 and 4 for the field $F_1^a(q)$.

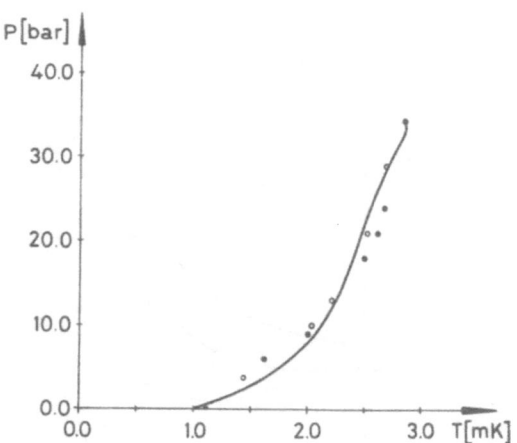

Figure 6. The second order phase transition line in the (P,T) thermodynamic plane as predicted by the SLJ functional modified by the interaction in Eq. (16). The coefficient δ_c in Eq. (15) takes the value 0.005. Experimental data are the same as in ref.[14].

is important to mention as well that with the present parameter set, we predict $^1S_0$ antipairing along the complete density interval.

SUMMARY

The present results indicate that energy density functionals originating in generalized Skyrme - type interactions of the form (16), whose major features are i) inclusion of both spin density - spin density and spin current density - spin current density couplings, ii) explicit inclusion of the finite range of the realistic atom - atom interaction as well as its long -scale behavior and iii) smoothing short scale inhomogeneities away by means of a coarse - grained averaging of the density, may be powerful tools to achieve satisfactory descriptions of most thermodynamical, spectral and scattering properties of liquid ^3He. In particular, we have shown that starting from a functional that has proven to behave satisfactorily in the spin symmetric channel[4], one may construct an extra interaction term of the Skyrme shape acting on the spin antisymmetric channel, with similar finite range characteristics as the former. The conjunction of both functionals gives rise to particle - particle scattering amplitudes capable of reproducing the Ginzburg - Landau instability in the (P, T) thermodynamic plane to an acceptable degree. The computation of transport coefficients remains a challenge to suministrate further support to the above conjecture.

ACKNOWLEDGEMENTS

We are pleased to acknowledge enlightening discussions with Jesús Navarro. One of us (E. S. H.) acknowledges warm hospitality at the University of Barcelona, where this research was initiated. This work was performed under partial finantial support from Consejo Nacional de Investigaciones Científicas y Técnicas of Argentina, grant PID 97/88 and from the DGICYT of Spain, grant PB89-0332.

REFERENCES

1. M. Barranco, M. Pi, A. Polls and X. Viñas, J. Low Temp. Phys. 80 (1990) 77.
2. S. Stringari and J. Treiner, Phys. Rev B36 (1987) 8369.
3. S. Weisgerber and P. G. Reinhard, Z Phys. D, (1992) in print.
4. C. García-Recio, J. Navarro, Nguyen Van Giai and L. L. Salcedo, Ann. Phys. 214 (1992) 293.
5. S. Weisgerber and P. G. Reinhard, Phys. Letters A158 (1991) 407.
6. S. Stringari, Phys. Letters 106A (1984) 267.
7. S. Weisgerber, P. G. Reinhard and C. Toepffer, *Spin Polarized Quantum Systems*, ed. S. Stringari, World Scientific, Singapore, 1988, p. 121.
8. Ll. Serra, J. Navarro and M. Barranco and Nguyen Van Giai, Phys. Rev. Lett. 67 (1991) 2311.
9. C. H. Aldrich III and D. Pines, J. Low Temp. Phys. 32 (1978) 689; D. Pines, Can. J. Phys. 65 (1987) 1357.
10. S. Stringari, Phys. Letters107A (1985) 36.
11. R. Scherm, K. Guckelsberger, B. Fåk, K. Sköld, A. J. Dianoux, H. Godfrin and W. G. Stirling, Phys. Rev. Lett. 59 (1987) 217.

12. J. Dupont-Roc, M. Himbert, N. Pavloff and J. Treiner, J. Low Temp. Phys. $\underline{81}$ (1990) 31.
13. P. Tarazona, Phys. Rev. $\underline{A31}$ (1985) 2672.
14. K. Bedell and D. Pines, Phys. Letters $\underline{78A}$ (1980) 281.
15. S. Greywall, Phys. Rev. $\underline{B27}$ (1983) 2747.
16. G. Baym y C. Pethick, *Landau Fermi Liquid Theory*, Wiley, New York, 1991.
17. K. Bedell and D. Pines, Phys. Rev. Lett. $\underline{45}$ (1980) 39.
18. K. S. Dy and C. J. Pethick, Phys. Rev. $\underline{185}$ (1969) 373.

QUANTUM THERMODYNAMIC PERTURBATION THEORY FOR FERMIONS

M.A. Solís[†], R. Guardiola and M. de Llano[‡]

Departamento de Física Atómica y Nuclear, Universidad de Valencia
Avda. Dr. Moliner 50, E–46100 Burjassot, Valencia, Spain

†On leave from Instituto de Física, U.N.A.M.,
Apartado Postal 20-364, 01000 México D.F.
‡On leave from Universidad Autónoma de Madrid, Spain
and from North Dakota State University, Fargo, ND 58105 USA

Abstract

The quantum version of classical thermodynamic perturbation theory is applied to the ground state of a fluid of spin-1/2 fermions interacting via the Aziz interatomic potential, as a model for liquid ^{3}He. Results from the rapidly-convergent sixth-order calculation about the unperturbed hard-sphere fluid for energy, density and sound velocity at the zero-pressure liquid equilibrium point, lie within a few percent of computer-simulation values and appreciably closer than the most elaborate recent variational calculation. The procedure explicitly avoids crossing phase boundaries and is relatively insensitive to varying the close-packing density up to a value somewhat below the maximum possible (*primitive hexagonal* packing) value. With the aid of a rigorous energy bound principle, and as also occurs with extrapolations of low-density computer-simulation energies, this suggests the existence of a more stable high density phase (presumably crystalline), barring a breakdown at very high densities of Padé-like extrapolants based on only four series coefficients or on only four computer-simulation datapoints.

Quantum thermodynamic perturbation theory (QTPT) is a scheme of successive approximations about an unperturbed N-fermion fluid state which is *not* the ideal Fermi gas as in the original [1] field-theoretic perturbation expansion, but rather a nontrivial fluid of repulsive fermions. The starting point is the low-density series for the ground-state (GS) energy-per-particle, E/N. The density is given by $\rho = N/\Omega = k_F^3/3\pi^2$, with N the number of particles, Ω the volume and k_F the Fermi sphere radius. The low density expansion is rigorously derived [2] with Feynman diagrammatic techniques from the above mentioned original perturbation expansion. This series takes the form [3]

$$
\begin{aligned}
E/N \;=\; & \frac{3\hbar^2 k_F^2}{10m}\Big\{1 + C_1(k_F a) + C_2(k_F a)^2 \\
& + \; [C_3 R_0/2a + C_4 A_1(0)/a^3 + C_5](k_F a)^3 + C_6(k_F a)^4 ln(k_F a) \\
& + \; [C_7 R_0/2a + C_8 A_0''(0)/a^3 + C_9](k_F a)^4 + o[(k_F a)^4]\Big\},
\end{aligned}
\tag{1}
$$

where the leading term is the (pure kinetic) energy of the ideal Fermi gas. Here, N particles of mass m interact pairwise via an arbitrary central potential $V(r)$, with r the relative coordinate of the pair, and manisfests itself solely [2] through the first few effective-range theory [4] two-body scattering parameters

$$
\begin{aligned}
a \equiv A_0(0) \;=\; & \int_0^\infty dr\, r V(r) u_0(r), \\
R_0 \;=\; & (2/a_0^2) \int_0^\infty dr\, [(r-a)^2 - u_0^2(r)], \\
A_1(0) \;=\; & (1/3) \int_0^\infty dr\, r^2 V(r) u_1(r), \\
A_0''(0) \;=\; & -(1/3) \int_0^\infty dr\, r^3 V(r) u_0(r),
\end{aligned}
\tag{2}
$$

where $u_0(r)$ and $u_1(r)$ are respectively the S-wave and P-wave reduced radial scattering wave functions [5]. Of the four parameters in Eq. (2) only $A_0''(0)$ is potential-shape dependent, as the other three parameter are related to scattering phase shifts alone. Thus, the $(k_F a)^4$ term in Eq. (1) can thus be interpreted as a three-body cluster contribution. The dimensionless coefficients C_i $(i = 1, 2, ..., 9)$ are given in Ref.[3] for both two- and four-species fermions, the former being the case for (spin-one-half) ^{3}He fermions to be treated here and for which $C_6 \equiv 0$. Hence, for two-species fermions we have a simple power series in k_F, at least through the order given in Eq. (1).

A realistic but otherwise arbitrary pair potential $V(r)$ will consist of some very repulsive short-ranged portion $V_{rep}(r)$, surrounded by a relatively weak, longer-ranged part $V_{att}(r)$ falling off to zero as r^{-6}. Introducing the dimensionless coupling parameter $0 \le \lambda \le 1$ of thermodynamic perturbation theory [6], we can write

$$
V(r) = V_{rep}(r) + \lambda V_{att}(r).
\tag{3}
$$

Ref.[7] summarizes and evaluates the various ways of decomposing $V(r)$ according to (3); we shall restrict ourselves here to the more common decompositions of Barker and Henderson (BH) and of Weeks, Chandler and Andersen (WCA). Substituting (3) into (2) one can expand (2) as power series in λ, namely

$$
\begin{aligned}
a \;=\; & a_0(1 + a_1\lambda + a_2\lambda^2 + \ldots), \\
R_0 \;=\; & r_0(1 + r_1\lambda + r_2\lambda^2 + \ldots), \\
A_1(0) \;=\; & t_0(1 + t_1\lambda + t_2\lambda^2 + \ldots), \\
A_0''(0) \;=\; & p_0(1 + p_1\lambda + p_2\lambda^2 + \ldots),
\end{aligned}
\tag{4}
$$

where the leading coefficients a_0, r_0, t_0 and p_0 pertain to the potential V_{rep} alone. The a_i, r_i, t_i and p_i $(i = 0, 1, 2, ...)$ coefficients have been determined [5] for several potential decompositions (3) including BH and WCA. Inserting (4) into (1) and defining $x \equiv k_F a_0$ leads to the fundamental QTPT expression for the GS energy-per-fermion

$$E/N \simeq \frac{3\hbar^2 x^2}{10ma_0^2} \Big[e_0(x) + \sum_{i=1}^{4} \sum_{j=1}^{\infty} f_{ij} x^i \lambda^j \Big], \tag{5}$$

where the dimensionless coefficients f_{ij} depend algebraically upon C_j $(j = 1, 2, ..., 9)$, and a_i, r_i, t_i, p_i $(i = 1, 2, ...)$ and are obtainable [8] via computer algebra, e.g., with the software package MACSYMA [9].

The $V_{rep}(r)$-interacting N-sphere fluid unperturbed state is described by the low-density expansion

$$\begin{aligned}
e_0(x) &= 1 + C_1 x + C_2 x^2 + [C_3 r_0/(2a_0) + C_4 t_0/a_0^3 + C_5] x^3 \\
&\quad + [C_7 r_0/(2a_0) + C_8 p_0/a_0^3 + C_9] x^4 \\
&\equiv 1 + D_1 x + D_2 x^2 + D_3 x^3 + D_4 x^4.
\end{aligned} \tag{6}$$

Instead of Eq. (6) it will be more practical in extracting the high-density behavior to deal with the expression

$$\begin{aligned}
e_0(x)^{-1/2} &= \Big(1 + D_1 x + D_2 x^2 + D_3 x^3 + D_4 x^4 \Big)^{-1/2} \\
&= 1 + F_1 x + F_2 x^2 + F_3 x^3 + F_4 x^4 + O(x^5),
\end{aligned} \tag{7}$$

where the F_i's depend algebraically on the D_i's in a simple manner. Likewise, the double summation in (5) will prove more convenient to write in the form

$$\sum_{i=1}^{4} \sum_{j=1}^{\infty} f_{ij} x^i \lambda^j = \sum_{j=1}^{\infty} f_{1j} x e_j(x) \lambda^j, \tag{8}$$

where

$$\begin{aligned}
e_j(x) &\equiv 1 + k_{1j} x + k_{2j} x^2 + k_{3j} x^3 + k_{4j} x^4, \\
k_{ij} &\equiv f_{i+1}/f_{1j} \quad (i = 1, 2, 3),
\end{aligned} \tag{9}$$

with the k_{4j} coefficients remaining unknown. Implementation of QTPT, to derive GS equations of state (EOS) for a given pair potential $V(r)$, can be reduced to two exercises in Padé approximants [10]: a) one of finding appropiate high (i.e., physical)–density extrapolants, say $\epsilon_0(x)$ and $\epsilon_j(x)$, respectively, for the low-density series $e_0(x)$ and $e_j(x)$ $(j = 1, 2, ...)$; and b) the other one of using these $\epsilon_0(x)$ and $\epsilon_j(x)$ in (5) and (8), instead of $e_0(x)$ and $e_j(x)$, respectively, and then extrapolating, if necessary, the resulting power series in λ from $\lambda = 0$ to the physical value $\lambda = 1$. Consequently we would then have

$$\begin{aligned}
E/N &= \frac{3\hbar^2 x^2}{10ma_0^2} \Big[\epsilon_0(x) + \sum_{j=1}^{\infty} f_{1j} x \epsilon_j(x) \lambda^j \Big] \\
&\equiv E_0 + \lambda E_1 + \lambda^2 E_2 + \dots,
\end{aligned} \tag{10}$$

217

which is precisely the QTPT scheme. Such a *two-path* procedure can be shown [2] to avoid crossing phase boundaries in the $T = 0$ gas-liquid $\lambda - \rho$ plane, and thus avoid uncertainties in converging to the wrong state.

In Ref.[11] moderately good results were obtained, when compared with computer simulations, for the GS energy-per-particle, equilibrium (zero pressure) density and sound velocity for a liquid of two-species fermions interacting with the Aziz interaction [12]. This potential has a *soft* core, which will be convenient to take as an effective *hard* core in order to fix high-density behavior better. The results [11] relied on the hypothesis that: a) the close-packing density for identical hard spheres of diameter a_0 of *any* statistics should be the same, since close-packing means perfect localization of all the system particles, which in turn implies *distinguishability*; and b) consequently, that the close-packing density $\rho_P = 0.371\rho_0$ of hard-sphere bosons extracted [13] from GFMC boson hard-sphere [14] data means that two-species *fermion* hard spheres should possess a second-order, uncertainty-principle pole at $x \equiv k_F a_0 = x_P$ given by

$$x_P = (3\pi^2 \rho_P)^{1/3} a_0 = \left[(3\pi^2\sqrt{2})(0.371) \right]^{1/3} \simeq 2.5, \tag{11}$$

if $\rho_0 \equiv \sqrt{2}/a_0^3$ (maximum, primitive hexagonal, geometric packing density) is used. Since the value (10) is somewhat less than the maximum possible x, namely $x_0 = (3\pi^2 \rho_0)^3 a_0 = (3\sqrt{2}\pi^2)^{1/3} \simeq 3.47$ (which is expected on physical grounds), we have lifted requirement (b) above in an attempt to improve upon the results of Ref.[11] as well as to extract a fermion hard-sphere EOS that may be compared against eventual computer simulations for this all-important many-particle system. The results of this effort will be reported here.

The extrapolants $\epsilon_0(x)$, $\epsilon_1(x)$, ... $\epsilon_6(x)$ found in Ref.[11] Figs. 3 and 4, were respectively $[0/4]^{-2}(x)$, $[2//2](x)$, $[4//0](x)$, $[2//2](x)$, $[0/3](x)$, $[0/3](x)$ and $[0/3](x)$. Here, $[L/M](x)$ is an ordinary Padé approximant of order $L + M$ to a power series of that same order, while $[L//M](x)$ is a *two*-point Pade approximat reproducing the series for small enough x through order $L + M$, *and* taking on a certain given value (see below) at $x = x_P$. Note that $\epsilon_j(x)$, $(j = 1, 2, 3)$ are all fourth order approximants, implying that the *fourth* (unknown) coefficients k_{4j}, $(j = 1, 2, 3)$ in (9) have been separately fixed. This was carried out in Ref.[11] in accordance with the Stell-Penrose [15] theorem stating that *classical* TPT for a fluid of hard-spheres surrounded by some attractive pair potential is exact at the close-packing density in 2D (and conjectured to be so also in 3D), which implies that $\epsilon_1(x_P) = $ *nonzero constant*, while $\epsilon_j(x_P) \equiv 0$ for $j \geq 2$. This latter condition was found to be well satisfied naturally only for $j \geq 4$ *without* the need to assume definite values for the corresponding k_{4j}'s, if an ordinary Padé approximant $[0/3](x)$ is constructed in each case to $e_j(x) = 1 + k_{1j}x + k_{2j}x^2 + k_{3j}x^3$.

Table 1 lists the energy-per-particle, at a specific moderate value of the density, of several Padé extrapolants to (7) and compares their value with the computer simulation kinetic energy [16]. The comparison between the repulsive-sphere *total* energy-per-particle E_0 and the pure kinetic energy of the fully-interacting fluid is not perfect and

Table 1. Padé energy-per-particle E_0 (in K) of soft-sphere fluid corresponding to the Aziz-BH potential, compared to kinetic energy per particle of 54 fermions interacting via the same potential, as determined by the mirror potential (MP) GFMC computer simulation reported in Ref. [16]. Here, $e_0^{-1/2}(x)$ is the 4th order polynomial (7), $[M/N]$ are ordinary Padé approximants to this, and the two-point approximants $[M//N](x)$, $M + N = 5$, are built to possess a second-order pole at $x = k_F a_0 = x_P = 3.13$, with $a_0 = 2.11166 \text{Å}$ being the S-wave scattering length of the soft-sphere repulsive pair potential $V_{rep}(r)$. All data are for $\rho\sigma^3 = 0.273$, with $\sigma = 2.556$ Å.

Padé to $e_0^{-1/2}(x)$	$E_0\,[K]$
[0/4]	11.47
[1/3]	12.76
[4//1]	21.13
[1//4]	7.29
[2//3]	17.80
[3//2]	11.47
GFMC-MP$< T >$	12.19 to 12.48

Table 2. QTPT, variational and computer-simulation energy-per-particle (in K), and sound velocity c (in m/s) at the (zero-pressure) saturation density ρ_s for Aziz fermions, all compared with the experimental results [22] for liquid ^{3}He. The length parameter $\sigma = 2.556\text{Å}$ is used to express density in dimensionless units. QTPT predictions refer to the $[6/0](\lambda)$ Padé approximant to the λ-series Eq.(10) with the soft-sphere fluid $e_0^{-1/2}(x)$ series (7) extrapolated by the $[3//2](x)$ two-point Padé approximant bearing a second-order pole at $x \equiv k_F a_0 = x_P = 2.5, 3.11, 3.12$ and 3.13, with $a_0 = 2.11166\text{Å}$ being the Aziz-BH soft-sphere potential S-wave scattering length. Column marked *Adj. min. GFMC* refers to values obtained by interpolation of data reported in Ref.[16].

		QTPT $[6/0](\lambda = 1)$			Variational	Adj. min.	Expt
		$x_P = 2.5$	3.11	3.12	3.13	Ref.[21]	GFMC	Ref.[22]
E/N [K]	(BH)	-3.17	-2.19	-2.26	-2.32	-2.47	-2.38±0.02	-2.47
	(WCA)	-2.33	-1.83	-1.88	-1.94			
$\rho_s\sigma^3$		0.326	0.266	0.270	0.274	0.277	0.267	0.267
		0.284	0.237	0.241	0.245			
c [m/s]		173	170	172	174	185	162±3	183
		151	156	158	160			

meant only as a rough guide to select the appropriate extrapolant. The ordinary approximant $[0/4](x)$ to (7) was employed in Ref.[11] and, together with $[1/3](x)$, brackets [3] from above and below the Jastrow Monte Carlo results [17] for the Bethe homework purely repulsive Yukawa potential. However, neither of these two extrapolants contain the required uncertainty-principle pole in E_0 at *any* density. Such a pole, determined to lie at $x_P = 3.13$, can be built in with the four remaining *two*-point Padé approximants in the Table. (Values of x_P larger than 3.13 gave worse agreement with the liquid en-

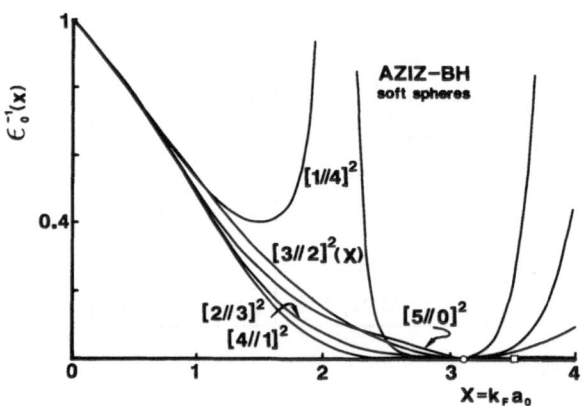

Figure 1. Different two-point Padé extrapolants to the square of low-density Aziz-BH soft-sphere fluid expression (7), as explained in text.

ergy, density, sound-velocity GFMC parameters). The $[1//4]$ approximant to $e_0^{-1/2}(x)$ has a pole around $x \simeq 2$ (see Figure 1), i.e., in the physical interval $0 \le x \le x_P = 3.47$, and so is discarded. Of the surviving approximants to $e_0^{-1/2}(x)$, namely $[4//1]$, $[2//3]$ and $[3//2]$, the first two give too large an energy E_0 compared to the GFMC kinetic energy and are also discarded. Only $[3//2]$ survives this analysis and will be used as the extrapolant for the repulsive-sphere fluid EOS. Figure 1 displays the behavior of the various two-point extrapolants just cited, the open circle on the x-axis marking the value $x_P = 3.13$, and the open square the value $x_0 = 3.47$ designating the ultimate density $\rho_0 \equiv \sqrt{2}/a_0^3$.

The $[3//2](x)$ approximant to (7) just adopted is further seen in Figure 2 to lie everywere rather close to the London low- and high-density interpolation formula [18], generalized to fermions [19], if the corresponding close-packing density is set at

$x_P = 3.13$. Also shown in the Figure is a rigorous *lower bound* to the GS $\epsilon_0^{-1/2}(x)$, obtained by using as trial wave function [20] a Slater determinant of mutually non-overlapping, localized single-particle orbitals. Employing the results of Ref.[20] the lower bound translates into

$$\epsilon_0^{-1/2} \geq \sqrt{3/10}\,(3/2\pi)^{1/3}(1 - x/x_0), \tag{12}$$

the pre-factor being $\simeq 0.428$, and x_0 defined as before just below (11). That this lower bound to the GS is violated by our $[3//2](x)$ approximant for x approaching x_0, as

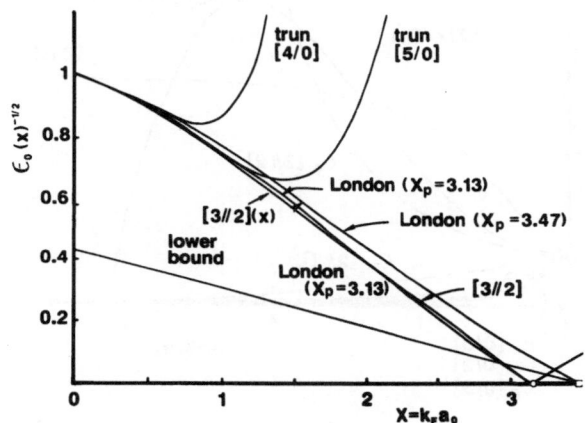

Figure 2. Same as Figure 1, but for some one- and two-point Padé approximants, for the London interpolant formula [19], ant the rigorous lower bound (12), as discussed in text.

can also be shown to happen for the boson hard-sphere fluid GFMC-selected Padé-like extrapolant [13], *could* mean that low-density-based extrapolants *may* extend naturally into a metastable (non GS) glassy-like branch at high density. Indeed, to obtain the presumably GS high-density *crystalline* phase in GFMC for boson hard spheres entirely different starting configurations were needed [14] in the simulations. However, a more plausible explanation is simply that such extrapolations break down at *some* density before reaching the geometric close-packing value.

Using the value $x_P = 3.13$ just cited, as well as the same technique to estimate the value of $\epsilon_1(x_P)$ given in Ref.[11], all approximants $\epsilon_0, \epsilon_1, ..., \epsilon_6$ to (6) and (9) through 6th order in (10) were constructed and are displayed in Fig. 3. Inserting these density extrapolants in (10), Table 2 gives the QTPT 6th order results (for both BH and WCA decompositions of the Aziz potential) for the minimum energy-per-particle E/N, equilibrium (or saturation) density ρ_s, and associated sound velocity c. These are given for $x_P = 3.13$ as well as for 3.12, 3.11 and 2.5, the last value corresponding to the results of Ref.[11]. Also listed in the Table are the variational results of Ref.[21], results based on the raw GFMC data reported in Ref.[16], and the experimental data [22] for liquid ^{3}He. The numbers labeled *adjusted minimum GFMC* are the results of a third-order

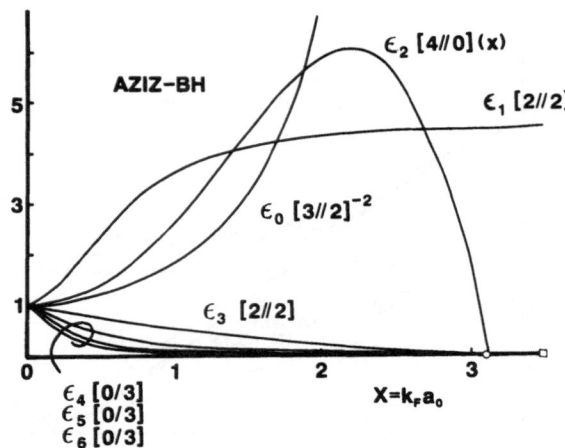

Figure 3: Behavior of extrapolants $\epsilon_j(x)$ for $j = 1, 2, ..., 6$ to (9), selected by criteria as discussed in text.

polynomial fit carried out on the raw GFMC data [16].

Figure 4 displays this *adjusted minimum GFMC* energy-density point with its uncertainty, the experimental EOS curve, and the QTPT results from several 6th order Padé approximants to the λ-series (10). The trivial approximant $[0/6](\lambda)$ is zeroless for all densities and so cannot give a negative energy, while the $[1/5](\lambda)$ is off scale. These approximants are all seen to be within about 0.01 K of each other, which is already a small fraction of the adjusted minimum GFMC uncertainty. Thus, the λ-series (10) is well-behaved and converges rapidly. In other words, the QTPT parameter λ is indeed a good *smallness* parameter (and the value $\lambda = 1$ is in fact *small*) for the description of the fermion liquid GS. Similar results were previously found [23] for the boson liquid of ^{4}He atoms interacting via the same Aziz potential.

Table 3. GS energy-per-particle for the two-species repulsive-sphere fluid, given in both dimensionless form, $mc^2E/N\hbar^2$, and in degrees K, E/N, for several relevant density values. The density is expressed in three different forms, as indicated. The two sets of energy-per-particle shown correspond with the second-order, uncertainty principle pole in the energy falling at $\rho/\rho_0 = 0.734$ and 1, respectively, with the former value being the one best fitting the liquid-^{3}He QTPT comparison with GFMC data.

Density ρ			$x_P = 3.13$ $(\rho/\rho_0 = 0.734)$		$x_P = 3.47$ $(\rho/\rho_0 = 1)$	
$x = k_F c^*$ $(\rho = k_F^3/3\pi^2)$	ρ/ρ_0 $(\rho_0 = \sqrt{2}/c^3)$	$\rho\sigma^3$ $(\sigma = 2.556\,\text{Å})$	$mc^2E/N\hbar^2$	E/N [K]	$mc^2E/N\hbar^2$	E/N [K]
0	0	0	0	0	0	0
0.25	0.0004	0.0009	0.021	0.075	0.021	0.075
0.50	0.003	0.007	0.095	0.342	0.95	0.342
0.75	0.010	0.025	0.254	0.916	0.251	0.906
1.00	0.020	0.060	0.556	2.01	0.541	1.95
1.25	0.047	0.117	1.11	4.00	1.05	3.78
1.49	0.080	0.200	2.09	7.55	1.90	6.85
1.58	0.094	0.237	2.62	9.44	2.33	8.41
1.66	0.109	0.273	3.18	11.5	2.79	10.0
1.67	0.110	0.277	3.25	11.75	2.84	10.24
1.71	0.120	0.300	3.65	13.2	3.15	11.4
1.76	0.130	0.327	4.15	15.0	3.54	12.8
1.85	0.150	0.377	5.19	18.7	4.31	15.6
2.00	1.91	0.479	7.88	28.4	6.19	22.3
2.25	2.72	0.682	16.5	59.4	11.4	41.2
2.50	0.373	0.936	39.8	143	22.3	80.6
2.75	0.497	1.245	133	479	49.1	177
3.0	0.645	1.617	1353	4880	137	494
3.47	1	–	–	–	∞	∞

* With $c = a_0 = 2.11166\,\text{Å}$ (Aziz-BH)

Finally, Table 3 gives *two* two-species fermion hard sphere EOS's based upon the approximant $[3//2](x)$ to (7), one with the second-order, uncertainty-principle energy pole at the value $x_P = 3.13$ ($\rho/\rho_0 = 0.734$) as arrived at in this paper based on Aziz fermion GFMC data [16], and one based on the maximum (geometric) close-packing value of $x_0 = 3.47$ (implying $\rho/\rho_0 = 1$). For convenience, the energy per particle is expressed in degrees kelvin (K) and in dimensionless units. Likewise, the density is expressed in three different ways: as $x \equiv k_F c$ (with $c \equiv a_0$), as ρ/ρ_0 with $\rho_0 \equiv \sqrt{2}/c^3$, and as $\rho\sigma^3$ (with $\sigma \equiv 2.556$ Å). It will be interesting to compare the results in this table with GFMC results for the two-species hard-spheres fluid in a not too distant future when such valuable benchmark numbers might hopefully be made available.

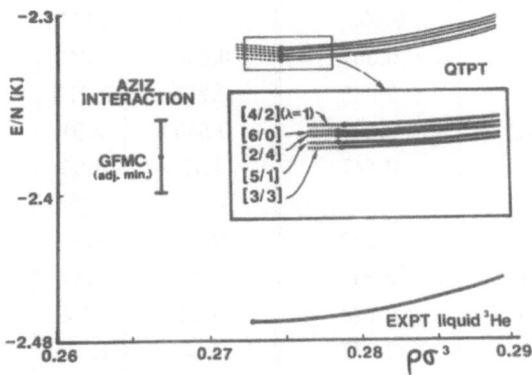

Figure 4. QTPT, GFMC and experimental ground-state energy for ³He modeled as a fluid of two-species fermions interacting with the Aziz-BH potential, as explained in text.

Acknowledgments

R. G. acknowledges support from DGICyT (*Dirección General de Investigación Científica y Técnica*) (Spain) under contract PB 90/417. M. de Ll. and R. G. acknowledge NATO (*North Atlantic Treaty Organization*) (Belgium) for support. M.A.S. thanks DGICyT (Spain) and DGAPA (*Dirección General de Asuntos del Personal Académico*), Universidad Nacional Autónoma de México (México) for financial support. M. de Ll. thanks the U.S. Army Research Office for a travel grant.

References

[1] A.L. Fetter and J.D. Walecka, *Quantum Theory of Many–Particle Systems* (Mc-Graw Hill, New York, 1971); D.J. Thouless, *The Quantum Mechanics of Many-Body Systems* (Academic, New York, 1972).

[2] G.A. Baker, Jr., *Rev. Mod. Phys.* **43**, 479 (1971).

[3] G.A. Baker Jr., L.P. Benofy, M. Fortes, M. de Llano, S.M. Peltier and A. Plastino, *Phys. Rev.* **A**, 3575 (1982).

[4] P. Roman, *Advanced Quantum Theory* (Addison-Wesley, Reading, MA, 1965).

[5] E. Buendía, R. Guardiola and M. de Llano, At. Data Nucl. Data Tables **42**, 293 (1989).

[6] D.A. McQuarrie and J.L. Katz, *J. Chem. Phys.* **44**, 2393 (1966); J.A. Barker and D. Henderson, *Rev. Mod. Phys.* **48**, 597 (1976); H.C. Andersen, D. Chandler, and J.D. Weeks, *Adv. Chem. Phys.* **34**, 105 (1976).

[7] D. Chandler, J.D. Weeks and H.C. Andersen, *Science* **220**, 787 (1983).

[8] V.C. Aguilera-Navarro, R. Guardiola, C. Keller, M. de Llano and M. Popovic, *Phys. Rev.* **A 35**, 3901 (1987).

[9] MACSYMA, 1976, 1984, Massachusetts Institute of Technology. Enhancements, 1984, Symbolics, Inc.

[10] G.A. Baker, Jr. and P. Graves-Morris, in *Encyclopedia of Mathematics and its Applications*, edited by G.C. Rota (Addison-Wesley, Reading, MA, 1981) Vols. **13** and **14**.

[11] Z. Hu, M. de Llano, E. Buendía, and R. Guardiola, *Phys. Rev.* **B 43**, 12827 (1991).

[12] R.A. Aziz, V.P.S. Nain, J.S. Carley, W.L. Taylor, and G.T. Mc Conville, *J. Chem. Phys.* **70**, 4330 (1979).

[13] V.C. Aguilera-Navarro, S. Ho and M. de Llano, *Phys. Rev.* **A 36**, 5742 (1987).

[14] M.H. Kalos, D. Levesque and L. Verlet, *Phys. Rev.* **A 9**, 2178 (1974).

[15] G. Stell and O. Penrose, *Phys. Rev. Lett.* **51**, 1397 (1983).

[16] R.M. Panoff, in *Recent Progress Many-Body Theories*, Vol. **2**, edited by Y. Avishai (Plenum, New York, 1990); R.M. Panoff and J. Carlson, *Phys. Rev. Lett.* **62**, 1130 (1989); R.M. Panoff (private communication).

[17] D. Ceperley, G.V. Chester, and M.H. Kalos, *Phys. Rev.* **B 16**, 308 (1977).

[18] F. London, *Superfluids*, vol. II (Dover, N.Y., 1964).

[19] M. de Llano and S.Z. Ren, *Eur. J. Phys.* **10**, 96 (1989).

[20] M. de Llano and S. Ramírez, *Ann. Phys. (NY)* **79**, 186 (1973).

[21] M. Viviani, E. Buendía, S. Fantoni, and S. Rosati, *Phys. Rev.* **B 38**, 4523 (1988).

[22] R.A. Aziz and R.K. Pathria, *Phys. Rev.* **A 7**, 809 (1973).

[23] C. Keller, M. de Llano, S.Z. Ren, E. Buendía, and R. Guardiola, *Phys. Rev.* **40**, 11070 (1989); M.A. Solís, V.C. Aguilera–Navarro, M. de Llano, and R. Guardiola, *Phys. Rev. Lett.* **59**, 2322 (1987).

NUCLEONIC SUPERFLUIDS

J. W. Clark, R. D. Davé, and J. M. C. Chen

McDonnell Center for the Space Sciences
and Department of Physics
Washington University, St. Louis, MO 63130 USA

INTRODUCTION

Nucleonic pairing inside finite nuclei and neutron stars gives rise to a rich variety of superfluids and superconductors. With energy gaps Δ on the nuclear scale of 1 MeV, these systems are characterized by incredibly high transition temperatures, of order $T_c = \Delta/k_B \sim 10^{10}$ K. In this paper, we shall focus on singlet pairing of like nucleons in neutron and neutron-star matter. The conventional picture of the hierarchy of pairing phenomena inside a neutron star is the following.

Going from the surface to the center of a cold neutron star, a very durable traveler would encounter several layers of neutral matter, distinguished by their particle content and their phase structure. In this journey the density increases from zero to several times the equilibrium density of symmetrical nuclear matter, $\rho_o = 0.17$ nucleons/fm$^3 = 2.8 \times 10^{14}$ grams/cm^3. The *outer crust* of the star is made up of bare nuclei arranged in a Coulomb lattice, interpenetrated by a gas of energetic electrons. The lattice ions grow ever more neutron rich as the depth increases, and eventually the neutron-drip line is reached and one enters the *inner crust*. In this region, background fluids of neutrons and relativistic electrons coexist with the lattice nuclei. In the background neutron sea, S-wave neutron-neutron collisions predominate, sampling the long-range attraction of the mutual $^1S_0$ potential. As a consequence there is (isotropic) pairing in the $^1S_0$ channel. The effect is maximal at a value of the neutron Fermi wave number around $k_{F_n} = 1$ fm^{-1} (about $\rho_o/5$) and persists until the density reaches something like half ρ_o ($k_F \simeq 1.4$ fm^{-1}). The colliding neutrons feel more and more of their mutual short-range repulsion, the effective pairing matrix element $V_{k_F k_F}$ turns positive, and the gap closes. As the density rises from $\rho_o/2$ to ρ_o, the material undergoes a transition from a crystalline structure to a fluid mixture of neutrons, protons, and electrons called the *quantum fluid interior*. By charge neutrality, there must be equal numbers of protons and electrons (plus negative muons), and the ratio of neutrons to protons is typically 10/1–20/1. At densities near ρ_o (i.e. in the vicinity of the crust-fluid transition) the proton Fermi wave number reaches values $\simeq 1$ fm^{-1} and the protons pair-condense into a $^1S_0$ superconductor, by the same mechanism responsible for $^1S_0$ neutron superfluidity in the inner crust. Proceeding into the quantum fluid interior, triplet pairing of neutrons, in the $^3P_2$-$^3F_2$ channel, will be favored by the noncentral components of the nucleon-nucleon (NN) interaction. In principle anisotropic, this type of pairing

Condensed Matter Theories, Vol. 8, Edited by
L. Blum and F.B. Malik, Plenum Press, New York, 1993

is expected to have maximum strength at a somewhat higher density than $^1S_0$ proton pairing. In the region $2\rho_o$ up to the central density, more exotic kinds of physics may come into play, variously involving hyperons, condensed pions or kaons, or individual quarks; pairing phenomena in this regime await concerted theoretical exploration.

Although the different kinds of nucleonic pairing exert relatively small effects on the gross energetics of neutron-star material, being hardly noticeable in the equation of state, they can have decisive effects (i) on the internal dynamics of neutron stars associated with the relaxation of the pulsar period following a "glitch" (sudden spin-up) and (ii) on the thermodynamic evolution of neutron stars. For a discussion of nucleonic superfluidity within the vortex-pinning–vortex-creep model of the rotational dynamics of neutron stars, see the review by Pines and Alpar.[1] The standard references for the theory of neutron-star cooling — with and without superfluidity — are Maxwell[2] and Nomoto and Tsuruta.[3] In quantitative studies of these macroscopic manifestations of superfluidity, the essential microscopic inputs are the pairing gaps at the Fermi surface, $\Delta \equiv \Delta_{k_F}$, of the relevant nucleonic superfluids. The reader may be acquainted with recent developments in neutron-star physics — and some of the observational implications of nucleonic superfluidity — through the proceedings volume of SENS90 and the recent review by Lamb.[4] The current state of microscopic calculations of nucleonic pairing is addressed in Chen et al.,[5] and a more comprehensive review of microscopic theories will be given by Davé, Clark, and Chen.[6]

To illustrate the importance of nucleon pairing to observable gross properties of neutron stars, we will cite just one example that is currently the source of some excitement. In the conventional view of the thermal history of a neutron star, energy loss by neutrino emission dominates the cooling at times up to about 10^5 years, after which photon mechanisms take over. In the neutrino-cooling phase, the direct Urca process is not considered to contribute, due to Pauli and kinematic constraints on the two reactions (viz. $n \rightarrow p + e^- + \bar{\nu}_e$; $p + e^- \rightarrow n + \bar{\nu}_e$). The star must then rely on the modified Urca process (involving an extra nucleon on the left and right sides of these reactions to meet the constraints), and on nucleon bremsstrahlung, which are not nearly so efficient. However, Lattimer et al.[7] have pointed out that recent calculations of neutron-star equations of state using advanced many-body techniques (e.g. Ref. 8) suggest that the proton concentration in the quantum fluid interior may rise to values that allow the direct Urca process to "go." If this does happen, the surface temperature of a young neutron star drops catastrophically after $\sim 10^2$ years, *provided* neither the neutrons nor protons participating are in a pair-condensed phase.[9] Pairing quenches the effect because the participating neutrons or the protons (or both) must be excited above the superfluid gap for the Urca reaction to occur. In the case that nucleonic superfluidity prevails over the bulk of the emitting region, the surface temperature will drop to a value determined by the superfluid transition temperature after $\sim 10^2$ years and decrease slowly for the next $\sim 10^5$ years. Thus, superfluidity can have a dramatic effect on the cooling scenario. In particular, at 10^3 years the surface temperature can be almost an order of magnitude lower than in the conventional picture if the Urca process is effective and superfluidity is turned off, but the conventional picture is largely restored if there is a strong and pervasive pairing effect. The situation is actually more complicated than this brief sketch would suggest. For example, a significant proton pairing effect that does *not* extend into the deep interior of the star (as found by Chao et al.[10]) has little effect on the Urca cooling process, whereas a density-independent critical temperature for $^1S_0$ proton pairing does gives a cooling curve not very different from the standard one corresponding to dominance of the modified Urca process. To put these ideas and arguments into perspective, it should be noted that satellite X-ray measurements (mostly upper limits) of the flux from compact objects indicate that some neutron stars (especially, the Vela pulsar) are too cold for consistency with the conventional cooling scenario. If the direct Urca process is viable, there is no need to invoke exotic cooling mechanisms involving pion or kaon condensates, quarks, or axions.

In this paper we will summarize the findings of recent studies of singlet-S neutron and proton pairing within the method of correlated basis functions[11-13] (CBF). The difficulties faced by microscopic theory in the evaluation of nucleonic gaps are twofold: (1) Due to the nature of the NN interaction, particularly the existence of a strong repulsive core, special care must be taken in solving the gap equation. (2) A far more serious obstacle is the proper treatment of higher-order medium effects on the pairing interaction, notably the induced interaction arising from exchange of density and spin-density excitations. Generically, these effects are called "polarization corrections."

As a counterpoint to these difficulties, the problem of the $^1S_0$ neutron gap in neutron-star matter admits certain simplifications, due principally to the low nucleon densities involved. First, it is permissible to work with pure neutron matter instead of neutron-star matter, since the interactions of a given neutron with the dilute proton admixture are relatively unimportant. Second, the gap will be insensitive to complicating details of the nucleon-nucleon interaction, since its noncentral character and behavior in higher states of angular momentum will not be probed over most of the density range where the pairing effect is strong. It will suffice to use a central, spin-dependent interaction that fits the low-energy NN data and the S-wave phase shift out to intermediate energies. Third, also because of the low density, evaluation of the required CBF matrix elements in leading cluster order should be an adequate approximation except at the higher densities where the gap closes.

After a brief description of the theoretical approach, we shall present some new results for the $^1S_0$ neutron gap in pure neutron matter, calculated at the variational level of CBF theory. These results will be supplemented by estimates of the gap based on medium-corrected versions of the pairing matrix elements and effective mass, which are determined by means of a scaling procedure that makes use of numerical data from the second-order CBF calculations of Ref. 14. We provide a quantitative commentary on the validity of essential ingredients of the variational calculation, including: (i) methods for numerical solution of the gap equation, (ii) the decoupling approximation (use of normal-state single-particle energies), (iii) the effective-mass approximation, and (iv) cluster expansion of the combinations of CBF matrix elements defining single-particle energies and pairing matrix elements. The new microscopic gap predictions are compared with older variational and CBF results,[15,14] with results based on bare pairing matrix elements,[16] and with estimates from the semi-microscopic polarization-potential approach of Ainsworth, Wambach, and Pines.[17] We also update an early evaluation of the proton gap in neutron-star matter.[10] The $^1S_0$ neutron and proton gaps are found to be significantly larger than in previous work. The increases are due either to an improved description of the geometrical two-body correlations induced by the short-range behavior of the nuclear forces, or to improved numerical treatment of the gap equation.

CBF THEORY OF PAIRING IN STRONGLY INTERACTING SYSTEMS

An exact approach to pairing in strongly interacting systems may be formulated in terms of correlated basis functions (CBF). In CBF approaches, the leading approximation takes the form of a variational treatment. To describe BCS-like pairing in the presence of bare two-body interactions $v(12)$ containing a strong or singular repulsive core, one may employ a trial ground-state ket

$$|\Psi_0^s> = \sum_{m,N} |\Psi_m^{(N)}> < \Phi_m^{(N)}| \text{ BCS } > \quad , \tag{1}$$

formed by linearly superposing the (normalized, but nonorthogonal) members

$$|\Psi_m^{(N)}> = F^{(N)}|\Phi_m^{(N)}> < \Phi_m^{(N)}|F^{(N)\dagger}F^{(N)}|\Phi_m^{(N)}>^{-1/2} \tag{2}$$

of a correlated basis of normal states, using the same coefficients with which the model states $|\Phi_m^{(N)}>$ enter the BCS ket

$$| \text{ BCS } > = \prod_k [(1 - h_k)^{1/2} + h_k^{1/2} a_{k\uparrow}^\dagger a_{-k\downarrow}^\dagger]| 0 > \quad . \tag{3}$$

In detail, $\{|\Phi_m^{(N)}>\}$ is a complete orthonormal set of plane-wave Slater determinants and $F^{(N)}$ is a symmetrical N-particle correlation operator having the cluster decomposition property.[18] The latter operator may be the familiar Jastrow factor, built as a product of state-independent pair correlations $f(r_{ij})$, or it may be something more complicated (e.g. with an operator structure to take account of important local spin/isospin-dependent correlations). Minimally, however, $F^{(N)}$ has the responsibility of describing the short-range geometrical correlations due to the repulsive cores. The function h_k appearing in Eq. (3) is assumed to be real and has the role of a pair occupation probability.

The next step in the variational-CBF theory of pairing is to evaluate the expectation value of $\hat{H} - \mu\hat{N}$ in the trial superstate (1), where $\hat{H}$ and $\hat{N}$ are the second-quantized Hamiltonian and number operators and μ is the chemical potential. Following Refs. 13,14, we introduce a simplifying approximation which is well justified for nucleonic superfluids. The required expectation value is evaluated to first order in $h_k - h_k^o$ [with $h_k^o = \Theta(k_F - k)$] and second order in $h_k^{1/2}[1 - h_k]^{1/2}$. This approximation corresponds to the "decoupling approximation" that is usually made in setting up the BCS gap equation (see below) and hence is given the same name. Upon invoking the extremum principle

$$\frac{\delta}{\delta h_k} < \Psi_0^s|\hat{H} - \mu\hat{N}|\Psi_0^s> = 0$$

and imposing the auxiliary condition $< \Psi_0^s|\hat{N}|\Psi_0^s> = A$, where A is the specified particle number, we are led to a nonlinear gap problem that is isomorphic to that of the familiar BCS scheme:

$$\Delta_k = -\frac{1}{2}\sum_{k'} \frac{V_{kk'}}{[(\epsilon_{k'} - \mu)^2 + \Delta_{k'}^2]^{1/2}}\Delta_{k'} \quad , \tag{4}$$

$$\epsilon_k = \frac{\hbar^2 k^2}{2m} + \sum_l h_l^o < k\sigma, l\sigma'|V(12)|k\sigma, l\sigma' - l\sigma', k\sigma > \quad , \tag{5}$$

$$h_k = \frac{1}{2}\left(1 - \frac{\epsilon_k - \mu}{[(\epsilon_k - \mu)^2 + \Delta_k^2]^{1/2}}\right) \quad . \tag{6}$$

In Eq. (5), h_l^o is the normal-state pair-occupation probability $\Theta(k_F - l)$ (implying single-particle energies ϵ_k of normal-state, Hartree-Fock form), while the chemical potential appearing in Eqs. (4) and (6) is to be determined from $\sum_{\mathbf{k}\sigma} h_k^o = A$ (thus $\mu = \epsilon_{k_F}$). Since we are specializing to singlet-S-wave pairing, the pairing matrix elements involved in the nonlinear gap equation (4) are given by

$$V_{kk'} = <\mathbf{k}\uparrow, -\mathbf{k}\downarrow |V(^1S_0)|\mathbf{k}'\uparrow, -\mathbf{k}'\downarrow> , \qquad (7)$$

where $V(^1S_0)$ is the $^1S_0$ component of the relevant two-body interaction (including the projector onto $^1S_0$ states).

In spite of the formal coincidence with the ordinary BCS treatment, there is the crucial distinction that the pairing matrix elements $V_{kk'}$ and single-particle energies ϵ_k that now appear are *renormalized* by the dynamical correlations incorporated through the $F^{(N)}$ operators. These renormalized quantities are specified in terms of an *effective* two-body potential $V(12)$ which assumes the role of the bare interaction $v(12)$ and is built from normal-state CBF matrix elements – i.e., from matrix elements of the Hamiltonian and the identity operators in the basis of correlated normal states (2) (see Refs. 12 and 13 for details). For a Jastrow choice of the correlation operator, Fermi hypernetted-chain (FHNC) techniques[18] are available for the accurate evaluation of the CBF pairing matrix elements and single-particle energies defining the gap equation.[12] For more complicated choices of the $F^{(N)}$, one must generally resort to cluster expansion of these quantities, which should be a satisfactory recourse for nucleon densities below ρ_o.

We stress that the great advantages of the decoupling approximation (relative, say, to the formulation of Fantoni[19]) are (i) the variational-CBF gap problem has *exactly* the same form as in the standard version of BCS theory and (ii) the renormalized pairing matrix elements and single-particle energies that enter may be evaluated entirely in terms of *normal-state* CBF matrix elements. These features actually persist to any order in CBF superstate perturbation theory, as formulated in Ref. 13. To develop the full CBF perturbative approach to the pair-condensed ground state, one first extends the trial correlated BCS ground state of the variational theory to a basis of correlated BCS states, by acting on $|\Psi_0^s >$ of Eq. (1) with all possible products of correlated Bogoliubov quasiparticle creation operators. Implementation of the extremum condition to any order within the corresponding CBF perturbation expression for the exact-ground state expectation value of $\hat{H} - \mu\hat{N}$ leads once again (in the decoupling approximation) to Eqs. (4)-(6), the inputs ϵ_k and $V_{kk'}$ now being dressed by CBF perturbative corrections to the given order; moreover, all the inputs for each order are determined by well-defined combinations of normal-state CBF matrix elements. The leading order of CBF superstate perturbation theory is just the variational treatment outlined above.

Obviously, variational-CBF pairing theory reduces to ordinary BCS theory upon replacement of the correlation operators $F^{(N)}$ by unity, in which case the operator V appearing in (5) and (7) becomes the bare NN interaction v. If the gap equation is solved in configuration space,[20] the strong or singular short-range repulsion present in v does not lead to any problems of principle, since the pairing correlations introduced by the BCS ansatz can adjust to suppress exploration of the core region. This statement is subject to the uncomfortable qualification that the single-particle energies ϵ_k are poorly behaved or undefined when computed from the bare interaction. Thus they must be renormalized before such a solution can proceed, which gives rise to an apparent inconsistency in the treatment of the inputs $V_{kk'}$ and ϵ_k. (The usual prescription is to replace the bare-potential ϵ_k by a Brueckner-Hartree-Fock spectrum, as in the recent work of Baldo et al.[16,21]) By contrast, in the CBF approach there is a division of labor between h_k and the F correlations, with the former taking care of the momentum-space correlations around the Fermi surface and the latter incorporating the short-range spatial correlations. Hence the key ingredients $V_{kk'}$ and ϵ_k are consistently renormalized in terms of the geometrical correlations contained in the $F^{(N)}$. In particular, at the variational-CBF level, the strong short-range repulsion of the bare potential v is largely eliminated, although there does remain a significant repulsive bump or spike at the core edge, corresponding in part to correlation kinetic energy. On the other hand, the variational-CBF version of ϵ_k behaves much like the single-particle energy of Brueckner-Hartree-Fock theory (at least in the sense of producing a similar value for the effective mass). Consequently, we can expect the gap values produced by variational CBF theory to be substantially larger than those found in by Baldo et al. using bare pairing matrix elements.

In closing this section on microscopic pairing theory, we note that the fully coupled BCS problem may be regained from Eqs. (4)-(6) by restoring h_k in place of h_k^o in (5) and determining μ from $\sum_{k\sigma} h_k = A$. Although not a complete test, some measure of the accuracy of the decoupling approximation of superfluid CBF theory may be obtained by comparing the gap parameters Δ_{k_F} produced by solution of the coupled and decoupled BCS problems for the variational CBF effective interaction V. In the cases of nucleonic pairing studied in the next section, the differences turn out to be tiny; with the scales used in Figs. 1-5, the results of the two procedures are almost indistinguishable, the Δ_{k_F} values for the decoupled problem being the smaller by a very narrow margin.

NEW RESULTS FOR $^1S_0$ NEUTRON AND PROTON GAPS

Two simplified nucleon-nucleon interactions have been employed in our numerical calculations. Both are central but spin-dependent, and both fit the low-energy NN data and (to reasonable accuracy) the $^1S_0$ phase shift out to intermediate energies (i.e. to lab energies of 250-300 MeV). Both involve strong repulsive cores. The older potential, the well-studied OMY4 interaction,[22,15,10,12] contains a hard core of radius 0.4 fm, surrounded (in some channels) by an exponential attractive well. In the $^1S_0$ neutron pairing problem this potential is considered to act in S waves only; whereas in the the treatment of proton pairing in neutron-star matter it is given a "mixed-Serber" exchange character[10] (with the hard core plus exponential well of the S-wave channel acting in all even-parity states, and only the hard core operating in odd partial waves). The newer potential, called Reid v_4, is a trimmed-down version of the famous Reid-soft-core interaction.[23] It is designed for use in pure neutron matter. In even states, this potential is identified with the (soft-core) interaction constructed by Reid for the $^1S_0$ channel, and in odd states it is given by the central portion of the Reid $^3P_2$-$^3F_2$ interaction. The use of these simplified interactions (and, in particular, the neglect or simulation of the noncentral portions of the realistic NN interaction) is justified in the case of $^1S_0$ neutron pairing in neutron (or neutron-star) matter by the low densities involved; on the other hand, our application of the OMY4 interaction to proton pairing, in neutron-star matter at density $\sim \rho_o$, can at best be expected to yield semi-quantitative results.

In conjunction with the Reid-v_4 choice, we adopt a spin-dependent two-body correlation operator[18] $F(12)$ of radial shape $f(r) = \exp\{-(b/r)^m \exp[-(r/b)^n]/2\}$, with parameters b, m, n that in general differ in singlet and triplet states (i.e. neutron-neutron states of even and odd parity). These parameters are determined by minimization of the normal-state energy expectation value, truncated at lowest – zeroth – order in a cluster expansion in the smallness parameter $\xi = \rho| \int [f^2(r) - 1]d^3r|$ (simplified from Ref. 18). The correlations chosen in association with the OMY4 interaction coincide with those used in the early estimates of the $^1S_0$ neutron gap by Yang and Clark[15] and of the $^1S_0$ proton gap by Chao et al.[10] In the former case, a two-body correlation function $F(12)$ of shape $f(r) = \Theta(r-c)[1 - \exp[-\gamma(r-c)]]$, vanishing for $r \leq c$, is present in (singlet) S waves, with $F(12) \equiv 1$ in all other channels. In the latter, the same form for $F(12)$ is employed in all partial waves. As in Refs. 15 and 10, the inverse-range parameter γ is chosen to minimize the leading cluster approximation to the normal-state variational energy, also defined in terms of the ξ ordering scheme. It is to be noted that, for the energy, evaluation to lowest order in ξ is equivalent to two-body cluster approximation. However, in the evaluation of variational-CBF pairing matrix elements, the two approximations differ, since the two-body cluster treatment contains some extra factorizable terms of first order in ξ. For details, see Refs. 15 and 12. We stress that either evaluation is effectively a low-density approximation. Some reassurance that such approximations are not unreasonable (even in application to the proton gap problem) is given by the slow, smooth variation of the "optimal" parameter μ (and of the "optimal" spin-dependent parameters b, m, and n in the Reid-v_4 case), over the density ranges of interest.

Again taking advantage of the low densities involved in the neutron-gap problem, the CBF effective pairing interaction $V_{kk'}$ and single-particle spectrum ϵ_k are likewise approximated to leading order in ξ (see Refs. 15 and 12 for explicit formulas). The same simplifications are implemented (with less confidence) in treating the proton gap.

Unless stated to the contrary, we follow standard practice and parametrize the normal-state single-particle energies (5) in terms of an effective mass m^* specified by $(d\epsilon_k/dk)|_{k=k_F} = \hbar^2 k_F/m^*$. This popular approximation has been thoroughly tested. For example, for the Reid-v_4 interaction, the variational $^1S_0$ neutron gap as evaluated in effective-mass approximation differs by less than 3% from the result obtained with the actual CBF-variational single-particle spectrum, over the range $k_F = 0.6 - 1.1$ fm^{-1}. At higher densities, the discrepancies are larger, but in this regime, other uncertainties (related to the interactions assumed and to cluster truncation) are more important. The effective-mass approximation tends to underestimate the gap, although only by about 1% near the peak of Δ_{k_F} versus k_F.

For pairing forces derived from realistic interactions — even when tamed with the correlations introduced in the CBF method — numerical solution of the gap problem becomes nontrivial in the sense that the integration over the intermediate momentum k' must be carried out at least to ~ 10 fm^{-1}. Our methods for solving the gap equation (based on straightforward iteration and on the linearization-eigenvalue method described in Ref. 14) take this fact into account and produce results of quite satisfactory accuracy. Older treatments, notably those applied in Refs. 10 and 15, do not, and accordingly yield results for Δ_{k_F} that are too small by 20-25% near its maximum. And, as will be seen, the omnipresent weak-coupling formula $\Delta = 2\epsilon_{k_F}\exp[-1/N(0)g]$ of BCS (with $g = -V_{k_F k_F}$ and the density of states factor $N(0)$ determined from m^*) is unreliable in the present context. For detailed critical assessment of procedures for numerical treatment of the gap equation, see Refs. 5, 6, and 16.

Some of our main results are displayed in Figs. 1–5. In considering $^1S_0$ neutron pairing in neutron-star matter, we disregard the presence of lattice nuclei in the relevant density regime and (as is customary) approximate neutron-star matter by pure neutron matter. Fig. 1 compares different evaluations of the $^1S_0$ neutron gap parameter $\Delta \equiv \Delta_{k_F}$ for the OMY4 potential. All three curves refer to the Yang-Clark S-wave choice of two-body correlations $F(12)$ (as specified above), with pairing matrix elements and single-particle potentials calculated in lowest cluster order. All three cases invoke decoupling and effective-mass approximations. This comparison demonstrates the good agreement of solutions of the gap equation (4) reached by straightforward iteration and by the linearization-eigenvalue procedure, when the cutoff k_c for the k' integration is chosen sufficiently large. It also documents the unreliability of the weak-coupling formula and the quantitative inaccuracy of the parametric method adopted by Yang and collaborators.[15,10] The accurate results from the former pair of methods indicate a peaking of the variational gap at about 0.75 fm^{-1}, with a maximum value of nearly 3.2 MeV. Fig. 2 shows a similar comparison based on the Reid-v_4 interaction and the associated choice of $F(12)$ (as specified above), leading to similar conclusions. The weak-coupling formula (which gives a result very similar to that for k'-cutoff $k_c = 2k_F$) underestimates the gap by roughly a factor two in the peak region, with larger errors in the wings. The variational

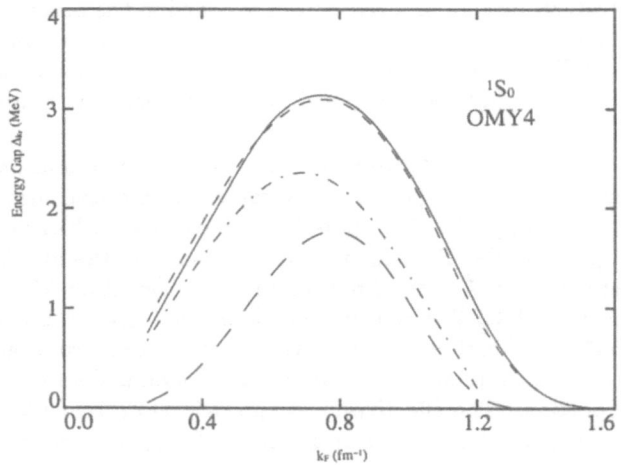

Fig. 1. Variational results for the $^1S_0$ neutron pairing gap Δ_{k_F} in pure neutron matter, based on the OMY4 potential (acting in S-waves only). Solid curve: straightforward iterative solution of gap equation (4) for k' cutoff $k_c = 16k_F$. Short dashes: linearization-eigenvalue solution ($k_c = 20k_F$). Dot-dashed curve: from Ref. 15, solution by parametric algorithm. Long dashes: weak-coupling formula. Correlation operator of Ref. 15 is used.

gap peaks at approximately 0.85 fm^{-1}, thus at a somewhat higher density than in the OMY4 case, and reaches a higher value, around 3.8 MeV. The stronger singlet-S pairing effect is due to the fact that the effective pairing interaction has a softer remnant repulsion when the Reid-v_4 potential is used as input.

Fig. 3 represents an attempt to characterize the range of validity of the lowest-order cluster treatment. The gap estimates appearing in this plot make use of the results of an earlier study by Krotscheck and Clark[12] of low-cluster-order and FHNC treatments of the CBF matrix elements entering the $^1S_0$ pairing problem in neutron matter. This study was based on the OMY4 potential, in its mixed-Serber (rather than pure-S-wave) version. A correlation operator of strict state-independent Jastrow type was assumed, with two-body correlations of the shape $f(r) = \Theta(r - c)[1 - \exp[-\gamma(r - c)]]$, the parameter γ being fixed (as above) by minimization of the variational normal-state energy in two-body cluster approximation. As improvements upon two-body cluster evaluation, the variational-CBF pairing matrix element $V_{k_F k_F}$ was calculated using two different approximations within FHNC theory, summing the important classes of cluster diagrams to all orders in ξ (or to all orders in the number of "bodies" involved). These integral-equation treatments will be referred to as the FHNC/C[W] and FHNC/C[P] approximations; the former differs from the latter in the inclusion of certain factorable diagrams that arise in the diagrammatic analysis of the pairing matrix elements. The FHNC/C[P] approximation is believed to be the more accurate. The effective mass computed in FHNC/C approximation was found to lie within about 1% of the two-body-cluster value until k_F reaches 1.1 fm^{-1} and is still within 5% at $k_F = 1.56$ fm^{-1}. At low densities, the two FHNC/C results for $V_{k_F k_F}$ are nearly coincident with the two-body

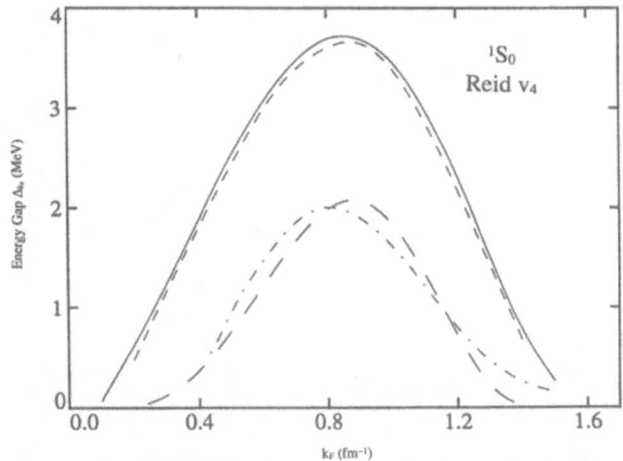

Fig. 2. Variational results for the $^1S_0$ neutron pairing gap Δ_{k_F} in pure neutron matter, based on the Reid-v_4 potential. Solid curve: straightforward iterative solution ($k_c = 32k_F$). Short dashes: linearization-eigenvalue solution ($k_c = 20k_F$). Dot-dashed curve: straightforward iterative solution ($k_c = 2k_F$). Long dashes: weak-coupling formula. Spin-dependent F correlations are used, with parameters optimized in the normal phase.

truncation, the discrepancy of this simple evaluation from either FHNC/C result remaining less than about 10% below $k_F = 1.1$ fm^{-1}. For k_F values corresponding to the peak region of Δ_{k_F}, the error in $V_{k_F k_F}$ due to omission of higher-order cluster contributions is quite small. For higher densities where the discrepancies become more significant, the FHNC/C[P] and FHNC/C[W] treatments bracket the two-body result. We would like to translate this information into statements about the gap parameter itself. It would be natural to apply the weak-coupling formula, but it has been demonstrated that this formula is not to be trusted. Thus some prescription is needed for extrapolating the FHNC/C results for $V_{kk'}$ to momenta k, k' away from the Fermi surface.

In setting up such a prescription, we shall – as simplicity dictates – ignore the distinction between cluster approximations to lowest (zeroth) order in ξ and to lowest order (two-body) in the number of correlated bodies. The difference between the specific OMY potentials used in our work and in Ref. 12 (pure S-wave versus mixed Serber character) affects m^* slightly, but the $V_{kk'}$ hardly at all. Using the data in Table 1 of Ref. 12, determine scaling factors $\beta_{m^*} = m^*(\text{FHNC/C})/m^*(\text{2-body})$ and $\beta_{V[\cdot]} = V_{k_F k_F}(\text{FHNC/C}[\cdot])/V_{k_F k_F}(\text{2-body})$, where $\cdot$ stands for P or W. The bottom curve [top curve] in Fig. 3 is generated by applying the factor $\beta_{V[P]}$ [respectively, $\beta_{V[W]}$] to *our* lowest-order $V_{kk'}$ values, multiplying *our* lowest-order m^* values by β_{m^*}, and solving the gap equation by the linearization-eigenvalue method with $k_c = 20k_F$. The middle curve is the corresponding lowest-order OMY4 result from Fig. 1

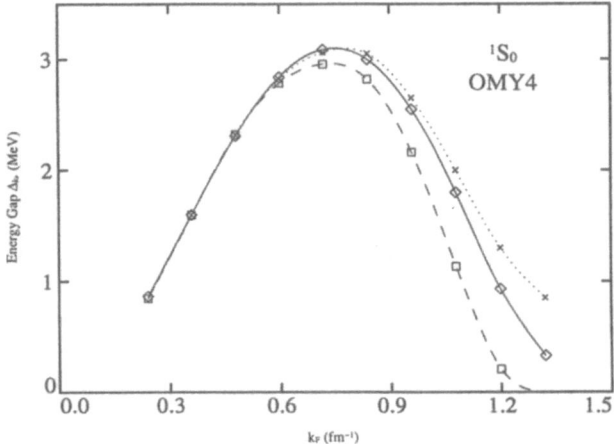

Fig. 3. Test of the efficacy of lowest-cluster-order evaluation of the inputs ϵ_k and $V_{kk'}$ of the gap equation in variational theory (see text for details). Comparison of lowest-order and FHNC results for the pairing gap is made for versions of the OMY4 potential. Curve with diamond data points: variational gap corresponding to truncation of cluster-expanded inputs at leading order in parameter ξ. Crosses: scaling results including higher-order cluster diagrams in FHNC/C[W] approximation. Squares: scaling results based on FHNC/C[P] approximation. Solutions of gap equation are obtained by the linearization-eigenvalue algorithm with $k_c = 20k_F$.

(the short-dashed curve, also obtained by the linearization-eigenvalue method with $k_c = 20k_F$). The lowest-ξ-order approximation for the gap is seen to be in good agreement with the scaled-FHNC/C estimates below and around the peak but begins to depart seriously from the scaled-FHNC/C[P] result as k_F increases beyond 1 fm^{-1}. Based on our extrapolation of the FHNC/C pairing matrix elements, it would appear that the lowest-ξ-order evaluation tends to overestimate the gap, and that such a treatment is unlikely to produce a quantitatively reliable value for the gap-closure density. A similar scaling procedure has been used to estimate FHNC-corrected gap values for the Reid-v_4 case, with qualitatively similar findings.

Fig. 4 brings together results from a number of calculations and estimates of the $^1S_0$ pairing gap in neutron matter. Included are: (a) the present work at the variational-CBF level, with the Reid-v_4 interaction as input, either spin-dependent F correlations [upper solid curve] or spin-independent Jastrow correlations [short-dashed curve], and no effective-mass approximation in the spin-dependent case, (b) the older variational-CBF calculation of Chen et al.[14] [upper dot-dashed curve], carried out for the Reid-v_6 interaction with density-independent Jastrow correlations, and (c) the predictions of Baldo et al.[16] based on the bare Paris and Argonne-v_{14} interactions [dotted curve for the Paris interaction, pluses for Argonne-v_{14}]. With regard to (a), it is interesting and reassuring that the spin-dependence of the geo-

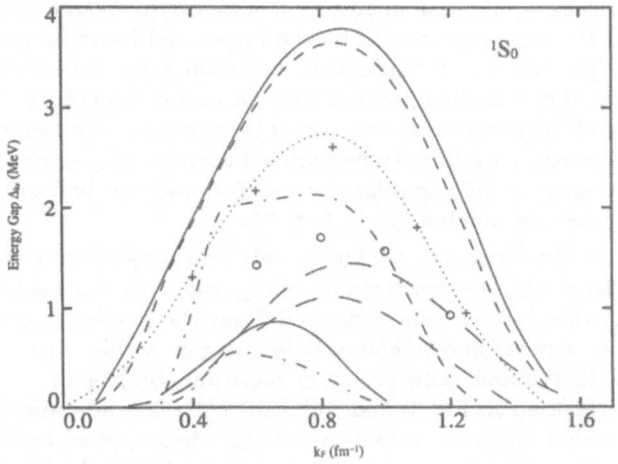

Fig. 4. Predictions for $^1S_0$ neutron pairing gap Δ_{k_F} in pure neutron matter. Upper solid curve: variational result for Reid-v_4 interaction, with optimized spin-dependent F correlations, and without effective-mass approximation (straightforward iterative solution with $k_c = 32k_F$). Short dashes: variational result for spin-independent correlations (straightforward iterative solution with $k_c = 16k_F$). Dotted curve [pluses]: bare-interaction result from Ref. 16, for Paris potential [for Argonne v_{14}]. Upper dot-dashed curve: Variational result from Ref. 14, for Reid v_6. Lower dot-dashed curve: second-order CBF result of of Ref. 14, for Reid v_6. Lower solid curve: new scaled-CBF prediction (see text). Upper and lower long-dashed curves: respective results from cases II and III of polarization-potential analysis of Ref. 17. Circles: alternative scaling estimate of polarization-corrected gap, based on the procedure of Ref. 24 (see text).

metrical correlations has little effect on the gap – the extra freedom allowed by spin-dependent parameters b, m, and n produces only a modest enhancement, mainly at the higher densities. The differences between the four potential models involved in (a)-(c) should not be of much consequence, at least for the behavior of Δ_{k_F} near its peak and at the lower densities. The large size of the current variational-CBF gaps compared to the corresponding results of Ref. 14 is due primarily to a considerable improvement in the spin-independent component of the two-body correlations $F(12)$. (Even so, the fact that Chen et al.[14] evaluated the CBF matrix elements within a

FHNC scheme, rather than in lowest cluster order, may be responsible for a significant share of the difference at the higher densities, and the current improvement in the accuracy of solution of the gap equation will have a noticeable effect). It is of course to be expected that the variational-CBF prediction is larger than that of Baldo et al., owing to the substantially less repulsive pairing interaction of the variational approach (which includes medium-induced modifications that would only be reached in higher orders of a conventional Green's function theory).

The remaining calculations represented in Fig. 4 take some account of higher-order medium effects, or polarization corrections, as defined within the pertinent theoretical frameworks. Plotted are: (d) the result obtained by Chen et al.[14] using second-order CBF superfluid perturbation theory, for the Reid-v_6 potential [lower dot-dashed curve], (e) the current prediction based on the Reid-v_4 interaction, with higher-order medium corrections estimated by a scaling procedure employing CBF results from Ref. 14 [lower solid curve], (f) two "limiting" predictions from the semi-microscopic polarization-potential analysis of Ainsworth, Wambach, and Pines[17] [their cases II and III being represented by the upper and lower long-dashed curves, respectively], and (g) alternative polarization-corrected gap estimates for the Reid-v_4 interaction, generated by a scaling procedure employing results of Ref. 17 for the "direct" interaction and Landau parameters. In all these cases, we observe the profound impact of the higher-order medium (polarization) corrections, which are responsible for a strong suppression of the gap relative to variational and bare-potential evaluations. This effect was first pointed out in Ref. 24.

As indicated in the foregoing, we have made two rough estimates of the influence of higher-order medium corrections on our present variational-CBF results for the Reid-v_4 case, which are of course derived from the leading approximations to the pairing matrix elements and single-particle energies within CBF superfluid perturbation theory. In the first (and probably more reliable) of these estimates, the numerical results reported in Fig. 2 and Table 1 of the second paper of Ref. 14 are used to form correction factors α_{m^*} and α_V which, when applied to the corresponding variational results of Ref. 14 for m^* and $V_{k_F k_F}$, yield the calculated second-order CBF approximations to these quantities. (Although neither factor deviates grossly from unity, the exponential sensitivity of the gap to $m^* V_{k_F k_F}$ leads to the very strong quenching of Δ_{k_F}.) Ignoring the difference between Reid-v_6 and v_4 interactions and (more seriously) between our F correlations and those adopted in Ref. 14, we have applied the *same* correction factors to the variational m^* and $V_{kk'}$ determined here for the Reid-v_4 interaction with spin-dependent correlations. Iteratively solving the gap equation for these scaled inputs, in decoupling and effective-mass approximations and with cutoff $k_c = 32k_F$, we arrive at the lower solid curve of Fig. 4. Relative to the very small CBF gaps of Ref. 14, the revised CBF results show improved agreement with the results of the polarization-potential scheme as reported in Ref. 17 and plotted as the long-dashed curves in Fig. 4. The improved agreement of microscopic and semi-empirical descriptions of $^1S_0$ neutron pairing is due simply to the fact that the variational gap, when properly calculated, is considerably larger than was previously believed.

In spite of the improved agreement, significant discrepancies between the two approaches are evident at the larger k_F values. More recent polarization-potential estimates, using putatively better inputs for the "direct" interaction of the theory, yield gaps that lie somewhat below the band defined by the long-dashed curves and indicate lower gap-closure densities, reducing the disagreement between the two approaches. The strengths and weaknesses of the current generation of CBF and polarization-potential gap estimates is discussed at some length in Ref. 5, with the essential conclusions that the scaled-CBF result (lower solid curve) in Fig. 4 represents an underestimate of the actual gap (due to an overestimate of the quenching effect of higher-order medium corrections), while there may be a significant cancellation of errors (due to overcounting and due to application of the weak-coupling formula) in the polarization-potential treatment.

Along with our estimation of higher-order medium corrections to the inputs m^* and $V_{kk'}$ by scaling the available CBF corrections, we have made a second determination of the suppression factor α_V by appealing to the simple Babu-Brown induced-interaction model[25] applied by Clark et al.[24] in the initial investigation of polarization quenching. The required inputs, which consist of "direct-interaction" components and Landau parameters, are taken from Table 1 of Ainsworth et al.[17] (selecting their intermediate case I as the most typical). At $k = 0.6$ fm^{-1} we find $\alpha_V = 0.78$, which is close to the value 0.74 quoted by Clark et al.[24] but appreciably smaller than the result from the CBF study of Chen et al.[14] (note that it is actually the dotted curve in Fig. 2 of Ref. 14 that corresponds to the Reid interaction). The gap equation was re-solved with Reid-v_4 variational pairing matrix elements attenuated by this factor, without any modification of the effective-mass parametrization of the single-particle energies. The procedure was repeated at higher densities. The resulting gaps are plotted as circles in Fig. 4. Considering all of the assumptions involved, the degree of agreement with the gap estimates of the polarization-potential approach is noteworthy.

Although further refinements and modifications of the various approaches to the $^1S_0$ neutron pairing problem will surely be forthcoming, it would appear that the most detailed studies are converging toward a common prediction that Δ_{k_F} reaches a maximum value of about 1 MeV at $k_F \simeq 0.7 - 0.8$ fm^{-1}. However, the (upper) gap-closure density, so critical to interesting aspects of neutron-star modeling (see e.g. Ref. 26), is still subject to great uncertainty, with current estimates of the corresponding Fermi wave number spanning the broad range $k_F \simeq 1.1 - 1.4$ fm^{-1}.

Finally, let us turn briefly to the evaluation of the $^1S_0$ proton gap in neutron-star matter. This is a more difficult problem than $^1S_0$ neutron pairing, since (i) the overall nucleon density is considerably higher (casting doubt on approximation to lowest cluster order and requiring more precision in modeling the NN interaction) and (ii) the nature of neutron-star matter as a mixture of different Fermi fluids becomes much more important (with the $^3S_1$-$^3D_1$ channel coming into play, and the background neutron sea giving rise to a substantial dispersive effect in the single-particle spectrum and very substantial polarization contributions to the pairing interaction). These matters are discussed in some qualitative and quantitative detail in Refs. 10, 21, and 27. Our immediate goal in this problem is modest: Adopting the model of neutron-star matter developed by Chao et al.[10] based on the OMY4 interaction with mixed-Serber exchange, we simply update the solution of the gap equation, using the original inputs for single-particle energies (effective masses) and pairing matrix elements. The (decoupled) gap equation is solved by straightforward iteration, with $k_c = 16k_F$, yielding the lower solid curve of Fig. 5. This result is to be compared with the original estimate of Chao et al.[10] (long-dashed curve in Fig. 5), derived from a relatively crude treatment of the gap equation (the "parametric" method). Although the parametric method does not lead to serious – i.e., qualitative – error, it is again found to fall short of quantitative reliability. Again the refined calculation leads to appreciably larger gaps.

For reference, we include in Fig. 5 the proton gap results obtained by Baldo et al.[21] with the bare Argonne-v_{14} potential as pairing interaction (short-dashed curve, with values computed in a constant-effective-mass approximation indicated by crosses). The (very rough) agreement of variational-CBF and bare-interaction gaps may be understood as follows. The hard core in the OMY4 potential leads to a more repulsive effective pairing force than would be the case if the Reid-v_4 potential (or other modern potential rooted in boson exchange) were used as input to the variational-CBF scheme; indeed, for the correlations assumed, this $V_{kk'}$ is – in effect – almost as repulsive as the bare Argonne-v_{14} pairing interaction. The four sets of results for the proton gap are put into perspective by the upper curves and data

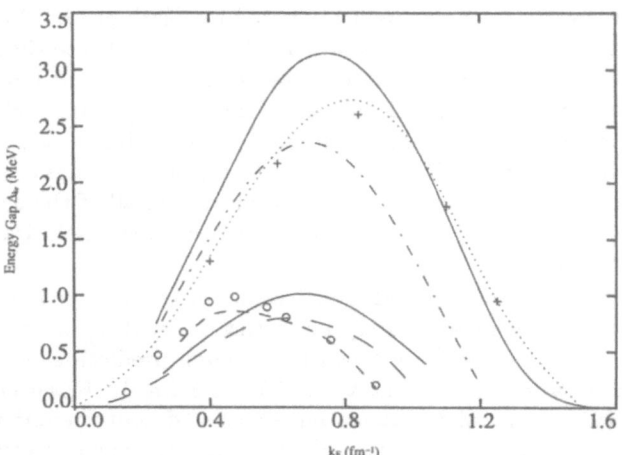

Fig. 5. Energy gaps Δ_{k_F} for $^1S_0$ neutron and proton pairing, plotted against the Fermi wave number k_F for the given nucleon type. Lower group of curves and points: results for proton gap in models of neutron-star matter. Upper group: neutron gap in pure neutron matter. Lower solid curve: variational proton gap from straightforward iterative solution of gap equation with $k_c = 16k_F$, based on mixed-Serber OMY4 inputs from Ref. 10. Long-dashed curve: corresponding variational proton gap of Ref. 10 from parametric method. Upper solid curve: similar update of the variational neutron gap for S-wave OMY4; same as solid curve of Fig. 1. Dot-dashed curve: corresponding neutron gap from Ref. 15; same as dot-dashed curve of Fig. 1. Short-dashed curve: bare-interaction $^1S_0$ proton-gap result of Ref. 21, for Argonne-v_{14} potential, with circles at values determined in a constant-effective-mass approximation. Dotted curve: bare-interaction $^1S_0$ neutron-gap result of Ref. 16 for Paris potential, with pluses indicating Argonne-v_{14} values.

points in Fig. 5, which represent selected results for the $^1S_0$ neutron gap in pure neutron matter. In particular, the upper solid curve is the current result obtained by accurate solution of the gap equation for OMY4 inputs (same as the solid curve in Fig. 1) and the dot-dashed curve is the original result of Yang and Clark.[15] The dotted curve traces the bare-gap prediction of Baldo et al.[21] for the Paris interaction, with pluses marking the corresponding Argonne-v_{14} values.

CONCLUSIONS

In this paper we have considered some of the pairing phenomena that can occur in nucleonic systems, with special attention to singlet S-wave pairing of neutrons and protons in neutron-star matter. The associated gap parameters Δ_{k_F} provide vital inputs to models that address observable features of the internal dynamics and cooling of neutron stars. We have indicated some of the difficulties and pitfalls in microscopic evaluation of these inputs. Improved calculations of the $^1S_0$ gaps at the variational level of CBF theory have been described. The new variational gaps are substantially larger than those found previously (due to improvements in the method of solution of the gap equation, or in the geometrical correlations employed); as expected, these new gaps are also significantly larger than the bare-interaction results, when input NN interactions with comparable inner repulsions are assumed. In the case of $^1S_0$ neutron pairing, rough estimates have been made of the quenching effect of higher-order medium corrections (including effects of exchange of density and spin-density fluctuations that cannot be described through energy-averaged propagators). The results are arguably consistent with recent findings of Ainsworth, Pines, and Wambach within the polarization-potential approach. However, quantitative microscopic evaluation of higher-order medium effects – along with implementation of FHNC techniques in the gap-closure regime, if the CBF approach is followed – will be necessary if the $^1S_0$ neutron gap is to be pinned down with the accuracy required to settle a number of key issues in neutron-star physics. The need for a more advanced microscopic treatment of singlet-S proton superfluidity is even more pressing; in this case the medium effects are much more complicated, and many-body cluster contributions cannot be ignored. Most difficult of all is the problem of $^3P_2$-$^3F_2$ neutron pairing in neutron-star matter at densities in the range $(1-4)\rho_o$, where complex many-body processes will abound and the usual technical simplifications are expected to fail. In spite of some recent progress in the numerical treatment of the relevant gap equation and the characterization of its solutions,[21] quantitative prediction of the $^3P_2$-$^3F_2$ gap remains a formidable challenge.

ACKNOWLEDGEMENTS

This research was supported in part by the National Science Foundation under Grant No. PHY-9002863. J. W. C. gratefully acknowledges travel support from the Army Research Office, Durham, through a grant to Southern Illinois University. We have benefited from numerous discussions and communications with T. Ainsworth, D. Pines, and J. Wambach. We thank V. V. Khodel for producing some of the data used in Fig. 4.

REFERENCES

1. D. Pines and M. A. Alpar, *Nature* **316**:27 (1985).
2. O. Maxwell, *Ap. J.* **231**:201 (1979).
3. K. Nomoto and S. Tsuruta, *Ap. J.* **312**:711 (1987).
4. D. Pines, R. Tamagaki, and S. Tsuruta, eds., *The Structure and Evolution of Neutron Stars, Proceedings of the U.S.-Japan Joint Seminar, Kyoto, November 6-10, 1990*, Addison-Wesley, New York (1992);
   F. K. Lamb, in *Frontiers of Stellar Evolution*, D. L. Lambert, ed., Astronomical Society of the Pacific (1991), p. 299.
5. J. M. C. Chen, J. W. Clark, R. D. Davé, and V. V. Khodel, submitted to *Nucl. Phys. A.*
6. R. D. Davé, J. W. Clark, and J. M. C. Chen, to be published.

7. J. M. Lattimer, C. J. Pethick, M. Prakash, and P. Haensel, *Phys. Rev. Lett.* **66**:2701 (1991).
8. R. B. Wiringa, V. Fiks, and A. Fabrocini, *Phys. Rev. C* **38**:1010 (1988).
9. D. Page and J. H. Applegate, Columbia University preprint (1992).
10. N.-C. Chao, J. W. Clark, and C.-H. Yang, *Nucl. Phys.* **A179**:320 (1972).
11. J. W. Clark and E. Krotscheck, in *Recent Progress in Many-Body Theories*, H. Kümmel and M. L. Ristig, eds., Springer-Verlag, Heidelberg (1984).
12. E. Krotscheck and J. W. Clark, *Nucl. Phys.* **A333**:77 (1980).
13. E. Krotscheck, R. A. Smith, and A. D. Jackson, *Phys. Rev. B* **24**:6404 (1981).
14. J. W. Clark, J. M. C. Chen, E. Krotscheck, and R. A. Smith, *Condensed Matter Theories* **1**, 313 (1986);
    J. M. C. Chen, J. W. Clark, E. Krotscheck, and R. A. Smith, *Nucl. Phys.* **A451**:509 (1986).
15. C.-H. Yang and J. W. Clark, *Nucl. Phys.* **A174**:49 (1971).
16. M. Baldo, J. Cugnon, A. Lejeune, and U. Lombardo, *Nucl. Phys.* **A515**:409 (1990).
17. T. L. Ainsworth, J. Wambach, and D. Pines, *Phys. Lett.* **B222**:173 (1989).
18. J. W. Clark, *Prog. Part. Nucl. Phys.* **2**:89 (1979).
19. S. Fantoni, *Nucl. Phys.* **A363**:381 (1981).
20. L. N. Cooper, R.L. Mills, and A. M. Sessler, *Phys. Rev.* **114**:1377 (1959).
21. M. Baldo, J. Cugnon, A. Lejeune, and U. Lombardo, *Nucl. Phys.* **A536**:349 (1992).
22. T. Ohmura, *Prog. Theor. Phys.* **41**:419 (1969).
23. R. V. Reid, Jr., *Ann. of Phys.* **50**:411 (1968).
24. J. W. Clark, C.-G. Källman, C.-H. Yang, and D. A. Chakkalakal, *Phys. Lett.* **B61**:331 (1976).
25. S. Babu and G. E. Brown, *Ann. of Phys.* **78**:1 (1973).
26. P. Haensel, V. A. Urpin, and D. G. Yakovlev, *Astron. Astrophys.* **229**:133 (1990).
27. J. Wambach, T. L. Ainsworth, and D. Pines, in *Neutron Stars: Theory and Observation*, J. Ventura and D. Pines, eds., Kluwer, Amsterdam (1991).

ENERGY DENSITY FUNCTIONAL FORMALISM AND NUCLEAR FISSION

Irwin Reichstein

School of Computer Science, Carleton University
Ottawa, Ontario, K1S 5B6, Canada

F. Bary Malik*

Physics Department, Southern Illinois University
Carbondale, Illinois 62901, U.S.A.

ABSTRACT

An energy-density functional theory is applied to deal with the nuclear fission process. The functional is found to reproduce observed nuclear masses using proper density distributions. The theory gives rise to an additional potential barrier between the saddle and scission points and accounts for observed half-lives with experimental total kinetic energies for ^{234}U, ^{236}U, ^{240}Pu, ^{244}Cm, ^{252}Cf and ^{258}Fm. It also accounts for the mass distributions. The theory accounts for the observed half-lives in the decay of ^{258}Fm to symmetric modes. The theory can also be applied to light-ion radioactivity and a preliminary calculation provides a good understanding of ^{14}C activity in the decay of ^{226}Ra.

I. INTRODUCTION

Nuclear fission in physics is an unique phenomenon where a large mass of matter, the parent, suddenly reorganizes itself and emits two separating heavy masses, usually called daughters. A parent nucleus usually decays to a series of daughter pairs which yields a mass as well as charge distribution which is different for different parent nuclei. The fission of a particular nucleus takes place neither to a symmetric mode, i.e., to a daughter pair, each having one half the mass of the parent nor to an asymmetric mode, i.e., one

---

*FBM thankfully acknowledges a travel grant from Army Research Office.

member of the daughter pair having about twice the mass of the other but produces a mass and charge distribution commensurate with the energy conservation given by the following equation:

$$M(A,Z)c^2 = M(A_iZ_i)c^2 + M(A_jZ_j)c^2 + Q_i \qquad (1)$$

with $A = A_i + B_j$ and $Z = Z_i + Z_j$ (i,j = 1...n). In the above M(A, Z) is the mass of a nucleus having baryon and atomic number A and Z, respectively. Q_i is the kinetic energy available to a particular daughter pair. In general daughter pairs are in various excited states and associated with a particular daughter pair, there is a most probable kinetic energy, known usually as TKE. Thus, there is also a TKE spectrum available for the fission of a parent nucleus.

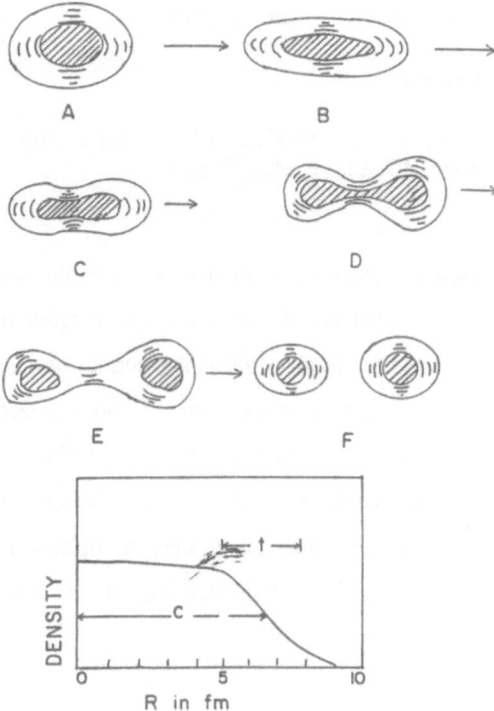

Fig. 1.   Schematic representation of the change of density distribution and shape in the fission process. A typical density distribution function of a nucleus is shown at the bottom.

Any theory of fission, therefore, must explain the proper mass distribution of daughter nuclei along with the associated values for TKE.

At the bottom of Fig. 1 we have depicted a typical mass density distribution of a parent nucleus, where the width of the nuclear surface, i.e., the distance where the central density falls from its 90% to its 10% value is between 2.5 to 3.5 fm. The constant density area typically has a radius of 5 to 7 fm. A simple estimate done later indicates that the mass of a nucleus in the surface area is comparable to that in the constant density zone. Fig. 1 depicts typical steps that a nucleus may go through in decaying to a daughter pair. Just prior to separation a daughter pair is likely to form a neck of densities considerable lower than the central density of a nucleus as depicted by configuration E in Fig. 1. Block et al. [1] in 1971 pointed out that such a configuration enroute to fission might seriously alter the barrier between the saddle point, approximately given by configuration B, and the scission point. Subsequent detailed calculations using an energy-density functional method [2,3,4] have established the existence of a barrier between saddle and scission points and this external barrier is critical in reproducing the observed mass distributions with the proper TKE spectrum. Using such an empirical external barrier, Hooshyar, Compani and Malik, [5-9] could explain the mass, charge and TKE spectra in spontaneous, induced and isomer fission.

In sections 2 and 3, we present, respectively, a schematic analysis of the consequence of the configurations like E of Fig. 1 for the fission barrier and a calculation of such a barrier using an energy-density formalism. In section 4, we present calculations of spontaneous fission half-lives, and discuss the mass distribution in the case of the fission of ^{240}Pu. We also present our calculation for the fission of ^{258}Fm where the dominant decay modes are symmetric and the decay of ^{226}Ra to ^{14}C, i.e., light-ion radioactivity, from the energy-density functional viewpoint.

2. SCHEMATIC CONSIDERATION

The number of nucleons in the surface area relative to those in constant density zone, i.e., in the interior can be easily obtained by using a trapezoidal function to represent the

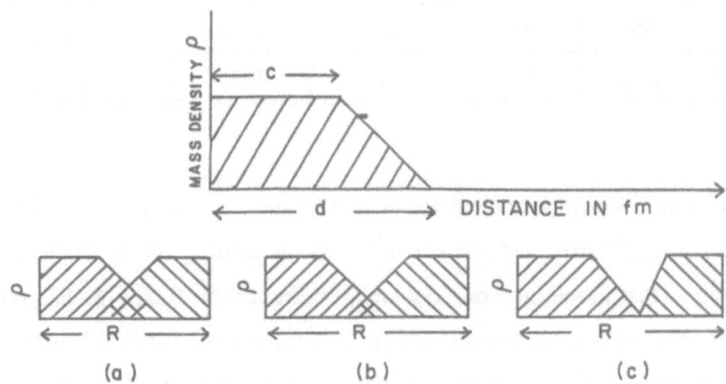

Fig. 2.   The trapezoidal function given by (2) is shown at the top . The density distribution change in a fission process is shown schematically by (a), (b), and (c).

nuclear mass density distribution. A typical case, shown at the top of Fig. 2, can be represented by

$$\rho(r) = \begin{cases} \rho_o & \text{for } r<c \\ \rho_o \dfrac{d-r}{d-c} & \text{for } c \leq r < d \\ 0 & \text{for } r \geq c \end{cases} \quad (2)$$

The total number of particles in a nucleus of mass number A, N(A) is simply given by

$$N(A) = (\pi \rho_o/3)(c+d)(c^2+d^2)$$

and the number of particles in a sphere of radius c, N(o) and in the surface region, N(s) are, respectively, given by

$$N(o) = (4/3)\pi \rho_o c^3 \quad (3a)$$

$$N(s) = (\pi \rho_o/3)[(c+d)(c^2+d^2) - 4c^3] \quad (3b)$$

Noting that $c \sim (1.2 \, A^{1/3} - s)$ where $s = d-c$, one may find N(A), N(o) and N(s) for a particular A.

Table 1. Fractions of total mass N(A) in units of $(\pi \rho_o/3)$ located at the center, N(o) and at the surface N(s) for surface thicknesses of 2.5 and 3.0 fm for mass number A=130 and 258.

A	d - c = 2.5 fm			d - c = 3.0 fm		
	N(A)	N(o)	N(s)	N(A)	N(o)	N(s)
130	937	451	486	1054	524	530
258	1853	926	927	2035	1043	992

Such a calculation is shown in table 1 for A = 130 and 258 and for s =2.5 and 3.0 fm in units of $(\pi \rho_o/3)$. Clearly, the number of nucleons residing on the nuclear surface is about the same as that in the interior. As such, configuration E of Fig. 1 is relevant for the fission process. In the bottom half of Fig. 2, we have represented schematically the density reorganization relevant to our estimation of the change in energy involved in going through configurations represented in E. Density is constant, ρ_o, everywhere in

Fig. 2(a) but in 2(b), the density in the overlapped region is less than ρ_o corresponding to the formation of a neck of low density-nuclear-matter. 2(c) represents the scission configuration. One may now use a nuclear matter calculation for the energy per nucleon as a function of density to estimate the change in energy as one goes from 2(c) to 2(a). The change is an attraction since a proper energy-density curve is negative and has a minimum at the saturation density ρ_o. Using the energy-density calculation for nuclear matter by Brueckner et al. [10], Block et al. [1] have estimated this attraction, D, to be about 57 MeV for the daughter pair $A_1 = 134$ and $A_2 = 92$.

Of course, Coulomb energy increases as one proceeds, from 2(c) to 2(a). Using a constant charge density, the increase in Coulomb energy, ΔE_c can be expressed as a ratio of interpenetration distance, p, and scission radius, R_{sc}.

$$\Delta E_s = E_{sc}(p/R_{sc} - (1/2)(p/R_{sc})^2) \tag{4}$$

where, $E_{sc} = Z_1 Z_2 e^2/R_{sc}$, is the Coulomb barrier height at the scission point which is typically about 200 MeV. For a ten percent penetration i.e., for $p/R_c = 0.1$ which is typified by Fig. 2(b), $\Delta E_s = 20$ MeV. Hence, the net attraction, V, as the density in the overlapped region changes from ρ_o to $(\rho_o/2)$,

$$V = -D + \Delta E_c = -27\ MeV \tag{5}$$

Thus, we expect a barrier of 25 to 40 MeV between the saddle and the scission points in the fission of actinides.

In the following section we improve upon this schematic estimation by using an energy-density functional formalism suitable for the fission process.

3. ENERGY-DENSITY FUNCTIONAL METHOD FOR FISSION

We present here an outline of the energy-density functional method developed by us [3, 4] to describe fission process. The starting point of the formalism is the following energy-density functional to represent total energy, $E(\rho)$, of each configuration shown in Fig. 1

$$E(\rho) = \int d^3r\, \epsilon[\rho(\vec{r})] \tag{6}$$

with

$$\epsilon[\rho(\vec{r})] = (3/5)(\hbar^2/2M)(3\pi^2/2)^{2/3}(1/2)[(1-\alpha)^{5/3} + (1+\alpha)^{5/3}]\rho^{5/3}$$
$$+ V(\rho,\alpha) + (e/2)\phi_c\,\rho - 0.739e^2\rho_p^{4/3} + (\hbar^2/8M)\eta\,(\nabla\rho)^2 \tag{7}$$

In (7) α is the neutron excess, M, ρ, and ρ_p are respectively, nucleonic mass, the nuclear matter density and proton charge density. The coefficient n in the density gradient term incorporates both the Weizsaecher correction to kinetic energy and an additional correction originating from correlation between nucleons that is not included in $V(\rho,\alpha)$ and taken empirically here to be 8 in all cases. $V(\rho,\alpha)$ is the average potential seen by a nucleon and is calculated in [10,11] using the Brueckner-Hartree-Fock approximation from the realistic two-nucleon Brueckner, Gammel and Thaler potential and its density dependence is given by

$$V(\rho_a) = b_1(1+a_1\alpha^2)\,\rho + b_2(1+a_2\alpha^2)\,\rho^{4/3} \\ + b_3(1+a_3\alpha^2)\rho^{5/3}. \tag{8}$$

The parameters a_i and b_i (i = 1,2,3,) have kindly been supplied to us by Dr. Buchler. Φ_c, the Coulomb potential is given by

$$\Phi_c = e\int d^3r'\rho_p(\vec{r}')/1\vec{r}-\vec{r}'1 \tag{9}$$

The term next to Φ_c in (7) is the correction to the classical Coulomb energy due to the Pauli principle [12].

The potential, V(R), between a daughter pair separated by a distance R is given by

$$V(R) = E(\rho_1(R),\rho_2(R)) - E(\rho_1(\infty)) - E(\rho_2(\infty)) \\ = E(R) - E(\infty) \tag{10}$$

In the above, $E(\rho_1(R), \rho_2(R))$ is the energy of the compound system calculated using (6) by superposing two densities $\rho_1(R)$ and $\rho_2(R)$, each of which varies continuously as a function of R. $E(\rho_1(\infty))$ and $E(\rho_2(\infty))$ are the densities of the daughter nuclei when they are far apart. One has further the auxiliary condition

$$A = \int \rho(r)d^3r \tag{11}$$

We further impose the conditions that (i) the density of the compound system at any point does not exceed the central density of the parent nucleus, (ii) at R = ∞ one reaches the appropriate density distributions of a particular daughter pair and (iii) at R = 0, one has the density appropriate to the parent nucleus.

Once the central densities, ρ_o, of the parent and a particular daughter pair are determined by (11) for the observed density distribution function, parameters characterizing density distribution function of the parent and the corresponding daughter pair such as half-density radius c and the surface thickness parameter t (i.e., the distance where the density drops from 90% of its value to 10%) may be determined by minimizing the energy $E(\rho_1(R), \rho_2(R))$. Such a point by point variation [3] is very time consuming and almost similar results may be obtained by using the following ansatz [3,13], called the special adiabatic approximation

$$C_d(R) = C_p \exp[\ell n(C_d(\infty)/C_p)] (R/R_{cut})^2 \tag{12a}$$

and

$$t_d(R) = t_d \exp[\ell n (ct(\infty)/t_p)] (R/R_{cut})^2 \tag{12b}$$

where the subscripts d and p stand for daughter and parent, respectively. R_{cut} in (12a) and (12b) is a scaling parameter which determines the distance at which the parameters of the daughter pair reach their final values $C_d(\infty)$ and $t_d(\infty)$. Beyond R_{cut} the density distributions of the daughter nuclei remain unchanged.

In the case of spheroidal density distributions, we have used eccentricities as inputs.

Calculations of half-lives requires a knowledge of the asymptotic kinetic energy, E_{kin} of a daughter pair. We have used the experimental data, whenever possible. In other cases, E_{kin} is determined from

$$Z_1 Z_2 e^2/R_{sc} = E_{kin} + f(A_p) \tag{13}$$

Classical physics demands $f(A_p) = 0$. However, there are uncertainties in the knowledge of scission radii, and the exact excitation of daughter pair etc. and $f(A_p)$ is an empirical way to correct for these energy losses

4. RESULTS AND DISCUSSION

The first test of the energy-density functional approach is to investigate whether or not it can reproduce the known masses. As noted earlier, this approach can, indeed, reproduce the observed masses with proper density distributions as well as those which are obtained from the standard mass formula based on the liquid drop model which, however, does not reproduce observed density distribution [3, 14-16]. In [15,16] we have compared our calculated masses with the observed ones and the agreement is excellent.

Table 2. Calculated and observed log of half-lives. Columns 1 through 7 are, respectively, the parent nucleus, the decay modes, eccentricity of the daughter pair for the spheroidal shape, kinetic energy used in calculation which is obtained from data, calculated half-lives for spherical shapes, calculated half-lives for spheroidal shapes and experimental half-lives [ref. 17 for Fm and ref. 20 otherwise].

Parent	Daughter	ϵ	E_{kin} (MeV)	logT/2 (th-s) years	log T/2 (th-d) years	logT/2 (exp) years
234 U 92	142 92 Xe+ Sr	.48	168.1	16.2	16.5	16.2
236 U 92	140 96 Xe+ Sr	.44	171.6	13.7	13.2	16.3
240 Pu 94	142 98 Ba+ Sr	.53	175.0	11.1	11.1	11.1
244 Cm 96	144 100 Ba+ Zr	.55	185.5	8.4	8.4	7.1
248 Cf 98	152 96 Sm+ Kr	.54	188.7	3.0	2.2	3.8
252 Cf 98	144 108 Xe+ Ru	.69	187.0	3.3	1.9	1.9
258 Fm 100	128 130 Sm+ Sn	0.00	215.0	-11.5	-	-10.9

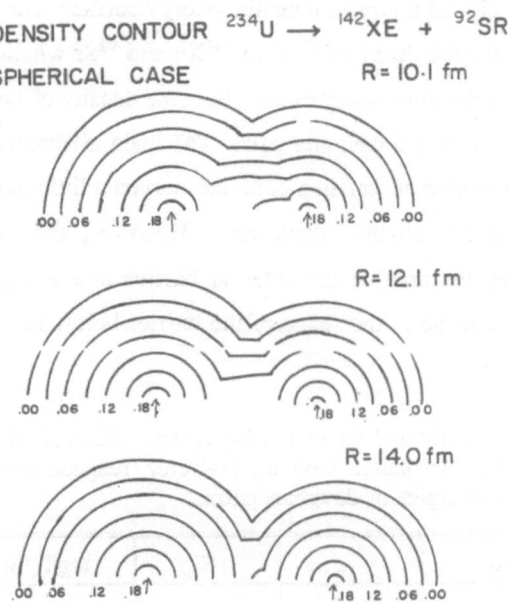

DENSITY CONTOUR ^{234}U ⟶ ^{142}XE + ^{92}SR

SPHERICAL CASE

R = 10·1 fm

R = 12.1 fm

R = 14.0 fm

Fig. 3.   A typical example of the evolution of the density contour as a function of seperation distance R for the case of spherical shape of a daughter pair.

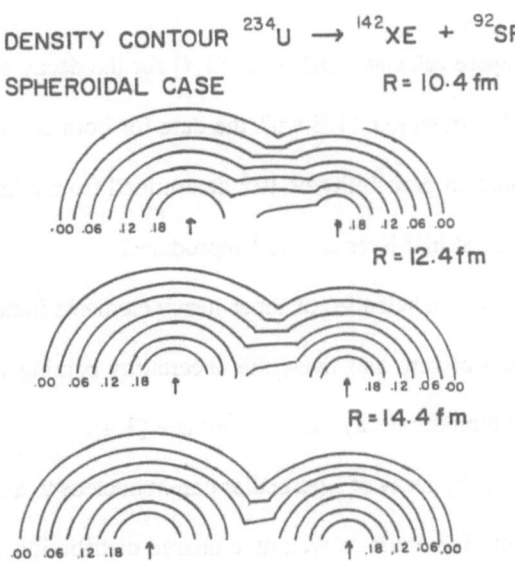

DENSITY CONTOUR ^{234}U ⟶ ^{142}XE + ^{92}SR

SPHEROIDAL CASE

R = 10·4 fm

R = 12.4 fm

R = 14.4 fm

Fig. 4.   The same as in Fig. 3, except each member of the daughter pair is spheroidal with an eccentricity ϵ of 0.48.

In Fig. 3 we have plotted a typical potential energy surface with density contours. This particular one is for the decay of ^{234}U to ^{142}Xe and ^{92}Sr which are assumed to be spherical. Fig. 4 shows the same decay mode where the density of the compound system is generated from the spheroidal daughter pair. The main differences between the two cases are that in the latter case the scission point has moved a little outward and the depth of the external potential is slightly shallower. However, the decay processes are essentially governed by the area of the external barrier above E_{kin}, the asymymtotic kinetic energy and the change in the shape of the barrier below E_{kin} does not affect the calculations of half-lives.

Table 3. Calculated, logT(th) and observed, logT(exp) [21] half-lives for the decay of ^{240}Pu to a number of daughter pairs. Column 1-4 refer, respectively, to mass numbers, eccentricity and kinetic energies of daughter pairs.

A_1	A_2	ϵ	E_{kin}	logT(th)	logT(exp)
92	148	0.66	168	11.3	11.5
98	142	0.62	175	11.2	11.1
100	140	0.62	178	12.1	11.1
106	134	0.62	186	11.0	11.0
108	132	0.57	186	11.6	11.2
118	122	0.72	168	16.3	14.0

In table 2, we compare calculated half-lives [3,4] for the decay of a few activides to their fastest modes using observed TKE with the data for both cases. The calculations are done for a preformation probability of 10^{-5} determined from a calculation using the shell model [1]. Observed half-lives are well reproduced.

Our calculations of fission half-lifes of super-heavy elements indicate that they could at best have a very short or zero half-lives, the uncertainty coming from the estimation of TKE and that the symmetric decay modes dominate [3,4].

The theory allows for the decay of a particular parent nucleus to a number of different daughter pairs i.e., there is a mass as well as a charge distribution for decay products which is a characteristic of fission. In table 3, we have compared the calculated half-lives with the data for the decay of ^{240}Pu in a number of modes using observed E_{kin}. The shape of the daughter pairs is assumed to be spheroidal, each member of a pair having the same eccentricity. The calculation confirms the dominance of the asymmetric modes of decay over the symmetric ones, as observed and reproduces the observed yields reasonably well.

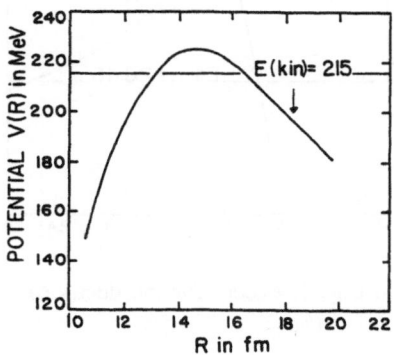

Fig. 5.   The external part of the potential energy surface for the decay of ^{258}Fm to ^{128}Sm and ^{130}Sn. Asymtotic kinetic energy, E(kin) refers to E_{kin} and is marked by a straight line. The arrow indicates scission point.

It has now been established that the dominat decay modes of ^{258}Fm [17] are symmetric, a fact that was explained by Hooshyar and Malik [6] using an empirical external barrier. In Fig. 5 we present the external part of our calculated potential surface for the decay of ^{258}Fm to ^{128}Sn and ^{130}Sn. A value of 215 MeV has been used for E_{kin} which is in line with the data [18]. The calculated half-life shown in table 2 is very short and is in reasonable agreement with the observed one.

The theory discussed here and in refs. [1-10] allows a parent nucleus to decay in all possible decay modes including alpha-particle and light nuclei. The decay mode may appear to be non-existent, if the half-lives are very long. In the last decade a number of activides has been found to emit light nuclei such as ^{14}C, ^{20}Ne [19]. We present here a preliminary calculation for the decay of ^{226}Ra to ^{212}Pb and ^{14}C. The external barrier for this case is shown in Fig. 6.

Using an $E_{kin} = 26.46$ MeV we get a half-life of 11.3 years which is the experimental number. However, $E_{kin} = 26.46$ MeV means that the daughter products are slightly excited.

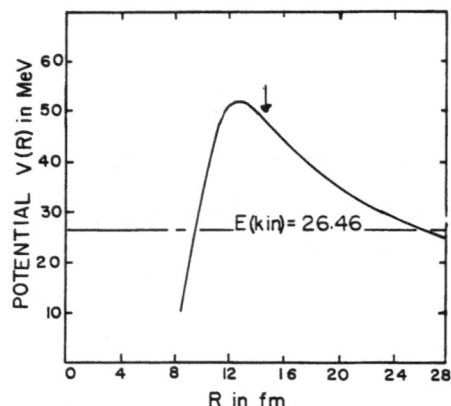

Fig. 6.   The same as in Fig. 5, except for the decay of ^{226}Ra to ^{212}Pb and ^{14}C.

5. CONCLUSION

The energy-density functional approach which has its roots in nuclear many body theory and a realistic two-nucleon potential can successfully explain half-lives and mass distributions observed in spontaneous fission, light-ion radioactivity and alpha-decay. It can also calculate the general features of the elastic scattering of two heavy-ions [13,15].

REFERENCES

1.   B. Block, J. W. Clark, M. D. High, R. Malmin and F. B. Malik, Ann. Phys. (N.Y.) 62, 464 (1971).

2.   I. Reichstein and F. B. Malik, Proc. Int'l. Conf. on Nucl. Phys. eds. de Boer (North - Holland Publ, Amsterdam 1973).

3.   I. Reichstein and F. B. Malik, Ann. Phys. (N.Y.) 98, 322 (1976).

4.   I. Reichstein and F. B. Malik, ed. M. A. K. Lodhi Super Heavy Elements (Gordon Bveach, 1978).

5.   M. A. Hooshyar and F. B. Malik, Phys. Lett. 38B, 495 (1972); ibid. 55B, 144 (1975).

6. M. A. Hooshyar and F. B. Malik, Helv. Phys. Acta 46, 720, 724 (1973); ibid, 45, 567 (1972).

7. B. Compani-Tabrizi, M. A. Hooshyar and F. B. Malik, Proc. Int'l. Conf. on the interaction with Nuclei, ed. E. Sheldon (ERDA publ. No. Conf. - 7607115-p1, 1976) p. 725.

8. B. Compani-Tabrizi, M. A. Hooshyar and F. B. Malik, Proc. V. Intl'l. Conf. on Nucl. Reaction Mechanism. ed. E. Gadioli (Univ. of Milan Press, 1988) p. 385.

9. B. Compani-Tabrizi, M. A. Hooshyar and F. B. Malik, 50 Years with Nuclear Fission eds. J. W. Behrens and A. D. Carlson (American Nuclear Society 1989) p. 643.

10. K. A. Brueckner, S. A. Coon and J. Dabrowski, Phys. Rev. 168, 1184 (1968).

11. K. A. Brueckner, J. R. Buchler, R. C. Clark and R. J. Lombard, Phys. Rev. 181, 1534 (1969).

12. D. C. Peaslee, Phys. Rev. 95, 717 (1959).

13. I. Reichstein and F. B. Malik, Phys. Lett. 37B, 344 (1971).

14. R. J. Lombard, Ann. Phys. (N.Y.) 77, 380 (1973).

15. I. Reichstein and F. B. Malik, Condensed Matter Theories, 1, 291 (1986).

16. I. Reichstein and F. B. Malik, Condensed Matter Theories, 6, 335 (1991).

17. W. John, E. K. Hulet, R. W. Longheed and J. J. Wesolowski, Phys. Rev. Lett. 27, 45 (1971).

18. E. K. Hulet, 50 Years with Nuclear Fission eds. J. W. Behrens and A. D. Carlson (American Nuclear Society 1989) p. 533.

19. For an overview see B. Price, Nuclear and Atomic Clusters eds. T. Loenroth, M. Brenner and F. B. Malik (Springer Verlag 1992) p. 273.

20. E. K. Hyde, The Nuclear Properties of heavy Elements (Prentice Hall, Englewood Cliffs, NJ. USA).

21. J. N. Neiler, E. J. Walter, and H. W. Schmitt, Phys. Rev 149, 894 (1966).

FHNC STUDY OF N=Z NUCLEI

Adelchi Fabrocini

Dept. of Physics, University of Pisa
and
INFN, Sezione di Pisa, I-56100 Pisa, Italy

Abstract: *Correlated Basis Function theory and Fermi HyperNetted Chain theory are used to study some ground state properties of N=Z nuclei. The ground state energy, the One Body Density Matrix and the Momentum Distribution are computed for $^4He,^{16}O$ and ^{40}Ca, using central nucleon-nucleon potentials and scalar correlations. The results compare satisfactorily with the available exact Monte Carlo calculations.*

1. Introduction

The non-relativistic nuclear many-body theory has achieved in the past years many successes in describing the properties of many nuclear systems, ranging from deuteron to nuclear matter, by means of the hamiltonian

$$H = T + V = \frac{-\hbar^2}{2\,m} \sum_i \nabla_i^2 + \sum_{i<j} v_{ij} + \sum_{i<j<k} v_{ijk}, \qquad (1.1)$$

where the two-body potential v_{ij} is required to fit the deuteron and the nucleon-nucleon scattering data, and the three-body potential v_{ijk} ensures a good description of both light nuclei and nuclear matter properties [1]

The many-body Schrödinger equation in the A=3,4 nuclei may be exactly solved for several nuclear potentials within the Faddeev [2] (A=3) and Green Function Monte Carlo (GFMC) [3] (A=3,4) theories and attempts have been made to use GFMC in ^{16}O, but only with a simplified interaction [4].

Good results in the description of both static and dynamical properties of light nuclei have been also obtained by means of variational methods [5,6], which appear to be the only ones able to succesfully deal with realistic interactions in heavy nuclear systems.

Within the variational theories, the hamiltonian (1.1) has been used, in conjunction with Correlated Basis Function (CBF) theory [7,8] to accurately study the properties of infinite systems of nucleons (neutron and nuclear matter) and light nuclei [5].

The CBF theory is based on a set of correlated A-body wave functions

$$\Psi_n(1,2...A) = F(1,2..A)\Phi_n(1,2...A), \qquad (1.2)$$

where $\Phi_n(1,2...A)$ is a generic mean field state and $F(1,2...A)$ is an A-body correlation operator acting on Φ_n and taking into account the effects on the wave function of the nuclear hamiltonian. The operatorial dependence of the correlation operator, in the most realistic CBF calculations, is shaped on that of the nucleon-nucleon interaction, having central, tensor and spin-orbit components. However, in the following, I will limit myself to simple scalar (Jastrow) correlations,

$$F(1,...,A) = \prod_{i<j} f(r_{ij}), \qquad (1.3)$$

as a first step toward more realistic cases. The function $f(r)$ usually depends upon some variational parameters, which are fixed by minimizing the ground state expectation value of the hamiltonian, $\langle H \rangle$.

With simple Jastrow-correlated wave functions, $\langle H \rangle$ and other quantities of interest may be calculated in finite systems by using Monte Carlo (MC) techniques to sample the necessary many-body integrals. This is not feasible in infinite nuclear and neutron matters, where Fermi HyperNetted Chain (FHNC) theory has been successfully applied [9,10]. Monte Carlo has been also used with realistic correlations and interactions in $A \leq 6$ nuclei, but the extension to heavier nuclei appears problematic. Recently Pieper et al. [11] have performed variational calculations of the binding energy of ^{16}O with realistic correlations using Monte Carlo techniques. In this calculation, the Jastrow part of the correlation has been exactly treated by MC, whereas the contribution of the non scalar parts has been approximated by considering up to four-body cluster terms.

Low order cluster expansions have been used by Guardiola et al. [12] to study nuclei up to ^{40}Ca, with simple central interactions and Jastrow correlations and Boscá and Guardiola [13] have applied similar techniques to light nuclei ($A \leq 16$), using realistic interactions and state dependent correlations.

Obviously, low order cluster expansions become more and more unreliable as A increases; however, microscopic studies of nuclei in the medium-heavy mass region are necessary to clearly assess the interplay between surface, shell and correlations effects. In this perspective, the application of the FHNC theory to finite nuclei appears particularly appealing since this approach could provide the tool to overcome the technical and numerical difficulties of the Variational Monte Carlo (VMC) and of the finite cluster expansion techniques.

Finite systems FHNC theory was originally developed by Fantoni and Rosati [14] for Jastrow correlations; later, Krothscheck et al. [15] implemented and applied the theory to study surface properties of quantum liquids.

In sect.2 of this paper, I will present a variational calculation of the binding energies of various N=Z nuclei, for the case of state-independent Jastrow correlation operators and semi-realistic nucleon-nucleon (NN) central interactions, in the framework of the FHNC theory. In sect.3, results for the One Body Density matrix (OBDM) and the Momentum Distribution (MD) will be given and discussed. Finally, short conclusions and possible future developments will be presented in sect.4.

2. Ground state energy

The FHNC calculation, in N=Z nuclei, of the expectation value on the ground state of an hamiltonian containing only two-body, central (no tensor components) interactions, and with a Jastrow correlated wave function, is described at length in ref.[16] (hereafter, referred to as I). Here I will briefly sketch the main features of the method.

The FHNC theory allows to calculate expectation values of operators by performing an expansion in power series of the function

$$h(r) = f^2(r) - 1. \tag{2.1}$$

The product form of the correlation operator generates *cluster terms* characterized by both the number of particles involved, and the number of the dynamical correlations $h(r_{ij})$ linking the particles. Moreover, each n-body cluster may contain *statistical correlations* generated by the Pauli principle and embedded in the mean field ground state wave function $\Phi_0(1, 2...A)$, which is taken to be the Slater determinant of single particle wave function $\phi_\alpha(i)$

$$\Phi_0(1, ...A) = \mathcal{A}(\phi_1(1), ...\phi_A(A)), \tag{2.2}$$

where $\mathcal{A}$ is the antisymmetrization operator and $\phi_\alpha(i)$ are eigenfunctions of the single-particle hamiltonian

$$h_{sp}(r) = -\frac{\hbar^2}{2m}\nabla_r^2 + U(r). \tag{2.3}$$

Key quantities entering the calculation of $\langle H \rangle$ are the one- and two- body distribution functions

$$\rho_1(\vec{r}_1) \equiv \rho(\vec{r}_1) \equiv \left\langle \Psi_0^* \sum_i \delta(\vec{r}_1 - \vec{r}_i)\Psi_0 \right\rangle, \tag{2.4}$$

$$\rho_2^{(n \leq 4)}(\vec{r}_1, \vec{r}_2) \equiv \left\langle \Psi_0^* \sum_{i \neq j} \delta(\vec{r}_1 - \vec{r}_i)\delta(\vec{r}_2 - \vec{r}_j)O^{(n)}(i, j)\Psi_0 \right\rangle, \tag{2.5}$$

where $O^{(n=1,4)}(i,j) = 1, \vec{\sigma}_i \cdot \vec{\sigma}_j, \vec{\tau}_i \cdot \vec{\tau}_j, (\vec{\sigma}_i \cdot \vec{\sigma}_j)(\vec{\tau}_i \cdot \vec{\tau}_j)$.

The clusters contributing to $\rho_2^{(n)}(\vec{r}_1, \vec{r}_2)$ are most conveniently represented by diagrams, that are grouped according to both the exchange character of the external points and their topology. Following this last criterion, the diagrams may be distinguished in nodal ($N_{\alpha\beta}$), composite ($X_{\alpha\beta}$) and elementary ($E_{\alpha\beta}$) (see I for a more complete discussion).

The sums, at all orders, of the nodal and composite diagrams can be expressed in a closed form through the FHNC integral equations, once the sum of the elementary diagrams is given. Unfortunately, no exact method to evaluate this contribution has been devised so far and one has to use approximations to estimate $E_{\alpha\beta}$. The simplest one consists in setting $E_{\alpha\beta} = 0$ (FHNC/0 approximation). FHNC/0 has been used in most of the calculations of the equations of state of nuclear and neutron matter [7], indicating that the elementary diagrams give very little contribution in the case of translational invariant nuclear systems. More sophisticated approximations have been also used in literature, as the FHNC/4 (only the lowest order elementary diagrams are considered), the scaling [17] and the interpolating [18] approximations.

In order to test the ability of the FHNC scheme of treating the strong nucleon-nucleon correlations, we have considered central potentials which are strongly repulsive at short distances and have been used in previous calculations on complex nuclei. The models that have been studied are: the spin-isospin independent Malfiet-Tjon (MT) potential [19] (v_{MT}), which has been used in Variational Monte Carlo, Green Function Monte Carlo [4] and FHNC/c [20] calculations of ^{16}O; the Brink-Boeker B1 central potential (v_{B1}) [21] and the semi-realistic central interaction S3 by Afnan and Tang [22](v_{S3}). MT does not saturate nuclear matter, leads to unphysically dense nuclei and has been considered only for the sake of comparison with previous calculations; B1 is an effective interaction that reproduces nuclear matter saturation properties and ^4He binding energy in Hartree-Fock theory; S3 reproduces the s-wave two-body scattering data up to roughly 60 MeV, it provides reasonable values of the binding energies and of the radii of the $A = 3, 4$ nuclei and an equation of state of nuclear matter which is not too different from those obtained with realistic two-body interactions. Since the original S3 potential is only defined for two-particle states of even parity, analogously to what has been done in ref.[12], we supplemented the potential in the odd channels with a repulsive interaction given by the repulsive term of the even channels.

In addition to the binding energies and the one-body density $\rho_1(\vec{r})$, particular attention has been also devoted to the normalization sum rules:

$$S_1 = \frac{1}{A} \int d^3r \, \rho_1(\vec{r}) = 1, \qquad (2.6.a)$$

$$S_2 = \frac{1}{A(A-1)} \int d^3r_1 d^3r_2 \, \rho_2^{(1)}(\vec{r}_1, \vec{r}_2) = 1, \qquad (2.6.b)$$

and

$$S_\sigma = \frac{1}{3A} \int d^3r_1 d^3r_2 \, \rho_2^{(n=2,3)}(\vec{r}_1, \vec{r}_2) = -1, \qquad (2.6.c)$$

which provide useful information on the accuracy of the approximations used to solve the FHNC equations.

The ground-state energy of ^{16}O interacting via the MT potential has been computed in FHNC/0 and compared with the VMC calculation of ref.[4]. The correlated wave function contains the same single particle wave functions of ref.[4] (generated by a Woods-Saxon potential with parameters $U_o = -207 MeV$, $a = 0.5 fm$ and $R = 1 fm$) and an Euler-like Jastrow factor $f_{Eul}(r)$ determined by minimizing the lowest order cluster expansion $\langle H_2 \rangle$ of $\langle H \rangle$, with healing distance $d = 2 fm$ (see I for more details). FHNC/0 provides $\langle H \rangle_{FHNC/0} = -987 MeV$, to be compared with $\langle H \rangle_{VMC} = -1024 \pm 5 MeV$; the sum rules S_1 and S_2 are satisfied within less than percent, whereas $S_\sigma = -1.030$. For the same potential, GFMC method gives $\langle H \rangle_{GFMC} = -1194 \pm 20 MeV$ [4] and, after a rough minimization on the Woods-Saxon parameters, we obtain $\langle H \rangle_{FHNC/0} = -1152 MeV$.

The above results seem to show that FHNC/0 is satisfactorily reliable, especially in view of the fact that, in the MT model, the central density is rather large $(\rho_1(0) = 1.21 fm^{-3})$ and the importance of the elementary diagrams is known to increase rapidly with the density of the system. On the other hand, one has to be careful in drawing this conclusion because the purely scalar character of the Malfiet-Tjon potential may mask the effects of the elementary diagrams of exchange type, as it is suggested by the fact the sum rule (2.6.c) is not very well satisfied. FHNC/0 cannot satisfy this sum rule since the elementary diagrams are disregarded, and, already at the first order in h, an elementary belonging to the class of the exchange diagrams appears. Its inclusion in the FHNC/0 scheme leads to the FHNC-1 approximation. FHNC-1 largely improves on FHNC/0, giving $S_\sigma = -0.998$ and leaving $\langle H \rangle$ unaffected.

A problem similar to the one encountered in the S_σ sum rule, may show up with potentials having strong exchange parts, as the B1 potential (a Wigner-Majorana admixture with a large Majorana component). The binding energies of ^4He, ^{16}O and ^{40}Ca have been computed with this potential and with harmonic oscillator single particle wave functions and correlation functions of gaussian form:

$$f_G(r) = 1 + \alpha \; exp(-\beta r^2), \tag{2.7}$$

In Table 1 the FHNC results are compared to the VMC energies. The FHNC-1 correction has been found negligible for the Wigner potential expectation value and the kinetic energy , which come out to be in good agreement with the VMC estimates. This result, together with the fact that the sum rules S_1 and S_2 are very well satisfied, makes one confident that the FHNC/0 treatment of the Jastrow correlations is rather accurate. However, for both S_σ and the expectation value of the Majorana part, a more accurate treatment of the statistical correlations is needed. Their FHNC-1 estimates are very close to the exact VMC results (S_σ is satisfied within less than 1%).

Finally, the S3 potential has been considered as an example of semi-realistic nucleon-nucleon interaction. No particular difference with respect to the previous cases has been found. The results for ^{16}O and ^{40}Ca are shown in Tab.2. It should be noticed that the FHNC-1 contribution is always of the same magnitude, irrespectively from the interaction used.

Table 1. *Binding energy of ^4He, ^{16}O and ^{40}Ca nuclei for the B1 potential, with harmonic oscillator single particle wave functions and gaussian correlations f_G. The parameters are given in ref.[16]. The expectation values of the Majorana and Wigner ($\langle v_M \rangle$ and $\langle v_W \rangle$) potentials, the kinetic energy ($\langle T \rangle$), the center of mass contribution T_{cm}, and the energy ($E = \langle H \rangle - T_{cm}$) are listed. All the energies are expressed in MeV. The VMC results for ^4He and ^{16}O are from Ref.[23] (the errors are $\sim \pm .2 MeV$); the VMC result for ^{40}Ca is from Ref.[24] (the error is $\sim \pm 1. MeV$).*

Nucleus		$\langle v_M \rangle$	$\langle v_W \rangle$	$\langle T \rangle$	T_{cm}	E
^4He	FHNC/0	−132.5	24.6	83.9	20.5	−44.5
	FHNC − 1	−125.6				−37.7
	VMC	−123.8	24.8	83.0		−36.4
^{16}O	FHNC/0	−421.6	−63.3	329.8	13.1	−168.2
	FHNC − 1	−403.8				−150.4
	VMC	−402.6	−62.3	327.1		−150.9
^{40}Ca	FHNC/0	−1065	−407	956	11.2	−516
	FHNC − 1	−1020				−482
	VMC					−481

Table 2. *Ground state energies of ^{16}O and ^{40}Ca nuclei for the S3 potential, with harmonic oscillator single particle wave functions and Euler correlations f_{Eul}. The parameters are given in ref.[16]. See Table 1.*

Nucleus		$\langle v \rangle$	$\langle T \rangle$	T_{cm}	E
^{16}O	FHNC/0	−479.1	378.7	13.2	−113.5
	FHNC − 1	−470.8			−105.3
^{40}Ca	FHNC/0	−1483	1124	11	−370
	FHNC − 1	−1463			−350

3. One Body Density Matrix and Momentum Distribution

The Momentum Distribution $n(k)$ in nuclei is known to be strongly affected by the correlations. In correlated nuclear matter, a fraction of the nucleons is promoted to momenta larger than the Fermi momentum k_F, depleting the MD respect to one below k_F and enhancing it respect to zero above. A similar picture holds in nuclei, where high momentum states are partially occupied because of the correlations.

VMC has been used to study MD in light nuclei [25] within CBF and, in conjunction with cluster expansions, in ^{16}O [26]. A low order cluster expansion has been used to compute MD, with realistic correlations, in ^{16}O and ^{40}Ca by Benhar and collaborators [27]. The FHNC method has been recently used to compute MD in N=Z nuclei, with Jastrow correlations [28], and its results will be shortly described in the following.

$n(k)$ is given by

$$n(k) = \frac{1}{A} \int d^3r_1 \int d^3r_{1'} \, \rho(\vec{r}_1, \vec{r}_{1'}) \, e^{i\vec{k}\cdot(\vec{r}_1 - \vec{r}_{1'})}, \tag{3.1}$$

where, $\rho(\vec{r}_1, \vec{r}_{1'})$ is the One Body Density Matrix, defined as

$$\rho(\vec{r}_1, \vec{r}_{1'}) = A \int d^3r_2 ... \int d^3r_A \, \Psi_o^\dagger(1, 2...A) \, \Psi_o(1', 2...A), \tag{3.2}$$

and whose diagonal part, times the degeneration number $\nu = 4$, gives the One Body Density (OBD) $\rho_1(\vec{r}_1)$.

MD has to be normalized

$$\frac{\nu}{(2\pi)^3} \int d^3k \, n(k) = 1, \tag{3.3}$$

and it is related to the kinetic energy $\langle T \rangle$ by the relation

$$\frac{A\nu}{(2\pi)^3} \frac{\hbar^2}{2m} \int d^3k \, k^2 \, n(k) = \langle T \rangle. \tag{3.4}$$

The FHNC equations for the OBDM in homogeneous infinite systems have been derived by Fantoni [29], and the extension of the theory to finite systems is discussed in ref.[28]. In strict analogy to what has been outlined in sect.2, a cluster expansion may be obtained for the OBDM, by developing it in powers of $h(r)$ and of the function $\eta(r) = f(r) - 1$. Points 1 and 1', whose coordinates are the arguments of the OBDM, can be reached only by the η dynamical correlation. Its presence increases the number of the types of diagrams, which, however, can still be grouped into nodal, composite and elementary ones; the first two classes may be summed via the FHNC equations, provided an approximation for the elementary diagrams is given.

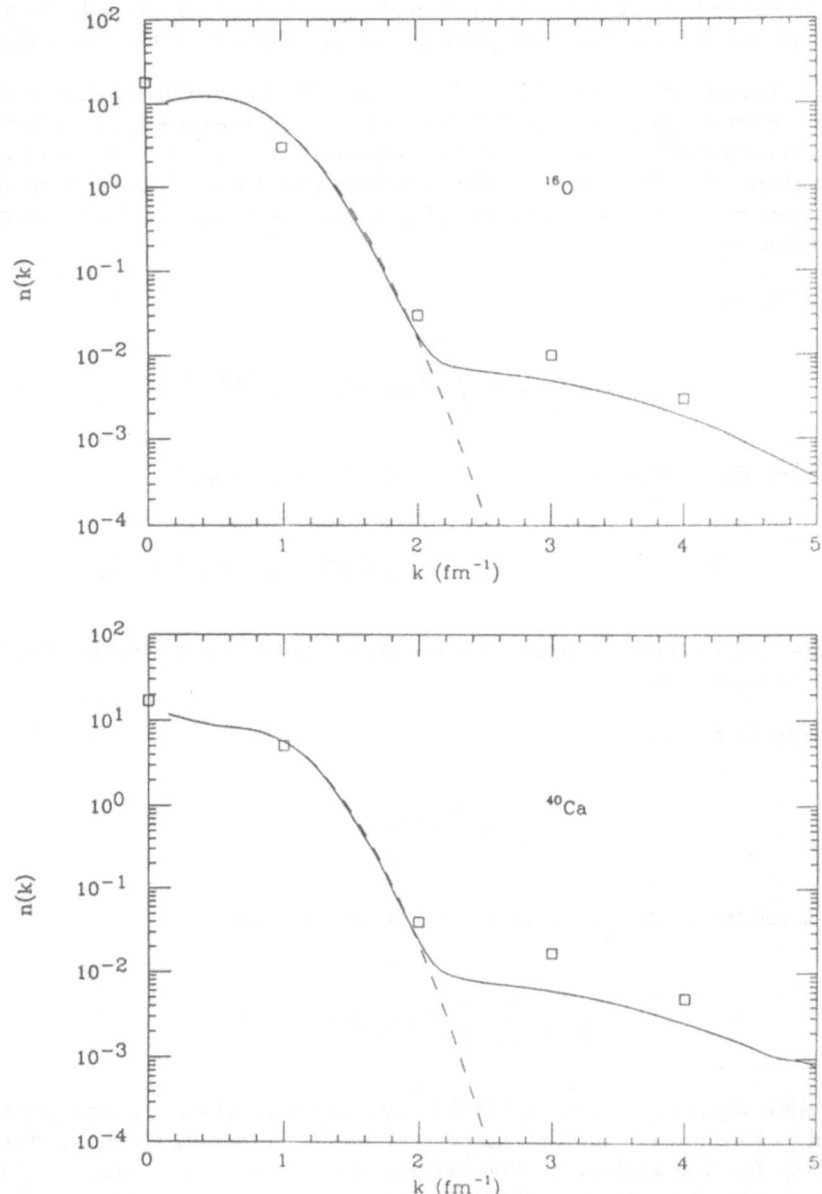

Fig. 1 *From ref.[28]. The momentum distributions of* ${}^{16}O$ *and* ${}^{40}Ca$. *The squares are from ref.[27]. See text.*

Again, the degree of violation of conditions (3.3) and (3.4) is a measure of the importance of the elementary diagrams. An additional, powerful condition is obtained by requiring that the diagonal part of the OBDM coincides with $\rho_1(\vec{r}_1)$ of eq.(2.4).

The OBDM has been computed for ^{16}O and ^{40}Ca in FHNC/0 and with the same wave functions adopted in sect.2 for the S3 potential. In both cases, condition (3.3) is satisfied to less than 0.1%. The kinetic energies, computed via eq.(3.4), turn out to be $384\,MeV$ in ^{16}O and $1140\,MeV$ in ^{40}Ca, to be compared with the results quoted in Table 2 ($379\,MeV$ for ^{16}O and $1124\,MeV$ for ^{40}Ca). Finally, the differences between the diagonal part of the OBDM and the OBD of eq.(2.4), are always less than 0.5 % in both nuclei.

The above results indicate that FHNC/0 is very accurate for the calculation of the OBDM and the MD. It is likely that the differences in the kinetic energies are to be ascribed to numerical difficulties in computing the veryh igh momentum tail of $n(k)$.

The momentum distributions of the two nuclei are shown in Figure 1. The solid lines give the FHNC/0 results; the dashed lines are the Independent Particle Model (IPM) momentum distributions and the squares are the low order cluster expansion results of ref.[27] with state-dependent correlations. The striking feature is the high momentum enhancement of $n(k)$ respect to the IPM results, due to the introduction of the correlations. By comparison with the results of ref.[27], it appears that the simple Jastrow choice slightly underestimates the net effect of the correlations and that the introduction of a more complicate operatorial dependence (especially tensorial) may provide further enhancement.

4. Conclusions

The work described in this paper intends to be a first step towards a microscopic description of medium-heavy nuclei with correlated wave functions of the type extensively used in nuclear matter.

Jastrow correlated wave functions and FHNC summation techniques have been used to compute the ground state energy, the One Body Density Matrix and the Momentum Distribution in N=Z nuclei, with model interactions, and compared with available Monte Carlo results.

A new FHNC scheme (named FHNC-1), that takes particular care of the exchange diagrams, has shown to be capable of treating at a satisfactory level of accuracy the state independent part of the nucleon-nucleon correlation. This is known to be the part of the correlation which needs to be treated with the highest possible accuracy. The FHNC-1 approximation is readily applicable to the case of state dependent correlation operators, and therefore to the use of realistic nucleon-nucleon interactions.

This last point, and the extension of the theory to $N \neq Z$ nuclei, are the next necessary steps to be taken. Our results appear to be encouraging and show that a microscopic, quantitative description of the ground state of heavy nuclei is at reach of CBF theory.

Acknowledgments

The results presented here have been obtained in close collaboration with Stefano Fantoni, GiamPaolo Co' and Isaac Lagaris. I am also very grateful to Omar Benhar, Rafael Guardiola and Vijay Pandharipande for many useful discussions. E.Buendia, A.M.Lallena, S.Pieper and M.Viviani are gratefully aknowledged for providing unpublished VMC results.

References

[1] V.R.Pandharipande, "Proceedings of Cargese summer School 1989", ed. J. Tran Tranh Van and J. Negele (Plenum Press, NY 1990).

[2] C.R.Chen, G.L.Payne, J.L.Friar and B.F.Gibson, Phys.Rev. **C33**(1986)1740;

A.Stadler, W.Glöckle add P.U.Sauer, Phys.Rev. **C44**(1991)2319.

[3] J.Carlson, Phys.Rev. **C36**(1987)2026.

[4] U.Helmbrecht and J.G.Zabolitzky, Nucl.Phys. **A442**(1985)109.

[5] J.Carlson, V.R.Pandharipande and R.Schiavilla, in "Modern Topics

in Electron-Scattering", ed. B.Frois and I.Sick (World Scientific, Singapore, 1991).

[6] M.Viviani, A.Kievsky and S.Rosati, Nuovo Cimento **A**(1992), in press.

[7] R.B.Wiringa, V.Ficks and A.Fabrocini, Phys.Rev. **C43**(1991)2605.

[8] O.Benhar, A.Fabrocini and S.Fantoni, in "Modern Topics in Electron-Scattering", ed. B.Frois and I.Sick (World Scientific, Singapore, 1991).

[9] V.R.Pandharipande and R.B.Wiringa, Rev.Mod.Phys. **51**(1979)821.

[10] S.Rosati, in "From nuclei to particles", Proc.Int.School E.Fermi, course LXXIX, ed. A.Molinari (North Holland, Amsterdam, 1982).

[11] S.C.Pieper,R.B.Wiringa and V.R.Pandharipande, Phys.Rev.Lett. **64**(1990)364.

S.Pieper, in "Many-Body Theories" ed. Y.Avishai, Vol.2 (Plenum Press ,NY 1990).

[12] R.Guardiola, A.Faessler, H.Müther and A.Polls, Nucl.Phys.**A371**(1981)79.

[13] M.C.Boscá and R.Guardiola, Nucl.Phys. **A476**(1988)471;

R.Guardiola and M.C.Boscá Nucl.Phys. **A489**(1988)45.

[14] S.Fantoni and S.Rosati, Nucl.Phys. **A328**(1979)478.

[15] E.Krotscheck, W.Kohn and G.-X.Qian, Phys.Rev. **B32**(1985)5693.

[16] G.Co',A.Fabrocini,S.Fantoni and I.E.Lagaris, Pisa preprint IFUP/TH 15/92.

[17] E.Manousakis et al., Phys.Rev. **B28**(1983)3770.

[18] M.Viviani et al., Nuovo Cimento **D8**(1986)561.

[19] R.A.Malfiet and J.A.Tjon, Nucl.Phys. **A127**(1969)161.

[20] E.Krotschek, Nucl.Phys. **A465**(1987)461.

[21] D.M.Brink and E.Boeker, Nucl.Phys. **A91**(1967)1.

[22] I.R.Afnan and Y.C.Tang,Phys.Rev. **175**(1968)1337.

[23] M.C.Boscá, E.Buendía, R.Guardiola, Phys.Lett.**198B**(1987)312
and "Condensed Matter Theories" ed. J.Arponen, R.Bishop and M.Manninen,
Vol.3 (Plenum Press, NY 1987).

[24] R.Guardiola, private communication.

[25] R.Schiavilla,V.R.Pandharipande and R.B.Wiringa, Nucl.Phys. **A449**(1986)219.

[26] S.Pieper, private communication.

[27] O.Benhar et al., Phys.Lett.**135B**(1986)135.

[28] G.Co',A.Fabrocini and S.Fantoni, Pisa preprint IFUP/TH 28/92.

[29] S.Fantoni, Nuovo Cimento **A44**(1978)191.

A COUPLED CLUSTER STUDY OF ABELIAN LATTICE GAUGE FIELD THEORIES

R.F. Bishop, A.S. Kendall, L.Y. Wong and Yang Xian

Department of Mathematics, UMIST
(University of Manchester Institute of Science and Technology)
P.O. Box 88, Manchester M60 1QD, England

1.   INTRODUCTION

It is widely accepted that the quantum chromodynamics (QCD) of interacting quarks and gluons is the best theory of the strong interaction, and hence of (sub-)hadronic structure, that is presently available. In particular, a full explanation of both the hadronic spectrum and quark confinement requires a careful treatment of this non-Abelian gauge field theory in the nonperturbative sector. A common approach to QCD is via the lattice formulation,[1,2] which involves an ultraviolet cutoff that both renders the theory well-defined and strictly preserves the important local gauge invariance.[3]

In view of the complicated structure of the non-Abelian QCD, it has seemed sensible to many people to study first various algebraically simpler Abelian gauge theories as a testing-ground for the theoretical tools to be used. In particular, both the continuous compact gauge group U(1) describing conventional quantum electrodynamics (QED),[2,4-6] and the discrete gauge groups {Z(N); N = 2,3,···}[7,8] have been suggested and widely studied on the lattice in this context. The latter groups, which are simply the sets of complex Nth roots of unity with ordinary multiplication as the group operation, are interesting for a number of reasons. In the first place, as N → ∞ the group Z(N) passes smoothly over to U(1). Secondly, for N = 2, the Z(2) lattice gauge theory reduces to a gauge-invariant generalization of the Ising model, first considered by Wegner[9] in the context of an investigation of condensed matter systems which exhibit phase transitions without possessing a local order parameter. Lastly, the Z(N) gauge theories are also closely connected to the spin-glass models of Edwards and Anderson.[10]

The close relationship of the Z(2) gauge model to the Ising spin model noted above is an example of a rather remarkable, more general set of relationships or analogies first noted by Migdal.[11] Thus, most (3+1)-dimensional gauge field theories are similar to their (1+1)-dimensional spin-chain system counterparts in a number of well-defined senses, as described in some detail by Fradkin and Susskind,[5] for example. Perhaps the principal difference between such spin models as the Ising model and such gauge models as Z(2) is the fact that in the latter a symmetry is incorporated at the local level which is present in the former only globally. Thus, the Ising model is symmetric with respect to a reversal of the direction of the spins on *all* lattice sites, whereas in the Z(2) model this symmetry is incorporated locally with respect to a definite site on the lattice. Thus, as explained in Sec. 2, the Z(2) model Hamiltonian is invariant with respect to reversing the sign of only the spins connected to that site. *All* gauge symmetries[1,2,7,9] are local symmetries in this sense, namely of involving an invariance with respect to an operation affecting only degrees of freedom localized near some point.

As we have already indicated, lattice techniques have become the principal tool for treating the interesting nonperturbative physical content of gauge theories. Much of the effort to date has come from large-scale Monte Carlo studies of various kinds,[1,12-16] and other finite lattice calculations.[17-20] Nevertheless, it is clearly also useful to apply such analytic tools as are available, in order to gain additional physical insight into the various gauge models. Amongst others, renormalization-group ideas, variational methods, weak- and strong-coupling expansions, and Padé-approximant analysis have all been employed. However, with one principal exception,[21] techniques from microscopic quantum many-body theory have largely been notable by their absence until now.

It is our contention that the application of *ab initio* many-body techniques to gauge field problems is especially timely for two distinct reasons. In the first place, the last decade has seen enormous progress in quantum many-body theory at both the formal level and in the range of successful applications made. Two methods in particular have proven themselves to be extremely versatile, able to achieve very high accuracy in practice, and capable of systematic improvement. These are the method of correlated basis functions[22] (CBF) and the coupled cluster method[23,24] (CCM). They are widely recognized as providing the two most powerful, microscopic formulations currently available for dealing with *ab initio* quantum many-body systems. Secondly, one of the founders of lattice gauge theory has recently voiced his strong pessimism and reservations about the developments still needed within lattice QCD before any really meaningful interaction with experiment can take place.[25] In particular, he quantifies the existing gap between theory and experiment by suggesting that an increase in computing power of at least a factor of 10^8 *and* equally powerful algorithmic or methodological advances are both needed.

It is also pertinent to note that as part of the above discussion Wilson recommends the lattice gauge community to look, for example, to the field of quantum chemistry for additional inspiration and new ideas, especially since both fields share a common concern with many-fermion systems interacting via long-range (unscreened) forces. What is valid for quantum chemistry in particular is clearly valid for many-body theory in general. Within this context we also note that one of the methods widely applied in modern *ab initio* quantum chemistry is the CCM. For example, the very high accuracy required nowadays for the calculation of parity violation in atoms, as well as for the calculation of molecular energy differences of chemical significance calls for extreme accuracy in the solution of the electron correlation problem.[26] The CCM is ideally suited for such applications. Indeed, it is now widely recognized as the method of first choice in terms of power and accuracy for calculations on, for example: ionization potentials, electron affinities, Auger spectroscopy, excitation energies, and energy gradients (for use, for example, in searching potential energy surfaces to predict vibrational spectra, or to locate transition states in decomposition reactions). A large number of atoms and molecules (including, e.g., LiH, H_2O, GaAs, benzene, etc.) has been studied, and state-of-the-art calculations are now being done on molecules with up to 80 active electrons.[26]

Finally, we note that the CCM has also recently been applied with considerable success to various spin-lattice problems. These include the solid phases of ^{3}He,[27] and various models exhibiting antiferromagnetism.[28] The latter includes the Heisenberg model on a two-dimensional square lattice, which is believed to be of relevance to the (undoped) ceramic cuprate materials exhibiting high-temperature superconductivity. In the light of all of the above comments, it seems especially worthwhile to attempt to apply similar CCM techniques to lattice gauge field theories. The purpose of the present paper is to outline our preliminary steps in this respect for the Abelian models Z(2) and U(1) in (2+1)-dimensions.

After outlining the two models themselves in Sec. 2, the formal aspects of the application of the CCM to them is described in Sec. 3. In order to implement the otherwise formally exact CCM in practice, approximation schemes must be introduced. The lowest-order member of a particular hierarchy of such truncation schemes which retains selected correlations of arbitrarily long range is described in Sec. 4 within the context of the U(1) model. A different, more localized, scheme is discussed in Sec. 5, where it is applied to the Z(2) model. In both cases comparison is made with the results of perturbation theory and other methods. Our own results are summarized and discussed in Sec. 6.

2. THE U(1) AND Z(2) MODELS

2.1. The U(1) Lattice Hamiltonian

The Hamiltonian for the free Maxwell field of electrodynamics in a (d+1)-dimensional continuum of d spatial dimensions and one time dimension is given in the axial or temporal gauge (where $A_0=0$) by the usual expression,

$$H = \tfrac{1}{2}\int d^d x[\vec{E}^2(x) + \vec{B}^2(x)], \tag{1}$$

in terms of the electric field $\vec{E} = \vec{E}(x)$ and magnetic field $\vec{B} = \vec{B}(x)$. The magnetic field in turn is given in terms of the vector potential $\vec{A} = \vec{A}(x)$ by the usual relation,

$$\vec{B} = \vec{\nabla} \times \vec{A}. \tag{2}$$

If the spatial continuum is now replaced by a d-dimensional hypercubic lattice, the discrete generalization of the vector potential field $\vec{A}(x)$ is the phase angle $A_\ell \in [-\pi,\pi]$ defined on the links $\{\ell\}$ joining nearest-neighbour lattice sites. In the usual Wilson[1] formulation each link now has associated with it an element $U_\ell \equiv \exp(iA_\ell)$ of the underlying group U(1). If we set the lattice spacing to be unity we have $A_\ell = g\vec{\ell}\cdot\vec{A}$ where $\vec{\ell}$ is the directed vector joining the two sites of the link, and g is the U(1) coupling constant. The electric field operators are also defined on the links, $\vec{\ell}\cdot\vec{E}(x) = E_\ell(x) \to E_\ell$, and the quantization of the Hamiltonian on the lattice is now associated with the commutation relation,

$$[A_\ell, E_{\ell'}] = i\delta_{\ell\ell'}. \tag{3}$$

Thus, the electric field operators $\{E_\ell\}$ are conjugate to the link phase angles, and can be represented as,

$$E_\ell = -i\frac{\partial}{\partial A_\ell} \ . \tag{4}$$

Just as in the continuum theory $\vec{A}(x)$ is not a gauge-invariant quantity, so nor are the equivalent link phases A_ℓ in the lattice case. However, we may introduce a gauge-invariant phase B_p associated with each plaquette p defined to be simply an elementary square formed of four links on the hypercubic lattice, by making a lattice generalization of Eq. (2). We thus define the group-element-valued plaquette variable U_p as an oriented product of the link group elements bordering the plaquette as shown in Fig. 1,

$$U_p \equiv \exp(iB_p) \equiv U_1 U_2 U_3^\dagger U_4^\dagger = \exp[i(A_1 + A_2 - A_3 - A_4)] \ . \tag{5}$$

The lattice generalization of the magnetic energy density, $\frac{1}{2}\vec{B}^2(x) = \frac{1}{2}|\vec{\nabla}\times\vec{A}(x)|^2$, in the continuum expression of Eq. (1) is now simply,

$$\frac{1}{2}\vec{B}^2(x) \rightarrow \frac{1}{g^2}\left[1 - \frac{1}{2}(U_p + U_p^\dagger)\right]$$

$$= \lambda(1 - \cos B_p), \tag{6}$$

where $\lambda \equiv g^{-2}$ is the non-negative dimensionless coupling constant. The strong-coupling limit is defined by $\lambda \rightarrow 0$ ($g \rightarrow \infty$), and the weak-coupling limit by $\lambda \rightarrow \infty$ ($g \rightarrow 0$). We thus have the U(1) lattice Hamiltonian analogue of Eq. (1),

$$H = -\frac{1}{2}\sum_\ell \frac{\partial^2}{\partial A_\ell^2} + \lambda \sum_p (1 - \cos B_p), \tag{7}$$

where the periodicity in the magnetic term plays a crucial role in maintaining the true phase structure of the theory.[4]

Since the physical states are gauge-invariant,[3] we may equivalently write the Hamiltonian of Eq. (7) wholly in terms of the plaquette phases $\{B_p\}$, rather than partially using the gauge-dependent link phases $\{A_\ell\}$, provided we restrict H to operate only on physical states. By making use of Eq. (5), we can thus easily express H in terms of the plaquette variables alone for any dimensionality d. We quote the result only for the d = 2 (square lattice case),

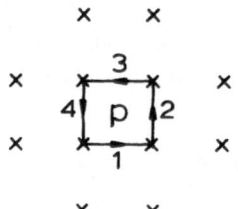

Fig. 1. Schematic definition of the plaquette variable p by the ordered product of the 4 links defining it.

$$H = \sum_p \left[-2\frac{\partial^2}{\partial B_p^2} + \lambda(1 - \cos B_p) \right] + \frac{1}{2}\sum_p \sum_\rho \frac{\partial^2}{\partial B_p \partial B_{p+\rho}}, \tag{8}$$

where ρ is a lattice vector connecting neighbouring plaquettes on the two-dimensional (2D) lattice. The Hamiltonian of Eq. (8) is invariant under the transformation $B_p \to B_p + 2n\pi$, with n integral. Hence the space of B_p is compact $(-\pi \le B_p \le \pi$, for all p), and the Hamiltonian is called that of compact (or periodic) QED.

We note that in the (2+1)-dimensional case studied here, compact QED is known to be asymptotically free and not to undergo a deconfining phase transition. Perturbative results are available in the strong-coupling regime,[29] whilst the weak-coupling limit has mainly been studied via the Villain[30] approximation as applied to the Euclidean lattice.[31,32] Monte Carlo techniques have also been applied to the 2D lattice.[14,33] Other treatments have included various variational approaches[34,35] and their systematic improvement via the Lanczos algorithm.[36,37] We shall make a comparison of our own results with those from some of these alternative methods in Sec. 4.

2.2. The Z(2) Lattice Hamiltonian

As in the U(1) case above, the Z(N) models are defined by a one-parameter (λ) family of Hamiltonians specified in terms of operators $\{P_\ell, Q_\ell\}$ associated with the links $\{\ell\}$ of a hypercubic lattice in d spatial dimensions. These are defined to obey the Z(N) algebra,

$$P_\ell^N = Q_\ell^N = 1, \qquad P_\ell^\dagger P_\ell = Q_\ell^\dagger Q_\ell = 1,$$

$$P_\ell Q_\ell = \exp(-2\pi i/N)Q_\ell P_\ell. \tag{9}$$

Operators on different links commute. It follows from Eq. (9) that P_ℓ and Q_ℓ are ladder operators with respect to one another. For example, in a representation in which P_ℓ is diagonalized,

$$P_\ell|L\rangle = \exp(-2\pi i L/N)|L\rangle , \tag{10}$$

the operator Q_ℓ has the following mode of action,

$$Q_\ell|L\rangle = |L+1\rangle ; \quad |L+N\rangle = |L\rangle. \tag{11}$$

Equivalently, just as in the U(1) case each link ℓ has associated with it an element U_ℓ of the group Z(N) defined by,

$$Z(N) = \{e^{in\delta} ; \quad \delta \equiv \frac{2\pi}{N} ; \quad n = 1,2,\cdots,N\} . \tag{12}$$

Clearly, for the case N = 2 considered here, the algebra of Eq. (9) can be realized by choosing P_ℓ and Q_ℓ to be any two of the usual Pauli matrices, $\sigma_i(\ell)$, i = 1,2,3, defined on the links $\{\ell\}$. We make here the (standard) choice of $\sigma_1(\ell)$ and $\sigma_3(\ell)$.

For reasons explained more fully elsewhere,[5,8,38] the quantum Z(N) Hamiltonian is taken to be,

$$H = \frac{1}{2} \sum_{\ell} (2 - P_{\ell}^{\dagger} - P_{\ell}) - \frac{1}{2}\lambda \sum_{P} (Q_1 Q_2 Q_3^{\dagger} Q_4^{\dagger} + \text{H.c.} - 2) , \qquad (13)$$

where the second term involves a sum over the elementary plaquettes of the oriented product of the link elements $\{Q_{\ell}\}$ bordering the plaquette, as shown in Fig. 1. The $N \to \infty$ limit is readily related to the U(1) model above, after a trivial re-scaling of the first term, by making the following representations of the operators P_{ℓ} and Q_{ℓ},

$$P_{\ell} = \exp\left(-\frac{2\pi i}{N} E_{\ell}\right) , \quad Q_{\ell} = \exp(iA_{\ell}), \qquad (14)$$

where E_{ℓ} and A_{ℓ} are dimensionless Hermitian operators associated with the link ℓ. They are the analogues of the electric field and vector potential operators, and they obey the commutation relation of Eq. (3). Further details are given in Ref. [8].

Henceforth, we restrict ourselves to the Z(2) model, which we can thus represent (after the omission of some trivial constant terms) in terms of spin variables in the form,

$$H = -\sum_{\ell} \sigma_1(\ell) - \lambda \sum_{P} \sigma_3(1)\sigma_3(2)\sigma_3(3)\sigma_3(4), \qquad (15)$$

in terms of the plaquette variables defined in Fig. 1. The Hamiltonian of Eq. (15) has a *local* gauge invariance as we now demonstrate. A local gauge transformation at a specific lattice site n is brought about by the application of the operator G(n) defined as,

$$G(n) = \prod_{i=1}^{2d} \sigma_1(\ell_{n;i}) , \qquad (16)$$

where the product is over all 2d links $\{\ell_{n;i} ; i = 1,2,\cdots,2d\}$ attached to the site n. The operator G(n) is unitary, and its mode of action on the spin operators is as follows,

$$G^{-1}(n)\sigma_1(\ell)G(n) = \sigma_1(\ell) ; \quad \text{all } \ell , \qquad (17)$$

$$G^{-1}(n)\sigma_3(\ell)G(n) = \begin{cases} -\sigma_3(\ell) ; \ \ell \text{ attached to n,} \\ \\ \sigma_3(\ell) ; \ \ell \text{ not attached to n.} \end{cases}$$

Thus the application of G(n) changes the sign of the σ_3 operators attached to site n, and leaves all other σ_1 and σ_3 operators unchanged. Clearly, the Hamiltonian of Eq. (15) commutes with all G(n), as it has been constructed from products of σ_3 operators taken around closed loops (i.e., elementary plaquettes), and hence H is left invariant under all G(n).

The ground state of the Hamiltonian of Eq. (15) may also be proven[3] to be invariant under these gauge transformations. Since we are interested only in gauge-invariant operators acting on the ground state, *all* of the states $\{|\Psi\rangle\}$ of physical interest are hence gauge-invariant,

$$G(n)|\Psi\rangle = |\Psi\rangle ; \text{ all n.} \qquad (18)$$

As a consequence, Eq. (18) immediately implies that the expectation value in any gauge-invariant state of any operator A which is not gauge-invariant [i.e., $G^{-1}(n)AG(n) \neq A$ for all n] must vanish identically. In particular, there can be no nonzero magnetization in any state satisfying gauge invariance, and no phase transition can lead to a magnetized phase. Nevertheless, as Wegner[9] first pointed out, the Z(2) lattice models for various values of the dimensionality d *do* exhibit phase transitions, as we discuss below.

The above is a particular example of Elitzur's theorem[3] that a local symmetry cannot break down spontaneously. In this respect such gauge field theories and their local invariance properties are very different from the global invariance properties of, for example, the related Ising model, whose Hamiltonian is invariant only under a change of the sign of all of the spin operators simultaneously. In that case, the ground state of the Ising model is doubly degenerate and the symmetry transformation takes one vacuum to the other. The difference in the stability properties of the spontaneously broken symmetry states in the two cases has especially been stressed by Fradkin and Susskind.[5]

Finally, we note that certain well-defined *duality relations* are known for the Z(2) model.[5,7-9] For the two-dimensional lattice (d = 2) the theory is dual to the Ising model in two spatial dimensions in a transverse magnetic field, and is thus a two-phase system which undergoes a second-order phase transition at some critical coupling, $\lambda = \lambda_c$. Estimates of λ_c have been obtained by various means, including perturbation-theory expansions,[39] 1/N-expansions,[40] finite lattice studies,[17,18] renormalization-group techniques,[41] and the application[42] of a perturbative-variational method due to Bessis and Villani.[43] The first two of these are among the most accurate. They give the respective values $\lambda_c \approx 3.08$ and $\lambda_c \approx 3.055$. By contrast, for d = 3 the Z(2) model is self-dual with a symmetry point at $\lambda = 1$. Free energy expansions[7] have indicated that the d = 3 system is a two-phase system with a first-order phase transition at $\lambda = 1$. Monte Carlo simulations[13] and all other subsequent work have provided confirmation of this.

3. A COUPLED CLUSTER TREATMENT OF THE U(1) AND Z(2) MODELS

Since the CCM has been well reviewed many times in the past, we refer the reader to the literature for both a general description of the method (and see, for example, Ref. [24] and the references cited therein) and its previous applications to spin lattice problems in particular.[28] We concentrate in this section on the specific applications to the two lattice gauge models described in Sec. 2.

3.1. The U(1) Model

In order to describe the correlations in a many-body system the CCM starts from a suitable reference state. This is usually (but not always) the properly symmetrized, non-interacting many-body state composed of some single-particle wave functions chosen to be the eigenstates of an appropriate one-body Hamiltonian. The correlations present in the physical system are then incorporated via many-body excitations from this reference state. In this respect the CCM is similar to perturbation theory. Nevertheless, the CCM is highly nonperturbative, since the correlation operator actually takes the form of an exponentiated function,[23,24] as is by now rather well known, and as we shall see in detail below.

The Schrödinger equation for the U(1) Hamiltonian of Eq. (8) reduces in the case of a single plaquette to the Mathieu equation,

$$-2 \frac{d^2}{dB^2} \psi_n(B) + \lambda(1-\cos B)\psi_n(B) = \varepsilon_n \psi_n(B); \quad -\pi \leq B \leq \pi. \tag{19}$$

In view of the fact that the Mathieu equation is well-known to be highly nonperturbative, it is pertinent to consider the simpler problem of its strong-coupling ($\lambda \to 0$) limit first. This (strong-coupling) unperturbed Hamiltonian,

$$H_0 = -2d^2/dB^2 ; \quad -\pi \leq B \leq \pi , \tag{20}$$

has two sets of eigenstates, namely {cos mB; $m = 0,1,2,\cdots$} with even parity, and {sin mB; $m = 1,2,\cdots$} with odd parity. The ground state of H_0 is clearly just a constant. We now use these simple strong-coupling single-plaquette wave functions to construct both the ket and bra ground states in terms of a CCM analysis.

The non-interacting reference state $|\Phi\rangle$ is hence simply a constant, in which all plaquettes are described by the ground state (g.s.) of H_0. The exact many-body ket g.s. of the U(1) Hamiltonian of Eq. (8) is then taken in the CCM form,

$$|\Psi_o(\{B_p\})\rangle = e^{S(\{B_p\})}|\Phi\rangle, \quad S(\{B_p\}) = \sum_{k=1}^{N_p} S_k(\{B_p\}) , \tag{21}$$

where N_p is the total number of plaquettes in the lattice and where the k-body correlation operators $S_k = S_k(\{B_p\})$ are decomposed as,

$$S_1 = \sum_{n=1}^{\infty} \sum_{p=1}^{N_p} s_p(n)\cos(nB_p) , \tag{22a}$$

$$S_2 = \frac{1}{2!} \sum_{n_1,n_2=1}^{\infty} \sum_{p_1,p_2=1}^{N_p} {}' \left[s_{p_1 p_2}^{(1)}(n_1,n_2) \cos(n_1 B_{p_1}) \cos(n_2 B_{p_2}) \right.$$
$$\left. + s_{p_1 p_2}^{(2)}(n_1,n_2) \sin(n_1 B_{p_1}) \sin(n_2 B_{p_2}) \right] , \tag{22b}$$

where the prime on the summation in Eq. (22b) excludes the term $p_1=p_2$; and similarly for the higher-order partitions S_k with k>2. For reasons which have been described many times in the past[24] the bra g.s., $\langle \tilde{\Psi}_o|$, in the CCM parametrization is not taken as the manifest Hermitian conjugate of $|\Psi_o\rangle$. Instead it is parametrized in the form,

$$\langle \tilde{\Psi}_o(\{B_p\})| = \langle \Phi|\tilde{S}(\{B_p\})e^{-S(\{B_p\})} , \quad \tilde{S} = 1 + \sum_{k=1}^{N_p} \tilde{S}_k(\{B_p\}) , \tag{23}$$

where $S = S(\{B_p\})$ is as given above in Eqs. (21)-(22), and $\tilde{S} = \tilde{S}(\{B_p\})$ is similarly defined as,

$$\tilde{S}_1 = \sum_{n=1}^{\infty} \sum_{p=1}^{N_p} \tilde{s}_p(n)\cos(nB_p), \tag{24a}$$

$$\tilde{S}_2 = \frac{1}{2!} \sum_{n_1,n_2=1}^{\infty} \sum_{p_1,p_2=1}^{N_p} \left[\tilde{s}_{p_1 p_2}^{(1)}(n_1,n_2) \, \cos(n_1 B_{p_1}) \, \cos(n_2 B_{p_2}) \right.$$
$$\left. + \tilde{s}_{p_1 p_2}^{(2)}(n_1,n_2) \, \sin(n_1 B_{p_1}) \, \sin(n_2 B_{p_2}) \right] , \tag{24b}$$

and similarly for the higher-order terms. We note that both $|\Psi_0\rangle$ and $\langle\tilde{\Psi}_0|$ are explicitly invariant under the transformation $B_p \rightarrow B_p + 2\pi$. As befits the ground state, they have also been constructed to have even parity under interchange of the sign of all of the coefficients $\{B_p\}$.

We define an inner product of wave functions $\langle\tilde{g}(\{B_p\})|$ and $|f(\{B_p\})\rangle$ as,

$$\langle\tilde{g}|f\rangle = \int_{-\pi}^{\pi} \prod_{p=1}^{N_p} \left(\frac{dB_p}{2\pi}\right) \tilde{g} \cdot f , \tag{25a}$$

and the expectation value of an operator $\theta = \theta(\{B_p\},\{d/dB_p\})$ with respect to the conjugate states $|f\rangle$ and $\langle\tilde{f}|$ as,

$$\bar{\theta} \equiv \langle\tilde{f}|\theta|f\rangle = \int_{-\pi}^{\pi} \prod_{p=1}^{N_p} \left(\frac{dB_p}{2\pi}\right) \tilde{f} \cdot \theta \cdot f . \tag{25b}$$

The normalization condition, $\langle\Phi|\Phi\rangle=1$, then gives that Φ is equal to unity. The CCM equations for the correlation coefficients of Eqs. (22) and (24) are now derived from the condition that the g.s. expectation value of the Hamiltonian of Eq. (8),

$$\bar{H} = \langle\tilde{\Psi}_0|H|\Psi_0\rangle = \langle\Phi|\tilde{S}e^{-S}He^S|\Phi\rangle , \tag{26}$$

should be stationary with respect to variations in both complete sets of bra- and ket-state coefficients. Thus, for example, the one-body coefficients $\{\tilde{s}_p(n)\}$ and $\{s_p(n)\}$ are derived from the variational principle,

$$\frac{\delta\bar{H}}{\delta\tilde{s}_p(n)} = 0 = \frac{\delta\bar{H}}{\delta s_p(n)} . \tag{27}$$

The equations for the two- and higher-body coefficients are similarly given by the corresponding stationarity conditions for $\bar{H}$.

Within the context of the above CCM parametrization there are clearly two distinct kinds of correlations in play. In the first place one has the mode-coupling terms between different modes $\{\cos(nB_p), \sin(nB_p)\}$ specified by the index n. Secondly, we have the more physical correlations between different plaquettes specified by the indices $\{p_i\}$. Clearly mode-coupling is included even at the one-plaquette level (i.e., as specified by S_1 and $\tilde{S}_1$ alone), whereas one needs to include at least S_2 and $\tilde{S}_2$ as well in order to describe plaquette correlations.

We also note that the CCM correlation operators of Eqs. (21)-(24) do not involve the usual creation and destruction operators as in the more conventional CCM.[24,28]   This distinct difference is related essentially to the fact that in the lattice gauge systems under consideration there are neither any real particles nor quasiparticles defined.   Whereas in more conventional many-body problems one is usually concerned with particle conservation, the primary concern in lattice gauge models is the gauge invariance.

3.2. The Z(2) Model

The strong-coupling $(\lambda \to 0)$ limit of the Z(2)-model Hamiltonian of Eq. (15) is

$$H_0 = - \sum_{\ell} \sigma_1(\ell) , \tag{28}$$

where the sum is over all links.   The model state $|\Phi\rangle$ of the CCM is now taken for this Z(2) gauge theory as the g.s. of Eq. (28), namely the state with all spins aligned with respect to the 1-axis, such that,

$$\sigma_1(\ell)|\Phi\rangle = |\Phi\rangle ; \text{ all } \ell . \tag{29}$$

Exactly as for the U(1) model, the ket and bra ground states $|\Psi_o\rangle$ and $\langle\tilde{\Psi}_o|$ of Eq. (15) are formed in terms of the correlation operators S and $\tilde{S}$, as before. We recall that these states must be gauge-invariant with respect to the transformations induced by the operators G(n) of Eq. (16).   Since $|\Phi\rangle$ is also gauge-invariant, hence so must S and $\tilde{S}$ be.   As explained in Sec. 2.2 the easiest way to ensure this is to construct them from products of σ_3 operators taken around closed loops on the lattice.   One way to achieve this is in terms of the elementary plaquette operators $\{U(p)\}$ defined as products around the links bordering the plaquettes as in Fig. 1,

$$U(p) = \sigma_3(1)\sigma_3(2)\sigma_3(3)\sigma_3(4) . \tag{30}$$

Hence, we may write for the ket g.s., by analogy with Eq. (21),

$$|\Psi_o\rangle = e^S|\Phi\rangle , \quad S = \sum_{k=1}^{N_p} S_k , \tag{31}$$

where N_p is the total number of plaquettes, and where the k-body partitions of the correlation operator are specified as,

$$S_k = \frac{1}{k!} \sum_{p_1=1}^{N_p}{}' \cdots \sum_{p_k=1}^{N_p}{}' s_{p_1 p_2 \cdots p_k} U(p_1) \cdots U(p_k) , \tag{32}$$

where the primes on the summations indicate that only such terms are included where no two or more of the indices $(p_1, p_2, \cdots, p_k)$ are equal.   The bra g.s. is similarly specified, by analogy with Eq. (23), in the form,

$$\langle\tilde{\Psi}_o| = \langle\Phi|\tilde{S}e^{-S} , \quad \tilde{S} = 1 + \sum_{k=1}^{N_p} \tilde{S}_k , \tag{33}$$

and where $\tilde{S}_k$ is parametrized exactly as in Eq. (32), except that $\{s_{p_1 \cdots p_k}\} \to \{\tilde{s}_{p_1 \cdots p_k}\}$.

4. THE SUB1 SCHEME FOR THE U(1) MODEL

In principle, the exact ket and bra ground states of our U(1) lattice gauge model are given by Eqs. (21)-(24), where *all* correlation operators $\{S_k, \tilde{S}_k; \ k=1,2,\cdots,N_p\}$ up to those between all N_p plaquettes should be included. However, in practice one clearly needs to approximate. The most common CCM truncation scheme is the so-called SUBn scheme,[24] in which only those correlations described by the cluster partitions $\{S_k, \tilde{S}_k\}$ with $k \leq n$ are included, and those with $k > n$ are set to zero. We focus preliminary attention here on the lowest-order SUB1 approximation, for which we therefore make the replacements $S \rightarrow S_1$, $\tilde{S} \rightarrow \tilde{S}_1$.

The retained coefficients $\{s_p(n), \ \tilde{s}_p(n)\}$ are then found as described in Sec. 3.1, and as given by Eqs. (26) and (27). In particular, the SUB1 g.s. ket-coefficients are clearly obtained by solving the coupled set of equations,

$$\langle \cos(mB_p) | e^{-S_1} H e^{S_1} | \Phi \rangle = 0 \ ; \quad m=1,2,\cdots, \tag{34}$$

where the matrix element is defined as in Eq. (25b) and where, as described in Sec. 3.1, the model state $|\Phi\rangle$ is simply a constant, chosen to be unity so that $\langle \Phi | \Phi \rangle = 1$ according to the inner product definition of Eq. (25a). Although the evaluation of the g.s. expectation value of an arbitrary operator in this SUB1 approximation would also require the bra g.s. coefficients $\{\tilde{s}_p(n)\}$, the g.s. energy, E_g, is clearly given, in view of Eqs. (23), (26) and (34), by the stationary value of $\bar{H}$, and hence purely in terms of the ket g.s. coefficients as,

$$E_g = \langle \Phi | e^{-S_1} H e^{S_1} | \Phi \rangle . \tag{35}$$

In terms of the parametrization of Eq. (22a), and using the explicit definition of the (2+1)-dimensional Hamiltonian in Eq. (8), it is not difficult to derive the explicit SUB1-approximation equations,

$$\sum_{p=1}^{N_p} \left[-\tfrac{1}{2}\lambda\delta_{m,1} + m^2 s_p(m) + \tfrac{1}{2} \sum_{n,n'=1}^{\infty} nn' s_p(n)s_p(n') \left(\delta_{m,n+n'} - \delta_{m,|n-n'|} \right) \right]$$
$$= 0 \ ; \quad m=1,2,\cdots, \tag{36}$$

$$E_g = \lambda N_p - \sum_{p=1}^{N_p} \sum_{n=1}^{\infty} n^2 s_p^2(n) . \tag{37}$$

In the limit of an infinite lattice $(N_p \rightarrow \infty)$ considered here the coefficients $s_p(n)$ are independent of plaquette index p by lattice translational invariance. Hence we may write,

$$a_m \equiv m s_p(m) \ ; \quad m = 1,2,\cdots . \tag{38}$$

Furthermore, if we extend the definition of the coefficients $\{a_m\}$ to non-positive integers by the relation,

$$a_{-m} = -a_m \; ; \quad m = 0,1,2,\cdots , \tag{39}$$

so that $a_0 \equiv 0$, we may readily rewrite Eqs. (36) and (37) as,

$$\left(\frac{E_g}{N_p} - \lambda\right)\delta_{m,0} + \tfrac{1}{2}\lambda(\delta_{m,1} + \delta_{m,-1}) - ma_m - \tfrac{1}{2}\sum_{n=-\infty}^{\infty} a_n a_{m-n} = 0 , \tag{40}$$

which is valid for all integers m. We may now define an odd function, $A = A(B)$, by the Fourier sum,

$$A(B) \equiv \sum_{m=-\infty}^{\infty} a_m \sin(mB) , \tag{41a}$$

whose inverse, taking into account the symmetry relation (39), is given by,

$$a_m = \frac{1}{2\pi} \int_{-\pi}^{\pi} dB \, A(B) \, \sin(mB) . \tag{41b}$$

It is easy to show that Eq. (40) is equivalent to the first-order Riccati equation,

$$\frac{dA}{dB} - \tfrac{1}{2}A^2 = E_g/N_p - \lambda(1-\cos B) . \tag{42}$$

Finally, the standard substitution,

$$A = -\frac{2}{\psi}\frac{d\psi}{dB} \tag{43}$$

reduces Eq. (42) to the Mathieu equation,

$$[-2 \, d^2/dB^2 + \lambda(1-\cos B)]\psi(B) = (E_g/N_p)\psi(B) , \tag{44}$$

which may be compared with the single-plaquette equation (19).

The fact that we regain the Mathieu equation in our SUB1 scheme is not too surprising since no multi-plaquette correlation effects have been included, and as we and others[15,44] have shown, the single-plaquette problem is equivalent to the Mathieu equation. Nevertheless, the many-body nature of our system has not been entirely lost, since the eigenvalue in Eq. (44) is the intensive quantity, E_g/N_p. Furthermore, one can without too much difficulty go beyond this SUB1 level to include 2-plaquette (and higher-order) correlations in our CCM formalism.

For present purposes, however, we remain at the SUB1 level. In order to solve Eq. (40) it is natural to define the so-called SUB1(n) sub-hierarchy in which one retains at the nth level of approximation only those coefficients a_m with which one retains at the nth level of approximation only those coefficients a_m with $|m| \leq n$, and sets the remainder with $|m| > n$ to zero. Thus, in the SUB(1) scheme, only the single independent coefficient $a_1(=-a_{-1})$

Table 1. The ground-state energy per plaquette, E_g/N_p, of the U(1) gauge field model in (2+1)-dimensions, for various values of the coupling constant λ. Results in various SUB1(n) sub-approximations are shown.

Method	λ					
	0.1	0.5	1.0	2.0	5.0	20.0
SUB1(1)	0.097500	0.437500	0.750000	1.000000	−1.25000	−80.0000
SUB1(2)	0.097503	0.439123	0.772689	1.249117	2.209359	5.525510
SUB1(3)	0.097503	0.439117	0.772425	1.242552	2.061439	0.553501
SUB1(4)	0.097503	0.439117	0.772431	1.243018	2.099494	4.273716
SUB1(6)	0.097503	0.439117	0.772431	1.243022	2.099907	4.302545
SUB1(8)	0.097503	0.439117	0.772431	1.243021	2.099974	4.335206
SUB1(20)	0.097503	0.439117	0.772431	1.243021	2.099977	4.343306

is retained. The solution is trivially given as,

$$a_1 = -\tfrac{1}{2}\lambda \ , \quad E_g/N_p = \lambda - \tfrac{1}{4}\lambda^2 \ ; \quad \text{SUB1(1)}. \tag{45a}$$

Similarly in the SUB1(2) approximation one retains only the two independent coefficients a_1 and a_2. The solution to Eq. (40) in this case is given by,

$$a_1 = \left[\sqrt{\lambda^2 + 64/27} + \lambda\right]^{1/3} - \left[\sqrt{\lambda^2 + 64/27} - \lambda\right]^{1/3} ,$$

$$a_2 = -\tfrac{1}{4}a_1^2 \ , \quad E_g/N_p = \lambda - a_1^2 - a_2^2 \ ; \quad \text{SUB1(2)}. \tag{45b}$$

Solutions for arbitrary n and λ to the general SUB1(n) approximation to Eq. (40) are easily obtained numerically on a micro-computer. Convergence of the iterated solution is rapid. An accuracy of 6 significant figures, for example, is usually obtained in less than about 5 iterations. The g.s. energy per plaquette, E_g/N_p, is given in Fig. 2 and Table 1 as a function of λ for several SUB1(n) schemes. The full SUB1 values, which are the *exact* solutions of the Mathieu equation (19), are actually well represented, to the level of accuracy shown, by the SUB1(20) results for the whole range of λ displayed. The convergence with index n of the SUB1(n) results to the SUB1 limit is clearly quite rapid, even in the large-λ (weak-coupling) limit.

It is also very interesting to compare our SUB1(n) scheme results with the corresponding results for the Mathieu problem from perturbation theory in the strong-coupling ($\lambda \to 0$) limit. Up to eighth-order in this limit, the g.s. eigenvalue of Eq. (19) is given by,[45]

$$\varepsilon_0 \xrightarrow[\lambda \to 0]{} \lambda - \frac{1}{4}\lambda^2 + \frac{7}{256}\lambda^4 - \frac{29}{4608}\lambda^6 + \frac{68687}{37748736}\lambda^8 + O(\lambda^{10}) . \tag{46}$$

We have explicitly verified that in this strong-coupling limit the SUB1(4) approximation for E_g/N_p exactly reproduces this series to the order shown. In general, the $\lambda \to 0$ limiting form of the g.s. energy in SUB1(n) approximation

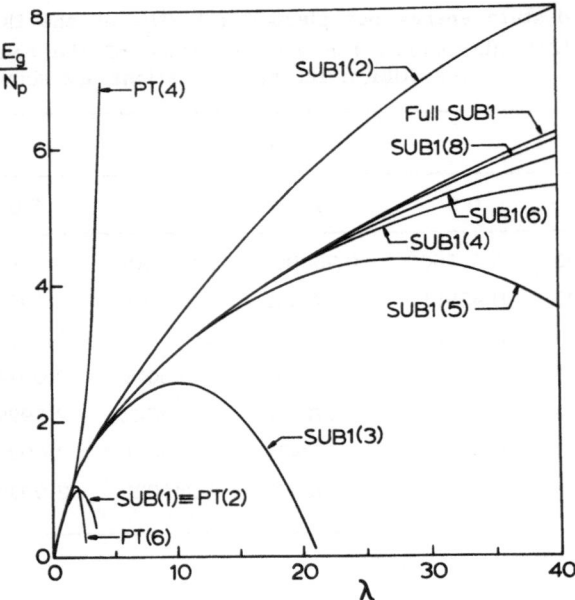

Fig. 2. Ground-state energy per plaquette, E_g/N_p, of the U(1) gauge field model in (2+1)-dimensions, as a function of coupling constant, λ. Our full SUB1 approximation and various SUB1(n) sub-approximations are shown as well as some results of nth-order strong-coupling ($\lambda \to 0$) perturbation theory, PT(n).

exactly reproduces the result from $(2n)^{th}$-order perturbation theory, PT(2n). Indeed, the SUB1(n) scheme actually provides a very efficient way to generate the coefficients of the terms in the perturbative treatment of the Mathieu problem! A more detailed comparison of the SUB1(n) and PT(2n) results for the g.s. energy is provided by Fig. 2, where we see clearly that the perturbative results are very poor for $\lambda \gtrsim 1$.

Clearly, the range of validity of the SUB1(n) results extends to values well above $\lambda=1$, which is an approximate natural boundary for the PT(2n) results, even for relatively low values of n. The accuracy of different orders of perturbation theory as a function of λ is shown more explicitly in Fig. 3. The accuracy, $\mathcal{A}$, of a quantity E is, roughly speaking, the number of significant figures of the approximate result, $E_{approx.}$, compared with its exact counterpart, E_{exact}. It is defined more precisely as,

$$\mathcal{A} \equiv \log_{10} \left| \frac{E_{exact}}{E_{exact} - E_{approx.}} \right|. \tag{47}$$

Figure 3 exhibits rather well the strongly non perturbative nature of the Mathieu problem, and the fact that the series of Eq. (46) has a finite radius of convergence.

By contrast, the accuracy of the comparable SUB1(n) results is displayed in the same way in Fig. 4. It is clear that these results represent a very natural extension of the PT(2n) approximations. They comprise, in effect, a well-defined analytic continuation or resummation of the PT(2n) results, within the context of a systematic hierarchy of approximations. In this sense, the

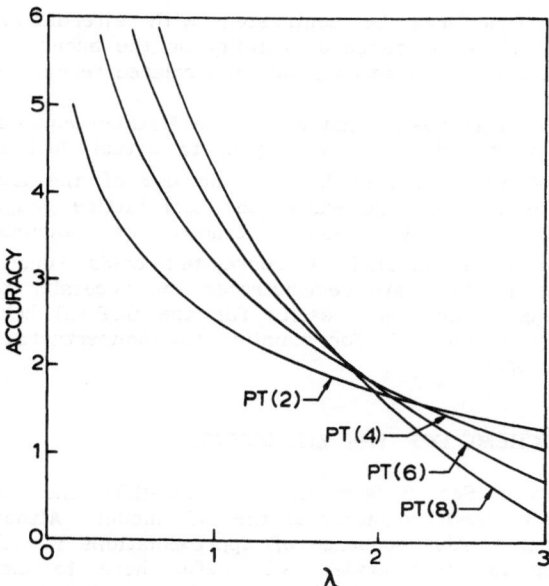

Fig. 3. The accuracy, defined in Eq. (47), as a function of coupling constant, λ, for the ground-state energy, ε_0, of the Mathieu equation (19) calculated in n^{th}-order strong-coupling ($\lambda \to 0$) perturbation theory, PT(n), for n = 2,4,6, and 8, compared to the exact result.

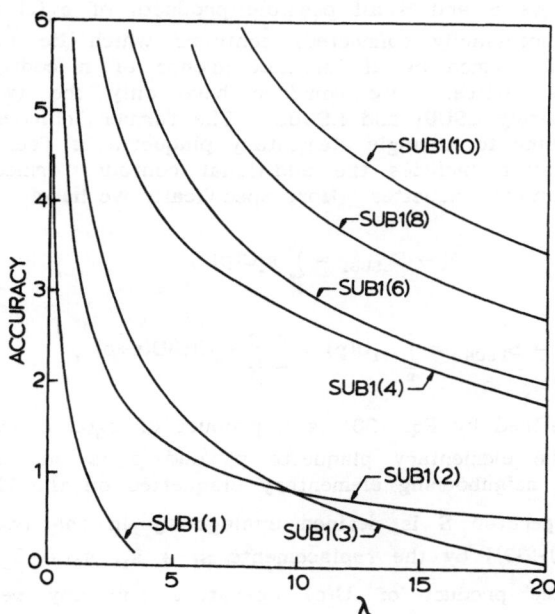

Fig. 4. The accuracy, defined in Eq. (47), as a function of coupling constant, λ, for the ground-state energy, ε_o, of the Mathieu equation (19) calculated in our SUB1(n) sub-approximation, for various values of n, compared to the exact result.

SUB1(n) scheme scheme may be contrasted with alternative rather *ad hoc* approaches for extending the range of validity or the accuracy of similar PT(n) sequences. These include Padé approximant and related techniques.

In fact, the nonperturbative nature of the Mathieu equation has long made it a testing-ground for different techniques to attack lattice gauge theories. These have included variational methods,[46] and use of the Lanczos algorithm[37] to improve upon them. More generally, the U(1) lattice gauge model has also been investigated by such other methods as various Monte Carlo techniques,[14,16,33] and the method of correlated basis functions.[21] Most of these alternative techniques have demonstrated the necessity to include higher-order mode couplings. Our own results for the SUB1(n) scheme have clearly proven the efficacy of the CCM for handling the nonperturbative sector of the U(1) lattice gauge model.

5. THE LSUBn SCHEME FOR THE Z(2) MODEL

We have given in Sec. 3 [and see Eqs. (31)-(33)] the (in principle) exact CCM forms for the ket and bra states of the Z(2) model. Although we could now in principle perform a SUBn sequence of approximations for the Z(2) model as outlined above for the U(1) model, we prefer here to employ a different hierarchy of approximations, which we call the LSUBn sequence. This approximation scheme, which has no simple counterpart for continuous extended systems, has been recently developed within the context of quantum spin lattice models,[28] where it has met with considerable success. It is particularly suited for treating lattice systems where the interaction forces are highly localized (i.e., short-ranged).

For our current 2D Z(2) model a given LSUBn approximation includes in the correlation operators S and $\tilde{S}$ all possible products of $\sigma_3(\ell)$ operators around closed (but not necessarily connected) contours which lie inside the closed, connected contours formed by all distinct groups of n contiguous elementary plaquettes on the lattice. We consider here only the two simplest such approximations, namely LSUB1 and LSUB2. The former includes only the single contour corresponding to a single elementary plaquette of four links as in Fig. 1, whereas the latter includes the additional contour formed by a pair of neighbouring elementary plaquettes. More specifically we have,

$$S \rightarrow S_{LSUB1} = \sum_p s_1 U(p) \, , \tag{48a}$$

$$S \rightarrow S_{LSUB2} = \sum_p s_1 U(p) + \sum_p \sum_\rho s_2 U(p) U(p+\rho) \, , \tag{48b}$$

where U(p), as defined by Eq. (30) is a product of $\sigma_3(\ell)$ operators around the links bordering the elementary plaquette p, and ρ is one of the 4 lattice vectors connecting neighbouring elementary plaquettes on the 2D square lattice. The correlation operator $\tilde{S}$ is defined analogously in the two approximations [and see Eqs. (31)-(33)] by the replacements $s_1 \rightarrow \tilde{s}_1$, $s_2 \rightarrow \tilde{s}_2$. We note that since $\sigma_3^2 = 1$, the product of U(p) operators for any set of contiguous elementary plaquettes is equivalent to a product of $\sigma_3(\ell)$ operators around the links forming their boundary.

The derivation of the CCM equations to determine the unknown cluster configuration coefficients now proceeds in the usual fashion. Thus, the g.s, energy expectation value, $\bar{H}$, of Eq. (26) is first evaluated. The ket g.s. cluster coefficients, for example, are then obtained as in Eq. (27), by requiring $\bar{H}$ to be stationary with respect to the corresponding bra g.s. coefficients. In order to evaluate $\bar{H}$ we need to calculate the similarity transform, $e^{-S}He^S$. As usual, we utilize the well-known nested-commutator expansion,

$$e^{-S}He^S = H + [H,S] + \frac{1}{2!} [[H,S],S] + \cdots . \tag{49}$$

However, whereas the infinite expansion of Eq. (49) usually terminates in practice for most standard many-body problems (and, more particularly, when S is composed only of creation operators with respect to the cyclic vector $|\Phi\rangle$, and H is a finite-order multinomial in the corresponding creation and destruction operators[24]), this is *not* the case here.

Nevertheless, for the Z(2) model under consideration, the non-terminating expansion of Eq. (49) is readily resummed. Thus, consider a general correlation operator S of the form,

$$S = \sum_p \sum_C s_C U_C(p) , \tag{50}$$

where $U_C(p)$ is a product of σ_3 operators around a particular lattice contour of shape and orientation specified by the index C, and where the index p labels some one particular plaquette in a specified ordering of the plaquettes inside C. We note that the c-number cluster amplitudes $\{s_C\}$ do not depend on the index p due to the lattice translational invariance. It is now straightforward to show that,

$$e^{-S}He^S = - \sum_{\ell} \left[\sigma_1(\ell) \cosh G - i\sigma_2(\ell) \sinh G\right] - \lambda \sum_p U(p) , \tag{51}$$

where the operator G is defined by,

$$G \equiv 2 \sum_{p \ni \ell} \sum_C s_C U'_C(p;\ell) , \tag{52}$$

wherein the sum over plaquettes is restricted to those which include the link ℓ, and where $U'_C(p;\ell)$ is identical to the product of σ_3 operators comprising $U_C(p)$ except that the single operator $\sigma_3(\ell)$ on the particular link ℓ is omitted and the ordering of the plaquettes inside C is relaxed. Both the terms cosh G and sinh G in Eq. (51) can then be expanded using the usual expansion rules for cosh(A+B) and sinh(A+B). Finally, using that $[U'_C(p;\ell)]^2 = 1$ for all C, p and ℓ, since $\sigma_3^2 = 1$, we have the relations,

$$\cosh[s_C U'_C(p;\ell)] = \cosh s_C ; \quad \sinh[s_C U'_C(p;\ell)] = U'_C(p;\ell)] \sinh s_C , \tag{53}$$

which enable us ultimately to rewrite Eq. (51) in a form in which the hyperbolic functions act only on the c-number amptitudes $\{s_C\}$.

We quote only the final result for $\bar{H}$ for the 2D Z(2)-model in the LSUB1 approximation,

$$\bar{H}_{LSUB1} = - \sum_{\ell} \left[\cosh^2(2s_1) + \tfrac{1}{2}\tilde{s}_1\{\lambda - 2\sinh(4s_1)\} \right] , \qquad (54)$$

where, in writing the result as a sum over link variables only, we have used that $N_\ell = 2N_p$ for the infinite 2D square lattice. By making $\bar{H}$ stationary with respect to $\tilde{s}_1$ and s_1, we readily find,

$$s_1 = \tfrac{1}{4} \sinh^{-1}(\tfrac{1}{2}\lambda) , \quad \tilde{s}_1 = \tfrac{1}{2}\lambda \left(1 + \tfrac{1}{4}\lambda^2\right)^{-1/2} ; \quad LSUB1, \qquad (55)$$

and the stationary value of the g.s. energy per link is given as,

$$E_g/N_\ell = - \tfrac{1}{2}\left(1 + \sqrt{1 + \tfrac{1}{4}\lambda^2}\right) ; \quad LSUB1. \qquad (56)$$

The comparable equations which determine the ket-state coefficients (s_1, s_2) and the g.s. energy per link in the LSUB2 approximation may also be derived after some algebra as,

$$\lambda + \sinh(2s_1)\cosh(2s_1)[12\sinh(2s_2) - 4\cosh(2s_2)]\cosh^5(2s_2)=0 , \qquad (57a)$$

$$\sinh^2(2s_1)[2\sinh^2(2s_2) + \cosh^2(2s_2)] - 6\cosh^2(2s_1)\sinh(2s_2)\cosh(2s_2)=0 , \qquad (57b)$$

$$E_g/N_\ell = - \cosh^2(2s_1)\cosh^6(2s_2) ; \quad LSUB2. \qquad (58)$$

Equations (57) are readily solved numerically and the solution for E_g/N_ℓ is shown in Fig. 5 together with the LSUB1 result from Eq. (56).

A priori, we expect the above LSUBn approximations to be most accurate in the strong-coupling ($\lambda \to 0$) regime, where we may compare with the known perturbation-theory results for the 2D Z(2)-model,[39]

$$\frac{E_g}{N_\ell} \xrightarrow{\lambda \to 0} - 1 - \frac{\lambda^2}{16} - \frac{\lambda^4}{3072} - \frac{\lambda^6}{196608} + O(\lambda^8) . \qquad (59)$$

Our LSUB1 result of Eq. (56) is clearly seen to agree with Eq. (59) to 2nd order. An analytic small-λ expansion of Eqs. (57)-(58) also shows that our LSUB2 result reproduces the result of 4th-order perturbation theory. Although we have not attempted a strict proof, we expect that (suitably constructed) LSUBn approximations with n > 2 will also reproduce the results of (2n)th-order perturbation theory, PT(2n).

It is also interesting to compare the weak-coupling ($\lambda \to \infty$) limits of our LSUB1 and LSUB2 results with the corresponding results from perturbation theory in this regime,[39]

$$\frac{E_g}{N_\ell} \xrightarrow{\lambda \to \infty} -\frac{1}{2}\lambda \left[1 + \frac{1}{2}\lambda^{-2} + \frac{5}{32}\lambda^{-4} + O(\lambda^{-6})\right] . \qquad (60)$$

As expected, we do not get agreement in this limit. Nevertheless, both LSUB1 and LSUB2 approximations give the correct leading asymptotic behaviour, namely linear in λ, albeit with an incorrect coefficient. More precisely, Eq. (56) trivially gives $E_g/N_\ell \rightarrow -0.25\lambda$ as $\lambda \rightarrow \infty$ in the LSUB1 approximation, whereas an analytic large-λ expansion of Eqs. (57)-(58) shows that the leading-order LSUB2 result is $E_g/N_\ell \rightarrow -k\lambda$ as $\lambda \rightarrow \infty$, where $k = (7+3\sqrt{7})/28 \approx 0.533$. This latter value is rather close to the exact value, $k = 0.5$.

Finally, we note that for the 2D Z(2)-model Cardy and Hamber[47] and also Suranyi[48] have proposed a trial variational g.s. wave function of precisely our LSUB1 form,

$$|\Psi\rangle = \exp\left[s_1 \sum_p U(p)\right]|\Phi\rangle . \tag{61}$$

It is not difficult to show that if the expectation value of the Hamiltonian of Eq. (15) is evaluated in this trial state in the usual fashion as $E_g = \langle\Psi|H|\Psi\rangle/\langle\Psi|\Psi\rangle$, where the bra state is now the explicit Hermitian adjoint of the ket state, we find

$$E_g/N_\ell = -\text{sech}^2(2s_1) - \tfrac{1}{2}\lambda \tanh(2s_1) . \tag{62}$$

This expression may be minimized with respect to s_1 to find the simple variational result,

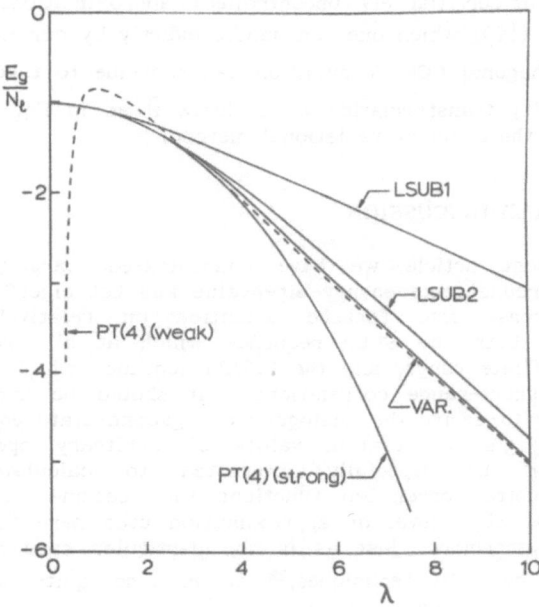

Fig. 5. Ground-state energy per link, E_g/N_ℓ, of the Z(2) gauge model on a 2D square lattice, as a function of the coupling constant, λ. Shown are our LSUB1 and LSUB2 results; the variational (VAR) result of Eq. (63); and the fourth-order perturbation theory, PT(4), results in both the strong-coupling ($\lambda\rightarrow 0$) and weak-coupling ($\lambda\rightarrow\infty$) limits, from Eqs. (59) and (60) respectively.

$$E_g^{var}/N_\ell = \begin{cases} - 1 - \frac{1}{16}\lambda^2 & ; \ \lambda \leq 4 \ , \\[2mm] - \frac{1}{2}\lambda & ; \ \lambda > 4 \ . \end{cases} \qquad (63)$$

Like our CCM LSUB1 approximation this also reproduces the 2nd-order strong-coupling ($\lambda \to 0$) perturbation theory result of Eq. (59). However, unlike our LSUB1 result, this variational treatment both gives the exact leading asymptotic behaviour in the weak-coupling ($\lambda \to \infty$) limit, and gives an indication of a (second-order) phase transition at a critical value $\lambda_c = 4$, which may be compared with the more precise value $\lambda_c \approx 3.1$.

The above simple variational approximation is clearly superior (at least so far as the g.s. energy is concerned) to our LSUB1 result. Interestingly, unlike the more general situation (including the U(1) case) where this variational approach is not easily extended to larger classes of trial wave functions without making further approximations in the evaluation of the energy expectation value (which then destroy the variational upper bound), this is not so for the Z(2) model on the 2D square lattice. The underlying simplicity of this model has, for example, enabled Dagotto and Moreo[49] to use larger classes of trial wave functions in exactly the spirit of our LSUBn scheme. For example, for our LSUB2 wave function of Eq. (48b) considered as a trial variational state, they now find a phase transition in the g.s. energy at a critical value, $\lambda_c = 3.85$. The interested reader is referred to Ref. [49] for further details.

We stress, however, that for more general lattice gauge field theories, one cannot evaluate the g.s. energy expectation value, $E_g = \langle \Psi | H | \Psi \rangle / \langle \Psi | \Psi \rangle$, without making further (largely uncontrolled) approximations, for the classes of wave functions $\{|\Psi\rangle\}$ which one can handle exactly by our CCM techniques. As usual,[24] the biorthogonal CCM formulation can continue to take advantage of the underlying similarity transformation to evaluate $\bar{H}$ as in Eq. (26), when this is not practicable in the ordinary variational method.

6. SUMMARY AND DISCUSSION

In the present article we have concentrated, largely for pedagogical reasons, on the ground-state energy eigenvalue and ket eigenfunction. We have, for similar reasons, also focused attention on relatively low-level CCM approximations in both the SUBn sequence which at all orders retains some correlations of infinite range, and the LSUBn sequence which is more geared to the inclusion of short-range correlations. It should be immediately apparent to the reader how to derive the analogous bra ground-state equations, and hence how to calculate g.s. expectation values of arbitrary operators. In this context we shall be especially interested to calculate the physically interesting 2-plaquette correlation functions that become meaningful when we proceed beyond the SUB1 level of approximation used here for the U(1) model, for example. Furthermore, just as in the particular case of the spin-lattice problems studied by CCM techniques,[28] it is also quite straightforward to implement the standard CCM technology to study the excited states[24] of our lattice gauge models.

For the 2D Z(2) model studied here we are particularly interested to study the phase transition which occurs at a critical coupling. The localized LSUBn sequence within our standard CCM approach shows no obvious sign of a transition in the g.s. energy although, interestingly, the variational approach based on the same class of wave functions does.[49] It will be of interest to study this matter further within the CCM framework, perhaps by requiring that the model state $|\Phi\rangle$ of Eq. (29) be redefined in terms of a product of more general maximum-overlap link orbitals involving an additional (rotation) parameter that can be chosen to maximize the overlap of the new model state $|\Phi'\rangle$ with the true state $|\Psi_0\rangle$. We note in this regard that Kümmel[50] has investigated the sufficiency condition for such maximum-overlap orbitals actually to maximize the overlap rather than simple to extremize it, as given by the usual necessary condition. He shows in particular how the extra sufficency condition can be utilized to identify within the CCM the onset of a critical "shape instability" which marks the transition to a different optimal basis, or configuration, as some appropriate internal parameter (here, λ) is varied.

The great attraction of the CCM is, as we have already remarked, that it is truly a universal method. Hence, motivated by the preliminary studies reported here, we intend to extend the application of CCM techniques to, for example: (i) the more general Z(N) model discussed in Sec. 2.2, particularly to investigate in detail the N→∞ limit to U(1); (ii) both the Z(2) and U(1) models considered here, in (3+1)-dimensions; and (iii) such non-Abelian gauge theories as SU(2). We hope to report on these calculations elsewhere.

ACKNOWLEDGEMENT

One of us (RFB) acknowledges support for this work in the form of a research grant from the Science and Engineering Research Council (SERC) of Great Britain.

REFERENCES

1. K.G. Wilson, Phys. Rev. D 10:2445 (1974).
2. J. Kogut and L. Susskind, Phys. Rev. D 11:395 (1975).
3. S. Elitzur, Phys. Rev. D 12:3978 (1975).
4. A.M. Polyakov, Phys. Lett. B72:477 (1978).
5. E. Fradkin and L. Susskind, Phys. Rev. D 17:2637 (1978).
6. S.D. Drell, H.R. Quinn, B. Svetitsky and M. Weinstein, Phys. Rev. D 19:619 (1979).
7. R. Balian, J.M. Drouffe and C. Itzykson, Phys. Rev. D 10:3376 (1974); 11:2098 (1975); 11:2104 (1975).
8. D. Horn, M. Weinstein and S. Yankielowicz, Phys. Rev. D 19:3715 (1979).
9. F.J. Wegner, J. Math. Phys. 12:2259 (1971).
10. S.F. Edwards and P.W. Anderson, J. Phys. F 5:965 (1975).
11. A.A. Migdal, Sov. Phys.-JETP 42:413, 743 (1976).
12. M. Creutz, "Quarks, Gluons and Lattices," Cambridge University Press (1983).
13. M. Creutz, L. Jacobs and C. Rebbi, Phys. Rep. 95:201 (1983).
14. D.W. Heys and D.R. Stump, Phys. Rev. D 28:2067 (1983).
15. S.A. Chin, J.W. Negele and S.E. Koonin, Ann. Phys. (NY) 157:140 (1984).
16. T. Barnes and D. Kotchan, Phys. Rev. D 35:1947 (1987).
17. H.H. Roomany and H.W. Wyld, Phys. Rev. D 21:3341 (1980).
18. A.C. Irving and A. Thomas, Nucl. Phys. B200 [FS4]:424 (1982).
19. A.C. Irving, J.F. Owens and C.J. Hamer, Phys. Rev. D 28:2059 (1983).
20. M.D. Kovarik, J.W. Darewych and R. Koniuk, Phys. Rev. D 33:3654 (1986).
21. A. Dabringhaus, M.L. Ristig and J.W. Clark, Phys. Rev. D 43:1978 (1991).
22. J.W. Clark and E. Feenberg, Phys. Rev. 113:388 (1959); H.W. Jackson and E. Feenberg, Ann. Phys. (NY) 15:266 (1961).
23. F. Coester, Nucl. Phys. 7:421 (1958); F. Coester and H. Kümmel, Nucl. Phys. 17:477 (1960).

24. R.F. Bishop, Theor. Chim. Acta $\underline{80}$:95 (1991).
25. K.G. Wilson, Nucl. Phys. B (Proc. Suppl.) $\underline{17}$:82 (1990).
26. R.J. Bartlett, Theor. Chim. Acta $\underline{80}$:71 (1991) -- and note also that issues 2-6 of this volume are entirely devoted to articles on the coupled cluster theory of electron correlations in many-electron systems.
27. M. Roger and J.H. Hetherington, Phys. Rev. B $\underline{41}$:200 (1990).
28. R.F. Bishop, J.B. Parkinson and Yang Xian, Phys. Rev. B $\underline{43}$:13782 (1991); $\underline{44}$:9425 (1991); Theor. Chim Acta $\underline{80}$:181 (1991).
29. V. Alessandrini, V. Hakim and A. Krzywicki, Nucl. Phys. $\underline{B200}$[FS4]:355 (1982).
30. J. Villain, J. Phys. (Paris) $\underline{36}$:581 (1975).
31. M. Göpfert and G. Mack, Commun. Math. Phys. $\underline{82}$:545 (1981).
32. J. Ambjørn, A.J.G. Hey and S. Otto, Nucl. Phys. $\underline{B210}$[FS6]:347 (1982).
33. G. Bhanot and M. Creutz, Phys. Rev. D $\underline{21}$:2892 (1980).
34. U.M. Heller, Phys. Rev. D $\underline{23}$:2357 (1981).
35. D.W. Heys and D.R. Stump, Nucl. Phys. $\underline{B257}$:19 (1985); $\underline{B285}$:13 (1987).
36. J.W. Choe, A. Duncan and R. Roskies, Phys. Rev. D $\underline{37}$:472 (1988).
37. E. Dagotto and A. Moreo, Phys. Rev. D $\underline{31}$:865 (1985).
38. J.B. Kogut, Phys. Rep. $\underline{67}$:67 (1980).
39. P. Pfeuty and R.J. Elliot, J. Phys. C $\underline{4}$:2370 (1971).
40. J. Kogut and D. Sinclair, Phys. Lett. $\underline{81A}$:149 (1981).
41. E. Fradkin and S. Raby, Phys. Rev. D $\underline{20}$:2566 (1979).
42. T. Banks and A. Zaks, Nucl. Phys. $\underline{B200}$[FS4]:391 (1982).
43. D. Bessis and M. Villani, J. Math. Phys. $\underline{16}$:462 (1975).
44. D. Robson and D.M. Webber, Z. Phys. C $\underline{7}$:53 (1980).
45. G. Blanch, in: "Handbook of Mathematical Functions," M. Abramowitz and I.A. Stegun (eds.), N.B.S. Appl. Math. Ser. $\underline{55}$, U.S. Govt. Printing Office, Washington, D.C. (1964), p.721.
46. W. Langguth, Z. Phys. C $\underline{23}$:289 (1984).
47. J. Cardy and H. Hamber, Nucl. Phys. $\underline{B170}$[FS1]:79 (1980).
48. P. Suranyi, Nucl. Phys. $\underline{B210}$:519 (1982).
49. E. Dagotto and A. Moreo, Phys. Rev. D $\underline{29}$:300 (1984).
50. H.G. Kümmel, Nucl. Phys. $\underline{A317}$:199 (1979).

PHYSICAL REGIONS OF AN EQUATION OF STATE

George A. Baker, Jr. and J. D. Johnson

Theoretical Division, Los Alamos National Laboratory
University of California, Los Alamos, N. M. 87545, USA

ABSTRACT

There are several key physical regions in any equation of state. We are concerned with astrophysical and similar such applications. The key regions are: (i) All the electrons have coalesced about the nuclei to form atoms (cold curve). Included here would be the further condensation into fluids and solids. (ii) There is a plasma of electrons and nuclei (hot curve). (iii) The low-density, partial-ionization region (ionization curve). The density and temperature go to zero together for this region, because for fixed, non-zero temperature, total ionization occurs in the low density limit. (iv) The intermediate region. Regions (i) and (ii) merge at high density. It is our thesis that region (iv) can be obtained from the hot curve, the ionization curve and the cold curve plus the low order perturbations about them by a suitable interpolation proceedure. By the "cold curve" we must also understand the pendant phase-diagram structure adjoining it. As in our previous work, we will use the Thomas-Fermi equation of state as a test for these ideas. We have previously developed the hot curve, the cold curve and their perturbations. Herein, we discuss the ionization curve. The virtue of our approach is that, if successful, the Thomas-Fermi approximations for these quantities (which have a great many well known defects) can be replaced by the more accurate results of quantum many-body theory to give a greatly improved equations of state.

1. INTRODUCTION AND SUMMARY

The system that we are interested in here is an electrically neutral system of electrons and a single species of nuclei with charge Z. The Coulomb potential energy, as is well known, is neither bounded from below (electron-nuclei part) nor from above (electron-electron part, for example). In addition, it is long-ranged so there is no such thing as a dilute system. Of course, Debye screening does occur, but this feature is a result of the theory, and can not be used *ab initio*. There are two important physical aspects to this problem. They are the quantum mechanical aspect, embodied by Planck's constant, h, and the Coulomb interaction, embodied by the electronic charge, e.

We know that if we take $h = 0$ and $e \neq 0$, then the electrons spiral into the nucleus with the emission of γ-rays, and the whole system collapses into a state of negatively infinite energy. Put otherwise, there is no classical limit for this system, and quantum mechanics is an absolutely essential feature. On the other hand, if we take $h \neq 0$ and $e = 0$, then we get a perfectly well defined system. This system is the ideal Fermi gas for the electrons. In the energy ranges we will be considering, the heavier, less numerous nuclei will behave pretty much like an ideal gas, however a better treatment should not be a particular problem if required. The many-body perturbation theory in e^2 about the $e^2 = 0$ system is, by the nature of the Coulomb potential, a singular one but not[1] a hopeless one.

In the second section, we examine the various limiting physical regions of our system and use this knowledge to construct a physical picture. We will conclude that the best (for our purposes at least) independent variables to use are not the usual temperature and density, but the de Broglie density and the density.

In the third section we discuss Thomas-Fermi theory as a model of the true system and recast it in terms of our independent variables. We review what is both known and useful in our work in this framework. We discuss the dilute-limit ionization curve for this model.

In the final section we show how the information about the various limiting physical regions can be combined in a simple functional form to give a good overall representation. Better information for the various physical regions than that given by the Thomas-Fermi or the Thomas-Fermi-Dirac theories should now be obtainable from many-body perturbation theory, from Saha theory and from various methods for the zero temperature pressure-density curve.

2. PHYSICAL REGIONS

The first physical region is that of the ideal Fermi gas, where $e \equiv 0$. As we will see later, it is best thought of for our purposes as the high-density region. The equation of state is given as,[2,3]

$$\frac{P\Omega}{ZNkT} = g(\zeta),$$

$$\zeta = \frac{ZN}{2\Omega} \left(\frac{h^2}{2\pi m k T} \right)^{\frac{3}{2}}, \tag{2.1}$$

$$\zeta = f_{\frac{3}{2}}(z), \quad g(\zeta) = \frac{f_{\frac{5}{2}}(z)}{f_{\frac{3}{2}}(z)},$$

where

$$f_n(z) = \frac{1}{\Gamma(n)} \int_0^\infty \frac{z y^{n-1} e^{-y} \, dy}{1 + z e^{-y}}. \tag{2.2}$$

In (2.1), z is determined as a parameter from ζ which in turn is determined from the temperature and density. Then the function $g(\zeta)$ is determined parametrically, and thus the equation of state. Here we use P for the electron pressure, Ω for the volume, m for the mass of the electron, Z for the nuclear charge, N for the number of atoms, T for the temperature, and k for the Boltzmann constant. As remarked above, in this region, the pressure due the the nuclei can just be added to that in (2.1) to give the total pressure.

By means of many-body perturbation theory, we can expand the pressure as,

$$\frac{P\Omega}{ZNkT} = \sum_{j=0} g_j(\zeta) y^j, \tag{2.3}$$

where

$$y = \left[\frac{Ze^2}{r_b kT}\right]^{\frac{1}{2}} = \left(\frac{\sqrt{\pi}}{4}\zeta\right)^{\frac{1}{3}}\sqrt{x}, \tag{2.4}$$

with the definitions

$$\frac{4\pi}{3}r_b^3 = \frac{\Omega}{N}, \quad x = \left(\frac{128Z}{9\pi^2}\right)^{\frac{1}{3}}\frac{me^2 r_b}{\hbar^2}. \tag{2.5}$$

The known terms of (2.3) are $g_0 = g$ of (2.1), $g_1 = 0$, g_2 is the exchange term and g_3 is the Debye-Hückel term. Note that the expansion parameter is proportional to e and not to e^2 because of the singular nature of the perturbation. This feature will not be reproduced in the Thomas-Fermi model theory to the extent it is known.

The other thing we know is that for $T = 0$, the pressure P is a function of the density alone. We expect, in order for this conclusion to hold in a straightforward manner for the $g_j(\zeta)$ of (2.3), that $g_j(\zeta) \propto \zeta^{(2-j)}$ as $\zeta \to \infty$. This feature does hold in Thomas-Fermi theory. Let us recast (2.5) as,

$$P = \left(\tfrac{2}{3}\right)^{\frac{5}{3}}\frac{128Z^{\frac{10}{3}}m^5 e^{10}}{\pi^3\hbar^8 x^5}\sum_{j=0}\left(\frac{\sqrt{\pi}}{4}\right)^{\frac{j}{3}}\zeta^{(2-j)/3}g_j(\zeta)x^{\frac{j}{2}}, \tag{2.6}$$

where account has been taken of (2.4). It is to be observed, since the density goes to infinity as $x \to 0$, that (2.6) is now a high-density expansion of the pressure at fixed de Broglie density ζ. If we next define

$$\hat{g}_j = \lim_{\zeta\to\infty}g_j(\zeta)\zeta^{(j-2)/3}, \tag{2.7}$$

then we can take the zero temperature (infinite ζ) limit of (2.6), which is

$$P = \left(\tfrac{2}{3}\right)^{\frac{5}{3}}\frac{128Z^{\frac{10}{3}}m^5 e^{10}}{\pi^3\hbar^8 x^5}\sum_{j=0}\left(\frac{\sqrt{\pi}}{4}\right)^{\frac{j}{3}}\hat{g}_j x^{\frac{j}{2}}. \tag{2.8}$$

This form of viewing the results of Thomas-Fermi theory suggest to us that if we can supply a low-density expansion for fixed ζ, then by use of two-point ($x = 0$ and $x = \infty$) approximation methods for each ζ, we should be able to give a good global representation for the equation of state. The low-density behavior corresponds to the knowledge of the dilute limit ionization curve, which may perhaps be approached through Saha theory.[4]

We now have the following physical picture. First at high ζ, we have the "cold curve" which is a function of density (or x) alone. At high densities this behavior merges into that of the ideal Fermi fluid system and here we have the many-body perturbation theory expansion. Since the $g_j(\zeta)$ are of order unity for small ζ, it is natural to expect that this expansion in $\sqrt{x}$ should have a region of validity of the order of $1/\sqrt[3]{\zeta}$. Thus, in accord with other ways of approaching the same question, we expect that the limiting behavior as $\zeta \to 0$ for fixed density (or x) will be a totally ionized system. Finally, for fixed ζ as $x \to \infty$ the behavior is governed by the dilute-limit, partial-ionization behavior of the system. The question we will be considering in the last section is, can we use our knowledge of the physics of these boundary regions to fill in, to decent accuracy, the equation of state over the whole region?

3. THOMAS-FERMI THEORY

In this section we illustrate the ideas of the previous section on the Thomas-Fermi[5-7] equation of state as a model theory. This theory, and more frequently its generalization the Thomas-Fermi-Dirac theory, is currently quite often used in practical applications, such as the astrophysics of stellar interiors, to give the equation of state over wide ranges of temperature and density.

In the finite-temperature, Thomas-Fermi, statistical theory of the atom, the electron density is given by

$$\rho(r) = \int_0^\infty \frac{2 \cdot 4\pi p^2 \, dp/h^3}{\exp[\{p^2/(2m) - eV(r)\}/kT + \eta] + 1}, \tag{3.1}$$

where $-eV(r)$ is the potential energy. Let us define,

$$I_n(\eta) = \int_0^\infty \frac{y^n \, dy}{e^{y-\eta} + 1}. \tag{3.2}$$

Then we can use Poisson's equation to determine $V(r)$,

$$\frac{1}{r}\frac{d^2}{dr^2}[rV(r)] = \frac{16\pi^2}{h^3} e(2mkT)^{\frac{3}{2}} I_{\frac{1}{2}}\left(\frac{eV(r)}{kT} - \eta\right). \tag{3.3}$$

Note that if the electronic charge is taken as $e = 0$, the right-hand side of (3.3) vanishes and the solution for $V(r)$ is just $a + b/r$ where a, b are constants. This solution corresponds to the ideal Fermi gas as described in the previous section.

The usual reduction of these equations to dimensionless variables[5] uses,

$$c = \left(\frac{h^3}{32\pi^2 e^2 m(2mkT)^{\frac{1}{2}}}\right)^{\frac{1}{2}} \propto T^{-\frac{1}{4}}. \tag{3.4}$$

Let $s = r/c$. Then (3.3) becomes,

$$\frac{d^2\beta}{ds^2} = sI_{\frac{1}{2}}\left(\frac{\beta}{s}\right), \tag{3.5}$$

where

$$\frac{\beta}{s} = \frac{eV(r)}{kT} - \eta, \tag{3.6}$$

since η is independent of r. As $r \to 0$, we must have $V(r) \to Ze/r$ to reproduce the nuclear charge. Therefore as one of the boundary conditions for (3.6) we must have,

$$\beta(0) = \alpha = \frac{Ze^2}{kTc} \propto T^{-\frac{3}{4}}. \tag{3.7}$$

In order to insure that the total number of electrons is Z for a neutral atom inside a sphere whose volume is the average volume per atom, Ω/N, a little manipulation[6] shows that,

$$\frac{d\beta}{ds} = \frac{\beta}{s} \quad \text{at } s = b, \tag{3.8}$$

where $r_b = cb$, see (2.5).

It is convenient for our purposes to re-express the Thomas-Fermi equations in terms of our chosen independent variables, ζ (2.1) and x (2.5). To this end let us make the further change of variables

$$s = \sigma b, \quad \beta = \gamma \alpha. \tag{3.9}$$

The defining equations (3.6-8) now become,

$$\frac{d^2\gamma}{d\sigma^2} = \frac{6\sigma}{\zeta\sqrt{\pi}} I_{\frac{1}{2}}\left(\left(\frac{\sqrt{\pi}}{4}\zeta\right)^{\frac{2}{3}}\frac{x\gamma}{\sigma}\right),$$

$$\gamma(0) = 1.0, \tag{3.10}$$

$$\frac{d\gamma}{d\sigma} = \frac{\gamma}{\sigma} \text{ at } \sigma = 1.0.$$

From our earlier analysis, one of the regions we need to study is the low density region, $x \to \infty$. To facilitate this study, we note that $I_{\frac{1}{2}}(\eta)$ is a continuously differentiable function that is asymptotically proportional to $\eta^{3/2}$ as $\eta \to +\infty$ and is asymptotically proportional to e^η as $\eta \to -\infty$. Thus, in the limit $x \to \infty$ (3.10) becomes,

$$\frac{d^2\gamma}{d\sigma^2} = \begin{cases} \sigma^{-\frac{1}{2}}(x\gamma)^{\frac{3}{2}}, & \gamma > 0, \\ \frac{3\sigma}{\zeta}\exp\left(\left(\frac{\sqrt{\pi}}{4}\zeta\right)^{\frac{2}{3}}\frac{x\gamma}{\sigma}\right), & \gamma < 0, \end{cases}$$

$$\gamma(0) = 1.0, \tag{3.11}$$

$$\frac{d\gamma}{d\sigma} = \frac{\gamma}{\sigma} \text{ at } \sigma = 1.0.$$

We can observe from (3.11) that when $\gamma < 0$ its second derivative goes exponentially to zero, and so in this region γ is a straight line in σ. The division of the regions between $\gamma > 0$ and $\gamma < 0$ correspond to an ion and the Fermi gas of electrons, respectively. From (3.11), γ starts positive at $\sigma = 0$. The region over which it remains positive can be worked out approximately analytically, or studied numerically. It turns out to tend to zero width as $x \to \infty$. It contains $N < Z$ electrons, and the outer region contains $Z_i = Z - N$ ionized electrons. A little manipulation, similar to that which lead to (3.8), shows that,

$$Z_i = -Z\sigma_0\gamma'(\sigma_0), \tag{3.12}$$

where σ_0 is defined by $\gamma(\sigma_0) = 0$. As remarked above, and as we confirm here, when $\zeta \to 0$ $(T \to \infty)$ we get the total ionization region $Z_i \to Z$. When $\zeta \to \infty$ $(T \to 0)$, $Z_i \to 0$, we get the region where all the electrons condense to form neutral atoms.

Let us now look more closely at the low-temperature region. First, for Thomas-Fermi theory, the pressure is given by,

$$\frac{P\Omega}{N} = \frac{2}{9}ZkT\left[\frac{r_b^3}{c^3\alpha}\right]I_{\frac{3}{2}}\left(\left(\frac{\sqrt{\pi}}{4}\zeta\right)^{\frac{2}{3}}x\gamma(1)\right), \tag{3.13}$$

which, in and near the low-temperature limit gives,

$$P = \frac{Z^2e^2}{10\pi\mu^4}\left[\frac{\phi_b}{x}\right]^{\frac{5}{2}}\left[1 + \left(\frac{3}{2}\right)^{\frac{4}{3}}\frac{5\pi^2x^2}{12\phi^2\alpha^{\frac{8}{3}}} + O\left(\frac{x^4}{\phi^4\alpha^{\frac{16}{3}}}\right) + O\left(\exp\left[-\frac{E\phi\alpha^{\frac{4}{3}}}{x}\right]\right)\right] \tag{3.14}$$

with E a numerical constant. The defining equations here are

$$\frac{d^2\phi}{d\xi^2} = \xi^{-\frac{1}{2}}\phi^{\frac{3}{2}}, \quad \phi(0) = 1.0, \quad \frac{d\phi}{d\xi} = \frac{\phi}{\xi} \text{ at } \xi = x. \tag{3.15}$$

Note is taken that since by (3.7), $\alpha^{4/3} \propto 1/T$, the last error term in (3.14) contributes no power series terms in T to the expansion about $T = 0$. For the large x (dilute) limit, Baker and Johnson[7] find that

$$\phi(x) \asymp \frac{287.40}{x^3}. \tag{3.16}$$

Thus in the dilute limit,

$$P \propto x^{-10} \left[1 + A(Tx^4)^2 + B(Tx^4)^4 + \cdots + O\left(e^{-C/(Tx^4)}\right)\right], \tag{3.17}$$

or in terms of ζ and x, we get,

$$P \propto x^{-10} \left[1 + A'(\zeta^{-\frac{2}{3}}x^2)^2 + B'(\zeta^{-\frac{2}{3}}x^2)^4 + \cdots + O\left(e^{-C'\zeta^{2/3}/x^2}\right)\right], \tag{3.18}$$

where A, A', B, B', C and C' are numerical constants. As the series is a function of $\zeta^{-\frac{2}{3}}x^2$ alone, in the dilute limit, the simplest hypothesis is that it takes the form

$$P \propto x^{-10} f\left(\zeta^{-\frac{2}{3}}x^2\right) \xrightarrow[x\to\infty]{} Kx^{-10}\left(\frac{\sqrt{x}}{\zeta^{\frac{1}{6}}}\right)^\gamma, \tag{3.19}$$

where K and γ are constants to be determined. Since we have failed to find them analytically, we have fit them to the numerical solutions of the Thomas-Fermi equations as solved for aluminum at a compression of 10^{-6} for a range to temperature from $0.25 - 20$ ev. We find that

$$\gamma \approx 7.0 \pm 0.2. \tag{3.20}$$

This result gives the general dilute limit asymptote for the Thomas-Fermi pressure. Needless to say, the other error term will modify the situation significantly away from this limit, and is expected to introduce a ζ-dependence in even the first term of the expansion off the large x limit.

4. GLOBAL REPRESENTATION OF THE THOMAS-FERMI PRESSURE

Since our goal is to create a "zipper" formula into which the separately computed or measured boundary-physical-region behaviors can be simply "zipped," and since the cold curve and its pendant phase diagram is far too complex for general treatment, we write,

$$P = \Delta P + P_{\text{cold curve}}. \tag{4.1}$$

Baker and Johnson[7] have previously computed for the Thomas-Fermi model the expansion of P about the "hot curve," or the high-density expansion as we saw in (2.6). It is,

$$P = 196.6889 Z^{\frac{10}{3}} x^{-5} \zeta^{-\frac{2}{3}} \left[g_0(\zeta) + \left(\frac{\sqrt{\pi}}{4}\zeta\right)^{\frac{2}{3}} g_2(\zeta)x + + \left(\frac{\sqrt{\pi}}{4}\zeta\right)^{\frac{4}{3}} g_2(\zeta)x^2 + \cdots\right], \tag{4.2}$$

where P is in megabars, the terms $g_1(\zeta) = g_3(\zeta) = 0$, and Baker and Johnson[7] give the representations, accurate to about 0.1%,

$$g_0(\zeta) \approx \left[\frac{1 + 0.61094880\zeta + 0.12660436\zeta^2 + 0.0091177644\zeta^3}{1 + 0.080618739\zeta} \right]^{\frac{1}{3}},$$

$$g_2(\zeta) = -\frac{3}{10},$$

$$g_4(\zeta) \approx \frac{97}{2880} \left[\frac{v_3(\zeta)}{u_5(\zeta)} \right]^{\frac{1}{3}}. \tag{4.3}$$

The definitions,

$$v_3(\zeta) = 1 + 0.17549205\zeta + 1.1833437 \times 10^{-2} + 3.0923597 \times 10^{-4}\zeta^3,$$

$$u_5(\zeta) = 1 + 1.2361522\zeta + 0.54327035\zeta^2 + 9.7985998 \times 10^{-2}\zeta^3, \tag{4.4}$$

$$+ 6.1912639 \times 10^{-3}\zeta^4 + 1.6191557 \times 10^{-4}\zeta^5,$$

have been used in (4.3). When we subtract the cold curve to form ΔP, we must subtract the large-ζ limiting behavior from the g_j's. When this step is done, we get

$$\Delta P = 196.6889 Z^{\frac{10}{3}} x^{-5} \left[\tilde{g}_0(\zeta) + \tilde{g}_4(\zeta) x^2 + \cdots \right], \tag{4.5}$$

where

$$\tilde{g}_0(\zeta) = g_0(\zeta)\zeta^{-\frac{2}{3}} - 0.48359758,$$

$$\tilde{g}_2(\zeta) \equiv 0, \tag{4.6}$$

$$\tilde{g}_4(\zeta) = 0.33782096 g_4(\zeta)\zeta^{\frac{2}{3}} - 0.041787498.$$

Since in the dilute limit, by (3.20),

$$\Delta P \propto x^{-\frac{13}{2}}, \tag{4.7}$$

then the [ ] in (4.5) should sum to something proportional to $x^{-3/2}$. After some experimentation, we have chosen the form,

$$\Delta P \approx \frac{196.6889 Z^{\frac{10}{3}} x^{-5} \tilde{g}_0(\zeta)}{\left[1 - 2\frac{\tilde{g}_4(\zeta)}{\tilde{g}_0(\zeta)} x^2 + \left\{ \frac{Z^{\frac{4}{3}}\zeta^{\frac{7}{6}} \tilde{g}_0(\zeta)}{d_0} \right\}^2 x^3 \right]^{\frac{1}{2}}}. \tag{4.8}$$

This form agrees with the series expansion in x, and if the asymptotic form (3.19) for $x \to \infty$ were exact, then d_0 would be a constant. Because of the necessary corrections to that asymptotic form we have chosen the approximate representation,

$$d_0 \approx \frac{1602.7288 + (19532.475\zeta^{\frac{1}{6}} - 12736.679)\zeta^{\frac{1}{6}}/\sqrt{x}}{1 + 35.483743\zeta^{\frac{1}{6}}/\sqrt{x}}. \tag{4.9}$$

This form corresponds to the asymptotic result (3.19) and two corrections to it. The form (4.9) is derivable as the [1/1] Padé approximant to series,

$$\Delta P \asymp 196.6889 Z^2 x^{-\frac{13}{2}} \zeta^{-\frac{7}{6}} [1602.7288 + (19532.475\zeta^{\frac{1}{6}} - 69607.496)\zeta^{\frac{1}{6}}/\sqrt{x}$$

$$+ (-693085.33\zeta^{\frac{1}{6}} + 2469934.6)\zeta^{\frac{1}{3}}/x + \cdots], \tag{4.10}$$

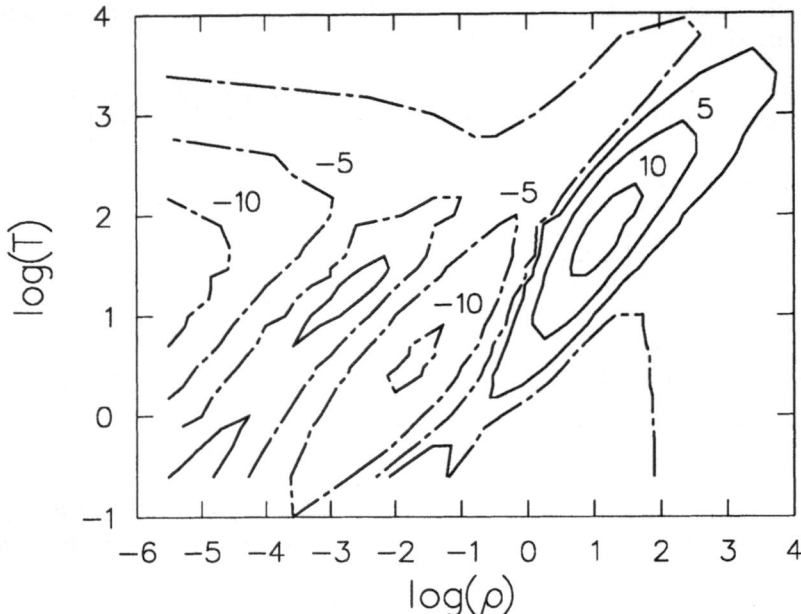

Figure 1. Error contours for the representation (4.1, 8, 11) of the Thomas-Fermi pressure. Positive errors, the fit larger than Thomas-Fermi, are solid lines while negative are chain-dashed. The 5 and 10 percent contours are explicitly labeled while the ±1 percent ones have no labels. Here the axes are the log, base 10, of the temperature in electron volts and the density in g/cc for aluminum. The conversion from ev to Kelvin is 11604.

which was derived by the expansion of (4.9). Form (4.9) is a rough but adequate representation of d_0 for our needs. Since by identification with (3.19), d_0 should be positive and since we are trying to keep our results simple, we found it to be computationally more efficient to simply cut-off d_0 rather than to generate a more elaborate form which doesn't need a cutoff. We use the maximum of the result of (4.9) and 50 for d_0. This number was choosen to be just a little lower than the smallest, well-determined value of d_0 found by a scan of direct Thomas-Fermi theory computation through our region of application. This value occurs at very high temperature and density.

In addition we have used Baker and Johnson's[7] representation for the cold curve pressure. It is based on the use of the $T = 0$ part of (3.14) and is,

$$
P_{\text{cold curve}} \approx 9.0549692 \left(\frac{Z\rho}{\mathcal{G}} \right)^{\frac{5}{3}} x^{\frac{5}{2}} \left[\frac{(1 + 1.59659 x^{-0.772} + 1.06595 x^{-1.544})^{\frac{1}{6}}}{1 + 0.2783436 x^{-0.772}} \right]^{9.715}
$$

(4.11)

where $P_{\text{cold curve}}$ is in megabars, the density ρ is in grams per cc, and $\mathcal{G}$ is the gram molecular weight.

We conclude by showing in Fig. 1 a contour map for the percent errors of our fit to Thomas-Fermi. The maximum deviation is about 12% and occurs in two islands along the "seam" between the high and low density regions. The error also rises for very low density, but we are not as interested in that. The rms error is about 4%. The representation is doing very well especially when we consider the large ranges over which we are working, about nine decades in density and almost five in temperature. We hope in future work to obtain representations of the internal energy and free energy. We also wish to include in our representation better physics than Thomas-Fermi.

ACKNOWLEDGEMENTS

One of the authors (G.B.) wishes to thank the USARO for partially supporting his attendance at the *XVI International Workshop on Condensed-Matter Theories*, where this paper was presented.

REFERENCES

1. D. J. Thouless. "The Quantum Mechanics of Many-Body Systems," Academic Press, New York (1961).
2. K. Huang. "Statistical Mechanics," John Wiley and Sons, New York (1963).
3. G. A. Baker, Jr. and J. D. Johnson, Thomas-Fermi equation of state-the hot curve, *in*: "Condensed Matter Theories, Volume 5," V. C. Aquilera-Navarro, ed., Plenum Press, New York (1990).
4. M. N. Saha, Article title, *Phil. Mag.* 40:472 (1920); D. Mihalas, "Stellar Atmospheres," W. H. Freeman & Co., San Francisco (1970).
5. R. P. Feynman, N. Metropolis and E. Teller, Equations of state of elements based on the generalized Fermi-Thomas theory, *Phys. Rev.* 75:1561 (1949).
6. R. D. Cowan and J. Ashkin, Extension of the Thomas-Fermi-Dirac statistical theory of the atom to finite temperatures, *Phys. Rev.* 105:144 (1957).
7. G. A. Baker, Jr. and J. D. Johnson, General structure of the Thomas-Fermi equation of state, *Phys. Rev. A* 44:2271 (1991).

CLUSTER EXPANSIONS AND VARIATIONAL MONTE CARLO IN MEDIUM LIGHT NUCLEI

E. Buendía
Departamento de Física Moderna, Universidad de Granada,
18071 Granada, Spain
and
R. Guardiola
Departamento de Física Atómica y Nuclear, Universidad de Valencia
Avda. Dr. Moliner 50, E-46100 Burjassot (Valencia), Spain

ABSTRACT

The B1 Brink-Boeker effective interaction is used to compute variational upper bounds for the ground state energy of nuclei from ^{16}O up to ^{40}Ca. The calculations are carried out by means of the Variational Monte Carlo method and with a multiplicative cluster expansion up to fourth order.

1 INTRODUCTION

After many years of concentrating on extended systems, many-body theories have turned their eyes back to finite systems, including nuclear systems. At present there exist reliable variational Monte Carlo calculations of light systems using realistic interactions [1], a very promising extension of the Green functions Monte Carlo Method for light systems [2] and variational calculations for ^{16}O using Monte Carlo method [3,4] for realistic interactions. On the other hand, the FHNC method of carrying massive summations of diagrams for finite systems, derived by Fantoni and Rosati [5] already in 1979, have been finally solved for ^{16}O, with model interactions [6,7].

The purpose of our work is to provide comparison results for the FHNC approach, as well as to test the validity of the simpler method of cluster expansions, particularly the FAHT cluster expansion of multiplicative nature. We will use again model interactions, the Malfliet-Tjon MTV interaction [8], which is a pure Wigner inter-action, and the Brink-Boeker B1 potential [9], which contains an important Majorana space exchange part. Of course, our calculations, as far as they involve model or effec-tive interactions, cannot be considered as a description of real nuclei. We also call the attention on the hyperspherical harmonics expansion method [10], which turns out at present to be a competitive description of light and medium-light nuclei.

2 MODEL INTERACTIONS AND NUCLEAR WAVE FUNCTIONS

We consider in this work two simple interactions. Firstly, the Malfliet-Tjon MTV interaction [8], which is a Wigner interaction defined as the average of s-wave singlet and triplet nucleon-nucleon interactions (i.e. the Wigner part of the MT I/III potential). This interaction will be used as if it were defined in all channels, as in [11], where it was applied to the variational and Green functions Monte Carlo for ^{16}O, and in [6] and [7], where the interaction was used in the framework of the FHNC theory for finite systems [5]. The interaction, which is quite good for the description of ^3He and ^4He, turns out to provide too much binding energy for ^{16}O, just because the mainly repulsive odd channels are not well described. Nevertheless, the comparison with previous work serves as a good check. Moreover, given that the system turns out to be more dense than physical nuclear matter, the test on approximate theories may be more selective.

The second interaction is the Brink and Boeker B1 interaction [9]. This is an effective interaction, and one should not expect to obtain a reliable description of nuclei when short range correlations are introduced in the wave function. The interest of this interaction is twofold. On one side, there is a large amount of work carried out with this interaction for light nuclei, medium nuclei and also nuclear matter, in the framework of several theories (cluster expansions [12,13,14], Brueckner theory [15], hyperspherical harmonics expansion [16], integro-differential description of the hyperspherical harmon-ics expansion expansion [17,18], etc). The B1 interaction is a Wigner-Majorana mixture, providing in this way a reasonable description of the odd channels. This Majorana com-ponent of the interaction is the second point of interest, because it may provide a test on the exchange part of the two-body distribution function, which is more sensitive to approximations of the various theories [19].

Because of the simplicity of the interactions, one may avoid the introduction of the spin and isospin degrees of freedom in the nuclear wave function. In particular, for a spin and isospin saturated system, the statistical part of the wave function may be written as the product of four Slater determinants, one for each fermionic species. We are going to use for the description of the nuclear states a Jastrow correlated shell model wave function. The single particle states will be eigenstates of a harmonic oscil-lator potential, characterized by a variational parameter $\alpha = \sqrt{m\omega/\hbar}$. The correlation function will be parametrized in a very simple way,

$$f(r) = 1 + ae^{-(r/b)^2} \tag{1}$$

Table 1. Configurations used to describe p-shell and sd-shell nuclei. The orbitals are represented in the cartesian basis (n_x, n_y, n_z). For $A > 16$ there is also a ^{16}O core.

Nucleus	Configuration	Shape
^8Be	$(000)^4 (001)^4$	Prolate
^{12}C	$(000)^4 (100)^4 (010)^4$	Oblate
^{16}O	$(000)^4 (100)^4 (010)^4 (001)^4$	Spherical
^{20}Ne	$(002)^4$	Prolate
^{24}Mg	$(101)^4 (002)^4$	Triaxial
^{28}Si	$(110)^4 (200)^4 (020)^4$	Oblate
^{32}S	$(110)^4 (101)^4 (200)^4 (020)^4$	Triaxial
^{36}A	$(110)^4 (101)^4 (011)^4 (200)^4 (020)^4$	Oblate
^{40}Ca	$(110)^4 (101)^4 (011)^4 (200)^4 (202)^4 (002)^4$	Spherical

and a and b will be again considered as variational parameters, controlling the depth and the range of the correlation, respectively.

For the closed shell nuclei, ^{16}O and ^{40}Ca, there is a natural configuration for the description of the shell model wave function. Between these two nuclei we have considered the $4n$ nuclei and have assumed to be spin and isospin saturated. The construction of shell model wave functions is not simple, and a good approximation may be achieved by considering the orbitals described in the cartesian basis and characterized by the three harmonic oscillator quantum numbers (n_x, n_y, n_z), and filling every orbital with four nucleons. This way of constructing the wave functions violates the rotational invariance and gives a deformed shape for all sd shell nuclei. The deformation related to the coupling may still be amplified by assuming different harmonic oscillator parameters for the three axes, generating in general triaxial shapes. In general, the deformation results in a gain of energy, and most of the systems prefer an axial symmetry. The configurations used in this work are listed in Table 1.

For the axially symmetric nuclei we also consider a deformation parameter d,

$$d = \alpha_z / \alpha, \tag{2}$$

in such a form that for $d > 1$ we have an oblate shape, and for $d < 1$ the shape is prolate. For the two cases where the configuration has already a triaxial shape, we tried also different harmonic oscillator parameters for each of the three axis. Only for ^{24}Mg we have obtained a gain in energy after the introduction of three different harmonic oscillator parameters.

The main advantage of using harmonic oscillator basis is the simplicity in removing the spurious center-of-mass motion. Once the gaussian part of the single particle orbitals is extracted from the Slater determinants, the latter are translationally invariant (they are actually a generalization of Vandermonde determinants) and the spurious center-of-mass coordinate only appears in the gaussian term $\exp(-\alpha^2 \sum_{i=1}^{A} r_i^2/2)$, or $exp[-\sum_{i=1}^{A}(\alpha_x^2 x_i^2 + \alpha_y^2 y_i^2 + \alpha_z^2 z_i^2)/2]$ for a triaxially deformed oscillator. As it is well known, this exponential may be rewritten as the product of two exponentials, one involving *only* the center-of-mass coordinate, and the other depending on interparticle distances r_{ij}. In terms of the parameter $\gamma = \alpha^2/2A$ and the hyperradius r defined as

$$r^2 = \sum_{i<j}^{A}(\mathbf{r}_i - \mathbf{r}_j)^2 \tag{3}$$

the shell model wave function in the harmonic oscillator basis is

$$\Psi = Ne^{-\gamma r^2} r^L \mathcal{Y}_L \tag{4}$$

i.e., the product of a specific hyperradial function times a hyperspherical harmonic function $\mathcal{Y}_L$, where L is the minimal degree of the harmonic polynomial compatible with the antisymmetry of the wave function.

The obvious way of generalizing this function is to replace the specific hyperradial function by a general functional form, and determine the new function by means of a minimization algorithm. This is the generalization of the harmonic oscillator basis to a Hartree-Fock like structure without center-of-mass motion spurious components. This way may be pursued by adding higher spherical harmonics and corresponding hyperradial functions, this procedure being equivalent to introducing $2p - 2h$, $3p - 3h$, ...excitations while still conserving the translational invariance.

The convergence of this expansion may be very slow, in particular when dealing with strongly repulsive cores. A way of improving the convergence is to use the Potential Basis (PB) approach, which is a Fadeev-like description of a many-body system which corresponds to a trial function [16]

$$\Psi_{PB} = \sum_{i<j}^{A} \mathcal{F}(\mathbf{r}, \mathbf{r}_{ij}). \tag{5}$$

In its lowest order approximation, the PB method is equivalent to the second order approximation to the translationally invariant coupled cluster theory [20,21], which in turn is equivalent to a configuration interaction expansion with all translationally invariant $2p - 2h$ excitations included. Higher orders of the PB formalism may be determined by solving an infinite set of coupled differential equations [16] or two coupled integro-differential equations (IDEA) [17]. Note that the IDEA method results from an approximation to the PB formalism and the variational character may be lost.

Another way of improving the convergence of the hyperspherical expansion is to include in the wave function Jastrow correlations, instead of the pair correlations included in the potential basis description. This method has been applied very successfully to light systems (up to four particles) [22]. The application of this method to medium-light nuclei will still require some technical improvements to deal with the complexities related to the Jastrow description.

3 VARIATIONAL MONTE CARLO CALCULATIONS

For the simple interactions we are dealing with, the expectation value of the hamiltonian for Jastrow correlated wave functions may be quite easily computed [23] by using the Monte Carlo method to carry out the integration over all coordinates of the nucleons, combined with the Metropolis [24] algorithm to generate the appropriate random positions of the particles. The method is an adaptation to the quantum many-body problem of the algorithm used in Classical Statistical Mechanics. An upper bound $\langle H \rangle$ to the ground state energy E_0 is obtained by computing the expectation value of the local energy $E_L(\mathbf{R})$ for a probability distribution function $P(\mathbf{R})$

$$\langle H \rangle = \int d\mathbf{R}\, P(\mathbf{R}) E_L(\mathbf{R}), \tag{6}$$

where $\mathbf{R}$ stands for the $3A$ coordinates of the system. The local energy is defined by the action of the hamiltonian on the trial wave function

$$E_L(\mathbf{R}) = \frac{1}{\Phi(\mathbf{R})} H\Phi(\mathbf{R}), \tag{7}$$

and the probability distribution function is the modulus squared of the (normalized) trial wave function

$$P(\mathbf{R}) = \frac{|\Phi(\mathbf{R})|^2}{\int d\mathbf{R}|\Phi(\mathbf{R})|^2}. \tag{8}$$

The Metropolis algorithm is specially adequate to generate random vectors $\mathbf{R}$ corresponding to this distribution function, particularly for boson systems. In the case of fermionic systems, the computation of the Slater determinants at each move of the particles may turn out to be very expensive in terms of computational effort. It is then convenient to use a technique involving the cofactor matrix of the Slater determinant to carry out efficiently the calculations [25].

It is also quite simple to compute the second moment of the hamiltonian $\langle [H - \langle H \rangle]^2 \rangle$. This only requires the evaluation of the integral

$$\langle H^2 \rangle = \int d\mathbf{R}\, P(\mathbf{R}) E_L^2(\mathbf{R}), \tag{9}$$

which may be carried out along with the evaluation of the energy expectation value.

The uncertainty of the energy in the quantum mechanical sense, i.e., the square root of the second moment of the hamiltonian with respect to the trial function $\Delta H = \langle [H - \langle H \rangle]^2 \rangle^{1/2}$, is a (quite loose) way of characterizing the goodness of the wave function. Let's call E_0 the (unknown) ground state energy, E_T the upper bound related to the trial function, and ϵ the amplitude measuring the difference between the trial wave function and the exact wave function,

$$\Psi_T = (\Psi_0 + \epsilon\Psi_1)/\sqrt{1 + \epsilon^2}, \tag{10}$$

where $\langle \Psi_1 | \Psi_0 \rangle = 0$. Then there is the lower bound condition

$$E_0 \geq E_T - |\epsilon|\Delta H. \tag{11}$$

The equation is not useful because the admixture parameter ϵ is unknown, and for this reason, the knowledge of the uncertainty ΔH is only a rough indication of the goodness or badness of the trial wave function.

4 CLUSTER EXPANSIONS

One of the oldest methods to calculate approximately expectation values of operators for Jastrow correlated wave functions is the expansion in clusters of the many-body matrix elements. Again, the various types of expansions have their roots in Classical Statistical Mechanics. The analogous to the partition function are the generalized subnormalization integrals

$$I_n = \langle \Psi | \prod_{i<j}^n f_{ij} e^{\beta H_n} \prod_{i<j}^n f_{ij} | \Psi \rangle, \ n = 1, \ldots, A, \tag{12}$$

where H_n is the hamiltonian of the n-particle subsystem,

$$H_n = \sum_{i=1}^n T_i + \sum_{i<j}^n V_{ij}. \tag{13}$$

The wave function $|\Psi\rangle$ which appears in this equation is the *uncorrelated* wave function. Its basic role is to incorporate into the system the statistical properties, to define the discrete quantum numbers and to confine the system, i.e., to localize the particles. Normally the uncorrelated wave function is a Slater determinant constructed with single particle orbitals from a central potential. Nevertheless, one may also include in this state long range correlations, such as deformations or configuration mixing. A total of A quantities $I_1, I_2, \ldots, I_A$ are defined in Eq.(12), and the quantity we are interested in is, obviously, I_A. The cluster expansion method is an extrapolation algorithm to approximate I_A, which involves A-body operators, in terms of the lowest subnormalization integrals $I_1, \ldots,$ up to I_n. This defines the n-th order approximation.

A complete description of the formal aspects of the cluster expansion method may be found in Ref.[26]. There are two basic types of expansions, additive (A) and multiplicative (F, from Factor), and also two ways of describing each of these two types, one singularizing the single particle orbitals, (Iwamoto-Yamada type, IY), and the other based on averages on the single particle orbitals (Aviles-Hartog-Tolhoek type, AHT).

A comparative analysis between several cluster expansions has been carried out by studying relations between the subnormalization integrals [27], i.e., the quantities I_n defined in Eq.(12) for $\beta = 0$, as well as by evaluating sum rules or model operator relations [19]. From these analyses it was concluded that the FAHT cluster expansion was more accurate than the other expansions. Calculations of energy expectation values for simplified interactions [12,15,28], as well as for realistic forces [4],[29]-[37], show the capabilities of the cluster expansions as well as its extension in the framework of the more general correlated basis functions theory [38]-[44].

The cluster integrals are defined as

$$I_n = \prod_{k=1}^{n} \mathcal{Y}_k^{\binom{n}{k}}, \quad n = 1, \dots, A \tag{14}$$

so that $\mathcal{Y}_n$ is defined in terms of I_n and all other clusters $\mathcal{Y}_k$, $k = 1, \dots, n-1$. Equations 14) are a hierarchy of definitions which become a set of extrapolation rules when all clusters $\mathcal{Y}_p$, from $p = N+1$ up to A are assumed to be zero. With this assumption and after computing the logarithmic derivative of I_A with respect to β at $\beta = 0$ one obtains the practical formulae to compute the expectation value of an operator. In particular, for the hamiltonian, there results

$$E = E^{(1)} + E^{(2)} + E^{(3)} + \cdots \tag{15}$$

where

$$E^{(k)} = \binom{A}{k} \left(\frac{d \log \mathcal{Y}_k}{d\beta} \right)_{\beta=0} \tag{16}$$

The lowest order contributions are

$$E^{(1)} = A \left[\frac{\langle \Psi | T_1 | \Psi \rangle}{\langle \Psi | \Psi \rangle} \right], \tag{17}$$

$$E^{(2)} = \binom{A}{2} \left[\frac{\langle \Psi | f_{12}(T_1 + T_2 + V_{12}) f_{12} | \Psi \rangle}{\langle \Psi | f_{12}^2 | \Psi \rangle} - 2 \frac{\langle \Psi | T_1 | \Psi \rangle}{\langle \Psi | \Psi \rangle} \right], \tag{18}$$

$$
\begin{aligned}
E^{(3)} = \binom{A}{3} &\left[\frac{\langle \Psi | f_{12} f_{13} f_{23}(T_1 + T_2 + T_3 + V_{12} + V_{13} + V_{23}) f_{12} f_{13} f_{23} | \Psi \rangle}{\langle \Psi | f_{12}^2 f_{13}^2 f_{23}^2 | \Psi \rangle} \right. \\
&\left. - 3 \frac{\langle \Psi | f_{12}(T_1 + T_2 + V_{12}) f_{12} | \Psi \rangle}{\langle \Psi | f_{12}^2 | \Psi \rangle} + 3 \frac{\langle \Psi | T_1 | \Psi \rangle}{\langle \Psi | \Psi \rangle} \right],
\end{aligned}
\tag{19}
$$

The most delicate point in the use of equations (17-19) is the important cancellations which occur between the terms in square brackets. For a saturating system, the total energy as well as each of the cluster contributions $E^{(k)}$ are proportional to A, the number of particles, but each of the quotients of expectation values appearing in eqs.(17-19) are $O(1)$ in the number of particles, so a strong cancellation must occur in each cluster contribution. The practical consequence is the necessity of computing the matrix elements with enough precision to have reliable results. Moreover, the FAHT cluster expansion cannot be directly applied to infinite systems, and some kind of renormalization is required [45].

On the other hand, the most remarkable property of the FAHT cluster expansion is its non-linearity. Due to its non-linear character all diagrams are counted (exactly or approximately), already from the second order of the expansion,. In the language of HNC theories, a diagram is characterized by a product of dynamical correlation factors $h_{ij} = f_{ij} - 1$ and also of statistical correlations related to the fermionic permutations. Then, even if at some order k of the expansion only k particles are involved, all other

Table 2. The ^{16}O ground state energy for the MTV potential computed with Jastrow correlated trial functions.

Method	Reference	S.P. wave functions	Correlation	Energy(MeV)
VMC	[11]	WS potential	2nd order EL	-1024 ± 5
FHNC	[6]	WS potential	FHNC EL	-1055
FHNC	[6]	FHNC EL	FHNC EL	-1059
FHNC	[7]	WS potential	2nd order EL	-1152
VMC	This work	HO potential	Eq.(1)	-1103 ± 1
GFMC	[11]			-1194 ± 20

diagrams involving a higher number of particles are approximated by combinations of products of diagrams involving less particles. Further details on this property may be seen in [27].

The only question regarding the validity of the FAHT cluster expansion, which is shared by any other theory involving expansions, is its convergence. In a previous work [13,14] a comparison of the FAHT expansion with variational Monte Carlo calculations was done for light nuclei, up to ^{16}O, confirming the rapid convergence of the expansion. Here the test is extended up to ^{40}Ca, and for these number of particles the expansion shows its asymptotic character.

5 SOME NUMERICAL RESULTS

Table 2 resumes several calculations of the ground state energy of ^{16}O with the MTV interaction. Our result is in the sixth row, and corresponds to $\alpha = 1.24 \, F^{-1}$, $a = -0.75$ and $b = 0.51 \, F$, with the parametrization described in Section 2. Is a variational Monte Carlo with two million samples. We obtain $1248 \pm 3 \, MeV$ for the expectation value of the kinetic energy and $-2350 \pm 4 \, MeV$ for the potential energy. The quantum uncertainty of the hamiltonian is $480 \pm 20 \, MeV$. In the other calculations the single particle orbitals correspond to a Wood Saxon potential or (fifth row) to fully optimized FHNC orbitals. The two-body correlations correspond to a constrained Euler-Lagrange optimization of the second order cluster expansion or to an Euler-Lagrange optimization of the FHNC functional of the energy. The last row is a Green Function Monte Carlo evaluation of the energy, and should be considered as the true ground state energy within the statistical error.

It is hardly difficult to understand why our calculation, which corresponds to a very simple wave function, is producing for the ground state energy an upper bound $50 \, MeV$ lower than very elaborated minimizations of the FHNC functional. We presume that the values of Ref.[6] do not include the center-of-mass correction. The amount of

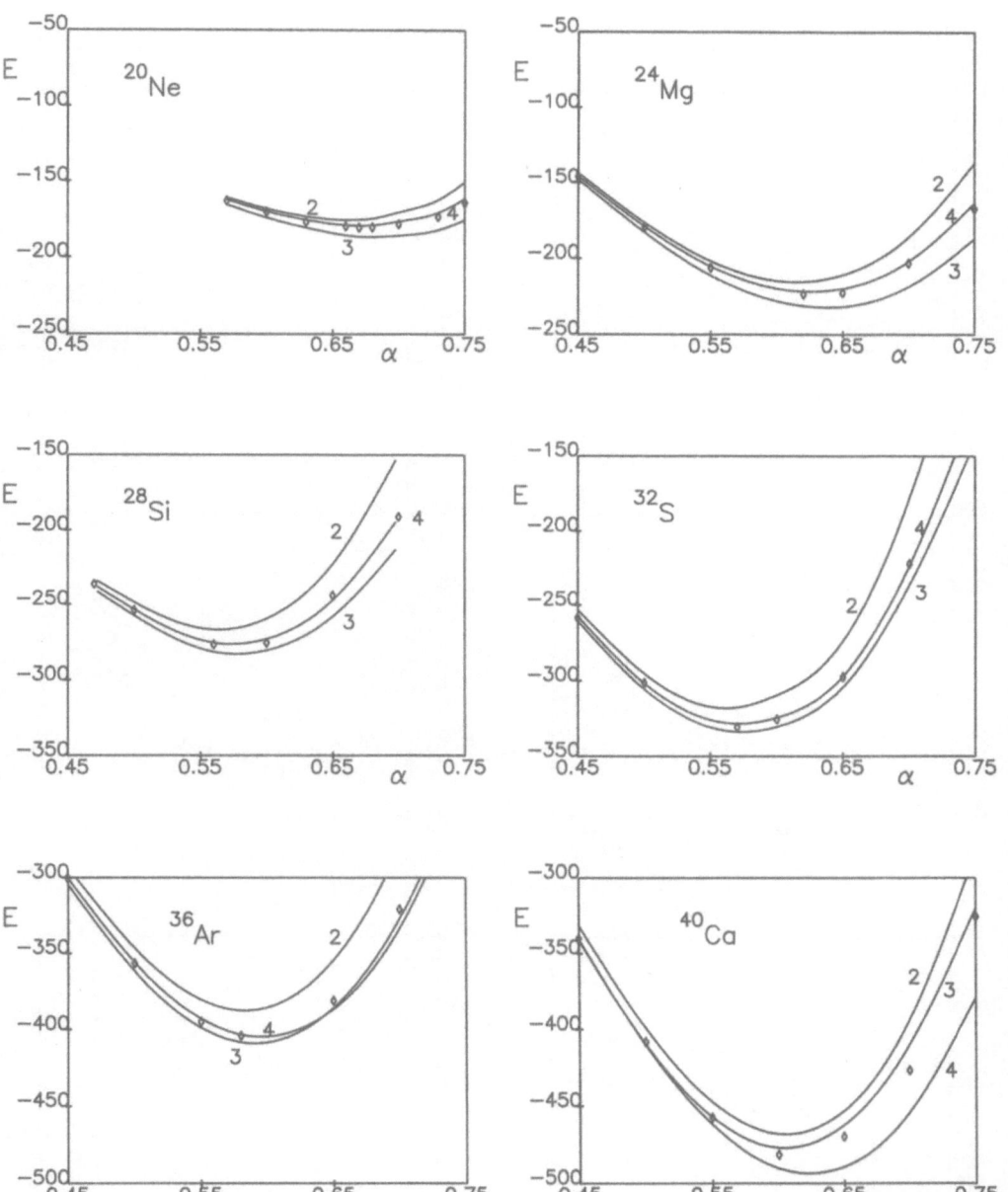

Figure 1. Comparison of the FAHT cluster expansion for the energy (in MeV) of sd-shell nuclei with the Monte Carlo calculation. The interaction is B1. Labels 2,3 and 4 refer to second, third and fourth order of the cluster expansion. The diamonds are the variational Monte Carlo result. The figure shows the variation with the harmonic oscillator parameter α, the other parameters of the wave function being those of Table 3.

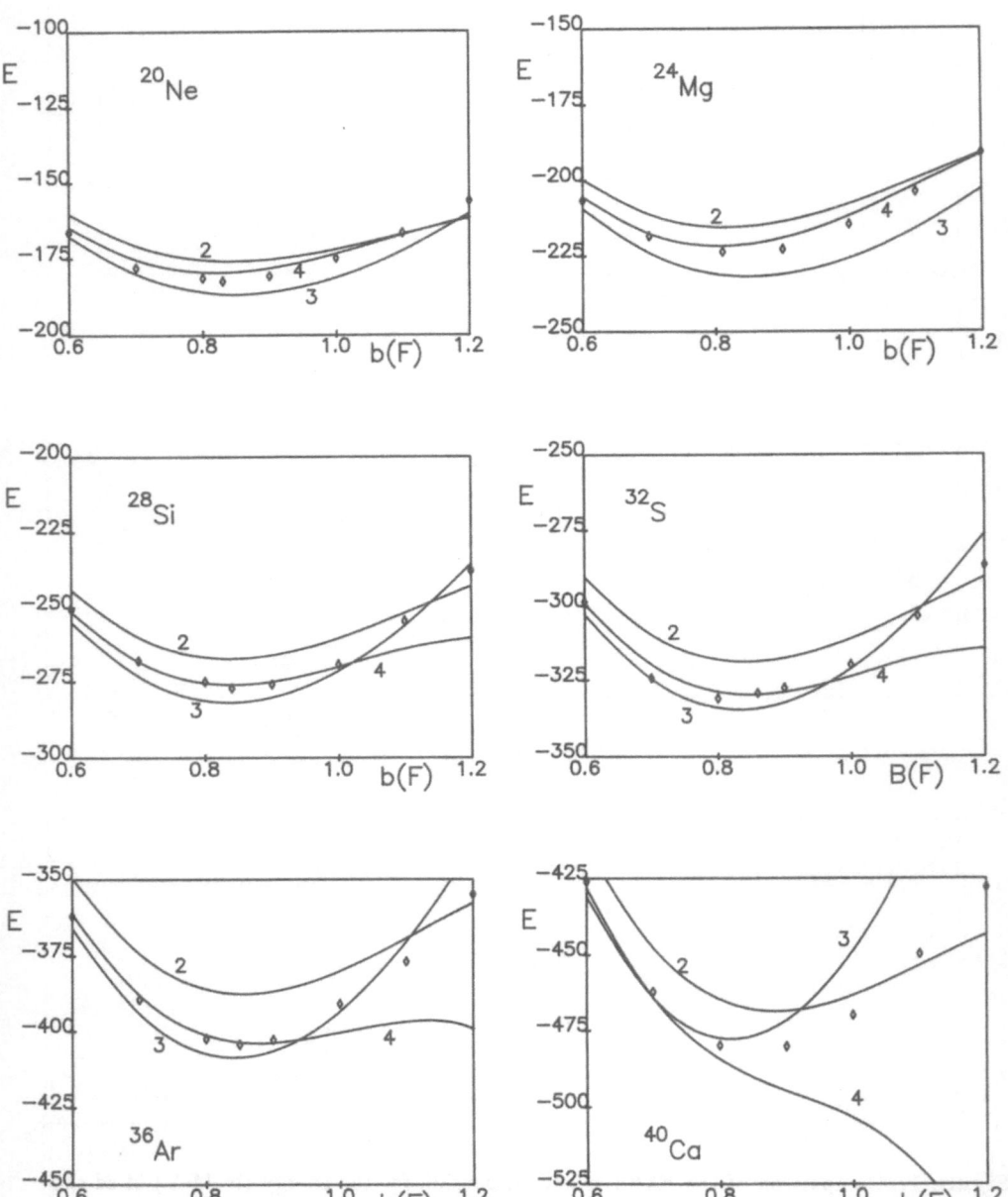

Figure 2. Same as Figure 1, but now showing the variation with the correlation length b. The other parameters of the wave function being those of Table 3.

Table 3. The p-shell and sd-shell upper bounds for the ground state energy computed with Jastrow correlated trial functions for the B1 potential. All cases correspond to spherical or axially symmetric functions, with the exception of Mg wich has a triaxial shape with $d_z = 0.87$ and $d_y = 1.09$.

Nucleus	$\alpha\,(F^{-1})$	a	$b\,(F)$	d	E(MeV)
^8Be	0.752	-0.46	0.726	0.71	-55.2 ± 0.2
^{12}C	0.600	-0.49	0.800	1.28	-92.3 ± 0.3
^{16}O	0.648	-0.51	0.800	1	-150.9 ± 0.3
^{20}Ne	0.666	-0.49	0.828	0.83	-181.3 ± 0.3
^{24}Mg	0.616	-0.49	0.814	0.87	-224 ± 1
^{28}Si	0.561	-0.50	0.842	1.22	-275.9 ± 0.3
^{32}S	0.566	-0.50	0.800	1.14	-328.6 ± 0.4
^{36}A	0.580	-0.51	0.850	1.10	-403.6 ± 0.5
^{40}Ca	0.600	-0.53	0.850	1	-483.0 ± 0.4

this correction, for our wave function, is of $3\hbar\omega/4 = 48\,MeV$. Our result still differs from the true ground state energy by about $90\,MeV$. By using eq.(11) one may get a minimum value for $\epsilon \simeq 0.2$, or a maximum value for the projection $\langle \Psi_T | \Psi_0 \rangle = 0.96$.

The variational Monte Carlo results corresponding to the B1 force are presented in Table 3 and in Figures 1 and 2. In Table 3 we list the parameters characterizing the trial wave function in the way described in Section 2, and the upper bound for the ground state energy. The values corresponding to Be, C and O are from [13,14]. Perhaps the only point to mention with regard to this table is that the first nucleus which is α-stable, accordingly with our variational results, is ^{28}Si. The Diffusion Monte Carlo binding energy for ^4He is $-38.32\pm0.01\,MeV$ [46], so that our results indicate that all nuclei with less than 28 particles would prefer to disintegrate in several α particles, or, more probably, that the trial function used is not good enough.

Figures 1 and 2 show the comparison of the FAHT cluster expansion with the Monte Carlo calculations, up to fourth order of the expansion. In Figure 1 we show the region around the minimum as a function of the harmonic oscillator parameter α, the remaining parameters being those listed in Table 5. Figure 2 shows the same comparison but in terms of the correlation length b. The conclusion from these figures is obvious: the FATH cluster expansion breaks down for the heavier systems. Nevertheless, the expansion works very well up to Ar, and the values at the fourth order cannot be distinguished from the Monte Carlo minimum in all cases but in Ca. In fact, the difference between the fourth order calculations and the Monte Carlo calculations is always less that $1\,MeV$, with the obvious exception of Ca. Finally, Figure 3 displays the one-body ρ_1 and the two-body ρ_2 densities for ^{16}O, normalized so that $\int_0^\infty r^2\rho(r)dr = 1$,

Table 4. A collection of results in MeV for ^{16}O and ^{40}Ca for the B1 interaction. The table also includes the value of the uncertainty ΔH related to our Jastrow correlated wave functions. The FHNC calculation uses the same wave functions as us. The FHNC-EL improves the Jastrow correlation by optimizing the contrained second order energy.

Method	Reference	^{12}C	^{16}O	^{40}Ca
VMC	This work	-92.3 ± 0.3	-150.9 ± 0.3	-483.0 ± 0.4
VMC-ΔH			83	151
FHNC	[7]		-150.4	-478
FHNC-EL	[7]		-152.4	-482
PB	[16]		-152.1	-468.14
IDEA	[18]	-80.1 to -79.6	-164.7 to -164.2	
BHF	[15]		-163.7	-507.2

corresponding to the uncorrelated function, the second-order cluster expansion and the Monte Carlo value. Again we observe that already the second order FAHT cluster expansion is giving a good description of the nuclear densities

We finally mention other results for two selected nuclei (O and Ca) obtained from other methods. We have collected several results in Table 4, as well as the value of the uncertainty related to our simple Jastrow-correlated wave function. The table confirms the validity of the FHNC method to compute the ground state energy, as well as the good performance of the hyperspherical harmonics expansion and the integro-differential approximation.

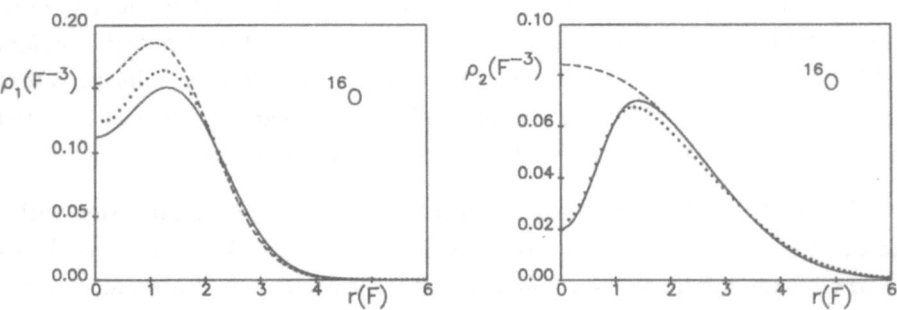

Figure 3. The one-body density (left) and the two-body density (right) of ^{16}O. The dashed line is the uncorrelated calculations, the continuous line is the FAHT/II approximation, and the dots correspond to the Monte Carlo calculation.

Acknowledgements The authors are supported by the *Dirección General de Investigación Científica y Tecnológica* of Spain. They are also grateful to Adelchi Fabrocini for showing the FHNC results prior to publication and to Jesús Navarro for many useful comments.

REFERENCES

[1] R. Schiavilla, V.R. Pandharipande and R.B. Wiringa, *Nucl. Phys.* **A449**, 219 (1986)

[2] J. Carlson, *Phys. Rev.* **C36**, 2026 (1987) and **C38**, 1879 (1988)

[3] J. Carlson and M.H. Kalos, *Phys. Rev.* **C32**, 2105 (1985)

[4] Steven C. Pieper, R.B. Wiringa and V.R. Pandharipande, *Phys. Rev. Lett.* **64**, 364 (1990)

[5] S. Fantoni and S. Rosati, *Nucl. Phys.* **A328**, 478 (1979)

[6] E. Krotscheck, *Nucl. Phys.* **A465**, 461 (1987)

[7] G. Co', A. Fabrocini, S. Fantoni and I.E. Lagaris, *Model calculations of doubly shell nuclei in CBF theory (I)*, Preprint IFUP-TH 15/95 (Pisa, 1992)

[8] R.A. Malfliet and J.A. Tjon, *Nucl. Phys.* **A127**, 161 (1969)

[9] D.M. Brink and E. Boeker, *Nucl. Phys.* **A91**, 1 (1967)

[10] M. Fabre de la Ripelle and J. Navarro, *Ann. Phys.* **123**, 185 (1979)

[11] U. Helmbrecht and J.G. Zabolitzky, *Nucl. Phys.* **A442**, 109 (1985)

[12] E. Buendía and R. Guardiola in *Recent Progress in Many-Body Theories*, J.G. Zabolitzky, M. de Llano, M. Fortes and J.W. Clark editors, *Lect. Not. Phys.* **142**, 348 (Springer Verlag, Berlin 1981)

[13] M.C. Boscá, E. Buendía and R. Guardiola, *Phys. Lett.* **B198**, 312 (1987)

[14] M.C. Boscá, E. Buendía and R. Guardiola, in *Condensed Matter Theories* 3, J.S. Arponen, R.F. Bishop and M. Manninen editors, p. 101. Plenum Press (New York, 1988)

[15] R. Guardiola, A. Faessler, H. Müther and A. Polls, *Nucl. Phys.* **A371**, 79 (1981)

[16] J.L. Ballot, M. Fabre de la Ripelle and J. Navarro, *Phys. Lett.* **143B**, 19 (1984)

[17] M. Fabre de la Ripelle, H. Fiedeldey and S.A. Sofianos, *Phys. Rev.* **C38**, 449 (1988)

[18] R.M. Adam, H. Fiedeldey, S.A. Sofianos anf M. Fabre de la Ripelle, *J. Phys.* **G17**, L157 (1991), and preprint IPNO/TH 91-72 (Orsay, 1991)

[19] R. Guardiola, A. Polls and J. Ros, *Il Nuovo Cimento* **59A**, 419 (1980)

[20] R.F. Bishop, M.F. Flynn, M.C. Boscá, E. Buendía and R. Guardiola, *Phys. Rev.* **C42**, 1431 (1980)

[21] R.F. Bishop, M.F. Flynn, E. Buendía and R. Guardiola, *J. Phys.* in press (1992)

[22] M. Viviani, A. Kiewski and S. Rosati, *Il Nuovo Cimento* in press (1992)

[23] D.M. Ceperley and M.H. Kalos, in *Monte Carlo Methods in Statistical Mechanics*, K. Binder editor. Springer Verlag (Berlin, 1979)

[24] N. Metropolis, A.W. Rosenbluth, M.N. Rosenbluth, AQ.H. Teller and E. Teller, *J. Chem. Phys.* **21**, 1087 (1953)

[25] D.M. Ceperley, G.V. Chester and M.H. Kalos, *Phys. Rev.* **B16**, 3081 (1976)

[26] J.W. Clark and P. Westhaus, *J. Math. Phys.* **9**, 131 (1968)

[27] R. Guardiola and A. Polls, *Nucl. Phys.* **A342**, 385 (1980)

[28] R. Guardiola *Nucl. Phys.* **A328**, 490 (1979)

[29] M.L. Ristig, W.J. Ter Low and J.W. Clark, *Phys. Rev.* **C3**, 1504 (1971)

[30] J.C. Owen, R.F. Bishop and J.M. Irvine, *Nucl. Phys.* **A274**, 108 (1976)

[31] J.C. Owen, R.F. Bishop and J.M. Irvine, *Ann. Phys.* **102**, 170 (1976)

[32] J.C. Owen, R.F. Bishop and J.M. Irvine, *Phys. Lett.* **61B**, 147 (1976)

[33] J.C. Owen, R.F. Bishop and J.M. Irvine, *Nucl. Phys.* **A277**, 45 (1977)

[34] R.F. Bishop, C. Howes, J.M. Irvine and M. Modarres, *J. Phys.* **G4**, 1709, L89 and L123 (1978)

[35] K.E. Kürten, M.L. Ristig and J.W. Clark, *Nucl. Phys.* **A317**, 87 (1979)

[36] V.R. Pandharipande and R.B. Wiringa, *Rev. Mod. Phys.* **51**, 821 (1979)

[37] M.C. Boscá and R. Guardiola, *Nucl. Phys.* **A476**, 471 (1988) and **A489**, 45 (1988)

[38] E. Feenberg and J.W. Clark, *Phys. Rev.* **113**, 388 (1959)

[39] E. Feenberg and C.W. Woo, *Phys. Rev.* **137**, 391 (1965)

[40] J.W. Clark and P. Westhaus, *Phys. Rev.* **141**, 833 (1966) and **149**, 990 (1966)

[41] J.W. Clark, L.R. Mead, E. Krotscheck, K.E. Kürten and M.L. Ristig *Nucl. Phys.* **A328**, 45 (1979)

[42] J.W. Clark, *Nucl. Phys.* **A 328**, 587 (1979)

[43] J. W. Clark, in *The Many-Body Problem: Jastrow Correlations versus Brueckner Theory*, R. Guardiola and J. Ros editors, *Lect. Not. Phys.* **138**, 184 (Springer Verlag, Berlin 1981)

[44] E. Krotscheck and J.W. Clark, *Nucl. Phys* **A328**, 73 (1979)

[45] P. Westhaus and J.W. Clark, *J. Math. Phys.* **9**, 149 (1968)

[46] R.F. Bishop, E. Buendía, M.F. Flynn and R. Guardiola, *J. Phys.* **G18**, L21 (1992)

QUANTUM STATISTICAL INFERENCE

R. N. Silver

Theoretical Division
MS B262 Los Alamos National Laboratory
Los Alamos, NM 87545

ABSTRACT. Can quantum probability theory be applied, beyond the microscopic scale of atoms and quarks, to the human problem of reasoning from incomplete and uncertain data? A unified theory of quantum statistical mechanics and Bayesian statistical inference is proposed. QSI is applied to ordinary data analysis problems such as the interpolation and deconvolution of continuous density functions from both exact and noisy data. The information measure has a classical limit of negative entropy and a quantum limit of Fisher information (kinetic energy). A smoothing parameter analogous to a de Broglie wavelength is determined by Bayesian methods. There is no statistical regularization parameter. A priori criteria are developed for good and bad measurements in an experimental design. The optimal image is estimated along with statistical and incompleteness errors. QSI yields significantly better images than the maximum entropy method, because it explicitly accounts for image continuity.

1. Introduction

Statistical inference may be defined as "the process of reasoning from incomplete and uncertain information". In physics, a successful theory of inference is quantum statistical mechanics, in which the goal is to determine the most probable macrostate of a many-body system from knowledge of only the Hamiltonian and a few constraints (e.g. average energy, number of particles, etc.). In statistics and information theory, the goal may be to determine the most probable density function (the *image*) and its reliability from a limited set of uncertain data (typically integral transforms of the density function). *Probability* is used here in the sense of Bayesian probability theory to represent a *degree of belief*. This is a more general definition than the common *frequency of occurence*. Similarly, *human ignorance* is a more general concept than *random variable*. The goal of probability theory should be to develop and apply universal principles of logical inference which systematically combine new data with prior knowledge. This approach is summarized by Jaynes' hypothesis: *probability theory as logic*.

From this perspective, the same principles of inference should apply to physics, statistics and information theory. Perhaps the success of quantum physics can provide a big clue toward finding these, as yet incompletely formulated, principles. However, there are notorious difficulties in finding a human language for quantum concepts (Jaynes, 1989) which are foreign to statistics and information theory. Despite a lack of mathematical rigor, quantum mechanics and statistical mechanics are the only two empirically validated probability theories. Despite a rigorous mathematical formulation, the probability theories commonly used in statistics and information theory lack comparable empirical support. This suggests that it should be possible to formulate an approach to statistics and information theory based on quantum statistical mechanics.

The practical need for a quantum approach to information theory and statistics can be established from experience with the maximum entropy (ME) class of Bayesian methods (Jaynes, 1983) for density function estimation. Although ME has been successfully applied to numerous data analysis problems, ME images often exhibit spurious artifacts and overly sharp structure (Skilling, 1989) The justifications for ME (see, e.g., Tikochinsky, 1984) require no prior correlations between points in an image. However, most applications of ME are to *continuous* density functions, which means that the first derivatives of the density function are finite. This prior knowledge can only be incorporated by additional image smoothing. The developers of the leading ME code have proposed this smoothing to be a user-chosen *pre-blur* of a *hidden ME image*.

The alternative approach to image smoothing using quantum probability theory is motivated by a series of observations. There exists a one-to-one mathematical analogy between ME and classical statistical mechanics. Quantum statistical mechanics is also a successful theory of inference. It has classical statistical mechanics as a limit. Quantum density functions are smoother than classical density functions. The quantum smoothing parameter is the de Broglie wavelength. The Schrödinger equation at the heart of quantum mechanics may be derived from a variational principle on Fisher information (kinetic energy) (Frieden, 1989). Variational principles are preferable to *ad hoc* formulations.

Hence, a natural way to improve upon ME would be to formulate a Bayesian density-function estimation method in one-to-one mathematical analogy with quantum statistical mechanics. *Quantum Statistical Inference* (QSI) is the realization of this approach. It introduces a broad new class of prior probabilities incorporating correlations between points in an image.

2. The Data Analysis Problem

The generic data analysis problem is to solve a Fredholm equation of the first kind,

$$\hat{D}_k = \int dy \hat{R}_k(y)\hat{f}(y) + \hat{N}_k \quad . \tag{1}$$

Here $\hat{D}_k$ are N_d data which are typically incomplete, $\hat{R}_k(y)$ is an integral transform which is often an instrumental resolution function, and $\hat{N}_k$ represents noise. The true $\hat{f}(y)$ which generated the data is usually termed the *object*. Finding a normalized density function, $\hat{f}(y)$, is an ill-posed inverse problem. The goal of a statistical inference method is to provide a best estimate of $\hat{f}(y)$, termed the *image*, and its reliability. These estimates should be based on both the data and any prior knowledge available. Success is measured by the fidelity of the image to the object.

In ME, the image is derived from a variational principle: maximize the Shannon-Jaynes entropy,

$$S_c \equiv - \int dy \hat{f}(y) \ln \left(\frac{\hat{f}(y)}{\hat{m}(y)} \right) \quad , \tag{2}$$

subject to the constraints of the data. Here, $\hat{m}(y)$ is a *default model* for $\hat{f}(y)$, so-named because it is the answer ME returns in the absence of data. It is most convenient to remove the explicit reference to the default model by introducing renormalized variables $x \equiv \int_{-\infty}^{y} dy'\hat{m}(y')$, $f(x) \equiv \hat{f}(y)/\hat{m}(y)$, and $R_k(x) \equiv \hat{R}(y)$. Since $\hat{m}(y)$ must also be a normalized density function, clearly $0 \leq x \leq 1$. The data analysis problem is then rewritten in vector notation,

$$\vec{D} = \int_0^1 dx\, \vec{R}(x) f(x) + \vec{N} \quad . \tag{3}$$

In the special case of noiseless data ($\vec{N} = \vec{0}$), the solution by the method of Lagrange multipliers is

$$f_C(x) = \frac{1}{Z_C} \exp\left(-\vec{\lambda}_C \cdot \vec{R}(x)\right) \quad Z_C \equiv \int_0^1 dx \exp\left(-\vec{\lambda}_C \cdot \vec{R}(x)\right) \quad . \tag{4}$$

The N_d Lagrange multipliers, $\vec{\lambda}_C$, are to be determined by the fits to the data. This is formally analogous to a density function in *classical* statistical mechanics (subscript "*C*") with the identifications $\vec{\lambda}_C \cdot \vec{R}(x) \Leftrightarrow V(x)/T$, where $V(x)$ is potential energy and T is temperature. Z_C is a classical partition function.

ME works best for problems, such as deconvolution, where $\vec{R}(x)$ is a broad function and $f(x)$ contains comparatively sharp features. However, for problems where $\vec{R}(x)$ is sharp and $f(x)$ broad, ME tends to produce spurious structure. An extreme example is the data interpolation problem, in which the goal is to infer a density function from knowledge of its values at a finite number of points. Then $\vec{R}(x)$ consists of a set of δ-functions, and the ME solution is nonsense. Figure 1 shows a test density function and three data analysis problems corresponding to interpolation (I), Gaussian deconvolution (G), and exponential deconvolution (E). The data are taken to be exact and measured at 32 equally spaced data points. The corresponding ME images are shown in Fig. 2 displayed in 128 pixels. The ME image for the G data is credible, but it exhibits overly sharp and occasionally spurious structure. The ME image for the E data exhibits sharp edges reflecting the sharp feature in the resolution function. The ME image for the I data equals the data for measured pixels and equals the default model value of 1.0 for unmeasured pixels. The I and E ME images are not credible because they clearly reflect how the data were measured.

3. Quantum Statistical Mechanics & Fisher Information

A general relation exists between the calculus of variations and eigenvalue problems, such that variational principles can be associated with most of the differential equations used in physics (see, e.g. Mathews, 1964). However, physicists are not generally aware that the variational functional appropriate to the Schrödinger equation has a significance in statistics (Frieden, 1989). It is the *Fisher information* originally introduced (Fisher, 1925) as a measure of the inverse uncertainty in determining a position parameter by maximum likelihood estimation. In information theory, Fisher information and entropy have been proven to be related by derivative and metrical relations (Dembo, 1991). Such a proposed relation between statistics, information theory and quantum mechanics concepts is very controversial among physicists. Nevertheless, a principle of *minimum Fisher information* (MFI) provides a convenient derivation of a Schrödinger-like wave equation for density function estimation.

Applying MFI to the data analysis problem defined by Eq. (3), extremize

$$Q_1 \equiv \frac{1}{4} \int_0^1 dx \frac{1}{f(x)} \left(\frac{\partial f(x)}{\partial x}\right)^2 + \vec{\lambda} \cdot \int_0^1 dx\, \vec{R}(x) f(x) - E \int_0^1 dx\, f(x) \quad . \tag{5}$$

The first term is Fisher information, the second term imposes the constraints due to the data with Lagrange multipliers $\vec{\lambda}$, and the third term is the normalization constraint with Lagrange multiplier E. Defining a *wave function* by $\psi(x) \equiv \pm\sqrt{f(x)}$, the Euler-Lagrange equation is

$$\frac{\partial Q_1}{\partial \psi(x)} - \frac{d}{dx}\left(\frac{\partial Q_1}{\partial (d\psi(x)/dx)}\right) = 0 \quad . \tag{6}$$

This results in a Schrödinger-like wave equation,

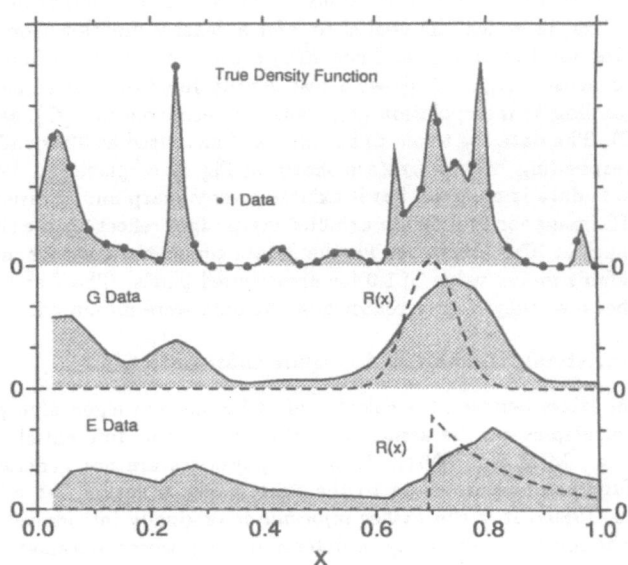

Fig. 1. Exact Data Example - Top curve is the test density function, and dots are the interpolation (I) data. Middle curves are the Gaussian (G) data (solid) and resolution function $R(x)$ (dashed ×0.35). Bottom curves are the exponential (E) data and resolution function. Data are exact and measured at **32** points. The Gaussian has a standard deviation of **0.044** and the exponential has a decay constant of **0.15**.

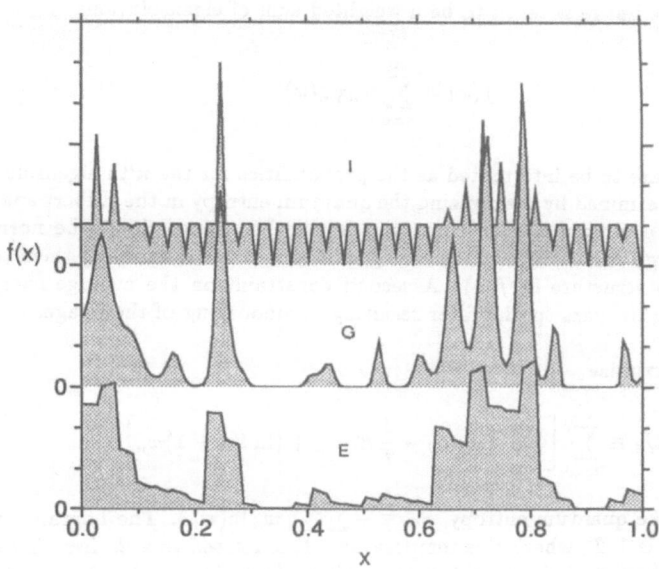

Fig. 2. Maximum Entropy Images - For exact data example in Fig. 1, displayed in 128 pixels.

$$-\frac{d^2\psi(x)}{dx^2} + \vec{\lambda} \cdot \vec{R}(x)\psi(x) = E\psi(x) \quad . \tag{7}$$

The data constraints $\vec{\lambda} \cdot \vec{R}(x)$ are analogous to potential energy $V(x)$, and E is analogous to total energy. Eq. (7) may also be written in matrix form as $\mathbf{H} \mid \psi >= E \mid \psi >$ using the *Heisenberg representation* $\psi(x) \equiv < x \mid \psi >$. The matrix $\mathbf{H}(x, x')$ is a *Hamiltonian*.

Solving the Schrödinger equation is an eigenvalue problem subject to a requirement that the solutions form a Hilbert space in $0 \le x \le 1$. For example, for $V(x) = 0$ a complete orthonormal set of solutions is $\psi(x) = \sqrt{2}\cos(k_n x); \sqrt{2}\sin(k_n x)$ where wave vectors are $k_n = 2\pi n$ and eigenenergies are $E_n^o = k_n^2$. This paper does not consider the possibility of complex solutions to the Schrödinger equation. For $V(x) \ne 0$, the $\psi(x)$ for the ground state is nodeless and would provide an image corresponding to the MFI principle.

However, the ME principle is certainly successful for many data analysis problems. Fortunately, it is not necessary to make a choice between these two variational principles, since quantum statistical mechanics (see, e.g., McQuarrie, 1986) provides a seamless interpolation between the MFI principle in the quantum limit and the ME principle in the classical limit. The image is taken to be a weighted sum of eigensolutions,

$$f(x) = \sum_{n=0}^{\infty} w_n \psi_n^2(x) \quad . \tag{8}$$

The weights, w_n, are to be interpreted as the probabilities for the n'th eigensolutions, and they should be determined by maximizing the quantum entropy in the Hilbert space subject to any applicable constraints. One constraint is that the image should be normalized to unity. Higher energy solutions have increasing numbers of nodes enabling them to describe increasingly sharp structure in $f(x)$. A second constraint on the average energy would, therefore, act as a low-pass spatial filter resulting in smoothing of the image.

Therefore, maximize

$$Q_2 \equiv \sum_{n=0}^{\infty} \left[-w_n \ln(w_n) - \frac{1}{T}E_n w_n + (\ln Z_Q + 1)w_n \right] \quad . \tag{9}$$

The first term is the quantum entropy, $S_Q \equiv -\sum_{n=0}^{\infty} w_n \ln(w_n)$. The Lagrange multiplier for average energy is $1/T$, where T is temperature. It is related to a *de Broglie wavelength*, Λ, which is the characteristic scale for smoothing of the image, by $\Lambda \equiv \sqrt{4\pi/T}$. (In ordinary quantum mechanics, $\Lambda = \sqrt{2\pi\hbar^2/mk_BT}$, so that QSI corresponds to setting $\hbar^2/2mk_B \to 1$.) The Lagrange multiplier for the normalization of the image is $\ln Z_Q + 1$. The result is the *canonical ensemble* of statistical mechanics,

$$w_n = \frac{1}{Z_Q}\exp\left(-\frac{E_n}{T}\right) \quad Z_Q \equiv \sum_{n=0}^{\infty}\exp\left(-\frac{E_n}{T}\right) = \exp\left(-\frac{F}{T}\right) \quad . \tag{10}$$

Z_Q is the quantum partition function, and F is the *Gibbs free energy*. The form chosen for the constraints was partially motivated by the many attractive mathematical properties of the canonical ensemble.

This theory may be written more compactly in the Heisenberg representation,

$$f(x) = \frac{1}{Z_Q} < x \mid e^{-\mathbf{H}/T} \mid x > \quad Z_Q = Tr\left(e^{-\mathbf{H}/T}\right) \quad . \tag{11}$$

The expectation value for any integral function, $O(x)$, of the image is termed an *observable*, and it is equivalent to a thermodynamic expectation value in statistical mechanics, i.e.

$$\ll O \gg \equiv \int_0^1 dx O(x) f(x) = \frac{1}{Z_Q} Tr \left(O e^{-\mathbf{H}/T} \right) \quad . \tag{12}$$

For example, setting $O(x) \to \delta(x - x_o)$ gives the image $f(x_o)$, and setting $O(x) \to R_k(x)$ gives the k'th fit to the data, $\ll R_k \gg$. The Lagrange multipliers, $\vec{\lambda}$, and the fits, $\ll \vec{R} \gg$, are conjugate variables in the sense of a Legendre transformation in statistical mechanics. The image is an implicit function of $\vec{\lambda}$ and Λ. In the classical limit ($\Lambda \to 0, T \to \infty$), the images reduce to the ME results given by Eq. (4) with the identifications $\vec{\lambda}_C \Leftrightarrow \vec{\lambda}/T$ and $Z_Q \Leftrightarrow Z_C/\Lambda$. In the quantum limit ($\Lambda \to \infty, T \to 0$), they reduce to the MFI images.

4. Bayesian Statistical Inference

The optimal $\vec{\lambda}$ and Λ are to be determined from the data by the application of Bayesian statistical inference (see, e.g., Loredo, 1990). Bayes theorem is

$$P[\vec{\lambda}, \vec{D}, \Lambda] = P[\vec{\lambda} \mid \vec{D}, \Lambda] \times P[\vec{D}, \Lambda] = P[\vec{D} \mid \vec{\lambda}, \Lambda] \times P[\vec{\lambda}, \Lambda] \quad . \tag{13}$$

In Bayesian terminology $P[\vec{\lambda} \mid \vec{D}, \Lambda]$ is termed the *posterior probability*, $P[\vec{D}, \Lambda]$ is the *evidence*, $P[\vec{D} \mid \vec{\lambda}, \Lambda]$ is the *likelihood function*, and $P[\vec{\lambda}, \Lambda]$ is the *prior probability*. The most probable $\vec{\lambda}$ is determined from the maximum of the posterior probability, and the most probable Λ is determined from the maximum of the evidence. The data are embodied in the likelihood function.

The prior probability is determined by an analogy between quantum statistical mechanics and statistical inference. In statistical mechanics, the system probability is proportional to the partition function, $Z_Q = \exp(-F/T)$, which counts the number of different eigensolutions the system can have within the constraints. In Bayesian inference the system probability is proportional to the product of the likelihood function and the prior probability. Since the data constraints have already been identified with the potential energy, the likelihood function may be identified as $\propto \exp(- \ll V \gg /T)$. That leaves a prior probability $P[\vec{\lambda}, \Lambda] \propto \exp(-I_Q - F_o/T)$. A subscript "o" will be used to denote a quantity evaluated at the default model value of $\vec{\lambda} = \vec{0}$. The quantum generalization of an information measure is

$$I_Q \equiv \delta \left[\frac{F}{T} - \frac{\ll V \gg}{T} \right] = \delta \left[\frac{1}{T} \ll -\frac{\partial^2}{\partial x^2} \gg -S_Q \right] \quad . \tag{14}$$

Here "δ" denotes a change in this quantity from the $\vec{\lambda} = \vec{0}$ value, so that $I_Q^o = 0$. The right hand side is a simple sum of the expectation values for Fisher information (kinetic energy) and quantum negentropy. In the classical ($\Lambda \to 0$) limit, I_Q reduces to the information measure in ME, $-S_C$.

To establish the properties of I_Q, consider the *susceptibility* (or *Hessian*) matrix,

$$\mathbf{K} \equiv -T\vec{\nabla}\vec{\nabla}F = -T\frac{\partial \ll \vec{R} \gg}{\partial \vec{\lambda}} \quad . \tag{15}$$

Here $(\vec{\nabla})_k \equiv \partial/\partial \lambda_k$. A susceptibility in statistical mechanics is the derivative of an observable (e.g. $\ll R_k \gg$) with respect to a *field* (e.g. $\lambda_{k'}$). The *fluctuation-dissipation theorem* (Kubo, 1959) of quantum statistical mechanics relates $\mathbf{K}$ to a correlation function, which may be written in two ways,

$$\mathbf{K}_{k,l} = \int_0^1 d\tau \ll \delta R_k e^{-\mathbf{H}\tau/T} \delta R_l e^{+\mathbf{H}\tau/T} \gg = \int_0^1 dx \int_0^1 dx' \delta R_k(x) \delta R_l(x') \Theta(x, x') \quad . \tag{16}$$

Here $\delta R_k(x) \equiv R_k(x) - \ll R_k \gg$.. Eq. (16) may be derived by second-order quantum perturbation theory. The first expression for $\mathbf{K}$ is the same as a susceptibility in the Kubo theory for linear response (see, e.g., Fick, 1990). The second expression introduces a *spatial correlation function* $\Theta(x, x')$, whose width characterizes prior correlations between points in the image. It is a strictly positive real symmetric function satisfying $\int_0^1 dx \Theta(x, x') = f(x')$. In the classical limit of $\Lambda \to 0$, $\Theta(x, x') \to f(x)\delta(x - x')$. The peak in $\Theta(x, x')$ broadens with increasing $\Lambda \neq 0$. In the default model limit of $\vec{\lambda} = \vec{0}$, the variance of the peak is given by $\Lambda/\sqrt{12\pi}$. In the quantum limit of $\Lambda \to \infty$, $\Theta(x, x') \to f(x)f(x')$. It follows that $\mathbf{K}$ must be a positive definite matrix. At fixed $\ll \vec{R} \gg$, the eigenvalues of $\mathbf{K}$ are maximal at $\Lambda = 0$, decrease monotonically with increasing Λ, and go to zero at $\Lambda = \infty$. This behavior of the eigenvalues will be important to determining the optimal Λ from the evidence.

Derivatives of I_Q with respect to observables may be evaluated using the chain rule,

$$\frac{\partial I_Q}{\partial \ll \vec{R} \gg} = -\frac{\vec{\lambda}}{T} \quad , \tag{17}$$

and

$$\frac{\partial^2 I_Q}{\partial \ll R_k \gg \partial \ll R_{k'} \gg} = \mathbf{K}^{-1}_{k,k'} \quad . \tag{18}$$

Since $\mathbf{K}$ is positive definite, Eq. (18) implies that I_Q must be a strictly positive convex function of the fits to the data, $\ll \vec{R} \gg$ and, therefore, $\vec{\lambda}$. In this paper, Gaussian approximations will be used to evaluate all integrals over these parameters. The normalized prior probability is then

$$P[\vec{\lambda}, \Lambda] = \frac{1}{T^{N_d}} \sqrt{\frac{\det(\mathbf{K})}{(2\pi)^{N_d}}} \exp\left(-I_Q - \frac{F_o}{T}\right) \qquad P[\ll \vec{R} \gg, \Lambda] = \frac{T^{N_d}}{\det(\mathbf{K})} P[\vec{\lambda}, \Lambda] \quad . \tag{19}$$

In the first relation, $\sqrt{\det(\mathbf{K})}$ is the effective *metric* for integrals over $\vec{\lambda}$. The second relation follows from Eq. (15), which states that $\det(\mathbf{K})/T^{N_d}$ is the Jacobian for the transformation from an integral over $\vec{\lambda}$ to an integral over $\ll \vec{R} \gg$. More generally, QSI satisfies geometrical properties related to those of classical statistical thermodynamics (Levine, 1986).

Consider the application of QSI to the exact data example considered in Figures 1,2. Then, the likelihood function is

$$P[\vec{D} \mid \vec{\lambda}, \Lambda] = \prod_{k=1}^{N_d} \delta(D_k - \ll R_k \gg) \quad . \tag{20}$$

The posterior probability is non-zero only for the $\vec{\lambda}$ such that $\vec{D} = \ll \vec{R} \gg$, which can be found by non-linear optimization methods such as Newton-Raphson. In principle, the image should be obtained by marginalizing Bayes theorem over Λ. In practice, a single Λ_{opt} may be used instead of marginalizing if the evidence, $P[\vec{D}, \Lambda]$, has a single sharp peak. The evidence for exact data is

$$P[\vec{D}, \Lambda] = \frac{1}{\sqrt{(2\pi)^{N_d} \det(\mathbf{K})}} \exp\left(-I_Q - \frac{F_o}{T}\right) \quad . \tag{21}$$

This separates into a product of three factors whose behavior may be understood using the dependence of the eigenvalues of $\mathbf{K}$ on Λ. The *Occam factor*, $1/\sqrt{\det(\mathbf{K})}$ (Bretthorst, 1988) favors the simpler model of larger Λ (or fewer quantum states contributing to the image). Eq. (18) implies that the *data factor*, $\exp(-I_Q)$, favors the more complex model of smaller Λ (or more quantum states contributing to the image). The factor $\exp(-F_o/T)$ may be regarded as a slowly varying Jeffrey's prior for Λ which is independent of $\ll \vec{R} \gg$.

Figure 3 shows the evidence, Occam factor and data factor for the interpolation problem as a function of Λ. The $\Lambda_{opt} \approx 0.6$, corresponding to a variance of the spatial correlation function given by $\Lambda_{opt}/\sqrt{12\pi} \approx 0.1$. Comparable values of Λ_{opt} are found for all three data analysis problems in Figure 1. Figure 4 shows the optimal QSI images for the data in Figure 1. They are clearly more credible than the corresponding ME images shown in Figure 2. Credibility is quantitatively expressed by the many orders of magnitude larger Bayesian evidences for the QSI images relative to the ME images.

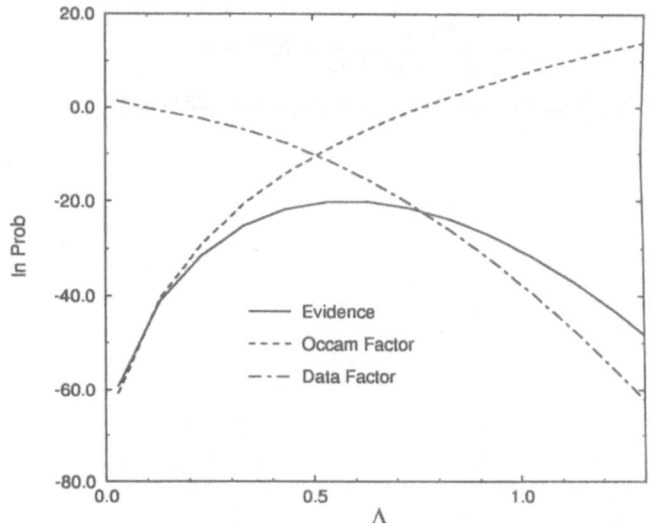

Fig. 3. Optimization of de Broglie Wavelength - The evidence, occam factor, and data factor for the interpolation problem in Fig. 1 as a function of Λ.

5. Noisy Data

Consider the extension of QSI to data subject to Gaussian independent noise defined by $E(\hat{N}_k) = 0$ and $Cov(\hat{N}_k \hat{N}_{k'}) = \delta_{k,k'} \hat{\sigma}_k^2$ in Eq. (1). It is most convenient to renormalize to new variables $D_k \equiv \hat{D}_k/\hat{\sigma}_k$ and $R_k(x) \equiv \hat{R}_k(y)/\hat{\sigma}_k$. The likelihood function is then

$$P[\vec{D} \mid \vec{\lambda}, \Lambda] = \frac{1}{\sqrt{(2\pi)^{N_d}}} \exp\left(-\frac{\chi^2}{2}\right) \quad \chi^2 = (\vec{D} - \ll \vec{R} \gg)^\dagger \cdot (\vec{D} - \ll \vec{R} \gg) \ . \quad (22)$$

Bayes theorem for noisy data becomes

$$P[\ll \vec{R} \gg \mid \vec{D}, \Lambda] \times P[\vec{D}, \Lambda] = \frac{1}{(2\pi)^{N_d}\sqrt{\det(\mathbf{K})}} \exp\left(-I_Q - \frac{\chi^2}{2} - \frac{F_o}{T}\right) \ . \quad (23)$$

Regarding $1/\sqrt{\det(\mathbf{K})}$ as a metric for integrals over $\ll \vec{R} \gg$, the most probable image is obtained from the minimum of

$$Q_3 \equiv I_Q + \frac{\chi^2}{2} \ . \quad (24)$$

In contrast to the traditional literature (Titterington, 1985) on regularizing ill-posed inverse problems including previous implementations of ME (Skilling, 1989), there is no statistical regularization parameter multiplying the regularizing functional I_Q, even in the classical limit where $I_Q \to -S_C$. Equivalently, the regularization parameter in quantum statistical mechanics is known *a priori* to be one when the density function is normalized to unity. The minimum is obtained from the first derivative of Q_3,

$$\frac{\partial Q_3}{\partial \ll \vec{R} \gg} = -\frac{\vec{\lambda}}{T} - \vec{D} + \ll \vec{R} \gg = \vec{0} \quad . \tag{25}$$

A sub(super)script "f" denotes a quantity evaluated at the *final* solution, $\vec{\lambda}^f$, to Eq. (25). The second derivative

$$\frac{\partial^2 Q_3}{\partial \ll \vec{R} \gg \partial \ll \vec{R} \gg} = \mathbf{K}^{-1} + \mathbf{1} \quad , \tag{26}$$

is strictly positive implying that a unique solution to Eq. (25) exists.

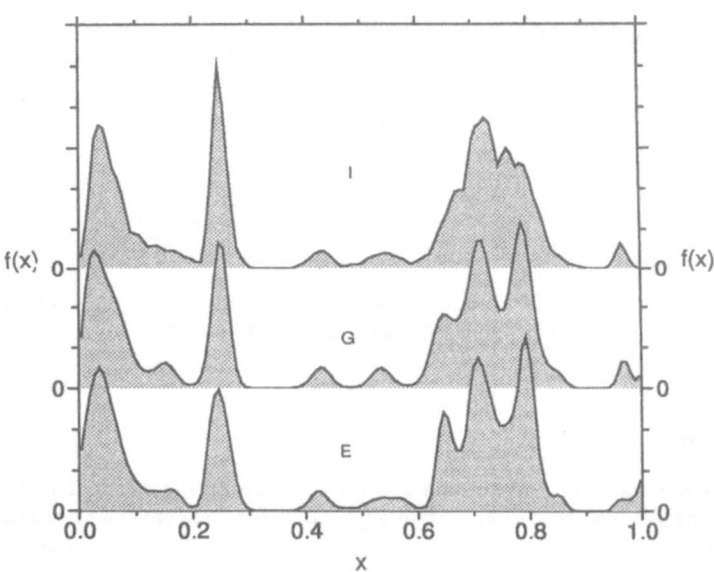

Fig. 4. Quantum Statistical Inference Images - For the exact data example in Fig. 1, displayed in 128 pixels.

For typical data analysis problems, the eigenvalue spectrum of $\mathbf{K}^f$ is a very steep function varying over many orders of magnitude. Independent measurements can be defined as eigenvectors of $\mathbf{K}^f$, and their quality can be ranked according to the size of their eigenvalues. *Good measurements* may be defined as eigenvectors of $\mathbf{K}^f$ whose eigenvalues are much greater than 1. The data for good measurements are dominated by signal, so that the typical values of $| \ll R_k \gg | \approx | D_k |$ are large compared to one. *Bad measurements* may be defined by eigenvectors of $\mathbf{K}^f$ whose eigenvalues are much less than 1. The data for bad measurements are dominated by noise so that the typical values of $|D_k|$ are of order 1 and the typical values for $|\ll R_k \gg|$ are much less than one. In other words, QSI fits the good measurements and ignores the bad ones. The *number of good measurements* may be defined by

$$N_g^f \equiv Tr\left[\mathbf{K}^f \cdot (1 + \mathbf{K}^f)^{-1}\right] \quad . \tag{27}$$

An error scaling consistency argument suggests that

$$\chi_f^2 \approx N_d - N_g^f \quad , \tag{28}$$

so that the χ^2 for the most probable image is less than the number of data. Numerical experiments suggest that N_g satisfies inequalities,

$$N_g^{\vec{\lambda} \neq \vec{0}, \Lambda \neq 0} \leq N_g^{o, \Lambda \neq 0} \leq N_g^{o, \Lambda = 0} \quad . \tag{29}$$

Therefore, the initial good measurements can be identified from the eigenvectors of the classical susceptibility, $\mathbf{K}^{o, \Lambda = 0}$. They can often be calculated *a priori* from the resolution functions and noise, before any data are acquired. Throwing out the initial bad measurements can dramatically reduce the computational and data acquisition tasks without affecting the results. Relations similar to Eqs. (27,28) were first found in the Classic MaxEnt formulation (Skilling, 1989), but they depended on the values of a statistical regularization parameter. Since the initial regularization parameter in Classic MaxEnt is infinite, the initial N_g in Classic MaxEnt is zero.

The evidence is given by

$$P[\vec{D}, \Lambda] = \frac{1}{(2\pi)^{N_d}\sqrt{\det(\mathbf{K}^f + 1)}} \exp\left(-I_Q^f - \frac{\chi_f^2}{2} - \frac{F_o}{T}\right) \quad . \tag{30}$$

Only good measurements contribute to the Occam factor $\propto 1/\sqrt{\det(\mathbf{K}^f + 1)}$. The data factor is now $\propto \exp(-I_Q^f - \chi_f^2/2)$. Again, the Occam factor favors small Λ, while the data factor favors large Λ.

Figure 5 shows noisy data sets for the test density function in Fig. 1 and the same Gaussian (G) and exponential (E) resolution functions. The data are measured at 64 equally spaced points with the noise levels indicated. Figure 6 shows the corresponding ME images obtained by setting $\Lambda = 0$ in QSI, although similar results would be obtained from using the MEMSYS routines (Skilling, 1989). The ME images clearly show excessive overfitting and noise artifacts. Figure 7 shows the corresponding QSI images for the same data, which are clearly superior to the ME images. Although χ^2 is larger for the QSI images than for the ME images, the evidence for them is greater because they correspond to the simpler model of larger Λ. The data sets are placed in order of increasing information content, as measured by the value of I_Q^f. This ranking also corresponds to the visual quality of the images in Figure 7 and to the values of N_g^f. Note that the lineshape can be more important than the statistical errors in determining the information content of the data, with sharper lineshapes generally yielding more information (everything else being equal).

The behavior of the evidence can be seen most readily by evaluating it in a quadratic approximation,

$$P[\vec{D}, \Lambda] \approx \frac{1}{(2\pi)^N \sqrt{\det(\mathbf{K}^o + 1)}} \exp\left(-\frac{1}{2}\delta\vec{D} \cdot (\mathbf{K}^o + 1)^{-1} \cdot \delta\vec{D} - \frac{F_o}{T}\right) \quad , \tag{31}$$

where the *discrepancy* between the data and the default model predictions is given by $\delta \vec{D} \equiv \vec{D} - \ll \vec{R} \gg_o$. Since the eigenvalues of $\mathbf{K}^o$ are monotonically decreasing functions of Λ, the Occam factor again prefers large Λ and the data factor again prefers small Λ. In this quadratic approximation, the image is given by a *quantum filter*,

$$f(x) \approx \int_0^1 dx' \Theta_o(x, x') \left[1 + \delta \vec{R}(x')^\dagger \cdot (\mathbf{K}^o + 1)^{-1} \cdot \delta \vec{D} \right] \quad . \tag{32}$$

This can be a computationally effecient approximation which is adequate for many data analysis problems, although it will not enforce the positivity of the image or achieve the full resolution of the non-linear QSI equations.

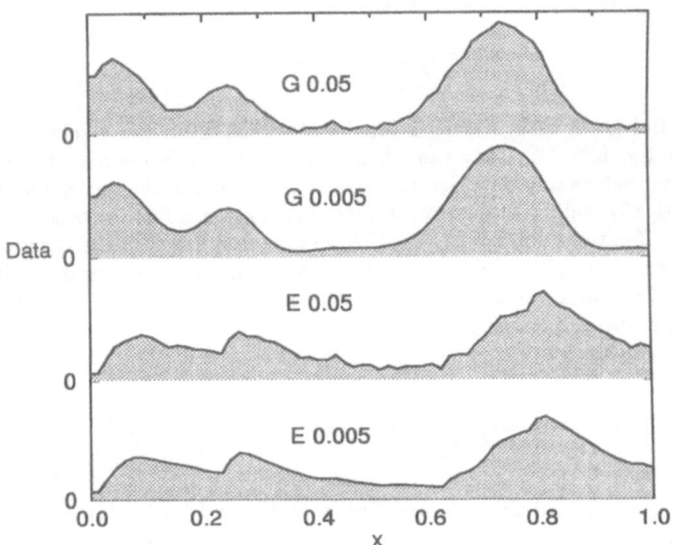

Fig. 5. Noisy Data Example - The test density function is convolved with the same resolution functions in Fig. 1 at 64 equally spaced data points, and then Gaussian independent noise is added. The notation "E 0.05" stands for the exponential resolution function with a noise standard deviation of 0.05.

6. Statistical and Incompleteness Errors

The goal of a statistical inference procedure is to use the data and prior knowledge to make predictions for unmeasured observables. This should include estimates of the reliability of those predictions. For any observable, $\ll O \gg$, these may be calculated from the conditional probability, $P[\ll O \gg | \vec{D}]$, using Bayes theorem. In the absence of data on that observable, the most probable value for $\ll O \gg$ is obtained at $\vec{\lambda}^f$ and $\lambda_O = 0$. This section summarizes the results for the errors on these estimates.

For two observables, $\ll A \gg$ and $\ll B \gg$, define generalized susceptibilities,

$$K_{A,B}^f \equiv \int_0^1 d\tau \ll \delta A e^{-\mathbf{H}\tau/T} \delta B e^{+\mathbf{H}\tau/T} \gg^f = \int_0^1 dx \int_0^1 dx' \delta A(x) \delta B(x') \Theta_f(x, x') \quad , \tag{33}$$

with $\delta A(x) \equiv A(x) - \ll A \gg_f$. Then the composite covariance $C[\delta \ll A \gg \delta \ll B \gg]$ is given by

$$C = K_{A,B}^f - \vec{K}_{A,R}^{f\dagger} \cdot (1 + \mathbf{K}^f)^{-1} \cdot \vec{K}_{R,B}^f \quad . \tag{34}$$

In the absence of measurements the covariance is maximal, given by $K_{A,B}^o$. The second term in Eq. (34) says that good measurements reduce the covariance for unmeasured quantities, as should be required of any valid statistical inference procedure.

Insight can be gained by separating the covariance into the sum of a *statistical covariance*,

$$C_S \equiv \vec{K}_{A,R}^{f\dagger} \cdot (\mathbf{K}^f + \mathbf{K}^f \cdot \mathbf{K}^f)^{-1} \cdot \vec{K}_{R,B}^f \quad , \tag{35}$$

and an *incompleteness covariance*,

$$C_I \equiv K_{A,B} - \vec{K}_{A,R}^{f\dagger} \cdot (\mathbf{K}^f)^{-1} \cdot \vec{K}_{R,B}^f \quad . \tag{36}$$

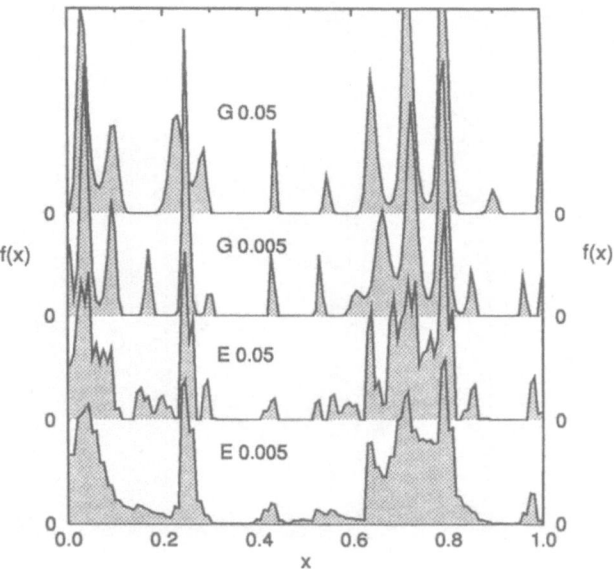

Fig. 6. Maximum Entropy Images - For the noisy data example in Fig. 5.

These definitions make sense in several respects. If the statistical errors on the data are zero, C_I remains finite while C_S goes to zero. If $\ll A \gg$ is one of the measurements, C_I goes to zero while C_S remains finite. Bad measurements are the dominant contribution to C_S, which cancel the reduction in C_I.

An example is the covariance on the fits to the data, $\delta \ll \vec{R} \gg^\dagger \cdot \delta \ll \vec{R} \gg$. Then $C_I = 0$ and $C_S = N_g^f$. Combined with Eq. (28), it implies that the average fit to the data satisfies $E(\chi^2) \approx N_d$, which corresponds to the intuitive expectation.

Another example is the covariance on points in the image. Define

$$\vec{\Gamma}(x) \equiv \int_0^1 dx' \Theta(x, x') \delta \vec{R}(x') \quad . \tag{37}$$

Then the covariance of two points in the image is

$$C[\delta f(x_1)\delta f(x_2)] = \Theta_f(x_1, x_2) - f(x_1)f(x_2) - \vec{\Gamma}(x_1)^\dagger \cdot (1 + \mathbf{K}^f)^{-1} \cdot \vec{\Gamma}(x_2) \quad . \qquad (38)$$

The case $x_1 = x_2$ would give the errors on points. The first term on the r.h.s. is the spatial correlation function. In the absence of data it indeed characterizes prior correlations between points in the image. In the classical ME limit of zero Λ, it approaches a δ-function, so that the errors on individual points diverge. For $\Lambda \neq 0$, the errors on points are finite and they decrease with increasing Λ. The second term preserves the normalization of the image. The third term states how good measurements reduce the covariance of points in the image.

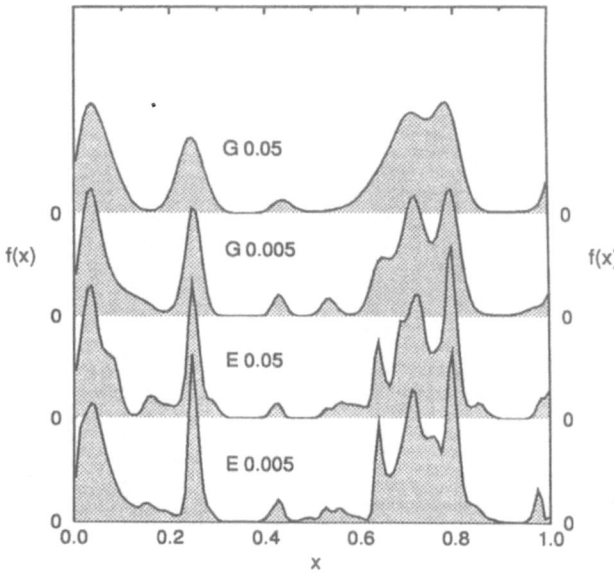

Fig. 7. Quantum Statistical Inference Images - For the noisy data example in Fig. 5.

Generally, the optimal QSI images are smoother and have smaller errors than the ME images for the same data. If the object consists of a small number of isolated δ-functions, $\Lambda_{opt} \approx 0$ (T large) so that QSI may reduce to ME with maximal errors. If the object is smooth but has sharper structure than the resolution function, the QSI image may have $\Lambda_{opt} \neq 0$ even though it may resemble a ME image at $\Lambda_{opt} = 0$; however, the errors on the QSI image will be smaller than errors on the ME image. If the data are close to the default model predictions, Λ_{opt} will be large ($T \approx 0$) and the errors will be small.

7. Conclusions and Discussion

QSI is the first explicit application of quantum theory to human reasoning. The compelling motivations for QSI include the unquestioned success of quantum statistical mechanics in the physical domain, the *probability-theory-as-logic* hypothesis, and the inadequacies of the maximum entropy method for a broad class of density function estimation problems. This paper has proposed that quantum mechanics is a theory of statistical inference for *continuous* density functions, it has demonstrated applications to the interpolation and deconvolution of data, and the results are superior to the maximum entropy method.

QSI can best be tested and developed by experience with diverse data analysis problems. So far QSI has used only a small fraction of the rich structure of quantum theory. Other quantum concepts may also apply to information theory and statistics. These may include pure quantum states, complex wave functions, path integrals, gauge fields, etc.

Did the founders of quantum mechanics, in solving the riddles of the atom, serendipitously discover how to reason from incomplete and uncertain information? The successful validation of QSI would do much more than provide a powerful new tool for data analysis. Lifting quantum theory out of its original physical context would aid in the development of a more rigorous mathematical approach. It would also support the probability-theory-as-logic hypothesis and, thereby, provide new perspectives on the interpretation of quantum mechanics.

ACKNOWLEDGMENTS: This research was supported by the U. S. Dept. of Energy. Travel grant provided by the U. S. Dept. of the Army.

REFERENCES

Bretthorst, G. L.: 1988, *Bayesian Spectrum Analysis and Parameter Estimation*, Springer-Verlag, Berlin.

Dembo, A.; Cover, T. M.; Thomas, J. A.: 1991, 'Information Theoretic Inequalities', *IEEE Trans. Info. Theory* **37**, 1501.

Fick, E.; Sauermann, G.: 1990, *The Quantum Statistics of Dynamic Processes*, Springer-Verlag, Berlin

Fisher, R. A.: 1925,'Theory of Statistical Estimation', *Proc. Cambr. Phil. Soc.* **22**, 700.

Frieden, B. R.: 1989, 'Fisher Information as the basis for the Schrödinger wave equation', *Am. J. Phys.* **57**, 1004.

Jaynes, E.T.: 1989, 'Clearing Up Mysteries – The Original Goal', in J. Skilling (ed.), *Maximum Entropy and Bayesian Methods*, Kluwer, Dordrecht.

Jaynes, E. T.: 1983, in R. D. Rosenkrantz, (ed.), *E. T. Jaynes: Papers on Probability, Statistics and Statistical Physics*, D. Reidel Publishing Co., Dordrecht.

Kubo, R.: 1959, 'Some Aspects of the Statistical-Mechanical Theory of Irreversible Processes', in Britten & Dunham, (eds.), *Lectures in Theoretical Physics*, **1**.

Levine, R. D.: 1986, 'Geometry in Classical Statistical Thermodynamics', *J. Chem. Phys.* **84**, 910.

Loredo, T.: 1990, 'From Laplace to Supernova SN 1987A: Bayesian Inference in Astrophysics',in P. Fougere (ed.), *Maximum Entropy and Bayesian Methods*, Kluwer, Dordrecht., 81

J. Mathews, R. L. Walker: 1964, *Mathematical Methods of Physics*, W. A. Benjamin, New York, p. 315

McQuarrie, D. A.: 1976, *Statistical Mechanics*, Harper & Row, New York.

Skilling, J.; Gull, S.: 1989, 'Classic MaxEnt', in J. Skilling (ed.), *Maximum Entropy and Bayesian Methods*, Kluwer, Dordrecht, 45-71.

Tikochinsky, Y.; Tishby, N. Z.; Levine, R. D.: 1984, 'Consistent Inference of Probabilities for Reproducible Experiments', *Phys. Rev. Letts.* **52**, 1357.

Titterington, D. M.: 1985, 'Common Structure of Smoothing Techniques in Statistics', *Int. Statist. Rev.*, *53*, 141.

INFORMATION THEORY-BASED VARIATIONAL APPROACH
IN CORRELATED FERMION LATTICES

L. Arrachea, N. Canosa, A. Plastino, R. Rossignoli

Departamento de Física, Univ. Nac. de La Plata
C.C.67, 1900 La Plata, Argentina

ABSTRACT

A variational approach based on Maximum Entropy considerations is used for the construction of trial ground state wave functions of correlated fermion systems. Results for the one-dimensional version of the Hubbard model show that the method allows for a sensibly better evaluation of the ground state energy than the one provided by the Gutzwiller approximation, being in a good agreement with the exact values.

1 INTRODUCTION

The study of strongly correlated fermion systems is one of the most exciting topics of research within the field of condensed-matter physics. Most of the areas of interest nowadays, such as heavy fermion systems and high-temperature superconductivity, are geared towards the understanding of this sort of systems. However, in spite of the great efforts undertaken in that direction, this complex quantum-mechanical many-body problem can by no means be regarded as solved (or even reasonably understood).

The Hubbard model [1], whose two-dimensional version is thought to explain the mechanism of high-Tc superconductivity [2], is the simplest Hamiltonian containing the basic ingredients of a strong correlation. This apparently simple model, has been exactly solved only for its one-dimensional version (in the thermodinamic limit) [3]. However, it is quite difficult to manage and most of the physics involved remains far from being understood. In this sense, progress has been done by the use of numerical (Monte Carlo) methods [4],[5] and variational wave functions [7]-[10]. The most extensively used of this variational wave functions was first proposed by Gutzwiller [6], who also introduced a suitable approximation in order to calculate the ground state energy.

In the present work we conserve the essence of the Gutzwiller treatment, in the sense of introducing correlations in the wave function via variational parameters. The procedure is performed starting from an interpretation of the ground state wave function within the conceptual framework of the Information Theory [14]-[17].

The work is organized as follows. In Sec. 2 the pertinent formalism is described and in Sec. 3 its application to the study of correlated lattice systems is discussed in relation to the Hubbard model. The pertinent results and comparisson with Gutzwiller approximation are dealt with in Sec. 4. Finally, some conclusions are drawn in Sec. 5.

2 GROUND STATE VARIATIONAL WAVE FUNCTIONS AND INFORMATION THEORY

Let us consider a quantum system in its ground state $|\Psi_0\rangle$, which is spanned by a given basis $\{|j\rangle, \; j = 1, \ldots, d\}$,

$$|\Psi_0\rangle = \sum_j a_j |j\rangle. \tag{1}$$

The set $\{|a_j|^2, \; j = 1, \ldots, d\}$ defines the probability distribution associated to the expansion of the wave function (in the given basis). It was found in [14]-[17] that, according to the information theory prescriptions, the pseudo (or quantal) entropy

$$S = -\sum_j |a_j|^2 \ln(|a_j|^2), \tag{2}$$

which provides a measure of the lack of information associated with the above probability distribution, is a quite useful tool in many circunstances.

Let us consider now that we posses incomplete knowledge about the system, consisting of a set of mean values of relevant observables:

$$O_i \equiv \langle \Psi_0 | \hat{O}_i | \Psi_0 \rangle. \tag{3}$$

If the operators $\hat{O}_i$ conform an abelian set, there exists a common basis in which they have a diagonal representation. In what follows, we shall select this basis for the expansion (1). The set of appropriate mean values is then written as

$$O_i \equiv \sum_{j=1}^{d} |a_j|^2 O_i(j) \quad i = 1, \ldots, m \tag{4}$$

with

$$O_i(j) \equiv \langle j | \hat{O}_i | j \rangle. \tag{5}$$

The Maximum Entropy criterium may be invoked in order to infer the probability distribution, i.e. we should look for the set $\{|a_j|^2, \; j = 1, \ldots, d\}$ that maximizes (2), with the constraints (4). This procedure is performed via the introduction of a Lagrange multiplier for each of the constraints (4), which leads to [18],[19]

$$|a_j|^2 = \exp\{-\lambda_0 - \sum_{i=1}^{m} \lambda_i O_i(j)\}, \tag{6}$$

where the normalization is impossed by introducing

$$\lambda_0 = \ln\{\sum_j^d \exp(-\sum_{i=1}^m \lambda_i O_i(j))\}, \qquad (7)$$

which satisfies

$$\frac{\partial \lambda_0}{\partial \lambda_i} = -O_i, \quad i = 1, \ldots, m. \qquad (8)$$

In the particular case that all the coefficients in the expansion (1) are real and posses the same sign, the above scheme allows for the inference of the a_j's from the available information, where only a global phase remains indetermined.

In most cases, however, we have to deal with the problem of predicting the mean values O_i rather than having them as available information. In this case, we may construct a trial wave function, assuming the dependence (6) for the coefficients, in which one includes those observables considered relevants. The Lagrange multipliers now play the role of *variational parameters* to be adjusted by minimizing the functional

$$E[\lambda_1, \ldots, \lambda_m] \equiv \langle \psi_0 | \hat{H} | \psi_0 \rangle, \qquad (9)$$

with $|\psi_0\rangle$ given by (1) and (6) and $\hat{H}$, the hamiltonian of the system, which is assumed known. This variational approach is the one to be employed in this work, see [14],[17] for additional details.

3 APPLICATION TO THE HUBBARD MODEL

We shall show that the ideas expounded in the previous section may be succesfully applied to the study of correlated lattice fermion systems.

Let us consider the Hubbard hamiltonian with nearest neighbour hopping

$$\hat{H} = -t \sum_{i,\delta,\sigma} c_{i,\sigma}^\dagger c_{i+\delta,\sigma} + U \sum_i \hat{n}_{i\uparrow} \hat{n}_{i\downarrow}, \qquad (10)$$

where i runs over the L lattice sites, δ over nearest neighbours and $\sigma = \uparrow, \downarrow$. We shall restrict ourselves to the one dimensional chain, with half band filling, i.e. the number of particles N equals the number of sites L.

According to Lieb and Mattis [11], the ground state is a singlet and thus belongs to the subspace of states with equal numbers of down and up particles. So, $N_\uparrow = N_\downarrow = N/2$. This subspace is $d = \binom{N}{N/2}^2$ dimensional and may be generated by:

$$|n_{1\uparrow}, n_{1\downarrow}, \ldots, n_{L\uparrow}, n_{L\downarrow}\rangle = \prod_{i \in \{k_l\}} \prod_{j \in \{k_s\}} c_{i\uparrow}^\dagger c_{j\downarrow}^\dagger |0\rangle = |l, s\rangle; \quad l, s = 1 \ldots K, \qquad (11)$$

where $\{k_l\}$ denotes a particular set of $N/2$ indices within the $K = \binom{N}{N/2}$ possible configurations corresponding to $N/2$ particles with "up" spin distributed over the $L = N$ sites and $\{k_s\}$, the corresponding to the configurations with "down" spin. The occupation numbers $n_{p\uparrow}$ and $n_{p\downarrow}$, with $p = 1, \ldots, L$ adopts values 0 or 1, depending on the configuration.

The ground state may thus be written as follows

$$|\Psi_0\rangle' = \sum_{(l,s)} a_{(l,s)} |l,s\rangle. \tag{12}$$

It may be seen that all the coefficients $a_{(l,s)}$ are real and have the same sign, (a detailed discussion may be found in [12] in relation with the Heisemberg antiferromagnet). We are now in a possition to apply the ideas of the previous section. The relevant diagonal observables, in this case, are taken to be related with the on-site and neighbouring-site correlations which are the ones involved in the dynamical behaviour of the system. The following selection is then made for the conmuting operators that appear in (6):

$$\hat{D} = \sum_i \hat{n}_{i\uparrow}\hat{n}_{i\downarrow}, \tag{13}$$

$$\hat{A}_\alpha = \sum_{i,\delta,\sigma} \hat{n}_{i,\sigma}\hat{n}_{i+(\delta)\alpha,-\sigma}, \tag{14}$$

$$\hat{B}_\alpha = \sum_{i,\delta,\sigma} \hat{n}_{i,\sigma}\hat{n}_{i+(\delta)\alpha,\sigma}, \tag{15}$$

where $\alpha = 1$ denotes nearest-neigbours, $\alpha = 2$, next nearest-neighbours, and so on, and $\delta = \pm 1$. The corresponding variational parameters are denoted by λ_D, $\lambda_{A\alpha}$ and $\lambda_{B\alpha}$, respectively.

The trial wave function is then written as

$$|\Psi_0\rangle = \sum_{(l,s)} a_{(l,s)}(\lambda_0,\lambda_D,\lambda_{A\alpha},\lambda_{B\alpha}) |l,s\rangle \tag{16}$$

with

$$a_{(l,s)}(\lambda_0,\lambda_D,\lambda_{A\alpha},\lambda_{B\alpha}) = \exp -\frac{1}{2}\{\lambda_0 + \lambda_D D(l,s) + \lambda_{A\alpha} A_\alpha(l,s) + \lambda_{B\alpha} B_\alpha(l,s)\}, \tag{17}$$

where

$$D(l,s) = \langle l,s| \hat{D} |l,s\rangle \ , \quad A_\alpha(l,s) = \langle l,s| \hat{A}_\alpha |l,s\rangle \ , \quad B_\alpha(l,s) = \langle l,s| \hat{B}_\alpha |l,s\rangle. \tag{18}$$

The minimization of the functional (9) leads to the pertinent values of the parameters.

The Gutzwiller wave function

$$|\Psi_G\rangle = \prod_i [1 - (1-g)\hat{n}_{i\uparrow}\hat{n}_{i\downarrow}] |\psi_{FS}\rangle \tag{19}$$

with $g = e^{-\eta}$; $0 \le g \le 1$, and $|\Psi_{FS}\rangle$ being the uncorrelated ground state in the reciprocal space, may be considered , within this context, as a particular case of a trial wave function of the sort studied here. In this scheme, the Gutzwiller wave function should correspond to the case in which only one relevant operator ,$\hat{D}$, is included, together with the "a-priori" probability distribution associated to the expansion of $|\Psi_{FS}\rangle$ over the direct space basis (11). The use of (19) for the evaluation of physical quantities is very difficult, and a certain approximation is necessary for it. Well known for the calculation of the ground-state energy is the Gutzwiller approximation (GA), which leads to the unphysical result that for half-band filling, the optimization of g

leads to $g = 0$ and, thus, to a vanishing value of the ground state energy for any strength of electron repulsion above the critical value $U_C = 8e_0$, where e_0 is the mean kinetic energy for the non-interacting system [6]. This phenomenon is known as the Brinkman-Rice-type transition [13], from a conducting state to a localized state, with single occupied lattice sites (metal-insulator transition). This feature, being a consequence of the GA, is not observed when the ground state energy is evaluated rigurously, and neither is it observed when evaluated with the present approach, in which case no approximation is necessary.

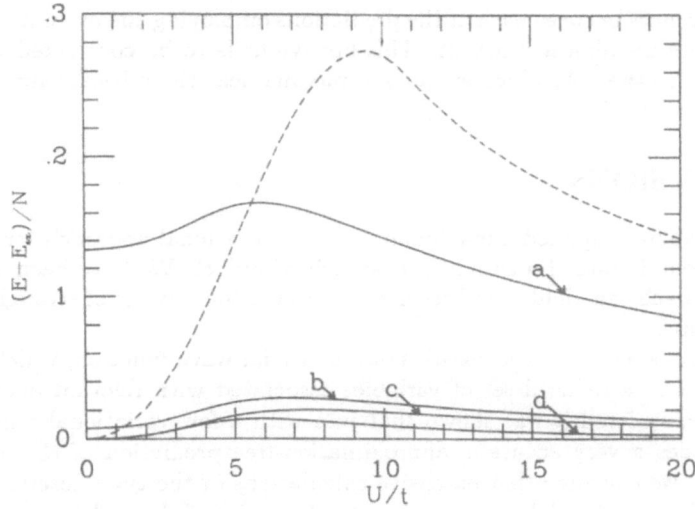

Figure 1. Differences between the inferred values of the ground state energy per particle and the exact value as a function of U/t. Curve (a) corresponds to inference with the inclusion of $\hat{D}$; curves (b), (c) and (d), to the addition of $\hat{A}_\alpha$, $\hat{B}_\alpha$, with $\alpha = 1, 2, 3$, respectively; whereas the curve in dashed lines correspond to the GA.

4 RESULTS

The evaluation of the parameters that leads to the minimun value of (9) has been performed numericaly for a finite chain of $L = 6$ lattice sites with periodic boundary conditions and half filling, $N = L$.

The ground state energy per particle $E/N = \langle \hat{H} \rangle /N$ has been computed as a function of U/t, taking into account the on-site correlation relevant operator, $\hat{D}$, and also including the correlations involving 1st, 2nd and 3th-neighbours, $\hat{A}_\alpha$, $\hat{B}_\alpha$, with $\alpha = 1, 2, 3$ respectively. The results are shown in the above figure, where the differences

$(E - E_{ex})/N$ between the corresponding predictions and the exact value of energy per particle E_{ex}/N are plotted.

The results corresponding to the present approach with the inclusion of both one-site and nearest-neighbour-site operators exhibit a very good quality for the whole range U/t considered. Moreover, this quality is not improved by considering additional neighbours. This fact is suggestive beyond the role of the correlations involved, and a consequence of studying a model in which the hopping takes place only between nearest neighbours. Another interesting feature is that even the prediction corresponding to the inclusion of only the on-site operator, which is the crudest possible approach within the present scheme, shows to be a better one than that obtained with GA, except for weak repulsion. It is worthwhile to remark that this case, which leads to a Gutzwiller-type trial wave function with a non "a-priori" probability distribution, allows for an approximation-free calculation, which thus becomes a tractable one. It can also be seen that the differences between each of the predictions employing the present approach and the exact value are almost constant. This behaviour is to be contrasted with the one corresponding to the GA, which exhibits a maxima near the critical value $U_C \approx 10$.

5 CONCLUSIONS

In this work we have applied a maximum entropy variational approach for the description of the ground state of a simple but not trivial model. We have been able to show that this approach can yield useful results concerning the physics of strongly correlated fermion systems.

The scheme is based on the construction of a trial wave function, which is parametrized in terms of a reduced set of variables associated with relevant observables. In the example considered it was shown that just with a few variational parameters the method provides a very accurate, approximation-free prediction of the ground state energy. Then, we conclude that extensive calculations of the type described here may indeed allow for useful insights concerning the dynamics of the rather complex systems studied in the present comunication.

ACKNOWLEDGEMENTS

L.A., N.C. and A.P. acknowledge support from Consejo Nacional de Investigaciones Científicas y Técnicas de la República Argentina. R.R. acknowledges support from Comisión de Investigaciones Científicas de la Provincia de Buenos Aires.

REFERENCES

[1] J. Hubbard, *Proc. R. Soc. London Ser. A* 276, 238 (1963).

[2] P. W. Anderson, *Science* 235, 1196 (1987).

[3] E. H. Lieb and F. Y. Wu, *Phys. Rev. Lett.* 20, 1445 (1968).

[4] J. E. Hirsch, R. L. Sugar, D. J. Scalpino and R. Blankenbecler, *Phys. Rev. B* <u>26</u>, *5033 (1982)*.

[5] J. E. Hirsch, *Phys. Rev. B* <u>28</u>, *4059 (1983)*; <u>31</u>, *4403 (1985)*; <u>34</u>, *3216 (1986)*; H. Q. Lin and J. E. Hirsch, *ibid*, <u>35</u>, *3359 (1987)*.

[6] M. C. Gutzwilller, *Phys. Rev. Lett.* <u>10</u>, *159 (1963)*.

[7] W. Metzner and D. Vollhardt, *Phys. Rev. B.* <u>37</u>, *7382 (1988)*.

[8] F. Gebhard and D. Vollhardt, *Phys. Rev. B.* <u>38</u>, *6911 (1988)*.

[9] M. L. Ristig, *Z. Phys B.* <u>79</u>, *351 (1990)*.

[10] K. Hashimoto, *Phys. Rev. B.* <u>31</u>, *7368 (1985)*.

[11] L. H. Lieb and D. C. Mattis, *Phys. Rev.* <u>125</u>, *164 (1962)*.

[12] E. Lieb, T. Schultz and D. C. Mattis, *Ann. Phys.* <u>16</u>, *407 (1961)*.

[13] W. F. Brinkman and T. M. Rice, *phys. Rev. B* <u>2</u> *4302 (1970)*.

[14] N. Canosa, A. Plastino, and A. Rossignoli, *Phys. Rev. A* <u>40</u>, *519 (1989)*.

[15] N. Canosa, R. Rossignoli and A. Plastino, *Nucl.Phys. A* <u>512</u>, *492 (1990)*.

[16] N. Canosa, R. Rossignoli and A. Plastino, *Phys.Rev. A* <u>43</u>, *1445 (1991)*.

[17] N. Canosa, R. Rossignoli, A. Plastino and H. G. Miller, *Phys. Rev. C* <u>45</u>, *1162 (1992)*.

[18] J. Jaynes, *Phys.Rev.* <u>106</u>, *620 (1957)*; <u>108</u>, *171 (1957)*.

[19] A.Katz, *Principles of Statistical Mechanics* (Freeman, San Francisco, 1967).

MAXIMUM ENTROPY VARIATIONAL APPROACH
TO COLLECTIVE STATES

N. Canosa, R. Rossignoli and A. Plastino

Departamento de Física, Universidad Nacional de La Plata
C.C. 67 (1900), La Plata, Argentina

Abstract

An approximation to the energy eigenstates of a many-body system, based on a previously introduced maximum entropy approach to the ground state, is developed and applied to a monopole fermion system. An excellent agreement with the exact eigenstates is obtained over the whole range of the pertinent coupling constant.

Introduction

The mean field method constitutes the basic approach to the many–fermion problem [1]–[3]. At the very least it yields the best zero–order wave function upon which to build up elementary excitations in order to describe low–lying excited states [2]. More recently, the maximum entropy principle derived from Information Theory (IT) [4]–[5] has proved to be a powerful tool to get insights into the complexity of the many–body problem. In a recent effort [6]–[10] an alternative, maximum entropy based approach to the description of many–body ground states has been introduced. By recourse to an appropiately defined *quantal* entropy that measures the lack of information concerning the probability distribution of a quantum state over an arbitrary basis, the method allows for a consistent theoretical picture of the ground state in terms of a small set of variables associated to relevant observables. It was shown that, just with a few one and two–body observables , diagonal in the given basis, this IT approximation yields ground state results in excellent agreement with the exact ones for a variety of many fermion models, for all values of the pertinent coupling constants, including transitional regions [6]–[10]. Here we wish to show how to build up elementary excitations upon such an IT based approximate ground state and, as an example, we tackle the description of the excited states of a many–body fermion model under a monopolar interaction.

The work is organized as follows: first, the IT approximation to the ground state is briefly reviewed followed by the extension of the formalism to excited states. Then, we illustrate our treatment with reference to an $SU(3)$ solvable model. Finally some conclusions are drawn.

Quantal Entropy and the Descrition of Ground States

We shall focus our attention upon systems described by a Hamiltonian of the form $\hat{H} = \hat{H}_0 + \hat{H}_{int}$, where $\hat{H}_0$ denotes the unperturbed term and $\hat{H}_{int}$ the corresponding interaction one. Let $\{\hat{O}_\alpha, \; \alpha = 1,\ldots,n\}$ be a set of relevant commuting operators that commute with $\hat{H}_0$ and which are thus diagonal in the appropriate common basis, in this case the unperturbed basis $\{|j\rangle, \; j = 1,\ldots,K\}$, formed by the eigenstates of $\hat{H}_0$. We shall consider a maximum entropy based exponential approximation [7, 8] to the ground state of the system, denoted by $|0\rangle$, of the form

$$|0\rangle = \sum_j C_j^{(0)}|j\rangle, \tag{1}$$

with

$$C_j^{(0)} = \exp\{-\tfrac{1}{2}[\lambda_0 + \sum_\alpha \lambda_\alpha O_\alpha(j)]\}, \tag{2}$$

where $O_\alpha(j) = \langle j|\hat{O}_\alpha|j\rangle$, $\{\lambda_\alpha = \lambda_\alpha^r + i\lambda_\alpha^i\}$ constitute a set of complex optimizable parameters and

$$\lambda_0 = \ln \sum_j \exp[-\sum_\alpha \lambda_\alpha^r O_\alpha(j)] \tag{3}$$

is the normalization constant (which can be taken as real). In this way the coefficients have the functional form typical of IT which *maximizes* the quantal entropy [6]–[7] in the common unperturbed basis, defined as

$$S = -\sum_j |C_j^{(0)}|^2 \ln |C_j^{(0)}|^2, \tag{4}$$

subject to the constraints

$$\langle \hat{O}_\alpha \rangle_0 \equiv \langle 0|\hat{O}_\alpha|0\rangle = O_\alpha. \tag{5}$$

We would like to remark here that the information entropy (4) is not the conventional thermodynamic one, which becomes zero for a pure state. The entropy (4) measures the lack of information concerning the probability distribution over the unperturbed basis, vanishing only in that special case in which $|0\rangle$ coincides with one of the eigenstates of $\hat{H}_0$. A smoothness criterium which is particularly suitable for ground states is obtained by means of the maximization of (4).

The formalism is able to yield both an *inference* scheme[7]–[8], in which the parameters λ_α are obtained according to the standard IT prescriptions, i.e. from the knowledge of the expectation values O_α (eqs. 5), and also can, alternatively, provide us with a pure *variational* treatment [7], [10]. The latter is the approach that we shall employ in this work. In the variational approximation the parameters λ_α result from the minimization of the ground state energy $\langle \hat{H} \rangle_0 = \langle 0|\hat{H}|0\rangle$, and in this case the appropriate relationships that define the set of general (complex) parameters λ_α are

$$-\partial\langle \hat{H} \rangle_0/\partial\lambda_\alpha^r = \tfrac{1}{2}\langle \hat{O}_\alpha\hat{H} + \hat{H}\hat{O}_\alpha \rangle_0 - \langle \hat{O}_\alpha \rangle_0\langle \hat{H} \rangle_0 = 0, \tag{6}$$

$$-\partial \langle \hat{H} \rangle_0 / \partial \lambda_\alpha^i = \tfrac{1}{2} \langle [\hat{H}, \hat{O}_\alpha] \rangle_0 = 0, \tag{7}$$

Eqs. (6) and (7) are together equivalent to the condition

$$\langle \hat{H}\hat{O}_\alpha \rangle = \langle \hat{O}_\alpha \hat{H} \rangle = \langle \hat{H} \rangle \langle \hat{O}_\alpha \rangle. \tag{8}$$

For the exact ground state, eqs. (6)–(7) are obviously satisfied for any operator $\hat{O}_\alpha$. Thus, it is apparent that convergence towards the exact ground state can be obtained by adding operators $\hat{O}_\alpha$ in the exponent of (2). The exact ground state coefficients can always be expanded in the form (2) if a complete set of diagonal operators $\hat{O}_\alpha$ is used [7].

If a prior estimate of the coefficients p_j is known, which can be either an approximate starting value or a multiplicity factor, i.e., a nonequal weight assigned to the unperturbed states the formalism can easily be extended to include this previous information. In this case one define the so called *surprisal* $I_j^2 = -\ln(|C_j^{(0)}|^2 p_j^2)$, (see for example Ref. [11]), so that the coefficients $C_j^{(0)}$ are now selected so as to maximize the *entropy defficienty*

$$\Delta S = \sum_j |C_j^{(0)}|^2 I_j^2, \tag{9}$$

i.e. the quantal entropy *relative* to the measure determined by p_j^2. The coefficients $C_j^{(0)}$ now acquire the appearance

$$C_j^{(0)} = p_j \exp\{-\tfrac{1}{2}[\lambda_0 + \sum_\alpha \lambda_\alpha O_\alpha(j)]\}. \tag{10}$$

The variational equations are obviously still given by (6)–(7). As we shall see, a good ansatz for the representation of excited states, which are characterized by the existence of nodes in the pertinent wave function, is available by recourse to suitable weight factors. This constitutes the central idea of the present work.

Quantal Entropy and Collective Excited States

We start the pertinent consideration by studying the states

$$|\alpha\rangle \equiv (\hat{O}_\alpha - \langle \hat{O}_\alpha \rangle_0)|0\rangle = \sum_j C_j^{(\alpha)}|j\rangle, \tag{11}$$

with

$$C_j^{(\alpha)} = C_j^{(0)}(O_\alpha(j) - \langle \hat{O}_\alpha \rangle_0) = -2\partial C_j^{(0)}/\partial \lambda_\alpha^r, \tag{12}$$

which can be regarded as maximum quantal entropy–defficiency coefficients. Here the prior knowledge entering the surprisal I_j is that provides by the measures $p_j = (O_\alpha(j) - \langle \hat{O}_\alpha \rangle_0)^2$. The states (11) are clearly *orthogonal* to the (approximate) ground state,

$$\langle 0|\alpha \rangle = 0 \tag{13}$$

and on account of the stability conditions (8), they verify,

$$\begin{aligned} \langle 0|\hat{H}|\alpha \rangle &= \langle \hat{O}_\alpha \hat{H} \rangle_0 - \langle \hat{O}_\alpha \rangle_0 \langle \hat{H} \rangle_0 \\ &= 0. \end{aligned} \tag{14}$$

Hence, the eqs. (6)–(7) imply that our approximate ground state is indeed stable against the excitations represented by the states $|\alpha\rangle$. This entails that for every operator included in the exponent of (2) there is no mixing between $|0\rangle$ and the states (11). Notice that these states are not orthogonal (and not normalized). For collective operators $\hat{O}_\alpha$, they can be regarded as furnishing a suitable basis for the description of collective excitations. As we can see, a first estimation of the low lying states can be obtained by diagonalizing $\hat{H}$ in this reduced space of dimension n. This can be accomplished by diagonalizing the pertinent overlap matrix, which coincides with the ground state covariance matrix,

$$\langle\alpha|\alpha'\rangle = \langle\hat{O}_\alpha\hat{O}_{\alpha'}\rangle_0 - \langle\hat{O}_\alpha\rangle_0\langle\hat{O}_{\alpha'}\rangle_0. \tag{15}$$

The hamiltonian matrix is of the form

$$\langle\alpha|\hat{H}|\alpha'\rangle = \langle\hat{O}_\alpha\hat{H}\hat{O}_{\alpha'}\rangle_0 - \langle\hat{O}_\alpha\rangle_0\langle\hat{O}_{\alpha'}\rangle_0\langle\hat{H}\rangle_0, \tag{16}$$

which is attained with the help of (8). The resulting approximate states can be cast as

$$|\beta\rangle \equiv (\hat{Q}_\beta - \langle\hat{Q}_\beta\rangle_0)|0\rangle, \tag{17}$$

where

$$\hat{Q}_\beta = \sum_\alpha b_{\alpha\beta}\hat{O}_\alpha, \tag{18}$$

and with the $b_{\alpha,\nu}$ arising from the eigenvector matrix of the system

$$hb = obh' \quad b^\dagger ob = I, \tag{19}$$

where o and h stand, respectively, for the overlap and hamiltonian matrices of elements (15) and (16), while $e_{\mu,\nu} = \delta_{\mu,\nu}E_\nu$ is the diagonal eigenvalue energy matrix. The orthogonal states (17) can thus be interpreted as *normal* modes.

In order to extend the formalism to higher excited states, we can construct in general the states

$$|\gamma\rangle \equiv (\prod_\alpha \hat{O}_\alpha^{n_\alpha} - \langle\prod_\alpha \hat{O}_\alpha^{n_\alpha}\rangle_0)|0\rangle, \tag{20}$$

where γ stands for $(n_1,\ldots,n_n)$, with $0 \le n_\alpha \le k_\alpha$, and diagonalize $\hat{H}$ in the ensuing reduced space. These states are in general not orthogonal, except with the approximate ground state ($\langle 0|\gamma\rangle = 0$), with an overlap matrix given by

$$\langle\gamma|\gamma'\rangle = \langle\prod_\alpha \hat{O}_\alpha^{n_\alpha+n'_\alpha}\rangle_0 - \langle\prod_\alpha \hat{O}_\alpha^{n_\alpha}\rangle_0\langle\prod_\alpha \hat{O}_{\alpha'}^{n_{\alpha'}}\rangle_0. \tag{21}$$

In order to avoid superposition, we should obviously exclude from (20) those operators $\hat{O}_\alpha$ which can be expressed as products of other $\hat{O}'_\alpha$s. We could also employ the operators $\hat{Q}_\beta$ instead of $\hat{O}_\alpha$ in (20), reducing in this way the number of non-vanishing elements in the ensuing overlap and energy matrices. The space spanned by the states (20) is similar to that generated by the states $|\gamma'\rangle = \sum_j C_j'^{(\gamma)}|j\rangle$, with

$$C_j'^{(\gamma')} = \partial^k C_j^{(0)}/\prod_\alpha \partial\lambda_\alpha^{n_\alpha}, \tag{22}$$

where $k = \sum_\alpha n_\alpha$.

It is expected that a considerable part of the corresponding collective space will be spanned with low values of k_α, so that an accurate prediction of the low lying energy states can be achieved with a hamiltonian matrix $\langle \gamma | \hat{H} | \gamma' \rangle$ of small dimension and hence, the parameters λ_α can still be obtained *before* diagonalization by solving equations (6)–(7).

As we shall see in the next section our formalism is able to yield, in the example considered, a very accurate description of the lowest energy levels.

Application to a Monopole Model

In order to illustrate our formalism, we shall examine a $U(n)$ model [12]. We are dealing with $N = 2\Omega$ fermions distributed among $n = 2\Omega$–fold degenerate single particle (sp) levels with unperturbed energy ε_i coupled by a monopole interaction. The sp states are denoted as $|p, i\rangle$, $i = 1, \ldots, n$, $p = 1, \ldots, 2\Omega$. We shall consider the Hamiltonian

$$\hat{H} = \sum_i \varepsilon_i \hat{G}_{ii} - \tfrac{1}{2} \sum_{i<j} V_{ij}(\hat{G}_{ij}^2 + \hat{G}_{ji}^2), \tag{23}$$

where $\hat{G}_{ij} = \sum_p c_{pi}^\dagger c_{pj}$ are collective operators satisfying a $U(n)$ algebra under commutation. We shall take as expansion basis the eigenvectors of the unperturbed Hamiltonian $\sum_i \varepsilon_i \hat{G}_{ii}$ and shall consider $V_{ij} > 0$.

In the case ($N = 2\Omega$) the ground state belongs to the completely symmetric representation $(N, 0, \ldots, 0)$ spanned by states $|m\rangle \equiv |m_1, \ldots, m_n\rangle$, $\sum_i m_i = N$, of dimension $\binom{N+2}{2}$, with m_i denoting the number of particles in the level i. The ground state $|\psi\rangle$ of (23) can be expanded as

$$|\psi_0\rangle = \sum_m C_m^0 |m\rangle, \tag{24}$$

where, due to the structure of this Hamiltonian, the sum can be restricted to states with even values of m_i, and all the ground state coefficients $C_m^{(0)}$ obviously possess the same phase.

We shall consider now the maximum quantal entropy approximation for the ground state (10). It has been shown that an excellent agreement with the exact ground state can be achieved employing just one and two–body diagonal operators in the exponent of the ground state coefficients, which amounts within the present context consider the operators $\hat{G}_{ii}$ and $\hat{G}_{ii}\hat{G}_{jj}$, $i \geq j > 1$. Thus, the ensuing approximate ground state coefficients are

$$C_m^{(0)} = p_m \exp[-\tfrac{1}{2}(\lambda_0 + \sum_{i>1} \lambda_i m_i + \sum_{i>j>1} \lambda_{ij} m_i m_j)], \tag{25}$$

with real λ_i and λ_{ij} determined by the set of equations (6)–(7). Here is interesting to remark that by setting $\lambda_{ij} = 0$ and employing a weight factor $p_m = N!/(\prod_i m_i!)$ in (25) we recover the *projected HF* coefficients [7]. Nevertheless, results with this factor and including $\lambda_{ij} \neq 0$ are of the same quality as those obtained from our maximum quantal entropy approximation [7] which, can obviously be obtained from (25) by setting $p_m = 1$.

Now let us turn our attention to the excited states of the symmetric representation. To this end we shall consider the approximate ground state (25) as a correlated vacuum upon which we shall generate collective excitations. Following the previous section we construct first the states

$$|\psi_\gamma\rangle = [\langle \prod_i \hat{G}_{ii}^{n_i+n_i'}\rangle_0 - \langle \prod_i \hat{G}_{ii}^{m_i}\rangle_0 \langle \prod_i \hat{G}_{ii}^{n_i'}\rangle_0]|m\rangle \qquad (26)$$

where the ensuing coefficients are given by

$$C_m^\gamma = C_m^{(0)}[\prod_{i>1} m_i^{n_i} - \langle \prod_{i>1} \hat{G}_{ii}^{m_i}\rangle_0] \qquad (27)$$

with an overlap matrix

$$\langle \psi_\gamma|\psi_{\gamma'}\rangle = \langle \prod_i \hat{G}_{ii}^{n_i+n_i'}\rangle_0 - \langle \prod_i \hat{G}_{ii}^{m_i}\rangle_0 \langle \prod_i \hat{G}_{ii}^{n_i'}\rangle_0. \qquad (28)$$

In this way the complete space correspond to the (projected) symmetric representation can be expanned, in the case $0 \leq m_i \leq k_i$ with $k_i = \binom{N+2}{2}$.

In order to construct the first excited states of the symmetric representation we shall consider the reduced basis formed by the states (25), with $0 \leq m_i \leq k_i$, and small decreasing values of k_i. In the case in which $0 \leq \sum_i m_i \leq 2$, we recover the states (11), the corresponding overlap and energy matrices (15)–(16) have the dimension $\binom{n+1}{2}$.

It is also possible to diagonalize first $\hat{H}$ in the space of dimension $n-1$ generated by the states

$$|i\rangle \equiv (\hat{G}_{ii} - \langle \hat{G}_{ii}\rangle)|0\rangle, \quad i \geq 2, \qquad (29)$$

taking $\sum_i n_i = 1$ in (27), and then work with the *normal* operators (17)

$$\hat{G}_{ii}' = \sum_{j \geq 2} B_{ji} \hat{G}_{jj}, \qquad (30)$$

with the matrix B determined by eqs. (19). We can construct higher excited states as in (27) employing the operators (30) instead of $\hat{G}_{ii}$.

As we shall see a very good agreement with the exact results can be obtained employing the previous formalism within the context of the present model. The figure is illustrative of the quality of the results obtained for the $SU(n)$ model with $n = 3$ (three–level case), for $N = 20$, $\varepsilon_i = (i-1)\varepsilon$ and $V_{ij} = (1 - \delta_{ij})v/(N-1)$. The three energy differences $\Delta E_i = E_i - E_0$; $i = 1, 2, 3$, corresponding to the four lowest lying levels of the pertinent energy spectrum are depicted. The approximate results were obtained after a 11×11 diagonalization in the reduced basis formed by states with coefficients (27). We note that in this model the HF approximation predicts second order ground state shape transitions in the classical limit ($N \to \infty$) with critical values at $v_{c1}/\varepsilon = 1$ and $v_{c2}/\varepsilon = 3$, corresponding to an spherical to deformed and to a deformed to deformed transitions [7]. Notice that the behaviour of the exact solution is obviously smooth for finite values of N, the same behaviour is obtained with our formalism. As we can see from the figure, our method is able to resolve even the tiny splitting between the second and the third excitation energies. The excellence of our approach is clearly appreciated, for the whole range of the coupling constant including transitional regions.

Conclusions

The present formalism provides one with a unified accurate description of the energy eigenstates, both in the low and strong coupling regimes, *including transitional* regions.

In a previous effort we had introduced a maximum quantal entropy approach which is able to yield an accurate description of the ground state of various fermion models in terms of a reduce set of variables associated with one and two-body relevant operators *diagonal* in a given unperturbed basis. In the present work we have extended the

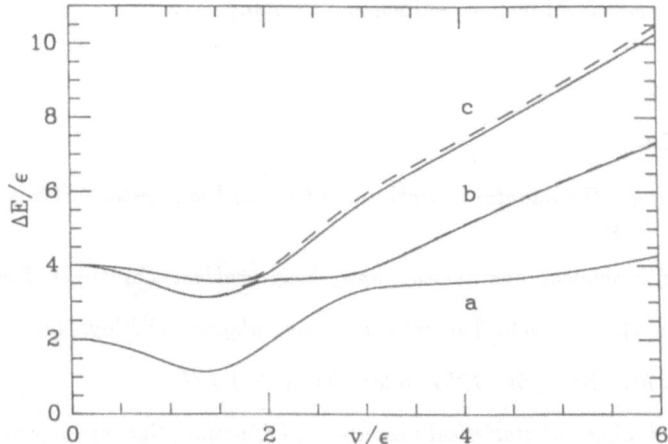

Figure 1. The first three excitation energies $\Delta E_i = E_i - E_0$, $i = 1, 2, 3$, curves (a), (b) and (c), respectively, corresponding to the four lying energy levels for $N = 20$ and $\varepsilon_i = (i-1)\varepsilon$ as a function of the coupling constant $V_{ij} = v/(N-1)$. Solid lines correspond to exact results and dashed lines to results obtained after a 11×11 diagonalization, indistinguishable in this scale.

scope of the previous general formalism in order to include the description of excited states. The approach provides a new simple scheme for generating collective excitations orthogonal to the maximum entropy ground state in terms of these diagonal operators.

In the example considered, results indicate that extremely accurate predictions of the lowest energy levels can be achieved in this way by means of the diagonalization of $\hat{H}$ in a basis of a quite *small* dimension. We can conclude that a new method for constructing general collective states on the basis of a suitably defined maximum quantal entropy correlated vacuum has been introduced.

Therefore, the present scheme may provide us with a useful way to generate collective spaces in many-body systems, in terms of a reduced set of variables. The excellent results should justify further works in more complex systems.

Acknowledgment

N.C. and A.P. acknowledge support from Consejo Nacional de Investigaciones Científicas y Técnicas de Argentina (CONICET), and R.R. acknowledges support from Comision de Investigaciones Científicas de la Provincia de Buenos Aires (CIC). One of us, R.R. also wishes to acknowledge support from CONICET for an external Fellowship. He is very indebted to Prof. Ring for his kind hospitality at the Physik Department der Technischen Universität München, München, Germany.

References

[1] D. J. Thouless, *The quantum mechanics of many body systems,* (Academic Press, New York, 1972).

[2] P. Ring and P.Schuck, *The nuclear many body problem,* (Springer, Berlin, 1980).

[3] A. de Shalit, H. Feshbach, *Theoretical nuclear physics,* (Wiley, New York, 1974).

[4] J. Jaynes, *Phys. Rev. 106 (1957) 620; 108 (1957) 171.*

[5] A. Katz, *Principles of statistical mechanics,* (Freeman, San Francisco, 1967).

[6] N. Canosa, A. Plastino and R. Rossignoli, *Phys. Rev. A 40 (1989) 519; Nucl. Phys. A 453 (1986) 457.*

[7] N. Canosa, R. Rossignoli and A. Plastino, *Nucl. Phys. A 512 (1990) 520.*

[8] N. Canosa, R. Rossignoli and A. Plastino *Phys. Rev. A 43 (1991) 1445.*

[9] N. Canosa, R. Rossignoli and A. Plastino, *Proceeding of Condensed Matter Theories XV (1992).* Ed. A. Protto and J. Aliaga (in press.)

[10] N. Canosa, A. Plastino, R. Rossignoli and H. G. Miller, *Phys. Rev. C 45 (1992) 1162.*

[11] J. Skilling, *Maximum Entropy and Bayesian Methods,* Kluwer Academic Publisher, Dordrecht, The Netherlands (1989).

[12] N. Meshkov, *Phys. Rev.C 3 (1971) 2214.*

TIME-DEPENDENT N-LEVEL SYSTEMS

J. Aliaga, J.L. Gruver and A.N. Proto

Grupo de Sistemas Dinámicos
Centro Regional Norte, UBA
C.C. 2, (1638) Vicente López, Argentina

INTRODUCTION

The Maximum Entropy Principle (MEP) formalism is applied in order to solve time-dependent N-level systems. The formalism, related to a cuasi-Lie algebra, allows us to find not only the dynamical equations of the relevant operators for any temporal dependence of the interaction between levels, but also to diagonalize the associated off-equilibrium (temperature dependent) density matrix. A set of generalized Bloch equations, in terms of the relevant operators, is given. The time-dependent two-level case is analyzed as a particular example.

MEP APPROACH TO TIME-DEPENDENT N-LEVEL SYSTEMS

The interaction of a two-level system or spin system with a time-dependent radiation field is of fundamental interest in several fields of physics like quantum optics, magnetic resonance, quantum electronics and chaotic behaviour of dynamical systems. The temporal evolution of the level population (or its difference) has been studied [1-3] for several applications, using different methods. Recently [4], S.V. Prants has used the method of Wei and Norman [5] in order to describe the interaction of a two-level, or spin system, with radiation. Departing from Bloch equations, and including relaxation parameters, the solution presented in ref.[4] have all the advantages of using Lie algebra's based methods.

The aim of this work is to present a general solution of an N-level system coupled to an arbitrary time-dependent interaction $S_{j,j+1}$ using the Maximum Entropy Principle (MEP) formalism [6]. In our approach, we depart from an N-level Hamiltonian, and Bloch equations [2,3] in a generalized version are recovered. The advantages of our approach become from three facts: a) a cuasi-Lie algebra can be defined; b) the connection with an off-equilibrium, temperature dependent density matrix can be straightforwardly done and; c) the generalized Bloch equations are reobtained in a Hamiltonian context. The MEP formalism can be summarized as follows [6]: Given the expectation values, O_j, of operators $\hat{O}_j$, the statistical operator $\hat{\rho}(t)$ is defined by:

$$\hat{\rho}(t) = \exp\left(\left(-\lambda_0\,\hat{I} - \sum_{i=1}^{L}\lambda_i\,\hat{O}_i\right)\right), \tag{1}$$

where the λ_i's, $L + 1$ of them, are Lagrange multipliers which will be determined to fulfill the set of constraints

$$< \hat{O}_j > = \text{Tr} \ [\hat{\rho}(t) \ \hat{O}_j] \ , j = 0,1,2,...,L \ , \tag{2}$$

($\hat{O}_0 = \hat{I} =$ identity operator) and the normalization condition in order to maximize the entropy, defined (in units of Boltzmann constant) by

$$S(\hat{\rho}) = - \text{Tr} \ (\hat{\rho} \ \ln \hat{\rho}) \ . \tag{3}$$

The time evolution of the statistical operator is given by

$$i \ \hbar \ \frac{d \ \hat{\rho}}{dt} = [\hat{H}(t), \hat{\rho}(t)] \ . \tag{4}$$

One should endeavour to find those (relevant) operators entering eq.(1) so as to guarantee not only that S is maximum, but also a constant of the motion. Introducing the natural logarithm of eq.(1) into eq.(4) it can be easily verified that the relevant operators are those that close a cuasi-Lie algebra under commutation with the Hamiltonian $\hat{H}$,

$$[\hat{H}, \hat{O}_j] = i\hbar \sum_{i=0}^{L} g_{ij} \hat{O}_i \ . \tag{5}$$

Equation (5) defines a G matrix, (which may depend upon the time if $\hat{H}$ is time dependent), and it constitutes the central requirement to be fulfilled by the operators entering in the density matrix. The Liouville equation (4) can be substituted by a set of coupled equations for the Lagrange multipliers λ_i, as follows:

$$\frac{d\lambda_i}{dt} = \sum_{j=0}^{L} \lambda_j \ g_{ij} \ , \ i = 0,....,L \ , \tag{6}$$

or alternatively [6],

$$\frac{d < \hat{O}_i >_t}{dt} = - \sum_{j=0}^{L} g_{ji} < \hat{O}_j >_t \ , \ i = 0,....,L \ . \tag{7}$$

The mean value of the operators and the Lagrange multipliers are related by

$$\frac{\partial \lambda_0}{\partial \lambda_i} = - < \hat{O}_i > \ . \tag{8}$$

The initial conditions (IC) play a crucial role in the dynamical behaviour of the system. As it was shown by A.N. Proto et al [7], even when the G matrix determines the character of the solution (dissipative or not, at least in the case of time independent Hamiltonians), *the initial conditions impose additional restrictions not contained explicitly in the dynamics of the problem*. For more details and applications we refer the reader to refs. [8-12].

In order to make a general treatment of the problem at hand, we begin with a Hamiltonian of the form:

$$\hat{H}(t) = \sum_{j=1}^{N} E_j \hat{a}_j^+ \hat{a}_j + \sum_{j=1}^{N-1} S_{j,j+1}(t)(\gamma_{j,j+1} \hat{a}_j \hat{a}_{j+1}^+ + \gamma_{j,j+1}^* \hat{a}_{j+1} \hat{a}_j^+), \tag{9}$$

$(\hbar = 1)$ where $\gamma_{j,j+1}$ is the interaction energy between the system and the external field, $\hat{a}_i^+$ $(\hat{a}_i)$ (boson operators) represents the creation (annihilation) of a particle in state i, $S_{j,j+1}(t)$ is any function of time and N is the number of levels. It can be easily demonstrated that the exact temporal evolution of the relevant operators for the Hamiltonian (9) depends only on the strength of $\gamma_{j,j+1}$ and the temporal behaviour of $S_{j,j+1}(t)$. As has been said, the MEP provides a formalism which enables us to construct a set of N^2 operators satisfying eq.(5) for arbitrary values of $S_{j,j+1}(t)$ and $\gamma_{j,j+1}$. The set of N^2 operators can be divided into three groups depending on their physical meaning. We call $\hat{N}_k$ the population of the k level

$$\hat{N}_k = \hat{a}_k^+ \hat{a}_k, \quad 1 \le k \le N. \tag{10}$$

$\hat{I}_{s,N-k}$ is the operator whose expectation value represents the interaction energy between the levels s and N-k and the external field,

$$\hat{I}_{s,N-k} = \left(\prod_{j=s}^{N-k-1} \gamma_{j;j+1} \right) \hat{a}_s \hat{a}_{N-k}^+ + \left(\prod_{j=s}^{N-k-1} \gamma_{j;j+1}^* \right) \hat{a}_{N-k} \hat{a}_s^+, \quad \begin{matrix} 0 \le k \le N-s-1 \\ 1 \le s \le N-1 \end{matrix} \tag{11}$$

and the expectation value of $\hat{F}_{s,N-k}$ gives the particle's current between the levels s and N-k,

$$\hat{F}_{s,N-k} = i\left[\left(\prod_{j=s}^{N-k-1} \gamma_{j;j+1} \right) \hat{a}_s \hat{a}_{N-k}^+ - \left(\prod_{j=s}^{N-k-1} \gamma_{j;j+1}^* \right) \hat{a}_{N-k} \hat{a}_s^+ \right], \quad \begin{matrix} 0 \le k \le N-s-1 \\ 1 \le s \le N-1 \end{matrix} \tag{12}$$

It can be proved that, for any time-dependent function $S_{j,j+1}$ connecting only adjacent levels, $\hat{F}_{s,N-k}$ and $\hat{I}_{s,N-k}$ (given by eqs. (11) and (12)) will be the relevant operators. The dynamical equations of these relevant operators (eqs. (10-12)) are given by eq. (7). These N^2 equations constitutes the *generalization* of the well-known Bloch equations [2,3] for the N=2 case. For the fermionic case, the same sets of relevant operators and dynamical equations are obtained. The only difference between this case and the bosonic one is that, in the former, the total population is limited by the number of levels. Up to here, we have briefly summarized the MEP's N-level system solution.

TWO-LEVEL SYSTEM

In order to compare our results with the previous ones, we particularize them for the well-known case of a two-level system. Enacting eq. (5) we arrive to the following set of relevant operators

$$\hat{O}_1 = \hat{a}_1^+ \hat{a}_1, \tag{13a}$$

$$\hat{O}_2 = \hat{a}_2^+ \hat{a}_2, \tag{13b}$$

$$\hat{O}_3 = i(\gamma_{12} \hat{a}_1 \hat{a}_2^+ - \gamma_{12}^* \hat{a}_2 \hat{a}_1^+), \tag{13c}$$

$$\hat{O}_4 = \gamma_{12}\,\hat{a}_1\hat{a}_2^+ + \gamma_{12}^*\,\hat{a}_2\hat{a}_1^+ \ . \tag{13d}$$

In Heisemberg representation we obtain

$$\frac{d\hat{O}_1}{dt} = i\,[\hat{H},\hat{O}_1] = S_{12}(t)\,\hat{O}_3 \ , \tag{14a}$$

$$\frac{d\hat{O}_2}{dt} = i\,[\hat{H},\hat{O}_2] = -S_{12}(t)\,\hat{O}_3 \ , \tag{14b}$$

$$\frac{d\hat{O}_3}{dt} = i\,[\hat{H},\hat{O}_3] = -2\,S_{12}(t)|\gamma_{12}|^2\,(\hat{O}_1 - \hat{O}_2) - (E_2 - E_1)\,\hat{O}_4, \tag{14c}$$

$$\frac{d\hat{O}_4}{dt} = i\,[\hat{H},\hat{O}_4] = (E_2 - E_1)\,\hat{O}_3 \ . \tag{14d}$$

Equivalently, the above equations can be rewritten using the generalized Ehrenfest theorem. This leads to the well-known Bloch equations in terms of the levels population, particle's current and interaction energy [2,3]

$$\frac{d<\hat{O}_1>}{dt} = S_{12}(t)<\hat{O}_3> \ , \tag{15a}$$

$$\frac{d<\hat{O}_2>}{dt} = -S_{12}(t)<\hat{O}_3> \ , \tag{15b}$$

$$\frac{d<\hat{O}_3>}{dt} = -2|\gamma_{12}|^2 S_{12}(t)(<\hat{O}_1> - <\hat{O}_2>) - (E_2 - E_1)<\hat{O}_4>, \tag{15c}$$

$$\frac{d<\hat{O}_4>}{dt} = (E_2 - E_1)<\hat{O}_3> \ . \tag{15d}$$

We consider two particular cases: a) $S_{12}(t) =$ cte, b) $S_{12}(t) =$ cte $\cos(\omega_1 t)\cos(\omega'_1 t)$. The case a) is shown in Fig. 1. We depict $<\hat{O}_1>$ (a), $<\hat{O}_2>$ (b), $<\hat{O}_3>/\omega$ (c), $<\hat{O}_4>/\hbar\omega$ (d) versus ωt for two different initial conditions: 1) $<\hat{O}_1>_0 = 2$, $<\hat{O}_2>_0 = 1$, $<\hat{O}_3>_0/\omega = <\hat{O}_4>_0/\hbar\omega = 0$ (continuous line), 2) $<\hat{O}_1>_0 = <\hat{O}_3>_0/\omega = <\hat{O}_4>_0/\hbar\omega = 2$, $<\hat{O}_2>_0 = 1$ (dashed line) with $\omega_2 - \omega_1 = 0$, $\omega = 1$ and $\gamma_{12} = S(t) = 1$. In Figure 2 we show the case b). In this Figure we depict the difference in the population level, A(t), versus ωt for $<\hat{O}_1>_0 = 1$, $<\hat{O}_2>_0 = 0$, $<\hat{O}_3>_0/\omega = <\hat{O}_4>_0/\hbar\omega = 0$ with $\omega_2 - \omega_1 = 0$, $\omega = 1$, $\gamma_{12} = 1$ and $S(t) = 5\cos(17711/28657\ t)\cos(4637/13313\ t)$. We consider these values because they were previously considered in Ref. 3. Thus, it can be seen that our results are in good agreement with those given in that Reference. If instead of N=2, we consider an N-level system, the above equations (obtained from eq. (7)) result in a set of *generalized Bloch equations*.

Alternatively, and in order to derive a thermodynamical approach to the problem at hand, we will write the density matrix including the Hamiltonian as a relevant operator. Then, the statistical operator can be written as

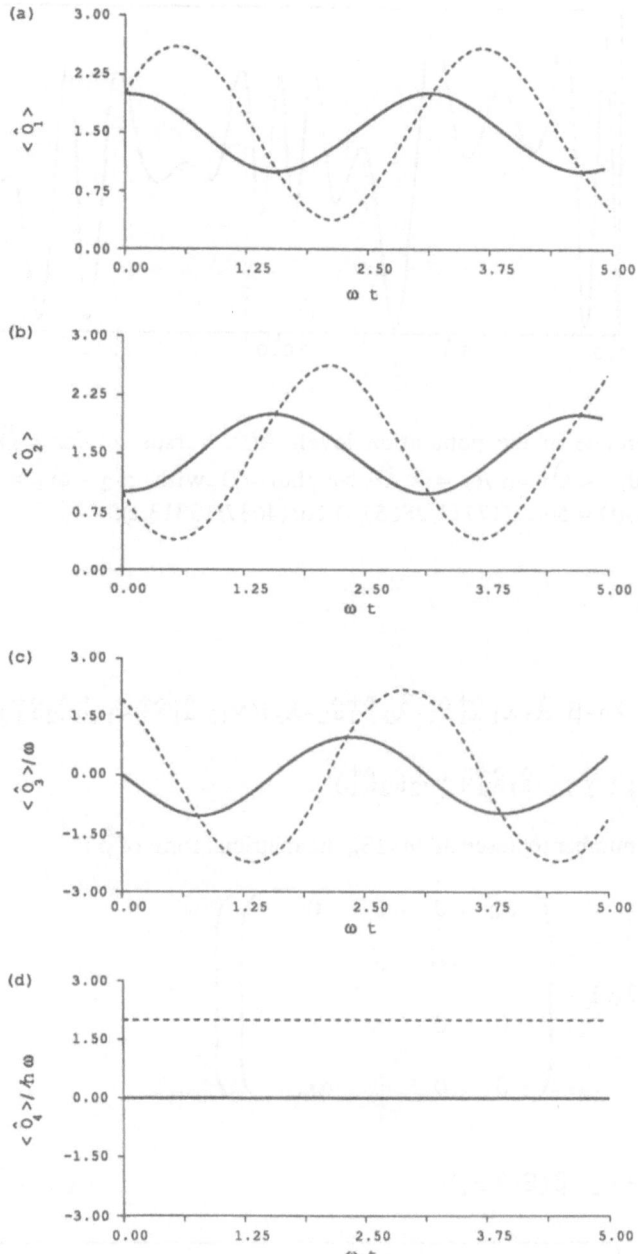

Fig. 1 $< \hat{O}_1 >$ (a), $< \hat{O}_2 >$ (b), $< \hat{O}_3 >/\omega$ (c), $< \hat{O}_4 >/\hbar\omega$ (d) versus ωt for two different initial conditions: 1) $< \hat{O}_1 >_0 = 2$, $< \hat{O}_2 >_0 = 1$, $< \hat{O}_3 >_0 /\omega = < \hat{O}_4 >_0 /\hbar\omega = 0$ (continuous line), 2) $< \hat{O}_1 >_0 = < \hat{O}_3 >_0 /\omega = < \hat{O}_4 >_0 /\hbar\omega = 2$, $< \hat{O}_2 >_0 = 1$ (dashed line) with $\omega_2 - \omega_1 = 0$, $\omega = 1$ and $\gamma_{12} = S(t) = 1$.

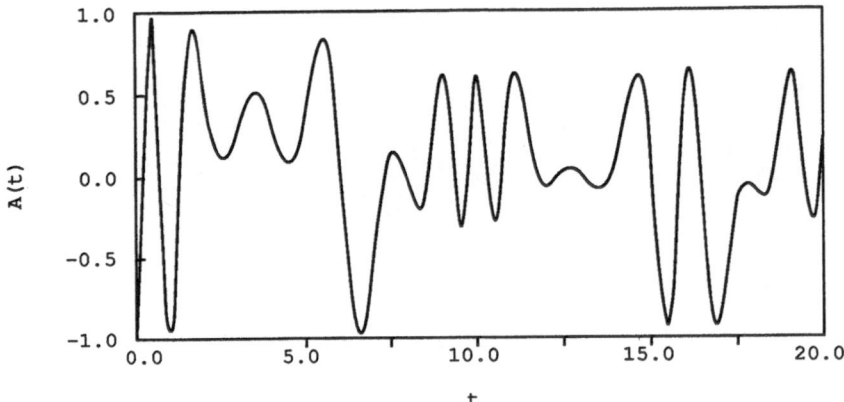

Fig. 2 Difference in the population level, A(t), versus ωt for $< \hat{O}_1 >_0 = 1$, $< \hat{O}_2 >_0 = 0$, $< \hat{O}_3 >_0 /\omega = < \hat{O}_4 >_0 /\hbar\omega = 0$ with $\omega_2 - \omega_1 = 0$, $\omega = 1$, $\gamma_{12} = 1$ and $S(t) = 5 \cos(17711/28657 \, t) \cos(4637/13313 \, t)$.

$$\hat{\rho}(t) = \exp[-\lambda_0 - \beta \hat{H} - \lambda_1 \hat{a}_1^+ \hat{a}_1 - \lambda_2 \hat{a}_2^+ \hat{a}_2 - \lambda_3 i(\gamma_{12} \hat{a}_1 \hat{a}_2^+ - \gamma_{12}^* \hat{a}_2 \hat{a}_1^+) -$$

$$- \lambda_4 (\gamma_{12} \hat{a}_1 \hat{a}_2^+ + \gamma_{12}^* \hat{a}_2 \hat{a}_1^+)] . \tag{16}$$

In the occupation number representation [13], the matricial form of $\hat{\rho}$ is

$$\hat{\rho} = \exp\left[-\lambda_0 \hat{I} - \begin{pmatrix} \alpha_0 & 0 & 0 & 0 \\ 0 & \alpha_1 & 0 & 0 \\ 0 & 0 & \ddots & \vdots \\ 0 & 0 & \vdots & \alpha_M \end{pmatrix} \right], \tag{17a}$$

where

$$\alpha_J = \frac{M}{2} [\lambda_1 + \lambda_2 + \beta (E_1 + E_2)] -$$

$$- \frac{M - 2J}{2} \sqrt{[\lambda_2 - \lambda_1 + \beta (E_2 - E_1)]^2 + 4|\gamma_{12}|^2 [\lambda_3^2 + (\lambda_4 + \beta S_{12}(t))^2]} \tag{17b}$$

$0 \leq J \leq M$, $< \hat{O}_1 > + < \hat{O}_2 > = M$ and λ_i are the Lagrange multipliers, which can take any value between $-\infty$ and ∞. Using eqs. (8) and (17), we can evaluate the expectation values of the relevant operators as a function of the Lagrange multipliers. We obtain

$$< \hat{O}_1 >_t = \frac{M}{2} + \cos\theta \, T(r) , \tag{18a}$$

$$<\hat{O}_2>_t = \frac{M}{2} - \cos\theta\, T(r) ,$$ (18b)

$$<\hat{O}_3>_t = -2|\gamma_{12}|\sin\theta\cos\phi\, T(r) ,$$ (18c)

$$<\hat{O}_4>_t = -2|\gamma_{12}|\sin\theta\sin\phi\, T(r) ,$$ (18d)

$$<\hat{H}>_t = \frac{M}{2}(E_2+E_1) - [(E_2-E_1)\cos\theta + 2|\gamma_{12}|\sin\theta\,\sin\phi]\,T(r) ,$$ (18e)

$$T(r) = \frac{\displaystyle\sum_{J=0}^{M} \frac{M-2J}{2} e^{-Jr}}{\displaystyle\sum_{k=0}^{M} e^{-kr}} ,$$ (18f)

where the following set of coordinates has been introduced

$$r\cos\theta = \lambda_2 - \lambda_1 + \beta\,(E_2 - E_1) ,$$ (19a)
$$r\sin\theta\cos\phi = 2|\gamma_{12}|\lambda_3 ,$$ (19b)
$$r\sin\theta\sin\phi = 2|\gamma_{12}|[\lambda_4 + \beta\,S_{12}(t)] .$$ (19c)

From eqs. (18), it is easy to see that the mean values are described by the Lagrange multipliers λ_1, λ_2, λ_3 and λ_4. It is important to notice that it is much easier to work in the space of Lagrange multipliers that in its dual space, i.e., the space of operators. Nevertheless, in order to obtain the explicit functional time dependence of eqs. (18), we need to know the particular choice of $S_{12}(t)$ and resolve eqs. (6). For some time-dependent couplings, $S_{12}(t)$, this set of differential equations, with the Lagrange multipliers as variables, can be analytically solved [4]. However, for every time-dependent interaction, a numerical solution of eqs. (6) can be obtained (see e.g. ref. [3]). Notice that, when S_{12} is time-independent, r is a constant of motion.

As can be seen from eqs.(19), r, θ and ϕ can take any value between $(-\infty,+\infty)$, $(0,\pi)$, $(0,2\pi)$ respectively. Using eqs. (19) it is easy to demonstrate that the expectation values of $\hat{O}_i$ are limited as follows:

$$0 \leq <\hat{O}_1>_t \leq M$$ (20a)

$$0 \leq <\hat{O}_2>_t \leq M$$ (20b)

$$-|\gamma_{12}|M \leq <\hat{O}_3>_t \leq |\gamma_{12}|M$$ (20c)

$$-|\gamma_{12}|M \leq <\hat{O}_4>_t \leq |\gamma_{12}|M$$ (20d)

$$ME_1 \leq <\hat{H}>_t \leq ME_2 + M|\gamma_{12}|$$ (20e)

Equations (20) give the set of compatible initial conditions (according with the dynamics of the system) for all the relevant operators and the energy of the system, in terms of M, E_1, E_2 and γ_{12}. These parameters can be freely chosen.

The method presented here, not only reproduces the original Bloch equations, but also allows us to give a thermodynamical treatment of a two-level system coupled to an external field by a general time-dependent function $S_{12}(t)$. It is important to notice that our solution for the two-level system is closely related to that developed by S.V. Prants [4], although the MEP solution is quite more general as it was developed for N-level systems. The same procedure briefly outlined here for $N=2$ can be withdraw for any other number of levels, in spite of the fact that the density matrix diagonalization's procedure becomes longer. Thus, our approach generalizes the results presented in ref. [4], allowing for a straightforward connection with the thermodynamical properties of the quantum system (i.e. a non-zero temperature method), via the density matrix formalism (eq. (16)).

CONCLUSIONS

Summarizing, we can conclude that our formalism presents the following advantages: a) it is a general formalism which works out for N-level Hamiltonians, b) it allows us to introduce the temperature of the quantal system, which provides a quantum thermodynamical approach to the problem, c) eq. (5) determines all the relevant operators needed in order to analyze the dynamical behaviour of the system, being these operators related by a cuasi-Lie algebra, d) The initial conditions of the relevant operators appears to be limited as a consequence of the quantum-statistical nature of the problem at hand (see eqs. (20)). As the initial conditions are related by eqs. (18), which are the consequence of eq. (8), it should be taken in mind that they can not be freely chosen. The crucial point in order to get a deeper insight into this class of systems is the *diagonalization of the density matrix*. This procedure allows to work in the dual space of Lagrange multipliers, turning the original non-commutative operator's structure into geometrical relationships (see eqs. (19)) [10-12].

ACKNOWLEDGMENTS

J. A. thanks the Argentine National Research Council (CONICET) for its support. A.N. P.(Member of the Research Career of CIC) acknowledges support from the Comisión de Investigaciones Científicas de la Provincia de Buenos Aires (CIC).

REFERENCES

[1]   H.A. Cerdeira and E.Z. Da Silva, Phys. Rev. **A 30** (1984) 1752.
[2]   P.W. Milonni, M.L. Shih and J. R. Ackerhalt, Chaos in Laser-Matter Interactions (World Scientific, Singapore, 1987).
[3]   Y. Pomeau, B. Dorizzi and B. Grammaticos, Phys. Rev. Lett. **56** (1982) 1636.
[4]   S.V. Prants, Phys. Lett. **A 144** (1990) 25.
[5]   J. Wei and E. Norman, Proc. Am. Math. Soc. 15 (1963) 327.
[6]   Y. Alhasid and R.D. Levine, J. Chem. Phys. **67** (1977) 4321; Phys. Rev. **A18** (1978) 89; A.N. Proto, Maximum Entropy Principle and Quantum Mechanics, Proc. International Workshop on Condensed Matter Theories, Valdir Aguilera-Navarro Ed. (Plenum Press, 1989) and references therein.
[7]   A.N. Proto, J. Aliaga, D. R. Napoli, D. Otero, and A. Plastino, Phys. Rev. **A39**, 4223 (1989);
       G. Crespo, D. Otero, A. Plastino, and A. N. Proto, Phys. Rev. **A 39**, 2133 (1989).
[8]   J. Aliaga, D. Otero, A. Plastino and A.N. Proto, Phys. Rev. **A37**, 918 (1988).
[9]   J. Aliaga, H. Cerdeira, A.N. Proto and D. Otero, Phys. Rev. **B40** (1989) 4375.
[10]   J. Aliaga and A.N. Proto, Phys. Lett. **A142** (1989) 63.
[11]   J. Aliaga, G. Crespo and A.N. Proto, Phys. Rev. **A42** 618 (1990).
[12]   J. Aliaga, G. Crespo and A.N. Proto, Phys. Rev. **A42** 4325 (1990).
[13]   C. Cohen - Tannoudji, B. Diu, F. Laloë, Quantum Mechanics, (John Wiley & Sons, Paris, 1977).

ATOMS IN MOLECULES*

Nikos Georgopoulos[†] and J. K. Percus[‡]

Physics Department
New York University
New York, NY 10003

ABSTRACT

We focus on the properties of a common species presented in weakly connected spatially distinct components. The classical two-phase interface serves as introduction. For the electron fluid of a system of atoms in a molecule, we introduce replica systems and then carry out a unitary transformation of the joint annihilation-creation fields to approximate a disjoint set of atoms. This approximation is coupled with variation of the unitary transformation to treat two primitive molecules. Possible correction procedures and extensions are discussed.

1. INTRODUCTION

Descriptions of physical systems are most often cast rewardingly in molecular form. Even a common species may pose as a set of weakly interconnected spatially distinct components; we then try to use firm knowledge of the *components to build up firm knowledge of the system.*

This decomposition may be largely one of convenience: two phases of the same

substance with a common interface have sufficiently different bulk properties that it may very well pay to regard them as distinct but interconvertible substances. Or the components may have essential structural distinctions, such as the electron fluid gathered into atoms in a molecule and here it seems folly to ignore the modular construction. But saying this and doing it are hardly equivalent. Moffitt,[1] forty years ago, worked in a basis of anti-symmetrized atomic state products to assemble a

---

* Supported in part by NSF grants.
† Current address: 6 Hadjimihali St., Neos Cosmos, Athens, Greece.
‡ and Courant Institute of Mathematical Sciences.

relatively small state space in which to carry out variational estimates for molecules. Girardeau,[2] starting thirty years ago, represented the electrons of an atom as a higher space entity with which to build up a molecule whose atoms were elementary particles. Practical considerations did not allow either of these ingenious approaches to go very far.

More recently, Bader[3] and collaborators have reexamined the conceptual basis of the division of molecules into atoms, and shown that one can define an electron charge distribution attached to an atom in such a fashion that it maintains its form under a variety of molecular environments. And the Harris[4] approximation that atomic charge densities simply overlap and add has been effective in a density functional viewpoint. Our intention in this paper is to examine a reformulation, intersecting those of Moffitt and Harris, which brings out salient features of the atoms-in-molecules picture at an early stage of the analysis, with fair quantitative success. Higher accuracy does not of course come cheaply in this or any other fashion.

2. CLASSICAL DECOMPOSITION

A single species classical fluid in thermal equilibrium at reciprocal temperature β and chemical potential μ is represented by the grand partition function

$$\Xi[\mu] = \sum_N \frac{1}{N!} \int \cdots \int \exp\left\{\beta \sum_1^N \mu(i)\right\} \exp\left\{-\frac{1}{2}\beta \sum_1^N {}'\phi(i,j)\right\} d(1\cdots N) \quad (2.1)$$

where $\mu(i) = \mu - u(i)$, u being the external potential, and $\phi(i,j)$ the pair interaction, $\sum'$ signifying that $i \neq j$. Suppose that the fluid has two qualitatively distinguishable forms, the first occurring in space with a weight $w_1(r) = e^{-\beta u_1(r)}$, the second with weight $w_2(r) = e^{-\beta u_2(r)}$. w_1 and w_2 can overlap in space, but must of course satisfy

$$w_1(r) + w_2(r) = 1 \quad (2.2)$$

Eq. (2.1) can then be rewritten as

$$\Xi[\mu] = \sum_N \frac{1}{N!} \int \cdots \int \prod_1^N \left(\exp\left\{-\beta u_1(i)\right\} + \exp\left\{-\beta u_2(i)\right\}\right)$$

$$\exp\left\{\beta \sum_1^N \mu(i)\right\} \exp\left\{-\frac{1}{2}\beta \sum_1^N {}'\phi(i,j)\right\} d(1\cdots N)$$

$$= \sum_N \frac{1}{N!} \int \cdots \int \sum_{\Lambda \subset (1,\dots,N)} \exp\left\{-\beta \sum_{i\in\Lambda} u_1(i)\right\} \exp\left\{-\beta \sum_{i\in\overline{\Lambda}} u_2(i)\right\}$$

$$\exp\left\{\beta \sum_1^N \mu(i)\right\} \exp\left\{-\frac{1}{2}\beta \sum_1^N {}'\phi(i,j)\right\} d(1\cdots N) \quad (2.3)$$

$$= \sum_N \frac{1}{N!} \binom{N}{S} \int \cdots \int \exp\left\{-\beta \sum_1^S u_1(i)\right\} \exp\left\{-\beta \sum_{S+1}^N u_2(i)\right\}$$

$$\exp\left\{\beta \sum_1^N \mu(i)\right\} \exp\left\{-\frac{1}{2}\beta \sum_1^S {}'\phi(i,j)\right\} \exp\left\{-\frac{1}{2}\beta \sum_{S+1}^N {}'\phi(i,j)\right\}$$

$$\exp\left\{-\beta \sum_1^S \sum_{S+1}^N \phi(i,j)\right\} d(1\cdots S)\, d(S+1\cdots N)$$

or finally as[5]

$$
\Xi[\mu] = \sum_{s,t} \frac{1}{s!t!} \int \cdots \int \exp\left\{ \beta \sum_1^s \mu_1(i) \right\} \exp\left\{ \beta \sum_1^t \mu_2(i') \right\}
$$

$$
\exp\left\{ -\frac{1}{2}\beta \sum_1^s {}'\phi(i,j) \right\} \tag{2.4}
$$

$$
\exp\left\{ -\frac{1}{2}\beta \sum_1^t {}'\phi(i',j') \right\} \exp\left\{ \beta \sum_{1,1}^{s\,t} \phi(i,j') \right\} d(1\cdots s)\, d(1'\cdots t') \ .
$$

In other words, the system has now been reinterpreted as a mixture of two fluids, under respective external fields $u(r) + u_1(r)$ and $u(r) + u_2(r)$, and with fluid-fluid interaction the same as between particles of a given fluid. Computing the densities as usual by $n_\alpha(r) = \frac{1}{\beta}(\delta\Xi/\delta\mu_\alpha(r))/\Xi$, it is clear that there is a hidden "sum rule"

$$
n_\alpha(r) = \exp\left\{ -\beta u_\alpha(r) \right\} n(r) \ , \tag{2.5}
$$

and similarly with higher distributions, but this is automatic, and only the fact that

$$
n(r) = n_1(r) + n_2(r) \tag{2.6}
$$

need ever be used. What is the virtue of (2.4)? Suppose for example that 1 and 2 represent liquid and gas phases. These will in practice yield to different suitable approximation schemes, and in (2.4) these can be used independently. As a very extreme instance, imagine the liquid and gas as initially being uncoupled, with intrinsic free energy functionals $\overline{F}_1[n_1]$, $\overline{F}_2[n_2]$, and then turn on the mutual Boltzman factor $\exp\{-\beta\phi(1,2)\}$. To first order, this yields the perturbation

$$
\Delta\overline{F}[n_1, n_2] = -\frac{1}{\beta} \int \int \left(\exp\left\{ -\beta\phi(1,2) \right\} - 1 \right) n_{1\,2}(r_1, r_2)\, dr_1\, dr_2 \ . \tag{2.7}
$$

Treating both free energies and $n_{1\,2}$ in mean field, then

$$
\overline{F}[n_1, n_2] = \int f_1(n_1(r))\, dr + \int f_2(n_2(r))\, dr
$$
$$
+ \int \int w(r_1, r_2)\, n_1(r_1)\, n_2(r_2)\, dr_1\, dr_2 \tag{2.8}
$$

which can be analyzed by standard techniques.

3. QUANTUM PROTOTYPE

The basic question that we address is how the reaction of electrons to several sources is related to the reactions to the separate sources. The most elementary situation in which this question is meaningful is that of a single electron, in one dimension, in the presence of two δ-function sources. Adopting the standard length scale in which $\hbar^2/2m = 1$, the ground state wave function thus satisfies

$$
-\psi''(x) - A(\delta(x+a) + \delta(x-a))\,\psi(x) = -k^2 \cdot \psi(x) \ , \tag{3.1}
$$

Figure 1.

i.e. the electron is free except for the continuity conditions

$$\psi'(x)\big|_{a+}^{a-} = A\psi(a), \qquad \psi'(x)\big|_{-a+}^{-a-} = A\psi(-a) \tag{3.2}$$

The ground state is even, and one finds at once

$$\psi(x) = \begin{cases} \alpha\cosh kx & |x| < a \\ \beta\, e^{-k|x|} & |x| > a \end{cases}, \tag{3.3}$$

where

$$\alpha^2 = 1/(a + \frac{1}{2k}(e^{2ka} - 1)), \quad \beta^2 = (e^{2ka} - 1)^2\alpha^2 \text{ and}$$

$$k = \frac{1}{2}A(1 - e^{-2ka}) \tag{3.4}$$

Our major interest lies in the "one-electron bond", i.e. the electron density at the midway point, which is found to be

$$\rho(0) = A/(Aa + e^{2ka}). \tag{3.5}$$

Now let us look at a single δ-function at $-a$. The ground state is trivial to solve, and we have (L means left)

$$\psi_L(x) = (A/2)^{1/2} \exp\left\{ -\frac{1}{2}A|x + a| \right\}$$

$$k = \frac{1}{2}A \tag{3.6}$$

$$\rho_L(0) = \frac{1}{2}A \exp\left\{ -Aa \right\}.$$

What is the relation between (3.4, 3.5) and (3.6)? With two identical strength sources, the ground state wave function will also satisfy $\psi'(0) = 0$ with a single source at $-a$. Hence if Aa is large enough that the reflected wave — identical with $\psi_A(x)$ continued from $x > 0$ — has decayed to near 0 by the time it gets to $-a$, we will have except for normalization

$$Aa \gg 1: \psi(x) = \psi_L(x) + \psi_R(x), \tag{3.7}$$

i.e. the LCAO wave function becomes exact.

However, we are concerned with chemical binding, in which the isolated atomic wave functions overlap the neighboring sources substantially. In the present context, many reflections would be required, so that a two source LCAO would not be great. And anyway, the two sources do not enter separately into computations of either energy or "bond" density. A possibility that at least satisfies the last criterion is simply that

$$\rho(0) = \rho_L(0) + 0 = 0 + \rho_R(0), \tag{3.8}$$

i.e. the sum of one electron bound by one source and none by the other. Numerically, • for ρ, o for ρ_L, this relationship is reasonably well verified, with some degree of constructive interference consistently underestimated. It is certainly a decent simple zeroth approximation.

Figure 2.

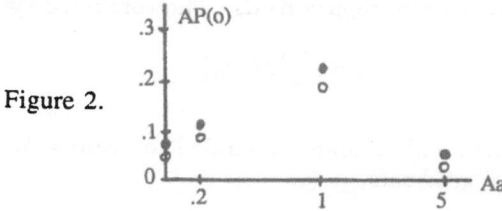

4. BASIC TRANSFORMATION

We have seen that in a classical grand ensemble, it is possible to separate particle density into nominally independent but interacting components, and that such a separation may lead to a reasonable starting approxiation in the multisource quantum domain. For Fermions in second quantized representation, this suggests that the annihilation field $\psi(r)$ — r denoting both space and spin degrees of freedom — be similarly decomposed into fields associated with N different atomic domains:

$$\psi(r) = \sum_1^N U_\beta \phi_\beta(r) \,. \tag{4.1}$$

Here U_α is an operator, but since the separation is clearest (and the computations simplest) in a spatial context. we will from the outset in this presentation assume U_α to be diagonal in space: $U_\beta(r)$. The condition that the Fermion fields ϕ_α produce in this way a valid Fermion field ψ: $[\psi(r), \psi(r')]_+ = [\psi^+(r), \psi^+(r')]_+ = 0$, $[\psi(r), \psi^+(r')]_+ = \delta(r - r')$, is seen to require only

$$\sum_1^N U_\beta^*(r) U_\beta(r) = 1 \,. \tag{4.2}$$

It will be extremely convenient[6] to extend $\psi(r)$ to a set of N independent Fermion fields, the other $N - 1$ auxiliary or "ghost" fields satisfying their own independent quantum mechanics. Then the Fermion to Fermion transformation

$$\psi_\alpha(r) = \sum_1^N U_{\alpha\beta}(r)\, \phi_\beta(r) \tag{4.3}$$

requires that the matrix $U(r)$ be unitary:

$$U^+(r)\, U(r) = I \,. \tag{4.4}$$

Now of course $\psi_1(r) = \psi(r)$ will contribute to the standard (non-relativistic, spin-independent $\cdots$) pair-interacting many-Fermion Hamiltonian

$$H[\psi] = \int (\hbar^2/2m)\, \nabla\psi^+(r) \cdot \nabla\psi(r)\, dr$$

$$+ \int \psi^+(r)\, \psi(r)\, v(r)\, dr, \tag{4.5}$$

$$+ \frac{1}{2} \int \int \psi^+(r) \psi^+(r') \psi(r') \psi(r) v(r,r') \, dr \, dr' \; .$$

Our assumption in this study will be that each ψ_α is controlled by the same Hamiltonian, so that we have a set of replica fields. The combined system then has the total Hamiltonian

$$H = \sum_\alpha H[\psi_\alpha] \; , \qquad (4.6)$$

or in condensed notation, with ψ and ϕ extended to vectors, but the dot product referring only to spatial gradients,

$$H[\phi] = (\hbar^2/2m) \int \nabla(\phi^+ U^+) \cdot \nabla(U\phi) + \int \phi^+ \phi v$$
$$+ \frac{1}{2} \int \sum_\alpha (\phi^+ U^+)_\alpha(r)(\phi^+ U^+)_\alpha(r')(U\phi)_\alpha(r')(U\phi)_\alpha(r) v(r,r') \; . \qquad (4.7)$$

What does (4.7) mean? We observe of course that

$$\rho(r) = \sum \langle \psi_\alpha^+(r) \psi_\alpha(r) \rangle$$
$$= \sum \langle \phi_\alpha^+(r) \phi_\alpha(r) \rangle \equiv \sum \rho_\alpha(r) \; . \qquad (4.8)$$

Now suppose that we arrange matters such that $\psi_2, \ldots, \psi_N$ in fact act on 0-particle states in the system state under study, i.e. are truly "ghost fields". Then we have

$$\langle \psi_\alpha^+(r) \psi_\alpha(r) \rangle = 0 \; , \qquad \alpha = 2, \ldots, N \; , \qquad (4.9)$$

and $\rho(r) = \langle \psi_1^+(r) \psi_1(r) \rangle$ will indeed have been decomposed into the sum of transformed system densities. In zeroth approximation we will write

$$H[\phi] \simeq \sum H_\alpha[\phi_\alpha] \; , \qquad (4.10)$$

so the ϕ_α will be uncoupled, giving us precisely a sum of atomic densities, although the uncoupled wave function inserted into $\langle H[\phi] \rangle$ for a variational estimate will include additional energy contributions in mean field form as well. The terms of the form $\phi_\alpha^+ \phi_\beta$ in (4.7) are precisely those that create a spatial region in which an atom β-type electron is transmuted to an α-type electron, thus maintaining the indistinguishability of electrons.

A few words on the nature of the unitary U, which in practice will be chosen for an energy minimum in any selected approximation scheme. The transformation $U(r) \to U(r) e^{i\delta(r)}$ for scalar δ affects only the kinetic energy in (4.7). Then

$$\nabla(\phi^+ U^+) \cdot \nabla(U\phi)$$
$$\to \nabla(\phi^+ U^+) \cdot \nabla(U\phi) + 2j(\phi,\phi) \cdot \nabla\delta + \phi^+ \phi |\nabla\delta|^2 \; , \qquad (4.11)$$
$$\text{where} \quad j(\phi,\psi) \equiv (\phi^+ \nabla\psi - \nabla\phi^+ \psi)/2i \; .$$

Since $\langle j(\phi_\alpha, \phi_\alpha) \rangle = 0$ for any state we use, (4.11) will have minimum expectation when $\nabla\delta = 0$, i.e. δ is a constant phase factor, which can be removed. Since we can always make $\det U(r) = 1$ by such a phase factor $\delta(r)$, we conclude that we may restrict $U(r)$ to be a member of the special unimodular unitary group, of determinant 1, and will do so.

5. KINETIC ENERGY MINIMUM — DIATOMIC CASE

If we have chosen a model polyatomic wave function, the model parameters and the functions $U_{\alpha\beta}(r)$ are to be determined by a Rayleigh-Ritz minimization of $\langle H[\phi]\rangle$. Let us at this stage inquire only as to how the kinetic energy in (4.7) is to be minimized. Since

$$
\begin{aligned}
K &\equiv \langle \nabla(\phi^+ U^+) \cdot \nabla(U\phi)\rangle \\
&= \langle \nabla\phi^+ \cdot \nabla\phi\rangle + \langle \phi^+ i\nabla U^+ U \cdot \frac{1}{i}\nabla\phi\rangle \\
&\quad + \langle i\,\nabla\phi^+ \cdot U^+ \frac{1}{i}\nabla U\phi\rangle + \langle \phi^+ i\nabla U^+ U \cdot U^+ \frac{1}{i}\nabla U\phi\rangle\,,
\end{aligned}
\tag{5.1}
$$

it is $U^+\frac{1}{i}\nabla U = i\nabla U^+ U$ (by virtue of $\nabla(U^+ U) = 0$) that we must find. Let us restrict our attention to the diatomic, $N = 2$, case where unimodular unitary transformations have the simple parametrization

$$
U = \begin{pmatrix} a & b \\ -b^* & a^* \end{pmatrix} / (a^* a + b^* b)^{1/2}\,.
\tag{5.2}
$$

Then we find

$$
U^+\frac{1}{i}\nabla U = \begin{pmatrix} j(a,a) - j(b,b) & 2j(a,b) \\ 2j(b,a) & j(b,b) - j(a,a) \end{pmatrix} / (\rho(a) + \rho(b))
\tag{5.3}
$$

$$
\text{where } j(a,b) = \frac{1}{2i}(a^*\nabla b - \nabla a^* b)\,, \quad \rho(a) = a^* a\,.
$$

If (5.3) is substituted into (5.1), and the analogous notation

$$
\begin{aligned}
j(\phi,\psi) &= \langle \frac{1}{2i}(\phi^+\nabla\psi - \nabla\phi^+\psi)\rangle \\
\rho(\phi) &= \langle \phi^+\phi\rangle
\end{aligned}
\tag{5.4}
$$

is adopted, one finds after a bit of algebra

$$
\begin{aligned}
K &= \langle \nabla\phi^+ \cdot \nabla\phi\rangle - |j(\phi_1,\phi_1) - j(\phi_2,\phi_2)|^2 / (\rho(\phi_1) + \rho(\phi_2)) \\
&\quad - 4|j(\phi_1,\phi_2)|^2 / (\rho(\phi_1) + \rho(\phi_2)) \\
&\quad + (\rho(\phi_1) + \rho(\phi_2)) \left| \frac{j(a,a) - j(b,b)}{\rho(a) + \rho(b)} - \frac{j(\phi_2,\phi_2) - j(\phi_1,\phi_1)}{\rho(\phi_1) + \rho(\phi_2)} \right|^2 \\
&\quad + 4(\rho(\phi_1) + \rho(\phi_2)) \left(\frac{j(a,b)}{\rho(a) + \rho(b)} + \frac{j(\phi_2,\phi_1)}{\rho(\phi_1) + \rho(\phi_2)} \right) \\
&\quad \cdot \left(\frac{j(b,a)}{\rho(a) + \rho(b)} + \frac{j(\phi_1,\phi_2)}{\rho(\phi_1) + \rho(\phi_2)} \right)
\end{aligned}
\tag{5.5}
$$

Minimizing with respect to a and b is not difficult, but if a and b can be found such that

$$
\rho(a) + \rho(b) = \rho(\phi_1) + \rho(\phi_2)\,, \quad j(b,a) + j(\phi_1,\phi_2) = 0\,,
$$
$$
j(a,a) + j(\phi_1,\phi_1) = j(b,b) + j(\phi_2,\phi_2) = 0
\tag{5.6}
$$

then it is trivial: the square terms vanish and, using $j(\phi, \phi) = 0$, the "kinetic energy density" (r implicit) is

$$K = \langle \nabla \phi^+ \cdot \nabla \phi \rangle - 4|j(\phi_1, \phi_2)^2 / (\rho(\phi_1) + \rho(\phi_2))| \tag{5.7}$$

When can (5.6) hold? Certainly if

$$\langle \phi_\alpha^+ \phi_\beta \rangle = \langle \phi_\alpha^+ \rangle \langle \phi_\beta \rangle \tag{5.8}$$

($\langle \phi_\beta \rangle \neq 0$ since the states will not conserve type $-\beta$ particle number separately), for then we need only choose

$$a = \langle \phi_1^+ \rangle , \quad b = \langle \phi_2^+ \rangle . \tag{5.9}$$

And, as a bonus, it is seen that then

$$\langle \psi_2^+ \psi_2 \rangle = 0 , \tag{5.10}$$

so that ψ_2 is indeed a "ghost" field.

6. PRIMITIVE APPLICATIONS

The prototypical realistic system, at its irreducibly simplest level, is the hydrogen molecule-ion, H_2^+, a one-electron system with

$$\begin{aligned} v(r) &= v_L(r) + v_R(r) \\ v_L(r) &= -e^2/|r - r_L| , \quad v_R(r) = -e^2/|r - r_R| ; \end{aligned} \tag{6.1}$$

here the spin degree of freedom is of no consequence. The corresponding left (L) and right (R) atoms have ground states

$$\begin{aligned} |u_L\rangle &= \phi_L^+|0\rangle , \qquad |u_R\rangle = \phi_R^+|0\rangle \\ \phi_L^+ &= \int C(r) \phi_1^+(r) , \qquad \phi_R^+ = \int D(r) \phi_2^+(r) , \end{aligned} \tag{6.2}$$

where $C(r)$ is the standard hydrogenic ground state wave function centered at r_L: $C(r) = (\pi a_0^3)^{-1/2} e^{-(r - r_2)a_0}$, and similarly with D. If $|00\rangle$ now denotes the common vacuum of type -1 and type -2 fields, (6.2) extends at once to the one-electron two-source states

$$|u_{L,0}\rangle = \phi_L^+|00\rangle , \qquad |0, u_R\rangle = \phi_R^+|00\rangle . \tag{6.3}$$

The states (6.3) are degenerate for the reference Hamiltonian, in obvious notation,

$$H^{(0)}[\phi] = H_L[\phi_1] + H_R[\phi_2] , \tag{6.4}$$

with energies $\mathcal{E}_0 = -e^2/2a_0$, and so the two-source reference state must be taken as

$$|u\rangle = \frac{1}{\sqrt{2}}(|u_L, 0\rangle + |0, u_R\rangle) . \tag{6.5}$$

Then indeed

$$\langle \phi_\alpha^+ \phi_\beta \rangle = \langle \phi_\alpha^+ \rangle \langle \phi_\beta \rangle , \tag{6.6}$$

leading at once to

$$\rho(\phi_1) = \frac{1}{2}C^2 , \qquad \rho(\phi_2) = \frac{1}{2}D^2$$

$$j(\phi_1, \phi_2) = \frac{1}{4i}(C\nabla D - \nabla C D) .$$

(6.7)

The kinetic energy is evaluated at once from (5.7), and on computing $\langle u|H[\phi]|u\rangle$, one finds the first order result

$$\mathcal{E}_0^{(1)} = -1.33 \text{ ev at } |r_R - r_L| = 2.43a_0$$

$$\rho(0) = 0.043 .$$

(6.8)

Comparing with the exact[7] $\mathcal{E}_0 = -2.8$ ev at $2.0a_0$, with $\rho(0) = 0.099$, one is not very impressed.

But H_2^+ is in fact a very demanding test case, with its forced transitions between 0-electron and 1-electron states. A fairer test would be the more complex hydrogen molecule H_2, where one can imagine each electron as being temporarily localized around one nucleus. It is easy enough to include spin, and now the degenerate singlet superposition for the combined independent atoms will be

$$|u\rangle = \frac{1}{2\sqrt{2}}[(\phi_{R\uparrow} + \phi_{L\uparrow})(\phi_{R\downarrow} + \phi_{L\downarrow})$$

$$- (\phi_{R\downarrow} + \phi_{L\downarrow})(\phi_{R\uparrow} + \phi_{L\uparrow})]\,|\,00\rangle .$$

(6.9)

Eq. (6.6) is still satisfied, so that the kinetic energy is evaluated as before; only the pair interaction, $v(r_1, r_2) = e^2/|r_1 - r_2|$, involves a new computation. Calculation of $\langle u|H[\phi]|u\rangle$ now leads to the first order result

$$\mathcal{E}_0^{(1)} = -4.35 \text{ ev} \quad \text{at} \quad 1.76\,a_0 ,$$

(6.10)

comparing quite acceptably with the exact -4.72 ev at $1.4a_0$.

7. LEADING CORRECTIONS

Although we understand the inadequacies of the leading order treatment of H_2^+, we clearly need as well an effective connection procedure, both for energy and density. Indeed, they involve very similar considerations. Suppose that

$$H_0|\phi\rangle = E_0|\phi\rangle$$

(7.1)

with normalized ϕ, is perturbed to

$$(H_0 + W)|\psi\rangle = E|\psi\rangle .$$

(7.2)

If the normalization $\langle \phi|\psi\rangle = 1$ is adopted, then we see that

$$E - E_0 = \langle \phi|W|\psi\rangle .$$

(7.3)

Furthermore, we have as well

$$|\psi\rangle = |\phi\rangle + \frac{1}{E_0 - H_0}(1 - |\psi\rangle\langle\phi|)W|\psi\rangle , \tag{7.4}$$

giving a first order correction

$$|\psi\rangle \simeq |\phi\rangle + \frac{1}{E_0 - H_0}(1 - |\phi\rangle\langle\phi|)W|\phi\rangle . \tag{7.5}$$

One consequence is the second order energy correction

$$\Delta E^{(2)} = \langle W\phi \mid \frac{1}{E_0 - H_0} \mid (W - \langle W\rangle)|\phi\rangle , \tag{7.6}$$

and another is the first order correction to the expectation of an arbitrary observable Z:

$$\Delta\langle Z\rangle^{(1)} = 2\,\mathrm{Re}\langle Z\phi \mid \frac{1}{E_0 - H_0} \mid (W - \langle W\rangle)|\phi\rangle . \tag{7.7}$$

Our task then is to evaluate matrix elements of

$$G_0 = (E_0 - H_0)^{-1} \tag{7.8}$$

between states with no ϕ-component. This has a long history — see Bazley and Fox,[8] Delves,[9] Spruch, ... For our purposes, it is most readily done by the variational principle

$$\langle \chi_1 \mid G_0 \mid \chi_2\rangle = \langle \xi_1 \mid \chi_2\rangle + \langle \chi_1 \mid \xi_2\rangle + \langle \xi_1 \mid E_0 - H_0 \mid \xi_2\rangle , \tag{7.9}$$

stationary with respect to variations of ξ_1 and ξ_2. An alternative that we have used takes advantage of a subsidiary maximum principle:[10]

$$\begin{aligned} \langle \chi_1 \mid G_0 \mid \chi_2\rangle &= \langle \chi_1 \mid \xi_1\rangle \\ \text{when } \xi \text{ maximizes } &\langle \xi \mid E_0 - H_0 \mid \xi\rangle + \langle \chi_2 \mid \xi\rangle + \langle \xi \mid \chi_2\rangle . \end{aligned} \tag{7.10}$$

The problem of course is the parametric choice of the auxiliary function ξ, and for theoretical as well as practical reasons, a linear combination of low excited states is suggested. Taking the most primitive form

$$|\xi\rangle = \int (C_1(r)\,\phi_1^+(r) + D_1(r)\,\phi_2^+(r))\,|\,00\rangle \tag{7.11}$$

for the H_2^+ problem, where C_1 and D_1 are the lowest excited s-states makes substantial improvements in both energy and density.

8. CONCLUSION

It is clear that even the zeroth order picture of *truly* independent atoms in a bound molecule is unexpectedly effective. This corresponds of course to some perturbation expansion resummation, and so its utility is not unlimited. In particular, strongly unbalanced electron distributions are not well handled. Even under these circumstances, bandaids in the form of correction procedures are quite useful, but only more structured "unperturbed states" appear to offer the hope of substantial improvement. In the next stage of our program, we will A) include bond transfer region terms of the form

$$\int w(r_1, r_2)\, \phi_1^+(r_1)\, \phi_2(r_2) + \cdots \qquad (8.1)$$

in the reference Hamiltonian, raising new issues which will be reported in the near future, and B) investigate the extent to which the reference atoms can be chosen in polarized or otherwise distorted form.

REFERENCES

1. W. Moffitt, Proc. Roy. Soc. London *210*, 245 (1951).

2. M. Girardeau, J. Math. Phys. *4*, 1096 (1963).

3. R. F. W. Bader, Acc. Chem. Res. *18*, 9 (1985).

4. J. Harris, Phys. Rev. *B 31*, 1770 (1985).

5. J. K. Percus, in "The Liquid State of Matter", ed. E. W. Montroll and J. L. Lebowitz, (North-Holland, 1982).

6. N. Georgopoulos, Ph.D. Thesis, Physics Dept., NYU (1991).

7. D. R. Bates, K. Ledsham, A. L. Stewart, Phil. Tr. Soc. London *A246*, 215 (1953).

8. N. W. Bazley, D. W. Fox, J. Math. Phys. *7*, 415 (1966).

9. L. M. Delves, Nuc. Phys. *41*, 497 (1963).

10. S. Aranoff, J. K. Percus, Phys. Rev. *166*, 1255 (1968).

DENSITY-FUNCTIONAL THEORY OF LARGE SYSTEMS:

A DIVIDE-AND-CONQUER APPROACH

Weitao Yang

Department of Chemistry
Duke University
Durham, NC 27706 USA

ABSTRACT

We describe here the divide-and conquer density-functional approach
to the calculation of electronic structure. This method has been
developed recently to treat large systems that are beyond the reach of
conventional Kohn-Sham and Hartree-Fock methods. There are two unique
features: (1) This method uses electron density as the basic computational variable; (2) it divides a large system into many subsystems and
then calculates electron densities of the subsystems separately. Progress of this method and its prospects are also discussed.

INTRODUCTION

Density-functional theory has become an important tool for calculating electronic structure of molecular and solid-state systems.[1-9] This
has been possible because of the ingenious formulation of Kohn and Sham[10]
and because of the algorithm development in solving the Kohn-Sham equations.[11-23]

In the Kohn-Sham formulation of the density-functional theory, the
total electron density is resolved into many single-electron orbitals
to make possible the accurate determination of the electronic kinetic
energy. This bypasses the need of constructing explicit kinetic energy
functional in terms of electron density, and turns the original many-
electron problem into a tractable noninteracting-electron problem.
Much progress has been made over the years in solving the Kohn-Sham
equations, with basis functions[12-21] or numerically with finite difference method.[22] The type of basis functions used include plane waves
in conjunction with pseudopotentials, muffin-tin orbitals, and gaussian,
Slater, and numerical atomic orbitals. It is now possible to do a Kohn-
Sham calculation for a system of about 100 atoms.

However, there remain two major tasks in density-functional theory.
The first one is the approximation of the exchange-correlation energy
functional in terms of electron density. This has to do with the
accuracy of the prediction from density-functional calculations, which
we do not intend to address in this paper. The second one has to do
with the efficiency of the theory and has its origin in the approximation of the kinetic energy functional. In the Kohn-Sham method, kinetic
energy is determined explicitly in terms of the Kohn-Sham orbitals. But

the electron density alone is the basic variable in density-functional theory and orbitals are just auxiliary quantities. The use of orbitals in Kohn-Sham method forces the computational effort to increase as N^3, where N is the number of electrons, and therefore makes it impossible to carry out calculations for very large systems, say, consisting of a few hundred or a few thousand atoms.

In this report, we describe a recently developed approach[24-26] which uses electron density as the basic computational variable in the spirit of the Thomas-Fermi theory, but takes the advantage of the Kohn-Sham method in treating the kinetic energy accurately.

THE DIVIDE-AND-CONQUER APPROACH

In the Kohn-Sham method, the electron density is expressed as,

$$\rho(\vec{r}) = 2 \sum_i^{N/2} \psi_i(\vec{r})\psi_i(\vec{r}) , \tag{1}$$

where ψ_i is the Kohn-Sham orbital satisfying the following orbital equation

$$\hat{H}\psi_i(\vec{r}) = (-\frac{1}{2}\nabla^2 + V_{eff}(\vec{r}))\psi_i(\vec{r}) = \varepsilon_i\psi_i(\vec{r}) , \tag{2}$$

with V_{eff} as the Kohn-Sham effective potential that depends in turn on the electron density. The total energy is given by

$$E[\rho] = 2 \sum_i^{N/2} \varepsilon_i + Q[\rho] + \sum_{a,b} Z_a Z_b/R_{ab}, \tag{3}$$

where

$$Q[\rho] = \int \rho[-\Phi(\vec{r})/2 - V_{xc}(\vec{r})]d\vec{r} + E_{xc}[\rho], \tag{4}$$

and

$$\Phi(\vec{r}) = \int \frac{\rho(\vec{r}')}{|\vec{r}-\vec{r}'|} d\vec{r}' . \tag{5}$$

In Eq (4), E_{xc} is the exchange-correlation functional and

$$v_{xc}(\vec{r}) = \delta E_{xc}[\rho]/\delta\rho(\vec{r}) .$$

The new method[24] is based on a well-known expression for the electron density[4]

$$\rho(\vec{r}) = 2 <\vec{r}|\eta(\varepsilon_F - \hat{H})|\vec{r}> , \tag{6}$$

were $\eta(x)$ is the Heaviside step function ($\eta(x) = 1$ for $x > 0$, and $\eta(x) = 0$ for $x \leq 0$), $\hat{H}$ is the Kohn-Sham Hamiltonian and ε_F can be any value between the highest occupied and the lowest unoccupied eigenvalues. Eqs. (1) and (6) are identical. But unlike Eq. (1), which uses orbitals, Eq. (6) focuses on the Hamiltonian and allow different schemes of approximation.

To use Eq. (6) to calculate the electron density, we have to approximate the Hamiltonian and turn it from a differential operator to certain finite projection of it. The projection of $\hat{H}$ to the space spanned by the linear combination of atomic orbitals is a _global_ approximation to $\hat{H}$, and leads to the usual LCAO approach for the Kohn-Sham equations. It is global because it uses the same LCAO finite projection of $\hat{H}$ for calculating the density at every point in the entire three-dimensional space.

A more efficient way is a _local_ approach in which the projection of

$\hat{H}$ depends on the point where the density is computed. To this end, we divide the system into subsystems in the physical space by the following smooth partition $1 = \Sigma_\alpha p^\alpha(\vec{r})$ where $p^\alpha(\vec{r})$ is a positive weighting function for the subsystem α. $p^\alpha(\vec{r})$ is large in the subspace where the subsystem α is and is small away from it. Then the total density can be exactly expressed as the sum

$$\rho(\vec{r}) = 2 \sum_\alpha p^\alpha(\vec{r}) \langle \vec{r} | \eta(\varepsilon_F - \hat{H}) | \vec{r} \rangle = \sum_\alpha \rho^\alpha(\vec{r}) , \tag{7}$$

where

$$\rho^\alpha(\vec{r}) = 2 p^\alpha(\vec{r}) \langle \vec{r} | \eta(\varepsilon_F - \hat{H}) | \vec{r} \rangle . \tag{8}$$

The preceding partition of the density is a generalization to subsystems of any size from what was first proposed by Hirshfeld[27] to define atoms in a molecule and later used by Becky[22] and Delley[14] to carry out three-dimensional integration for molecules. Here the smooth partition of the density in Eqs. (7) and (8) allows a local approximation--we can make different approximations to $\hat{H}$ for different subsystems. Thus, we introduce the following approximation

$$\tilde{\rho}^\alpha(\vec{r}) = 2 p^\alpha(\vec{r}) \langle \vec{r} | f_\beta(\varepsilon_F - \hat{H}^\alpha) | \vec{r} \rangle , \tag{9}$$

where $f_\beta(x)$ is the Fermi function, $f_\beta(x) = [1 + \exp(-\beta x)]^{-1}$, and $\hat{H}^\alpha$ is the subspace approximation of the Kohn-Sham Hamiltonian operator. Two separate approximations are involved: the step function $\eta(x)$ by Fermi function $f_\beta(x)$ and $\hat{H}$ by $\hat{H}^\alpha$. The Fermi function in Eq. (9) is a convenient choice to make the value of ε_F unique; the uniqueness of ε_F is necessary as will be shown below.

We now let $\hat{H}^\alpha$ be the projection of the original Kohn-Sham Hamiltonian operator $\hat{H}$ to the space spanned by the non-orthogonal basis functions $\{\phi_j^\alpha(\vec{r})\}$ that are localized in the subsystem α:

$$\hat{H}^\alpha = \sum_{jk\ell m} |\phi_j^\alpha\rangle \, (\underset{\approx}{S^\alpha})^{-1}_{jk} \, (\underset{\approx}{H^\alpha})_{k\ell} \, (\underset{\approx}{S^\alpha})^{-1}_{\ell m} \langle \phi_m^\alpha| \tag{10}$$

$$= \sum_i |\psi_i^\alpha\rangle \, \varepsilon_i^\alpha \, \langle \psi_i^\alpha| ,$$

where $(\underset{\approx}{S^\alpha})^{-1}_{jk}$ is the (j,k) element of the inverse matrix of the overlap matrix $\underset{\approx}{S^\alpha}$, and $(\underset{\approx}{H^\alpha})_{k\ell}$ the (k,ℓ) element of the Hamiltonian matrix $\underset{\approx}{H^\alpha}$, with

$$(\underset{\approx}{S^\alpha})_{ij} = \langle \phi_i^\alpha | \phi_j^\alpha \rangle ; \quad (\underset{\approx}{H^\alpha})_{ij} = \langle \phi_i^\alpha | \hat{H} | \phi_j^\alpha \rangle . \tag{11}$$

In Eq. (10), the first equality is a standard projection, while the second expresses the projected Hamiltonian $\hat{H}^\alpha$ in terms of its eigenvalues $\{\varepsilon_i^\alpha\}$ and eigenfunctions $\{\psi_i^\alpha\}$. The eigenfunctions $\{\psi_i^\alpha\}$ are obtained as the linear combinations of the basis functions $\{\phi_j^\alpha\}$:

$$\psi_i^\alpha(\vec{r}) = \sum_j c_{ji}^\alpha \phi_j^\alpha(\vec{r}) , \tag{12}$$

where the linear coefficients are the solutions of the following generalized eigenvalue equation derived from the Rayleigh-Ritz variational principle,

$$(\underset{\approx}{H^\alpha} - \varepsilon_i^\alpha \underset{\approx}{S^\alpha}) \, \underset{\sim}{c_i^\alpha} = 0 , \tag{13}$$

where the matrices involved are given in Eq. (11). With $\{\psi_i^\alpha, \varepsilon_i^\alpha\}$ given by Eqs. (12) and (13), one can easily verify the equivalence of the two expressions for $\hat{H}^\alpha$ in eq. (10).

Using the spectral resolution of $\hat{H}^{\alpha}$ in Eq. (10), we can evaluate the subspace density ρ^{α} by Eq. (9). Then by Eq. (7), we obtain the expression for the direct calculation of the total electron density

$$\tilde{\rho}(\vec{r}) = 2 \sum_{\alpha} p^{\alpha}(\vec{r}) \sum_i f_{\beta}(\varepsilon_F - \varepsilon_i^{\alpha}) |\psi_i^{\alpha}(\vec{r})|^2 , \tag{14}$$

where the value of ε_F is determined by the normalization constraint

$$N = \int \tilde{\rho}(\vec{r}) d\vec{r} = 2 \sum_{\alpha} \sum_i f_{\beta}(\varepsilon_F - \varepsilon_i^{\alpha}) \langle\psi_i^{\alpha}|p^{\alpha}|\psi_i^{\alpha}\rangle . \tag{15}$$

To guarantee a unique solution of ε_F for a given N, it is necessary to keep a finite β so that the right hand side of Eq. (15) is a continuous monotonic function of ε_F. The value of β can be choosen such that its increase does not significantly change the total energy.

We also need to determine the eigenvalue summation in the expression for the total energy, Eq. (3); namely,

$$\varepsilon = 2 \sum_i^{N/2} \varepsilon_i = 2 \int d\vec{r} \langle\vec{r}|\hat{H} \eta(\varepsilon_F - \hat{H})|\vec{r}\rangle . \tag{16}$$

Now make an approximation in a similar fashion as in Eq. (19),

$$\tilde{\varepsilon} = 2 \int d\vec{r} \sum_{\alpha} p^{\alpha}(\vec{r}) \langle\vec{r}|\hat{H}^{\alpha}f_{\beta}(\varepsilon_F - \hat{H}^{\alpha})|\vec{r}\rangle$$
$$= 2 \sum_{\alpha} \sum_i f_{\beta}(\varepsilon_F - \varepsilon_i^{\alpha}) \varepsilon_i^{\alpha}\langle\psi_i^{\alpha}|p^{\alpha}(\vec{r})|\psi_i^{\alpha}\rangle , \tag{17}$$

which leads to the approximate total energy

$$\tilde{E} = \tilde{\varepsilon} + Q[\tilde{\rho}] + \sum_{a,b} z_a z_b/R_{ab}, \tag{18}$$

where $Q[\tilde{\rho}]$ can be evaluated by three-dimensional integrations.

To summarize, the procedure to calculate $\tilde{\rho}(\vec{r})$ from a given $V_{eff}(\vec{r})$ is as follows: (i) Choose a partition function $p^{\alpha}(\vec{r})$ and a localized basis set $\{\phi_i^{\alpha}\}$ for each subsystem α; (ii) for each α, calculate the matrices S^{α} and H^{α} of Eq. (11) and then solve Eq. (13), which gives $\{\psi_i^{\alpha}, \varepsilon_i^{\alpha}\}$; (iii) determine ε_F by solving Eq. (15) and then $\tilde{\rho}(\vec{r})$ by Eq. (14). This procedure is coupled with the equation for $V_{eff}(\vec{r})$ to achieve self-consistency. Finally, the total energy is given by Eq. (18).

In this new approach, we employ a divide-and-conquer strategy. Not attempting the global approximation of the N/2 Kohn-Sham orbitals, we divide the electron density into contributions from subsystems using partition functions $p^{\alpha}(\vec{r})$ via Eq. (7), and then determine each contribution using local basis functions $\{\phi_i^{\alpha}\}$ via Eq. (9). It is conceptually appealing that the determination of structure of a molecule can be based on its division into its constituent atoms, or chemical bonds, or functional groups, or fragments (example of suitable partition functions will be given below). Computationally, the advantage of this approach is obvious: no construction nor diagonization of the global Hamiltonian matrix is needed. Instead, the diagonization of the Kohn-Sham Hamiltonian for each subsystem as described in Eq. (13) can be carried out separately and concurrently. This is ideal for parallel processors. The coupling between each subsystem is minimal--only through the local potential and the value of ε_F.

It is easily seen that the Kohn-Sham theory is a limit of the present theory: In Eq. (9), if we let $\beta \to \infty$ and let $\hat{H}^{\alpha}$ be the same for all the subsystems, the usual LCAO Kohn-Sham Hamiltonian, for example, then the theory becomes the Kohn-Sham theory, independent of the choice of partition functions. Division of a system per se does not introduce error, the projection of the original Kohn-Sham Hamiltonian to the space of the local basis set for each subsystem contains all the approximation. This suggests that the local basis set for each subsystem should at its limit approach a complete set for the present theory to approach the Kohn-Sham theory. The partition functions should be designed to make Eq. (14) an accurate approximation to the density without going to this limit. We have found the following prescription works well.

We define the general normalized partition function for subsystem α as

$$p^\alpha(\vec{r}) = g^\alpha(\vec{r}) / \sum_{\alpha'} g^{\alpha'}(\vec{r}) , \qquad (19)$$

with

$$g^\alpha(\vec{r}) = \sum_{A \in \alpha} [\rho_o{}^A (|\vec{r} - \vec{R}_A|)]^2 . \qquad (20)$$

The summation in Eq. (19) is over all the subsystems, while the summation in Eq. (20) is over all the atoms in subsystem α. $\rho_o{}^A(|\vec{r}-R_A|)$ is the spherical atomic density for atom A located at R_A; accurate Kohn-Sham atomic densities are used in our work.

Thus each atom has a weight in the three-dimensional space proportional to the square of its spherical atomic density; the weight of a subsystem is just the summation of the weights of all atoms in the subsystem. Such continuous partition ensures that the total electron density obtained through Eq. (14) is continuous everywhere in the space, even though each subsystem density is calculated with different local projection of the Hamiltonian.

PROGRESS AND PROSPECTS

We have implemented the new method for general molecular computations and shown that the method is capable of attaining the accuracy of the Kohn-Sham method. The tests so far include energetics in bond stretching of benzene molecule[25] and in the internal molecular rotation of a tetrepeptide.[26] We have also calculated the density of electronic states for the tetrapeptide. In all the tests, the prediction of the divide-and-conquer method invariably converges to that of the Kohn-Sham method, as the sizes of the local basis increase.

In particular, we have found that to make the divide-and-conquer method comparable in accuracy to the Kohn-Sham method, we need to include in a local basis set the atomic orbitals from the nearest-neighbor atoms for correct prediction of bond lengths and from the first, second and third nearest-neighbor atoms for the correct prediction of molecular internal dihedral angles. This reflects the geometric requirement that two atoms are needed to defined a bond length and four atoms are need to defined a dihedral angle.

The main advantage of the divide-and-conquer method over the Kohn-Sham method is in its efficiency. It does not diagonizes nor constructs the global Hamiltonian matrix and hence eliminates the N^3 requirement on computation time and the N^2 requirement on storage space. It in principle can scale as N in computational effort for large systems. We are currently making progress in the following directions: (1) calculating forces to incorporate the method with molecular dynamics and molecular geometry optimization, (2) generalizing the method to extended solid-state systems, including crystallines, defects and absorptions, and (3) applying the method to large biological molecules that are beyond calculations by the conventional Kohn-Sham or Hartree-Fock method.

ACKNOWLEDGEMENT

The Research is funded by the National Science Foundation (CHE-9109156). Participation in this workshop has been made possible by a travel grant from the U.S. Army Research Office.

REFERENCES

1. S. Lundquist and N.H. March, editors, *Theory of the Inhomogeneous Electron Gas*, Plenum, New York, 1983.
2. D. Langreth and M. Suhl, editors, *Many-Body Phenomena at Surfaces*, Academic Press, Orlando, 1984.
3. N.H. March, and B.M. Deb, editors, *The Single-Particle Density in Physics and Chemistry*, Academic Press, New York, 1987.
4. R.G. Parr and W. Yang, *Density-Functional Theory of Atoms and Molecules*, Oxford University Press, New York, 1989.
5. E.S. Kryachko and E.V. Ludeñā, *Energy Density Functional Theory of Many-Electron Systems*, Kluwer, Boston, 1990.
6. S.B. Trickey, editor, *Density Functional Theory of Many-Fermion Systems*, (*Adv. Quantum Chem. Vol. 21*), Academic Press, Orlando, 1990.
7. M. Levy and J. Perdew, *Density Functional Methods in Physics*, edited by R.H. Dreizler and J. da Providencia, Plenum, New York, 1985.
8. D.R. Salahub, Adv. Chem. Phys. 69 (1987) 447.
9. R.O. Jones and O. Gunnarsson, Rev. Med. Phys. 61, (1989) 689.
10. W. Kohn and L.J. Sham, Phys. Rev. 140A (1965) 1133.
11. K.H. Johnson, Adv. Quantum Chem. 7 (1973) 143.
12. E.J. Baerends, D.E. Ellis and P. Ros, Chem. Phys. 2 (1973) 41.
13. P.M. Boerrigter, G. Te Velde and E.J. Baerends, Int. J. Quantum Chem. 33 (1988) 87.
14. B. Delley, J. Chem. Phys. 92 (1990) 508.
15. H. Sambe and R.H. Felton, J. Chem. Phys. 62 (1975) 1122.
16. B.I. Dunlap, J.W.D. Connally and J.R. Sabin, J. Chem. Phys. 71 (1979) 3396.
17. M.L. Skriver, *The LMTO Method*, Springer-Verlag, Berlin, 1984.
18. F.W. Averill and G.S. Painter, Phys. Rev. B 39 (1989) 8115.
19. R. Car and M. Parrinello, Phys. Rev. Lett. 55 (1985) 2472.
20. M.P. Teter, M.C. Payne and D.C. Allan, Phys. Rev. B 40 (1989) 12255.
21. E. Wimmer, H. Krakauer, M. Weinert and A.J. Freeman, Phys. Rev. B 24 (1981) 864.
22. A.D. Becke, Int. J. Quantum Chem., Symposium 23 (1989) 599.
23. D. Heinemann, B. Fricke and D. Kolb, Phys. Rev. A 38 (1988) 4994.
24. W. Yang, Phys. Rev. Lett. 66 (1991) 1438.
25. W. Yang, Phys. Rev. A44 (1991) 7823.
26. C. Lee and W. Yang, J. Chem. Phys. 96 (1992) 2408.
27. F.L. Hirshfeld, Theoret. Chim. Acta (Berl.) 44 (1977) 129.

THE EXCHANGE-ONLY SELF-CONSISTENT FIELD PROCEDURE IN THE LOCAL-SCALING VERSION OF DENSITY FUNCTIONAL THEORY: SOME THEORETICAL AND PRACTICAL CONSIDERATIONS

Eugene S. Kryachko

Centre for Advanced Studies, Research
and Development in Sardinia, CRS4
Piazza del Carmine 22, Cagliari, Italy
and
Institute for Theoretical Physics
Kiev, 25130, USSR

E.V. Ludeña, R. Lopez-Boada and J. Maldonado

Chemistry Center, Venezuelan Institute
for Scientific Research, IVIC,
Apartado 21827, Caracas 1020-A, Venezuela

ABSTRACT

The independent particle model of electronic structure is considered in the context of the local-scaling transformation version of density functional theory. In particular, the intra-orbit energy optimization is discussed in terms of density, square-root of the density and orbital variations. The decomposition of the exchange-only correlation potential within a given orbit is elucidated and an approximate scheme based on its replacement by the exchange-energy density functional is advanced. Finally, a procedure based on combined position-momentum optimizations of the energy represented by functionals of the one-particle densities in position and momentum spaces is introduced for the purpose of carrying out inter-orbit optimizations.

I. INTRODUCTION

The one-particle equations appearing in the context of the independent particle model of quantum mechanics are in general coupled partial integro-differential equations for which, due to their complexity, it is not possible to obtain analytical solutions. For this reason, they are treated approximately by means of iterative procedures where the "field" or effective potential in which a particle moves, is optimized at each iteration step until, at the final stage self-consistency is attained among these fields. In the Hartree approximation [1], the field is an explicit functional of the one-particle density $\rho(\vec{r})$ and comprises the external potential and the Coulomb potential exerted by the charge distribution. In the Hartree-Fock approximation [2], an additional contribution to the effective potential arises due to the exchange among electrons which comes from the antisymmetry condition on the N-particle wavefunction. This additional term, however, is neither a simple nor an explicit functional of the one-particle density.

Density functional theory and the self-consistent field method came together when Slater [3] proposed to approximate the exchange potential by means of an expression depending on $\rho(\vec{r})$. This approach has been further extended by Gáspár [4], Kohn and Sham [5] and by many other authors [6-12]. Of particular importance in this respect is the exchange-only approximation within the Hohenberg-Kohn formulation of density functional theory. But, as it has been discussed elsewhere [13,14], in all the above approximations the functional N-representability problem is not properly dealt with and as a consequence, the strict upper-bound character of the ordinary self-consistent field methods is lost.

The local-scaling transformation version of density functional theory [13,15-22] has been developed with the aim of accounting properly for functional N-representability. In this sense, it is a variational formulation whose basic variable is the one-particle density $\rho(\vec{r})$. For this reason, it is strictly equivalent to the variation principle from which the Schrödinger equation ensues. In a previous paper [21], we have considered the application of the local-scaling version of density functional theory to the exchange-only case for many-electron systems. In the present paper, we further analyze this situation from both theoretical and practical standpoints. In Section II we briefly review this formulation and discuss intra-orbit optimization in terms of density variation as well as of orbital variation. These are contrasted with the previously discussed variation [21] given in terms of the square root of the density. In Section III, we analyze the terms contributing to the density-dependent exchange potential and propose an approximate scheme for the calculation of the one-particle density. Finally, in Section IV, a scheme is presented for inter-orbit jumping which is based on a combined position- and momentum-space intra-orbit optimization.

II. THE SELF-CONSISTENT FIELD IN LOCAL-SCALING DENSITY FUNCTIONAL THEORY

Let us denote by $\mathcal{L}_N$ the antisymmetric N-particle Hilbert space, and by $\mathcal{S}_N \subset \mathcal{L}_N$ the class of single Slater determinants. Let us call $\mathcal{N}_B^N$ the set of N-representable one-particle densities defined by

$$\mathcal{N}_B^N \equiv \left\{ \rho(\vec{r}) \mid \quad \rho(\vec{r}) \equiv N \int d^3\vec{r}_2 \cdots \int d^3\vec{r}_N \Phi_\rho^*(\vec{r}_1,\cdots,\vec{r}_N)\Phi_\rho(\vec{r}_1,\cdots,\vec{r}_N); \quad \Phi_\rho \in \mathcal{L}_N \right\} \tag{1}$$

one can easily show [25] that any $\rho(\vec{r}) \in \mathcal{N}_B^N$ may also be generated from the class $\mathcal{S}_N \subset \mathcal{L}_N$ of single Slater determinants. Furthermore, there is a many to one correspondence between wavefunctions $\Phi_\rho \in \mathcal{S}_N$ and one-particle densities $\rho(\vec{r}) \in \mathcal{N}_B^N$ [13]. It follows from this fact that the class $\mathcal{S}_N$ can be broken up into disjoint subclasses $\mathcal{O}_S^{[i]}$, the so-called "orbits", such that $\mathcal{S}_N = \bigcup_{i=1} \mathcal{O}_S^{[i]}$. On the other hand, within an orbit $\mathcal{O}_S^{[i]}$ any wavefunction $\Phi_\rho^{[i]} \in \mathcal{O}_S^{[i]}$ is in a one to one correspondence with its one-particle density $\rho(\vec{r}) \in \mathcal{N}_B^N$ [13,16]. Let us stress that this one to one correspondence is essential for setting up the energy as a well-defined functional of the one-particle density. An orbit $\mathcal{O}_S^{[i]}$ can be obtained by applying a local-scaling transformation $\hat{F}_{g,\rho} \equiv \prod_{i=1}^N \hat{f}_{g,\rho}(\vec{r}_i)$ to an arbitrary orbit-generating wavefunction [13] $\Phi_g(\vec{r}_1,\cdots,\vec{r}_N)$:

$$\Phi_\rho(\vec{r}_1,\cdots,\vec{r}_N) = \hat{F}_{g,\rho}\,\Phi_g(\vec{r}_1,\cdots,\vec{r}_N) \tag{2}$$

where the density transformation operators $\hat{f}_{g,\rho}(\vec{r})$ satisfy the equation

$$\rho(\vec{r}) = J(\vec{f}_{g,\rho}(\vec{r});\vec{r})\rho_g(\vec{f}_{g,\rho}(\vec{r})) \tag{3}$$

In the above equation, $J(\vec{f}_{g,\rho}(\vec{r}); \vec{r})$ is the Jacobian of the local-scaling transformation.

Within an orbit $O_S^{[i]}$, the energy is given by

$$E[\Phi_\rho^{[i]}] \equiv \mathcal{E}\left[\rho(\vec{r}); \Phi_g^{[i]}\right] = T_W[\rho(\vec{r})] + T_{NW}\left([\rho(\vec{r})]; \Phi_g^{[i]}\right) + E_{Coulomb}[\rho(\vec{r})] + E_X\left([\rho(\vec{r})]; \Phi_g^{[i]}\right) + E_{ext}[\rho(\vec{r})] \tag{4}$$

where T_W and T_{NW} stand for the Weizsäcker and non-Weizsäcker kinetic energy terms, respectively [26]. These components are defined by

$$T_W[\rho(\vec{r})] = \frac{1}{8} \int d^3\vec{r} \frac{(\nabla_{\vec{r}} \rho(\vec{r}))^2}{\rho(\vec{r})}$$

$$T_{NW}\left([\rho(\vec{r})]; \Phi_g^{[i]}\right) = \frac{1}{2} \int d^3\vec{r} \, \rho(\vec{r})\left[\nabla_{\vec{r}} \nabla_{\vec{r}'} \tilde{\gamma}_g^{[i]}\left(\vec{f}(\vec{r}), \vec{f}(\vec{r}')\right)\right]_{\vec{r}=\vec{r}'}$$

$$E_{Coulomb}[\rho(\vec{r})] = \frac{1}{2} \int d^3\vec{r} \int d^3\vec{r}' \frac{\rho(\vec{r})\rho(\vec{r}')}{|\vec{r} - \vec{r}'|} \tag{5}$$

$$E_X\left([\rho(\vec{r})]; \Phi_g^{[i]}\right) = \int d^3\vec{r} \, \rho(\vec{r}) \, \mathcal{E}_{X,g}^{[i]}\left([\rho(\vec{r})]; \Phi_g^{[i]}\right)$$

$$E_{ext}[\rho(\vec{r})] = \int d^3\vec{r} \, \rho(\vec{r}) \, v(\vec{r})$$

In Eq.(3), the non-Weizsäcker term $T_{NW}\left([\rho(\vec{r})]; \Phi_g^{[i]}\right)$ contains the non-local part $\tilde{\gamma}_g^{[i]}(\vec{f}(\vec{r}), \vec{f}(\vec{r}'))$ of the 1-matrix [23]. Similarly, the exchange energy density is

$$\mathcal{E}_{X,g}\left([\rho(\vec{r})]; \Phi_g^{[i]}\right) \equiv \frac{1}{2} \int d^3\vec{r}' \frac{\rho(\vec{r}')}{|\vec{r} - \vec{r}'|} f_{X,g}^{[i]}\left(\vec{f}(\vec{r}), \vec{f}'(\vec{r}')\right) \tag{6}$$

where $f_{X,g}^{[i]}(\vec{f}(\vec{r}), \vec{f}(\vec{r}'))$ is the exchange correlation factor [24] evaluated at the transformed vectors $\vec{f}(\vec{r})$ and $\vec{f}(\vec{r}')$. Let us notice that because $E[\Phi]$ for $\Phi \in S_N$ attains its minimum value when $\Phi = \Phi^{HF}$, it follows that the absolute minimum of the energy density functional is $E[\Phi_{\rho_{HF}}^{[HF]}] \equiv \mathcal{E}[\rho_{HF}(\vec{r}); \Phi_g^{[HF]}]$. Hence, there exists a particular orbit, namely the Hartree-Fock orbit, which contains the exact Hartree-Fock wavefunction [21]. Since it is within this orbit that the energy attains its minimum value any density optimization within an arbitrary orbit $O_S^{[i]} \neq O_S^{[HF]}$ yields necessarily an energy value above the Hartree-Fock one. As a consequence, total optimization requires that in addition to density or intra-orbit optimization, inter-orbit optimization should also be performed [21]. This fact is ilustrated in Fig. 1.

Intra-orbit optimization

A. Variation with respect to the one-particle density

In what follows we consider the derivation of the Euler-Lagrange equations for the energy functional $\mathcal{E}[\rho(\vec{r}); \Phi_g^{[i]}]$ subject to the normalization condition $\int d^3\vec{r} \, \rho(\vec{r}) = N$, when the variation is performed with respect to the one-particle density $\rho(\vec{r})$ (for other one-particle density equations, see [27-30]). For this purpose, we introduce the auxiliary functional

$$\Omega[\rho(\vec{r}); \Phi_g^{[i]}] \equiv \mathcal{E}[\rho(\vec{r}); \Phi_g^{[i]}] - \mu^{[i]}\left(\int d^3\vec{r} \, \rho(\vec{r}) - N\right) \tag{7}$$

where $\mu^{[i]}$ is the Lagrange multiplier accounting for the normalization of the density. Setting

$$\frac{\delta\Omega[\rho(\vec{r}); \Phi_g^{[i]}]}{\delta\rho(\vec{r})} = 0, \tag{8}$$

one is led [13] to the following integro-differential equation for the one-particle density:

$$-\frac{1}{8}\left[\frac{\nabla\rho(\vec{r})}{\rho(\vec{r})}\right]^2 - \frac{1}{4}\frac{\nabla^2\rho(\vec{r})}{\rho(\vec{r})} + v_{T,g}^{[i]}([\rho(\vec{r})]; \vec{r}) + v(\vec{r}) + v_H([\rho(\vec{r})]; \vec{r}) + v_{X,g}^{[i]}([\rho(\vec{r})]; \vec{r}) = \mu^{[i]} \tag{9}$$

where $v_{T,g}^{[i]}$ is the potential arising from the non-local part of the kinetic energy

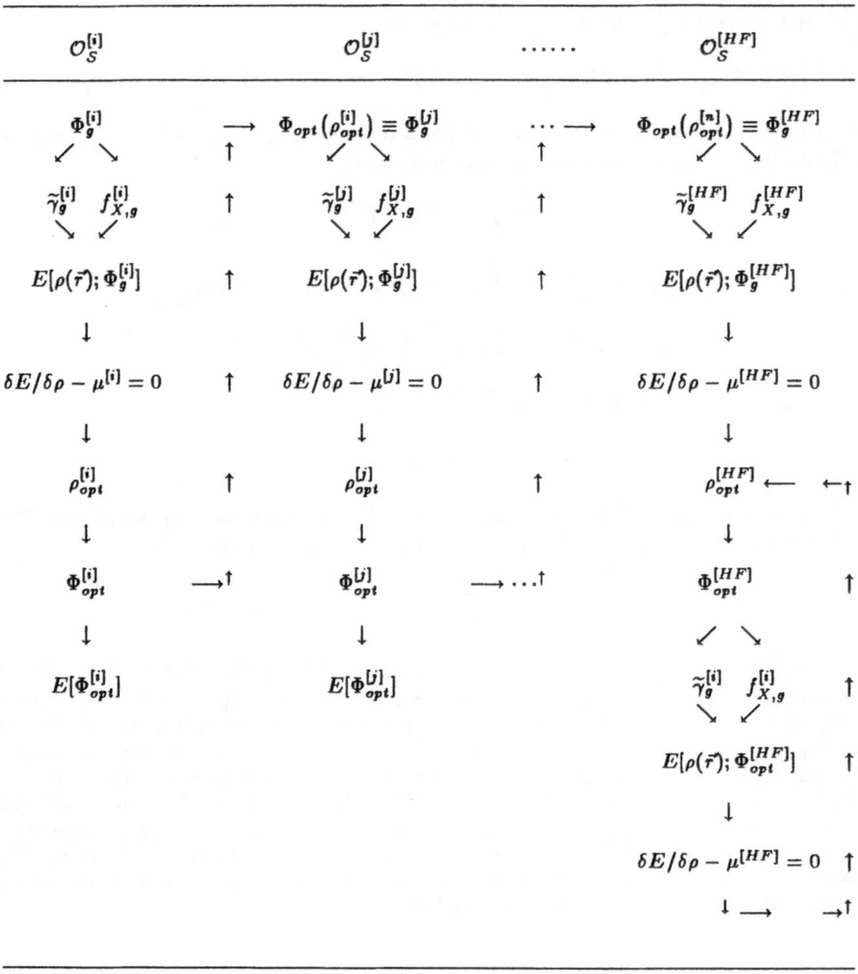

Fig. 1. Schematic representation of intra- and inter-orbit optimization in position space.

$$v_{T,g}^{[i]}\left([\rho(\vec{r})];\vec{r}\right) = \frac{1}{2}\left\{ \left[\nabla_{\vec{r}}\nabla_{\vec{r}'}\tilde{\gamma}_g^{[i]}\left(\vec{f}(\vec{r}),\vec{f}(\vec{r}')\right)\right]_{\vec{r}'=\vec{r}} + \rho(\vec{r})\frac{\delta}{\delta\rho(\vec{r})}\left(\left[\nabla_{\vec{r}}\nabla_{\vec{r}'}\tilde{\gamma}_g^{[i]}\left(\vec{f}(\vec{r}),\vec{f}(\vec{r}')\right)\right]_{\vec{r}'=\vec{r}}\right) \right\} \quad (10)$$

$v_H([\rho(\vec{r})];\vec{r}) = \int d^3\vec{r}\,\rho(\vec{r})|\vec{r} - \vec{r}'|^{-1}$ is the Hartree potential and $v_{X,g}^{[i]}$, the exchange-only potential

$$v_{X,g}^{[i]}\left([\rho(\vec{r})];\vec{r}\right) = \mathcal{E}_{X,g}\left([\rho(\vec{r});\Phi_g^{[i]}];\vec{r}\right) + \rho(\vec{r})\frac{\delta\mathcal{E}_{X,g}\left([\rho(\vec{r});\Phi_g^{[i]}];\vec{r}\right)}{\delta\rho(\vec{r})} \quad (11)$$

B. Variation with respect to the square-root of the one-particle density

Instead of considering $\rho(\vec{r})$ as the basic variable, we may deal with the "shape-wave-function" [31,32] $u(\vec{r})$ and its complex conjugate $u^*(\vec{r})$ which are related to the density through $\rho(\vec{r}) = u(\vec{r})u^*(\vec{r})$. In this case, the auxiliary variational functional is [21]

$$\Omega[u(\vec{r})u^*(\vec{r});\Phi_g^{[i]}] \equiv \mathcal{E}[u(\vec{r})u^*(\vec{r});\Phi_g^{[i]}] - \mu^{[i]}\left(\int d^3\vec{r}\,u(\vec{r})u^*(\vec{r}) - N\right) \quad (12)$$

Noticing that the Weizsäcker term takes the form

$$T_W[\rho(\vec{r})] = \frac{1}{2} \int d^3\vec{r}\, \left(\nabla_{\vec{r}} u^*(\vec{r})\right)\left(\nabla_{\vec{r}} u(\vec{r})\right) \tag{13}$$

the variation of the auxiliary energy functional with respect to the shape wavefunctions, that is

$$\frac{\delta\Omega[u(\vec{r})\,u^*(\vec{r});\,\Phi_g^{[i]}]}{\delta u^*(\vec{r})} + c.c = 0 \tag{14}$$

leads to the equation

$$\left[-\frac{1}{2}\nabla^2 + v_{T,g}^{[i]}([u^*(\vec{r}),u(\vec{r})];\vec{r}) + v(\vec{r}) + v_H([u(\vec{r}),u(\vec{r})];\vec{r}) + v_{X,g}^{[i]}([u^*(\vec{r}),u(\vec{r})];\vec{r})\right] u(\vec{r}) = \mu u(\vec{r}) \tag{15}$$

plus the corresponding complex conjugate equation for $u^*(\vec{r})$ where

$$
\begin{aligned}
v_{T,g}^{[i]}([u^*(\vec{r}),u(\vec{r})];\vec{r}) &= \nabla_{\vec{r}}\nabla_{\vec{r}'}\tilde{\gamma}_g^{[i]}([u^*(\vec{r}),u(\vec{r})];\vec{r},\vec{r}')\big|_{\vec{r}=\vec{r}'} \\
&+ u^*(\vec{r})\frac{\delta}{\delta u^*}\left[\nabla_{\vec{r}}\nabla_{\vec{r}'}\tilde{\gamma}_g^{[i]}([u^*(\vec{r}),u(\vec{r})];\vec{r},\vec{r}')\big|_{\vec{r}=\vec{r}'}\right]
\end{aligned} \tag{16}
$$

is the contribution to the potential arising from the non-local part of the kinetic energy, $v(\vec{r})$ is the external potential, $v_H^{[i]}([u^*(\vec{r}),u(\vec{r}),];\vec{r})$ is the ordinary Hartree potential

$$v_H([u^*(\vec{r}),u(\vec{r}),];\vec{r}) = \int d^3\vec{r}'\,\frac{u^*(\vec{r}')u(\vec{r}')}{|\vec{r}-\vec{r}'|}, \tag{17}$$

and $v_{X,g}^{[i]}([u^*(\vec{r}),u(\vec{r})];\vec{r})$ is the exchange potential

$$
\begin{aligned}
v_{X,g}^{[i]}([u^*(\vec{r}),u(\vec{r})];\vec{r}) &= \int d^3\vec{r}'\,\frac{u^*(\vec{r}')u(\vec{r}')}{|\vec{r}-\vec{r}'|}f_{X,g}^{[i]}([u^*(\vec{r}),u(\vec{r})];\vec{r},\vec{r}') \\
&+ u^*(\vec{r})\int d^3\vec{r}'\,\frac{u^*(\vec{r}')u(\vec{r}')}{|\vec{r}-\vec{r}'|}\frac{\delta}{\delta u^*}\left[f_{X,g}^{[i]}([u^*(\vec{r}),u(\vec{r})];\vec{r},\vec{r}')\right].
\end{aligned} \tag{18}
$$

C. Variation with respect to N local-scaled orthonormal orbitals

Let us consider the following expansion of the one-particle density in terms of N orthonormal orbitals

$$\rho(\vec{r}) = \sum_{k=1}^{N} |\phi_{\rho,k}^{[i]}(\vec{r})|^2 \quad \text{with} \quad \int d^3\vec{r}\,\phi_{\rho,k}^{*[i]}(\vec{r})\phi_{\rho,l}^{[i]}(\vec{r}) = \delta_{kl} \tag{19}$$

We assume that these orbitals have been obtained from the generating orbital set $\{\phi_{g,k}^{[i]}(\vec{r})\}$ by means of a local-scaling transformation:

$$\phi_{\rho,k}^{[i]}(\vec{r}) = \hat{f}_{g,\rho}\phi_{g,k}^{[i]}(\vec{r}) = [J(\vec{f}_{g,\rho};\vec{r})]^{1/2}\phi_{g,k}^{[i]}(\vec{f}_{g,\rho}(\vec{r})) \tag{20}$$

The single Slater determinant

$$\Phi_\rho^{[i]}(\vec{r}_1,\cdots,\vec{r}_N) \equiv \frac{det}{\sqrt{N!}}[\phi_{\rho,1}^{[i]}(\vec{r}_1)\cdots\phi_{\rho,N}^{[i]}(\vec{r}_N)] \tag{21}$$

constructed from these one-particle orbitals yields the kinetic energy term

$$T_S[\{\phi_{g,k}^{[i]}(\vec{r})\}] = \sum_{k=1}^{N}\frac{1}{2}\int d^3\vec{r}\,\nabla_{\vec{r}}\phi_{\rho,k}^{[i]}(\vec{r})\nabla_{\vec{r}'}\phi_{\rho,k}^{[i]}(\vec{r}')\big|_{\vec{r}'=\vec{r}} \tag{22}$$

The auxiliary variational functional becomes in the present case

$$\Omega[\{\phi_{\rho,k}^{[i]}(\vec{r})\};\Phi_g^{[i]}] \equiv \mathcal{E}[\{\phi_{\rho,k}^{[i]}(\vec{r})\};\Phi_g^{[i]}] - \sum_{k=1}^{N}\sum_{l=1}^{N}\lambda_{kl}\left(\int d^3\vec{r}\,\phi_{\rho,k}^{*[i]}(\vec{r})\phi_{\rho,l}^{[i]}(\vec{r}) - \delta_{kl}\right) \tag{23}$$

377

where the $\{\lambda_{kl}\}$ are the Lagrange multipliers enforcing orbital orthonormality. The energy functional is given by

$$\mathcal{E}[\{\phi^{[i]}_{\rho,k}(\vec{r})\}; \Phi^{[i]}_g] = T_S\left[\{\phi^{[i]}_{g,k}(\vec{r})\}\right] + E_{Coulomb}[\rho(\vec{r})] + E_X\left([\rho(\vec{r})]; \Phi^{[i]}_g\right) + E_{ext}[\rho(\vec{r})] \qquad (24)$$

Carrying out the variation of $\Omega[\{\phi^{[i]}_{\rho,k}(\vec{r})\}; \Phi^{[i]}_g]$ with respect to the one-particle orbital, we have

$$\frac{\delta T_S\left[\{\phi^{[i]}_{g,k}(\vec{r})\}\right]}{\delta\phi^{[i]}_{\rho,k}(\vec{r})} + \frac{\delta}{\delta\rho(\vec{r})}\left(E_{ext}[\rho(\vec{r})] + E_{Coulomb}[\rho(\vec{r})] + E_X\left([\rho(\vec{r})]; \Phi^{[i]}_g\right)\right)\frac{\delta\rho(\vec{r})}{\delta\phi^{[i]}_{\rho,k}(\vec{r})}$$

$$-\sum_{k=1}^{N}\sum_{l=1}^{N}\lambda_{kl}\frac{\delta}{\delta\phi^{[i]}_{\rho,k}(\vec{r})}\left(\int d^3\vec{r}\,\phi^{*[i]}_{\rho,k}(\vec{r})\phi^{[i]}_{\rho,l}(\vec{r}) - \delta_{kl}\right) = 0 \qquad (25)$$

from where one obtains the single particle equations

$$\left[-\frac{1}{2}\nabla^2 + v(\vec{r}) + v_H\left([\rho(\vec{r})]; \vec{r}\right) + v^{[i]}_{X,g}\left([\rho(\vec{r})]; \vec{r}\right)\right]\phi^{[i]}_{\rho \circ \rho t, k}(\vec{r}) = \sum_{l=1}\lambda_{kl}\phi^{[i]}_{\rho \circ \rho t, l}(\vec{r}) \qquad (26)$$

Notice that $v(\vec{r})$, $v_H\left([\rho(\vec{r})]; \vec{r}\right)$ and $v^{[i]}_{X,g}\left([\rho(\vec{r})]; \vec{r}\right)$ appearing in Eq.(26) are the same as those of Eq.(9). Notice also the absence of a potential arising from the kinetic energy term. The latter, however, is just a peculiarity which occurs in the exchange-only case, where the total wavefunction is given by Eq.(21). Clearly when the total wavefunction is given by a linear combination of Slater determinants, then T_S is not equal to $T_W + T_{NW}$. This gives rise, in the general case to a kinetic energy contribution to the one-particle potential [33].

III. DENSITY-DEPENDENT LOCAL EXCHANGE POTENTIAL

The exchange potential appearing in the Hartree-Fock equations is a non-local potential. However, in the exchange-only version of the Hohenberg-Kohn-Sham density functional theory, this potential is replaced by a local one, which, needs not, of course, be a local functional of the one-particle density [34,35]. Furthermore, a distinction must be made between the local exchange potential - the Talman potential - which minimizes the energy [36-38], and the exact exchange-only potential of the Kohn-Sham equations [39-41]. It is only the latter which corresponds to density functional theory proper.

In the local-scaling version of density functional theory, there arises also a local exchange potential $v^{[i]}_{X,g}([\rho(\vec{r})]; \vec{r})$ which is defined by Eq.(11). Note that the exchange energy density $\mathcal{E}_{X,g}([\rho(\vec{r})]; \Phi^{[i]}_g)$, appearing in Eq.(11), is given by Eq.(6). An important distinction between the Hohenberg-Kohn-Sham version and the present one concerns the possibility of rigorously representing in the latter an N-representable local exchange potential in each one of the orbits $O^{[i]}_S \subset \mathcal{L}_N$. It is clear, that the exact exchange-only potential of the Kohn-Sham equations occurs in the Hartree-Fock orbit $O^{[HF]}_S$; however, in order to guarantee N-representability, it must be constructed according to the specific prescriptions stipulated by Cioslowski [42]. But, as we have shown elsewhere [22], this is equivalent to applying local-scaling transformations in order to maintain the N-representability of the energy functional.

For a generating wavefunction corresponding to a single Slater determinant formed by the one-particle functions $\{\phi^{[i]}_{g,k}(\vec{r})\}^N_{k=1}$, the exchange-factor $f^{[i]}_{X,g}(\vec{f}(\vec{r}), \vec{f}(\vec{r}'))$ is equal to

$$f^{[i]}_{X,g}(\vec{f}(\vec{r}), \vec{f}(\vec{r}')) = -\frac{\gamma^{[i]}_g(\vec{f}(\vec{r}), \vec{f}(\vec{r}'))\gamma^{[i]}_g(\vec{f}(\vec{r}'), \vec{f}(\vec{r}))}{\rho_g(\vec{f}(\vec{r}))\rho_g(\vec{f}(\vec{r}'))} \qquad (27a)$$

where $\gamma_g^{[i]}(\vec{f}(\vec{r}), \vec{f}(\vec{r}'))$ is the first-order density matrix evaluated at the local-scaled vectors $\vec{f}(\vec{r})$ and $\vec{f}(\vec{r}')$. Expanding the first-order density matrix in terms of the generating one-particle orbitals, we obtain

$$f_{X,g}^{[i]}(\vec{f}(\vec{r}), \vec{f}(\vec{r}')) = \frac{1}{\rho_g(\vec{f}(\vec{r}))\rho_g(\vec{f}(\vec{r}'))} \sum_{k=1}^{N} \sum_{l=1}^{N} \phi_{g,k}^{*[i]}(\vec{f}(\vec{r}))\phi_{g,l}^{[i]}(\vec{f}(\vec{r}))\phi_{g,k}^{[i]}(\vec{f}(\vec{r}'))\phi_{g,l}^{*[i]}(\vec{f}(\vec{r}')) \quad (27b)$$

This expression, however, may be rewritten in terms of the transformed one-particle functions and the transformed density; in this case, Eq.(27 b) becomes:

$$f_{X,g}^{[i]}(\vec{f}(\vec{r}), \vec{f}(\vec{r}')) = -\frac{1}{\rho(\vec{r})\rho(\vec{r}')} \sum_{k=1}^{N} \sum_{l=1}^{N} \phi_{\rho,k}^{*[i]}(\vec{r})\phi_{\rho,l}^{[i]}(\vec{r})\phi_{\rho,k}^{[i]}(\vec{r}')\phi_{\rho,l}^{*[i]}(\vec{r}') \quad (28)$$

Using Eq.(28), we can rewrite Eq.(6) as follows

$$\mathcal{E}_{X,g}([\rho(\vec{r})]; \Phi_g^{[i]}) = -\frac{1}{2\rho(\vec{r})} \sum_{k=1}^{N} \sum_{l=1}^{N} \phi_{\rho,k}^{*[i]}(\vec{r})\phi_{\rho,l}^{[i]}(\vec{r}) \int d^3\vec{r}' \frac{\phi_{\rho,k}^{[i]}(\vec{r}')\phi_{\rho,l}^{*[i]}(\vec{r}')}{|\vec{r} - \vec{r}'|} \quad (29)$$

Carrying out the variation of Eq.(29) with respect to $\rho(\vec{r})$ and multiplying times $\rho(\vec{r})$, we obtain for the second term of the right-hand side of Eq.(11) the following expression:

$$\rho(\vec{r}) \frac{\delta \mathcal{E}_{X,g}([\rho(\vec{r}); \Phi_g^{[i]}]; \vec{r})}{\delta \rho(\vec{r})} = -\mathcal{E}_{X,g}([\rho(\vec{r}); \Phi_g^{[i]}]; \vec{r}) - \frac{1}{2} \frac{\delta}{\delta\rho(\vec{r})} \sum_{k=1}^{N} \sum_{l=1}^{N} \phi_{\rho,k}^{*[i]}(\vec{r})\phi_{\rho,l}^{[i]}(\vec{r}) \int d^3\vec{r}' \frac{\phi_{\rho,k}^{[i]}(\vec{r}')\phi_{\rho,l}^{*[i]}(\vec{r}')}{|\vec{r} - \vec{r}'|}$$
$$= -\mathcal{E}_{X,g}([\rho(\vec{r}); \Phi_g^{[i]}]; \vec{r}) + v_{X,g}^{[i]}([\rho(\vec{r})]; \vec{r}) \quad (30)$$

This result allows us to write the exchange potential as the following identity:

$$v_{X,g}^{[i]}([\rho(\vec{r})]; \vec{r}) = \mathcal{E}_{X,g}([\rho(\vec{r}); \Phi_g^{[i]}]; \vec{r}) + \left[-\mathcal{E}_{X,g}([\rho(\vec{r}); \Phi_g^{[i]}]; \vec{r}) + v_{X,g}^{[i]}([\rho(\vec{r})]; \vec{r}) \right] \quad (31)$$

It is clear, therefore, that the first contribution to the exchange potential in Eq.(11), that is, the term $\mathcal{E}_{X,g}([\rho(\vec{r}); \Phi_g^{[i]}]; \vec{r})$ is completely eliminated by a contribution arising from the second term. Let us notice, however, that if $\mathcal{E}_{X,g}([\rho(\vec{r}); \Phi_g^{[i]}]; \vec{r})$ approximates closely the exchange potential $v_{X,g}^{[i]}([\rho(\vec{r})]; \vec{r})$, then the contribution corresponding to $\rho(\vec{r})\frac{\delta \mathcal{E}_{X,g}([\rho(\vec{r}); \Phi_g^{[i]}]; \vec{r})}{\delta \rho(\vec{r})}$ becomes small. This opens to way to modelling the exchange potential by means of just $\mathcal{E}_{X,g}([\rho(\vec{r}); \Phi_g^{[i]}]; \vec{r})$.

Let us consider, therefore, the situation where the exchange potential is approximated by the exchange energy density:

$$v_{X,g}^{[i]}([\rho(\vec{r})]; \vec{r}) \simeq \mathcal{E}_{X,g}([\rho(\vec{r}); \Phi_g^{[i]}]; \vec{r}) \quad (32)$$

and let us discuss its implications on the variational procedure of intra-orbit optimization. Clearly, one cannot obtain by this method the $\rho^{opt}(\vec{r})$, namely, the optimal density corresponding to the optimal wavefunction $\Phi_{\rho,\rho^t}^{[i]}$ within orbit $O_S^{[i]}$. Assume, however, that the optimal one-particle density compatible with Eq.(32) is $\tilde{\rho}^{opt}(\vec{r})$. This density can be calculated by solving the equations

$$\left[-\frac{1}{2}\nabla^2 + v(\vec{r}) + v_H([\rho(\vec{r})]; \vec{r}) + \mathcal{E}_{X,g}([\rho(\vec{r}); \Phi_g^{[i]}]; \vec{r}) \right] \tilde{\phi}_{\rho^{opt},k}^{[i]}(\vec{r}) = \sum_{l=1} \tilde{\lambda}_{kl} \tilde{\phi}_{\rho^{opt},l}^{[i]}(\vec{r}) \quad (33)$$

and by computing

$$\tilde{\rho}^{opt}(\vec{r}) = \sum_{k=1}^{N} |\tilde{\phi}_{\rho^{opt},k}^{[i]}(\vec{r})|^2 \quad (34)$$

Now, in order to compute an energy value which is an upper bound to $E[\Phi_{\rho,\text{opt}}]$, we carry out a local-scaling transformation between the orbit generating density $\rho_g(\vec{r})$ corresponding to the orbit generating wavefunction $\Phi_g^{[i]}$ and the density $\tilde{\rho}^{opt}(\vec{r})$ of Eq.(34). This is done by solving the implicit first order differential equation

$$\frac{d\,f([\tilde{\rho}^{opt}]; r, \theta_o, \phi_o)}{d\,r} = \frac{r^2}{f([\tilde{\rho}^{opt}]; r, \theta_o, \phi_o)^2} \frac{\tilde{\rho}^{opt}(r, \theta_o, \phi_o)}{\rho_g(f([\tilde{\rho}^{opt}]; r, \theta_o, \phi_o), \theta_o, \phi_o)} \tag{35}$$

for each fixed direction θ_o, ϕ_o and spanning over all directions in order to generate $f(r, \theta, \phi)$. Of course, for a spherical symmetric atom, all fixed directions are equivalent.

Using this transformation, one can then obtain the new set (cf. Eq.(19))

$$\phi_{\rho,k}^{[i]}([\tilde{\rho}^{opt}], \vec{r}) = \left[\frac{\tilde{\rho}^{opt}(\vec{r})}{\rho_g(\vec{f}([\tilde{\rho}^{opt}]; \vec{r}))} \right]^{1/2} \phi_{g,k}^{[i]}(\vec{f}([\tilde{\rho}^{opt}]; \vec{r})) \tag{36}$$

and construct from it the transformed single Slater determinant

$$\Phi_\rho^{[i]}([\tilde{\rho}^{opt}]; \vec{r}_1 \cdots, \vec{r}_N) \equiv \frac{det}{\sqrt{N!}} [\phi_{\rho,k}^{[i]}([\tilde{\rho}^{opt}], \vec{r}_1) \cdots \phi_{\rho,k}^{[i]}([\tilde{\rho}^{opt}], \vec{r}_N)] \tag{37}$$

whose energy value is given by

$$E[\Phi_\rho^{[i]}([\tilde{\rho}^{opt}])] = \mathcal{E}[\tilde{\rho}^{opt}(\vec{r}); \Phi_g^{[i]}]. \tag{38}$$

The explicit expression for the different terms appearing in the expansion of Eq.(38) is given by (cf. Eq.(4)):

$$\mathcal{E}[\tilde{\rho}^{opt}(\vec{r}); \Phi_g^{[i]}] = T_W[\tilde{\rho}^{opt}(\vec{r}); \Phi_g^{[i]}] + T_{NW}[\tilde{\rho}^{opt}(\vec{r}); \Phi_g^{[i]}] + E_{Coulomb}[\tilde{\rho}^{opt}(\vec{r})]$$
$$+ E_X[\tilde{\rho}^{opt}(\vec{r}); \Phi_g^{[i]}] + E_{ext}[\tilde{\rho}^{opt}(\vec{r})] \tag{39}$$

Notice that the following inequalities are satisfied by the above energy expressions:

$$E[\Phi_\rho^{[i]}([\tilde{\rho}^{opt}])] \geq E[\Phi_{\rho,\text{opt}}^{[i]}] \geq E[\Phi_{\rho,\text{opt}}^{[HF]}] \tag{40}$$

Let us remark at this point that the approximate procedure which we advocate uses the correct energy expression, given as a functional of the one-particle density. The difference lies in that $E[\Phi_\rho^{[i]}([\tilde{\rho}^{opt}])]$ and $E[\Phi_{\rho,\text{opt}}^{[i]}]$ are calculated using different densities. Furthermore, the approximate density $\tilde{\rho}^{opt}$ is calculated from one-particle orbitals which are obtained from an approximate one-particle equation where the exact exchange potential $v_{X,g}^{[i]}([\rho(\vec{r})]; \vec{r})$ is replaced by the simpler local potential $\mathcal{E}_{X,g}([\rho(\vec{r}); \Phi_g^{[i]}]; \vec{r})$.

IV. INTER-ORBIT OPTIMIZATION THROUGH COMBINED POSITION - MOMENTUM ENERGY DENSITY FUNCTIONALS

As it is evident from Fig. 1, in addition to intra-orbit optimization, in order to reach the Hartree-Fock orbit, one must also carry out "orbit-jumping". In a previous work [21] we have discussed how this step may be accomplished by means of pseudo one-particle equations. In what follows we show how it is also possible to go from one orbit to another by coupling intra-orbit optimizations in coordinate and momentum spaces [43,44].

Let us consider the Fourier transformation of the orbit generating wavefunction $\Phi_g^{[i]}(\vec{r}_1, \cdots, \vec{r}_N) \equiv \Phi_g^{[i]}(r)$:

$$\Phi_g^{[i]}(\vec{p}_1, \cdots, \vec{p}_N) = \widehat{FT} \Phi_g^{[i]}(\vec{r}_1, \cdots, \vec{r}_N) \tag{41}$$

where the Fourier transformation operator is

$$\widehat{FT} = (2\pi)^{-3N/2} \int \prod_{i=1}^{N} d^3 \vec{r}_i \, exp\left(-i\sum_{k=1}^{N} \vec{p}_k \cdot \vec{r}_k\right) \tag{42}$$

The momentum density corresponding to the Fourier-transformed wavefunction $\Phi_g^{[i]}(\vec{p}_1,\cdots,\vec{p}_N) \equiv \Phi_g^{[i]}(p)$ is $\pi_g^{[i]}(\vec{p})$ and is defined by

$$\pi_g(\vec{p}) \equiv \pi_{\Phi_g^{[i]}(p)}(\vec{p}) = N \int d^3\vec{p}_2 \cdots \int d^3\vec{p}_N \Phi_g^{*[i]}(\vec{p}_1,\cdots,\vec{p}_N)\Phi_g^{[i]}(\vec{p}_1,\cdots,\vec{p}_N) \tag{43}$$

It is important to notice that $\pi_g^{[i]}(\vec{p})$ and the position one-particle density $\rho_g^{[i]}(\vec{r}) \equiv \rho_{\Phi_g^{[i]}(r)}(\vec{r})$ arising from the position N-particle wavefunction $\Phi_g^{[i]}(\vec{r}_1,\cdots,\vec{r}_N)$ are not connected to each other by means of a Fourier transformation. This fact has deep implications with regard to inter-orbit optimization.

Starting from an orbit generating wavefunction $\Phi_g^{[i]}(\vec{r}_1,\cdots,\vec{r}_N)$ in position space, we may compute, as was indicated in the previous Sections, the optimal one-particle density $\rho_{opt}^{[i]}(\vec{r}) \equiv \rho_{\Phi_{opt}^{[i]}(r)}(\vec{r})$ by optimizing the energy functional $\mathcal{E}[\rho(\vec{r});\Phi_g^{[i]}]$ subject to a normalization condition on the density (cf. Eq.(7)). By means of a local-scaling transformation of the orbit-generating wavefunction, one obtains the optimal wavefunction $\Phi_{opt}^{[i]}(\vec{r}_1,\cdots,\vec{r}_N)$ within the position orbit $\mathcal{O}_S^{[i]}$. Now, when a Fourier transformation is applied to this wavefunction, one obtains a new wavefunction, say, $\widetilde{\Phi}_{opt}^{[i]}(\vec{p}_1,\cdots,\vec{p}_N)$. The tilde indicates that although this wavefunction is obtained by Fourier-transforming the optimal position wavefunction in orbit $\mathcal{O}_S^{[i]}$, it does not necessarily follow that $\widetilde{\Phi}_{opt}^{[i]}(\vec{p}_1,\cdots,\vec{p}_N)$ is the optimal momentum wavefunction within the momentum orbit $\mathcal{P}_S^{[i]}$. In fact, the optimal wavefunction $\Phi_{opt}^{[i]}(\vec{p}_1,\cdots,\vec{p}_N)$ in momentum space, is the extremum solution of the variational minimization of the energy functional $\mathcal{E}[\pi(\vec{p});\Phi_g^{[i]}]$ subject to $\int d^3\vec{p}\,\pi(\vec{p}) = N$.

The proposed scheme for inter-orbit optimization is precisely based on the fact that upon variation the functional

$$\Omega\left[\pi_{\widetilde{\Phi}_{opt}^{[i]}(p)}(\vec{p});\Phi_g^{[i]}\right] = \mathcal{E}\left[\pi_{\widetilde{\Phi}_{opt}^{[i]}(p)}(\vec{p});\Phi_g^{[i]}\right] - \mu_p^{[i]}\left(\int d^3\vec{p}\,\pi_{\widetilde{\Phi}_{opt}^{[i]}(p)}(\vec{p}) - N\right) \tag{44}$$

attains its extremum

$$\frac{\delta}{\delta\pi(\vec{p})}\Omega\left[\pi(\vec{p});\Phi_g^{[i]}\right] = 0 \qquad \text{at} \qquad \pi(\vec{p}) = \pi_{\Phi_{opt}^{[i]}(p)}(\vec{p}) \tag{45}$$

If we now take the inverse Fourier transformation of $\Phi_{opt}^{[i]}(\vec{p}_1,\cdots,\vec{p}_N)$, namely, if we perform the operation described by the following equation

$$\Phi_g^{[j]}(\vec{r}_1,\cdots,\vec{r}_N) = \widehat{IFT}\Phi_{opt}^{[i]}(\vec{p}_1,\cdots,\vec{p}_N) \tag{46}$$

we obtain a wavefunction $\Phi_g^{[j]}(\vec{r}_1,\cdots,\vec{r}_N)$ in position space which must necessarily belong to an orbit $\mathcal{O}_S^{[j]}$ different from $\mathcal{O}_S^{[i]}$ (where $[i] \neq [j]$) because $\Phi_{opt}^{[i]}(\vec{p}_1,\cdots,\vec{p}_N)$ was obtained by solving the additional variational problem described by Eq.(44). The specific connection between the original orbit generating wavefunction in position space $\Phi_g^{[i]}(\vec{r}_1,\cdots,\vec{r}_N)$ and the resulting orbit generating wavefunction $\Phi_g^{[j]}(\vec{r}_1,\cdots,\vec{r}_N)$ is given as follows:

$$\Phi_g^{[j]}(\vec{r}_1,\cdots,\vec{r}_N) = \int \prod_{k=1}^{N} d^3\vec{r}_k\, T(\vec{r}_k;\vec{r}_k')\Phi_{opt}^{[i]}(\vec{r}_1,\cdots,\vec{r}_N)$$

$$= \int \prod_{k=1}^{N} d^3\vec{r}_k\, Q(\vec{r}_k;\vec{r}_k')\Phi_g^{[i]}(\vec{f}_{g,opt}(\vec{r}_1),\cdots,\vec{f}_{g,opt}(\vec{r})) \tag{47}$$

where

$$T(\vec{r}_k; \vec{r}_k') = (2\pi)^{-3} \int d^3\vec{p}_k \left[J\big(\vec{s}^{[i]}_{opt,opt}(\vec{p}); \vec{p}\big) \right]^{1/2} exp\left(-i\big[\vec{p}_k \cdot \vec{r}_k - \vec{s}^{[i]}_{opt,opt}(\vec{p}) \cdot \vec{r}_k'\big]\right) \qquad (48)$$

and

$$Q(\vec{r}_k; \vec{r}_k') = (2\pi)^{-3} \int d^3\vec{p}_k \left[J\big(\vec{s}^{[i]}_{opt,opt}(\vec{p}_k); \vec{p}_k\big)\right]^{1/2} \left[J\big(\vec{f}^{[i]}_{g,opt}(\vec{r}_k'); \vec{r}_k'\big)\right]^{1/2}$$
$$\times \ exp\left(-i\big[\vec{p}_k \cdot \vec{r}_k - \vec{s}^{[i]}_{opt,opt}(\vec{p}) \cdot \vec{r}_k'\big]\right) \qquad (49)$$

In Eq.(49), the local-scaling vector function $\vec{f}^{[i]}_{g,opt}(\vec{r}_k')$ in position space corresponds to transformation of the spatial density $\rho_g(\vec{r})$ into $\rho_{opt}(\vec{r})$. Similarly, $\vec{s}^{[i]}_{opt,opt}(\vec{p}_k)$ is the local-scaling vector function in momentum space for the transformation which carries the momentum density $\tilde{\tau}_{opt}(\vec{p})$ into $\pi_{opt}(\vec{p})$.

Acknowledgements

The authors would like to acknowledge very fruitful discussions with Prof. T. Koga. E.S.K would like, in addition, to express his gratitude to Dr. E. Clementi for his kind hospitality at CRS4.

REFERENCES

[1]   D.R. Hartree, The Calculation of Atomic Structures. Wiley, New York, (1957).

[2]   J.C. Slater, Quantum Theory of Atomic Structure, Vol. II. McGraw-Hill, New York (1960).

[3]   (a) J.C. Slater, Phys. Rev. **81**, 385 (1951); **Ibid.**, **82**, 538 (1951); **Ibid.**, J. Chem. Phys. **43**, S228 (1965).
(b) J.C. Slater, in Adv. Quantum Chem., P.-O. Löwdin (Ed.), Academic, N.Y., 1972, vol. 6, p. 1.

[4]   R. Gáspár, Acta Phys. Hung. **3**, 263 (1954).

[5]   W. Kohn and L.J. Sham, Phys. Rev. **140A**, 1133 (1965); B.Y. Tong and L.J. Sham, Phys. Rev. **144**, 1 (1966).

[6]   O. Gunnarsson and B.I. Lundqvist, Phys. Rev. **B13**, 4274 (1976); O. Gunnarsson, M. Jonson and B.I. Lundqvist, Phys. Rev. **B20**, 3136 (1979); O. Gunnarsson and R.O. Jones, Phys. Scr. **21**, 394 (1980).

[7]   U. von Barth, in The Electronic Structure of Complex Systems, P. Phariseau and W.M. Temmerman (Eds.), NATO ASI Ser. B: Physics, Plenum, N.Y., 1984, vol. 113, p. 67.

[8]   C.-O. Almbladh and U. von Barth, in Density Functional Methods in Physics, R.M. Dreizler and J. da Providencia (Eds.), NATO ASI Ser. B: Physics, Plenum, N.Y., 1985, vol. 123, p. 209.

[9]   W. Kohn and P. Vashishta, in Theory of the Inhomogeneous Electron Gas, S. Lundqvist and N.H. March (Eds.), Plenum, N.Y., 1983, p. 79.

[10] L. Hedin and S. Lundqvist, in Solid State Physics, F. Seitz, D. Turnbull and H. Ehrenreich (Eds.), Academic, N.Y., 1969, vol. 23, p. 1.

[11] D.C. Langreth and J.P. Perdew, Phys. Rev. **B21**, 5469 (1980); D.C. Langreth and M.J. Mehl, Phys. Rev. Lett. **47**, 446 (1981); **Ibid.**, Phys. Rev. **B28**, 1809 (1983).

[12] J. Callaway and N.H. March, in Solid State Physics, H. Ehrenreich and D. Turnbull (Eds.), Academic, N.Y., 1984, vol. 38, p. 135.

[13] E.S. Kryachko and E.V. Ludeña, Phys. Rev. **A 43**, 2179 (1991).

[14] E.S. Kryachko and E.V. Ludeña. In: Condensed Matter Theories **7**, 229 (1992).

[15] E.S. Kryachko and E.V. Ludeña, Phys. Rev. **A35**, 957 (1987).

[16] E.S. Kryachko and E.V. Ludeña, Energy Density Functional Theory of Many-Electron Systems. Kluwer, Dordrecht (1990).

[17] (a) I.Zh. Petkov, M.V. Stoitsov, E.S. Kryachko, Int. J. Quantum Chem. **29**, 149 (1986); E.S. Kryachko, I.Zh. Petkov, M.V. Stoitsov, Int. J. Quantum Chem. **32**, 467 (1987); **Ibid.** **32**, 473 (1987); **Ibid.** **34**, 305(E)(1988).
(b) I.Zh. Petkov, M.V. Stoitsov, Nuclear Density Functional Theory. Oxford Univ. Press, New York (1991).

[18] E.S. Kryachko and E.V. Ludeña, Phys. Rev. **A 43**, 2194·

[19] E.S. Kryachko, E.V. Ludeña and T. Koga, J. Math. Chem. (in press).

[20] E.S. Kryachko, E.V. Ludeña and V. Mujica In: Condensed Matter Theories 6, 161 (1991), ibid, Int. J. Quantum Chem. 40, 589 (1991).

[21] E.S. Kryachko and E.V. Ludeña, Int. J. Quantum Chem. (in press).

[22] E.S. Kryachko and E.V. Ludeña, J. Chem. Phys. 95, 9054 (1991).

[23] J.L. Gázquez and E.V. Ludeña, Chem. Phys. Lett. 83, 145 (1981).

[24] R. McWenny, Proc. Roy. Soc. Ser. A 253, 242 (1959).

[25] This may be accomplished using a similar argument as that following Eq.(36) of Ref. [22].

[26] E.V. Ludeña, J. Chem. Phys. 76, 3157 (1982).

[27] N.H. March, Int. J. Quantum Chem. Quantum Biol. Symp. 13, 3 (1986).

[28] M.R. Nyden and R.G. Parr, J. Chem. Phys. 78,4044 (1983).

[29] K.A. Dawson and N.H. March, J. Chem. Phys. 82, 323 (1985).

[30] G. Hunter and C.C. Tai, Int. J. Quantum Chem., S19, 173 (1985).

[31] A. Tachibana, Int. J. Quantum Chem. 34, 309 (1988).

[32] M. Levy, J.P. Perdew and V. Sahni, Phys. Rev. A30, 2745 (1984).

[33] C.-O. Almbladh and A.C. Pedroza, Phys. Rev. A 29, 2322 (1984).

[34] P. Hohenberg and W. Kohn, Phys. Rev. 136, B684 (1964); P.C. Hohenberg, W. Kohn and L.J. Sham, Adv. Quantum Chem. 21, 7 (1990).

[35] P. W. Payne, J. Chem. Phys. 71, 490 (1979).

[36] J.D. Talman and W.F. Shadwick, Phys. Rev. A 14, 36 (1976).

[37] J.D. Talman, P.S. Ganas and A.E.S. Green, Int. J. Quantum Chem., S13, 67 (1979).

[38] K. Aashamar, T.M. Luke and J.D. Talman, Phys. Rev. A 19, 6 (1979).

[39] J.B. Krieger, Y. Li and G.J. Iafrate, Phys. Rev. A 45, 101 (1992).

[40] Y. Li and J.B. Krieger, Phys. Rev. A 39, 992 (1989).

[41] F. Aryasetiawan and M.J. Stott, Phys. Rev. B 38, 2974 (1988).

[42] J. Cioslowski, Adv. Quantum Chem. 21, 303 (1990).

[43] E.S. Kryachko and T. Koga, J. Chem. Phys. 91, 1108 (1989).

[44] T. Koga, Y. Yamamoto and E.S. Kryachko, J. Chem. Phys. 91, 4758 (1989).

THE FOCK-SPACE COUPLED CLUSTER METHOD
EXTENDED TO HIGHER SECTORS[*]

S. R. Hughes and Uzi Kaldor

School of Chemistry, Tel Aviv University
69978 Tel Aviv, Israel

INTRODUCTION

The $\exp(S)$ or coupled cluster (CC) method[1-4] has been used increasingly in recent years for atomic and molecular calculations. The single-reference method, applicable when the state of interest may be described approximately by a single determinant, is used routinely (see Bartlett[5] for a review). Our contributions to three previous volumes in the present series document the development and implementation of the multireference Fock-space method to atomic and molecular systems. The first of these[6] includes a detailed presentation of the methodology and pilot applications. The second[7] emphasizes the problem of intruder states and incomplete model spaces (using the Be $2p^2$ ^1S resonance as example) and describes the first calculation of molecular potential functions. Large scale molecular applications, which include the calculation of 35 vertical excitation energies of N_2, accurate potential functions for the nine lowest states of Li_2, and the first calculation of a chemical reaction path, are described in the most recent contribution.[8] The method has since been applied widely and successfully,[9,10] but its scope has so far been restricted (with a single exception[11]) to states which may be reached from a closed-shell system by the addition or ionization of two electrons at most. This constitutes a severe limitation on the usefulness of the method. It excludes most atomic states, and allows the description of stretching and breaking molecular single bonds[12] but not double or triple bonds. It is therefore desirable to extend the method to higher sectors of the Fock space. Our adventures in the higher sectors are described below.

---

[*]Supported in part by the U.S. – Israel Binational Science Foundation.

METHOD

The basic method follows Lindgren's[13] choice of a normal-ordered wave operator,

$$\Omega = 1 + S + \frac{1}{2}\{S^2\} + \cdots, \quad . \tag{1}$$

S is the excitation operator describing *connected* single, double, ... excitations

$$S = S_1 + S_2 + \cdots = \sum_{ij}\{a_i^\dagger a_j\}s_j^i + \frac{1}{2}\sum_{ijkl}\{a_i^\dagger a_j^\dagger a_l a_k\}s_{kl}^{ij} + \cdots \tag{2}$$

where $a_i^\dagger$ and a_j are creation and annihilation operators, s_j^i, s_{kl}^{ij}, ... , are excitation amplitudes, and the curly brackets denote normal order with respect to a reference (core) determinant. The summation is carried out over connected terms only. The equations determining the excitation amplitudes in a complete model space may be derived from the generalized Bloch equation[13]

$$[S, H_0] = (QV\Omega - \Omega V_{eff})_{conn} \tag{3}$$

$$V_{eff} = PV\Omega P \tag{4}$$

where P and Q are the usual projection operators, and H_0 and V result from the partitioning of the Hamiltonian,

$$H = H_0 + V \quad . \tag{5}$$

The Fock-space method is designed to handle open-shell states. A closed-shell reference determinant is selected, and the excitation operator for states with n valence holes (electrons removed from the reference) and m valence particles (electrons added) is denoted by $S^{(n,m)}$. The total S operator may be written as

$$S = \sum_n \sum_m S^{(n,m)} \quad . \tag{6}$$

Haque and Mukherjee[14] have shown that this partitioning allows for partial decoupling of the Fock-space CC equations. The equations for $S^{(i,j)}$ involve only $S^{(k,l)}$ elements with $k \leq i$ and $l \leq j$, and the very large system of non-linear coupled equations separates therefore into smaller subsystems which are solved consecutively. The equations for $S^{(0,0)}$ are first iterated to convergence, then $S^{(0,1)}$ and $S^{(1,0)}$ are solved using the known $S^{(0,0)}$, and so on. This separation is exact, and reduces the computational effort significantly. It should not be confused with the expansion (2) of S in terms of the number of virtual excitations, i.e. the *total* number of electrons excited from the P-space determinant. Each term of (6) has an expansion similar to (2), e.g.

$$S^{(0,2)} = S_2^{(0,2)} + S_3^{(0,2)} + \cdots \tag{7}$$

where $S_j^{(n,m)}$ describes a j-body excitation from a state obtained by removing n electrons from the reference determinant and adding to it m electrons.

Diagrams

The working equations used in the computations were obtained from the generalized Bloch equation (3) in a topological, rather than algebraic, fashion. The full set of diagrams in the (0,0) sector when S is truncated at the CCSD (coupled cluster, singles and doubles) level

$$S \simeq S_1 + S_2 \tag{8}$$

is used routinely in closed-shell CC work, and may be found in Cullen and Zerner.[15] The corresponding diagrams in other (n, m) sectors, $n + m \leq 2$, are obtained from their $(0, 0)$ counterparts by bending down n outgoing particle lines and turning them into valence holes, at the same time bending down m incoming hole lines which become valence particles. This should be done in all ways generating distinct (n, m) diagrams.

When the CCSD approximation (8) is used, only (n, m) sectors with $n + m \leq 2$ have equations for $S^{(n,m)}$ which are solved iteratively. However, H_{eff} matrices for higher sectors may also be constructed and diagonalized to give state energies. The H_{eff} diagrams are similar to the S diagrams, except that all incoming and outgoing lines must be valence lines. The diagrams for sectors with one or two valence particles (and/or holes) may be easily obtained from Cullen and Zerner.[15] Lee, Kucharski and Bartlett[16] give S_3 skeletons, which may generate the S_3 diagrams upon drawing the arrows in all possible ways. Figure 1 shows the CCSD H_{eff} diagrams for four valence electrons, and five- and six-electron diagrams are given in figure 2.

Haque and Kaldor[17] noted that the CCSD approximation does not give a good description of sectors with more than two valence electrons. They developed the CCSD+T method, where triple virtual excitations were included noniteratively, computed from the converged CCSD excitation amplitudes. The *full* CCSDT approximation is given by

$$S \simeq S_1 + S_2 + S_3 . \tag{9}$$

Bartlett and coworkers[16,18] have proposed a hierarchy of approximations to (9), where S_3 is partially included. The first of the series, the CCSDT1 scheme, uses the full CCSDT definitions of S_1 and S_2, but keeps only $V S_2$ terms in the definition of S_3. The various CCSDT approximations have so far been applied to single reference CC only. We have adapted the CCSDT1 scheme to the Fock space method, and applications are described below.

APPLICATIONS

The (0,n) sectors: Fluorine and its Ions

The first application presented here aims at calculating the ionization potentials and excitation energies of the fluorine atom and its ions. The basis consisted of $(19s14p4d3f)$ Gaussian-type orbitals, contracted to $(8s7p4d3f)$. The first 7 s and 6 p GTOs were taken from table LXXXI of Partridge's tables,[19] and we added one diffuse s and p, as well as four d and three f functions. Spherical (5 d and 7 f components), rather than Cartesian (6 d, 10 f components) Gaussians were used. Starting from the closed shell F^{+5}, the $2p$ and $3s$ orbitals were defined as valence particles, so that the P space comprises all states with electrons in these orbitals. The CCSD and CCSDT1 coupled cluster equations for the $(0, n)$ Fock space sectors with $0 \leq n \leq 6$ were solved, and the H_{eff} matrices were diagonalized to yield the desired transition energies.

The results are shown and compared with experiment[20] in table 1. The quality of the CCSD values deteriorates slowly upon going from the (0,1) sector (F^{+4}) to the (0,4) sector (F^+), and becomes poor for the higher sectors. The effect of including S_3

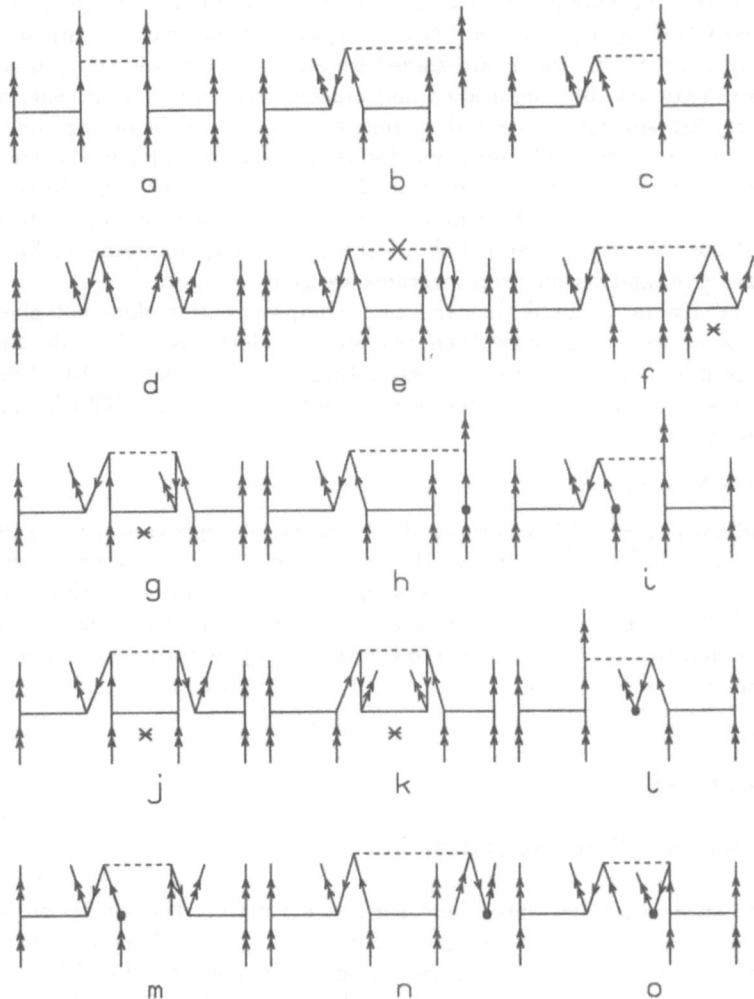

Figure 1. Four-body H_{eff} diagrams in the CCSD scheme.

Table 1. Ionization potentials and excitation energies of fluorine and its ions (eV). Errors given in parentheses.

			Exp.[20]	CCSD		CCSDT1	
F^{+4}	IP	^{2}P	114.21	114.00	(-0.21)	113.95	(-0.26)
	$2s^23s$	^{2}S	65.06	64.81	(-0.25)	64.80	(-0.26)
	Average absolute error				(0.23)		(0.26)
F^{+3}	IP	^{3}P	87.23	86.86	(-0.37)	86.92	(-0.31)
	$2s^22p^2$	^{1}D	3.13	3.19	$(+0.06)$	3.26	$(+0.13)$
		^{1}S	6.64	6.22	(-0.42)	6.57	(-0.07)
	$2s^22p3s$	^{3}P	51.63	51.33	(-0.30)	51.43	(-0.20)
		^{1}P	52.52	52.20	(-0.32)	52.31	(-0.21)
	Average absolute error				(0.29)		(0.18)
F^{+2}	IP	^{4}S	62.65	62.27	(-0.38)	62.59	(-0.06)
	$2s^22p^3$	^{2}D	4.23	4.36	$(+0.13)$	4.41	$(+0.18)$
		^{2}P	6.39	6.17	(-0.22)	6.39	(0.00)
	$2s^22p^23s$	^{4}P	39.26	38.96	(-0.30)	39.16	(-0.10)
		^{2}P	40.23	39.96	(-0.27)	40.05	(-0.18)
		^{2}D	42.65	42.41	(-0.24)	42.71	$(+0.06)$
		^{2}S	46.20	45.48	(-0.72)	45.99	(-0.21)
	Average absolute error				(0.32)		(0.11)
F^+	IP	^{3}P	34.98	34.07	(-0.92)	35.25	$(+0.27)$
	$2s^22p^4$	^{1}D	2.59	2.63	$(+0.04)$	2.59	(0.00)
		^{1}S	5.57	5.05	(-0.52)	5.20	(-0.37)
	$2s^22p^33s$	^{5}S	21.90	21.47	(-0.43)	21.97	(-0.07)
		^{3}S	22.67	22.25	(-0.42)	22.57	(-0.10)
		^{3}D	26.27	25.93	(-0.34)	26.52	$(+0.25)$
		^{1}D	26.66	26.32	(-0.34)	26.82	$(+0.16)$
		^{3}P	28.17	27.76	(-0.41)	28.46	$(+0.29)$
		^{1}P	28.46	28.15	(-0.31)	28.82	$(+0.37)$
	Average absolute error				(0.41)		(0.21)
F	IP	^{2}P	17.42	15.54	(-1.88)	19.03	$(+1.61)$
	$2s^22p^43s$	^{4}P	12.70	15.54	(-0.67)	13.74	$(+1.04)$
		^{2}P	12.98	12.09	(-0.89)	13.78	$(+0.80)$
		^{2}D	15.37	14.63	(-0.74)	16.35	$(+0.98)$
	Average absolute error				(1.04)		(1.11)
F^-	IP	^{1}S	3.45	-0.82	(-4.31)	6.60	$(+3.15)$

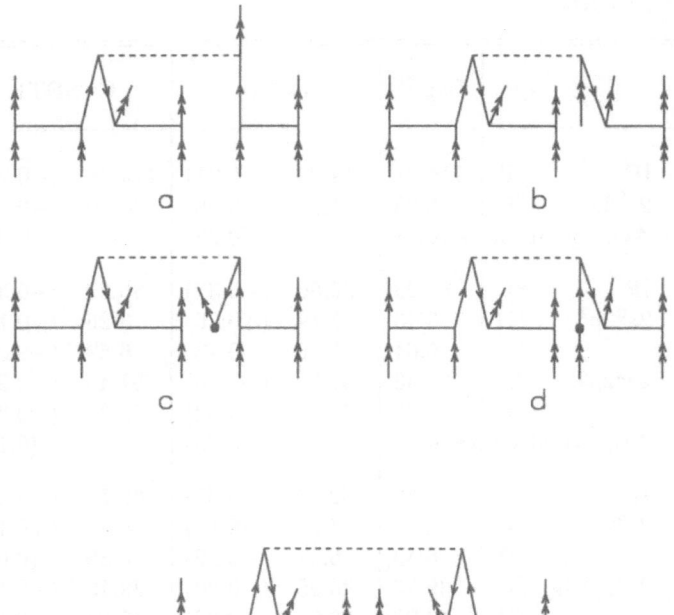

Figure 2. Five-body (a-d) and six-body (e) H_{eff} diagrams.

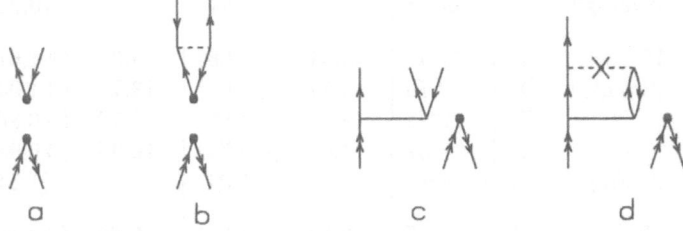

Figure 3. Unlinked diagrams in the $S_2^{(1,1)}$ (a,b) and the $S_2^{(1,2)}$ (c,d) expansion.

Table 2. Ionization potential, electron affinities and excitation energies of aluminum and its ions (eV). Errors given in parentheses.

		Sector	Exp.	CCSD	
1st IP		(0,1)	5.98	5.90	(−0.08)
2nd IP		(1,0)	18.82	18.63	(−0.19)
EA		(0,2)	0.46	0.39	(−0.07)
Al$^+$ EE		(1,1)			
3s3p	^{3}P		4.64	4.58	(−0.06)
3s3p	^{1}P		7.42	7.50	(+0.08)
Al$^-$ EE		(0,2)			
3s^{2}3p^2	^{1}D		0.34	0.37	(+0.03)
Al$^+$ EE		(1,1)			
2s3p	^{3}P		4.64	4.58	(−0.06)
	^{1}P		7.42	7.50	(+0.08)
Al EE		(1,2)			
3s3p^2	^{4}P		3.60	3.54^{a,b}	(−0.06)
	^{2}D		4.02	4.26^a	(+0.24)
				5.35^b	(+1.33)
	^{2}S		6.42	6.54^{a,b}	(+0.12)
	^{2}P		7.02	7.18^{a,b}	(+0.16)
Average absolute error, excluding ^{2}D					(0.10)

^{a}With $S_2^{(1,2)}$
^{b}Without $S_2^{(1,2)}$

increases with the number of valence particles. This effect is moderate for the first four sectors, and the CCSDT1 results are quite satisfactory. S_3 effects are large for the (0,5) and (0,6) sectors, and the CCSDT1 results for them are poor. The approximate inclusion of S_3 overshoots the mark, giving errors of opposite sign but about the same magnitude as CCSD. The indication is that the CCSDT1 approximation to full CCSDT is inadequate when S_3 effects are large.

The (1,2) Sector: Excited States of Aluminum

The (1,1) and higher sectors present problems not encountered in the (0, n) or (n, 0) sectors. The model spaces in these sectors are *always* incomplete,[21-23] due to the presence of deexcitation operators like $S_1^{(1,1)}$. This is demonstrated in figure 3a. The $S_1^{(1,1)}$ operator at the bottom half of 3a represents a transition from a (1,1) P-space determinant to the reference determinant, which for the (1,1) sector belongs to the Q space. Thus, 3a is a legitimate, though unlinked, $S_2^{(1,1)}$ diagram, and will lead to further unlinked diagrams such as 3b. The problem is not too severe for the (1,1) sector, as the offending terms may be ignored completely if one is only interested in obtaining energies.[22] The unlinked diagrams do appear in the Ω expansion for the (1,1) and higher sectors (see figure 3c and 3d), and for the higher terms they contribute to the energy too.

A special type of incomplete model space, proposed by Lindgren,[24] is the quasicomplete model space. It includes all determinants which may be constructed by distributing valence electrons in groups of valence orbitals, so that the number of electrons in each group is constant. The spaces used here are constructed so that the (n, m) P space will include determinants with all possible distributions of n electrons in valence particle orbitals and m holes in the valence hole orbitals. This clearly forms a quasicomplete space, and it can be shown that (i) the subduced $(k, 0)$ and $(0, l)$ spaces obtained by deleting an appropriate number of valence holes and/or particles from the (n, m) space are complete, and (ii) only linked diagrams appear in the energy expression. Details will be given elsewhere.[25]

The Al atom served as one of the test systems for the (1,2) Fock-space method. A $(19s/14p/4d/3f)$ set of Gaussian-type orbitals, taken from table XXX of Partridge[26] with the addition of diffuse s and p orbitals, as well as d and f functions, was contracted to $(13s/8p/4d/3f)$. The closed shell system Al^+, $1s^2 2s^2 2p^6 3s^2$ served as reference. The $3s$ orbital was designated valence hole, and the $3p$ — valence particle. Thus, the excited $3s3p^2$ states could be obtained. On the way, the lower sectors provided ionization potentials and electron affinities for the system.

The results are collected and compared with experiment[20,27] in table 2. A problem arose with the $3s3p^2$ ^2D state of the atom, which made the $S_2^{(1,2)}$ difficult to converge. The problem is caused by the excitations

$$2s2p^2\ ^2D \rightarrow 2s^2nd\ ^2D,$$

some of which are associated with small energy denominators and large two-electron integrals, giving rise to so-called "intruder states".[28,29] $S_2^{(1,2)}$ has a large effect on the ^2D state, but its effect on other excitations is less than 0.01 eV.

Other sectors converged easily, and all results (with the exception of the ^2D state) are in good agreement with experiment. Of particular interest is the excitation energy of Al^-, as very few anions have excited states.[27]

The (2,1) Sector: Bromine and its Ions

Several systems were calculated in the (2,1) sector. The last application described here involves the transition energies for the bromine atom and its ions. A $(16s12p5d3f)$ basis was used, contracted to $(11s10p4d3f)$. Br^- served as reference; the $4p$ orbital was valence hole, and the $5s$ and $5p$ — valence particles. It should be mentioned that the largest $S_2^{(2,1)}$ amplitude was obtained for the deexcitation

$$4s^2 4p^4 5s \rightarrow 4s^2 4p^5.$$

Results are given in table 3. Agreement with experiment[20,27] is good, with an average error of 0.17 eV.

SUMMARY AND CONCLUSION

Extension of the Fock-space coupled cluster method to hitherto unexplored sectors has been presented. Previous applications involved (n, m) sectors with $n + m \leq 2$, and the scope has now been broadened to include $(0, n)$ and $(n, 0)$ sectors with n as

Table 3. Transition energies in bromine (eV). Errors given in parentheses.

	Sector	Exp	CCSD	
EA	(1,0)	3.36	3.36	(0.00)
IP	(2,0)	11.84	11.67	(−0.17)
Br$^+$ EE	(2,0)			
$4s^24p^4$ $\quad$ ^{1}D		1.41	1.41	(0.00)
$4s^24p^4$ $\quad$ ^{1}S		—	3.76	
Br EE	(2,1)			
$4s^24p^4(^3P)5s$ $\quad$ ^{4}P		7.86	8.36	(+0.50)
$4s^24p^4(^3P)5s$ $\quad$ ^{2}P		8.33	8.39	(+0.03)
$4s^24p^4(^1D)5s$ $\quad$ ^{2}D		9.58	9.75	(+0.17)
$4s^24p^4(^3P)5p$ $\quad$ ^{4}P		9.26	9.57	(+0.31)
$4s^24p^4(^3P)5p$ $\quad$ ^{4}D		9.36	9.72	(+0.36)
$4s^24p^4(^3P)5p$ $\quad$ ^{2}D		9.73	9.75	(+0.03)
$4s^24p^4(^3P)5p$ $\quad$ ^{4}S		9.75	9.92	(+0.17)
$4s^24p^4(^3P)5p$ $\quad$ ^{2}P		9.78	9.89	(+0.11)
$4s^24p^4(^3P)5p$ $\quad$ ^{2}S		9.90	9.89	(−0.01)
$4s^24p^4(^1D)5p$ $\quad$ ^{2}P		10.97	11.38	(+0.41)
$4s^24p^4(^1D)5p$ $\quad$ ^{2}F		10.99	11.07	(+0.08)
$4s^24p^4(^1D)5p$ $\quad$ ^{2}D		11.13	11.25	(+0.12)
Absolute average error				(0.17)

big as 6, and to (1,2) and (2,1) sectors as well. In general, results are quite satisfactory, particularly where triple excitations are included in the CC iterations, albeit approximately. Applications to molecular systems are in progress.

REFERENCES

1. J. Hubbard, *Proc. Roy. Soc. (London)* A240:539 (1957); *ibid.* A243:336 (1958).
2. F. Coester, *Nucl. Phys.* 7:421 (1958); F. Coester and H. Kümmel, *Nucl. Phys.* 17:447 (1960); H. Kümmel, K. H. Lührmann, and J. G. Zabolitzky, *Phys. Rept.* 36:1 (1978).
3. J. Čížek, *J. Chem. Phys.* 45:4256 (1966).
4. J. Paldus, J. Čížek, and I. Shavitt, *Phys. Rev. A* 5:50 (1972); J. Paldus, *J. Chem. Phys.* 67:303 (1977); B. G. Adams and J. Paldus, *Phys. Rev. A* 20:1 (1979).
5. R. J. Bartlett, *Ann. Rev. Phys. Chem.* 32:359 (1981); *J. Phys. Chem.* 93:1697 (1989).
6. U. Kaldor, *in:* "Condensed Matter Theories", vol. 3, J. Arponen, R. F. Bishop, and M. Manninen, eds., Plenum, New York (1988).
7. U. Kaldor, *in:* "Condensed Matter Theories", vol. 4, J. Keller, ed., Plenum, New York (1989).
8. U. Kaldor, *in:* "Condensed Matter Theories", vol. 5, V. C. Aguilera-Navarro, ed., Plenum, New York (1990).
9. U. Kaldor, *Chem. Phys. Lett.* 166:599 (1990); *ibid.* 170:17 (1990); *Intern. J. Quantum Chem.* S24:291 (1990); *Chem. Phys. Lett.* 185:131 (1991); S. Roszak, U. Kaldor, D. A. Chapman, and J. J. Kaufman, *J. Phys. Chem.* 96:2123 (1992).
10. C. M. L. Rittby and R. J. Bartlett, *Theor. Chim. Acta* 80:469 (1991); M. Barysz, H. J. Monkhorst, and L. Z. Stolarczyk, *ibid.* 40:483 (1991).

11. A. Haque and U. Kaldor, *Chem. Phys. Lett.* 52:347 (1985).
12. S. Ben Shlomo and U. Kaldor, *J. Chem. Phys.* 89:956 (1988); U. Kaldor, *Chem. Phys.* 140:1 (1990).
13. I. Lindgren, *Intern. J. Quantum Chem.* S12:33 (1978); S. Salomonsson, I. Lindgren, and A.-M. Mårtensson, *Phys. Scr.* 21:351 (1980); I. Lindgren and J. Morrison, "Atomic Many Body Theory", Springer, Berlin (1982).
14. A. Haque and D. Mukherjee, *J. Chem. Phys.* 80:5058 (1984); *Pramana* 23:651 (1984).
15. J. Cullen and M. Zerner, *J. Chem. Phys.* 77:4088 (1982).
16. Y. S. Lee, S. A. Kucharski, and R. J. Bartlett, *J. Chem. Phys.* 81:5906 (1984).
17. A. Haque and U. Kaldor, *Chem. Phys. Lett.* 120:261 (1985).
18. M. Urban, J. Noga, S. J. Cole, and R. J. Bartlett, *J. Chem. Phys.* 83:4041 (1985); J. Noga, R. J. Bartlett, and M. Urban, *Chem. Phys. Lett.* 134:126 (1987).
19. H. Partridge, *J. Chem. Phys.* 90:1043 (1989); NASA Technical Memorandum 101044 (1989).
20. C. E. Moore, "Atomic Energy Tables", NBS circular 467, Washington, D.C. (1949).
21. D. Mukherjee, *Chem. Phys. Lett.* 125:207 (1986).
22. D. Sinha, S. Mukhopadhyay, and D. Mukherjee, *Chem. Phys. Lett.* 129:369 (1986).
23. B. H. Brandow, *Adv. Quantum Chem.* 10:187 (1977).
24. I. Lindgren, *Phys. Scr.* 32:291 (1985); ibid. 32:611 (1985).
25. S. R. Hughes and U. Kaldor, to be published.
26. H. Partridge, *J. Chem. Phys.* 87:6643 (1987); NASA Technical Memorandum 89449 (1987).
27. H. Hotop and W. C. Lineberger, *J. Phys. Chem. ref. Data* 4:539 (1975).
28. T. H. Schucan and H. A. Weidenmüller, *Ann. Phys. (NY)* 73:108 (1972); 76:483 (1973).
29. U. Kaldor, *Phys. Rev. A* 38:6013 (1988).

LIQUID-STATE THEORY FOR SOME NON-EQUILIBRIUM PROCESSES

James A. Given and George Stell

Department of Chemistry
State University of New York at Stony Brook
Stony Brook, New York 11794-3400

ABSTRACT

Recent advances in liquid-state theory permit the calculation of thermodynamic quantities and correlation functions for systems in which some of the degrees of freedom are quenched, or frozen in place, while the rest are annealed. Basic examples include models for porous media, crystals containing quenched impurities, and spin glasses. We further extend these methods to treat materials constructed in layers, each layer being added to the system and allowed to equilibrated, then frozen in place before the next layer is added. We discuss sequentially adsorbed systems as an important class of examples.

INTRODUCTION

Amorphous matter is ubiquitous in nature.[1] Indeed, if an amorphous system is defined as a system lacking long-range positional order of its particles, but possibly with ordering among the other degrees of freedom that characterize a particle, then matter generically is amorphous, and the ordered solid state is seen to be rather specialized! Under such a definition, amorphous systems include complex biological fluids, such as blood and milk, complex construction materials, such as cement and plastics, and even solid state materials such as glasses and amorphous alloys. Research in recent years[2,3] has shown that one can extend to the general category of amorphous materials the techniques developed during the past several decades to study the properties of simple, equilibrated fluids. These techniques include virial expansions for the bulk properties, integral equations for the correlation functions, and sequences of upper and lower bounds for all of these objects.

Extension of equilibrium liquid-state methods to a general class of disordered materials presents one immediate problem: one can no longer assume the positions of particles are determined by the same equilibrium distribution as the other degrees of freedom characterizing those particles. Attempts to extend liquid-state methods to the class of disordered or partly quenched systems[4-10] must confront the fact that expressions for the free energy and correlation functions of such systems are not in the form familiar from equilibrium statistical mechanics, that is, a single average over states of the system, with each state weighted by its Boltzmann weight. Rather the free energy and correlation functions take the form of double averages: one first calculates these quantities by averaging over all values of the annealed degrees of freedom, keeping the quenched degrees of freedom constant.[6] The quantities thus obtained are then averaged over all values of the quenched degrees of freedom. In this paper we will discuss a recently developed approach to this problem, namely the "liquid-state" version of the replica method. This method allows us to rewrite the double averages just described, as single averages of a type familiar from equilibrium statistical mechanics. We shall also discuss a complementary approach that is an extension of a very old method first used by Boltzmann[11] in treating equilibrium problems. This second method yields a set of Kirkwood-Salsburg equations that facilitate the exploration of rigorous results. This approach also yields a class of approximate integral equations for the pair distribution function.

The replica method was originally developed to treat model systems on a lattice[12] in which the exchange interaction strengths between lattice sites were chosen from a fixed random distribution. It involved integrating out the random exchange interactions to yield a non-random effective interaction between sites. The liquid-state replica method,[4,5,9] on the other hand, is applied to systems in which some of the particles have been quenched or frozen in position with a distribution corresponding to a different temperature, and a different pairwise interaction, from that characterizing the other particles in the system. Application of the replica method to liquid-state systems, rather than averaging out the quenched degrees of freedom, places all degrees of freedom on an equal footing by mapping a partly quenched system onto a limiting case of a fully equilibrated system, to be called the replicated system. The replica method, when applied to a partly quenched system, thus "erases" the distinction between quenched and annealed particles, allowing us to deal with a fully equilibrated system containing both kinds of particles. The standard machinery of equilibrium liquid-state theory can then be applied to this replicated system.

Of course, not all spatially disordered system can be simply idealized in the way just described, as a mixture of two fractions of particles, one of them "quenched" and one "annealed." However, many systems, both of practical and theoretical interest, can be described as being constructed in layers,[4] which are added successively to the system volume, each being brought into an equilibrium with the layers already present, then quenched in place, before the next layer is added. We emphasize that each layer of particles may have its own effective pairwise interaction, and its own effective temperature.[13] The liquid-state replica method can be applied repeatedly to such a layered system, each time "erasing" one of the boundaries between layers, until a fully equilibrated system results. Disordered systems that accord with the picture just described include Eden clusters, diffusion-limited aggregates,[14] chemical reaction models,[15] porous materials,[16] sequential adsorption processes,[4,17,18] and sequential polymerization.[5] The description along these lines, of certain irregular porous materials such as Berea sandstones,[14,16] also relies on a picture of successive aggregation. As can be seen from the description of the "layering" common to the formation of all the systems we can conveniently describe by replication, an absence of development in time of the structure of the system, once it is formed, is also a common feature, despite the nonequilibrium character of such systems. In this sense, such systems are all "preparation procedure" and no "dynamics".[18]

In this paper, we will describe the continuum replica method and develop the effective equilibrium picture that results from applying it to a two-layer system. We extend this scheme in two directions. First, we use chemical association techniques to apply our method to a model spin glass, in which the quenched and annealed degrees of freedom belong to the same set of particles. Second, we sketch the features of a model in which each particle is a separate layer, namely the class of sequentially adsorbed systems. For the case of sequential adsorption, we shall also introduce the Kirkwood-Salsburg equations to illustrate some of the features those equations that are common to both equilibrated and non-equilibrated systems, as well as the way the features that differentiate such systems from one another appear in the Kirkwood-Salsburg equations.

LIQUID-STATE THEORY FOR PARTLY QUENCHED SYSTEMS

In this section we develop the liquid state replica method[4] in its most natural setting, namely a system in which some of the particles are quenched or frozen in place and the others are annealed, or allowed to equilibrate. The result of applying this method is to map a partly quenched system onto a limiting case of a fully equilibrated system with additional "replicated" degrees of freedom. This mapping in turn yields Mayer expansions for the free energy and correlation functions of partly quenched systems. It also yields a set of integral equations, the replica Ornstein-Zernike (ROZ) equations, satisfied by the correlation functions.

We consider then a mixture of two species, one quenched and one annealed, which we denote species '0' and species '1', respectively. Species 0 particles are quenched in place with the spatial correlations corresponding to a temperature T and pairwise interaction $v_{00}(ij)$. Species-1 particles have pairwise interactions $v_{10}(ij)$, $v_{11}(ij)$ with the quenched particles and with each other, respectively. Here the arguments i and j represent the configurations of the particles labelled i and j. The average free energy of our two-species system is

$$-\beta F = \ln Z_{TOTAL} = \frac{1}{Z'} \int e^{-\beta H_{00}} \ln Z_1 d\vec{0} \tag{2.1}$$

with

$$Z' \equiv \int e^{-\beta H_{00}} d\vec{0} \tag{2.2}$$

and

$$Z_1 \equiv \int e^{-\beta[H_{01}+H_{11}]} d\vec{1} \tag{2.3}$$

Here we write H_{ij} for the sum of all pairwise interactions between particles of species i and species j. Also, we write $d\vec{0}$, $d\vec{1}$, to represent integration over all the positions of particles of species 0 and 1, respectively. The average in (2.1) is difficult to treat analytically because of the presence, under the integral sign, of the logarithm. We thus make use of the replica trick, which consists of replacing the logarithm with an exponential by using the identity

$$\ln Z \equiv \lim_{n \to 0} \frac{1}{n}[Z^n - 1] \tag{2.4}$$

Substituting (2.4) into (2.1) gives, for the total partition function

$$\ln Z_{TOTAL} = \lim_{n \to 0} \frac{1}{n} \int \left\{ \exp\left[-\beta \sum_{i=1}^{n} [H_{01}^{(i)} + H_{11}^{(i)}] \right] - 1 \right\} \exp[-\beta H_{00}]\{d\vec{1}\} d\vec{0} \tag{2.5}$$

The variables describing species 1 have been replicated and now appear in n copies, in accordance with (2.4). The notation $\{d\vec{1}\}$ indicates an integration over the n sets of position variables corresponding to these particles.

We first note that the expression on the RHS of (2.5) is, in fact, a limiting case of the equilibrium partition function for a particular system, namely the system with Hamiltonian

$$H = \sum_{<i,j>} v_{00}(x_{ij}) + \sum_{\alpha=1}^{n} \sum_{<i,j>} v_{0\alpha}(x_{ij}) + \sum_{\alpha,\beta=1}^{n} \sum_{<i,j>} v_{\alpha\beta}(x_{i,j})\delta_{\alpha\beta} \tag{2.6}$$

This system is a mixture of a one-component fluid (the quenched species) with an n-component fluid, given by n identical copies or replicas of the annealed species. Each pair of particles has the same pairwise interaction in this replicated system as in the partly quenched model from which it was derived **except** that a pair of annealed particles from different replicas has **no** interaction (because of the Kronecker delta $\delta\alpha\beta$ in the last term on the RHS). Thus the system of type 0 particles, i.e., the original quenched phase, can be thought of as a "solvent" which mediates interactions between the different type 1 components. This analogy is apt in the sense that a quenched phase, like a solvent, can induce effective interactions that are both long-range and many-body.

We now find it conceptually useful to again recast the problem, describing the n-replica fluid of type 1 particles as being instead a one-component fluid whose particles have a discrete internal degree of freedom which we call 'spin'. The spin of a type 1 particle i, which we write σ_i, is just another name for what we previously called its replica index. The conceptual change here is the reverse of that originally used by Onsager[20] in his treatment of liquid crystals: Onsager treated molecules with different orientations as members of different species; we are treating particles from different replicas as members of a single species differing only by values of a fictitious internal coordinate. This description is equivalent to the original one, in which the number of particles in each replica was held constant, because we work in the grand canonical ensemble.

Thus in considering Mayer expansions for the properties of this system, each internal vertex, or root point, associated with a type 1 particle will be accompanied by both a spatial integration over its position and a summation over its spin state. Also, because of the spin-dependence of the last term in the Hamiltonian (2.6), each pair of root points of type 1 that are connected by a Mayer bond must be in the same spin state.

We pause to comment upon the peculiar spin-dependent potential, given by the last term in eqn. (2.6), which acts between pairs of type 1 particles; we have previously[2-4,21-23] described this as a generalized Widom-Rowlinson interaction. The original Widom-Rowlinson model was introduced as a model of phase separation;[21] it involved a mixture of particles of two different species (or 'spin states') with a repulsive interaction between particles in **different** species. However, n-species generalizations such as that used here occur frequently in the theory of random media.[27,28] They are actually continuum generalizations of the ferromagnetic Potts model.[22] A basic insight of the work described here is that the continuum version of the well-studied replica method, used for treating many quenched random systems, falls within the same framework. It is a limiting case of an **anti-Widom Rowlinson model**, so-called because repulsive interactions are present between particles in the same species, rather than those in different species. This model is a continuum generalization of the **anti-ferromagnetic** Potts model. The intriguing notion suggested by these mappings, namely that there are two basically different kinds of models for randomly disordered materials, is still under study. It is also important to note that both of these well-studied classes of non-equilibrium systems are isomorphic to Hamiltonian models with pair interactions that define **non-additive** particle diameters; this is seen to be a general trait of such systems.

Here we note that the quenched two-phase system discussed in this section is the $n \to 0$ limit of the mixture just described, in the sense implied by eqn. (2.5). This relation provides explicit formulae for the coefficients in the virial expansions of the basic physical quantities. The Kronecker delta occurring in the interaction (2.6) ensures that any pair of species-1 particles connected by a Mayer bond must be in the same spin state. Since each group of such particles in a Mayer graph that are connected by bonds into a cluster will be weighted by a factor of n by the summation over spin states, the only graphs making a non-zero contribution in the $n \to 0$ limit will be those having all the species-1 particles connected directly or indirectly by Mayer bonds. By a similar argument, the graphs contributing to the correlation functions are precisely those in which each species-1 particle is connected, directly or indirectly, to a root point by a chain of Mayer bonds passing only through species-1 particles.

The lowest-order example of such functions is the density, for which the corresponding Mayer graphs are those with a single root point. A set of Ornstein-Zernike equations has been developed[9] by exploiting the description just given for the structure of the Mayer graphs that contribute to the correlation functions.

We can derive these equations, called the replica Ornstein-Zernike (ROZ) equations[9] satisfied by the correlation functions of a partly quenched system.[24] To do this, we write down the OZ equations for the replicated $(n+1)$-species system with Hamiltonian (2.6), isolate the n-dependence of these equations (by grouping together identical terms), and take the $n \to 0$ limit. The resulting ROZ equations are:

$$h^{00} = c^{00} + \rho c^{00} \otimes h^{00} \tag{2.7}$$

$$h^{01} = c^{01} + \rho c^{00} \otimes h^{01} + \rho c^{01} \otimes h^{11} \\ - \rho c^{01} \otimes h^{12} \tag{2.8}$$

$$h^{10} = c^{10} + \rho c^{10} \otimes h^{00} + \rho c^{11} \otimes h^{10} \\ - \rho c^{12} \otimes h^{10} \tag{2.9}$$

$$h^{11} = c^{11} + \rho c^{01} \otimes h^{01} + \rho c^{11} \otimes h^{11} \\ - \rho c^{12} \otimes h^{12} \tag{2.10}$$

$$h^{12} = c^{12} + \rho c^{01} \otimes h^{01} + \rho c^{11} \otimes h^{12} \\ + \rho c^{12} \otimes h^{11} - 2\rho c^{12} \otimes h^{12} \tag{2.11}$$

The symbol $\otimes$ denotes both a spatial convolution and an integral over the spin of the intermediate particle. That is, we have

$$f \otimes g \equiv \int d^3 x_2 \frac{d\Omega_2}{\Omega} f(x_1, s_1, x_2, s_2) g(x_2, s_2, x_3, s_3) \tag{2.12}$$

The integral $d\Omega_2$ is over the orientation of the vector spin s_2. The quantity Ω is a normalizing factor $\Omega = \int d\Omega_2$ which we have extracted from the fugacity of a 'spin'; it turns the integral over spin orientation into an average over this quantity.

We prefer to rewrite the ROZ equations in a way that emphasizes their graphical structure. To do this, it is useful to identify the function we shall denote here as c_b, which is represented by the subset of c_{11} graphs such that all paths between the two white species-1 vertices pass through at least one ρ_0-vertex. We further identify the function represented by the corresponding subset of h_{11} graphs as h_b and write

$$c_{11}(12) = c_c(12) + c_b(12) \tag{2.13}$$

$$h_{11}(12) = c_c(12) + h_b(12) \tag{2.14}$$

so that between the two white vertices of every c_c and h_c graph there is at least one unbroken path free of ρ_0-vertices (the subscripts 'c' and 'b' denote 'connected' and 'blocking', respectively.) We can then rewrite the ROZ equations as

$$h_{00} = c_{00} + c_{00} \rho_0 \otimes h_{00} \tag{2.15}$$

$$h_{10} = c_{10} + \rho_0 c_{10} \otimes h_{00} + \rho_1 c_c \otimes h_{10} \tag{2.16}$$

$$h_{11} = c_{11} + \rho_0 c_{10} \otimes h_{01} + \rho_1 c_c \otimes h_{11} + \rho_1 c_b \otimes h_c \tag{2.17}$$

$$h_c = c_c + \rho_1 c_c \otimes h_c \tag{2.18}$$

where, by symmetry, $c_{01} = c_{10}$ and $h_{01} = h_{10}$. Here the symbol $\otimes$ denotes a convolution.

An alternative equation for h_{01} that can be derived from using (2.12-13) in (2.19) is

$$h_{01} = c_{01} + \rho_0 c_{00} \otimes h_{01} + \rho_1 c_{01} \otimes h_c. \tag{2.19}$$

When c_{00}, c_{01}, c_{11}, c_c and the $\{\rho_i\}$ are prescribed, (2.15-18) are a closed set of equations for h_{00}, h_{01}, h_{11}, and h_c. For some $v_{\alpha\beta}$ a reasonable approximation in the ROZ equations is given by the assumption

$$c_b(12) = 0. \tag{2.20}$$

This approximation is implied by the Percus-Yevick (PY) closure, as well as the mean-spherical approximation (MSA) closure. We shall refer to the approximation (2.20) as the Madden-Glandt (MG) approximation. There is a class of approximate closures of the ROZ equations that imply $c_b(x) = 0$; within the context of such a closure the Madden-Glandt approximation is exact. This class of closures includes the Percus Yevick (PY) closure, $c_{ij} = f_{ij}y_{ij}$, where f_{ij} is the Mayer function, and the cavity function y_{ij} is defined by the equation

$$h_{ij} + 1 \equiv (f_{ij} + 1)y_{ij} \tag{2.21}$$

It also includes the mean-spherical approximation,

$$h_{ij} = -1, \qquad x_{ij} < R_{ij} \tag{2.22}$$

$$c_{ij} = \beta v_{ij}, \qquad x_{ij} > R_{ij} \tag{2.23}$$

Here v_{ij} is the pair potential, x_{ij} is the distance between particle centers and R_{ij} is the average of hard core diameters. However, the HNC approximation is not among this class of closures.

A fundamental fact about the MG approximation is this: in a sense to be made precise, the only difference between the MG OZ equations and the equilibrium OZ equations is in the quenched system input. In a quenched system, the functions h_{00} and c_{00} are supplied "externally", i.e., they are not influenced by the behavior of the annealed degrees of freedom. If we take these functions as input and proceed to solve both the MG OZ equations and the equilibrium OZ equations for the functions h_{01}, h_{11} (using the same closure) the results will be identical. This shows that the MSA (which implies the MG approximation) is "blind" to the difference between partly quenched and annealed systems.

This raises the question: what systems will exhibit appreciable "c_b effects," that is, effects that go beyond the effect of the quenched system input? The discussion above implies that such effects should be most pronounced for partly quenched systems in which the most natural closure lies outside the class discussed above, for which the MG approximation is exact. We discuss here a class of example for which the HNC closure is natural. Specifically, we consider the class of systems in which the quenched species consists of randomly centered particles ($f_{00} = 0$, ρ_0 spatially uniform) that freely overlap each other but are impenetrable to the annealed particles. Such a partly quenched system gives a natural model for a fluid adsorbed in a porous medium, the annealed and quenched particles representing, respectively, the fluid atoms, and the partly overlapping inclusions that make up the porous matrix. For this system, one can calculate c_b exactly in the limit $\rho_1 \to 0$ by summing the Mayer graphs that contribute to this quantity. Such Mayer graphs consist of two or more ρ_1-vertices, each connected to both of the root points. The sum of such graphs is easily obtained. It is

$$c_b(12) = \exp[\rho_0 O^\circ] - 1 - \rho_0 O^\circ \tag{2.24}$$

where $O^\circ(12)$ is the "overlap volume" integral

$$O^\circ(12) = \int d3\, f_{10}(13)f_{01}(32) \tag{2.25}$$

Physically, this contribution to c_b represents a screening or interference between the inclusions that make up the porous matrix; such screening is very substantial, even at low density of adsorbed fluid, for the intermediate to high densities of matrix inclusions that characterize many microporous materials. Another limit in which the RHS gives an exact result for $c_{11}(x)$ is that of an ideal fluid of noninteracting atoms, again adsorbed in a rigid matrix of freely overlapping inclusions. In this case we also have $c_{11} = c_{12}$, $c_{01} = f_{01}$, (with f_{01} the Mayer bond) and $c_{00} = 0$.

We are now studying simulations of such a porous medium. For low fluid density but moderate matrix density, the approximation given by (2.24) is superior to the MG approximation. However, both approximations underestimate the contact values of correlation functions. At higher fluid particle densities, the two approximations become very similar. We are trying to improve (2.24) by using a reference-system closure, in which the contributions to $\{c_{ij}\}$ which are set equal to zero in simpler closures, are instead evaluated within the reference system. Here we use as the reference system the pure species-0 system into which a single pair of species-1 particles are inserted. We represent the direct correlation function c_{ij} as follows:

$$c_{ij} = f_{ij} y_{ij} + d_{ij} \tag{2.26}$$

The function d_{11} is zero in the PY approximation. We note that this function contains all the terms in c_b because, by definition, it contains all the terms in c_{11} that lack a Mayer bond between the roots and all contributions to c_b, by definition, have this property. We can thus incorporate the screening effects described above by approximating d_{11} by c_b, and using for the latter the reference-system value (2.24). which takes us beyond the MG result and is exact in the limit $\rho_1 \to 0$. These approximations are still under study.

SPIN GLASS

We consider[10] each atom in a spin glass[25] to be a tightly associated pair of "pseudoatoms", one of which we call a 'spin', the other of which we call an 'atom'. We will relax the association between these pseudoatoms, use the replica method described in Section 2 to write the free energy and correlation functions of the spin glass in terms of the corresponding quantities for an equilibrium mixture of particles, then impose the constraint of complete association.

The interaction $v_{11}(ij)$ between a pair of spins i and j is taken to be the sum of a classical Heisenberg interaction and a spin-independent excluded-volume interaction (taken here to be of hard-sphere type):

$$v_{11}(ij) = v_{HS}(x_{ij}) + J(x_{ij})(\vec{s}_i \cdot \vec{s}_j) \tag{3.1}$$

where the dot product between the vector spins located at atoms i and j is multiplied by the separation-dependent exchange coupling $J(x_{ij})$. In many materials a reasonable form for $J(x)$ is provided by the sinusoidally modulated Yukawa coupling

$$J(x) = \frac{\exp[-\kappa x]}{x^p} \sin(\alpha x + \theta) \tag{3.2}$$

The interatomic interaction $v_{00}(x)$ is taken to be hard-core at short range; the long-range tail $v_{LR}(x)$ can be left arbitrary

$$v_{00}(x_{ij}) = v_{LR}(x_{ij}) + v_{HS}(x_{ij}) \tag{3.3}$$

Finally, the interaction v_{01} is taken to be an extremely strong, short-range attractive force binding a spin to the center of an atom. The presence of a hard-core component in both the spin-spin and atom-atom potentials ensures that no complexes containing more than one atom and one spin may form.

The continuum replica method provides an explicit mapping of the spin glass onto a limiting case of an equilibrium Hamiltonian system, to be called the replicated system. Specifically, this is an $(n + 1)$-species mixture, consisting of one copy of the quenched species, i.e., of the atoms, together with n copies of the annealed species, i.e., the spins. Here quenched and annealed particles have the same pairwise interactions that they have in the partly quenched system being studied, **except** that annealed particles from different replicas do not interact. In the complete association limit, the replicated system is entirely composed of tightly bound complexes, each consisting of an atom, together with one spin from **each** of the replicas.

We discuss two complementary approaches to developing integral equations for the correlation functions of the spin glass. The first is a version, for partly quenched systems of the RISM approximation.[26,27] It has the advantage of simplicity, and aids both intuition (in the formation of closures) and the process of numerical solution. It has the disadvantage of not providing a direct correlation function that can be expressed as a resummation of Mayer perturbation theory, i.e., of not being "proper," in the language of interaction site theory.[28,29] The second approach remedies that difficulty; it uses the chemical association[30] formalism of Wertheim[31,32] to develop a set of four coupled integral equations for the different terms contributing to h_{11}. The theory involves four direct correlation functions that correspond to these terms.

We first develop a RISM-like equation for the spin glass. It is natural to use site-site notation, with subscripts '0' and '1' for sites that are atoms and spins, respectively. The site-site formalism allows us to treat atoms and spins as separate particles, eg., for the purpose of applying the replica method, and yet be able to separate out the delta-function contributions to correlation functions from spins and atoms that are bound together. It is thus natural to use the SSOZ equations for spin glass correlation functions. However, because of the high degree of symmetry posessed by the Heisenberg spin glass, the distinction between the OZ and SSOZ equations will not be important here. This is discussed below.

It is useful to write for any pair correlation function $a(12)$

$$a(12) = \bar{a}(12) + \tilde{a}(12) \tag{3.4}$$

where $\bar{a}(12)$ is an average over the orientation Ω_i of both particles

$$\bar{a}(12) = \int a(12)d\Omega d.\Omega_2/\Omega^2, \Omega = \int d\Omega_i \tag{3.5}$$

In the complete association limit, the positions of atoms and spins must coincide. Thus, we have for a general interaction

$$\bar{h}_{11} = \bar{h}_{10} = h_{00} \tag{3.6}$$

Also, for interactions having the symmetry of the Heisenberg interaction, we have

$$\tilde{h}_{10} = \tilde{h}_b = 0 \tag{3.7}$$

where our notation is the same as in (2.15-2.18).

Using eqns. (3.6-7) in eqn. (2.17), we can write a simple Ornstein-Zernike (OZ) equation for $\tilde{h}_{11}$

$$\tilde{h}_{11} = \tilde{c}_{11} + \rho\tilde{c}_{11} \otimes \tilde{h}_{11} \tag{3.8}$$

Adding eqns. (3.6) and (2.7) gives an OZ equation for $h(12)$ itself,

$$h = c + \rho c \otimes h, \tag{3.9}$$

where $h = h_{00} + \tilde{h}_{11}$ and $c \equiv c_{00} + \tilde{c}_{11}$. It follows that in the mean-spherical approximation, the spin-spin correlation function for a spin fluid and a spin glass are identical, assuming the two have identical atom-atom correlation functions. This was surmised in [33].

Appropriate closures of PY and HNC (hypernetted chain) type for the spin glass[33] have been developed. We do not derive these here, but simply present them. For the PY-type closure one has

$$\tilde{c} = f(g-c) - \overline{f(g-c)} \tag{3.10}$$

where $g = h + 1$. For long-range $J(x_{ij})$, one might expect the analogous HNC-type closure to prove more accurate. This is given by

$$\tilde{c} = h - \ln y - \overline{(h - \ln y)} \tag{3.11}$$

where $g = (f+1)y$.

In order to extend this treatment in a systematic manner, it is important to develop a proper integral-equation treatment of the spin glass. This involves classifying the Mayer graphs that contribute to h_{11} into four types, according to whether they have neither, one, or both of the root points (representing spins) in such a graph connected directly to type 0 vertices (representing atoms.) The corresponding decomposition of h_{11} is

$$h_{11} = h_{11}^{00} + h_{11}^{01} + h_{11}^{10} + h_{11}^{11}. \tag{3.12}$$

The quantities $\{h_{11}^{ij}\}$ obey a coupled set of OZ equations which generalize (3.8)

$$\tilde{h}_{11}^{ij} = \tilde{c}_{11}^{ij} + \tilde{c}_{11}^{ik} \otimes \rho^{k\ell}\tilde{h}_{11}^{\ell j}. \tag{3.13}$$

Here we have $\rho^{00} = \rho^{01} = \rho^{10} = \rho$ and $\rho^{11} = 0$.

SEQUENTIAL ADSORPTION PROCESSES

The adsorption of large molecules and molecular aggregates, such as polymers, colloids, and proteins, to membranes and surfaces is frequently associated with a large binding energy so that the time needed for the surface involved to become saturated with particles is small compared to a typical desorption time. In such cases, the correlation functions are substantially different from those associated with equilibrium. It has been shown to be a good approximation, at least in some cases, to use for such processes an idealized model known as **random sequential adsorption** (RSA).[18] In this model, hard, i.e., non-overlapping particles are placed on a surface, one at a time, each in such a way that it does not overlap those already in place. Once placed, a particle is quenched, i.e., frozen in that location. This is the simplest RSA model. We generalize this model in two ways: first, by allowing a longer-range potential interaction between particles; second, by allowing for a finite probability of rearrangement. The former generalization is a natural way to allow the adsorption of charged particles; the latter gives a model that smoothly interpolates between an equilibrium hard-sphere system and a sequential adsorption system. As described in the Introduction, we can develop liquid-state theory for this process by viewing it as an extreme case of a system constructed in layers: here, each particle constitutes a separate component or layer, and the graphical rules developed in Section 2 for the two-layer system must be generalized.[4] Some reflection shows that we get prescriptions for the graphs contributing to both the free energy and the correlation functions if we replace the term "annealed path" in the prescriptions for the two-layer system by the term "uphill path", where an **uphill path** is defined to be a sequence of vertices, each successive pair connected by a Mayer bond, such that in traversing the path from field point to root point, the species labels of the vertices encountered increase monotonically. (Here, every particle in RSA is considered to belong to a different species, and successive particles are given species labels in the order in which they are introduced.)

We map the general RSA model just described onto a Hamiltonian system. Consider a system of particles bearing spins σ_i which can take any of n discrete values, such that particles i and j, $i < j$, have interaction

$$V_{ij}(x) = v(x)\delta_{\sigma_i,1} + \phi(x) \tag{4.1}$$

We consider this system to be a multi-species mixture, with each particle belonging to a distinct "species" labelled by its particle number. In the $n \to 0$ limit, the thermodynamic quantities of this model become the basic physical quantities that describe RSA. For example, we have the formulas for the adsorption rate Φ:

$$\Phi = \lim_{n \to 0} \frac{\rho}{z} \tag{4.2}$$

with z the fugacity and ρ the density of the mixture (4.1).

The two potential functions in this model, $v(x)$ and $\phi(x)$, can be chosen independently to give a variety of interesting models. If the potential $v(x)$ in (4.1) is zero, the interaction will give an equilibrium system with potential $\phi(x)$. If the potential $\phi(x)$ is zero, it gives various models of sequential adsorption. Choosing $v(x)$ to be a hard-sphere interaction gives the model usually called RSA. Choosing $v(x)$ to be the sum of a hard-sphere interaction and a Coulomb potential gives a model for the correlated but irreversible adsorption of charged, hard particles. This system could be realized experimentally by placing a static charge on small latex spheres, then allowing them to adsorb strongly onto a surface. We note that unless the potential $v(x)$ has a hard core, the resulting RSA model will have no jamming density. Nevertheless, one can ask whether the asymptotic behavior of the adsorption rate will still be a power law in the elapsed time, as it is in naive RSA.

We have two paths that lead to integral equations for the correlation functions in RSA. The first is to derive the ROZ equations by using the Hamiltonian mapping defined by eqn. (4.1). The second is to write the Kirkwood-Salsburg hierarchy for this process, again using the Hamiltonian (4.1) and use a relation between the LHS of its first two equations to derive a Percus-Yevick equation. We will sketch both of these.

We define the two-point correlation function of the Hamiltonian system of Section 2 as follows: $g_{ij}^{\sigma_i\sigma_j}(x)$ is the probability density associated with finding a particle of species i in spin state σ_i and a particle of species j in spin state σ_j, the two of them separated by a distance x. The $n \to 0$ limit of this function gives, depending on the spin indices σ_i and σ_j, two correlation functions useful in describing RSA. In this limit, the "species" of our original model become particle labels, describing the sequential order in which particles are placed. We will use the particle labels i and j, with the convention that $i < j$. We define two RSA correlation functions as follows:

$$g_{ij}^{11}(x) \to_{n \to 0} g_{ij}(x) \equiv g_{ij}^b(x) + g_{ij}^c(x) \tag{4.3}$$

$$g_{ij}^{i1}(x) \to_{n \to 0} g_{ij}^b(x) \tag{4.4}$$

The function $g_{ij}(x)$ is the probability density associated with finding, in a realization of RSA, the ith and j^{th} particle separated by a distance x. The Mayer series for this function is easily constructed using the Hamiltonian (4.1): it consists of all labelled, two-rooted graphs satisying the uphill constraint. A labelled Mayer graph is interpreted as an ordering on the particle labels; the field points in a labelled graph must be summed over all sets of particle labels that preserve this ordering. The function $g_{ij}^b(x)$ is a related function defined as the sum of the subset of Mayer graphs contributing to $g_{ij}(x)$ such that there exists **no** uphill path joining the two root points. Similarly, we have $g_{ij}^c(x)$ defined as the sum of the subset of Mayer graphs contributing to $g_{ij}(x)$ such that there exists **at least one** uphill path joining the two root points. Combining eqns. (4.3) and (4.4) gives

$$[g_{ij}^{11}(x) - g_{ij}^{i1}(x)] \to_{n \to 0} g_{ij}^c(x) \tag{4.5}$$

This function occurs in the topological reduction of the Mayer expansion.

By virtue of the isomorphism we establish onto the many body system defined by the Hamiltonian (4.1), the RSA system has the structure of a polydisperse system. In such systems, each particle has, in addition to its position, an extra parameter characterizing it. In the standard examples of polydisperse systems,[23,24] this parameter is the particle radius or orientation. Here, the extra parameter characterizing the ith particle in the system is ρ_i, the density of the system at the time that particle was added. This density is given by

$$\rho_i = i/V \tag{4.6}$$

In the thermodynamic limit, each intermediate summation over the particles in the system occurring e.g. in the Mayer expansions can be replaced by an integration over the extra parameter ρ_i. The integration measure for such an integration is trivial because each particle added to the system corresponds to a distinct value of ρ_i.

We henceforth assume this limit to be taken, and use as limits the densities ρ_i instead of particle labels i, density integrations instead of summations over particles, etc. Each specific correlation function $g_{ij}(x)$ is actually a function of three densities: ρ_i and ρ_j, the extra parameters characterizing the two root points, and ρ, the final particle density of the RSA system in which the correlation functions are computed. However, because each particle in a realization of RSA is independent of the particles placed **after** it, we have

$$g_{ij}(x,\rho) = g_{ij}(x,\rho_j) \tag{4.7}$$

for any $\rho > \rho_j$. In this equation and what follows, we will adopt the convention, unless stated otherwise, that $j > i$.

The ROZ equations satisfied by the functions $h_{ij}^b \equiv g_{ij}^b - 1$ and $h_{ij}^c \equiv g_{ij}^c - 1$ can be obtained by writing down the Ornstein-Zernike equations for the correlation functions of a polydisperse system with pairwise interaction (4.1), and taking the RSA limit $n \to 0$. We rewrite these immediately in terms of the definitions (4.3 - 4.5):

$$h_{ij}^c = c_{ij}^c + \int_{\rho_i}^{\rho_j} \left[c_{ik}^c \otimes h_{kj}^c \right] \tag{4.8}$$

$$h_{ij}^b = c_{ij}^b + \int_0^{\rho_i} \left[c_{ki}^c \otimes h_{kj}^c + c_{ki}^c \otimes h_{kj}^b + c_{ki}^b \otimes h_{kj}^c \right] d\rho_k + \int_{\rho_i}^{\rho_j} \left[c_{ik}^b \otimes h_{kj}^c \right] d\rho_k \tag{4.9}$$

The functions $\{g_{ij}^b(x)\}$ and $\{g_{ij}^c(x)\}$ are specific RSA correlation functions, i.e., they give the probability density associated with finding a specific pair of particles at a separation x. These functions, like the correlation functions of any polydisperse system, depend upon an additional parameter, i.e., an extra density. One must solve (4.8-9) and then form the generic correlation function, or probability density associated with finding any two particles with separation x. It would be of great value to be able to rewrite the ROZ equations (4.8-9) directly in terms of generic correlation functions. This we have not been able to do. However, we have obtained a formally exact equation for the generic correlation functions by using the Kirkwood-Salsburg equations for RSA. We will define the generic correlation functions for RSA, give the Kirkwood-Salsburg equations for them, and then relate these functions to the specific functions that satisfy (4.8-9).

Consider the function $\Phi_n(x_1; x_2 \ldots x_n)$ that represents the probability density associated with being able to insert at position x_1 an extra particle into a system consisting of $N + n$ particles, with $n - 1$ of them found at $x_2, \ldots, x_n$, respectively. [We shall write $\Phi_1(x)$ for Φ_n in the absence of the $n - 1$ particles at prescribed locations; in a spatially homogenous system in which $\Phi_1(x_1)$ is independent of x_1 we shall simply write Φ for $\Phi_1(x_1)$.]

There is a way of representing such functions as a series that goes back to Boltzmann,[11] who considered the first few terms in such a representation of Φ_1 and Φ_2 for a hard-sphere system in equilibrium. Subsequently Kirkwood and Salsburg considered the full series of the Φ_n for arbitrary pair potentials in an equilibrated system.

We can derive the Kirkwood-Salsburg equations for RSA by considering RSA as the $n \to 0$ limit of the process with Hamiltonian (4.1). The result is

$$\Phi_n(x_1; x_2 \ldots x_n) =$$
$$e_n(x_1; x_2 \ldots x_n) \sum_{s=0}^{\infty} \frac{1}{s!} \int dx_{n+1} \ldots dx_{n+s} \prod_{s=1}^{n} f(x_i x_{n+s}) \rho_{n+s-1}(x_2 \ldots x_{n+s}) \qquad (4.10)$$

Here

$$e_n(x_1; x_2 \ldots x_n) = \prod_{i=2}^{n} [1 + f(x_1 x_i)] \qquad (4.11)$$

and the $\rho_n(x_1 \ldots x_n)$ are **generic** n-particle probability density functions associated with finding n unspecified particles at positions $x_1 \ldots x_n$. On the other hand the Φ_n have an intrinsically **specific** quality in the sense that they specifically refer to the extra particle, which must be the **last** of the particles being inserted into the system. We could replace Φ_n by a fully specific probability density by referring to specified (i.e., labelled) particles at $x_2 \ldots x_n$. This would define a function that differs from Φ_n only through a trivial change in normalization; Φ_n already has the symmetry properties of its specific version. Strictly speaking, however, it is neither fully specific or fully generic. We shall refer to such functions as "mixed." In RSA the most important mixed functions are those, like the Φ_n, that refer to one **last** particle, with the rest of the particles unspecified.

In a system of particles at equilibrium that are identical except for labelling, the entire difference between specific and generic functions is just a trivial one of normalization, because particle labelling can be done in any order — order is of no consequence. This implies that Φ_n is very simply related to ρ_n. One has

$$\frac{\rho_n(x_1 \ldots x_n)}{\rho_1(x_1)} = \frac{\Phi_n(x_1; x_2 \ldots x_n)}{\Phi_1(x_1)} \qquad (4.12)$$

as the closure of (4.10). In the spatially uniform case this reduces to

$$\rho_n(x_1 \ldots x_n)/\rho = \Phi_n(x_1; x_2 \ldots x_n)/\Phi_n \qquad (4.13)$$

In the case of RSA, on the other hand, the ordering of the particles constitutes the whole problem! As a result the generic ρ_n are related to the Φ_n via a combination transformation that takes this ordering into account. This can be carried out term by term in the density expansion of ρ_n and summarized in terms of a single differentiation with respect to ρ, which yields the coefficients that are found in the expansion of Φ_n, which the latter are appropriately symmetrized. Considering only the spatially uniform case for simplicity, we have

$$\frac{\partial \rho_n(x_1 \ldots x_n)}{\partial \rho} = \sum_{k=1}^{n} \Phi_n(x_1; x_1 \ldots x_{k-1}, x_{k+1} \ldots x_n)/\Phi \qquad (4.14)$$

We shall find it convenient to introduce the probability densities $\Psi_n(x_1/x_2 \ldots x_n)$ associated with being able to insert a particle into the system at position x_1 in the presence of $n - 1$ particles at positions $x_2, \ldots, x_n$, respectively. One has

$$\Phi_n(x_1; x_2 \ldots x_n) = \Psi_n(x_1/x_2 \ldots x_n) \rho_{n-1}(x_2 \ldots x_n) \qquad (4.15)$$

In particular

$$\Psi_2(x_1/x_2) = \Phi_2(x_1; x_2)/\rho_1(x_2) \qquad (4.16)$$

and in a uniform system

$$\Psi_2(x_1/x_2)/\Phi = G(x_1/x_2) = G(x_{12}) \qquad (4.17)$$

where $G(x_{12})$ is a radial distribution function. In a uniform equilibrium system we would have simply

$$G(x_{12}) = \rho(x_1 x_2)/\rho^2 \tag{4.18}$$

but from (4.14) and (4.17) we have instead

$$2G(x_{12}) = \frac{1}{\rho}\frac{\partial\rho(x_1 x_2)}{\partial\rho} \tag{4.19}$$

or

$$G(x_{12}) = g(x_{12}) + \frac{\rho}{2}\frac{\partial g(x_{12})}{\partial\rho} \tag{4.20}$$

where $\rho(x_1 x_2) = \rho^2 g(x_{12})$, so $g(x_{12})$ is the generic two-particle distribution function.

Thus we have both a mixed cavity function $Y(x) = G(x)/[1 + f(x)]$ and a generic cavity function $y(x) = g(x)/[1 + f(x)]$. It is the former that appears in a more fundamental way in our theory. In particular we have the zero-separation condition for hard-particle RSA

$$\Phi Y(0) = 1 \tag{4.21}$$

from (4.10) with $n = 2$ as well as its obvious generalization to $n > 2$.

We now relate the generic and specific correlation functions. Thus, from its definition, we can express $G(x_1/x_2)$ as

$$\rho G(x_1/x_2) = \sum_{i=1}^{j} \rho_i g_{ij}(x) \tag{4.22}$$

In the limit of large N, this can be rewritten in terms of an integral over density:

$$\rho_j G(x_1/x_2, \rho_j) = \int_0^{\rho_j} g_{mj}(x_1, x_2, \rho_m, \rho_j) d\rho_m \tag{4.23}$$

The generic distribution function $g(x)$ is the probability density associated with finding any two particles at positions x_1 and x_2; it can be expressed in terms of the $\{g_{ij}(x)\}$ as

$$\begin{aligned}\rho^2 g(x_1, x_2) &= \sum_{j=1}^{N-1} \rho_j[G(x_1/x_2, \rho_j) + G(x_2/x_1, \rho_j)] \\ &= \sum_{i,j=1}^{N-1} \rho_i \rho_j g_{ij}(x_1, x_2, \rho_i, \rho_j)\end{aligned} \tag{4.24}$$

This can also be rewritten in terms of an integral over density:

$$\rho^2 g(x_1, x_2) = \int_0^{\rho} [G(x_1/x_2, \rho_i) + G(x_1/x_2, \rho_i)] d\rho_i \tag{4.25}$$

From (4.10) one can also obtain a number of interesting rigorous results for non-negative pair potentials, for which the remainder terms in the series alternate in sign, yielding rigorous upper and lower bounds on Φ_n.

For hard particles, the function $\Psi_n(x_1/x_2 \ldots x_n)$ has the significance of being the volume available to the center of the particle to be inserted at x_1, given particles fixed at $x_2, \ldots x_n$, divided by V, the volume of the box in which the system is contained. It is also useful to consider the volume available to the center of a cavity of the same size as a particle. This is just VA_n, where

$$A_n(x_1/x_2 \ldots x_n) = \frac{\Psi_n(x_1/x_2 \ldots x_n)}{e_n(x_1; x_2 \ldots x_n)} \tag{4.26}$$

with $A_1(x_1) = \Phi_1(x_1)$. We have, in a uniform system,

$$A_1(x_1/x_2) = \Phi Y(x_{12}) \tag{4.27}$$

To generate an exact functional expansion that yields a Percus-Yevick equation upon truncation, we can consider $A_1(x_1/x_2)$ expanded around its value $\Phi_1(x_1)$ when the particle at x_2 is turned off. We have

$$A_1(x_1/x_2) = \Phi_1(x_1) + \int \frac{\delta\Phi_1(x_1)}{\delta\rho(x_3)}[\rho(x_3 x_2) - \rho(x_3)]dx_3$$
$$+ \frac{1}{s!}\int \frac{\delta^s \Phi_1(x_1)}{\prod_{i=3}^{s+2}\delta\rho(x_i)}\prod_{i=3}^{s+2}\left[(\rho(x_i x_2) - \rho(x_i))dx_i\right] + \cdots \tag{4.28}$$

The mixed direct correlation functions $c_n(x_i; x_2 \cdots x_n)$ are generated by $\ln\Phi(x_i) = c_1(x_i)$

$$\frac{\delta^s c(x_1)}{\delta\rho(x_2)\ldots\delta\rho(x_{s+1})} = c_{s+1}(x_1; x_2 \ldots x_{s+1}) \tag{4.29}$$

Using (4.27) and (4.28), yields, in the uniform case,

$$Y(x_{12}) = 1 + \rho \int c_2(x_1; x_3)h(x_3 x_2)dx_3$$
$$+ \frac{\rho^2}{2}\int \left[c_3(x_1; x_3 x_4) - c_2(x_1; x_2)c_2(x_1; x_3)\right]h(x_3 x_2)h(x_4 x_2)dx_3 dx_4 \cdots \tag{4.30}$$

Truncation after the first non-trivial terms yields a Percus-Yevick approximation suitable for RSA

$$Y(x_{12}) = 1 + \rho \int c_2(x_1; x_3)h(x_3 x_2)dx_3 \tag{4.31}$$

In equilibrium, where $Y(x) = y(x)$, this relation defines a unique approximation when combined with the OZ equation which identifies the correlation terms as $h - c_2$, yielding $y = 1 + h - c_2$. For hard particles in equilibrium, this yields a good approximation $c_2 = 0$ outside the hard core, where $y = 1 + h$. Inside the core it yields the result $y = c_2$. This represents a very poor approximation for y; however inside the core, the PY approximation is very useful in situations in which knowledge of y inside the core is not needed. In the case of RSA, the OZ equations do not have as simple a form. In particular, the convolution on the LHS of (4.30) is not equal to $h - c_2$. Instead of obtaining a unique approximation via the OZ equations, it is more convenient to obtain one directly by introducing the closure

$$c_2(x_1; x_2) = f(x_{12})[G(x_1/x_2) - c_2(x_1; x_2)] \tag{4.32}$$

For hard particles, this simply says that $G = 0$ inside the core, which is identically true, and that

$$c_2(x_1; x_2) = 0 \tag{4.33}$$

outside the core, which is likely to be a good approximation for the same reasons that the PY approximation is good for hard particles in equilibrium. Equations (4.31), (4.32), and (4.19) yield a unique approximation for soft- as well as hard- core particles.

CONCLUSIONS

We have developed very general techniques for mapping a class of nonequilibrium problems into equilibrium statistical mechanics. We are now studying various approximations and closures of the equations presented here. Still before us is the task of incorporating phase transition behavior[34] into our account of systems with quenched disorder.

ACKNOWLEDGMENTS

J.G. is grateful to the National Science Foundation for their support of this work. G.S. gratefully acknowledges financial support from the Division of Chemical Sciences, Office of Basic Energy Sciences, Office of Energy Research, U.S. Department of Energy.

REFERENCES

1. J.M. Ziman, "Models of Disorder", Cambridge Univ. Press, Cambridge (1979).
2. G. Stell, A.M.S., Statistical mechanics applied to random-media problems, in: "AMS-SIAM Workshop on Random Media", G. Papanicoleau, ed., New York, AMS, (1991).
3. J.A. Given and G. Stell, Towards a realistic, general, continuum theory of clustering, in: "On Clusters and Clustering: from Atoms to Fractals", P. Reynolds ed., North Holland, New York (1991).
4. J.A. Given, Liquid-state methods for random media: I. random sequential adsorption, *Phys. Rev.* A 45:816 (1991).
5. J.A. Given, Liquid-state methods for random media: II. spin glasses, *J. Chem. Phys.* 96:2287 (1992).
6. W.A. Madden and E.D. Glandt, Distribution functions for fluids in random media, *J. Stat. Phys.* 51:537 (1988).
7. W.A. Madden, Fluid distributions in random media: arbitrary matrices, *J. Chem. Phys.* 96:5422 (1992).
8. L.A. Fanti, E.D. Glandt, and W.G. Madden, Fluids in equilibrium with disordered porous materials: integral equation theory, *J. Chem. Phys.* 51:537 (1988).
9. J.A. Given and G. Stell, Comment on: fluid distributions in two-phase random media: arbitrary matrices, *J. Chem. Phys.* 96: xxxx (1992).
10. J.A. Given and G. Stell, A realistic model for a spin glass, in: "Complex Fluids: MRS Symposium Proceedings", Materials Research Society, Pittsburg, 1991.
11. G. Stell, Mayer-Montroll equations (and some variants) through history for fun and profit, in: "The Wonderful World of Stochastics", M.F. Schlesinger and G.H. Weiss eds., North Holland, Amsterdam (1985).
12. M. Mezard, G. Parisi, and M.A. Virasoro, "Spin Glass Theory and Beyond", World Scientific, Singapore (1987).
13. R. Kraichnan and S.Y. Chen, Is there a statistical mechanics of turbulence?, *Physica* D37:160 (1989).
14. H.E. Stanley and N. Ostrowsky, "Random Fluctuations and Pattern Growth: Experiments and Models", Kluwer Academic, Norwell MA (1988).
15. K. Kang, S. Redner, P. Meakin, and F. Leyvraz, Long-time crossover phenomena in coagulation kinetics, *Phys. Rev.* A33:1171 (1986).
16. See e.g. E. Guyon, C.D. Mitescu, J.P. Hulin, and S. Roux, *Physica* D38:172 (1989).
17. G. Tarjus, P. Schaaf, and J. Talbot, Random sequential addition: a distribution function approach, *J. Stat. Phys.* 63:167 (1991).
18. J.W. Evans, RSA review.
19. P.L. Garrido, J.L. Lebowitz, C. Maes, and H. Spohn, Long-range correlations for conservative dynamics, *Phys. Rev.* A 42:1954 (1990).
20. L. Onsager, Ann. N.Y. Acad. Sci. 51:627 (1949).

21. B. Widom and J.S. Rowlinson, New model for the study of liquid-vapor phase transitions, *J. Chem. Phys.* 52:1670 (1970).
22. J.A. Given and G. Stell, The continuum Potts model and continuum percolation, *Physica* A 161:152 (1989).
23. J.A. Given, I.C. Kim, S. Torquato, and G. Stell, Comparison of analytic and numerical results for the mean cluster density in continuum percolation, *J. Chem. Phys.* 93:5128 (1990).
24. R. Fisch and A.B. Harris, Critical behavior of random resistor networks near the percolation threshold, *Phys. Rev.* B18:416 (1978).
25. K. Binder and A.P. Young, Spin glasses: experimental facts, theoretical concepts and open questions, *Rev. Mod. Phys.* 58:801 (1986).
26. D. Chandler, H.C. Anderson, Optimized cluster expansion for classical fluids III molecular fluids, *J. Chem. Phys.* 57:1930 (1972).
27. D. Chandler and L. Pratt, Statistical mechanics of chemical equilbrium and the intramolecular structure of non-rigid molecules in condensed phases, *J. Chem. Phys.* 65:2925 (1976).
28. P.T. Cummings, G.P. Morriss, and G. Stell, Solution of the site-site Ornstein-Zernike equation for non-ideal dipolar spheres, *J. Phys. Chem.* 86:1696 (1982).
29. L. Pratt, Connection between central-force model treatment of polyatomic molecular liquids and the interaction-site cluster expansion, *Mol. Phys.* 43:1163 (1981).
30. H.C. Andersen, Cluster expansions for hydrogen-bonded fluids II. dense liquids, *J. Chem. Phys.* 61:4985 (1974).
31. M.S. Wertheim, Fluids with highly directional attractive forces IV. equilibrium polymerization, *J. Stat. Phys.* 42:477 (1986).
32. M.S. Wertheim, Integral equation for the Smith-Nezbeda model of associated fluids, *J. Chem. Phys.* 88:1145 (1988).
33. J.S. Høye and G. Stell, Configurationally disordered spin systems, *Phys. Rev. Lett.* 36:1569 (1976).
34. S.H. Adachi, A.E. Panson, and R.M. Stratt, The effect of an unusual type of quenched disorder on phase transitions: illustration in a mixed-valence system, *J. Chem. Phys.* 88:1134 (1988).

NEW FREE ENERGY MODEL FOR NON-UNIFORM FLUIDS

Yaakov Rosenfeld
Physics Department
Nuclear Research Center - Negev
P.O.Box 9001, Beer-Sheva
Israel 84190

I. INTRODUCTION

Adsorption, wetting, catalysis, filtration, and freezing are fluid-solid interfacial phenomena of considerable practical importance, and of fundamental theoretical interest . The understanding of the physical properties of the solid-fluid interface has improved significantly during the past decade[1] as the result of the development of increasingly more refined theoretical techniques coupled to extensive computer simulations. In particular, a comprehensive picture of the complex thermodynamic behavior of fluids in confined geometries is now emerging[2]. The density functional method, a quite general approach to the equilibrium properties of non uniform (i.e. of spatially varying density) fluids, has proven to be one of the more successful and widely applicable approaches to a variety of interfacial and bulk phenomena[3]. In particular, the more sophisticated versions of the density functional theory, namely those based on non-local entropy functionals[4], provide an accurate description of the microscopic structure of fluids in the vicinity of solid surfaces, and of the oscilatory behavior of the density profiles. The central quantity in the density functional theory for nonuniform fluids is the excess (over "ideal gas" contributions) free energy which originates in interparticle interactions. It is a unique functional of the spatially varying one particle density $\rho(\mathbf{r})$, from which many equillibrium properties of the fluid can be derived, but which is in general unknown. The non-local free energy functionals employ weighted (coarse grained) densities and are constructed to fit available structural and thermodynamic properties of the homogeneous (bulk) fluid[4].

The major present day approxomate theories of the structure and thermodynamics of simple liquids[5] interpolate between the "ideal gas" and the asymptotic limit of the hypernetted-chain integral equation (denoted the Onsager limit) which for all practical purposes plays the role of the "ideal liquid"[6]. This Onsager limit corresponds to an exact lower bound for the potential energy of the system, and provides the needed reference "ideal" state for developing a systematic theory of liquid structure. This reference state corresponds analytically to an unphysical state where the spherical hard-cores fill all the volume. Yet the expansion around this state is fastly convergent at liquid densities, involving mathematical constructs (i.e. liquid-like <u>basis functions</u>) that enable analytic connection to functions described by low order diagrams[6]. The scaled-particle[7] - Percus-Yevick[8] description of the homogeneous hard sphere system is one example[9]. Other examples[10-11] are given by analytic and semi-analytic theories for the classical electron gas (or the one component plasma) and for charged hard particle fluids. When considering a model free energy for the inhomogeneous fluid, these new results enable to abandon the practice of refering to the liquid pair structure as a numerically given entity from a complicated liquid state calculation, and use instead a description in terms of "natural" <u>basis functions</u>, which are anyway <u>implicitly assumed</u> even in the numerical representations.

Following this general approach, a new type of free energy functional for the inhomogeneous hard sphere fluid mixture was derived recently[12-15] which is based on the idea of interpolation between (i) the limit of high density (eventually reaching analytically the unphysical packing fraction, $\eta = 1$) where the pair direct correlation function is dominated by convolutions of single particle geometries, i.e. overlap volume and overlap surface area, and (ii) the limit of low density where it is given by the pair exclusion volume. The formal structure of the theory, which thus also contains the exact solution of the Percus-Yevick equation[8] as a special case, requires to expand the pair exclusion as a convolution of weight functions which are characteristic functions for the geometry of a single particle. In comparison with other weighted - density approximations proposed in the recent litterature[4], the formal structure of the new free energy functional is advantageous in treating mixtures and pure components on equal footing, making the application to multicomponent fluids straightforward. The weight functions are purely geometric and density independent, and this reduces the computational effort needed for solving the Euler-Lagrange equations for the free energy minimum. The new theory also yields simple closed-form convolution expressions for the m-body direct correlation functions. In particular, the analytic, simple expression for the three particle direct correlation function[12], $c^{(3)}$, is found to agree well with extensive simulations[13]. The model also yields excellent results for the density profile of hard spheres in contact with steep potential-walls[16-18]. The predictions of this new theory have been recently tested very successfuly against computer simulations for a variety of situations where size or packing effects play an important role: adsorption of fluids in micropores, selective adsorption from mixtures at constant pressure, and adsorption of electrolytes onto highly charged surfaces[16-18]. As demonstrated for hard disks[14], this scaled-particle type geometric theory is also capable of handling analytically hard particle systems in an even number of dimensions. The new model is exact in one dimension (hard-rods[4a]).

The main puposes of this lecture are : (1) To outline the general features of the model for hard sphere mixtures along with specific free energy functionals for three dimensional spheres. (2) To describe certain new results and tests of the model[18], mainly in order to correct and complete the picture as emerging from recent extensive applications[16,17] of our new theory. (3) To present the extention of the model to charged-hard-spheres with Coulomb or Yukawa potentials, and for general interactions.

II. GENERAL PROPERTIES OF THE FREE ENERGY MODEL FOR HARD-SPHERE FLUID MIXTURES

II.1. Global Features

For the inhomogeneous fluid mixture of hard spheres in D-dimensions, characterized by the set of one particle densities $\{\rho_i(r)\}$, follow the exact result in one dimension[4a] and (A) postulate the following general excess (over ideal gas contribution) free energy functional

$$F_{ex}[\{\rho_i(r)\}]/k_BT = \int dx \Phi[\{n_\alpha(x)\}] \tag{1}$$

where (B) it is assumed that the excess free energy density Φ is a function of only the system averaged fundamental geometric measures of the particles,

$$n_\alpha(x) = \Sigma_i \int \rho_i(x')\omega_i^{(\alpha)}(x-x')dx \tag{2}$$

(C) The weighted densities $n_\alpha(x)$ are dimensional quantities with dimensions $[n_\alpha] = (\text{volume})^{(\alpha-D)/D}$ where $0 \leq \alpha \leq D$, and provide a functional basis set for expanding the function Φ which has dimension $(\text{volume})^{-1}$. The "weight functions" $\omega_i^{(\alpha)}$ are characteristic functions for the geometry of the particles. This general "fundamental geometric measure" description clearly originates[9] from the scaled particle theory of Riess et al.[7] for the uniform fluid. Thus, (D) the uniform fluid limit of this description should feature a scaled particle equation of state which contains a minimal set of D+1 reduced densities,

$$\{n_\alpha(x)\}_{\text{uniform fluid}} \rightarrow \{n_q ; q=0,1,...D \} \tag{3}$$

where n_q is proportional to the system averaged q-th power of the particle's radius, $\Sigma_i\rho_iR_i$. In particular, the uniform fluid limit of n_D corresponds to the total packing fraction of the hard particles. The minimal number of dimensional weighted densities is clearly D+1, but the minimal function space required for the description, i.e. the number of linearly independent functions $\omega_i^{(\alpha)}$ can be smaller, as demonstrated[12] for D=3. The complexity of the specific

413

fundamental-measures free energy model is determined by the number of linearly independent weight-functions, and not by the number of weighted-densities[13].

II.2. Specific Features

(a) Despite remaining freedom to choose the weight functions, there is no ambiguity concerning the characteristic scalar volume (index D) and surface (index (D-1)) functions for a sphere of radius R_i

$$\omega_i^{(D)}(r) = \theta(r - R_i) \tag{4a}$$

$$\omega_i^{(D-1)}(r) = |\nabla\theta(r - R_i)| = \delta(r - R_i) \tag{4b}$$

or the surface vector function

$$\omega_{v,i}^{(D-1)}(r) = \nabla\theta(r - R_i) = (r / r)\delta(r - R_i) \tag{4c}$$

$\theta(x)$ is the unit step function, $\theta(x>0)=0$, $\theta(x\leq0)=1$, ($\theta(x)=H(-x)$ where $H(x)$ is the standard Heaviside step function). $\theta(r-(R_i+R_j))$ represents the Mayer f-bond that characterizes the pair exclusion of particles i and j.

(b) In order to be able to reproduce exactly the lowest order (2 - particle) diagram in the diagramatic expansion, the set of weight functions must be able to decompose the pair exclusion function of two particles in a convolution expansion form

$$\theta(|r_i - r_j| - (R_i+R_j)) = \Sigma_{\alpha,\gamma}\omega_i^{(\alpha)}\otimes\omega_j^{(\gamma)} \tag{5}$$

with combinations having the correct dimension, $[\omega^{(\alpha)}]+[\omega^{(\gamma)}]=(volume)^{-1}$, where the symbol $\otimes$ denotes the convolution with implied "dot" product between vectors.

$$\omega_i^{(\alpha)}\otimes\omega_j^{(\gamma)} = \int\omega_i^{(\alpha)}(r_i - x)\omega_j^{(\gamma)}(r_j - x)dx \tag{6}$$

(c) In order to comply with the scaled-particle theory interpolation idea concerning the work required to introduce an additional small or large particle into the system, the "volume" ($\alpha=D$) coefficient of the chemical potential, $\mu_\alpha=\partial\Phi/\partial n_\alpha$ must equal the pressure. Thus the uniform fluid free energy density should obey the "scaled particle" differential equation :

$$(P/k_BT =) \qquad -\Phi+\Sigma_\alpha n_\alpha\partial\Phi/\partial n_\alpha + n_0 = \partial\Phi/\partial n_D \tag{7}$$

which (d) we impose on the general non-uniform case.

II.3. General Properties of the Direct Correlation Functions

The m-th order direct correlation function $c^{(m)}$ is given by functional derivatives of the excess free energy functional. In particular the pair (m=2) direct correlation function is:

$$k_B T c_{ij}^{(2)}(\mathbf{r}_1, \mathbf{r}_2) = - \delta^2 F_{ex}[\{\rho_i(\mathbf{r})\}]/\delta\rho_i(\mathbf{r})\delta\rho_j(\mathbf{r}) \tag{8a}$$

Because the weight functions characterize the individual hard particles, $c^{(m)}$ is non-zero only for tight-configurations of the particles, i.e. when the intersection of all core-overlaps is non-zero. The Fourier transforms of the homogeneous $c^{(m)}$ is simply a linear combination of products of the weight function transforms, $\omega_i^{(\alpha)}(k)$. This process of differentiation is easily automated once the free energy density function Φ is given. In view of (4) and (8a) we find that the leading terms for the uniform fluid pair-direct correlation function are ($r=|\mathbf{r}_1-\mathbf{r}_2|$)

$$c^{(2)}_{ij}(r) = (\partial\mu_D/\partial n_D)\Delta V_{ij}(r) + (\partial\mu_{D-1}/\partial n_D)\Delta S_{ij}(r) + \ .....\tag{8b}$$

where $\Delta V_{ij}(r)$ and $\Delta S_{ij}(r)$ are, respectively, the overlap volume and the overlap surface area of the two D-dimensional fused spheres at distance r. Similar expressions for the high order direct correlation are easily obtained[13]. This result[(8b)] is in agreement with previous general analysis of integral equation theories for liquid structure[6,10,11]. On the basis of our general analysis of the hypernetted-chain integral equation for the D-dimensional uniform fluid as well as the Percus-Yevick integral equation for D-dimensional hard spheres we also expect the pair direct correlation function $c_{ij}^{(2)}$ to be dominated at high densities by terms corresponding to the overlap volume of the two spheres[6,10,11].

III. HARD SPHERES IN THREE DIMENSIONS

III.1. Free Energy Functional

The <u>unique minimal function set</u> in three dimensions, composed exactly of the <u>three</u> functions (4a-4c), thus provides[12] a unified derivation of the scaled-particle free energy and the Percus-Yevick pair direct correlation functions. The other functions are proportional to these three, and are given by $\omega_i^{(1)}(r) = \omega_i^{(2)}(r)/4\pi R_i$, $\omega_i^{(0)}(r) = \omega_i^{(2)}(r)/4\pi R_i^2$, $\omega_{v,i}^{(1)}(r) = \omega_{v,i}^{(2)}(r)/4\pi R_i$. The scalar weights have the property $\omega_i^{(\alpha)}(k=0)=R_i^{(q)}$ where $R_i^{(q)}$ $=1, R_i, S_i, V_i$ for q=0,1,2,3 respectively (S_i, V_i denote the surface area and the volume of the sphere), while the Fourier transforms of the vector type weights vanish, $\omega_{v,i}^{(\alpha)}(k=0)=0$. Specifically, the Fourier transforms are given by (denoting $t=kR_i$)

$$\omega_i^{(q)}(k)/R_i^{(q)} = \sin(t)/t \qquad , \text{ for } q=0,1,2 \tag{9a}$$

$$\omega_i^{(q)}(k)/R_i^{(q)} = 3[\sin(t) - t\cos(t)]/t^3 \ , \quad \text{for } q=3 \tag{9b}$$

$$\omega_{V,i}^{(2)}(k) = -(-1)^{1/2}k\omega_i^{(3)}(k) \tag{9c}$$

The resulting[12] excess free energy density can be written as the sum of scalar and vector contributions in the form $\Phi = \Phi_S + \Phi_V$, where

$$\Phi_S = -n_0\ln(1-n_3) + n_1n_2/(1-n_3) + n_2^3/(1-n_3)^2/(24\pi) \tag{10a}$$

is easily recognized (with $\{n_q\} \to \{\xi_q\}$, i.e. $n_3 \to \eta$) as the scaled particle theory excess free energy density of the uniform hard sphere mixture. Recall that $\xi_0 = \Sigma_i\rho_i = \rho$, $\xi_1 = \Sigma_i\rho_iR_i$ $\xi_2 = \Sigma_i\rho_i4\pi R_i^2$, $\xi_3 = \Sigma_i\rho_i(4\pi/3)R_i^3 = \eta$. The vector contributions are

$$\Phi_V = n_1 \cdot n_2/(1-n_3) + n_2(n_1 \cdot n_2)/(1-n_3)^2/(8\pi) \tag{10b}$$

The pair direct correlation function for the homogeneous fluid takes the form $(r = |r_i - r_j|)$:

$$- c_{ij}^{(2)}(r) = \chi^{(3)}[\omega_i^{(3)} \otimes \omega_j^{(3)}] + \chi^{(2)}[\omega_i^{(3)} \otimes \omega_j^{(2)} + \omega_i^{(2)} \otimes \omega_j^{(3)}] + \chi^{(1)}[\omega_i^{(3)} \otimes \omega_j^{(1)}$$

$$+\omega_i^{(1)} \otimes \omega_j^{(3)} + (\omega_i^{(2)} \otimes \omega_j^{(2)} + \omega_{V,i}^{(2)} \otimes \omega_{V,j}^{(2)})/(4\pi)] + \chi^{(0)}[\omega_i^{(3)} \otimes \omega_j^{(0)} + \omega_i^{(0)} \otimes \omega_j^{(3)}$$

$$+(\omega_i^{(1)} \otimes \omega_j^{(2)} + \omega_i^{(2)} \otimes \omega_j^{(1)} + \omega_{V,i}^{(1)} \otimes \omega_{V,j}^{(2)} + \omega_{V,i}^{(2)} \otimes \omega_{V,j}^{(1)}] \tag{11}$$

where $\chi^{(q)} = \partial^2\Phi_S[\{\xi_m\}]/\partial\xi_3\partial\xi_q$. The result is identical, by each $\chi^{(q)}$ term, to the Percus-Yevick direct correlation function[8] written as[9]

$$- c_{ij}^{(2)}(r) = \chi^{(3)}\Delta V_{ij}(r) + \chi^{(2)}\Delta S_{ij}(r) + \chi^{(1)}\Delta R_{ij}(r) + \chi^{(0)}\theta(r-(R_i+R_j)) \tag{12}$$

For two spheres R_i and R_j at distance r : $\Delta V_{ij}(r)$ is the overlap volume, $\Delta S_{ij}(r)$ is the overlap surface area, $\Delta R_{ij}(r) = \theta(r-(R_i+R_j))[R_i+R_j -$ (mean radius of the convex envelope of the union of the two spheres)$] = \Delta S_{ij}(r)/[4\pi(R_i+R_j)] + [R_iR_j/(R_i+R_j)]\theta(r-(R_i+R_j))$.

The minimal set (4), i.e. (9), is composed of weight functions with <u>definite geometric meaning</u>. Our geometric approach remains conceptually the same for <u>non-spherical</u> convex hard particles, the main difficulty for application being the "addition theorem" (eq.5) representing the pair exclusion in terms of the individual particles geometry.

It is easy to observe that the minimal weight function set (9) is not the only one that can reproduce the Percus-Yevick direct correlation function. This is the general[13] situation for D>1. In particular, consider[16,17] the <u>four</u>-function set for which the weight functions $\omega_i^{(3)}$ and $\omega_i^{(2)}$ remain the same, and the single vector function $\omega_{V,i}^{(2)}(k)$ is replaced by two scalar functions (denoting $t = kR_i$)

$$\omega_i^{(1)}(k)/R_i^{(1)} = [\sin(t) + t\cos(t)]/(2t) , \quad \omega_i^{(0)}(k)/R_i^{(0)} = [\cos(t) + t\cos(t)/2] , \qquad (13a)$$

corresponding to derivatives of the delta function:

$$\omega_i^{(1)}(\mathbf{r}) = (1/8\pi) \, \delta'(R_i - r) , \quad \omega_i^{(0)}(\mathbf{r}) = -(1/8\pi) \, \delta''(R_i - r) + (1/2\pi r) \, \delta'(R_i - r) \qquad (13b)$$

With this 4-function set, the excess free energy density Φ is given by Φ_S above, $\Phi=\Phi_S$. This variant as well as other possibilities for enlarging the weight-function set in order to systematically improve the agreement with simulations, have been considered within the general framework of the theory[13-15].

III.2. Performance Tests

(1) **General comments**. It should be noted that any computer program written for one set of "fundamental measure" weight functions requires only trivial modifications in order to perform for another. This easy exploration of different weight functions is one of the favorable features that are originally built into the theory. The theory is so structured that the computational complexity of the model is determined by the number of different <u>functions</u> involved, and not by their labling. Indeed, as long as the density variation for the three dimentional spheres is not very extreme, the numerical predictions of the model are dominated by the common scalar functions (eq.4a,b), while the other weight-functions, in <u>any</u> fundamental-measure representation, play only a secondary role[18]. In particular, the results from the 4-function scalar description and the 3-function original description are almost identical in all the applications of the model so far[16-18]. However, when it comes to the description of narrow cylindrical pores, namely one dimensional limit for the three dimensional spheres, the weighted densities $n_0(r)$ and $n_1(r)$ corresponding to eq.13 have nonintegrable singularities[17] making a useless functional, while the original model with the geometric 3-function set continues to perform exceptionally well[18]. In addition, the scalar weights (eq.13) do not guarantee positive weighted densities. The introduction of derivatives of the delta function, and in particular $\omega_i^{(0)}$ in (13b) which does not have simple geometric meaning, does not allow a straightforward generalization of the functional <u>as is</u> to non-spherical convex particles.

The main reason for the special attention here to the 4-function scalar description is that it was specifically introduced[16] as a "simplified version" of the original theory. A series of extensive applications[16-17] followed which provide an excellent test of the performance of our new method. Unfortunately, these applications of our theory were published[16-17] only for the so-called "simplified version", they do not present a balanced and complete picture regarding its general capability, and may even lead to wrong conclusions. The main results from the thus required supplementary comparative calculations[18] are described below.

(2) Bulk uniform fluid in three dimensions. The "fundamental measure" free energy model provides the first unified derivation[12] of the of the most comprehensive, and highly accurate, analytic description of the hard sphere thermodynamics and pair-structure, as given by the scaled particle and Percus-Yevick theories. Until recently scaled particle and Percus-Yevick theories were considered as independent! Extensive computer simulations[13] performed specifically to test the new theory have shown the overall high accuracy of the 3-particle direct correlation functions, $c^{(3)}$, predicted by our model. As unticipated[13], the fully scalar 4-function representation based on (13) is in a slightly better[16] agreement with the simulations than the minimal 3-function set (9). Both are more accurate[19] than the weighted-density-approximation[4e].

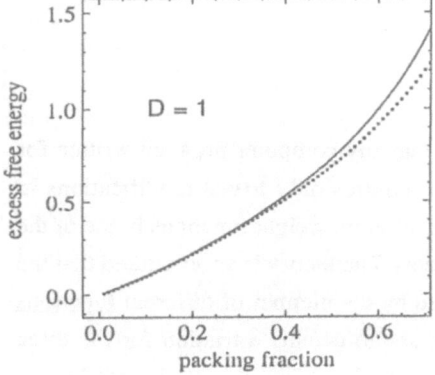 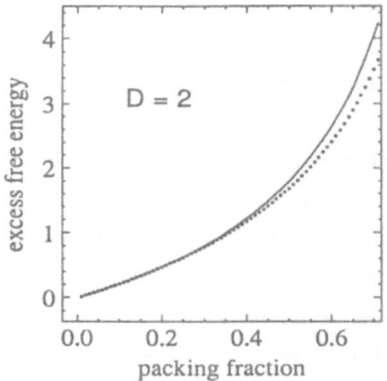

Fig.1. Excess free energy density from eq.10 for the uniform hard- rod fluid, compared with the exact result (dots).

Fig.2. Excess free energy density from eq.10 for the uniform hard-disk fluid, compared with the simulation data (dots).

(3) One- and two- dimensional limits for the 3D functional. Fluids strongly adsorbed at a planar surface, or confined in narrow slits, resemble a uniform fluid in two dimensions (D=2), while fluids confined in narrow cylindrical micropores resemble one dimensional (D=1) uniform fluids. The performance of the three dimensional density functional in these situations can be inferred from its ability to provide an accurate description of the uniform one- and two- dimensional fluids. For the D=2 limit consider the three dimentional densities of the form $\rho_i(r) = \rho_i^{(2D)}\delta(z)$, while for the D=1 limit consider $\rho_i(r) = \rho_i^{(1D)}\delta(x)\delta(y)$, where $\rho_i^{(2D)}$ and $\rho_i^{(1D)}$ are the two- and one- dimensional uniform densities, respectively. The results for the corresponding excess free energies per particle, $\Phi^{(2D)}/\rho^{(2D)}$ and $\Phi^{(1D)}/\rho^{(1D)}$, (which are obtained analytically by elementary integrations) along with $\Phi^{(3D)}/\rho^{(3D)}$ for the single component fluids are presented in Figures 1-3, where they are compared with exact or simulation results. For D=3 the functionals based on the 3-function or 4-function sets, (9) or (13), are of course identical, while for D=2 they are indistinguishable on the scale of the plot. For D=1 the weighted densities from (13) have non-integrable singularities giving an infinite excess free energy, while the result of the minimal-set (9) continue to agree very well with the exact result. The smoothed-density-approximation[4c] is much less accurate[20] than the present model.

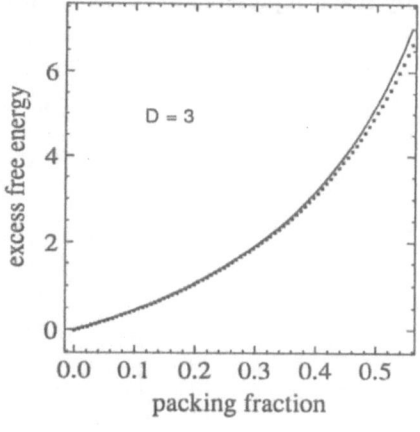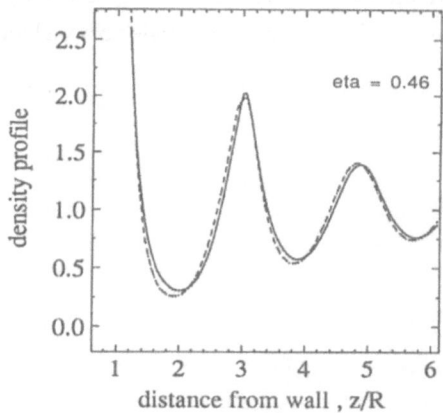

Fig.3. Excess free energy density from eq.10 for the uniform hard- sphere fluid , compared with the simulation data (dots).

Fig.4. Density profile for a hard sphere fluid of packing fraction $\eta=0.46$, near a hard wall. 3-function set compared with 4-function set (broken line).

(4) Hard sphere fluid mixtures near a hard wall. The density profiles for the fluid subject to external potentials $u_i(\mathbf{r})$ which couple to the particles of type i is obtained by solving the Euler-Lagrange equations

$$\delta\Omega[\{\rho_i(\mathbf{r})\}]/\delta\rho_i(\mathbf{r}) = 0 , \quad i=1,2,.... \tag{14}$$

which correspond to the minimization of the grand potential $\Omega[\{\rho_i(\mathbf{r})\}]$, given by

$$\Omega[\{\rho_i(\mathbf{r})\}] = F_{id}[\{\rho_i(\mathbf{r})\}] + F_{ex}[\{\rho_i(\mathbf{r})\}] + \Sigma_i \int d\mathbf{r}\rho_i(\mathbf{r})[u_i(\mathbf{r}) - \mu_i] \tag{15}$$

where μ_i are the chemical potentials. The excess free energy is known only approximately, e.g. eqs.1,10, while the ideal-gas free energy is given by the exact relation

$$F_{id}[\{\rho_i(\mathbf{r})\}] = k_BT \Sigma_i \int d\mathbf{r}\rho_i(\mathbf{r})\{\ln[\rho_i(\mathbf{r})\lambda_i^3] - 1\} \tag{16}$$

where λ_i are the de Broglie wavelengths. In the special case of a hard sphere fluid in a slit defined by two hard walls at z=0 and z=L, the external potentials are: $u_i(\mathbf{r}) = 0$ for $R_i \leq z \leq L-R_i$; $u_i(\mathbf{r}) = \infty$, otherwise. In practice the symmetry with respect to z=L/2 is employed, and when L is large enough (formally L=∞) one obtains the profiles of the hard sphere fluid near a wall at z=0. The calculation requires to evaluate integrals of the general form

$$f_i^{(\alpha)}(\mathbf{r}) = \int h(\mathbf{r}')\omega_i^{(\alpha)}(\mathbf{r}-\mathbf{r}')d\mathbf{r}' = \int h(z')W_i^{(\alpha)}(z,z')dz' \tag{17}$$

for functions which vary only along the z-axis $h(\mathbf{r})=h(z)$. The minimal-set (9) requires the expressions $W_i^{(3)}(z-z')=\pi[R_i^2 - (z-z')^2]$, $W_i^{(2)}(z-z')=2\pi R_i$, $\mathbf{W}_i^{(2)}(z,z')=2\pi(z'-z)(\mathbf{z}/z)$ where the integrals are from $(z-R_i)$ to $(z+R_i)$. The fully scalar representation employs two expressions[16] instead of the single vectorial term.

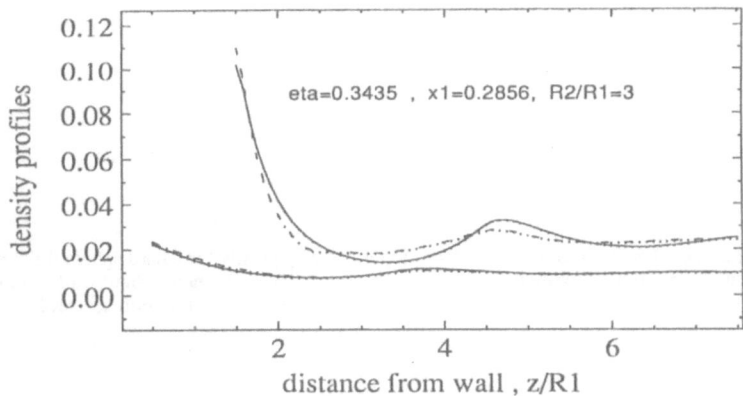

Fig.5. Density profiles for a hard sphere mixture of packing fraction $\eta=0.3435$ near a hard wall. The size ratio is $R_2/R_1=3$, and the relative concentration of the smaller spheres is $x_1=0.2856$. Results using the 3-function set compared with those for the 4-function set (broken line).

The results of our calculations using the fundamental-measure excess free energy functional based on the minimal 3-function set and on the 4-function scalar set are presented in Figures 4,5 , for densly packed hard sphere systems near a hard wall at $z=0$. The two functionals yield almost the same density profiles for the one component system, in excellent agreement with the computer simulations (Fig.4 is designed for comparison with Fig.2 of ref.16), eventhough they cannot strictly satisfy $\rho(0) = (P/k_BT)_{exact}$. Such high accuracy is reached by no other density functional. The results for the mixture exhibit a larger difference, with the results based on the 4-function set (13) in comparable accuracy to the modified-weighted-density-approximation[21] for $c^{(1)}$, while those based on the 3-function set (9) are in better agreement with the simulations[22] (see Fig.4 in ref.21)

(5) **Bulk pair-correlation functions by the test particle method.** When the external potential is obtained by fixing a test particle of type k at the origin, $u_i(\mathbf{r}) = \phi_{ik}(\mathbf{r})$, where $\phi_{ik}(\mathbf{r})$ is the corresponding pair potential between particles of types i and k in the fluid, then the density profiles normalized to unity at large r, correspond to the pair distribution functions in the bulk uniform fluid $g_{ik}(r) = \rho_i(r)/\rho_i(r=\infty)$. The pair distribution function,

g(r), thus obtained for the single component hard sphere fluid at high packing fraction, and the corresponding structure factor, $S(k)=1+\rho\int[g(r)-1]e^{i\mathbf{k}\cdot\mathbf{r}}d\mathbf{r}$, are presented in Figures 6,7. The results compare well with the Percus Yevick results which are also obtained from our theory via the direct correlation function (11) by using the Ornstein Zernike relation, $S(k) = [1-\rho c^{(2)}(k)]^{-1}$. Again, the results based on the 3-function and on the 4-function sets are indistinguishable on the scale of the plots, both improve significantly on the Percus-Yevick result near contact, and are altogether very close to the simulation results (compare Fig.6 with Fig.4 in ref.13). To our knowledge, no other density functional exhibits this high degree of consistency between the two alternative routes to the pair correlations.

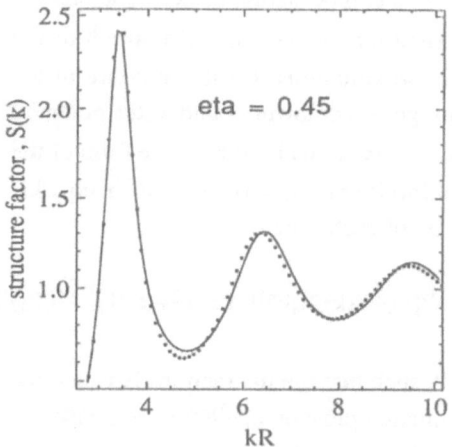
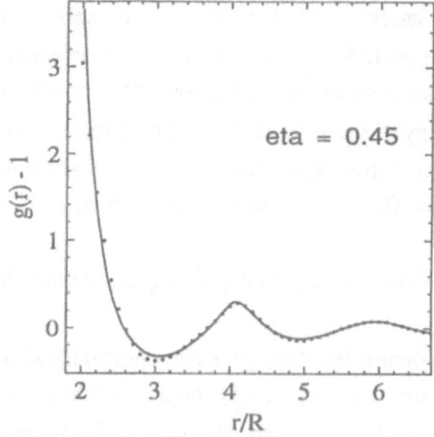

Fig.6. Bulk structure factor for a hard spheres at packing fraction $\eta=0.45$. Test particle limit compared with the Percus-Yevick result (dots).

Fig.7. Bulk pair correlation function for a hard sphere fluid at packing fraction $\eta=0.45$. Test particle limit compared with the Percus-Yevick result (dots).

(6) Conclusion. Considering together the accuracy of the theory under diverse physical circumstances, its easy implementation for mixtures, and its relative ease in calculations, it seems that the original fundamental-measure free energy model, based on the minimal-set of geometric weighted densities (eqs.3), is the overall most adequate presently available free energy functional for the hard sphere fluid. Its structure provides a starting point for systematic improvements based on simulation data or higher order expansions[13]. Its inability to stabilize a solid[12] is related to its derivation by interpolation of the fluid between the ideal-gas and ideal-liquid limits, bypassing the solid. This result led to calculations which question[23] the density functional theory of freezing as based today on a free energy model for the fluid. The present functional is not adequate for describing solid-like (e.g. narrow Gaussians on a lattice) density distributions.

IV. FREE ENERGY MODEL FOR CHARGED YUKAWA MIXTURES AND GENERAL INTERACTIONS

Consider an arbitrary mixture of charged hard spheres interacting via $\phi_{ij}(r) = \phi^{HS}_{ij}(r) + \phi^C_{ij}(r)$, where $\phi^{HS}_{ij}(r)$ ($= \infty$ for core overlap configurations, $= 0$ otherwise) is the hard sphere interaction, $\phi^C_{ij}(r)/k_BT = Z_iZ_j\Sigma_\alpha\Gamma_\alpha e^{-\alpha r}/r$ is the corresponding "Charge" contribution, and the Γ_α are independent coupling parameters. The special case of a single Yukawa in the limit of $\alpha=0$ corresponds to charged hard spheres with no screening. We consider the multi-Yukawa interaction because with free parameters α and Γ_α we can model even a Lennard-Jones potential[24]. We first consider the uniform fluid, and distinguish between two cases: (1) Total charge neutrality, $<\rho Z> = \Sigma_i\rho_iZ_i = 0$. (2) no charge neutrality, $<\rho Z> \neq 0$, which for the Coulomb case, $\alpha=0$, requires a compensating background charge. The analytic solution of the mean-spherical-approximation, as well as the solution of the hypernetted chain or modified hypernetted-chain approximations, for this mixture in the strong coupling limit, $\Gamma_\alpha \rightarrow \infty$, correponds to the Onsager exact lower bound to the potential energy of the system[6,10,11]. The direct correlation functions in this limit have the form of the original interaction but with the charges smeared uniformly on the surface (m=2 in eq.18 if $<\rho Z> = 0$) or in the volume (m=3 in eq.18 if $<\rho Z> \neq 0$) of each sphere:

$$c_{ij}^{(2)}(k) \rightarrow \Sigma_\alpha\Gamma_\alpha Z_iQ_{i,\alpha}^{(m)}Z_jQ_{j,\alpha}^{(m)}[\omega_i^{(m)}(k)/R_i^{(m)}][\omega_j^{(m)}(k)/R_j^{(m)}][4\pi/(k^2+\alpha^2)] \qquad (18)$$

The quantities $Z_iQ_{i,\alpha}^{(m)}$ are renormalized charges[25], such that the interactions between the smeared charges satisfy the mean spherical approximation closure, $c_{ij}^{(2)}(r) = -\phi^C_{ij}(r)/k_BT$, outside core overlap. For the Coulomb case ($\alpha=0$) $Q_{i,\alpha}^{(m)}=1$, by the Newton-Gaus law. Thus, the fundamental measure weight functions connect well with the Onsager smearing idea as related to the solution of liquid state integral equations. Moreover, the full analytic solution of the mean spherical approximation for this system[26] is exactrly of the analytic form of (18) with[6,11,18] m=3. It can be completely expanded in terms of convolutions of the fundamental measure weight functions[6,11,18], and is dominated by the asymptotic result even for weak coupling. The generalization of (10) that <u>manifestly</u> reproduces the analytic solution of the mean spherical approximation may eventually be found[11c].

Without loss of generality separate the excess free energy into the Hard-Sphere (HS) and Charge (C) parts:

$$F_{ex}[\{\rho_i(\mathbf{r})\}] = F_{ex}^{HS}[\{\rho_i(\mathbf{r})\}] + F_{ex}^C[\{\rho_i(\mathbf{r})\}] \qquad (19)$$

with the ensuing separation for the direct correlation functions. Consider a functional Taylor expansion of the Charge contribution to the free energy for the non-uniform fluid starting at its uniform limit. The Onsager energy bound which dominates in strong coupling, is <u>quadratic</u> in weighted densities for both uniform and non uniform systems. Thus, in the strong coupling limit this Taylor expansion is practically terminated at <u>second order</u>. In turn, a second order expansion of the charge part should be effective in weak coupling. These limiting behaviors provide the rationale for a free energy functional based on a second order expansion for the Charge part, in powers of $\Delta\rho_i(\mathbf{r})=\rho_i(\mathbf{r}) - \rho_i$, where ρ_i is the uniform fluid reference density, e.g. the density of the fluid far from the surface:

$$F_{ex}^{C}[\{\rho_i(\mathbf{r})\}] = F_{ex}^{C}[\{\rho_i\}] - k_BT\Sigma_i\, c_i^{(1),C}[\{\rho_i\}]\!\int\! d\mathbf{r}\Delta\rho_i(\mathbf{r})$$

$$- (k_BT/2)\Sigma_{ij}\!\int\!\!\int\! d\mathbf{r}d\mathbf{r}'c_{ij}^{(2),C}[\{\rho_i\};(|\mathbf{r}-\mathbf{r}'|)]\Delta\rho_i(\mathbf{r})\Delta\rho_i(\mathbf{r}') \tag{20}$$

Here $c_i^{(1),C}[\{\rho_i\}]$ ($= \mu_{i,ex}^{C}[\{\rho_i\}]/k_BT$, the excess chemical potential) and $c_{ij}^{(2),C}[\{\rho_i\};r]$ denote the Charge part of the one- and two-particle direct correlation functions in the uniform fluid limit. The Euler-Lagrange equations for the density profiles read (denote $h_i(\mathbf{r}) = \rho_i(\mathbf{r})/\rho_i - 1$):

$$h_i(\mathbf{r})+1= \exp[\{-u_i(\mathbf{r})+\mu_{i,ex}^{HS}[\{\rho_i\}]-\mu_{i,ex}^{HS}[\{\rho_i(\mathbf{r})\}]\}/k_BT$$

$$+\Sigma_i\rho_i\!\int\! d\mathbf{r}'c_{ij}^{(2),C}[\{\rho_i\};(|\mathbf{r}-\mathbf{r}'|)]h_i(\mathbf{r}')] \tag{21}$$

Note the analogy between this form and the <u>modified</u> hypernetted chain approximation[5c] for the pair correlation function in the uniform fluid. By expanding also the Hard-Sphere part to second order , eq.21 yields the hypernetted chain approximation for the density profile.

We can model a Lennard-Jones type potential by a linear combination of Yukawa potentials, and use the same argument as above for truncating the "attractive" contributions after the second order in the expansion. The case with a soft repulsive potential can be treated in analogy with the uniform fluid by mapping the "repulsive" contributions on the hard spheres and considering the radii of the particles as adjustable (e.g.variational) parameters, asuming additivity of effective hard cores. This perturbation-theory approach, with the hard spheres reference system, for treating non-uniform fluids seems to be promising[27], and is particularly successful[17,18] when the present theory is used for the hard spheres. The second order perturbation expansion for the "attractive" contributions is justified by the Onsager limit for charged Yukawa mixtures as the asymptotic strong coupling limit for the solution of the modified hypernetted equation[25]. Considering (21) in the test particle limit (III.2.(5), Figure 7) and recalling the universality of the bridge functions for the density profiles[28], this second order theory is expected to be of the accuracy of the modified hypernetted chain integral equation for the density profile. Another alternative[18], which also follows from our argument for the validity of the second order expansion based on the asymptotic strong coupling behavior, is to expand the full excess free energy functional to second order around a reference non-uniform hard sphere system, and to optimize (e.g. by variation) the reference system parameters.

V. CONCLUSION

Compared to other nonlocal density functional theories as applied to describing packing effects at the solid-liquid interface and to the study of adsorption of simple fluid mixtures at substrates, the new theory seems to be the most accurate while rquiring a significantly lower computational effort. The new model provides a paradigm for a new class of free energy functionals.

REFERENCES

1. (a) R. Evans, Adv.Phys. 28:143 (1979). (b) J.S. Rowlinson and B. Widom, "Molecular Theory of Capilarity", Clarendon Press, Oxford (1982). (c) D. Nicholson and N.G. Parsonage, "Computer Simulation and the Statistical Mechanics of Adsorption", Academic,NewYork (1982). (d) D.E. Sullivan and M.N. Telo da Gama, in: "Fluid Interfacial Phenomena", edited by C.A. Croxton,Wiley, Justin , Elsevier, New York (1989).

2. R. Evans, J.Phys.Condens.Matter 2: 8989 (1990).

3. (a) J.K. Percus, in: "The Liquid State of Matter: Fluids Simple and Complex", edited by E.W.Montroll and J.L.Lebowitz , North- Holland (1982). (b) M. Baus, J.Phys.Condens.Matter 2:2111 (1990); (c) M.Baus, J.Stat.Phys. 48: 1129 (1987). (d) A.D.J. Haymet, Ann.Rev.Phys.Chem. 38:89 (1987). (e) J.K. Percus, J.Stat.Phys. 52:1157 (1988). (f) David W. Oxtoby, in: "Liquids, Freezing and the Glass Transition", Les Houches Session 51, edited by J.P. Hansen, D. Levesque, and J. Zinn-Justin, Elsevier, New York (1990). .

4. (a) J.K. Percus,J.Stat.Phys. 15:505 (1976); A. Robledo, J.Chem.Phys. 72:1701 (1980); J.K. Percus, J.Chem.Phys. 75:1316 (1981); A. Robledo and C. Varea, J.Stat.Phys. 26:513 (1981); J. Fischer and U. Heinbuch, J.Chem.Phys. 88:1909 (1988). (b) S. Nordholm, M. Johnson, and B.C. Freasier, Aust.J.Chem. 33:2139 (1980); B.C. Freasier and S. Nordholm, J.Chem.Phys. 79:4431 (1983); Mol.Phys. 54:33 (1986). (c) P. Tarazona, Mol.Phys.52:81 (1984); P. Tarazona and R. Evans, Mol.Phys. 52:847 (1984); P. Tarazona, Phys.Rev.A 31:2672 (1985); P. Tarazona, U. Marini Bettolo Marconi, and R. Evans, Mol.Phys. 60:573 (1987). (d) T.F. Meister and D.M. Kroll, Phys.Rev. A 31:4055 (1985); R.D. Groot and J.P. van der Eerden, Phys.Rev.A 36:4356 (1987); S. Sokolowski and J. Fischer, Mol.Phys. 68:647 (1989).(e) W.A. Curtin and N.W. Ashcroft, Phys.Rev. A 32:2909 (1985); (f) A.R. Denton and N.W. Ashcroft, Phys.Rev.A 39:426,4701 (1989). (g) T.K. Vanderlich, L.E. Scriven, and H.T. Davis, J.Chem.Phys. 90:2422 (1989). (h) D.M. Kroll and B.B. Laird, Phys.Rev.A 42:4806 (1990). (i) J.F.Lutsko and M.Baus, Phys.Rev.Lett. 64:761(1990). (j) X.C.Zeng and D.W.Oxtoby, Phys.Rev.A 41:7094 (1990).

5. (a) J.A.Barker and D.Henderson, Rev.Mod.Phys. 48:587 (1976). (b) J.P. Hansen and I.R. McDonald, "Theory of Simple Liquids", 2'nd edition , Academic Press, London (1986). (c) Y. Rosenfeld and N.W. Ashcroft, Phys.Rev.A 20:1208 (1979), and see J.Talbot, J.L.Lebowitz, E.M.Weisman, D.Levesque, and J.J.Weis, J.Chem.Phys. 85:2187 (1986).

6. Y. Rosenfeld, Phys.Rev.A 32:1834 (1985); Phys.Rev.A 33:2025 (1986).

7. H. Reiss, H. Frisch, and J.L. Lebowitz, J.Chem.Phys. 31:369 (1959); H. Reiss, in: " Statistical Mechanics and Statistical Methods in Theory and Applications", edited by U. Landman, Plenum, New-York (1976); For more recent applications ,see e.g. H.Reiss and P.Schaaf, J.Chem.Phys. 91:2520 (1989) ; P.Schaaf and H.Reiss , J.Chem.Phys. 92:1258 (1990).

8. (a) J.K. Percus and G.J. Yevick, Phys.Rev. 110:1 (1958). (b) M.S. Wertheim, Phys.Rev.Lett.10:321 (1963); E. Thiele, J.Chem.Phys. 39:474 (1963). (c) J.L. Lebowitz,Phys.Rev.A 133:895 (1964); J.L. Lebowitz and J.S. Rowlinson, J.Chem.Phys. 41:133 (1964).

9. Y. Rosenfeld, J.Chem.Phys. 89:4272 (1988).

10. Y.Rosenfeld, Phys.Rev.A 35:938 (1987); ibid, 37:3403 (1988); Y.Rosenfeld, D.Levesque and J.J.Weis, Phys.Rev.A 39:3079 (1989); see the review, Y.Rosenfeld, in: "High-Pressure Equation of State : Theory and Applications", edited by S. Eliezer and R. Rici, International School of Physics Enrico Fermi, Course CXIII (1989), North-Holland, Amsterdam (1991).

11. (a) Y. Rosenfeld and W.M. Gelbart, J.Chem.Phys. 81:4574 (1985); (b) Y. Rosenfeld and L. Blum, J.Chem.Phys.85:1556 (1986); (c) L. Blum and Y.Rosenfeld, J.Stat.Phys. 63:1177(1991).

12. Y. Rosenfeld, Phys.Rev.Lett. 63:980 (1989).

13. Y. Rosenfeld, D.Levesque and J.J.Weis, J.Chem.Phys. 92:6818 (1990).

14. Y. Rosenfeld,Phys.Rev.A 42:5978 (1990).

15. Y. Rosenfeld, J.Chem.Phys. 93:4305 (1990)

16. E. Kierlik and M.L. Rosinberg, Phys.Rev.A 42:3382(1990)

17. E. Kierlik and M.L. Rosinberg, Phys.Rev.A 44:5025 (1991).

18. Y. Rosenfeld, to be published; Y. Rosenfeld and L. Blum, to be published.

19. W.A. Curtin, J.Chem.Phys. 93:1919 (1990).

20. B.K. Peterson, K. Gubbins, G.S. Heffelfinger, U. Marini Bettolo Marconi, and F. van Swol, J.Chem.Phys. 88:6487 (1988).

21. A.R. Denton and N.W. Ashcroft, Phys.Rev.A 44:8242 (1991).

22. Z. Tan, U. Marini Bettolo Marconi, F. van Swol, and K.E. Gubbins, J.Chem.Phys. 90:3704 (1989).

23. Y. Rosenfeld, Phys.Rev.A 43:5424 (1991); S.J. Smithline and Y. Rosenfeld, ibid 42:2434 (1990).

24. S.M. Foiles and N.W. Ashcroft, J.Chem.Phys. 75:3594 (1981). The Yukawa potential is reviewed by J.S. Rowlinson, Physica A 156:15 (1989).

25. Y. Rosenfeld, submitted to Phys.Rev.Lett. (1992).

26. E. Waisman, Mol.Phys. 25:45 (1973); J.S. Hoye and L. Blum, J.Stat.Phys. 16:300 (1977); L. Blum and J.S. Hoye, J.Stat.Phys. 19:317 (1977); K. Nizeki, Mol.Phys. 43:251 (1981).

27. L. Mier-y-Teran, S.H. Suh, H.S. White, and H.T. Davis, J.Chem.Phys. 92:5087 (1990); Z. Tang, L.E. Scriven, and H.T. Davis, J.Chem.Phys. 95:2659 (1991).

28. Y. Rosenfeld and L. Blum, J.Chem.Phys. 85:2197 (1986); A recent verification by simulations for the Lennard-Jones system is presented by M. Llano-Restrepo and W.G. Chapman, submitted to J.Chem.Phys. (1992).

INTEGRAL EQUATIONS FOR INHOMOGENEOUS FLUIDS

Douglas Henderson

Utah Supercomputing Institute/IBM Partnership
and Department of Chemistry
University of Utah
Salt Lake City, UT 84112

INTRODUCTION

Useful theories of inhomogeneous fluids can be obtained either from the singlet Ornstein-Zernike (OZ1) equation (Henderson et al, 1976),

$$h_1 = c_1 + \rho \int h_2 c_{12}^B d\underset{\sim}{r}_2 ,$$ (1)

where $\rho = N/V$ is the density (N and V are the number of molecules and volume), or from the pair Ornstein-Zernike (OZ2) equation,

$$h_{12} = c_{12} + \int \rho_3 h_{13} c_{23} d\underset{\sim}{r}_3 .$$ (2)

In Eqs. (1) and (2), $h_1 = h(\underset{\sim}{r}_1) = g(\underset{\sim}{r}_1) - 1$ and $h_{12} = h(\underset{\sim}{r}_1, \underset{\sim}{r}_2) - 1$ are the singlet and pair *total correlation functions* and the analogous c's are the pair and singlet *direct correlation functions*. The argument $\underset{\sim}{r}_i$ is the position of molecule i. The distribution functions, g_1 and g_{12}, give the probability of finding a singlet molecule at $\underset{\sim}{r}_1$ or a pair of molecules at $\underset{\sim}{r}_1$ and $\underset{\sim}{r}_2$, normalized so that g_1 and g_{12} become unity for large argument. The function $\rho_3 = \rho(\underset{\sim}{r}_3) = \rho g(\underset{\sim}{r}_3)$ is the density profile of the inhomogeneous fluid.

For a homogeneous or bulk fluid, $\rho(\underset{\sim}{r}) = \rho$ and Eq. (2) becomes the usual Ornstein-Zernike (OZ) equation. If the intermolecular potential is central, the correlation functions of a homogeneous fluid have the form $h_{12} = h(r_{12})$, where $r_{12} = |\underset{\sim}{r}_1 - \underset{\sim}{r}_2|$. If the inhomogeneity is planar, as will be assumed hereinafter, $h(\underset{\sim}{r}_1) = h(x_1)$ and $h(\underset{\sim}{r}_1, \underset{\sim}{r}_2) = h(R_{12}, x_1, x_2)$, etc., where x_i is the normal distance of molecule i from the plane of the inhomogeneity and R_{12} is the projection of r_{12} onto the plane of the inhomogeneity. For clarity and emphasis, a superscript B has been placed on c_{12}^B in Eq. (1). It is the pair direct correlation function of the homogeneous (or bulk) fluid. It is *not* the inhomogeneous direct correlation function, c_{12}. Of course, c_{12} becomes c_{12}^B when both molecules are in the homogeneous region.

Condensed Matter Theories, Vol. 8, Edited by
L. Blum and F.B. Malik, Plenum Press, New York, 1993

THEORY

Equations (1) and (2) are just definitions of c, and c_{12}. Quite formally

$$h_1 - c_1 = \ln y_1 + B_1 \tag{3}$$

and

$$h_{12} - c_{12} = \ln y_{12} + B_{12}, \tag{4}$$

where

$$y_1 = g_1 \exp[\beta V_1], \tag{5}$$

and

$$y_{12} = g_{12} \exp[\beta u_{12}]. \tag{6}$$

The function $V_1 = V(z_1)$ is the external potential and $u_{12} = u(r_{12})$ is the pair potential. The functions B_1 and B_{12} are called *bridge functions*. Approximations for the bridge functions give integral equations. Popular approximations are the *hypernetted chain (HNC) approximation*,

$$B_1 = 0, \tag{7}$$

$$B_{12} = 0, \tag{8}$$

and the *Percus-Yevick (PY) approximation*,

$$B_1 = y_1 - 1 - \ln y_1, \tag{9}$$

$$B_{12} = y_{12} - 1 - \ln y_{12}, \tag{10}$$

Equation (1), coupled with Eqs. (7) or (9), gives an integral equation which can be used to obtain the density profile, if c_{12}^B is known (say from the solution of the bulk OZ equation in some approximation). In contrast, Eq. (2), even when coupled with Eqs. (8) or (10), is incomplete unless ρ_3 is specified. For a homogeneous fluid ρ is constant and there is no problem. One possibility for an inhomogeneous fluid would be to use the solution of OZ1. It is less convenient, but more accurate, to use the equation (Lovett et al, 1976; Wertheim, 1976)

$$\frac{\partial \ln y_1}{\partial z_1} = \int \frac{\partial \rho_2}{\partial z_2} c_{12} d\mathbf{r}_2. \tag{11}$$

Other equations can be used in place of Eq. (11). However, they will not be considered here.

The use of Eq. (1) with Eq. (2), and some approximation, will give results for the pair correlation functions but, quite obviously, can only produce the density profile of Eq. (1). The density profile obtained from Eqs. (2), (11), and some approximation, is generally better. As an example, if $B_{12} = 0$ is assumed, the bridge function contribution to g_1, and c_1, is *exact*, to order ρ^2, and is better than the PY or HNC approximations, Eqs. (7) or (8), for high order ρ.

For a fluid near a hard wall, force balance considerations yield the contact condition

$$\rho(x_0) = \beta p,\tag{12}$$

where p is the pressure of the bulk fluid at an infinite distance from the wall, $\beta = 1/kT$, k being the Boltzmann constant and T being the temperature, and x_0 is the distance of closest approach of a molecule to the wall. If the HNC approximation is used with Eq. (1) for a fluid near a hard wall,

$$\rho(x_0) = \frac{1}{2}\,\rho\left[1 + \beta\,\frac{\partial p}{\partial\rho}\right].\tag{13}$$

A similar result is not known for the PY approximation. However, for hard spheres near a hard wall, the PY approximation gives

$$\rho(x_0) = \sqrt{\beta\,\frac{\partial p}{\partial\rho}}\tag{14}$$

To be precise $\beta\partial p/\partial\rho$ in Eqs. (13) and (14) is the compressibility of the bulk fluid as calculated from the compressibility equation in the HNC or PY approximation, respectively. For most applications, Eqs. (13) and (14) are poor approximations to Eq. (12).

At this stage it is with worth noting that setting $c_{12} = c_{12}^B$ in Eq. (11) yields Eq. (1) in the HNC approximation. As a result, Eq. (11), with $c_{12} = c_{12}^B$, yields Eq. (13) for a fluid near a hard wall. There are no other known contact value formulae, for uncharged hard walls, resulting from Eq. (11) except that, of course, if no approximation is made, Eq. (12) results.

The above observation that setting $c_{12} = c_{12}^B$ in Eq. (11) yields Eq. (1) in the HNC approximation, or some similar observations, has lead to occasional suggestions that c_{12}^B in Eq. (1) should be replaced by c_{12} for the inhomogeneous fluid and the resulting (incorrect) version of Eq. (1) be solved together with some relation specifying c_{12}, say Eq. (2). Such a procedure would be more convenient, if it were correct, than a coupling of Eqs. (2) and (11) because y_1, rather than its derivative, appears. However, this incorrect procedure yields incorrect results. For example, the first bridge diagram would have a coefficient of unity instead of one half. It is the presence of the derivative in Eq. (11) which yields the correct coefficient of one half after integration.

SOME RESULTS

Three applications of Eqs. (1) and (2) will be considered. They are hard spheres near a hard wall, Lennard-Jones molecules near a hard wall, and charged hard spheres near a charged hard wall.

First consider hard spheres near a hard wall. The PY approximation is the preferred approximation for hard particles. The results of the PY version of Eq. (1), the PY1 profile, and the PY version of Eqs. (2) and (11), called the PY2 profile, for $\rho(x)$ are plotted in Fig.1. The PY1 results (Henderson et al, 1976), which can be given by explicit formula (Henderson and Smith, 1978), are in quite good agreement with the simulation results (Snook and Henderson, 1978). The error is only significant near contact. Although incorrect, the PY1 results are at least qualitatively correct because βp and $\sqrt{\beta \partial p / \partial \rho}$ are both large at this high density. The PY2 results (Plischke and Henderson, 1986b) are an improvement. They can be made even better by modifying Eq. (11) slightly, in light of Eq. (12). We refer to Plischke and Henderson (1990) for details. Kjellander and Sarman (1988) have made similar calculations. The first solutions of the PY2 equation for hard spheres at a hard wall was made by Sokolowski (1980) who obtained results at low densities.

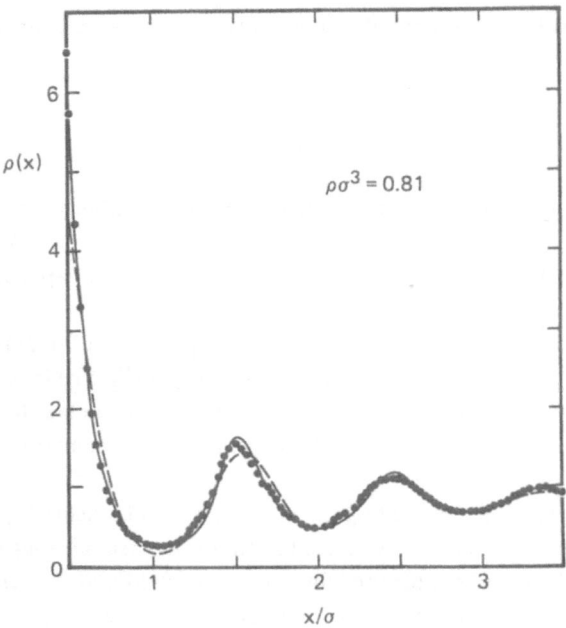

Figure 1. Density profile for hard spheres at a hard wall (in units of σ^3). The circles give the computer simulation results while the solid and broken curves give the PY2 and PY1 results, respectively. The parameter σ is the hard sphere diameter.

Plischke and Henderson (1986a) have given PY1 and PY2 results for Lennard-Jones molecules near a hard wall. At high temperatures the results are similar to those of Fig. 1. However, as is seen in Fig. 2, the situation changes at low temperatures and high densities. The PY1 results are rather poor near contact. In contrast, the PY2 results are in quite good agreement with simulations. The withdrawal of the molecules from the wall which is seen in the PY2 results and the simulations is referred to as drying. This drying of a wall by a liquid is the inverse of the more familiar wetting of a wall by a gas.

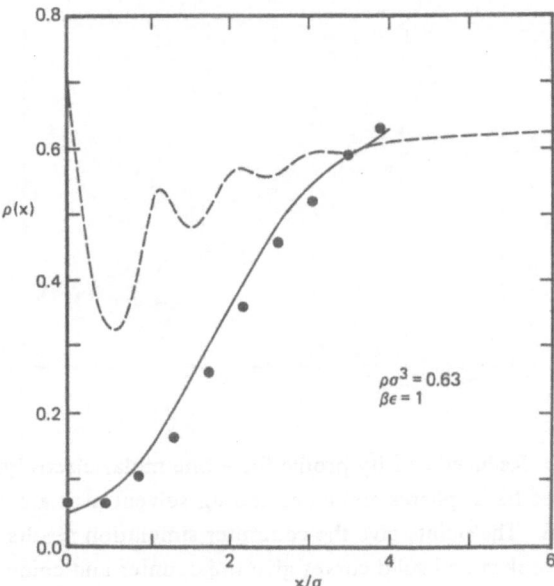

Figure 2. Density profile of Lennard-Jones molecules at a
hard wall curves have (in units of σ^3). The circles and
the same meaning as in Figure 1. The parameters ε and σ are the
usual Lennard Jones energy and distance parameters.

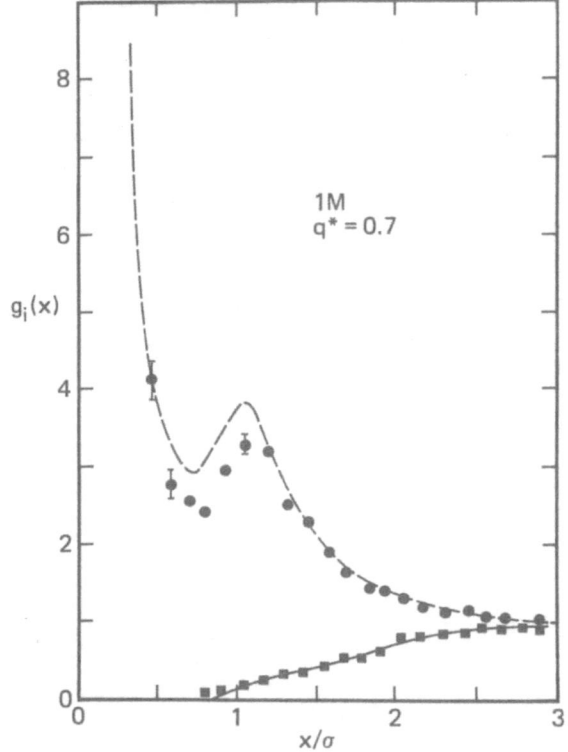

Figure 3. Reduced density profile for a one molar electrolyte
of charged hard spheres and a continuum solvent near a charged
hard wall. The points give the computer simulation results
and the broken and solid curves give the counter and coion
HNC2 results. The parameter q* is the reduced charge on the wall.

Finally, in Fig. 3 some of the HNC2 results of Plischke and Henderson (1988) for charged hard spheres near a charged wall are shown. The HNC approximation is the preferred approximation for charged particles. The HNC1 results are not shown but are monotonic. The HNC2 results are also monotonic at low charge on the wall, but at high charge the counterion profile is nonmontonic because of the formation of a second layer of counterions. This second layer is also seen in the simulations of Torrie and Valleau (1980).

SUMMARY

Integral equations, particularly those based on the pair OZ equation, Eq. (2), provide a useful and reliable theory of inhomogeneous fluids. Further details about integral equation approaches for inhomogeneous fluids, as well as other approches, can be found in a forthcoming monograph (Henderson, 1992).

ACKNOWLEDGEMENTS

The author is grateful to IBM/Salt Lake City for its support. He is also grateful to IBM/Dallas and the organizers and sponsors of this conference and for the financial assistance which made his attendance at the workshop possible.

REFERENCES

Abraham, F. F., 1978, *J. Chem. Phys.* 68:3713.

Henderson, D., 1992, *"Fundamentals of Inhomogeneous Fluids"*, Dekker, New York.

Henderson, D. and Smith, W. R., 1978, *J. Stat. Phys.* 19:191.

Henderson, D., Abraham, F. F., and Barker, J. A., 1976, *Mol. Phys.* 31:1291.

Kjellander, R., and Sarman, S., 1988, *Chem. Phys. Lett.* 149:102.

Lovett, R., Mou, C., and Buff, F. P., 1976, *J. Chem Phys.* 65:570.

Plischke, M., and Henderson, D., 1986a, *J. Chem. Phys.* 84:2846.

Plischke, M., and Henderson, D., 1986b, *Proc. Roy. Soc. (London)* A404:323.

Plischke, M., and Henderson, D., 1988, *J. Chem. Phys.* 88:2712.

Plischke, M., and Henderson, D., 1990, *J. Chem. Phys.* 93:4489.

Snook, I. K., and Henderson, D., 1978, *J. Chem. Phys.* 68:2134.

Sokolowski, S., 1980, *J. Chem. Phys.* 73:3507.

Torrie, G. M., and Valleau, J. P., 1980, *J. Chem. Phys.* 73:5807.

Wertheim, M., 1976, *J. Chem. Phys.* 65:2377.

FLUIDS OF HARD CONVEX MOLECULES

M. S. Wertheim

Physics Department
Michigan Technological University
Houghton, MI   49931

ABSTRACT

Recent work on fluids of hard convex molecules uses differential and integral geometry to reformulate the graphical expansion of the Helmholtz free energy. The reformulated expression involves only one-body geometry rather than the two-body geometry implicit in the Mayer f-function. The price paid for this reduction is the appearance of sets of n-point measures of one-body geometry. For the ring graphs, only 2-point measures are required; this makes it feasible to evaluate these graphs, previously obtained by Monte Carlo integration, by fast and accurate direct integration. The calculations involved in going beyond 2-point measures appear formidable. We present an overview of the theory, omitting technical details discussed in the original papers.

INTRODUCTION

Fluids of hard molecules are a continuing challenge to theory. Conceptually, they are among the simplest of systems. The interaction potential $\emptyset(A,B)$ of two hard bodies A and B is $+\infty$ if they overlap, and 0 otherwise. In terms of the Mayer f-function, $f(A,B) = \exp[-\emptyset(A,B)/kT] - 1$, where k is Boltzmann's constant and T the temperature, we have

$$f(A,B) = \begin{cases} 0 & \text{for } A \cap B = \emptyset \\ -1 & \text{for } A \cap B \neq \emptyset \end{cases} \qquad (1)$$

where $\emptyset$ denotes the empty set.

In contrast to the simplicity of the interaction is the richness of the phase diagram for highly aspherical hard molecules, as demonstrated in the Monte Carlo study of prolate and oblate spheroids by Frenkel et al.[1-2] These authors found four phases: isotropic, nematic, plastic solid, and solid. The nematic phase appears for $\lambda > 2.75$ and $\lambda < 1/2.75$, where λ is the ratio

Condensed Matter Theories, Vol. 8, Edited by
L. Blum and F.B. Malik, Plenum Press, New York, 1993

of the diameter corresponding to the symmetry axis to the doubly degenerate diameter.

Nearly all our knowledge of fluids of aspherical hard molecules beyond the level of the second virial coefficient B_2 derives from computer simulation. Solutions of the Percus-Yevick (PY) and hypernetted chain (HNC) integral equations have been obtained for moderately elongated spheroids[3] and spherocylinders.[4-5]

ROLE OF CONVEXITY

Liquid crystalline behavior of aspherical hard particles does not depend on convexity. The reason for the restriction to convex bodies is essentially mathematical: it enables us to apply methods based on differential and integral geometry.[6,7] The classical result of this type is the remarkable result for the second virial coefficient for two convex molecules A and B:

$$B_2 = (1/8\pi)[V(A)G(B) + S(A)M(B) + M(A)S(B) + G(A)V(B)], \tag{2}$$

where V is the volume, S is the surface, and M and G are integrals of the mean and Gaussian curvatures, respectively:

$$M(A) = \int_{\delta A} \overline{K} dS, \qquad G(A) = \int_{\delta A} K_G dS \tag{3}$$

where $\overline{K}$ is the mean curvature and K_G the Gaussian curvature. For any convex body we have $G = 4\pi$.

Eq. (2) as a result in mathematics is due to Blaschke and Santalo.[6] The application to B_2 by Isihara[8] stimulated many attempts to generalize this result to the higher B_n. These generalizations, ingenious and successful for moderately aspherical convex bodies, are either frankly semi-empirical or draw on ideas from the scaled particle model for aspherical molecules.[9] For example, Kihara and Miyoshi[10] proposed the semi-empirial formula

$$B_3 = (1/24\pi)\left[\sum_3 V(i)V(j)G(k) + \sum_6 V(i)S(j)M(k) + \sum_3 S(i)S(j)M^2(k)\right], \tag{4}$$

where i,j,k are permutations of the particle species. References to a large body of semi-empirical work are given in ta review by Boublik and Nezbeda.[11]

If we go to highly aspherical molecules, the semi-empirical schemes must be expected to fail. With increasing elongation, the sign of B_4 changes from + to - as the negative ring graph becomes more and more dominant. No hint of this is seen in the semi-empirical theories. Moreover, the asymmetry of the curvature does not appear explicitly, although an elongation parameter takes its place.

The standard way of dealing with nonspherical molecules is the expansion of the Mayer f-function in complete sets of functions of the orientations Ω_1 and Ω_2 of the two bodies. Applied to a Mayer f-function that is -1 or 0, this is likely to require enormous sets of basis functions. In addition, the number of f-functions for M species is $\frac{1}{2}M(M+1)$. This is a serious burden for large M, and probably an insuperable one for poly-dispersed mixtures of convex shapes.

Representation of f-function

Our reformulation[12-13] is based on reduction of the 2-body geometry implicit in the Mayer f-function to 1-body geometry. The starting point is the following. We note that for convex bodies we have

$$f(A \cap B) = -G(A \cap B)/4\pi. \tag{5}$$

The validity of this is restricted to convex bodies: the intersection of two convex bodies is necessarily a single convex body, so that $G(A \cap B) = 4\pi$. As soon as one of the bodies is not convex, there are configuration such that the surface $\delta(A \cap B)$ consists of more than one closed surface.

The integral Gaussian curvature on the right hand side of Eq. (5) may be expressed as a sum of two types of integrals:

$$G(A \cap B) = G_S(A \cap B) + G_L(A \cap B) \tag{6}$$

$G_S(A \cap B)$ consists of integrals over the parts of the surface of body A that lie inside B, and similarly over the parts of δB that lie inside A:

$$G_S(A \cap B) = \int_{\delta A \cap B} K_G(A)dS_A + \int_{\delta B \cap A} K_G(B)dS_B. \tag{7}$$

One of the terms is absent if one molecules engulfs the other.

$G_L(A \cap B)$ is the contribution to $G(A \cap B)$ from the the intersection curve(s) $\delta A \cap \delta B$. There may be more than one intersection curve; the maximum number that can appear depends on the geometry of the two convex bodies. The case of no intersection curve occurs when one body engulfs the other.

The contribution to $G(A \cap B)$ from the intersection curve(s) can be expressed in terms of the outward unit normals n_A and n_B, and the curvatures K_A and K_B:

$$G_L(A \cap B) = \int_{\delta A \cap \delta B} \left[(1 - \mathbf{n}_A \cdot \mathbf{n}_B)(\overline{K}_A + \overline{K}_B) - \right.$$

$$\left. - \frac{\mathbf{n}_A \cdot \mathbf{\Delta}_B \cdot \mathbf{n}_A + \mathbf{n}_B \cdot \mathbf{\Delta}_A \cdot \mathbf{n}_B}{1 + \mathbf{n}_A \cdot \mathbf{n}_B} \right] \frac{ds}{|\mathbf{n}_A \times \mathbf{n}_B|} . \tag{8}$$

The second rank tensor $\mathbf{\Delta}$ is the curvature deviator, related to the curvature $\mathbf{K}$ by

$$\mathbf{K} = \overline{K}(\mathbf{I} - \mathbf{nn}) + \mathbf{\Delta} , \tag{9}$$

where $\mathbf{I}$ denotes the unit tensor. The result for $G(A \cap B)$ as expressed by Eqs. (6-9) is equivalent to the sum over all connected pieces of surface of the surface integral of K_G plus the integral of the geodesic curvature of the boundary curves. This verifies consistency with the Gauss-Bonnet theorem.[12]

When this representation of the f-function is applied to the Mayer graph consisting of a single line, the terms containing $\mathbf{\Delta}$ do not contribute, and we recover the classical result of Eq. (2).

Effect on higher graphs

When the f-function is represented by $- G/4\pi$, we introduce one additional integration per line. We then interchange integrations and integrate over positions and orientations of the molecules. The terms free of $\mathbf{\Delta}$ are of the form of sums of products of terms which refer to only one molecule of the pair. In the terms with $\mathbf{\Delta}$, this property is spoiled by the denominator $1 + \mathbf{n}_A \cdot \mathbf{n}_B$ and can be recovered only by expansion.

Integration over a molecule with n incident lines than produces a sum of products of internal and external factors. The external factors are tensors formed from the relative vectors between pairs of the n space-fixed points arising from the representatiuon of the f-function. The external factors are universal, and do not refer in any way to the geometry of the molecule. The internal factors are scalar n-point measures of the molecular geometry for a given configuration of the n points. In particular, 2-point measures are functions only of the distance r between the two points, and 3-point measures are functions only of the three relative distances.

The reformulation may be represented in graphical expansion as follows: starting from the irreducible graphs in the expansion of the excess Helmholtz free energy, every point with n incident line is replaced by an n-gon, which represents an n-point measure. Every line of of the Mayer graphs is replaced by a point shared by two n-gons.

We note the following obvious corollaries. The Mayer graphs and the reformulated graphs are in one to one correspondence. In the reformulated graphs every point is shared by exactly two n-gons. Let N denote the number of points and n the number of lines of a Mayer graphs. Then $n \geqslant N$, with equality only for the ring graphs. For the number N_R of points of the re-formulated graphs we have

$$N_R = n \geqslant N \tag{10}$$

with equality only for the ring graphs. The complications of dealing with 3-point measures are severe. A very large set of functions is required, and the calculation of 3-point measures is harder than the case of 2-point measures. The greatest complication, however is the final integration of the reformulated graph. The practical feasibility of calculations involving 3-point measures is an open question at this time.

RING GRAPHS AND 2-POINT MEASURES

The ring graphs are unique in having the property that Mayer rings go into reformulated rings. In the reformulated rings, the lines represent 2-point measures of the geometry of one convex body. Calculationally, this represents a great simplification because fast and highly accurate computa-tion of 2-point measures is easily achieved for such standard models as spheroids and spherocylinders. In the 2-point measures, each point may be an interior (volume) point or a surface point. Surface points may enter un-weighted or with the weights

$$\overline{K}, \ K_G, \ \overline{Kn}, \ n^S, \ \Delta n^S \quad (s = 0, 1, 2....) \tag{11}$$

where the weights appear in the form of invariants formed from the weights at the two points and the unit vector $e = r/r$, where r is the vector which connects the two points.

For long or flat shapes a very large set of 2-point measures may be re-quired, but the calculation of the measures is very fast and accurate, orders of magnitude easier than the expansion of the f-function in (generalized) spherical harmonics. Some other advantages of the 1-body formulation may be noted. In the absence of symmetry which defines a best choice, the expansion of the f-function depends on the choice of molecular centers. The 1-body formulation is invariant; no reference to a molecular center is made. Finally, for a polydispersed mixture of shapes, the 1-body quantities are simple the sums, weighted by number densities, of the corresponding quanti-ties for each species. Thus the complexity scales as M, the number of spe-cies, rather than as $\frac{1}{2}M(M+1)$.

The properties of 2-point measures are intricate enough to warrant an entire article for their exposition.[13] In summary, the following features generate welcome simplification. First of all, some measures vanish, though not trivially. Some non-vanishing measures do not enter the calculation of the ring graphs. Finally, there are relations between measures that facilitate their calculation. For example, all measures involving volume points can be calculated from measures containing two surface points.

Most important of all are relations of two types. Both the integral and the small r behavior of 2-point measures are given in terms of 1-point measures. For models such as spheroids and spherocylinders there are known analytic expressions for these 1-point measures. This is extremely valuable

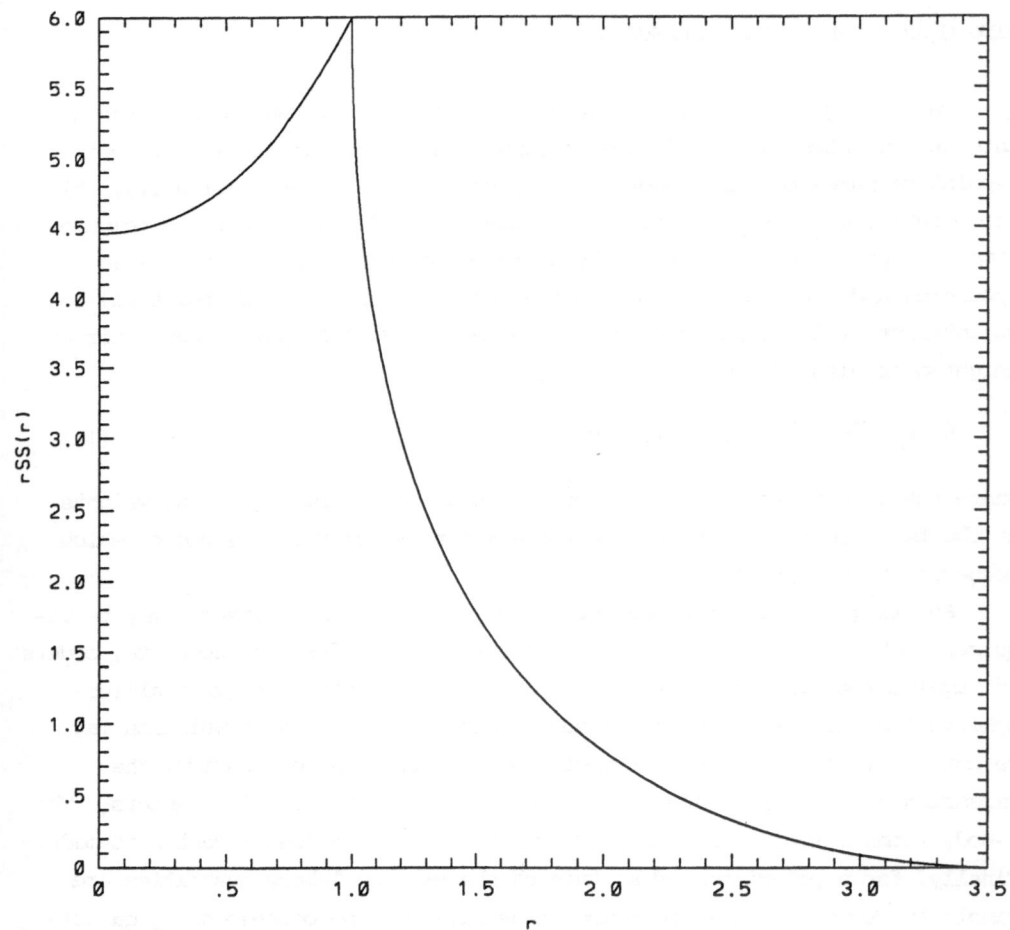

Fig.1. 2-point measure for unweighted surface points of a prolate spheroid with axis ratio 3.5

in checking the five digit accuracy which we are maintaining in our ongoing calculations of 2-point measures.

For example, the 2-point measure for two unweighted surface points satisfies

$$\lim_{r \to 0} rf_{SS}(r) = \tfrac{1}{2}S, \qquad \int f_{SS}(r)d^3r = S^2, \tag{12}$$

where S is the value of the surface area.

Fig. 1 exhibits $rf_{SS}(r)$ for a prolate spheroid with axis ratio $\lambda = 3.5$. The minor diameter is taken as the unit of length. As the cusp at $r = 1$ is approached from $r > 1$, the slope tends to ∞. In the case of spherocylinders, there is an integrable logarithmic singularity rather than a cusp. Dealing with cusps and integrable singularities requires some care to avoid loss of accuracy in the subsequent convolutions that are performed in the calculations of the ring integrals.

An obligatory exercise is the verification that the method of 2-point measures recovers the third virial coefficient of three hard spheres of arbitary diameters A, B, and C. The execution of this calculation was found to be exceedingly tedious, between two and three orders of magnitude longer than the direct calculation of B_3. However, this is so precisely because the calculations are completely _analytic_. In the realistic case where the calculation of the ring integral is _numerical_, only very simple and fast programs are needed.

OUTLOOK

The reformulation enables us to carry out a direct calculation of all ring graphs in the expansion of A/kT, where A is the Helmholtz free energy. Previously these graphs could be obtained only by Monte Carlo integration. For both long and flat shapes, this can be used to obtain accurate third virial coefficients for both one- and multi-component systems. This will produce a welcome check on the values obtained by Monte Carlo integration for prolate spheroids[14] and spherocylinders.[15]

Apart from an accurate B_3, the reformulation appears to be relevant primarily to systems of _long_ molecules. For long molecules, each additional f-bond inserted into a ring is a severe restriction on the configurations that contribute. It is natural to conjecture that in the very long limit the rings become the dominant graphs beyond B_2, which yields the Onsager limit.[16] If this is so, the sum of the rings should provide the leading correction to the Onsager result. When can higher terms be expected to become negligible? For spherocylinders, B_4 changs sign at a length to width ratio of about 10. This indicates that terms beyond rings are not small until we go to very long shapes.

In the opposite limit of very flat shapes, B_4 remains positive, and B_5 becomes negative.[17] This indicates that the ring sum is not appropriate in this case.

So far we have considered only the isotropic phase, and only completely integrated graphs, leading to the Helmholtz free energy. Generalizations to anisotropic phases and pair distribution functions are of interest, but should probably await a clarification of the feasibility of going beyond ring graphs.

REFERENCES

1. D. Frenkel, B. M. Mulder and J. P. McTague, Phase Diagram of a System of Hard Ellipsoids, Phys. Rev. Lett. 52:287 (1984).
2. D. Frenkel and B. M. Mulder, The hard ellipsoid-of-revolution fluid. I. Monte Carlo simulation, Mol. Phys. 55:1171 (1985).
3. A. Perera, P. G. Kusalik and G. N. Patey, the solution of the hyper-netted chain and Percus-Yevick approximations for fluids of hard non-spherical particles. Results for hard ellipsoids of revolution. J. Chem. Phys. 87:1295 (1987); erratum 89:5969 (1988).
4. A. Perera and G. N. Patey, The solution of the hypernetted chain and Percus-Yevick equations for fluids of hard spherocylinders, J. Chem. Phys. 89:4349 (1988).
5. S. Lago and P. Sevilla, Solution of the Percus-Yevick equation for hard spherocylinders. I. The entire pair correlation function, J. Chem. Phys. 91:4912 (1987).
6. W. Blaschke, Vorlesungen uber Integralgeometrie, Teubner, Leipzig and Berlin, 1936-7; reprinted Chelsea, New York, 1949.
7. L. A. Santaló, Integral Geometry and Geometric Probability, volume 1 of Encyclopedia of Mathematics and its Applications, Addison-Wesley, Reading, 1976.
8. A. Isihara, J. Chem. Phys. 18:1446 (1950).
9. R. M. Gibbons, The scaled particle theory for particles of arbitrary shape, Mol. Phys. 17:81 (1969).
10. G. Kihara and K. Miyoshi, Geometry of Three Convex Bodies Applicable to three-Molecule Clusters in Polyatomic Gases, J. Stat. Phys. 13:337 (1975).
11. T. Boublik and I. Nezbeda, P-V-T Behavior of Hard Body Fluids. Theory and Experiment, Czechoslovak Chem. Commun. 51:2301 (1986).
12. M. S. Wertheim, Fluids of hard convex molecules. I. Reformulation in terms of 1-body geometry, to be published.
13. M. S. Wertheim, Fluids of hard convex molecules. II. Properties of 2-point measures, to be published.
14. D. Frenkel, Computer simulation of hard-core models for liquid crystals, Mol. Phys. 60:1 (1987).
15. D. Frenkel, Onsager's Spherocylinders Revisited, J. Phys. Chem. 91:4912 (1987).
16. L. Onsager, Ann. New York Acad. Sci. 51:627 (1949).
17. W. R. Cooney, S. M. Thompson and K. E. Gubbins, Virial coefficients for the hard oblate spherocylinder fluids, Mol. Phys. 66:1272 (1989).

STATISTICAL MECHANICS OF FLUIDS AT HIGH DIMENSIONS

H.L. Frisch

Department of Chemistry
State University of New York
Albany, NY 12222

High spatial dimensionality affects profoundly the qualitative, statistical mechanical behavior of many-body systems. The mean spherical nature of lattice gases of unbounded coordination number has been pointed out some time ago[1] while more recently significant new results have been obtained for the Hubbard model of very high dimensions.[2] Here we shall limit ourselves to reviewing the dramatic structural simplification of purely repulsive finite range continuum fluids in the high dimensionality limit.[3-11] Our review is not exhaustive (we apologize in advance for slighting authors of any missing citations) but we focus instead on stating what can be achieved by this paradigm and point out, where relevant, what open problems need be addressed.

The infinite dimensional hard sphere fluid[3] (of diameter a) provides a natural example of how straight forward geometrical considerations dominate the statistical mechanics because of volume scaling. The natural scale of volume is $v = \pi^{D/2} a^D / \Gamma[1 + (1/2)D]$, the D dimensional volume of a sphere of radius a, As $D \to \infty$ not only does $v/a^D \to 0$ but also the overlap cross section between spheres vanishes rapidly. As a consequence[3-5] the Mayer series for the grand partition function in the limit as $D \to \infty$ has as the only non-vanishing contribution, the one resulting from tree diagrams.[3,5] The mechanical equation of state for the pressure, p, of this hard sphere fluid truncates after the second virial coefficient[3-5]. The profound effect of the surface to volume ratio considerations as $D \to \infty$ are mirrored further though in this system by a departure from the first order nature of the fluid-to-solid phase transition expected in lower dimensions.[4] (The first order nature of this transition in three dimensions is inferred from numerical simulations).

The high dimensionality paradigm thus can provide three kinds of insight for model systems which have some bearing to real systems in three dimensions:

Type 1 - Exact results (of high physical plausibility if not complete mathematical rigor) for high dimensional analogs whose phase transition(s) are different from that expected in three or two dimensions. Examples of such systems are the hard sphere,[3-8], hard hypercube[6] and probably spherical, purely repulsive finite-range systems.[4]

Type 2 - Exact result for high dimensional analogs, but these are systems whose phase transitions are the same as expected in three dimensions. Examples are the first order, orientational phase transition of rodlike hypercyliners from an isotropic fluid to a nematic[10] and similar liquid crystal transitions in high dimensions.

Type 3 - Tests of well-defined approximations to the statistical behavior of model systems which should be (or not) particularly sensitive to the limiting surface to volume ratio. An example[5] is provided by the scaled particle theory of hard spheres originally developed by Reiss et. al.[12] Below we shall comment briefly on these types of results.

The simplest examples of Type 1 results are that the pressure p of the hard sphere fluid (as c) is given exactly[3,5] by (ρ the number density, $\beta = 1/k_B T$)

$$\beta p/\rho = 1 + 1/2 \ v\rho \ , \tag{1}$$

while for arbitrary, finite range repulsive interactions $1/2 \ v$ has to be replaced[4] by the second virial coefficient of the model, B. Both the free energy and non-uniform profile can be obtained[4] in the $D \to \infty$ limit. The reaction of the equilibrium fluid to an external field (whose potential energy is u(r)) is given[4] by the nonlinear Debye-Hückel equation for the number density n(r) and chemical potential μ,

$$\beta[\mu - u(r)] = \ln n(r) - \int f(r-r')n(r')d^D r' \tag{2}$$

with f(r) the Mayer function $f(r) = \exp[-\beta\phi(r)]-1$

with $\phi(r)$ the interparticle potential which is nonnegative, translation invariant and of finite range. One sees from eq. (2) that for u =o there is a uniform fluid state which becomes metastable to excitations at some wave vector if

$$n \geq n_0 = \min_k [1/f(k)] \quad , \tag{3}$$

where $f(k)$ is the Fourier transform of $f(r)$, suggesting a phase transition at n_0. This is the Kirkwood instability.[13] Indeed the Kirkwood instability leads continuously (i.e. not first order) to an ordered lattice, whose free energy is lower that that of the uniform fluid.[4]

The kinetic equation of the infinite dimensional hard sphere fluid, as obtained from the binary-collision expansion of the one-particle distribution function is precisely the modified Euskog theory, providing the initial many particle distribution function can be factorized into a product of one particle distribution functions.[9] From this kinetic equation one can deduce the macroscopic, Navier-Stokes hydrodynamic equations in the limit of small gradients. In the dilute (Boltzmann) limit all transport coefficients are scaled by a single parameter (which can be taken to be the kinematic viscosity) reflecting the fact that a single, grazing collision, mechanism governs all transport processes.

Certainly one of the outstanding problems of these investigations is to provide the necessary mathematical underpinnings for a rigorous theory. These questions involve the justification of interchange of thermodynamic and $D \rightarrow \infty$ limits; the convergence of series e.g. the binary-collision expansion which has never been proven to converge in any dimension; or the proof of the physical realizability of n_0, etc. To validate these conclusions reached on the model systems associated with the $D \rightarrow \infty$ limit (in particular to estimate an upper critical dimension), it will be necessary to find consistent asymptotic large D expansions.

The quantum mechanical, $D = \infty$, hard sphere boson fluid behaves like an ideal Bose gas and undergoes a Bose-Einstein condensation.[5] Jensen and Percus[14] have demonstrated that the analogous fermion fluid exhibits a complex transition with a "superconducting" like gap separating the ground state.

As a final example of a Type 1 result we mention the behavior at high dimensions of binary non-additive hard-sphere mixtures.[11] Subject to suitable scaling the classical system of such interacting particles truncates at second virial terms. A binary mixture of non-additive hard spheres with sufficiently repulsive interactions between unlike particles decomposes at sufficiently high density into two coexisting phases. The region around the critical density behaves classically[15], unlike the two dimensional result.[16]

The best example of a Type 2 result is provided by the high dimensional limit $(D \to \infty)$ of hard hypercylinders of length L and radius R.[10] In that limit the free energy per particle truncates to the second virial coefficient (cf. the usual Onsger theory[17]). For such spatially uniform fluids, there are two phases: The lower density isotropic phase and a higher density anisotropic phase characterizaed by an orientational distribution function of the form $\exp[-\alpha(1-\gamma^2\cos2\theta)^{1/2}]$, with $\theta = 0$ the polar axis of the nematic liquid crystal, and α and γ explicitly known functions of the number density ρ. The first order transition (which occurs well below the Kirkwood transition) connects the two phases at $\beta p = 5.700\ D^2/B$, $\beta\Delta\mu = 3.376D$ and reduced densities of the anisotropic phase $B\rho/D = 3.376$, and isotropic phase $B\rho/D = 3.376$, with B the limiting second virial coefficient

$$B = L^2(2R)^{D-2}S_{D-2}/(D-2); \quad S_D = D\pi^{D/2}/(D/2)! \qquad (4)$$

These values of the state variables characterizing the first order transition are quite comparable with the best numerical estimates in two and three dimensions. We are seeking[18] a suitable asymptotic theory for correcting these results to lower dimensions as part of an ongoing program to assess the validity of the $D \to \infty$ limit. Conservative error estimates on the free energy (for finte D) can be gotten from knowledge of third virial terms. It is because of this that we are confident that first order transitions, which occur at relatively low density, maintain validity as D is lowered, but can make no seuch claim for transitions such as Kirkwood transitions.

The use of various spatial dimensions to test theoretical predictions is a well established procedure. The $D \to \infty$ limit employed in Type 3 results to test approximate theories has two possible advantages: a) The approximation may be carried through exactly and b) then can be tested against results for suitable models that can be solved exactly in this limit (i.e. Type 1 or 2 results). Thus the simplest version of scaled particle theory, as expected from surface to volume considerations, while giving a correct second virial coefficient gives an incorrect, finite third virial coefficient. This points out the importance of recent corrections to this theory of Reiss and collaborators; rather than its intrinsic weakness. The conjecture of Barboy and Gelbart[19] that the hard sphere equation of state can be written in the Y form in D dimensions

$$p = \sum_{i=1}^{D} a_i Y^i , \qquad (5)$$

$$Y = y(1-y)^{-1} \quad ; \quad y = \rho v / 2^D ,$$

is verified in the $D \to \infty$ limit as it is for $D = 1$. Thus Cuesta et. al.[20] used this paradigm to test the density functional theory, within the effective-liquid approximation, for the isotropic - nematic transition. The test of the mean spherical approximation for high dimensional, suitably scaled model hard core polar and charged fluids would be an interesting example of these results.

Much remains to be done in this field: We believe not only questions of rigor, but substantive questions remain of the physical behavior in the $D \to \infty$ limit of the Coulomb system, transport in dense quantum (as well as classical) versions of these models, the existence of metastable "glassy" states, to mention only a few of the outstanding challenges.

ACKNOWLEDGEMENT

This work was supported by the National Science Foundation Grant DMR 90233541. The author's travel expenses to the XVI Workshop on Condensed Matter Physics in San Juan, P.R. were supported by the U.S. Army Research Office.

REFERENCES

(1) H.E. Stanley, Phys. Rev. 176, 718 (1968).

(2) I. Ichinose and T. Matsui, Phys. Rev. B, 45, 9976 (1992); M. Grilli, C. Castellani, and G. Kotliar, Phys. Rev. B., 45, 805 (1992) and references cited therein.

(3) H.L. Frisch, N. Rivier, and D. Wyler, Phys. Rev. Lett. 54, 2061 (1985).

(4) H.L. Frisch and J.K. Percus, Phys. Rev. A 35, 4696 (1987).

(5) D. Wyler, N. Rivier and H. L. Frisch, Phys. Rev. A, 36, 2422 (1987).

(6)   T.R. Kirkpatrick, J. Chem. Phys. 85, 3515 (1986).

(7)   J.K. Percus, Commun. Pure Appl. Math. XL, 449 (1987).

(8)   J.K. Percus, in "Simple Models of Equilibrium and Nonequilibrium Phenomena", ed. by J.L. Lebowitz, Elsevier Science Publishers, Amsterdam, 1987.

(9)   Y. Elskens and H.L. Frisch, Phys. Rev. A, 37, 4351 (1988).

(10)   H.-O. Carmesin, H.L. Frisch and J.K. Percus, Phys. Rev. B, 40, 9416 (1989).

(11)   H.-O. Carmesin, H.L. Frisch and J.K. Percus, J. Stat. Phys., 63, 791 (1991).

(12)   H. Reiss, H.L. Frisch, and J.L. Lebowitz, J. Chem. Phys., 31, 369 (1959).

(13)   W. Klein and H.L. Frisch, J. Chem. Phys., 84, 2 (1986).

(14)   K. Jensen and J.K. Percus, unpublished results.

(15)   Thus high spatial dimenionality of a system joins the presence of long-range forces and increasing number of internal degrees of freedom which enter the interparticle coupling as causes of mean field behavior.

(16)   D. Ruelle, Phys. Rev. Lett., 27, 1040 (1971).

(17)   L. Onsager, Ann. N.Y. Acad. Sci., 51, 627 (1949).

(18)   J.K. Percus and H.L. Frisch, unpublished results.

(19)   B. Barboy and W.M. Gelbart, J. Chem. Phys., 71, 3053 (1979).

(20)   J.A. Cuesta, C.T. Tejero, and M. Baus, Phys. Rev., (in press).

MOLECULAR SIMULATION OF VAPOR-LIQUID EQUILIBRIUM IN MIXED SOLVENT ELECTROLYTE SOLUTIONS

Peter T. Cummings

Department of Chemical Engineering
Thornton Hall, University of Virginia
Charlottesville, Va 22903-2442

ABSTRACT

We review Gibbs ensemble Monte Carlo simulations of vapor–liquid equilibrium in pure water, water/NaCl, water/methanol and water/methanol/NaCl mixtures conducted in our laboratory. The intermolecular potentials used are taken directly from the published literature. For the cases in which interaction potentials between unlike species were not available in the literature, these were constructed using the Berthelot mixing rules. The Ewald sum method with tinfoil boundary conditions was used to take into account long range forces.

INTRODUCTION

Electrolyte solutions play an important role in both naturally occurring systems (biological and geological) and industrial processes. Their importance has led to the development of many empirical or semi–empirical [1]–[7] correlations for their physical properties. While adequate for fitting existing experimental data and for interpolating the thermodynamic properties of mixed solvent electrolyte systems (systems consisting of water, another non–electrolyte species, and a dissociating salt) not deviating markedly from the data sets used in regressing the model parameters, their predictive power is very limited. Almost all of these models treat the solvent as a continuum [8], and thus are limited to low salt concentrations. We are in the process of developing a fundamentally based model for mixed solvent electrolytes based on statistical mechanics that would be applicable over a wide range of salt concentrations. This requires a model that explicitly includes the solvent, thus introducing ion–solvent and solvent–solvent interactions as well as the ion–ion interactions.

Recently, we have performed a series of molecular simulations of vapor–liquid equilibrium in a prototypical pure water [9, 10] and in a prototypical water/alcohol/salt system, water/methanol/NaCl [11]. We used the Gibbs ensemble Monte Carlo [12, 13, 14] method using the SPC model [15] for the water–water interaction, the Haughney et al. [16] model for the methanol–methanol interaction, the Fumi-Tosi potentials[17, 18] for the ion–ion interactions and the Chandresakhar et al. [19] model for the ion–water interactions. Cross interactions between unlike species were determined by the usual Berthelot rules, which were also used to deduce the ion–methanol potential.

For pure water, we found that the phase envelope of pure water was predicted reasonably well by the standard SPC model. The predictions for the liquid and vapor densities, and the critical temperature, are improved considerably by taking into account the difference between the effective dipole moment in the liquid and vapor phases. For the mixtures, we found that using these potentials taken from the literature and without introducing any adjustable constants in the cross interactions leads to accurate results for the water/methanol vapor–liquid equilibrium (VLE) and qualitatively accurate water/methanol/NaCl VLE. In particular, as is observed experimentally, the methanol is salted out of the methanol/water mixture (i.e., its relative volatility increases in the presence of salt).

The salting out phenomenon is of particular interest industrially since it forms the basis of salt effect extractive distillation. When a nonvolatile salt is added to a nonelectrolyte mixture, the relative volatility of the nonelectrolyte species is altered. For example, suppose a strong electrolyte (3) such as potassium acetate or sodium chloride (which dissociate to form $K^+/C_2H_3O_2^-$ and Na^+/Cl^- ions respectively) is added to an ethanol(2)–water(1) solution. The conventional qualitative picture is that the ions preferentially complex with the water molecules creating a high molecular weight species which, compared to the salt–free case, has lower vapor pressure relative to the ethanol. Thus, the relative volatility of ethanol, defined by

$$\alpha_2 = \frac{y_2/x_2}{y_1/x_1} \tag{1}$$

where x_i and y_i are respectively the liquid and vapor phase compositions of species i at equilibrium, is increased by the addition of salt. Even at very small concentrations of these salts, this increase in volatility is sufficient to eliminate the azeotrope at 89.4 mole % ethanol, suggesting that salt may be a useful mass separating agent for extractive distillation. There are many other important azeotropic systems systems that can be separated via extractive distillation using salts as the separating agent [20, 21] and several authors [22, 23] have evaluated the advantages and disadvantages of using salts as separating agents for aqueous nonelectrolyte systems. Discussion of these pros and cons is beyond the scope of the present paper; it is sufficient to note that where solubility relationships permit, extractive distillation by salt effect at least offers a viable alternative to presently used methods for difficult separations. The challenge in designing such separations processes is developing thermodynamic models applicable to mixed solvent electrolytes systems given the complex nature of the systems involved. We are attempting to meet this challenge through a combination of molecular simulation and statistical mechanical theory.

In the next section, the details of the intermolecular potentials and the simulation methodology are briefly described. Then, in separate sections, results for pure water, water/NaCl mixtures, water/methanol mixtures and water/methanol/NaCl mixtures are presented. Finally, we conclude with a discussion of future work.

INTERMOLECULAR POTENTIALS AND SIMULATION METHODOLOGY

The SPC potential model is a site–site model for the water–water intermolecular potential which contains three sites developed by Berendsen *et al.* [15] as a modification of Jorgensen's TIPS3 [24] potential. The SPC potential has a Lennard–Jones interaction between the oxygens (regarded as the centers of mass in this model) with

Table 1. Parameters in the SPC and Haughney H1 models for the intermolecular potentials of water and methanol.

Property	SPC Water		Property	H1 Methanol
			r_{CO} (Å)	1.4246
r_{OH} (Å)	1.0		r_{OH} (Å)	0.9451
$\angle HOH$ (deg)	109.47		$\angle COH$ (deg)	108.33
σ_{OO} (Å)	3.1656		σ_{CC} Å	3.861
ϵ/k_B	78.2 K		ϵ_{CC}/k_B	91.2 K
q_O	-0.82		σ_{OO} Å	3.083
q_H	0.41		ϵ_{OO}/k_B	87.9 K
			q_O	-0.728
			q_H	0.431
			q_C	0.297

9 other site–site charge–charge interactions making a total of 10 site–site interactions. The potential is given by

$$u_{WW}(\vec{r}_{12}, \vec{\omega}_1, \vec{\omega}_2) = \sum_{a=1}^{3} \sum_{b=1}^{3} \frac{q_1^a q_2^b e^2}{r_{12}^{ab}} + 4\epsilon_{OO} \left[\left(\frac{\sigma_{OO}}{r_{OO}} \right)^{12} - \left(\frac{\sigma_{OO}}{r_{OO}} \right)^6 \right] \tag{2}$$

where e is the electrostatic charge, q_i^a is the charge on site a of molecule i in units of the electronic charge, e, r_{OO} is the distance between the centers of the oxygen atoms in the two molecules, ϵ_{OO} and σ_{OO} are the Lennard–Jones parameters. This expression gives the potential between molecule 1 (located at position $\vec{r}_1$ in orientation $\vec{\omega}_1$) and molecule 2 (located at position $\vec{r}_2$ in orientation $\vec{\omega}_2$), $\vec{r}_{12} = \vec{r}_1 - \vec{r}_2$ is the vector joining the centers of mass of the two molecules, and $r_{12}^{ab} = |\vec{r}_1^a - \vec{r}_2^b|$ is the distance and u_{ab} the pair potential between site a on molecule 1 and site b on molecule 2. Note that $\vec{r}_i^a$ denotes the position of site a in molecule i. The parameters characterizing SPC are given in Table 1 in which the angle $\angle HOH$ is the angle subtended by the two hydrogen sites at the oxygen site. The distance r_{OH} is the distances from the oxygen site to the hydrogen site. The point charges for the molecule are q_O for oxygen and q_H for hydrogen.

Haughney's *et al.* [16] modification of Jorgensen's [24] methanol–methanol potential is used in this work. The model consists of three sites: a hydrogen, an oxygen, and a methyl group, with a point charge at each site and a Lennard–Jones interaction at the oxygen site and at the methyl group site. The parameters for this model are given in Table 1 in which $\angle COH$ is the angle made between the carbon site, the oxygen site, and the hydrogen site next to the oxygen site. The distances r_{CO} and r_{OH} are the distances from the carbon site to the oxygen site and from the oxygen site to the neighboring hydrogen site, respectively. The point charges for the molecule are q_O for the oxygen, q_C for the carbon, and q_H for the hydrogen site attached to the oxygen site. There are two Lennard–Jones sites in methanol with σ_{CC} and ϵ_{CC}, and σ_{OO} and ϵ_{OO} being the parameters for the carbon and oxygen sites, respectively.

The salt studied was sodium chloride modeled by the interionic potentials of Fumi and Tosi [17, 18], which are fitted to solid state properties of the alkali halides, . These potentials were used by Pettitt and Rossky [25] in their integral equation work and on this basis appear to be acceptably accurate. The model consists of a single point charge and an exponential–six potential for the ion–ion interactions and a single point charge and a Lennard–Jones potential for the ion–water interactions. The potential function

Table 2. Parameters of the intermolecular potential for the ion–ion interactions.

	Na^+-Na^+	Na^+-Cl^-	Cl^--Cl^-
B/k_B	4.92×10^8 K	1.46×10^7 K	4.04×10^7 K
ρ (Å)	0.317	0.317	0.317
C (Å$^6/k_B$)	1.22×10^4 K	8.11×10^4 K	8.47×10^5 K

is of the Huggins–Mayer form

$$U_{ij}(r) = \frac{q_i q_j}{r} + B_{ij}e^{-r/\rho_{ij}} - \frac{C_{ij}}{r^6} \tag{3}$$

The parameters are summarized in Table 2. For the ion–solvent interactions the potential model is 1–6–12 potential

$$U_{Wj}(r) = \sum_{i=1}^{3} \frac{q_i^{SPC}q_j e^2}{r} + 4\epsilon_{Wj}\left[\left(\frac{\sigma_{Wj}}{r_{Oj}}\right)^{12} - \left(\frac{\sigma_{Wj}}{r_{Oj}}\right)^{6}\right] \tag{4}$$

of Chandrasekhar and Jorgenson with the oxygen site regarded as the center of the water molecule. The parameters for the water–Na$^+$ interaction are σ_{WNa^+}=1.89744 Å and $\epsilon_{WNa^+} = 1.11697 \times 10^{-20} Joules$ and for water–Cl$^-$ are σ_{WCl^-}=4.41724 Å and $\epsilon_{WCl^-} = 8.18635 \times 10^{-22} Joules$.

All the cross–interactions must be calculated. The Berthelot mixing rules

$$\sigma_{ij} = (\sigma_i + \sigma_j)/2 \qquad \epsilon_{ij} = (\epsilon_i\epsilon_j)^{1/2} \tag{5}$$

are used for characteristic distances and dispersion energies; the interaction between charges is determined by electrostatics.

The Gibbs ensemble Monte Carlo (GEMC) method was designed by Panagiotopou-los [12, 13, 14] to simulate directly coexisting phases. Two systems, one liquid and one vapor each subject to periodic boundary conditions, at fixed *total* molecule number and *total* volume are simulated simultaneously so that the systems come to configurational equilibrium while exchanging molecules and volume in order to equilibrate to the same chemical potential for each species and the same pressure. [For mixtures, one can specify the pressure externally and the the volume exchanges of the two systems then take place with a volume reservoir rather than with each other.] We have used this simulation technique to calculate vapor–liquid equilibria for pure water [9, 10], water/NaCl, water/methanol and water/methanol/NaCl systems [11]. To treat the long range forces adequately, we utilize the Ewald summation method [26, 27, 28] with the so-called tinfoil boundary conditions corresponding to the macroscopic system of spherical symmetry defined by the simulation being surrounded by a conductor. For the mixtures considered in this study, the Ewald summation is applied to all the charge–charge interactions between all the species, including partial charges in the neutral molecules (water and methanol). Using an Ewald sum method on each phase (vapor and liquid) results in negligible system size dependence [9, 10] in contrast with the finding of de Pablo and Prausnitz [29] for water simulations using a spherically truncated potential.

The details of the simulation methods can be found in the original papers to which the interested reader is directed. In this paper, we will discuss only the results.

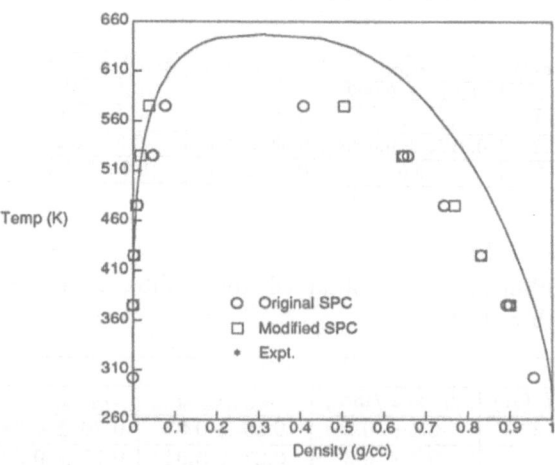

Figure 1. GEMC simulation results for SPC water with the same dipole moment in each phase (○) [9] and with different dipole moments in each phase (□)[10] compared with experimental data.

GEMC SIMULATION OF WATER

Using the GEMC simulation method, de Pablo *et al.* [9] obtained the coexistence curve of SPC water from 25 to 300°C. At each temperature simulated, the liquid densities obtained in the simulation were consistently lower than the corresponding experimental quantities and the vapor densities consistently higher. By fitting the simulation results to a Wegner expansion for the difference between simulated liquid and vapor orthobaric densities, the estimated critical temperature was $T_c = 587$ K and the estimated critical density was $\rho_c = 0.27$ gm/cm^3. This differs substantially from the experimental quantities values of $T_c = 647.3$ K and $\rho_c = 0.32$ gm/cm^3.

One of the features of the SPC model which makes it accurate at ambient (liquid) conditions is that it has a dipole moment of 2.24 D. Since the bare dipole moment of a water molecule is 1.8 D, it is clear that the higher dipole moment of SPC is an effective dipole moment that includes the effect of polarizability that is important at high density. However, in GEMC one of the phases is a vapor phase where it seems more appropriate to use the bare dipole moment of 1.8 D rather than the higher effective value. In a subsequent publication [10], Strauch and Cummings reported GEMC simulation with a modified SPC potential. The SPC potential used in the liquid phase was the usual SPC; in the vapor phase, the partial charges were reduced by a factor of 0.804 so that the molecules in the vapor phase had the bare dipole moment of water. The liquid and vapor phase densities are in better agreement with experiment. Using the Wegner expansion, we find that the critical point of the modified system can be estimated from the simulation results to be $T_c = 606$ K and $\rho_c = 0.27$ gm/cm^3, which is considerably closer to the experimental results. The simulation results are compared with experimental data in Figure 1.

Table 3. Phase equilibrium results obtained from GEMC simulation of pure water and water/NaCl systems.

T (K)	U_l (kcal/mol)	x_{NaCl}	ρ_l (gm/cc)	U_v (kcal/mol)	ρ_v (gm/cc)
375	-9.4 ± 0.3	0.017 ± 0.002	0.81 ± 0.04	-0.1 ± 0.1	0.00036 ± 0.00007
373	-8.9 ± 0.4	0.0	0.89 ± 0.02	-0.1 ± 0.9	0.00065 ± 0.0001
473	-8.1 ± 0.2	0.0172 ± 0.0004	0.63 ± 0.03	-0.3 ± 0.2	0.006 ± 0.001
473	-7.4 ± 0.2	0.0	0.75 ± 0.01	-0.85 ± 0.2	0.011 ± 0.002

Table 4. Phase equilibrium results obtained from Gibbs ensemble Monte Carlo simulation of water/methanol systems.

Liquid Phase			
T (K)	U_l (kcal/mol)	$x_{Methanol}$	ρ_l (g/cc)
344	-8.1 ± 0.2	0.56 ± 0.02	0.70 ± 0.03
351	-8.7 ± 0.3	0.27 ± 0.01	0.79 ± 0.03
360	-8.5 ± 0.3	0.144 ± 0.007	0.84 ± 0.01

Vapor Phase			
T (K)	U_v (kcal/mol)	$y_{Methanol}$	ρ_v (g/cc)
344	-0.11 ± 0.03	0.80 ± 0.02	0.0011 ± 0.0001
351	-0.02 ± 0.01	0.66 ± 0.01	0.0010 ± 0.0001
360	-0.09 ± 0.07	0.48 ± 0.01	0.00086 ± 0.00007

GEMC SIMULATION OF WATER/NACL MIXTURES

The GEMC simulations for water/NaCl were carried out using 108 molecules in the initial liquid phase and 32 molecules in the initial vapor phase [11] at temperatures of 375 K (5.6 million configurations) and 473 K (7.8 million configurations). The internal energy U, mole fraction, and density of each phase were computed and are given in Table 3. The mole fraction of NaCl was fixed in the vapor phase at zero, since experimentally it is known that the low volatility of the salt means that no salt will be present in the vapor phase. The simulations were run at constant total volume. It would be preferable to simulate this system at constant pressure, but fixing the vapor phase mole fraction eliminates the degree of freedom which would ordinarily permit us to fix the pressure [30]. Notice that the densities for of the vapor phase for the NaCl system are approximately one half those of the pure water simulations. This implies that the equilibrium pressure for the simulations is different. Figure 2 shows the convergence of the liquid and vapor densities over the course of the simulation. For a more detailed discussion of the results, the reader is referred to the original paper [11].

GEMC SIMULATION OF WATER/METHANOL MIXTURES

The constant pressure [30] GEMC simulations of water/methanol mixtures were carried out using 108 molecules in the initial liquid phase and 108 molecules in the initial vapor phase. Three simulations of water/methanol mixtures were performed at a pressure of 1 atm and at temperatures of 351 K (7 million configurations), 344 K (4.8 million) and 360 K (5 million). The internal energy, U, mole fractions, and densities of both the vapor and liquid phases are reported in Table .

At constant pressure, it is common to report results in the form $x - y$ (where x is the liquid mole fraction and y is the vapor mole fraction) and $T - x - y$ plots. We

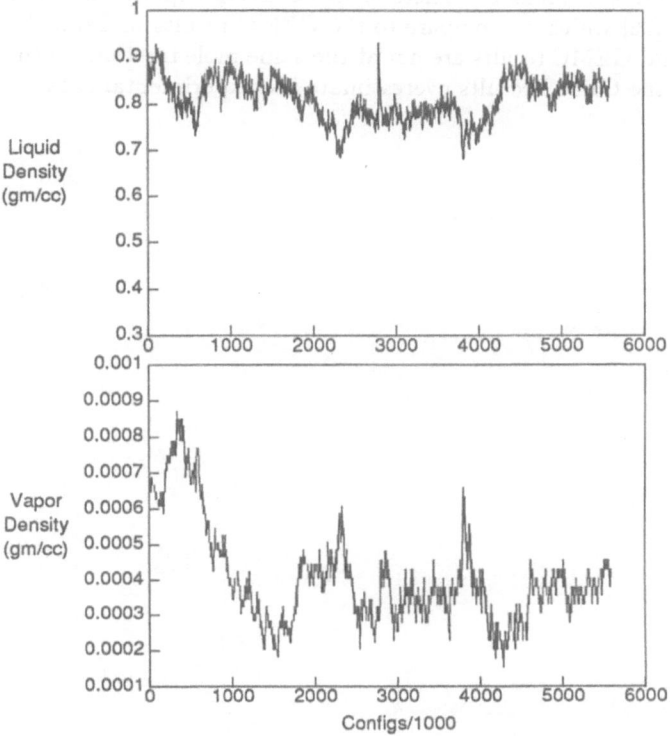

Figure 2. Convergence of the liquid and vapor densities for the NaCl/water mixture at $T = 375$ K.

present these in Figures 3 and 4 along with experimental data [31]. We see that we get excellent results for both curves.

GEMC SIMULATION OF WATER/METHANOL/NACL MIXTURES

Each constant pressure (1 atm) water/methanol/NaCl GEMC simulation was begun with 108 molecules in each phase with two sodium and two chloride ions in the liquid phase. As in the water/NaCl simulations, the salt concentration in the vapor phase was fixed at zero. The thermodynamic and phase equilibrium properties obtained are reported in Table for three temperatures, 344 K (9.4 million configurations), 351 K (5 million) and 360 K (8.8 million). Figures showing the approach to steady state are reported in Ref. [11].

In Figure 5 we plot the $x - y$ diagram of our water/methanol and water/methanol/NaCl mixtures. We see that as expected the water/methanol/NaCl $x - y$ curve simulation is moved to the left of the water/methanol $x - y$ curve. This at least qualitatively predicts the salt's effect on the water/methanol system. We can calculate the relative volatilities, α, of alcohol (A) to water (W),

$$\alpha_{A/W} = \frac{y_A x_W}{y_W x_A} \tag{6}$$

of the water/methanol/NaCl simulation and the water/methanol simulation at the same temperature and pressure. There is also existing experimental data collected in our laboratory [32] which we compare to the GEMC results in Table 6. Although the experimental and GEMC results are not at the same mole fractions in the liquid phase, it is clear that the GEMC results overestimate the experimental data.

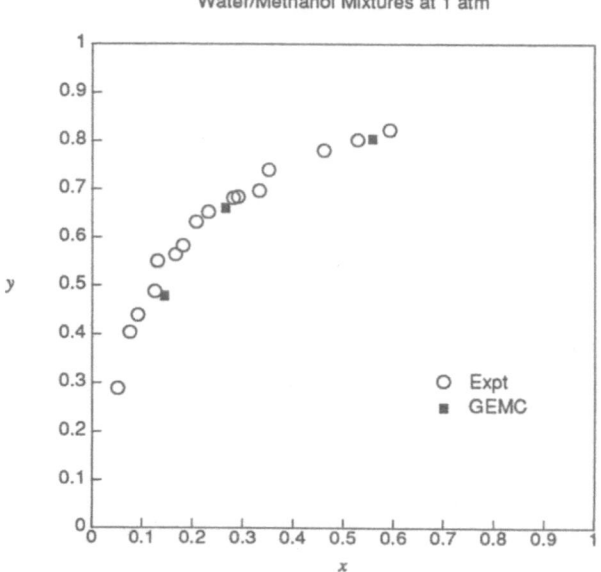

Figure 3. $x - y$ Diagram for water/methanol mixtures at $P = 1$ atm. Mole fractions x and y are of methanol.

CONCLUSIONS

In this paper, we have reviewed our GEMC simulation work on water, water/salt, water/alcohol and water/alcohol/salt. All the pair potentials used are taken or directly inferred without adjustable constants from the existing published literature. Quantitatively accurate predictions for the vapor–liquid equilibrium in water/methanol mixtures at 1 atm are obtained. For water/methanol/NaCl, the qualitatively correct behavior–i.e., the salting out of the methanol–is predicted by the simulations. Compared to experiment, however, the salt effect is overestimated.

Obtaining qualitatively correct, and in some cases quantitatively correct, predictions of phase equilibria in complex mixtures such as water/methanol/NaCl using published intermolecular potentials with no adjustable constants is very encouraging and suggests that other intermolecular potential models, or slight variations of them, might lead to quantitatively accurate predictions of the salt effect. We are currently investigating extensions of the research reported here.

Table 5.   Phase equilibrium results obtained from GEMC simulation of water/methanol/NaCl systems.

Liquid Phase				
T (K)	U_l (kcal/mol)	$x_{Methanol}$	ρ_l (g/cc)	x_{salt}
344	-9.2 ± 0.2	0.55 ± 0.03	0.69 ± 0.02	0.016 ± 0.001
351	-9.5 ± 0.1	0.25 ± 0.03	0.73 ± 0.05	0.0176 ± 0.0008
360	-9.6 ± 0.2	0.15 ± 0.02	0.82 ± 0.02	0.015 ± 0.001

Vapor Phase			
T (K)	U_v (kcal/mol)	$y_{Methanol}$	ρ_v (g/cc)
344	-0.15 ± 0.07	0.85 ± 0.03	0.0011 ± 0.0001
351	-0.08 ± 0.02	0.68 ± 0.03	0.0001 ± 0.0001
360	-0.07 ± 0.05	0.58 ± 0.03	0.0009 ± 0.0001

Table 6. Water/methanol/NaCl results from simulation and experiment at a temperature of 351 K and pressure of 1 atm.

System	x_{salt}	$x_{Methanol}$	$y_{Methanol}$	α
Expt.	0.0	0.307	0.688	4.978
Expt.	0.015	0.304	0.714	5.592
Expt.	0.030	0.302	0.737	6.198
GEMC	0.0	0.265	0.661	5.408
GEMC	0.0174	0.251	0.740	8.252

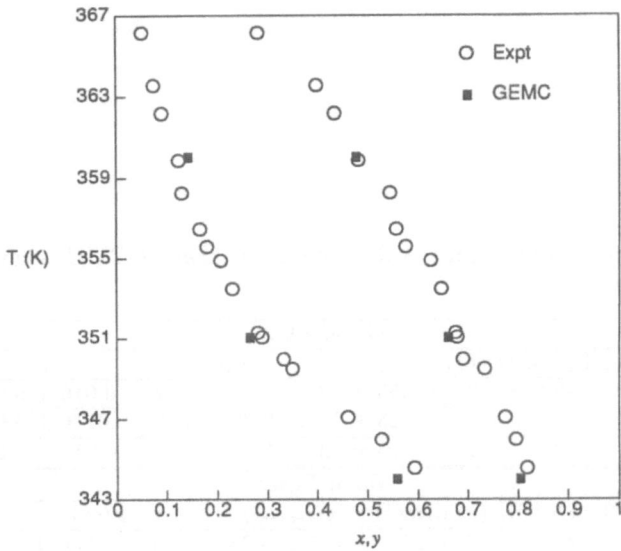

Figure 4. $T - x - y$ Diagram for water/methanol mixtures at $P = 1$ atm. Mole fractions x and y are of methanol.

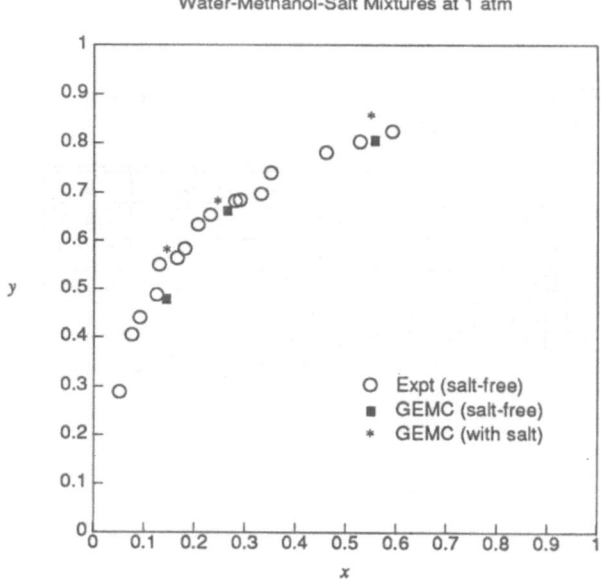

Figure 5. $x - y$ diagram for water/methanol/salt and water/methanol mixtures at $P = 1$ atm. Mole fractions x and y are of methanol.

ACKNOWLEDGEMENTS

The work reviewed here was primarily performed by Dr. H. J. Strauch as part of his dissertation work at the University of Virginia. The author gratefully acknowledges support of this research by the Division of Chemical Sciences, Office of Basic Energy Sciences, Office of Energy Research, U.S. Department of Energy. The presentation of this research at the XVIth International Workshop on Condensed Matter Theory was partially supported by a travel grant from the Army Research Office.

References

[1] P. Debye and E. Hückel. *Physik Z.*, 24:185, 1923.

[2] E. A. Guggenheim. *Phil. Mag.*, 19:588, 1935.

[3] K. S. Pitzer. *J. Phys. Chem.*, 77:268, 1973.

[4] K. S. Pitzer and G. Mayorga. *J. Phys. Chem.*, 77:2300, 1973.

[5] A. I. Johnson and W. F. Furter. *Can. J. Chem. Eng.*, 38:78, 1960.

[6] P. Debye and J. McAuley. *Physik Z.*, 26:22, 1923.

[7] D. Jaques and W. F. Furter. *A.I.Ch.E. Journal*, 18:343, 1972.

[8] H. L. Friedman. *Ionic Solution Theory.* Interscience Publishers, 1962.

[9] J. J. de Pablo, J. M. Prausnitz, H. J. Strauch, and P. T. Cummings. *J. Chem. Phys.*, 93:7355, 1991.

[10] H. J. Strauch and P. T. Cummings. *J. Chem. Phys.*, 96:864, 1992.

[11] H. J. Strauch and P. T. Cummings. *Fluid Phase Equil.*, 1991. accepted for publication.

[12] A. Z. Panagiotopoulos. *Molec. Phys.*, 61:813, 1987.

[13] A. Z. Panagiotopoulos. *Molec. Phys.*, 62:701, 1987.

[14] A. Z. Panagiotopoulos. Molecular simulation of phase equilibria: Simple, ionic and polymeric fluids. Presented at the 11th International Symposium on Thermophysical Properties, Boulder, CO, June 23–27, 1991.

[15] H. J. C. Berendsen, J. P. M. Postma, W. F. von Gunsteren, and J. Hermans. Interaction models for water in relation to protein hydration. In B. Pullman, editor, *Intermolecular Forces*, page 331. Reidel, Dordrecht, 1981.

[16] M. Haughney, M. Ferrario, and I. R. McDonald. *Mol. Phys.*, 58:849, 1986.

[17] F. G. Fumi and M. P. Tosi. *J. Phys. Chem. Solids*, 25:31, 1964.

[18] M. P. Tosi and F. G. Fumi. *J. Phys. Chem. Solids*, 25:45, 1964.

[19] J. Chandrasekhar, D. C. Spellmeyer, and W. L. Jorgensen. *J. Am. Chem. Soc.*, 106:903, 1984.

[20] W. F. Furter and R. A. Cook. *Int. J. Heat Mass Transfer*, 10:23, 1967.

[21] W. F. Furter. *Can. J. Chem. Eng.*, 55:229, 1977.

[22] W. F. Furter. *The Chemical Engineer*, No. 219:CE173, 1968.

[23] R. A. Cook and W. F. Furter. *Can. J. Chem. Eng.*, 46:119, 1968.

[24] W. L. Jorgensen. *J. Am. Chem. Soc.*, 103:335, 1981.

[25] B. M. Pettitt and P. J. Rossky. *J. Chem. Phys.*, 84:5836, 1986.

[26] S. W. de Leeuw, J. W. Perram, and E. R. Smith. *Proc. Royal Soc. London*, A373:27, 1980.

[27] S. W. de Leeuw, J. W. Perram, and E. R. Smith. *Proc. Royal Soc. London*, A373:57, 1980.

[28] S. W. de Leeuw, J. W. Perram, and E. R. Smith. *Proc. Royal Soc. London*, A388:177, 1983.

[29] J.J. de Pablo and J. M. Prausnitz. *Fluid Phase Equil.*, 53:177, 1989.

[30] A. Z. Panagiotopoulos, N. Quirke, M. Stapleton, and D. J. Tildesley. *Mol. Phys.*, 63:527, 1988.

[31] J. Gmehling, U. Onken, and W. Arlt. *Vapor-Liquid Equilibrium Data Collection*. Dechema, 1977.

[32] J. C. Morrison, J. F.and Baker, H. C. Meredith, K. E. Newman, T. D. Walter, R. L. Massie, J. D.and Perry, and P. T. Cummings. *J. Chem. Eng. Data*, 35:395, 1990.

SOLVATION DYNAMICS IN A STOCKMAYER FLUID

Lalith Perera and Max L. Berkowitz

Department of Chemistry
University of North Carolina
Chapel Hill, North Carolina, 27599

INTRODUCTION

The dynamic response of the solvent to the sudden change of the charge distribution in a solute molecule has been the subject of many investigations, experimental and theoretical[1]. A rough sketch of what happens in the experiment is given in Figure 1, where we display the free energies of the ground and excited states of the solute as a function of the generalized solvent coordinate. Optical excitation of the solute from the ground to excited state produces, according to the Franck-Condon principle, a non-equilibrium configuration of the solvent which relaxes to equilibrium. This relaxation of the solvent can be observed as a time dependent fluorescence spectral shift in the solute emission and it is usually used to quantify solute dynamics.

While the experimental data are obtained from the applications of fast laser pulse techniques to rather complicated systems[2,3], the majority of analytical theories use rather simplified models to describe the solutions[4-10]. At the same time the computer simulations are mostly performed on models that use a simple description of the solute and a rather realistic description of the solvent[11-15]. Thus, recent simulations were performed to study the response in solvents such as water[11,12], methanol[13] and acetonitrile[14]. To compare the results of the simulations with the analytical type theories, it is desirable to perform a simulation, where both solute and solvent are described by simple potentials. Therefore, we present here the results from a set of molecular dynamic simulations[16], where we studied the dynamics of ionic solvation in a Stockmayer fluid. The same model, but with a different choice of interaction parameters,

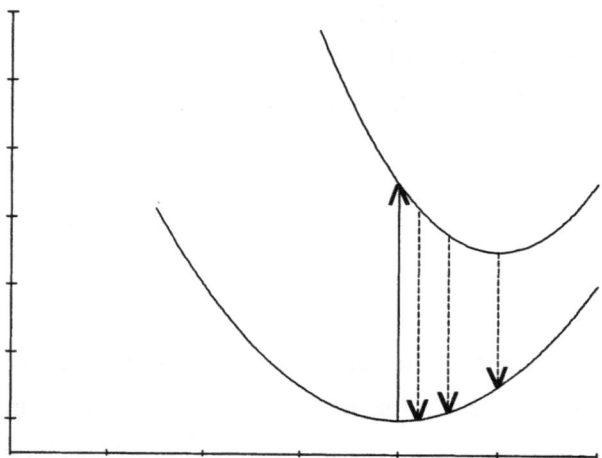

Figure 1. Free energies of the solute as a function of generalized solvent coordinate.

was simultaneously studied by Neria and Nitzan[17]. Here we report the results of our studies and compare them with the results from the existing analytical theories, such as dielectric continuum model (DCM)[4], the dynamical mean-spherical approximation (MSA)[6-8] and theories based on the Smoluchowski-Vlasov Equation (SVE)[8-10]. The observations reported by Neria and Nitzan[17] are qualitatively similar to ours.

MODEL AND METHODOLOGY

Our system consists of one solute particle dissolved in a Stockmayer solvent[18]. The choice of this particular model is primarily due to the fact that analytical theories based on the dynamical extension of the MSA model[6-8] and on the SVE[8-10] are available for a model very similar to ours (the theories are done for hard spheres with embedded dipoles). In addition the detailed knowledge of the dielectric properties of this solvent is available from molecular dynamics simulations[19-21]

The interaction energy for a pair of Stockmayer particles is given by the following expression:

$$U_{ij}(r) = 4\epsilon[(\sigma/r)^{12}-(\sigma/r)^6] + \mu_i \cdot T_{ij} \cdot \mu_j \tag{1}$$

where r is the distance between particles i and j, μ_i is the dipole moment of particle i, μ_j is the dipole moment of particle j, ϵ and σ are the Lennard-Jones parameters. The dipole tensor T_{ij} has its usual form:

$$T_{ij} = 1/r_{ij}^3[I-3r_{ij}r_{ij}/r_{ij}^2] \tag{2}$$

where I is the unit tensor.
The solute-solvent interaction is represented by the term:

$$U_j(r,\mu_j) = 4\epsilon[(\sigma/r)^{12}-(\sigma/r)^6]-\mu_j \cdot E \tag{3}$$

where E is the electric field due to the solute charge, and is given by an equation $E = qr_{ij}/r_{ij}^3$. For the neutral solute, this quantity becomes zero and the last term in Equation (3) vanishes. The equations of motion for our system are obtained from the Lagrangian

$$L = \sum_i \frac{1}{2}mv_i^2 + \frac{1}{\mu^2}\sum_i \frac{1}{2}I\mu_i^2 - \frac{1}{2}\sum_{i \neq}\sum_j \mu_i \cdot T \cdot \mu_j$$
$$-\frac{1}{2}\sum_i\sum_j u_{ij}^{LJ} + \sum_i \lambda_i(\mu_i^2-\mu^2) + \sum_i \mu_i \cdot E \tag{4}$$

where the first two terms represent translational and rotational kinetic energies of the solvent molecules, the third term represents the dipole-dipole interaction energy of the solvent, the fourth term is representing the Lennard-Jones interaction energy of the system. To preserve the magnitude of the dipole moment of the solvent we use the method of constraints[22]. Therefore the fifth term of the Lagrangian contains a constraint variable with the corresponding Lagrange multiplier. The last term in the Lagrangian is due to the solute-solvent electrostatic interaction. The solvent is characterized (in reduced units) by a dipole moment $(\mu^*)^2 = 3.0$, a moment of inertia $I^* = 0.025$, a density $\rho^* = 0.822$ and a temperature $T^* = 1.15$. The Lennard-Jones parameters for solute-solvent interaction were taken to be the same as for solvent-solvent interaction. The time step in reduced units was chosen to be equal to 0.0025. To facilitate the comparison of our results with the results from other simulations, note, that this time step is equal to 5.4 fs if the argon parameters ($\epsilon = 119.8K$ and $\sigma = 3.405Å$) are chosen for solvent particles.

The simulations were carried out in a cubic box, which contained 511 solvent molecules and one solute molecule. For simplification of the treatment, the solute particle was considered to be immobile. Equations of motion for solvent molecules were integrated using the leapfrog algorithm and the rotational part of the motion was handled by the methods of constraints. A good description on the Hxplementation of this method is found in Ref. [23]. Periodic boundary conditions along with the spherical cut-off convention were implemented in the calculations and the reaction field technique was used to account for the long range dipole-dipole interactions. In the reaction field calculations we used the value of the bulk dielectric constant of the Stockmayer fluid, which is 66.1 for the values of the reduced parameters given above[20].

We have performed three equilibrium simulations and four sets of non-equilibrium simulations. The first equilibrium simulation (ESI) was performed for the system with the uncharged solute, while the solute was charged ($q = 1e$) in the second simulation (ESII). In the first two simulations the reduced moment of inertia I^* of the solvent molecule was assigned the value of 0.025, in the third equilibrium simulation the value of I* was increased hundred times. By increasing the moment of inertia one achieves slow rotational motion of the solvent and thus amplifies the effects of translational motion on the overall dynamics. In all the equilibrium simulations the trajectories consisted of 20 ps equilibration period followed by 100 ps production run. The average reduced temperature of each trajectory was maintained at $T^* = 1.15$ by occasional rescaling of the velocities.

To mimic the solvation phenomena we use the non-equilibrium simulation technique. In these simulations the solute charge distribution is instantaneously changed and the response of the solute is measured. To quantify the solvation dynamics one calculates in the simulation the normalized response function or relaxation function

$$S(t) = \frac{\delta E(t) - \delta E(\infty)}{\delta E(0) - \delta E(\infty)} \qquad (5)$$

In equation (5) $\delta E(t)$ is the fluctuation in the solvent-solute interaction energy at a given time t, averaged over all the trajectories in a particular set of simulations. The equilibrium trajectories of the system with the neutral solute are used to generate initial configurations for the non-equilibrium trajectories. Therefore, we have selected 50 configurations (separated by 2 ps each) from the equilibrium simulation SEI and at $t = 0$ we placed a charge of $+1e$ on the solute. To compare the results of our simulations with some analytical theories, such as MSA, in which the relaxational mechanism is due to the rotational motion only, we froze the translational motion of the molecules in the first set of 50 trajectories. This first set of 50 non-equilibrium trajectories we call NESI (non-equilibrium set I.) To freeze the translations we imposed very heavy masses on the solvent molecules (10^{10} times the actual mass). Also, the initial linear velocities were set equal to zero to ensure that particles do not move initially. After the instantaneous charge jump, the trajectories from the molecular dynamics were monitored for 5.4 ps. (The equilibration has been achieved within 2.0 ps).

It has been emphasized that translational motion of the solvent molecules plays a substantial role in the overall mechanism of the relaxation. This was confirmed recently by Bagchi and Chandra[10], who extended the SVE treatment to include the effects of the translational motion. They demonstrated that if translational diffusion is considerably large, the Onsager picture[24] of the solvent relaxation occurring in a "snowball" fashion is no longer acceptable[25]. We tried to address this issue in two other sets of non-equilibrium simulations. In one set of simulations (NESII), consisting again of 50 trajectories, we started the trajectories from the same initial positions as in set NESI. The initial velocities of the solvent molecules, which were set equal to zero in the NESI, were now assigned from the equilibrium trajectories. The lengths of the trajectories were restricted to 2.7 ps since the equilibration was attained within 2 ps. The third

RESULTS AND DISCUSSION

Using the data from equilibrium simulations, we first consider the structural properties of solvent around the solute molecule. The solute-solvent

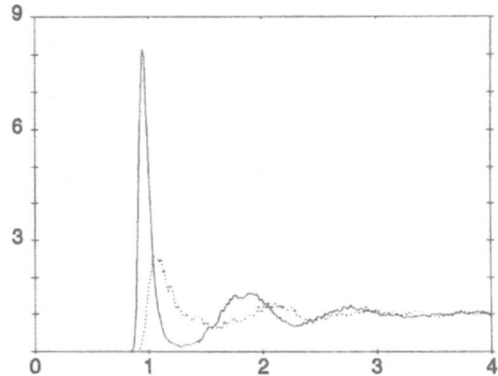

Figure 2. The solute-solvent radial distribution functions g(r) vs distance, obtained from ESI and ESII. The solid line corresponds to the ion-solvent rdf, the dotted line to the neutral solute-solvent rdf.

set of non-equilibrium trajectories (NESIII) was started from 50 points of ESIII trajectory, where the moment of inertia of the solvent was 100 times larger than that used in ESI. Trajectories were continued for 22 ps after the initial charge jump. The fourth set of non-equilibrium simulations (NESIV) consisted of 50 trajectories which had the same starting configurations as the trajectories from the NESI set. The trajectories in NESIV set were calculated with solvent-solvent interactions turned off.

Table 1. A summary of information on solvation shells. The values in the table are obtained from ESI and ESII simulations.

a) From an equilibrium simulation with a neutral solute

shell	radius in reduced units	number of molecules	$<\cos(\theta)>$	$<$density$>$ in reduced units
1	1.52	11.8	-0.0061	0.798
2	2.52	41.7	-0.0036	0.800
3	3.40	80.6	-0.0004	0.825
bulk-like		376.9	0.0003	0.822

b) From an equilibrium simulation with a charge solute

shell	radius in reduced units	number of molecules	$<\cos(\theta)>$	$<$density$>$ in reduced units
1	1.27	9.5	0.889	1.226
2	2.27	33.3	0.336	0.824
3	3.24	76.4	0.170	0.817
bulk-like		391.8	0.054	0.816

radial distribution functions (rdfs) obtained from ESI and ESII are shown in Figure 2. The curves on this figure demonstrate the existence of the well defined solvation shell structures around the solute, regardless of the charge on the solute. In Table 1 we summarize the information extracted from the rdf's along with the other relevant data. As the rdfs in Figure 2 indicate we distinguish three solvation shells around the solute. The rest of the solvent is "bulk-like". The boundaries of the shells are given by the locations of the minima of the rdfs. The properties shown in Table 1 are averages over the shells defined by such boundaries. The orientational distribution of the angle between the solvent dipole and a vector connecting the center of a solute and the center of the solvent is shown in Figure 3.

From Figures 2 and 3 and Table 1 we can see that the incorporation of the charge on the solute site results in the increase of the sharpness of the first peak of the rdf, while the peak position shifts inward. The size of the first shell decreases and as a result the number of nearest neighbors decreases. The same is true for the second shell. At the same time, note, that the density of solvent in the first shell around the ion is larger than the density of the solvent in the first shell around the neutral solute. The orientational distribution is homogeneous in the case of solvated neutral solute, but displays a strong orientational preference in the case when the solute is charged. As we can see in this case the orientational preference holds for all 3 shells around the solute. The rdfs and orientational distribution functions given by Figures 2 and 3 represent the initial and final states of our non-equilibrium simulations.

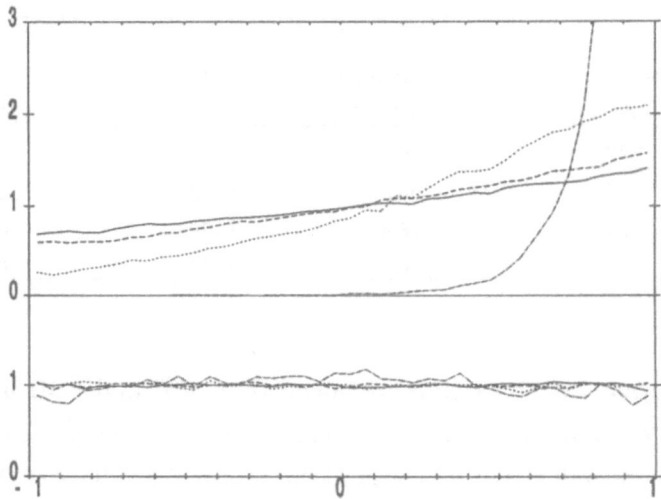

Figure 3. Orientational distribution for different solvation shells from ESI (bottom) and ESII (top); first shell (-.-), second shell (---), third shell (...) and bulk-like (—).

As we have already mentioned, the response of the solvent to the instantaneous change of the solute charge distribution is measured in a molecular dynamics computer experiment by the normalized response function (see equation 5). In Figure 4 we display the results for such a response function S(t) obtained from the NESI simulation along with the results obtained from continuum dielectric model[4], dynamical MSA theory[7] and SVE type theory[10]. The functional forms of the relaxation function S(t) obtained in the framework of

MSA and SVE and used here are presented in the Appendix. Since MSA theory does not take into account the translational mode of the relaxation, the comparison of this theory with the data from NESI simulation is most appropriate. The SVER curve was obtained by solving SVE where only the rotational mode of the relaxation was included.

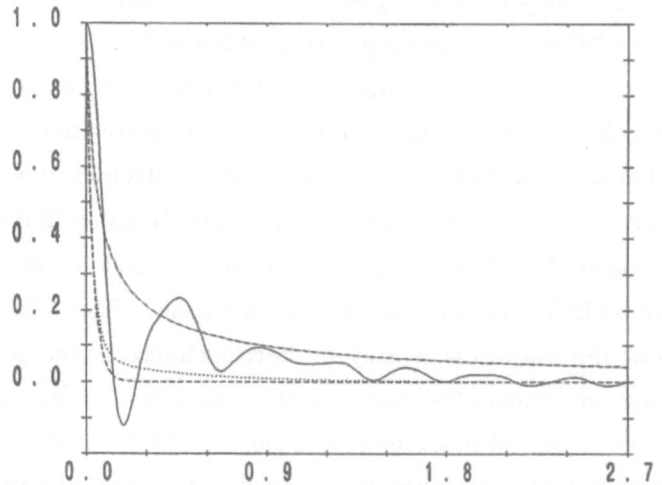

Figure 4. Response functions S(t) vs time (ps) obtained from NESI (—), dynamical MSA (-.-), SVER (...) and the DCM (---).

As we observe in Figure 4 the relaxation curve corresponding to NESI displays a fast initial decay (~0.1ps) followed by a relatively slow decay (~2.0ps) modulated by periodic oscillations. We also observe that after the initial response of the solvent the energy of the system is even lower than its equilibrium value (i.e. S(t) becomes negative). This may be attributed to a quick initial reorientation of dipoles towards the ion, which results in the lowering of the ion-dipole energy. But the system does not remain in this configuration with the lowest ion-dipole energy due to the presence of dipole-dipole interactions and thermal motion. The observed oscillations in the response function are the results of the timely adjustment of the dipoles to the field of the freshly created ion and the field emanating from the neighboring dipoles.

Let us now compare the relaxation behavior predicted by the DCM, MSA and SVE theories with the molecular dynamics results. Perhaps we should mention here, that our solvent is quite accurately described by the continuum Debye model[19]. For such a solvent the DCM predicts that the relaxation is exponentially decaying with a time constant corresponding to the longitudinal relaxation time $\tau_l = \tau_d * \epsilon_\infty / \epsilon_0$. Substitution into this formula of the values 2.15ps, 66.1 and 1.0 of the Debye relaxation time (τ_d), static(ϵ_0) and optical(ϵ_∞) dielectric constants respectively[20], predicts in our case a relatively short relaxation time (0.033ps). A more sophisticated theory that takes the solvent structure into account is the dynamical MSA theory[7]. It predicts a multi-exponential relaxation behavior with relaxation times ranging between τ_l and τ_d. Although the dynamical MSA theory considers the molecular character of the solvent, it does it in a very average way and as a result no oscillatory behavior of the relaxation is predicted by it. Figure 4 also shows, that the initial behavior of S(t) predicted by DCM and MSA theories is incorrect. From MD simulations we conclude that the initial decay may be better characterized by a gaussian function than by an exponential one. Such a gaussian decay of the initial response has been observed in other simulation works[15]. In addition we also observe from Figure 4, that the time to approach the asymptotic limit (S(t)=0) is much longer in the framework of MSA than what MD shows. The relaxation curve obtained from the theory based on the use of SVE[10] is close in our case to the predictions of the simple continuum model. As we can see, the curve from the molecular dynamics displays a different behavior at the initial time but its asymptotic behavior is the same as predicted by models based on SVE and continuum description. In general we can say that the molecular dynamics displays a behavior somewhere between that predicted by the continuum theory, MSA and SVE.

Let us now study how the inclusion of solvent translational motion affects its relaxation dynamics. In Figure 5 we compare the results obtained from a simulation in which the translational motion is included in the dynamics (NESII) with that of the previous simulation (NESI). Except for certain details discussed below, the two curves display similar behavior. This leads us to a conclusion that for this particular system, the rotational motion of the solvent essentially governs the overall relaxation mechanism. To study the relative contributions of translational and rotational diffusional modes into the total relaxation process a parameter p' defined by an equation

$$p' - D_t/(2D_r\sigma^2) \tag{6}$$

is introduced[10]. For our Stockmayer solvent which has a relatively large rotational diffusion coefficient (D_r = 1.167/ps) and a translational diffusion coefficient (D_t = 0.42Å^2/ps) the value of p' is rather small (p'=0.012). As was pointed out by Bagchi and Chandra this value of p' should not change the relative importance of translational and rotational modes[10].

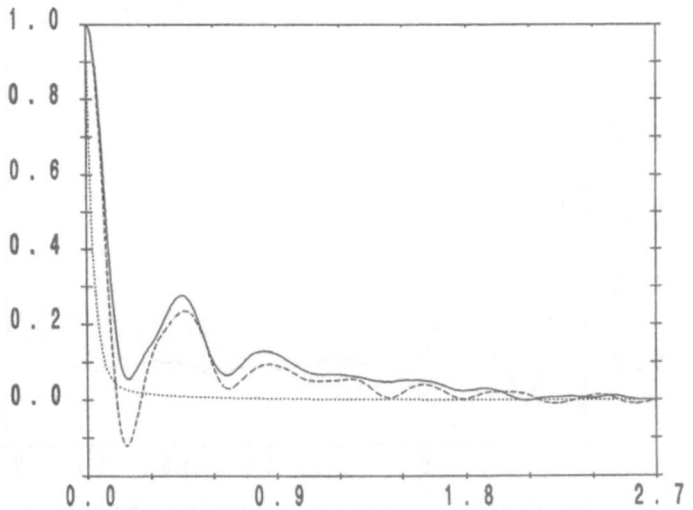

Figure 5. Response functions obtained from NESI (---), NESII (—) and SVERT (...).

Nevertheless translational motion does play some role in the relaxation, as we can observe from Figure 5, where we display the relaxation curves obtained from NESI and NESII sets. The main difference in the behavior of these is seen around their first minimum. This is due to the translational motion that results in the reconstruction of the solvent shells around the ion (the electrostriction effect). This reconstruction of the shell structure brings the molecules in the first two shells much closer and thus, causes the solvent molecules to feel the repulsions from their neighbors somewhat stronger than in the case when the translation is frozen. The reconstruction also results in an earlier activation and stronger dipolar interactions between solvent molecules, which is visible from the curves in the region around 0.18ps. However, the reconstruction does not seem to affect the long time relaxation behavior. Figure 5 also includes a curve labeled SVERT. This curve was obtained from the solution of SVE[10] where the effect of rotational and translational diffusion were included. Again only long time behavior of the SVERT curve is correct, as expected.

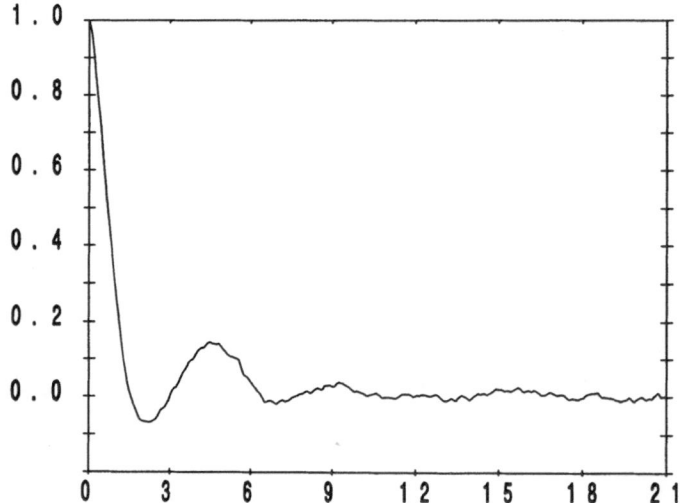

Figure 6. Response function obtained from NESIII.

To study how the translational motion of the solvent influences the relaxation dynamics at larger values of p′, we suppressed the fast rotations of solvent molecules by increasing their moment of inertia by a factor of 100 and performed an equilibrium (ESIII) and a series of non-equilibrium simulations (NESIII) for this new system. The equilibrium simulations show that the translational diffusion coefficient remains near that of the original system

(0.38Å²/ps), but the rotational diffusion coefficient decreases by a factor of 38 (to the value 0.031/ps). For this new Stockmayer solvent p' = 0.42, and therefore we expect to see a more pronounced effect of translation on the relaxation. If we would freeze out completely the translational motion of the solvent, the dynamics in NESIII would be scaled up by a factor of 10. Consequently, one would expect to see a complete relaxation of the solvation around the ion to take place after around 17-18 ps. However, as Figure 6 indicates, a complete relaxation is achieved after around 7 ps. This points to an appreciable contribution of translational motion to the total relaxation process. Another effect that translational motion has on the relaxation is related to the Onsager "snowball" conjecture and is discussed below.

As we have seen from our simulation and other simulations, the relaxation of the system can be described by a fast initial decay and a subsequent slow relaxation. The existence of these two time scales are also confirmed by experiments[26]. Therefore the main questions we should ask are: what are these two (or more) time scales observed in solvation dynamics and how do they correlate with the dynamics of solvent molecules? In the previous simulations, the short time decay was assigned to the inertial motion of the solvent molecules[11,13,14]. To find out if this inertial motion has a collective or independent character we performed the set NESIV. Figure 7 displays the behavior of the relaxation function S(t) obtained from the sets NESI and NESIV. As we can see from this figure, up to ~0.09 ps (at this time the energy relaxed ~70%) the decay functions obtained from NESI and NESIV are practically indistinguishable. We also know from our previous work and from other simulations that the main contribution into relaxation comes from the rotational motion of molecules in the first solvation shell (the difference in S(t) for the NESI and NESII is very small). Therefore we conclude that the initial relaxation in our system is due to the independent inertial rotational motion of solvent molecules.

The nature of the long time decay is also clarified in Figure 7, where the relaxation function S(t) is compared to the single dipole autocorrelation function $\Phi(t)$, defined as:

$$\Phi(t) = \frac{<\mu(t)\cdot\mu(0)>}{<\mu^2>} \tag{7}$$

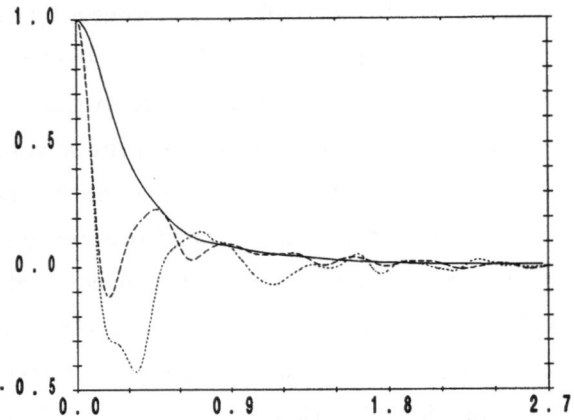

Figure 7. Response functions obtained from NESI (dashed line) and NESIV (dotted line). The single particle dipole autocorrelation function (solid line) is also shown.

The autocorrelation function $\Phi(t)$ was calculated from a separate simulation performed on 512 Stockmayer particles. As we can see from **Figure 7** the long time decay of S(t) is characterized by the same relaxation time as the decay of $\Phi(t)$. That does not mean that the long-time dynamics of a dipole is determined by solving the dynamical equations for a single particle, since to find a single particle correlation function in the many-body problem one has to solve the many-body problem. The agreement in long-time behavior of S(t) and $\Phi(t)$ provides a reasonable conformation to the idea that the long time decay of the relaxation is due to the adjustment of the dipole orientation to the resulting field from the rest of the system.

At this stage we want to return to the discussion of the Onsager "snowball" effect. Onsager once commented that the relaxation of solvent around a newly created charge distribution will proceed from outside towards the solute - like a "snowball"[24]. In an indirect way the Onsager's conjection was corroborated by the dynamic MSA theory[6-8], which included only the rotational motion of the solvent. Using the SVE treatment Bagchi and Chandra concluded that when translational motion is important, the relaxation occurs in an opposite manner--from inside to outside[10]. We can check Onsager's conjecture, since molecular dynamics simulations allow us to examine the contributions to the relaxation emanating

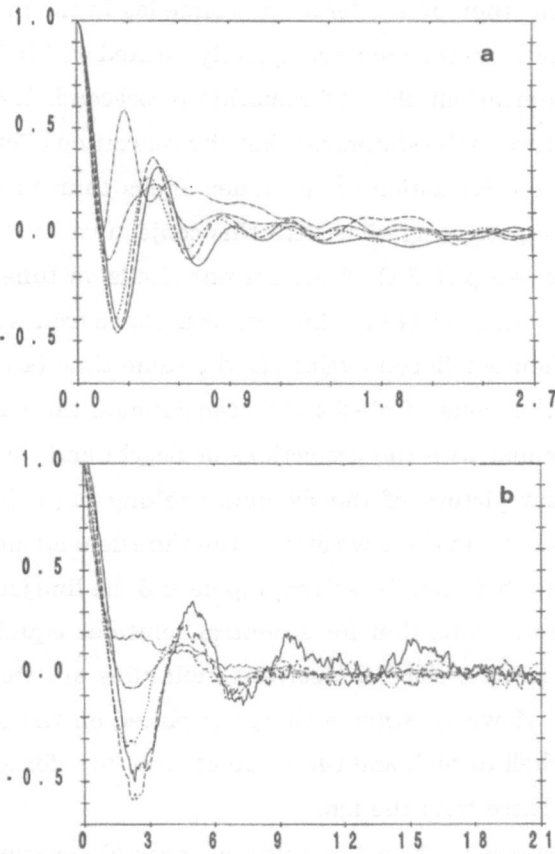

Figure 8. Responce functions [-..-] and their solvent shell components [first shell (-.-), second shell (...), third shell (---) and bulk (—)] from a) NESI and b) NESII.

from the different solvent shells around the ion. Thus in Figure 8a we show the plots of S(t) vs time for each solvent shell for NESI. Initially, all the shells seem to respond similarly to the instantaneous change in the charge on the solute and this response follows a gaussian behavior. However, the first shell molecules begin to react to the field exerted by the neighboring dipoles earlier than the molecules in other shells. Since this shell contributes more than ~70% to the overall response function, one expects to observe the characteristics of the 1st shell relaxation in the overall response function. The initial relaxation behavior of the solvent molecules in the subsequent shells does not show any appreciable difference from each other, except for a small time lag in the propagation of the response. These shells are seen to be completely relaxed within 1 ps time period, but the first shell remains unrelaxed for another picosecond. If one understands the Onsager conjecture as the statement that the relaxation of outer shells takes shorter time than the relaxation of the inner shells than in this respect the emerging picture is quite compatible with this conjecture.

In Figure 8b, we plot S(t) of the solvent shells vs time for the case in which solvent is modified to have $I^* = 2.5$. One can notice from the figure that all the curves reach their equilibrium values at the same time (around 7 ps). This indicates that for the value of $p' \sim 0.4$ the translational motion influences the relaxation in agreement with the predictions of Bagchi and Chandra[25].

A rather clear picture of the dynamics taking place in our system is obtained from Figures 9. In these we display the time dependence of the average cosine of the angle (θ) between the solvent dipole and the line joining the centers of solute and solvent. Note, that for a neutral solute at equilibrium with the solvent, $<\cos(\theta)>$ is zero, since the solvent molecules are randomly oriented around the solute. However, when a charge is placed on the solute, $<\cos(\theta)>$ may change from shell to shell and the value of $<\cos(\theta)>$ for a particular shell depends on the distance from the ion.

Figure 9a represents how the value of $<\cos(\theta)>$ varies with time for different solvent shells when the translational motion is absent. At $t = 0$, almost all the shells have their $<\cos(\theta)>$ values close to zero showing that the initial configurations of the non-equilibrium trajectories are selected from a rather good random distribution. With the time proceeding, $<\cos(\theta)>$ evolves towards values obtained from the equilibrium simulations.

As we see from Figure 9a the molecules in the first shell reorient fast towards the ion and even "overshoot" the equilibrium angle. Subsequently, due to interactions with the neighbors and the ion the average angle oscillates on its way towards equilibrium. For the molecules in other shells the picture is similar, but due to the smaller perturbation from the ion the initial deviation in the angle is smaller; as a result the relaxation proceeds faster to the equilibrium. Not surprisingly, the values of the time taken to complete relaxation obtained from Figure 9a agree quite well with those evaluated from S(t) plots.

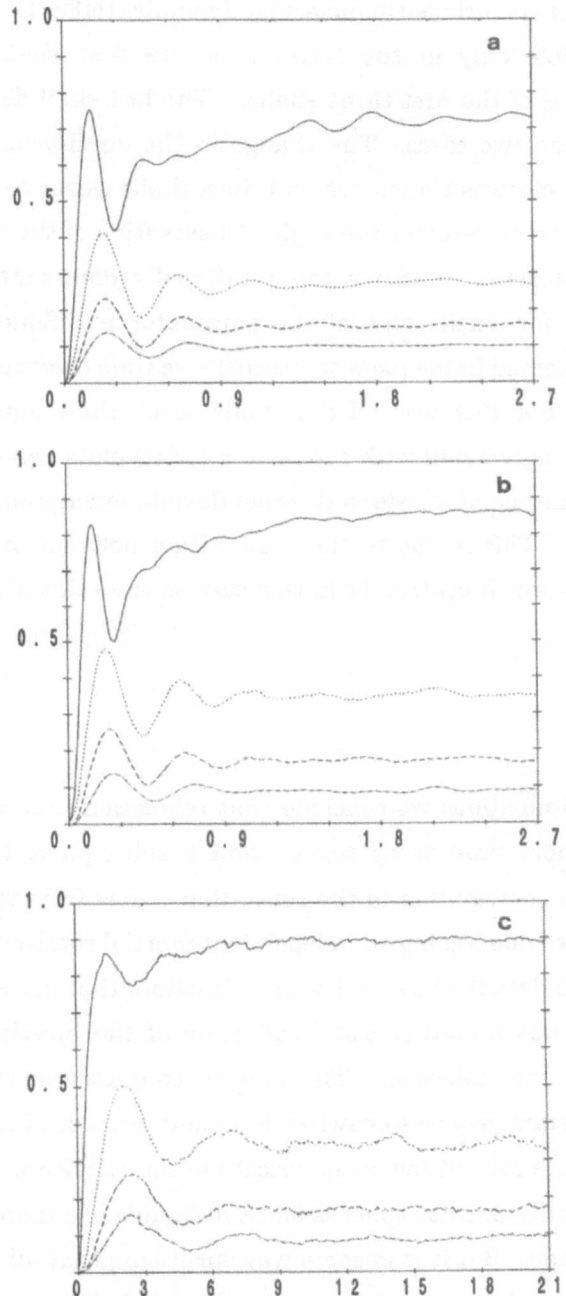

Figure 9. Variation of <cos(θ)> with time for different solvation shells [first shell (—), second shell (...), third shell (---) and bulk (-.-) from a) NESI, b) NESII and c) NESIII.

477

In Figure 9b, we present a similar set of plots for the case when the solvent translations are included in molecular dynamics (NESII). Inclusion of the translation is visible only in the features of the first shell and the final (equilibrium) values of the first three shells. The last shell displays quite the same results for the two cases. The change in the equilibrium values of the angle is due to the electrostriction, which brings shells closer to the ion and the distance from the ion determines the angle. Observation of the similar behavior of the angular relaxation confirms the rotationally dominant nature of the solvent relaxation for small value of the parameter p'. Somewhat different features can be observed in the plots of $<\cos(\theta)>$ vs time (Figure 9c) for the case in which $I^* = 2.5$. For this case all the shells reach their equilibrium values around 7ps in good agreement with the results of S(t) plots. As we see from the figure the first minimum of $<\cos(\theta)>$ does not deviate much from its equilibrium value in this case. This is due to the competition between translational and rotational motions which contribute in this case to the relaxation.

CONCLUSIONS

From our simulations we conclude that relaxation proceeds in two time regimes: initial short time decay regime and a subsequent long time decay regime. The main contribution to the relaxation comes from the first shell, of which molecules are undergoing an independent inertial rotational motion. The initial decay can be described by a relaxation function, that has a gaussian form, which is in some way a mathematical reflection of the inertial nature of the motion of the solvent molecules. The inertial character of the initial decay observed in our system may be somewhat dominant because of the nature of the model describing the solvent (small spherical molecules with point dipoles). In real life the molecules are not spheres and the liquids are more likely to be in the overdamped limit. But it is in some way surprising that all the simulations up to date show the dominant character of the inertial motion even for such solvents as water[11].

Our simulations also show that the long time decay of the relaxation function S(t) has the same character as the decay of the correlation function of the single dipole. It can, therefore, be characterized by an exponential relaxation with the relaxation time τ_s, which is known to be of the same order as τ_d[20]. This is why we think of this time regime as a diffusional regime.

We have also investigated the role of translational and rotational motion in the relaxation. We have observed that when only the rotational mode is participating in the relaxation the outer shells relax before the inner shells do. When the translational mode participates in the dynamics of the solvent we observed that all the shells relaxed simultaneously. It is possible that in some cases the translational mode plays a dominant role in the relaxation[27].

We observed that although our model fluid can be described quite well by a simple Debye-type continuum model, the relaxation process can not be described by a continuum theory. Even more sophisticated theories that we used to compare with our results from the simulations, (theories based on dynamical MSA or on SVE) are not performing that well either. Very recently Bagchi and Chandra presented a non-Markovian theory which included the viscoelastic behavior of the solvent responsible for the initial inertial response[28]. Numerical studies revealed that the relaxation function displays short-time oscillations, followed by a slow long-time decay[28]. Comparison of this new theory with the computer simulation performed on the same model will be very interesting.

Acknowledgments

This work was supported by a grant from the Office of Naval Research. The simulations were performed on the Cray YMP at the North Carolina Supercomputing Center and on the Convex C240 at UNC.

APPENDIX

The Functional Form of S(t) in the Framework of the Dynamical MSA

The dynamical MSA theory was proposed by Wolynes[6] and solved exactly by Rips, Klafter and Jortner (RKJ)[7] and by Nichols and Calef[8]. Here we present the final expression for the relaxation function S(t) that we use in order to get the corresponding MSA curves. For the details on the development of the MSA the reader is referred to the original references [6-8].

According to RKJ we can write an analytical expression for the Laplace transform of the relaxation function S(p)

$$S(p) = [\chi(p)-\chi(0)]/p[\chi(\infty)-\chi(0)] \tag{A1}$$

where the complex admittance, $\chi(p)$, of the solvent within the MSA has the form:

$$\chi(p) = \chi_{MSA}(p) = (1-1/\epsilon(p))/2r_i[1+\Delta(p)] \tag{A2}$$

The dynamic correction factor $\Delta(p)$ is given by the expression:

$$\Delta(p) = (3r_s/r_i)\{[f(p)]^{1/3}+[f(p)]^{-1/3}-2\}^{-1} \tag{A3}$$

where

$$f(p) = g(p)-\{[g(p)]^2-1\}^{1/2} \tag{A4}$$

and

$$g(p) = 1+54[\epsilon(p)]^{1/2} \tag{A5}$$

r_i and r_s are the solute and solvent radii respectively. In equations (A2), A(5) the dielectric susceptibility of the medium has the Debye form:

$$\epsilon(p) = \epsilon_\infty+(\epsilon_0-\epsilon_\infty)/(1+p\tau_d) \tag{A6}$$

Substitution of the equations (A2)-(A6) into equation (A1) results in the functional form for the relaxation function S(p). To get the desired function S(t) the inverse Laplace transform was performed with the help of numerical techniques[29].

The Functional Form of S(t) in the Framework of SVE

The Smoluchowski-Vlasov Equation (SVE) was originally proposed by Wolynes and Calef[9] to describe a structural relaxation of the polarization through diffusion in the mean field. Subsequently Nicols and Calef[8] solved it for the case when only rotational diffusion of the molecules was considered. Bagchi and Chandra[10b] extended the solution to include the translational diffusion of the molecules.

The functional form of S(t) in the framework of SVE is given by the equation (19) from ref. 10b, which is:

$$S(t) - 2\frac{r_i}{\pi}\int_0^\infty dk \exp[-t/\tau_L(k)](\int_{kr_i}^\infty dx \frac{\sin(x)}{x})^2 \tag{A7}$$

where $\tau_L(k)$ is given by the following expression:

$$1/\tau_L(k) - 2D_R\{1 + p'(k\sigma)^2 - 1/3\rho[1 + p'(k\sigma)^2](C_A + 2C_D)\} \tag{A8}$$

ρ is the density of the solvent, D_R is the solvent rotational diffusion coefficient, σ is the solvent diameter and p' is the dimensionless solvent parameter defined by equation (7). In equation (A8) C_A and C_D are components of the direct correlation function $C(k)$. In linear theories $C(k)$ can be separated into two parts, i.e.

$$C(k) - C_A I + C_D D \tag{A9}$$

where

$$D - 3kk - I \tag{A10}$$

If MSA is used to find $C(k)$, the analytical expressions for C_A and C_D exist, and for the linear combination that appears in (A9) we get:

$$C_A + 2C_D - 6KC_{PY}(k;2K\rho) \tag{A11}$$

where $C_{PY}(k;2K\rho)$ is the Percus-Yevick direct correlation function for hard spheres at density $2K\rho$. K is a parameter obtained from the relationships:

$$K = \xi/\eta \tag{A12}$$

and

$$\eta = \pi\rho\sigma^3/6 \tag{A13}$$

where ξ is the solution of the equation

$$\frac{(1+4\xi)^2(1+\xi)^4}{(1-2\xi)^6} = \epsilon_0 \tag{A14}$$

The solution of this equation and the form of $C_{PY}(k;2K\rho)$ are given in references [8] and [30]. Note that in our numerical work we used $r_i = r_s = \sigma/2$.

REFERENCES

1. For recent reviews and references therein see, for example: P.F. Barbara and W. Jarzeba, Adv. Photochem, **15**, 1 (1990); M. Maroncelli, J. MacInnis, and G.R. Fleming,Science, **243**, 1674 (1989); B. Bagchi, Annu. Rev. Phys. Chem. **40**, 115 (1989); B. Bagchi, and A. Chandra, Adv. Chem. Phys. **LXXX**, 1 (1991); J.D. Simon, Acc. Chem. Res. **21**, 128 (1988).

2. M. Maroncelli, and G.R. Fleming, J. Chem. Phys. **86**, 6221 (1987); E.W. Castner, M. Maroncelli, and G.R. Fleming, J. Chem. Phys. **86**, 1090 (1987).

3. W. Jarzeba, G.C. Walker, A.E. Johnson, M.A. Kahlow, and P.F. Barbara, J. Phys. Chem. **92**, 7039 (1988); M.A. Kahlow, W. Jarzeba, T.J. Kang, and P.F. Barbara, J. Chem Phys. **90**, 151 (1989).

4. C.J.F. Bottcher, and P. Bordewijk, Theory of Electric Polarization. Vol 2, Elsevier, Amsterdam, (1978).

5. G. van der Zwan, and J.T. Hynes, J. Phys. Chem, **89**, 4181 (1985).

6. P.G. Wolynes, J. Chem. Phys. **86**, 5133 (1987).

7. I. Rips, J. Klafter, and J. Jortner, J. Chem. Phys. **88**, 3246 (1988); ibid. **89**, 4288 (1988).

8. A.L. Nichols III., and D.F. Calef, J. Chem. Phys. **89**, 3783 (1989).

9. D.F. Calef, and P.G. Wolynes, J. Chem. Phys. **78**, 4145 (1983).

10. a). A. Chandra, and B. Bagchi, Chem. Phys. Lett. **151**, 47 (1988); b). ibid. **155**, 533 (1989).

11. M. Maroncelli, and G.R. Fleming, J. Chem. Phys. **89**, 5044 (1988).

12. O.A. Karim, A.D.J. Haymet, M.J. Banet, and J.D. Simon, J. Chem. Phys. **92**, 3391 (1988).

13. T. Fonseca, and B.M. Ladanyi, J. Phys. Chem. **95**, 2116 (1991).

14. M. Maroncelli, J. Chem. Phys. **94**, 2084 (1991).

15. E.A. Carter, and J.T. Hynes, J. Chem. Phys. **94**, 5961 (1991).

16. L. Perera, and M.L. Berkowitz, J. Chem. Phys. **96**, 3092 (1992).

17. E. Neria, and, A. Nitzan, J. Chem. Phys. **96**, 5433 (1992).

18. W.H. Stockmayer, J. Chem. Phys. **9**, 398, 863 (1941).

19. M. Neumann, Mol. Phys. **50**, 841 (1983).

20. M. Neumann, O. Steinhauser, and G.S. Pawley, Mol. Phys. **52**, 97, (1984).

21. E.L. Pollock, and B.J. Alder, Physica, **A 102**, 1 (1980); E.L. Pollock, and B.J. Alder, Phys. Rev. Lett. **46**, 950 (1981).

22. J.P. Ryckaert, G. Ciccotti, and H.J. Berendsen, J. Comp. Phys. **23**, 327 (1977).

23. S.H. Lee, J.C. Rasaiah, and J.B. Hubbard, J. Chem. Phys. **85**, 5232 (1986); ibid. **86**, 2383 (1987).

24. L. Onsager, Can. J. Chem. **55**, 1819 (1977).

25. A. Chandra, and B. Bagchi, J. Chem. Phys. **91**, 2594 (1989).

26. S.J. Rosenthal, X. Xie, M. Du, and G.R. Fleming, J. Chem. Phys. **95**, 4715 (1991).

27. R.M. Levy, D.B. Kitchen, J.T. Blair, and K. Krogh-Jespersen, J. Phys. Chem. **94**, 4470 (1990).

28. A. Chandra, and B. Bagchi, Chem. Phys. **156**, 323 (1991).

29. User Manual, MATH/Library, Vol II (version 2.0) 756 (1987).

30. C.G. Gray, and K.E. Gubbins, Theory of Molecular Fluids. Vol I, Clarendon Press, Oxford, (1984).

POLAR / NONPOLAR FLUID MIXTURES

Simon W. de Leeuw

Laboratory for Physical Chemistry
University of Amsterdam
Nieuwe Achtergracht 127
1018 WS Amsterdam, The Netherlands

1. INTRODUCTION

Polar fluid mixtures form a technologically important class of systems, whose thermodynamic and structural properties are at present poorly understood. Reliable prediction of equations of state and phase behavior is of crucial importance for a range of technological problems. In addition mixtures of polar fluids play an important role in chemistry. Often mixed solvents of different polarity are used as reaction media in order to control the dielectric properties of the medium through changes in the composition[1]. The consequences of doing this are not well understood at a molecular level. In particular the relation between molecular interactions parameters, such as polarity, polarizability on the one hand and macroscopic behavior, such as thermodynamic, structural and dielectric properties and phase behavior, is not clear.

Mixtures of fluids consisting of polar molecules having spherical cores provide a convenient model to study the effects of polarity and polarizability on the microscopic structure and thermodynamics of these systems. Gubbins and Twu have applied thermodynamic perturbation theory[2,3] to these systems to investigate the mixing behavior. They extended an approach, proposed by Stell and Narang[4,5], based on a Padé approximant for the free energy expansion, to mixtures of polar fluids. Using this approach they investigated the phase behavior for a wide range of interaction parameters in these mixtures and showed that these mixtures exhibited a rich diversity in phase behavior. More recently integral equation theories have been applied to study structural and

Condensed Matter Theories, Vol. 8, Edited by
L. Blum and F.B. Malik, Plenum Press, New York, 1993

dielectric properties of binary dipolar mixtures. Cummings and Blum[6] studied the dielectric properties of polar hard sphere mixtures using the mean spherical approximation. Morris and Isbister[7] used an integral equation to study mixtures of polar and nonpolar hard dumbbells and predicted phase separation into an almost pure nonpolar component and a mixture rich in polar component. Lee and Ladanyi[8] studied dipolar hard sphere mixtures with the reference hypernetted chain theory. Very recently Chen et al showed that hypernetted chain theory predicts demixing in mixtures of polar and nonpolar hard spheres[9,10].

In this paper I will discuss results of computer simulation studies of mixtures of Stockmayer molecules and polarizable Lennard-Jones molecules. The thermodynamic properties are studied as a function of dipolar strength and polarizability and various excess properties of mixing are obtained.

The free energy of mixing for mixtures of Stockmayer and Lennard-Jones molecules can be obtained readily through a thermodynamic integration procedure. Our results indicate strongly that demixing into an almost pure nonpolar phase and a phase rich in polar component occurs in these systems for sufficiently large dipole moments of the Stockmayer molecules, in qualitative agreement with perturbation[2,3] and integral equation[9,10] theories. Increasing the polarizability leads to an increase in miscibility for these mixtures, as expected.

The phase boundaries of these mixtures can be determined directly using Gibbs ensemble Monte Carlo[11]. I will discuss some prelimi- nary results, using this technique , which confirm the analysis based on free energies.

In the last part I will discuss structural and dielectric properties. In particular it will be shown that the pair correlation functions display a marked asymmetry with respect to composition. At low concentrations of the polar component a strong increase in orientational ordering of the polar molecules is observed. Also the main peak in the center of mass (COM) distribution function for pairs of polar molecules attains a sharp first maximum, reflecting the onset of phase separation. Such behavior has also been observed in Monte Carlo calculations of mixtures of quadrupolar molecules with spherical cores[12]. Again an increase in polarizability of the Lennard-Jones diminishes these effects. A qualitative explanation, based on a simple frustration model, is proposed for these effects.

2. Computational Details

The systems studied consist of N_s nonpolarizable Stockmayer molecules with permanent dipole moment μ and N_{LJ} polarizable Lennard-Jones molecules with (isotropic) polarizability α. The potential energy of the system is given by:

$$V(\vec{R}_1,..,\vec{R}_N;\vec{\mu}_1,..,\vec{\mu}_N) = \frac{1}{2}\sum_{i=1}^{N}\sum_{j\neq1}^{N}\varphi_{LJ}(R_{ij}) + V_{dd}(\vec{R}_1,..,\vec{R}_N;\vec{\mu}_1,..,\vec{\mu}_N) \qquad (2.1)$$

Here the first term describes the Lennard-Jones interaction between the particles. In all simulations discussed here the well depth ε and diameter σ of the Lennard-Jones and Stockmayer molecules are identical. The second term describes the dipolar interactions between the molecules and is given by:

$$V_{dd}(\vec{R}_1,..,\vec{R}_N;\vec{\mu}_1,..,\vec{\mu}_N) = -\frac{1}{2}\sum_{i=1}^{N}\sum_{j\neq1}^{N}\vec{\mu}_i\,\vec{\vec{T}}(\vec{R}_{ij})\,\vec{\mu}_j + \frac{1}{2}\sum_{i=1}^{N}(\vec{\mu}_i^{ind})^2/\alpha \qquad (2.2)$$

The $\vec{\vec{T}}$ - tensor is given by:

$$\vec{\vec{T}}(\vec{R}) = (3\,\vec{R}\vec{R} - \vec{\vec{I}})/R^3 \qquad (2.3)$$

The Stockmayer molecules are nonpolarizable, so that $\vec{\mu}_i$ corresponds to the permanent dipole moment vector and the induced dipole moment $\vec{\mu}_i^{ind} = 0$. For Lennard-Jones molecules the permanent dipole moment vanishes and

$$\vec{\mu}_i = \vec{\mu}_i^{ind} = \alpha\,\vec{E}_i \qquad i = N_{LJ}+1, ...,N \qquad (2.4)$$

Henceforth we shall employ reduced units defined in the usual manner. Thus all lengths are measured in units of the Lennard-Jones size parameter σ and energies in units of the Lennard-Jones well depth ε. The unit of time is given by $\tau = \sqrt{(m\sigma^2/\varepsilon)}$ and the dipole moment is given in units of $\mu/\sqrt{(\varepsilon\sigma^3)}$. The polarizability is given in dimensionless units α/σ^3.

2.1 MOLECULAR DYNAMICS

Molecular dynamics simulations were carried out for mixtures of Lennard-Jones and Stockmayer fluids in which the concentration x_s and the dipolar strength μ of the Stockmayer molecules and the polarizability α of the Lennard-Jones molecules was varied. All simulations were carried out at a reduced density $\rho = 0.822$ and reduced temperature $T = 1.15$, in the middle of the liquid regime. Both the pure Stockmayer system and the Lennard-Jones system has been studied extensively at this state point[13-15]. The temperature of the system was held fixed with Nosé's technique[16]. The dipolar energy was evaluated with Ewald sums, using 'tinfoil' boundary conditions[17].

The evaluation of the induced dipole moments $\vec{\mu}_i^{ind}$ for nonvanishing polarizabilities of the Lennard-Jones molecules is a highly non-linear many-body problem, which requires iterative procedures to determine the induced dipole moments self consistently[18,19]. The calculation can be speeded up considerably by a suitable starting value of $\vec{\mu}_1^{ind}$. A predictor method, using

$$\vec{\mu}_1^{ind}(t+\Delta t) = 3\vec{\mu}_1^{ind}(t) - 3\vec{\mu}_1^{ind}(t-\Delta t) + \vec{\mu}_1^{ind}(t-2\Delta t) \tag{2.5}$$

yields an accurate starting value. We found, that on the average 2.3 iterations were required to determine the induced dipole moments with an accuracy of 1 part in 10^3.

In all molecular dynamics simulations we used 108 or 256 particles. Salient details of the simulations are given elsewhere[20,21].

2.2 FREE ENERGY COMPUTATION

The free energy of mixing of the mixtures of Stockmayer and Lennard-Jones described above can be computed in a simple and elegant manner, which does not require knowledge of the free energy of the pure components. The technique is based on a standard expression for the free energy of a system, whose hamiltonian depends parametrically on a coupling parameter λ[22]. If the hamiltonian is given by:

$$H = H_0 + \Delta H(\lambda), \tag{2.6}$$

where H_0 describes a reference system and ΔH a perturbation, then the free energy can be obtained as:

$$A = A_0 + \int_0^\lambda < \frac{\partial H}{\partial \lambda'} >_{\lambda'} d\lambda' \tag{2.7}$$

where A_0 denotes the energy of the reference system, described by the hamiltonian H_0. For the systems described above two coupling parameters can be identified, namely the dipolar strength μ^2 of the Stockmayer molecules and the polarizability α of the Lennard- Jones molecules. Application of eq. 2.7 to this system then leads to:

$$\Delta a(x_s) = a(x_s) - x_s a(1) - (1-x_s)a(0)$$

$$= T \left[x_s \ln x_s + (1-x_s) \ln (1-x_s) \right] + \Delta a_{dd}(x_s) + \Delta a_{pol}(x_s) \tag{2.8}$$

Here $\Delta a(x_s)$ is the free energy of mixing of a mixture, which contains a mole fraction x_s of Stockmayer molecules; $a(x_s)$ the total free energy of this mixture. Note that $x_s = 1(0)$ corresponds to pure Stockmayer (Lennard-Jones) fluids. The first term in the rhs of eq. 2.8 describes ideal mixing.

$$\Delta a_{dd}(x_s) = \int_0^{\mu^2} \langle \frac{\partial v_{dd}(\mu'^2)}{\partial \mu'^2} \rangle_{\mu'^2} \, d\mu'^2 = \int_0^{\mu^2} \langle v_{dd}(\mu'^2) \rangle_{\mu'^2} \frac{d\mu'^2}{\mu'^2} \qquad (2.9)$$

This term describes the reversible work needed to charge the dipoles to the required value of the dipolar strength μ^2. Similarly:

$$\Delta a_{pol}(x_s) = \int_0^{\alpha} \langle \frac{\partial v_{dd}(\alpha')}{\partial \alpha'} \rangle_{\alpha'} \, d\alpha' = -\frac{1}{2} \int_0^{\alpha} \langle \vec{E}^2(\alpha') \rangle_{\alpha'} \, d\alpha' \qquad (2.10)$$

Eq. 2.10 describes the reversible work required to polarize the Lennard-Jones atoms. $\vec{E}$ denotes the local field acting on a polarizable Lennard-Jones molecules. Both $\vec{E}$ and v_{dd} are readily obtained during a simulation, so that $\Delta a(x_s)$ can be computed readily.

2.3 GIBBS ENSEMBLE MONTE CARLO

An alternative way to determine phase equilibria was proposed about 5 years ago by Panagiotopoulos[11]. This technique is particularly useful for studying liquid-gas and liquid-liquid phase equilibria. Panagiotopoulos recognized, that one of the central problems in studying phase-equilibria directly in computer simulations was related to the relatively large surface area between the different phases. Therefore two separate simulation boxes are introduced in this technique and particles are allowed to move from one box to the other. The transition probabilities for moving from one box to the other are chosen such temperature and pressure in the two boxes are the same and that chemical potentials of identical particles are the same in the two boxes. These are precisely the conditions of phase equilibrium between two phases. Hence if a system is simulated in a two-phase region it will attain its lowest free energy by forming one phase in the first simulation box and another phase in the second. Note that this technique avoids the formation of interfaces, which lead to an increase in the free energy of the system and consequently makes phase separation difficult to observe directly in computer simulations of small finite systems. The technique has been introduced by Panagiotopoulos on an intuitive basis, using a combination of three different (conventional) ensembles. A rigorous treatment has been given by Smit and Frenkel[23,24]. They considered the system consisting of two noninteracting subsystems as a new ensemble, the "Gibbs ensemble". They showed that (1) this ensemble is equivalent to conventional ensembles and (2) separate phases will form in the boxes in the two-phase region. Moreover, they showed that finite size effects near the critical temperature partly cancel, so that we may expect these to be much smaller than in other ensembles.

In this section a brief expose is given of the Gibbs ensemble Monte Carlo technique for liquid–liquid equilibria. Consider a system of two components A and B in a total volume V at temperature T. The system contains N_A particles of component A and N_B particles of component B. The system is divided into two non-interacting subsystems I and II, as indicated in fig. 1. The particles are distributed over the two subsystems in such a way that the total number of particles of each species is constant. The volume of each subsystems may vary such that the total volume remains constant.

V_I	$V - V_I$
n_A, n_B	$N_A - n_A$, $N_B - n_B$
μ_A, μ_B	μ_A, μ_B
T, P	T, P

Fig.1 Gibbs ensemble for a two-component system.

The partition function $Q(N_A, N_B, V, T)$ of the system is given by[25]:

$$Q(N_A, N_B, V, T) = \frac{1}{\Lambda_A^{3N_A} \Lambda_B^{3N_B} N_A! N_B!} \sum_{n_A=0}^{N_A} \sum_{n_B=0}^{N_B} \binom{N_A}{n_A} \binom{N_B}{n_B}$$

$$\times \int_0^V V_I^{(n_A+n_B)} \left(V - V_I \right)^{(N_A-n_A+N_B-n_B)}$$

$$\times \int_I d\vec{\xi}_A^{(n_A)} d\vec{\xi}_B^{(n_B)} \exp(-\beta U_I) \int_{II} d\vec{\xi}_A^{(N_A-n_A)} d\vec{\xi}_B^{(N_B-n_B)} \exp(-\beta U_{II}) \qquad (2.11)$$

Here n_A (n_B) denotes the number of A (B) particles in subsystem I, V_I the volume of subsystem I; $\vec{\xi}_A^{(n_A)}$ ($\vec{\xi}_B^{(n_B)}$) denotes the set of all scaled coordinates of A (B) particles in subsystem I ($\vec{\xi} = \vec{r}/V^{1/3}$) and U_I its energy. A similar notation is used for subsystem II. Λ_A (Λ_B) is the thermal wavelength of A (B) particles and $\beta = 1/k_B T$ the reciprocal temperature. Note that the integrand corresponds to a Boltzmann distribution with pseudo-Boltzmann factor

$$\exp(-\beta W)= \exp\left[\ln\binom{N_A}{n_A}\binom{N_B}{n_B} + (n_A+n_B)\ \ln\ V_I\ +\right.$$

$$\left. (N_A-n_A+N_B-n_B)\ \ln\ (V-V_I)\ -\ \beta(U_I+U_I)\right] \tag{2.12}$$

Starting from the pseudo-Boltzmann factor acceptance rules for generating a Markov chain of configurations in the Gibbs ensemble is straightforward to derive. The following moves are allowed:

1. random displacement of a particle in subsystem i (= I or II). The change in the pseudo-Boltzmann factor is:

$$\Delta W = \beta(U_I'' - U_I') \tag{2.13}$$

Here the superscript "(') refers to the new (old) configuration.

2. Volume change of V_I by ΔV. In this case the pseudo-Boltzmann factor changes by:

$$\Delta W = (U_I''- U_I') + (U_{II}''- U_{II}') - k_BT(n_A+n_B)\ \ln\left(\frac{V_I+\Delta V}{V_I}\right) -$$

$$(N_A-n_A+N_B-n_B)\ \ln\left(\frac{V-V_I+\Delta V}{V-V_I}\right) \tag{2.14}$$

3. Moving of a particle from one subsystem to another. Suppose a particle of species A is moved from subsystem II to subsystem I. Then the change in pseudo-Boltzmann factor is:

$$\Delta W = (U_I''- U_I') + (U_{II}''- U_{II}') + k_BT\ \ln\left(\frac{(V- V_I)(n_A+1)}{V_I(N_A-n_A)}\right) \tag{2.15}$$

4. Interchanging a particle of one species in a subsystem with a particle from another species in the other subsystem. Suppose for definiteness that a particle of species A in subsystem I is interchanged with a particle of species B in subsystem II. Then the change in the pseudo-Boltzmann factor is given by:

$$\Delta W = (U_I''- U_I') + (U_{II}''- U_{II}') + k_BT\ \ln\left[\frac{(n_A+1)(N_B-n_B+1)}{(N_A-n_A)\ n_B}\right] \tag{2.16}$$

The simplest way to generate trial configurations for the Markov chain is to perform the simulation in cycles. One cycle consists (on the average) of N_1 attempts to displace a particle, N_v attempts for a volume change, N_m attempts to move particles from one subsystem to the other and N_1 attempts to interchange a particle from one species in a subsystem with a particle from the other species in the second subsystem. For each trial the pseudo-Boltzmann factor is computed and the configuration is accepted with probability $\exp(-\beta\Delta W)$.

3. THERMODYNAMIC PROPERTIES AND PHASE BEHAVIOR

Extensive discussions on the thermodynamic properties of polar fluid mixtures can be found in the literature[22,23], so that only a brief summary will be given here. These simplest quantity, which can be computed directly from a simulation, is the excess energy Δv_{exc}. The excess energy is found to be positive, with a maximum at concentrations low in polar component. This is similar to what has been observed experimentally, in eg. mixtures of ethanol/n-hexane[25].As the polarity of the Stockmayer molecules increases the excess energy is seen to saturate. The excess energy can be reduced by making the Lennard–Jones molecules polarizable, as expected. In fig. 2 the free energy of mixing and the resulting phase diagram for mixtures of nonpolarizable Lennard–Jones and Stockmayer molecules is plotted. In computing the free energy the dipolar energy was fitted to a functional form, given by:

$$\langle v_{dd}(x_s;\mu^2)\rangle = -Ax_s^2 \mu^4/(1 + C(x_s)\mu^2) \qquad (3.1)$$

For small values of μ the energy is proportional to μ^4, in accordance with perturbation theory. For large values of the dipolar strength the dipolar energy increases as μ^2, as suggested by Onsager[26]. Perturbation theory yields $A = 1.70$ at the density and temperature considered. $C(x_s)$ has been treated as a fitting parameter. Good accuracy was obtained for:

$$C(x_s) = 0.876 x_s - 0.134 \qquad (3.2)$$

The free energy is no longer convex for $\mu^2 > 3.2$ and phase separation occurs. Fig. 4 shows that the mixture separates into a phase rich in polar component and an almost pure Lennard–Jones fluid. In calculating the phase diagram in fig. 4 we have neglected volume effects.The phase diagram is similar to that obtained from perturbation theory[2,3]. The onset of phase

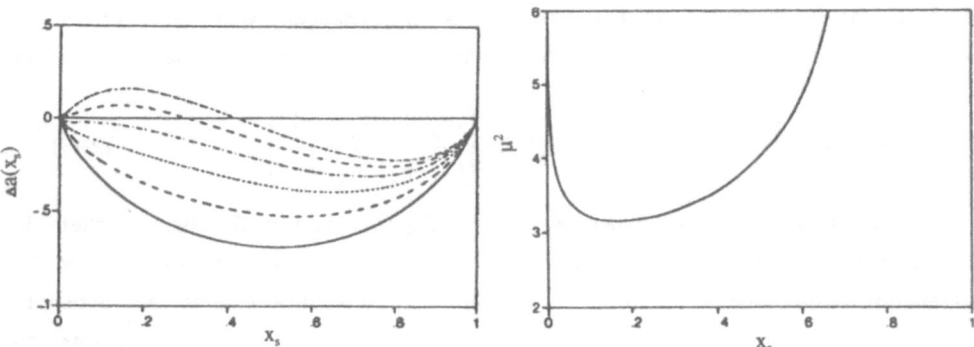

Fig 2 (a) Helmholtz free energy of mixing for mixtures of Stockmayer and Lennard–Jones fluids. ——: $\mu^2 = 1$; − −: $\mu^2 = 2$; ---; $\mu^2 = 3$; −.−: $\mu^2 = 4$; − −: $\mu^2 = 5$; −..−; $\mu^2 = 6$; (b) Phase diagram obtained from eq. 3.1 and 3.2.

separation occurs however at larger values of μ than predicted by perturbation theory. A similar phase diagram has been proposed by Chen et al. for mixtures of polar and neutral hard spheres on the basis of integral equation calculations[9,10].

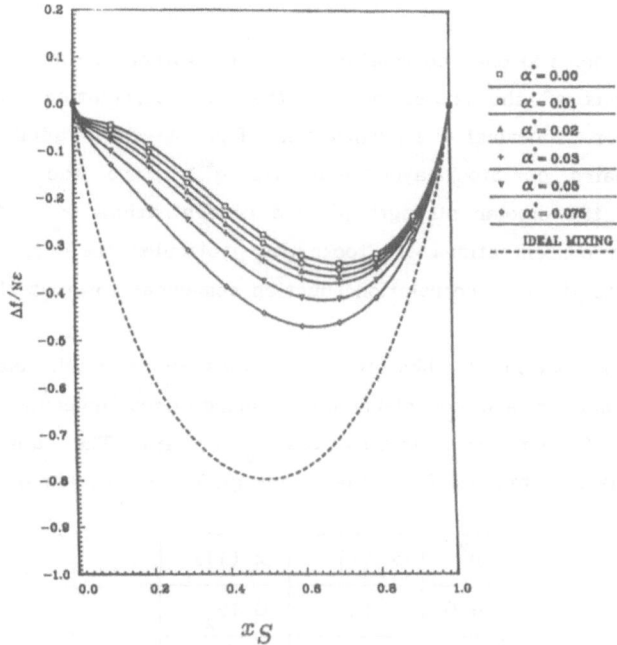

Fig. 3 Helmholtz free energy of mixing for mixtures of polar/polarizable fluids. The dipolar strength of the Stockmayer nolecules is $\mu^2 = 4$.

Fig. 3 shows the free energy of mixing of mixtures of polarizable Lennard-Jones and Stockmayer molecules for various values of the polarizability α. The dipolar strength of the Stockmayer molecules is $\mu^2=4$. Clearly as the polarizability increases $\Delta a(x_s)$ becomes more negative and the miscibility increases. For $\alpha \geq 0.05$ the two fluids become completely miscible again.

Perturbation theory predicts values of the energy and free energy in reasonable agreement with simulation results. However, the agreement is less for the excess energy and the compressibility factors $Z = PV/Nk_BT$

Gibbs ensemble Monte Carlo simulations have been carried out for 216 particles distributed over two subsystems as described in section 2.3. Here we give some preliminary results, which confirm, at least qualitatively, the phase behavior discussed above. A series of simulations was carried out using an overall mole fraction $x_s = 1/6$ of Stockmayer molecules. Phase separation is observed for $\mu^2 > 4$, which is somewhat higher than obtained on the basis of free energy calculations described above. For $\mu^2 = 4$ the simulation doesn't yield a clear answer and further analysis is needed. In table I the compositions of the phases in the two subsystems are collected.

4. STRUCTURAL PROPERTIES

The spatial and orientational correlations in the system are most conveniently discussed in terms of the projections of the pair correlation functions onto the lowest order spherical harmonics. In fig. 4 the radial distribution functions for pairs of Stockmayer molecules $g^{ss}(r)$ are shown for various values values of the dipolar strength μ^2 and concentrations $x_s = 0.125$ and $x_s = 0.875$. At high concentrations of Stockmayer molecules the increase in dipole moment only shifts the pair-correlation function somewhat towards the center.

Table I Composition of subsystems in Gibbs Monte Carlo calculations of mixtures of Stockmayer and nonpolarizable Lennard-Jones molecules. The overall mole fraction of Stockmayer molecules was $x_s = 0.167$. The subscribts denote the standard deviation obtained from block averaging over the simulation.

μ^2	$x_s(I)$	$x_s(II)$
4.0	0.14_2	0.19_2
4.5	0.07_3	0.33_5
5.0	0.02_2	0.44_3
6.0	0.02_1	0.49_3

At low concentrations we observe a strong increase in the first peak of the pair correlation function, showing a strong tendency for polar molecules to cluster. This behavior is also reflected in the radial distribution function for unlike pairs $g^{sl}(r)$. An increase in the dipolar strength at low Stockmayer concentrations results in a decrease in the main peak of the pair correlation function, showing that pairs of unlike particles effectively repel each other.

A similar behavior is observed for the orientational correlation functions. A strong increase in local ordering is observed at low concentrations of the polar component, as shown in fig. 5 for the projection

$h_\Delta(r)$ onto the angular function $\Delta(1,2)=\cos\gamma$, where γ is the angle between two dipole orientations. Similar effects are observed in the other angular correlation functions.

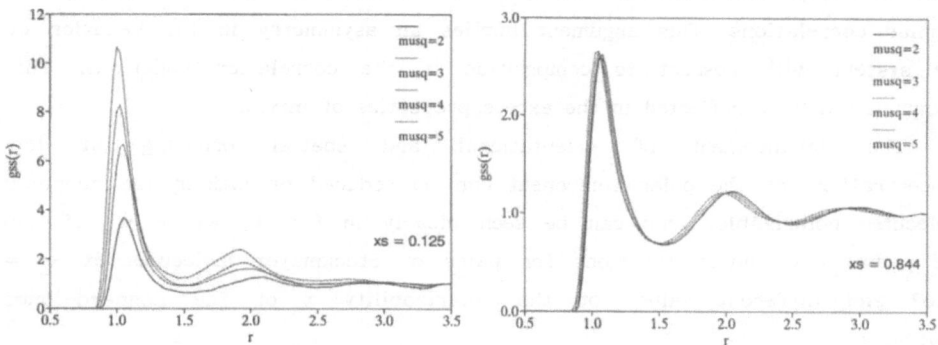

Fig. 4 Radial distribution functions for pairs of Stockmayer molecules at two different concentrations.

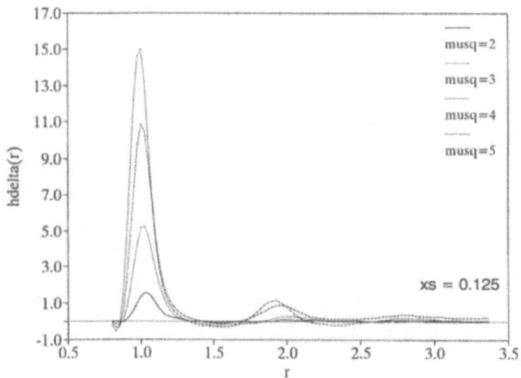

Fig. 5 Angular correlation function $h_\Delta(r)$ at x_s = 0.125 for various values of the dipolar strength.

The strong increase in orientational ordering can be understood on the basis of a simple frustration model. Consider a highly dilute solution, so that only pairs of dipoles occur. In that case the polar molecules are free to align themselves in the energetically most favorable orientation, i.e. head-to-tail for Stockmayer molecules. This leads to a large value of the angular functions, such a Δ(1,2) and consequently a large value of the corresponding angular correlation functions. For larger concentrations more dipoles are likely to be in the vicinity of another and an optimal configuration can no longer be achieved for all polar interactions. Therefore a compromise must be achieved, reducing the alignment of Stockmayer molecules. Thus at large concentrations there is a "frustration" effect, reducing the angular correlations. This argument implies an asymmetry in the behavior of the system with respect to composition in the correlation functions. This asymmetry will be reflected in the excess properties of mixing.

The enhancement of orientational and spatial ordering at low concentrations of the polar component can be reduced by making the nonpolar molecules polarizable. This can be seen clearly in fig. 6, where we display the radial distribution functions for pairs of Stockmayer molecules at x_s = 0.167 and different values of the polarizability α of the Lennard–Jones molecules.

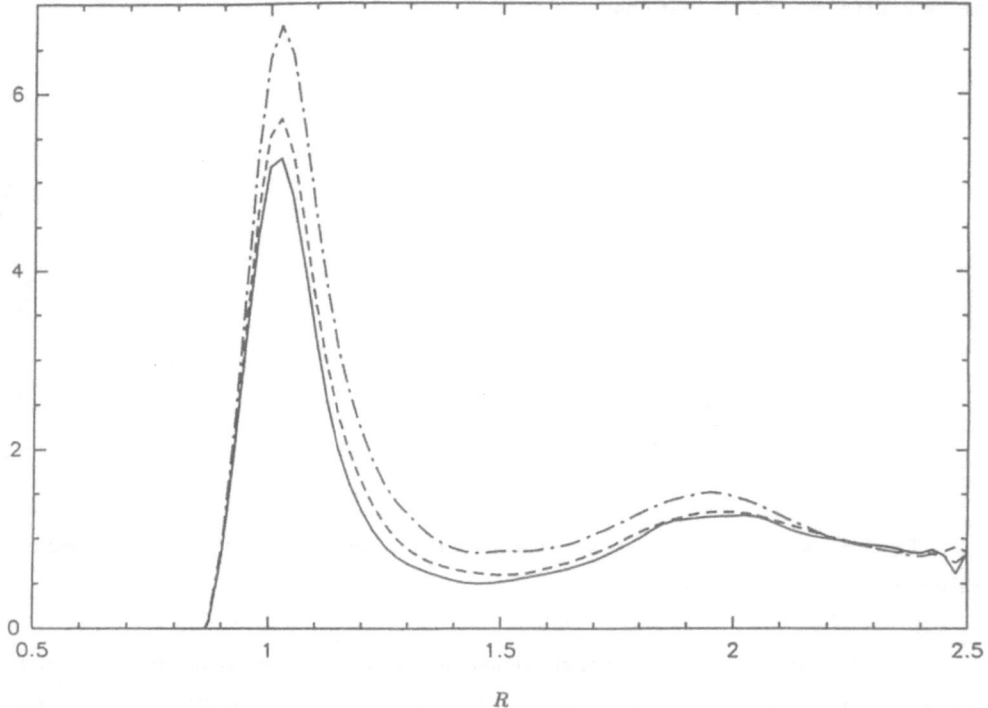

Fig. 6 Radial distribution functions for pairs of Stockmayer molecules for different polarizabilities of the nonpolar molecules. x_s = 0.167; μ^2 = 4; ——: α = 0.05; – – –: α = 0.03; ......: α = 0.0.

5. DIELECTRIC BEHAVIOR

The dielectric constant is obtained from the simulations through the relation:

$$\varepsilon = \varepsilon_\infty + 3y \frac{\langle \vec{M}^2 \rangle}{N\mu^2} \tag{5.1}$$

where $y = 4\pi\beta\rho\mu^2/9$ and μ is the permanent moment of the Stockmayer molecules. $\vec{M}$ is the fluctuating dipole moment of the simulation cell:

$$\vec{M} = \sum_{i=1}^{N} \vec{\mu}_i$$

and ε_∞ is obtained from the polarizability through the Clausius-Mossotti relation[27]. In fig. 7 the dielectric constant is plotted for a mixture of polarizable Lennard–Jones and Stockmayer molecules. The lines denote a theoretical approximation due to Looyenga[28,29] on the basis of a macroscopic electrostatic model. This simple model is seen to give a reasonable approximation to the curves and in fact leads to better results than for example the mean spherical approximation. Similar behavior has been noted by Cummings and Blum in their study of mixtures of dipolar hard spheres.

Note that the polarizability only has a small effect on the dielectric constant of these mixtures.

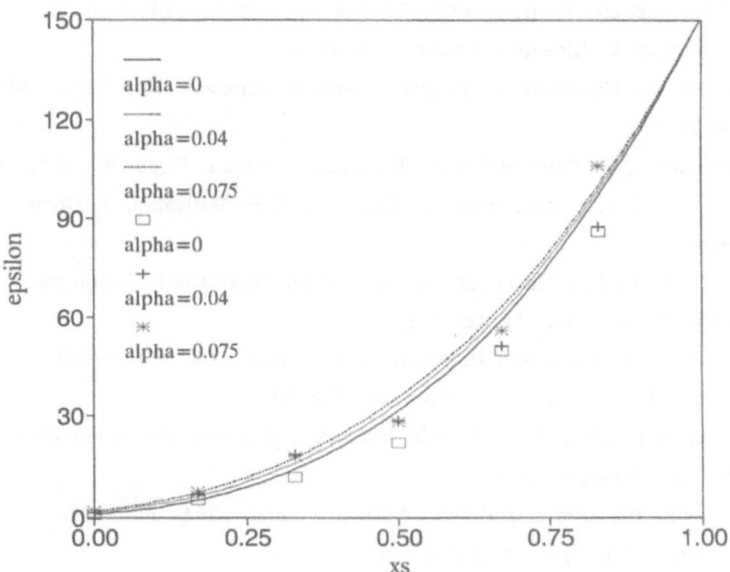

Fig. 7 Dielectric behavior of mixtures of polarizable Lennard–Jones and Stockmayer fluids. The curves are based on a macroscopic model due to Looyenga. The points are simulation results. $\mu^2 = 4$.

6. REFERENCES

1. see e.g. Electron Transfer in Inorganic, Organic and Biological Systems, James R. Bolton, N. Mataga and G. McLendon, Eds. Adv. Chem. Ser. **228**, Washington, 1991

2. K.E. Gubbins and C. H. Twu, Chem. Eng. Sci. **33**, 863 (1977)

3. K.E. Gubbins and C. H. Twu, Chem. Eng. Sci. **33**, 879 (1977)

4. G. Stell, J.C. Rasaiah and H. Narang, Mol. Phys. **23**, 393 (1972)

5. G. Stell, J.C. Rasaiah and H. Narang, Mol. Phys. **27**, 1393 (1974)

6. P.T Cummings and L. Blum, J. Chem. Phys. **85**, 6658 (1986)

7. G. Morriss and D. Isbister, Mol. Phys. **51**, 911 (1986)

8. P.H. Lee and B. M. Ladanyi, J. Chem. Phys. **91**, 7063 (1989)

9. X. S. Chen, M. Kasch and F. Forstmann, Phys. Rev. Lett. **67**, 2674 (1991)

10. X. S. Chen and F. Forstmann, preprint

11. A.Z. Panagiotopoulos, Mol. Phys. **61**, 813 (1987)

12. M. Wojczik and K. E. Gubbins, Mol. Phys. **51**, 911 (1986)

13. M. Neumann, O. Steinhauser and G.S. Pawley, Mol. Phys. **52**, 97 (1984)

14. D.J. Adams and E.M. Adams, Mol. Phys. **42**, 907 (1981)

15. M. Neumann, Mol. Phys. **50**, 841 (1983)

16. S. Nosé, J. Chem. Phys. **81**, 511 (1984)

17. S.W. de Leeuw, J.W. Perram and E.R. Smith, Ann. Rev. Phys. Chem. **37**, 240 (1986)

18. C.J.F. Böttcher and P. Bordewijk, Theory of Electric Polarization, Vol I, Chapter V, Elsevier, Amsterdam, 1978

19. P Ahlström, A. Wallqvist, S. Engström and B. Jönsson, Mol. Phys. **70**, 963 (1990)

20. S.W. de Leeuw, B. Smit and C.P. Williams, J. Chem. Phys. **93**, 2207 (1990)

21. G.C.A.M. Mooij, S.W. de Leeuw, B. Smit and C.P. Williams, J. Chem. Phys. (in press)

22. see eg. D. Chandler, "Introduction to Modern Statistical Mechanics", OUP, New York, 1987, chapter VII

23. B. Smit, Ph. de Smedt and D. Frenkel, Mol. Phys. **68**, 931 (1989)

24. B. Smit and D. Frenkel, Mol. Phys. 68, 951 (1989)

25. J.S. Rowlinson, Liquids and Liquid Mixtures, Butterworth Scientific Publications, London, 1959

26. E.L. Pollock, B.J. Alder and G.N. Patey, Physica **108A**, 14 (1981)

27. H. Looyenga, Mol. Phys. 9, 501 (1965)

28. L.D. Landau and E. M. Lifshitz, Electrodynamics of Continuous Media, Pergamon Press, Oxford, 1975, Chapter 3

TOPOLOGICAL FIELD THEORY OF POLYMER ENTANGLEMENTS

Arkady L. Kholodenko

375 H. L. Hunter Laboratories
Clemson University
Clemson, SC 29634-1905

INTRODUCTION

Some time ago Delbrück[1] had formulated the following conjecture. Let $\tilde{Q}_N$ be a set of *closed* polymer configurations (for the polymer chain containing N subunits) which may contain knots of any type and let Q_N° be a subset of *closed* polymer configurations which are knotedless (or unknotted). Then, for $N \to \infty$, the ratio $\zeta_N = Q_N^\circ / \tilde{Q}_N$ is anticipated to approach zero, i.e. for $N \to \infty$ practically all cosed (i.e. circular) polymers are knots.

Although the above conjecture, if it is true, has some biological applications[1] we shall not discuss them here. Instead, we shall be mainly interested in proving that the above conjecture is correct. The attempt to prove numerically the above conjecture was made by several authors. Among them, most notably, by Michels and Weigel[2] and by Windwer[3]. In order to understand their results, several comments are in place. To generate loops (knots) on the lattice by computer, a number of requirements must be met. First, there should be a routine which generates closed *selfavoiding* (SAW) loops. Second, there should be another routine which allows to distinguish between the different topologies. Third, in order to achieve an adequate statistical accuracy the number of closed polymer configurations of each topological type should be large to account for all possibilities of putting a knot of a given type on the lattice. The authors of Ref. 2 have obeyed the above requirements, except the requirement of selfavoidance, while Windwer had accounted for SAW constraint. He obtained the ratio with $\alpha = 0$, $\tilde{\mu} = 0.9949$ and $\tilde{c} = 1.2325$.

$$\zeta_N = \tilde{c}\tilde{\mu}^N N^\alpha \tag{1}$$

Eq. (1) can be conveniently rewritten in the equivalent form:

$$\zeta_N = \tilde{c}(\frac{1}{\tilde{c}})^{\frac{N}{\ell_T}}$$ (2)

where the topological persistence length ℓ_T is determined by the equation

$$\zeta_{\ell_T} = 1 = \tilde{c}\mu^{\ell_T}$$ (3)

Using values of $\tilde{\mu}$ and $\tilde{c}$ given above we obtain $\ell_T = 41$. The topological persistence length ℓ_T has the meaning similar to the usual persistence length ℓ for random walk models of polymers.[4] In the case of usual random walks on the lattice ℓ can be chosen of order of lattice size (i.e. there are *no* walks *smaller* than the lattice size). In the same time, on the square lattice the minimal number of steps to form a loop is 4. On a cubic lattice, evidently, there is a minimal number of steps to form a simplest knot, while if the lattice is of different geometry, the number of steps should, in general, be different. Whence, ℓ_T is *lattice-specific* quantity as well as ℓ. With these remarks it is obvious that ℓ_T cannot be universal. In the same time, $\tilde{c}$ is universal as we shall demonstrate. Using the results of knot theory[5] combined with that of Chern-Simons (CS) field theory[6] we have obtained

$$\tilde{c} = q^{\frac{1}{2}} \exp\{-\frac{\pi}{6}c\},$$ (4)

where q is the number of states of q-state two-dimensional Potts Model while c is the central charge of the Potts Model. We have regorously demonstrated[7] that q = 4 which implies c = 1. For the above values of q and c we obtain, using Eq. (4), $\tilde{c}$ = 1.1847696 which differs from the result, Eq. (1), by 0.047. We shall provide an explanation of the above discrepancy below.

STATISTICAL MECHANICS OF KNOTS

Knots are *intrinsically* 3-dimensional objects. Any kind of knot can be associated with some kind of embedding of a circle into 3-dimensional space. Closely related to knots are links. A link is some union of circles (connected or not). Because of this definition, the mathematical treatment of links parallels that of knots[5] and therefore, we shall discuss mainly knots below.

Mathematicians have reduced the problem of description of knots to that of planar graphs. The idea of the above reduction lies in the following. A given knot can be projected into an arbitrary plane so that the shadow of the knot produces 4-valent planar graph. The graph is 4-valent because each entanglement, when projected into the plane, produces the corresponding 4-vertex so that *the difference between different knots lies in the convention associated with the way 4-vertex is drawn*, specifically, 4-vertex can be oriented[5] (or not) so that different orientations would correspond to different knots or, if it is unoriented, one line of 4-vertex can pass under (over) another. In the both cases we have just two states associated with each vertex. Whence, if 4-valent graph contains n 4-vertices it is possible, in principle, to have 2^n *different* states of such planar graph. The question arises about the relationship between a given knot and the set of these 2^n configurations. At the present time, no unique relationship is known, that is, *the same* knot can be described in a number of possible *formally nonequivalent* ways and there is no known way to describe *all* possible knots. Different approaches differ by the number of nonequivalent knots they are able to describe.

In spite of the differences between the approaches to describe knots, they all contain the same type of elements as will be illustrated below. First, we would like to notice that *not all* of n crossings represent true entanglements. This is so because the plane into which a given knot is projected is arbitrary, i.e. if another plane is chosen this could, in principle, result in different number n. There is, however, some *minimal* n intrinsic to a given knot. This minimal n can be obtained via so called Reidemeister moves.[5] These moves (in the plane) serve to correct the arbitrariness in the choice of a plane, i.e. they force the projection of a given knot to be topologically the*same* (irrespective to the initial choice of a plane). Once this is accomplished, knots can be described recursively so that different methods differ by the way the recursive algorithm is designed. Consider, for example, the recursive procedure for the HOMFLY polynomial G_K. It is described by the set of axioms:

Axiom 1: if knots K and $\tilde{K}$ are equivalent (in the sense of Reidemeister moves), i.e. $K \sim \tilde{K}$, then $G_K = G_{\tilde{K}}$.

Axiom 2: if $K \sim 0$, where 0 is unknot, then $G_K = 1$.

Axiom 3: for some α and z the skein (recurrence) relationship between three knots K, $\tilde{K}$ and L which respectively differ *locally* by just one crossing: under (K), over ($\tilde{K}$), or no crossing at all (L), is given by

$$\alpha G_K - \alpha^{-1} G_{\tilde{K}} = z G_L. \tag{5}$$

The above three axioms allow us to unknot a given knot in expense of creating a polynomial in terms of α and z so that ideally *different knots should have different polynomials*. In reality, however, HOMFLY polynomial can "recognize" many but not all knots. Closely related to HOMFLY are Alexander-Conway ∇_K and Jones V_K polynomials. For ∇_K we have to put $\alpha = 1$, while for V_K we have to put $z = \dfrac{1}{\sqrt{\alpha}} - \sqrt{\alpha}$.

As it was demonstrated by Jones,[8] V_K is directly connected with 2-dimensional Potts model. Independently, Witten[6] had demonstrated how skein relation, Eq. (5), can be obtained field-theoretically using the nonabelian CS field theory. This kind of field theory is also called "topological" because it is independent of the metric of the embedding place and because it is able to generate, in addition to the well-known Gauss linking number,[5] other topological invariants for which the analytical form was unknown. In addition to G_K, V_K and ∇_K, there are other polynomials built according to similar works.

The connection between V_K and the Potts model discovered by Jones[8] allowed us to apply the methods of statistical mechanics to knots, one one hand, and to provide an interpretation of statistical mechanical models in terms of knot theory on the other. Specifically, a given knot can be made of braids (for which there are analogs of Reidemeister moves called Markov moves) a trivial braid can be visualized as a ladder of n steps while a nontrivial bride is made of trivial by "mixing" m and m + 1st steps, etc. A trivial link can be formed from a trivial braid if the ends of each ladder step are connected outside the ladder. Accordingly, the nontrivial links (knots) are formed by connecting the corresponding ends of nontrivial braids. An analogue of skein relation, Eq. (5), for the braid is the Temperley-Lieb (TL) algebra[9] given by where e_i are the so called braid generators (responsible for all of the above mixing) and q is the number of states

$$e_i e_j = e_j e_i, \quad |i - j| \ge 2,$$

$$e_i^2 = e_i,$$

$$e_i e_{i+1} e_i = \frac{1}{q} e_i,$$

(6)

in the Potts model. Jones had demonstrated how V_K is related to TL alegbra while Temperley and Lieb much earlier had demonstrated[9] how TL is related to the Potts model. Kauffman[5] found a much easier way to establish the above connection. In the next section, we shall use Kauffman approach.

ANALYTICAL PROOF OF DELBRÜCK CONJECTURE

Following Kauffman, we define new polynomial $\langle K \rangle$ via the set of axioms given below:

Axiom 1: $\langle O \rangle = 1$, where O is unknot

Axiom 2: $\langle O U K \rangle = \tilde{d} \langle K \rangle$, where U is usual set theory summation sign and $\tilde{d}$ is some number

Axiom 3: $\langle \asymp \rangle = A \langle \times \rangle + B \rangle (\rangle$

Axiom 4: $\langle \times \rangle = B \langle \times \rangle + A \langle) (\rangle$

with A, B and $\tilde{d}$ being, in general, arbitrary and independent numbers. To make $\langle K \rangle$ invariant with respect to action of Reidemeister moves, it is necessary to require $B = A^{-1}$ and $\tilde{d} = -A^2 = A^{-2}$. Thus corrected expression is given then by

$$f_K = \alpha^{-w(K)} \langle K \rangle$$

(7)

where $\alpha = -A^3$ and w(K) is writhe number connected with the torsion of the curve.[9,10] Kaufmann had shown that

$$f_K(t^{-\frac{1}{4}}) = V_K(t).$$

(8)

The polynomial $\langle K \rangle$ can be presented in a form of statistical mechanics grand partition function as

$$\langle K \rangle = \sum_{\{s\}} A^{n_A(s)} B^{n_B(s)} \tilde{d}^{|s|-1} \equiv Z \tag{9}$$

where summation $\{s\}$ takes place over 2^n states of 4-valent planar graph with n crossings, $n_A(s)$ and $n_B(s)$ correspond to the splice of A-type, i.e. $\asymp \Rightarrow \times$ and B-type, ie. $\asymp \Rightarrow)($, respectively, needed to obtain s disjoint closed curves from the graph corresponding to 3d knot.

Using the rules of statistical mechanics, we obtain:

$$\langle n_A \rangle = A \frac{\partial}{\partial A} lnZ \Big|_{B=A^{-1}} \tag{10}$$

$$\langle n_B \rangle = B \frac{\partial}{\partial B} lnZ \Big|_{B=A^{-1}}, \tag{11}$$

$$\langle s \rangle = \tilde{d} \frac{\partial}{\partial \tilde{d}} ln(\tilde{d}Z) \Big|_{B=A^{-1}}, \tag{12}$$

with additional condition: $\tilde{d} = -(A^2 + A^{-2})$. Comparison between Eq.s (7) and (10)-(12) leads us to conclusion that the above results will not change even if $\langle K \rangle$ is replaced by f_K and, whence, in view of Eq. (8), by V_K. Moreover, evidently, it is sufficient to assign only $\langle s \rangle$ in order for $\langle n_A \rangle$ and $\langle n_B \rangle$ to be determined.

ζ_N defined in the Introduction, can be written now as

$$\zeta_N^{-1} = \frac{lim}{n \to \infty} Z \Big|_{B=A^{-1}}, \quad \tilde{d} = -(A^2 + A^{-2}). \tag{13}$$

The above equation should be considered along with the condition given by Eq. (12) for n $\to \infty$. It is obvious that n=n(N) but the specific form of this N-dependence is generally unknown, except the fact that $n(N) < \frac{n}{\ell}$ where ℓ is the persistence length discussed in the Introduction. Combining Eq.s (8), (9)-(13) with the results of Jones it is rather easy to understand the physical meaning of Eq. (1) (for $\alpha = 0$). Indeed, if $Z \propto Z_{Potts}$, and $Z_{Potts} \equiv \exp\{n(N)f\}$, where f is free energy for the Potts model, then we obtain:

$$\zeta_N = const \ \exp\{n(N)f\} \tag{14}$$

which has the same functional form as Eq. (1). Additional arguments given in Ref. 7 produce the final result, Eq. (2), where the term $\exp\{-\frac{\pi}{6}\}$ comes from finite size corrections.[11] The discrepancy by the factor of 0.047 mentioned in the Introduction could originate from two independent sources. First, Windwer[3] apparently was not able to detect all knots of length N by the computer routine he had used, or second, the Jones polynomial V_K is unable to describe all possible knots of length N (which is, of course, always true).

DISCUSSION

Obtained results might have several immediate applications. First, in polymer physics, we have recently proposed new topological mechanism of reptation[10] which essentially uses the concepts and results of knot theory. The key experimental evidence in favor of topological mechanism lies in the fact that *both* ring and linear polymers exhibit almost identical visco-elastic behaviour.[12] In particular, for linear polymers the viscosity η in reptation regime scales as $\eta \propto N^{3.4}$ while for ring polymers the above exponent is slightly higher, e.g. 3.5-3.8. This higher exponent can be explained, based on Eq. (2), by the fact that practically all ring polymers are knotted. The degree of knottedness is given by Eq. (14) or by the explicit dependence of n on N. Because it is anticipated[5] that n depends on N rather weakly, this explains slightly higher exponents for rings (for more details see Ref.s. 10,12). Second, recently some papers have appeared[13] which claim that the presence of topological constraints such as links, in the case of percolation, affects the very nature of the percolation threshold Pc. Monte Carlo results which these authors have presented indicate that Pc-Pe $\leq 2.3 \cdot 10^{-7}$ where Pe is the so called entanglement threshold. Thus result (even though small) has been recently theoretically analyzed in Ref. 14 where it was shown that Pc is affected by the presence of entanglements.

The role of topological invariants in solid state physics in connection with dislocations, disclinations and point defects was recently discussed in 15. In Ref. 16, the topological invariants other than Gaussian linking numbers were discussed in connection with superconductivity, in Ref. 17, in connection with hydrodynamics, while in Ref. 18 in connection with fractional Hall effect.

ACKNOWLEDGEMENTS

The author gratefully acknowledges partial support by U. S. Army Research Office.

REFERENCES

1. M. Delbrück, Knotting problems in biology, in "Mathematical Problems in Biological Sciences", American Mathematical Society, Rhode Island (1962).
2. J. Michels and F. Weigel, Proc. Roy. Soc. London A403: 269 (1986).
3. S. Windwer, J. Chem. Phys. 93: 765 (1990).
4. P. de Gennes, "Scaling Concepts in Polymer Physics", Cornell U. Press, Ithaca (1979).
5. L. Kauffman, "Knots in Physics", World Scientific, Singapore (1991).
6. E. Witten, Comm. Math. Phys. 121: 351 (1989).
7. A. Kholodenko, J. Phys. A. (1992) to be published.
8. V. Jones, Bull. of the Amer. Math. Soc. 12: 103 (1985).

9. R. Baxter, "Exactly Solved Models in Statistical Mechanics", Academic Press, New York (1982).

10. A. Kholodenko, Phys. Lett. A159: 437 (1991).

11. M. Karowski, Nucl. Phys. B300: 473 (1988).

12. G. McKenna and B. Hostet her, Macromolecules 22: 1834 (1989).

13. Y. Kantor and G. Hassold, Phys. Rev. A40: 5334 (1989).

14. M. Azenman and G. Grimmet, J. Stat. Phys. 63: 817 (1991).

15. A. Holz, J. Phys. A25: L1 (1992).

16. T. Kephart, Phys. Rev. B32: 7583 (1985).

17. V. Arnold and B. Khesin, Ann. Rev. Fluid Mech. 24: 145 (1992).

18. Y. Hatsugai, M, Khomoto and Y. Wu, Phys. Rev. B43: 2661 (1991).

ELASTICITY OF A SELF-AVOIDING POLYMER

K.A. Dawson and D. Bratko*

Department of Chemistry
University of California
Berkeley, California 94720

Abstract

Elongation and concomitant lateral contraction of a dissolved macromolecule under traction is studied. A simple first order perturbative analysis is applied to a model polymer made up of repulsive beads connected by harmonic springs. The reference system consists of an anisotropic chain with transverse elasticity differing from the longitudinal one. The observed response to the tension conforms with the predictions from scaling arguments and agrees well with the simulation results for short homopolymers. Extensions to more complex systems are discussed.

Introduction

Understanding the elastic response of a macromolecule in solution is important in interpretations of a variety of phenomena ranging from protein folding to rubber elasticity or adhesion. While several features of polymer stretching can be explained by analogies with the random coil behavior, all interactions among the beads must be considered in more detailed studies of a real polymer. An important role is played by the excluded volume effect which by-and-large determines the dimensions of an unperturbed chain. The influence of the steric forces depends on the strength of the tension. Distinct stretching regimes have been predicted in the theory [1,2] and observed in the simulation [3]. At weak force f, with $\eta = \beta f R_o < 1$, where $\beta = 1/kT$ is the reciprocal thermal energy, $f=|f|$, $R_o =< h^2 >^{1/2} \cong aN^\alpha$ is the square root of the mean square end to end distance of the chain and N the degree of polymerization, one finds the linear response [1,2]:

$$< R_f > \cong \frac{\eta R_o}{3} \tag{1}$$

---

*Also affiliated to the Institute Josef Stefan, Ljubljana, Slovenia

Condensed Matter Theories, Vol. 8, Edited by
L. Blum and F.B. Malik, Plenum Press, New York, 1993

$<R_f> = <z_N-z_0>$ being the mean extension in the direction of the force **f** acting on the opposite ends of the chain. At stronger stretching, $\eta \geq 2$, the elasticity of the chain is reduced according to the relation

$$< R_f > \sim R_o \eta^\lambda$$

with   (2)

$$\lambda \equiv \frac{1}{\alpha} - 1 \leq 1$$

The value $\alpha = 3/5$ is widely accepted [2-8] to describe the swelling of polymers in good solvents. The strong-force power law of the form $R_f \propto f^{2/3}$ is therefore suggested. Different values of λ are expected with polymer models, characterized by other values of α [9-10]. A simple model of a self-avoiding macromolecule satisfying the Flory relation $\alpha \cong 3/5$ [4] was used in a study by Edwards and Singh [11], where polymer stretching was also briefly considered. A perturbative approach was applied to a model polymer consisting of N beads with effective volume u_2, connected by harmonic springs of spring constant equal to $3kT/2\ell^2$, ℓ being the average bond length in the absence of the excluded volume interaction. The reference system consisted of an ideal Gaussian coil with modified spring constant $3kT/2\ell\ell_1$, $\ell_1 > \ell$ being the average length of a bond in the actual polymer. The model gave the scaling law $R_0^2 = N\ell\ell_1 \propto N^{6/5}$ in agreement with other methods [2-8]. It was also proposed for use in studies of stretching at weak force when the deformation of the coil is not significant. In the present note, a generalization of the above method is described, whereby the difference in the polymer swelling along the two directions, parallel and perpendicular to the stretching force is taken into account. The reference Hamiltonian H_o consisting of two harmonic terms, corresponding to the extension in the two directions is therefore introduced. Distinct spring parameters ℓ_1 and p_1 determine the configuration fluctuations in these two directions. The extension $<R_f>$ along the direction of the force **f**, and the reduction of the lateral spread $< R_\perp^2 > = 3N\ell p_1 / 2$, are studied as functions of the force f. Numerical results for $<R_f>$ and $<R_\perp^2>$ are compared with Monte Carlo experiment [3] on a similar model. A fair agreement with the simulation results and with the scaling relations, Eqs. (1) and (2) for simple homopolymer systems is found. Extensions to more general models involving arbitrary interactions among the monomers in the chain are also discussed.

Analysis

A dissolved macromolecule is represented by a necklace of N equal beads connected by harmonic springs. The beads interact through excluded volume potential of the form $u_2\delta(\mathbf{r}_i - \mathbf{r}_j)$. The configuration dependent part of the Hamiltonian $H = \overline{H} - \mathbf{f}(\mathbf{r}_N - \mathbf{r}_o)$ is given by

$$\beta\overline{H} = \frac{3}{2\ell^2}\sum_{m=0}^{N-1}(\mathbf{r}_{m+1} - \mathbf{r}_m)^2 + \frac{u_2}{2}\sum_m\sum_{m'}\delta(\mathbf{r}_m - \mathbf{r}_{m'})$$ (3)

where r_m is the position of the bead m at given configuration and u_2 the apparent volume of the beads. The canonical average of any dynamic observable 0 is calculated using the probability distribution $P = e^{-\beta H_o}$, H_o being the reference Hamiltonian

$$\beta H_o = \frac{1}{2\ell} \sum_{m=0}^{N-1} \left\{ \frac{1}{p_1} \left[(x_{m+1} - x_m)^2 + (y_{m+1} - y_m)^2 \right] + \frac{1}{\ell_1} (z_{m+1} - z_m)^2 \right\} \tag{4}$$

and the direction z is taken to coincide with the direction of the force f acting on the ending units m=0 and m'=N. The difference between the elasticity parameters ℓ_1 and p_1 of the reference system implies the anisotropy of the polymer under the tension f. The quantities of our interest are the z projection of the end to end distance $<z_N-z_0>$ and lateral spread $< R_\perp^2 > = < (t_N - t_o)^2 > = < (x_N - x_o)^2 > + < (y_N - y_o)^2 > = 2N\ell p_1 / 3$. The mean extension

$$<z_N - z_o> = \frac{\text{Tr}\left[e^{-\beta H} (z_N - z_o) \right]}{\text{Tr } e^{-\beta H}} \tag{5}$$

can be determined by performing the averaging in the force-free ensemble, characterized by the Hamiltonian $\overline{H} = H + f \cdot (z_N - z_o)$, where $f(z_N-z_0)$ is the potential energy associated with the tension f:

$$<z_N - z_o> = \frac{\text{Tr}\left[e^{-\beta H} e^{\beta f(z_N-z_o)} (z_N - z_o) \right]}{\text{Tr}\left[e^{-\beta H} e^{\beta f(z_N-z_o)} \right]}$$

$$= \frac{< e^{\beta f(z_N-z_o)} (z_N - z_o) >_{\overline{H}}}{< e^{\beta f(z_N-z_o)} >_{\overline{H}}}$$

$$= \frac{\partial}{\partial f} \ln < e^{\beta f(z_N-z_o)} >_{\overline{H}} \tag{6}$$

Above, the brackets $<>$ denote the average at the true Hamiltonian H and $<>_{\overline{H}}$ corresponds to the averaging at the effective, force-free Hamiltonian $\overline{H}$. Replacing $\overline{H}$ by the reference Hamiltonian H_o, Eq. (4), that imposes a Gaussian distribution $P_o = e^{-\beta H_o}$, Eq. (6) simplifies to

$$<z_N - z_o> \cong \frac{\partial}{\partial f} \ln < e^{\beta f(z_N-z_o)} >_{H_o} = \frac{\partial}{\partial f} \ln \left[e^{\frac{1}{2}\beta^2 f^2 <(z_N-z_o)^2>_{H_o}} \right]$$

$$= \frac{1}{2} \frac{\partial}{\partial f} \left[\beta^2 f^2 < (z_N - z_o)^2 >_{H_o} \right] \tag{7}$$

$< (z_N - z_o)^2 >_{H_o}$ being the mean variance of z_N-z_0 in the reference system with Hamiltonian

H_o. It remains to determine the parameters ℓ_1 and p_1 of eq. (4) corresponding to the best agreement with the actual Hamiltonian H. We begin by noting that, for an arbitrary observable quantity 0, we may write

$$<0>=\frac{\text{Tr}\left[e^{-\beta(H-H_o)}e^{-\beta H_o}0\right]}{\text{Tr}\left[e^{-\beta(H-H_o)}e^{-\beta H_o}\right]}=\frac{<e^{-\beta(H-H_o)}0>_{H_o}}{<e^{-\beta(H-H_o)}>_{H_o}} \tag{8}$$

Expansion in cumulant series gives

$$<0>=<0>_{H_o}+\beta<0(H-H_o)>_{H_o}-\beta<0>_{H_o}<H-H_o>_{H_o}+\beta^2 0[(H-H_o)^2]+\cdots(9)$$

The average $<0>_{H_o}$ will represent a correct estimate of $<0>$ within the leading order of $(H-H_0)$, when H_0 satisfies the factorization [11,12]

$$<0(H-H_o)>_{H_o}-<0>_{H_o}<H-H_o>_{H_o}=0 \tag{10}$$

Being interested in calculation of extension $<z_N-z_0>$, (Eq. (7)), and lateral variance $<R_\perp^2>=<(t_N-t_0)^2>$, we are concerned with the generating functions $0_1=e^{\beta f(z_N-z_1)}$, $0_2=e^{\beta f(z_N-z_0)}(t_N-t_0)^2$. The reference Hamiltonian H_0 should therefore satisfy the relations

$$<e^{\beta f(z_N-z_o)}(\overline{H}-H_o)>_{H_o}-<e^{\beta f(z_N-z_o)}>_{H_o}<\overline{H}-H_o>=0 \tag{11}$$

$$<e^{\beta f(z_N-z_o)}(t_N-t_o)^2(\overline{H}-H_o)>_{H_o}-<e^{\beta f(z_N-z_o)}>_{H_o}<(t_N-t_o)^2>_{H_o}<\overline{H}-H_o>_{H_o}=0 \quad (12)$$

where the absence of direct coupling between the longitudinal and transverse terms of the reference Hamiltonian was taken into account. Eqs. (11) and (12) determine the values of the parameters ℓ_1 and p_1 at given force f. Combining Eqs. (11-12) and (4), we obtain

$$<e^{\beta f(z_N-z_o)}\left[\frac{1}{2\ell}\left(\frac{1}{\ell}-\frac{1}{\ell_1}\right)\sum_m(z_{m+1}-z_m)^2+\frac{u_2}{2}\sum_m\sum_{m'}\delta(\mathbf{r}_m-\mathbf{r}_{m'})\right]>_{H_o}$$

$$-<e^{\beta f(z_N-z_o)}>_{H_o}<\frac{1}{2\ell}\left(\frac{1}{\ell}-\frac{1}{\ell_1}\right)\sum_m(z_{m+1}-z_m)^2+\frac{u_2}{2}\sum_m\sum_{m'}\delta(\mathbf{r}_m-\mathbf{r}_{m'})>_{H_o}=0 \tag{13}$$

and

$$<e^{\beta f(z_N-z_o)}(t_N-t_o)^2[\frac{1}{2\ell}\left(\frac{1}{\ell}-\frac{1}{\ell_1}\right)\sum_m(z_{m+1}-z_m)^2+\frac{1}{\ell}\left(\frac{1}{\ell}-\frac{1}{p_1}\right)\sum_m(t_{m+1}-t_m)^2]$$

$$+\frac{u_2}{2}\sum_m\sum_{m'}\delta(\mathbf{r}_m-\mathbf{r}_{m'})]>_{H_o}-<e^{\beta f(z_N-z_o)}>_{H_o}\cdot<(t_N-t_o)^2>_{H_o}\cdot<\frac{1}{2\ell}\left(\frac{1}{\ell}-\frac{1}{\ell_1}\right)\sum_m(z_{m+1}-z_m)^2$$

$$+\frac{1}{\ell}\left(\frac{1}{\ell}-\frac{1}{p_1}\right)\sum_m (t_{m+1}-t_m)^2 + \frac{u_2}{2}\sum_m\sum_{m'}\delta(\mathbf{r}_m-\mathbf{r}_{m'})>_{H_o}=0 \tag{14}$$

In order to proceed, we expand the δ function in the plane wave and take advantage of the Gaussian form of P_o. Using relations

$$<(z_m-z_{m'})^2>_{H_o}=\frac{1}{3}\ell\ell_1|m-m'| \tag{15}$$

$$<(t_m-t_{m'})^2>_{H_o}=\frac{2}{3}\ell p_1|m-m'| \tag{16}$$

the average of $\delta(\mathbf{r}_m\text{-}\mathbf{r}_{m'})$ can be expressed as

$$<\delta(\mathbf{r}_m-\mathbf{r}_{m'})>=\left(\frac{3}{2\pi\ell}\right)^{\frac{3}{2}}\frac{1}{\ell_1^{1/2}p_1|m-m'|^{3/2}} \tag{17}$$

The averages of the terms containing $e^{\beta f(z_m-z_{m'})}$ take the form

$$<e^{\beta f(z_m-z_{m'})}>_{H_o}=e^{\frac{\beta^2 f^2}{2}<(z_m-z_{m'})^2>}=e^{\frac{\beta^2 f^2\ell\ell_1|m-m'|}{6}} \tag{18}$$

and

$$<e^{\beta f(z_N-z_o)}(z_m-z_{m'})^2>_{H_o}$$

$$=\lim_{g\to 0}\frac{\partial^2}{\partial g^2}<e^{\beta f(z_N-z_o)+\beta g(z_m-z_{m'})}>_{H_o}=e^{\frac{\beta^2 f^2}{2}<(z_N-z_o)^2>_{H_o}}\times \tag{19}$$

$$\left\{<(z_m-z_{m'})^2>_{H_o}+\frac{\beta^2 f^2}{4}\left[<(z_m-z_o)^2>_{H_o}+<(z_N-z_{m'})^2>_{H_o}-<(z_{m'}-z_o)^2>_{H_o}-<(z_N-z_m)^2>_{H_o}\right]^2\right\}$$

$$=e^{\frac{\beta^2 f^2\ell\ell_1 N}{6}}\left\{\frac{\ell\ell_1}{3}|m-m'|+\frac{\beta^2 f^2\ell\ell_1}{9}|m-m'|^2\right\}$$

and similar procedures [11] can be applied to other terms in Eqs. (13) and (14). One finally obtains:

$$\frac{N\beta^2 f^2\ell_1^2\ell}{6}\left[\frac{1}{\ell}-\frac{1}{\ell_1}\right]+\frac{u_2}{2}\left(\frac{3}{2\pi\ell}\right)^{\frac{3}{2}}\frac{1}{\ell_1^{\frac{1}{2}}}\cdot\frac{1}{p_1}\sum_m\sum_{m'}\frac{e^{\frac{\beta^2 f^2\ell\ell_1|m-m'|}{6}}-1}{|m-m'|^{\frac{3}{2}}}=0 \tag{20}$$

and

$$\frac{2Np_1^2}{3}\left[\frac{1}{\ell}-\frac{1}{p_1}\right]-\frac{u_2}{3\ell_1^{\frac{1}{2}}}\left(\frac{3}{2\pi\ell}\right)^{\frac{3}{2}}\sum_m\sum_{m'}\frac{e^{\frac{\beta^2 f^2\ell\ell_1|m-m'|}{6}}}{|m-m'|^{\frac{1}{2}}}=0 \tag{21}$$

511

At weak force f when the coil is virtually isotropic, $\ell_1 \to p_1$, Eq. (20) becomes a discretized analog of Eq. (B9) of ref [11]. In general cases, however, the longitudinal and the transverse stretching correspond to different spring parameters of the reference Hamiltonian. Eqs. (20) and (21) represent a system of equations to be solved to determine ℓ_1 and p_1 as functions of the force f. Newton-Raphson method has been found to provide a stable and rapid solution typically involving 10-15 iterations.

Results and Discussion

We present numerical results for the chains at two degrees of polymerization $N=10^4$ and $N=80$. The former value is typical for a real macromolecule and the latter was used for the sake of comparison with available Monte Carlo data [3]. The interaction parameter $u_2 = 4\pi\ell^3 / 3$ was used, thus making the excluded volume potential consistent with the mean field free energy in a gas of hard spheres of diameter ℓ. A series of calculations of at very weak force f, where $\ell_1 \to p_1$, was also carried out over a broad range of N. The scaling law

$$R_o^2 = aN^{2\alpha} \tag{22}$$

with a $\sim 1.44\ell^2$ and $\alpha = 0.596 \pm 0.003$ was found valid within the range $60 < N < 10^4$. The small deviation from the anticipated result $\alpha = 0.6$ [4,11] may probably serve as a measure of the accuracy of the numerical procedure. In Table 1, the calculated values of R_f, $< R_\perp^2 >$ and the exponents $\lambda = \partial \ln R_f / \partial \ln f$ and $\lambda' = \partial \ln < R_\perp^2 > / \partial \ln f$ at $N = 10^4$ are

Table 1. The elongation $R_f = < z_N - z_o >$ and lateral spread $< R_\perp^2 >$ as functions of the reduced tension $\eta = \beta f R_o$, at $N = 10^4$, $R_o = 290.5\ell$, $< R_{\perp,o}^2 > = 2R_o^2 / 3$, $\lambda = \partial \ln R_f / \partial \ln f$ and $\lambda' = \partial \ln < R_\perp^2 > / \partial \ln f$.

η	R_f / R_o	$< R_\perp^2 >/< R_{\perp,o}^2 >$	λ	$-\lambda'$
0.02905	0.00968	1.000	1.000	0.00026
0.2905	0.0968	0.9987	0.9997	0.026
1.453	0.482	0.965	0.992	0.059
2.905	0.950	0.891	0.954	0.182
5.81	1.762	0.738	0.824	0.333
14.53	3.527	0.536	0.726	0.344
29.05	5.752	0.424	0.692	0.333
58.1	9.260	0.337	0.676	0.332
145.3	17.14	0.247	0.666	0.356
290.5	26.83	0.189	0.634	0.399

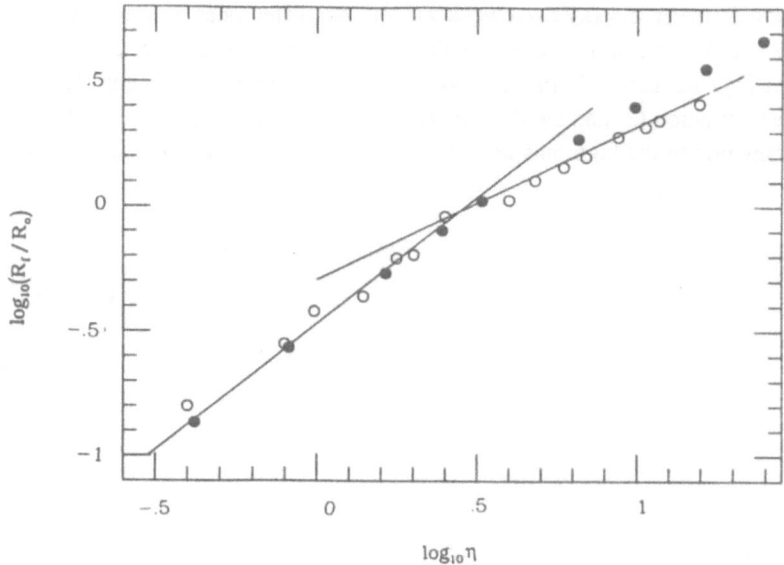

Figure 1. Relative elongation R_f / R_o due to the reduced tension $\eta = \beta R_o f$ at $N=80$. Open symbols denote the results of simulation [3] and solid circles are from the present calculation. Straight lines correspond to the slopes $R_f \propto f$ and $R_f \propto f^{2/3}$

collected. The range from vanishingly weak tension to strong stretching where the elongation approaches the contour length of the chain is considered. At weak force f, the linear response, Eq. (1), is observed. At elongations, comparable to the equilibrium end-to-end distance R_o, however, the reduction of the excluded volume effect due to the lower density of interacting beads becomes visible. As a result, the slope λ falls below unity and a lateral contraction takes place. The transition to the strong stretching regime with $\lambda \approx 2/3$ and $\lambda' \approx -1/3$ [1,2] is, however, not as sharp as found in the simulation [3]. At very strong tension $f > 1 / \beta \ell$, the scaling arguments leading to the above values of λ and λ' are no more valid [1-3]. At elongation $R_f \approx N\ell$, the calculated exponent λ passes through the minimum value close to 1/2 and then slowly increases towards unity, i.e. the value characteristic of a fully stretched chain obeying Hook law. The lateral spread $< R_l^2 >$, in turn, assymptotically approaches the value $3N\ell^2 / 2$, characteristic of the chain without excluded volume interaction. Hence, λ' vanishes at large elongation. Numerical results for these nonuniversal conditions are not presented since the model can no more describe the behavior of a real macromolecule.

In Figures 1 and 2 we compare the theoretical predictions and the results of Monte Carlo simulation [3]. The simulated chains were made up of 40 or 80 beads separated by links of fixed length ℓ. The beads interacted through a truncated Lennard-Jones potential [3] with the effective diameter $\sigma = 0.7\ell$. At weak tension, the differences between the Hamiltonian used in the simulation and that applied in the present work do not seem to affect the relative elongations R_f / R_o shown in Fiure. 1. An almost quantitative agreement between our model, the simulation [3] and the theoretical relation, Eq. (1), is observed below $\eta \approx 1.5$. A sharp transition to the high-tension law, Eq. (2), at $\eta \approx 2$ is seen in simulation.

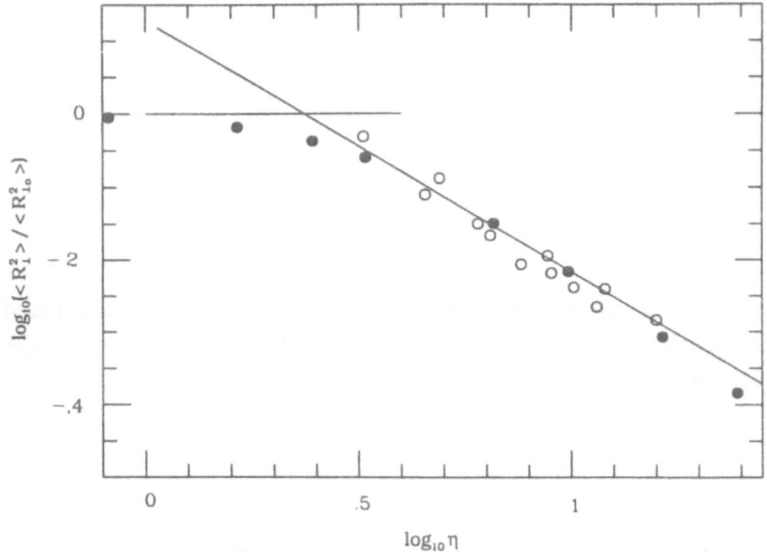

Figure 2. The lateral spread of the stretched chain $< R_\perp^2 >/< R_{\perp_o}^2 >$ as a function of the tension $\eta = \beta R_o f$. Open symbols represent simulation results for N=40 and 80 [3] and solid circles are from the present theory for N=80. Straight lines represent the relations $< R_\perp^2 >=$ const. and $< R_\perp^2 > \propto f^{-1/3}$.

A broader crossover towards the slope $\lambda \cong 2/3$ is predicted by the theory. It is not clear whether this difference can be rationalized by specific details of the models or by the approximations introduced in the perturbative analysis of this work.

Figure 2 illustrates the effect of the tension on the lateral spread $< R_\perp^2 >$. Except at very strong traction $\eta > 20$, the calculated results for N=80 show a good agreement with the Monte Carlo data as well as with the theoretical prediction $\lambda' \approx -1/3$ [1,2]. The deviations from this law observed beyond $\eta > 25$ reflect the limitations discussed in the preceding paragraph.

On the whole, the comparisons made in Figures 1 and 2 show the present perturbative method, combined with a simple polymer model to provide a reasonable description of the stress-strain behavior for simple polymers. Extensions to more complex and more realistic models are also possible. With systems characterized by arbitrary interactions among the bead pairs, the reference Hamiltonian H_o will, in general, contain NxN couplings but the periodicity in the bead-bead potential may substantially reduce the number of unknown spring constants V_{mm} . Transition to the Fourier space representation is convenient in these cases, since H_o then becomes a sum of independent contributions, $H_o = \Sigma \Delta V_q |r_q|^2$ [8,10,12]. Here, ΔV_q and r_q represent discrete Fourier transforms of coupling constants $V_{mm'}$ and positions r_m. A system of equations, analogous to Eq. (10) can be derived to calculate the components of ΔV_q in longitudinal and transverse directions. An application of this approach to collapsed polymer rings and helix-forming macromolecules will be described in a separate report.

Acknowledgment

D. B. is grateful to Professors L. Blum and F. B. Malik for support which made his participation and the presentation of this lecture at the XVI. International Workshop on Condensed Matter Theories possible. This work was supported by a David and Lucille Packard Foundation Award to K. A. D.

References

1. P. Pincus, *Macromolecules* 9:386 (1976).
2. P. G. de Gennes, "Scaling Concepts in Polymer Physics", Cornell University Press, Ithaca (1988).
3. I. Webman, J. L. Lebowitz and M. H. Kalos, *Phys. Rev. A* 23:316 (1981).
4. P. J. Flory, *J. Chem. Phys.* 17:303 (1949).
5. C. Tanford, "Physical Chemistry of Macromolecules", Wiley, New York (1969).
6. M. K. Kosmas and K. F. Freed, *J. Chem. Phys.* 68:4878 (1978).
7. M. Doi and S. F. Edwards, "The Theory of Polymer Dynamics", Oxford Science Publishers, New York (1989).
8. J. des Cloizeaux and G. Jannink, "Polymers in Solution. Their Modelling and Structure", Clarendon Press, Oxford (1989).
9. H. Reiss, *J. Chem. Phys.* 47:186 (1967).
10. J. des Cloizeaux, *J. Phys. Soc. Japan* 26:42 (1969); *J. Phys.* 31:715 (1970).
11. S. F. Edwards and P. Singh, *J. Chem. Soc., Faraday Transactions II* 75:1001 (1979).
12. D. Bratko and K. A. Dawson, in press.

DEFECT DIFFUSION WITH DRIFT AND THE KINETICS OF REACTIONS

C.A. Condat and E. Ulloa

Department of Physics
University of Puerto Rico
Mayagüez, PR 00681

INTRODUCTION

A considerable amount of work has been devoted to the study
of biased random walks. The effects of a random bias and of
random jump rates have been taken into account. Random
distributions of traps have also been added.[1-7] However, aside
from some calculations for walks on finite segments,[3,7] the
systems considered have been, in general, homogeneous on the
average. The purpose of this paper is to present results for
one-dimensional biased random walks where this "average
homogeneity" has been destroyed by the introduction of a
partially absorbing boundary. The boundary may represent a
defect-mediated relaxation process, as in the Glarum model of
dielectric relaxation,[8-10] or a chemical reaction occurring at
the surface of a sample if one of the reactants is allowed to
diffuse in the sample interior. It may also be a special site
in a protein where a process is triggered by the arrival of a
diffusing catalyst.[11-14] The sensitivity of ionic channels in
cell membranes to applied voltages is currently the object of
intense study.[15,16] Biological systems are of particular
interest, because of the many spatial constraints they impose
on particle motion.[17] A different application would be to the
problem of particle diffusion in a liquid when gravity effects
are nonnegligible.

The bias originates in a field that favours the hopping of
the walker in a given direction. The effects of even a small
field are dramatic. For instance, we will see that if the field
favours motion away from the reaction site, there is a finite
probability that the reaction will never occur. This is also
true if we let a random distribution of non-interacting walkers
compete to reach the boundary and give rise to the reaction. We
will also obtain the time dependence of the reaction rates,
whose asymptotic form is markedly different from that found for
the unbiased problem.

In the following sections we first present the model to be
studied and then analyze separately the one- and many-walker
problems.

The Master Equation and its Solution

We consider a walker on a semi-infinite one-dimensional lattice. Upon reaching site n = 1, the walker may go into an absorbing state with a rate α. The time units are chosen so that the jump rates to the left and the right are unity and Γ, respectively. If the bias is due to a static, uniform, field E and the jumps are thermally activated, the jump rate to the right is given by,

$$\Gamma = \exp(E/k_B T). \tag{1}$$

Here T is the temperature and k_B is Boltzmann's constant. The rate α may or may not depend on the applied field. This asymmetric random walk with an absorbing (or reacting) boundary is described by the master equation,

$$\frac{dp_{ns}}{dt} = p_{n+1,s} + \Gamma p_{n-1,s} - (1+\Gamma) p_{ns} \qquad (n>1), \tag{2a}$$

$$\frac{dp_{1s}}{dt} = p_{2s} - (\alpha+\Gamma) p_{1s} \tag{2b}$$

with the initial condition $p_{ns}(t=0) = \delta_{ns}$. Here $p_{ns}(t)$ is the probability that a walker at site s at time t=0 is at site n at time t.

The master equation can be symmetrized using the transformation,[18]

$$q_{ns} = p_{ns} (\sqrt{\Gamma})^{s-n} \exp[(\sqrt{\Gamma}-1)^2 t] \tag{3a}$$

$$\tau = \sqrt{\Gamma} t \tag{3b}$$

$$\beta = \frac{(\alpha-1)}{\sqrt{\Gamma}} + 1 \tag{3c}$$

We solve the resulting symmetrized master equation using an eigenfunction expansion.[19,20] We obtain,

$$p_{ns}(t) = (\sqrt{\Gamma})^{n-s} [C(t) + S(t)] \exp - (\sqrt{\Gamma}-1)^2 t, \tag{4}$$

where C(t) is the contribution of a continuum of eigenstates,

$$C(t) = \int_0^\pi \varphi_n \varphi_s e^{-\lambda t} dq, \tag{5}$$

and $S(t)$ is the contribution of the "surface state",[19,20]

$$S(t) = [\theta(\beta-2) + \theta(-\beta)] \chi_n \chi_s e^{-\mu t}. \tag{6}$$

Here $\theta(x)$ is the Heaviside step function and,

$$\lambda = 2\sqrt{\Gamma}(1-\cos q), \tag{7a}$$

$$\mu = \frac{(\alpha-1+\sqrt{\Gamma})^2}{\alpha-1}, \tag{7b}$$

$$\varphi_n = \sqrt{(2/\pi)} \frac{\sin(qn) + (\beta-1)\sin(q(n-1))}{\sqrt{1+2(\beta-1)\cos q + (\beta-1)^2}}, \tag{7c}$$

and,

$$\chi_n = \frac{\sqrt{\beta(\beta-2)}}{(1-\beta)^n}. \tag{7d}$$

Applications

Next we use Eq. (4) as a starting point to investigate the main properties of the diffusion-reaction process when a drift is present.

We assume that the walker departs from site s at t=0. The population of the absorbing state at time t, i.e., the probability that the reaction has occurred by the time t, is given by,

$$P_s^*(t) = \alpha \int_0^t p_{1s}(t') dt'. \tag{8}$$

The reaction rate can now be computed as,

$$f_s(t) = -\frac{dP_s^*}{dt}. \tag{9}$$

The function $P_1^*(t)$ is plotted in Fig. 1 as a function of time for $\alpha = 2$ and several values of Γ. While for $\Gamma<1$ (drift towards the reaction site) $P_1^*(t)$ increases very fast towards unity, for $\Gamma>1$ (drift away from the reaction site) $P_1^*(t)$ goes asymptotically to a value less than unity. This indicates that

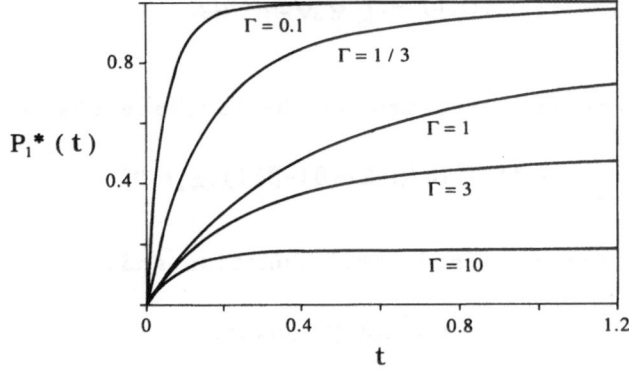

Figure 1. Probability that the reaction has occurred by time t if the walker departed from s=1 at t=0, for several values of Γ. Here α = 2.

there is a finite probability that the reaction will never occur. If Γ>1, the asymptotic form of $P_s^*(t)$ is given by,

$$P_s^*(\infty) = \frac{\alpha \Gamma^{1-s}}{\Gamma + \alpha - 1}.$$ (10)

Obviously, $P_s^*(\infty) = 1$ if Γ≤1: If the drift is towards the reaction site, the reaction will certainly occur. The probability that the transition has occurred by the time t=∞ is shown in Fig. 2 for s=1 and several values of the parameter α. We see that, if Γ>1, the decrease is quite fast for small values of α.

Using Eq. (9) we can obtain the reaction rate. Here we present only its long-time forms,

$$f_s \sim t^{-1.5} \exp(-(\sqrt{\Gamma}-1)^2 t) \qquad (\alpha > 1 - \sqrt{\Gamma}),$$ (11a)

and,

$$f_s \sim \exp\left(-\frac{\alpha(\alpha + \Gamma - 1)}{(\alpha - 1)} t\right) \qquad (\alpha < 1 - \sqrt{\Gamma}).$$ (11b)

Equation (11a) clearly reduces to the well-known power law $f_s = t^{-1.5}$ if there is no bias. The asymptotic form in Eq. (11a) arises from the band states [see Eq. (5)], while (11b) is due to the contribution of the "surface state", Eq.(6). We remark that an exponential form is obtained both for drift towards and away from the reaction site. The power law occurs only for the completely unbiased problem.

Using Eq. (1) we can write the reaction rate in terms of the field. The low and high field results are:

$$f_s \sim \exp[-(E/2k_B T)^2 t] \qquad (E \ll k_B T),$$ (12a)

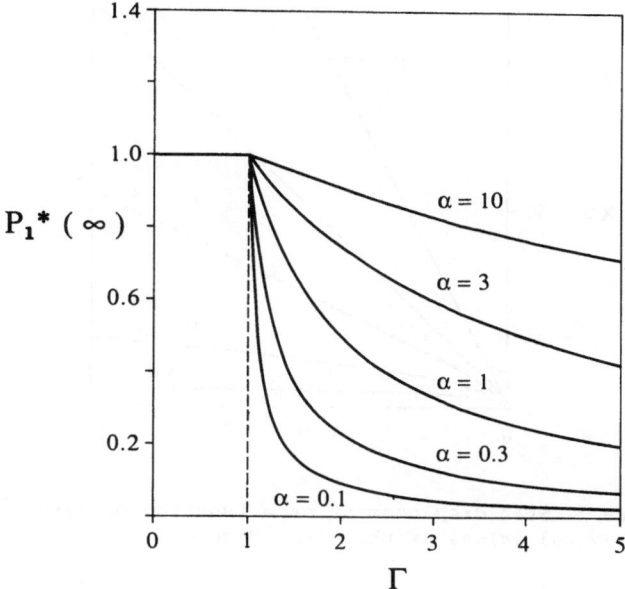

Figure 2. Probability that the reaction has taken place by $t = \infty$ as a function of the bias Γ for the indicated values of α.

and,

$$f_s \sim \exp\left[-\exp\left(E/k_BT\right) t\right] \qquad (E \gg k_BT) . \qquad (12b)$$

With equation (4) we can also calculate the moments. Assuming that the walker departs from site s=1,

$$\langle x^j \rangle = \frac{\sum_{n=1}^{\infty} (n-1)^j p_{n1}}{\sum_{n=1}^{\infty} p_{n1}} . \qquad (13)$$

The numerical results for the mean displacement and the variance are displayed as functions of the time in Figs. 3 and 4, respectively. We chose $\alpha = 1$ and several values of Γ. The long-time results for $\Gamma > 1$ are,

$$\langle x \rangle \sim v_d t, \qquad (14)$$

where $v_d = \Gamma - 1$ is the drift velocity. The variance is,

$$(\Delta x)^2 = \langle x^2 \rangle - \langle x \rangle^2 = 2Dt, \qquad (15)$$

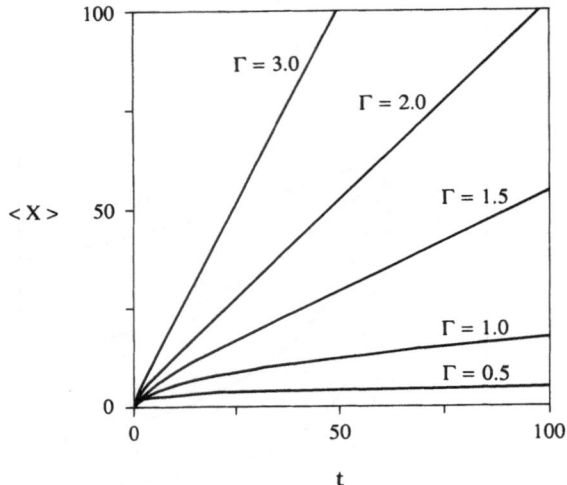

Figure 3. Mean displacement as a function of time for several values of the bias. Here $\alpha = 3$.

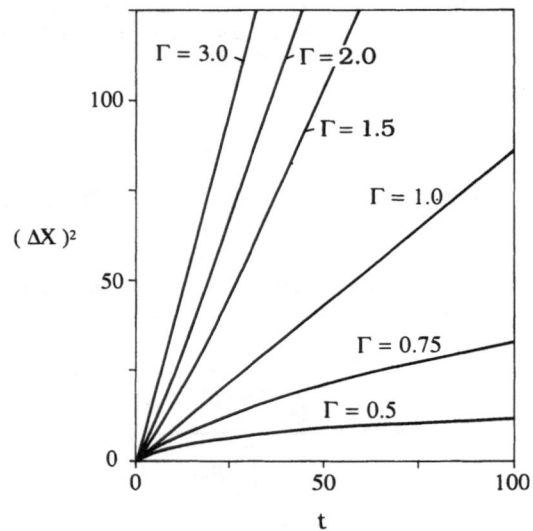

Figure 4. Variance as a function of time for $\alpha = 3$ and the indicated values of Γ.

with $D=(\Gamma+1)/2$ being the diffusion coefficient. For these leading terms in the asymptotic forms the absorption is irrelevant and v_d and D are identical to the results obtained in the absence of absorption.[2] Equations (14) and (15) can be obtained very simply if we start from the continuum limit of Eqs. (2). The asymptotic results for $\Gamma>1$ are evident in Figs. 3 and 4.

Using Eqs. (4) and (5), it is also easy to see that Einstein's relation between diffusion coefficient and mobility is obeyed only for low values of the field ($E<<k_BT$).

In this section we will assume that there is an initial concentration c of randomly distributed walkers ("scavengers"). These walkers compete to reach the target and induce the reaction. We are interested in the survival probability $\Phi(t)$ of the target, i.e. the probability that the reaction has not occurred by the time t; its time derivative is equal to minus the reaction rate. The function $\Phi(t)$ can be calculated using the equation,[21,22]

$$\phi(t) = \exp[-cQ(t)], \qquad (16)$$

where $Q(t)$ is given by,

$$Q(t) = \sum_{s=1}^{\bullet} P_s^*(t). \qquad (17)$$

Let us first review the well-known result obtained in the absence of drift.[9,20,22] At long times,

$$Q(t) \sim \frac{2\sqrt{t}}{\sqrt{\pi}}. \qquad (18)$$

The long-time fractional exponential form is due to those configurations where no scavenger is initially in the neighbourhood of the target. Note that the coefficient of the leading term in $Q(t)$ does not depend on α.

For $\Gamma > 1$ we find,

$$Q(t) = Q(\infty) + A(\alpha,\Gamma) t^{-1.5} \exp[-(\sqrt{\Gamma}-1)^2 t]. \qquad (19)$$

Here A is a not very illuminating function of α and Γ, and $Q(\infty)$ is given by,

$$Q(\infty) = \frac{\alpha\Gamma}{(\alpha+\Gamma-1)(\Gamma-1)}. \qquad (20)$$

The drift breaks the delicate balance existing in the unbiased case, where the absence of α in the leading term for $Q(t)$ is due to the exact compensation between a higher annihilation rate and a higher depletion of the survival probability.[10] The probability $\Phi(\infty)$ that the reaction will never occur is plotted in Fig. 5 as a function of the drift for several values of α. We see that the probability that the reaction will never occur is quite high, even for a small positive bias. Of course, the weakest the absorption rate α, the highest is $\Phi(\infty)$, since for small α the scavengers may reach many times the reaction site without triggering any response. If $\Gamma < 1$ the reaction must sooner or later occur.

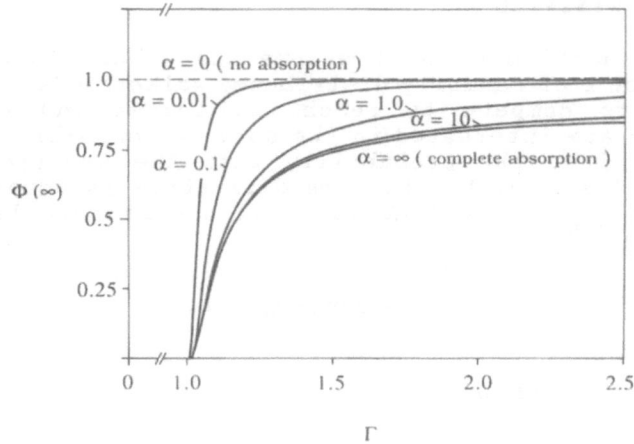

Figure 5. Probability that the reaction never occurs if we start with a random distribution of walkers whose concentration is c = 0.1. Results for several values of α are shown.

Let us finally make a remark concerning the "trapping" model, where a defect diffuses amidst a random distribution of traps. In this case, if a field is present, the fractional exponential $\exp(-at^{1/3})$ giving the long time form of the survival probability is replaced[4] by the simple exponential $\exp(-Kt)$.

CONCLUSIONS

We have computed exactly the properties of a diffusion-controlled reaction process in the presence of a uniform drift. The reaction can occur with a certain probability only when a walker or defect reaches a specific lattice site. The one- and many-walker problems were considered. We found explicit long-time results which are very different from those obtained in the absence of bias.

ACKNOWLEDGEMENT

This research was supported by an award from Research Corporation.

REFERENCES

1. J.W. Haus and K.W. Kehr, Diffusion in regular and disordered lattices, *Phys. Rep.* 150:263 (1987).
2. J. Bernasconi and W.R. Schneider, Random walks in one-dimensional random media, *Helv. Phys. Acta* 58:597 (1985).
3. V.V. Bryksin, Diffusion in an applied field in a one-dimensional disordered model of a broken network, *Sov. Phys. Solid State* 27:1083 (1985).
4. A. Aldea, M. Dulea, and P. Gartner, Long-time asymptotics in the one-dimensional trapping problem with large bias, *J. Stat. Phys.* 52: 1061 (1988).

5. C. Aslangul, N. Pottier, and D. Saint-James, Velocity and diffusion coefficient of a random asymmetric one-dimensional hopping model, *J. Phys. France* 50:899 (1989).

6. H. Weissman and M.J. Stephen, Drift and diffusion in a one-dimensional disordered system, *Phys. Rev. B* 40:1581 (1989).

7. A. Onipko, Charge-carrier-trapping kinetics in a chain with chaotically distributed traps and broken bonds: biased random walk model, *Phys. Rev. B* 43:13528 (1991).

8. S.H. Glarum, Dielectric relaxation of isoamyl bromide, *J. Chem. Phys.* 33:639 (1960).

9. P. Bordewijk, Defect-diffusion models of dielectric relaxation, *Chem. Phys. Lett.* 32:592 (1975).

10. C.A. Condat, Solution to the Glarum model with a finite relaxation rate, *Z.Phys. B* 77:313 (1989).

11. P. Läuger, Internal motions in proteins and gating kinetics of ionic channels, *Biophys. J.* 53:877 (1988).

12. C.A. Condat and J. Jäckle, Closed-time distribution of ionic channels, *Biophys. J.* 55:915 (1989).

13. R.E. Oswald, G.L. Millhauser, and A.A. Carter, Diffusion model in ion channel gating, *Biophys. J.* 59:1136 (1991).

14. W. Nadler and D.L. Stein, Biological transport processes and space dimension, *Proc. Natl. Acad. Sci. USA* 88:6750 (1991).

15. S. Cukierman, Asymmetric electrostatic effects on the gating of rat brain sodium channels in planar lipid membranes, *Biophys. J.* 60:845 (1991).

16. C.A. Vanderberg and F. Bezanilla, A sodium channel gating model based on single channel, macroscopic ionic, and gating currents in the squid axon, *Biophys. J.* 60:1511 (1991).

17. H. Qian, M.P. Sheetz, and E.L. Elson, Single particle tracking, *Biophys. J.* 60:910 (1991).

18. N.G. van Kampen. "Stochastic Processes in Physics and Chemistry," North-Holland, Amsterdam, (1981).

19. N.G. van Kampen and I. Oppenheim, Expansion of the master equation for one-dimensional random walks with boundary, *J. Math. Phys.* 13:842 (1971).

20. C.A. Condat, Defect diffusion and closed-time distributions for ionic channels in cell membranes, *Phys. Rev. A* 39:2112 (1989).

21. M. Tachiya, On the multi-step tunneling model for electron scavenging in low-temperature glasses, *Radiat. Phys. Chem.* 17:447 (1981).

22. M.F. Shlesinger and E.W. Montroll, On the Williams-Watts function of dielectric relaxation, *Proc. Natl. Acad. Sci. USA* 81:1280 (1984).

FLEXIBILITY OF POLYSACHARIDES USING MOLECULAR DYNAMICS

J.Raul Grigera[1,2], Cristina Donnamaría[1], and
Eduardo I. Howard[1]

[1]Instituto de Física de Líquidos y Sistemas Biológicos
(IFLYSIB) (CONICET-UNLP)
[2]Departamento de Ciencias Biológicas
Facultad de Ciencias Exactas, Universidad Nacional de La Plata
c.c. 565, 1900 La Plata, Argentina

INTRODUCTION

Polysacharides are part of the cell wall structure of plants and bacteria and have an important role as energetic reservoir in all living beings. Besides their participation from an energetic and structural point of view, they play a relevant part in the field of molecular recognition. In this field, knowledge of conformation and flexibility is becoming increasingly important. Proteins and nucleic acid are, in a sense, better known and their structure and dynamics have been widely studied by theoretical, experimental and simulation methods. The number of building blocks (nucleotides, aminoacids and monosacharides for nucleic acids, proteins and polysacharides, respectively) is very different for the three families. While there are only four nucleotides and about 20 aminoacids, monosacharides form a huge family. This is, probably, one of the drawbacks for a comprehensive study of these macromolecules. For a long time carbohydrates have been considered rigid and behaving in water as ideal solutions. In recent times, due to the new techniques introduced, these ideas are changing (Franks and Grigera 1989), although much work is still necessary.

Flexibilty can be studied by different experimental techniques, such as nuclear magnetic resonance, dielectric relaxation, viscosity, etc. Molecular dynamics computer simulation is also a valuable tecnique to be used, provided a good model and parametrization is implemented. In spite of the large amount of work done in the study of proteins by molecular dynamics simulation (MD), aplications of this technique to carbohydrates are relatively new (Brady 1986, Koelher *et al.*, 1988, Grigera 1988), although explosive.

The purpose of this article is to present some new data of mobility of trehalose in vacuum and in aqueous solution studied by molecular dynamics. Also, minimum energy calculations have been done with a force field different from that used in MD, to evaluate the sensitivity of the system to the model.

METHODS

Molecular dynamic simulation was performed using the GROMOS package (Biomos n.v., Groningen-Zurich) running in a VAX/750 and a VAX-Station/2000. Vacuum simulations were done on a free volume while the solution was simulated in a periodic box at contant pressure, constant temperature. Molecular mechanics calculations were performed with MM2(87) in a 386 personal computer with mathematical co-processor. Plots were done in a Houston Inst. Plotter or in a personal computer using PLOT88 (Plotworks Inc. Ramona).

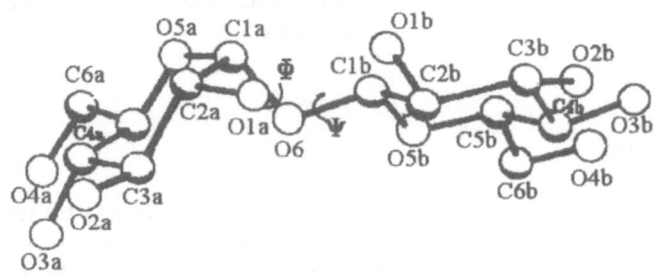

Figure 1. Schematic diagram of trehalose. All hydrogen atoms have been omited.

MODEL

Trehalose (α - D - glucopyranosil - (1 - 1) - α - D - glucopyranose) consists in two hexapyranose rings connected *via* 1-1 glycosidic linkage. In the molecular dynamics simulation the rings were kept rigid in 4C1 conformation by applying improper torsional potentials that avoid transitions between internal conformations. Improper torsion potentials were also used to keep the terahedral geometry of carbon atoms. No torsion potential was used between atoms involved in the glycosidic linkage. This point is important to remark because the conformation and mobility may be affected by the hindrance produced by torsional potentials. In our case it comes out from atom-atom interaction and solvent effects.

Force field parameters were used as given in GROMOS. Two sets of charges were used. In one of them (set 1) the atomic partial charges were taken as 1/10 of the value used by Sookee *et al.* (1988) for glucose, and it was used for vacuum simulation. The vacuum value considers an effective dielectric constant of 100 -- if one accepts the glucose charges as correct for each trehalose ring. The second set was computed with CNINDO/2RF and reduced. 1/10 for the vacuum calculation. For the molecular mechanics calculation the force field provieded by the MM2(87) program Allinger *et al.* (see MM2(87) Manual 1991) and the charges corresponding to set 1 were used.

The main difference with the force field used for MD is that in this case rings are no longer rigid and the glicosidic linkage is affected by torsion potentials.

Figure 1 shows an schematic diagram of threhalose. Hydrogens were omited.

The water model

Water was taken as the SPC/E model (Berendsen *et al.* 1984). It consist of an effective potential with three point charges. The point with negative charge (oxygen) has a repulsion-atraction potential of Lenard-Jones type. The other two (positive) charges occuppy the positions of the protons and do not have London-type forces acting on them. The model, though simple, reproduces the oxygen-oxygen radial distribution function acceptably, gives the right density for the liquids, and the computed diffusion coefficient agrees very well with experimental data.

RESULTS AND DISCUSSION

We have taken the rotation around the glycosidic bond as a measure of the overall mobility of trehalose. It is clear that individual rings also have their own mobility, but looking from the point of view of a linear molecule, the relevant movement will be the flexibility in the joining points of monomers. Following this criterium we record the trajectories of the two torsion angles of the glycosidic bond, namely C_{2a}-C_{1a}-O_6-C_{1b} (Φ) and C_{1a}-O_6-C_{1b}-C_{2b} (ψ). Figure 2 shows such trajectories for different cases. In figure 2a we see the trajectory of Φ and ψ angles for trehalose in vacuum as given by molecular dynamics at T=300 °K for charge set 1. Both angles fluctuate around almost the same mean value. The average values of 100 ps simulation time are for Φ, 224.9 (RMS=14.8) and 224.7 (RMS=15.3) for ψ.

These angles correspond to an almost fully extended conformation (for the definition used for the angles the ideal fully extended form corresponds to 180 for both angles).

It is seen that during 100 ps only one conformation has been accesed (besides the obvious fluctuation). This may be due to the very existence of only one probable conformation or to a very low transition probability. We have explored this situation in two ways. First by making the simulation at a higher temperature and secondly by calculating the minimum energy map.

Figure 2b shows the angles trajectories for 100 ps at T=3000 °K. To mantain the stability we had to reduce the time step to 0.0006 ps and increase the restoring force of angles to 4000 Kcal/mol/rad^2. (See Note 1)

The curves of figure 2b show less noise because of the larger sampling intrerval, otherwise no difference appears. This results induce to think that no other accessible conformation is present. Early studies of oligosaccharide conformation were done by using molecular mechanics (MM) with rigid ring models, as in our case. In most cases the energy surface shows more than one minimum, but the height of the barriers precludes the accesibility to most of them. The inclusion of ring flexibility produces a relaxed map allowing transition between substates. To test the situation we have made MM calculation with flexible rings.

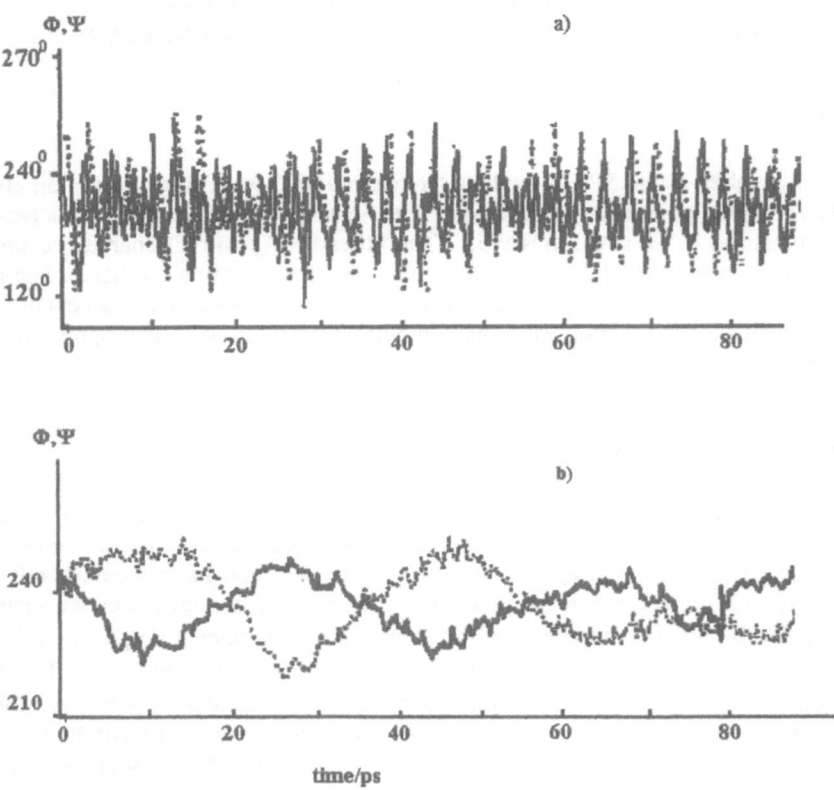

Figure 2. Trajectories of dihedral angles of trehalose during 100 ps. in vacuum with 'effective' dielectric constant at 300 °K. b) The same case at 3000 °K.

The minimum energy surface is shown in figure 3. It should be understood that the minimum energy map does not represent exactly the same situation that the molecular dynamics simulation does. While in the later we are dealing with the structure and dynamics at a certain temperature, in which the equilibrium will be determined by the free energy, in the former we consider the minimum potential energy at 0 K.

In the energy surface there is but one global minimum. The surface is relatively smooth, with no local minima. The global minimum corresponds to Φ=200.4 and ψ=198.2 with E=21.57 Kcal/mol. This confirms the presumption that only one conformation is present in the system.

Figure 4 shows the contour map of the same data of figure 3, with a number of angle trajectories of MD superimposed. It can be seen that the MD trajectories are close to the global minimum obtained by MM. In a sense this is striking since we are dealing with a different force field. The sensitivity of MD to the force field., reflecting the mobility of oligosaccharides, has been reported (Brady 1991). The main point stressed by Brady is the

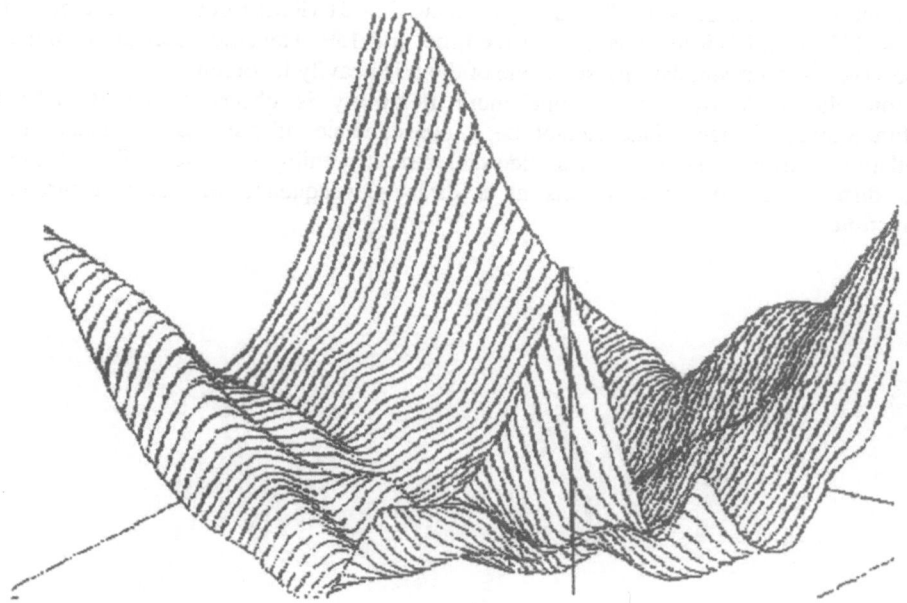

Figure 3. Energy surface for the glycosidic bond of trehalose as obtained with MM2.

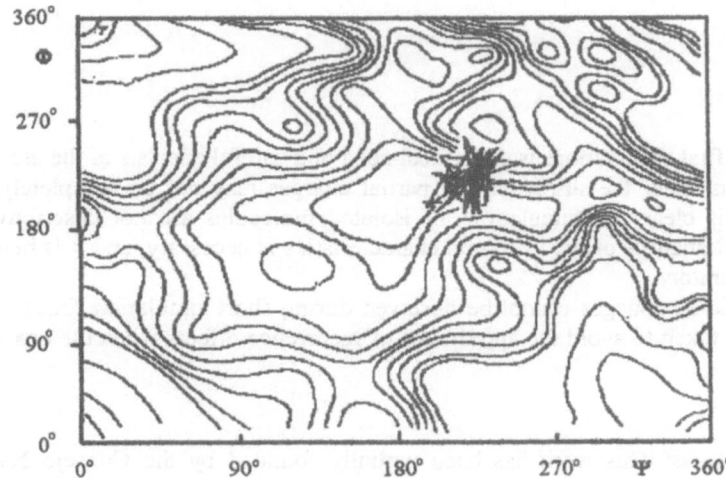

Figure 4. Contour plot corresponding to the energy surface of Figure 3.

assignment of charges. The present results indicate that, using the same charges, the remaining properties of force field are not critical.

The dihedral angle trajectories are shown in figure 5. We can see two different conformations. The dihedral Φ is always around 234.01 (RMS) but ψ stays around two values: 255, in which it remains most of the time, and 159. Transition betwen them cannot be observed in short simulations, since one of them is heavily favoured.

From the angle trajectories some more flexibility is observed, appart from the conformational changes. This cannot be a consequence of the charges, since in the simulation of isolated molecules a pseudo-dielectric screening is applied. The accesiblity of a different conformation results as a direct consequence of the water-trehalose interaction.

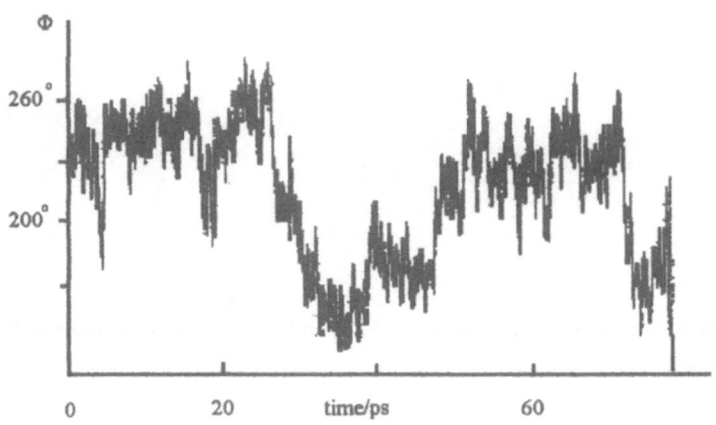

Figure 5a. Trajectories of dihedral angles for the simulation of trehalose in water at 300 K.

CONCLUSION

One of the first observations is the critical need of a careful analysis of the assignment of charges. Even under the suspicion that partial charges may not be completely correctly computed, it is clear that simulations of isolated molecules are not descriptive of real systems. A full analysis of the hydration characteristics is necessary, and it is being carried out in our laboratory.

Conformational changes cannot be observed during short simulation times. However, care has to be taken to avoid the appearance of recurrence effects when the box size is not large enough.

Acknowledgments: This work has been partially founded by the Consejo Nacional de Investigaciones Científicas y Técnicas (CONICET) of Argentina. JRG and CD are members of the Carrera del Investigador of CONICET and Comisión de Investigaciones

Vacuum molecular dynamics simulation was repeated for the second set of charges. In that case also one conformation is observed, but the equilibrium angles are different. We get $\Phi = 271$ (RMS=12.6), almost as in the previous case, and $\psi = 229$ (RMS=8.8). This change can be only attributed to the different charge set. The importance of solvents has allways been recognized. Moreover, it seems that, at least for some polyalcohols, the particular characteristics of solvent molecules may make large differences. That is the case for mannitol (Grigera 1988) for which the presence of water keeps it elongated, while in vacuum or in a non-polar solvent it becomes sickled. The MD simulation including water explicitly was imperative. It is worth to note in passing that most of the calculations 'in vacuo' are not completely in such state. Most force fields are based in data from aqueous solutions, which means that they are 'equivalent potentials'. However, they lack the specific interaction between water and the different groups of the molecule under study. The example of mannitol is illustrative, but certainly not the only one.

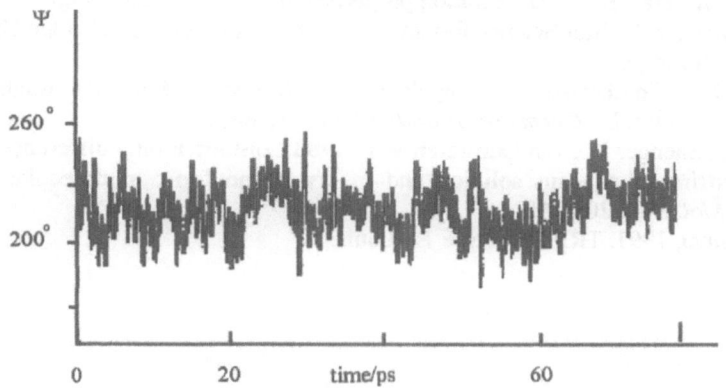

Figure 5b. Trajectories of dihedral angles for the simulation of trehalose in water at 300 K(continuation).

Simulation with solvent

One molecule of trehalose and 234 SPC/E water molecules were simulated in a rectangular box ($1.786 \times 1.813 \times 2.283$ nm^3) using the charges obtained by CNDO. From the point of view of solute-solute interaction such a solution can be considered as being at infinite dilution. However, water molecules are interacting with the threhalose of the box and its images, which may produce some indirect effects. From this point of view the solution can be taken as being of 0.237 mol/l.

ACKNOWLEDGEMENTS

Científicas of the Province of Buenos Aires, Argentina, respectively. EIH is fellow of CONICET. We wish to thank Prof. Blum for his interest in the work and help as editor.

Note 1. This computation can also be performed by increasing the atomic masses by the same factor than the temperature. The procedure is sometimes called 'weighted molecular dynamics'. It may allow to access conformations that cannot be reached at lower temperatures, but *it is not* a different method and there is no reason to give an special name.

REFERENCES

Berendsen H.J.C., Grigera J.R., Straatsma T., 1987, The missing term in effective pair potentials. *J.Phys. Chem.* **91**, 6269.

Brady J.W., 1986, Molecular dynamics simulation of α-D-glucose. *J.Am. Chem. Soc.***108**: 8153.

Brady J.W., 1991, Theroretical studies of oligosaccharides structure and conformational dynamics. *Current Opinion in Struc. Biol.* **1**, 711.

Franks F., and Grigera J.R., 1990, Solution properties of low molecular weight polyhydroxy compounds., in " Water Science Review" vol 5, F. Franks, ed., Cambridge UniversitY Press. Cambridge.

Grigera J.R. 1988, Conformation of polyols in water. Molecular-dynamics simulation of mannitol y sorbitol. *J.Chem.Soc. Faraday Trans. 1*, **84**, 2603.

Koelher J.E.H., Saenger W., van Gunsteren W.F.,1988,Conformational differences between -ciclodextrine in aqueous solution and in crystalline form: a molecular dynamics study. *J.Mol. Biol.* **203**:241.

MM2(87) Manual, 1991. TRIPOS Assoc. St. Louis.

SIMPLE MODELS OF THE INTERMOLECULAR POTENTIAL

FOR THE CONDENSED PHASES OF C_{60}

Z. Gamba* and M. L. Klein**

*División Física del Sólido, Comisión Nacional de Energía Atómica
Av. del Libertador 8250, 1429 Buenos Aires, Argentina
**Deparment of Chemistry, University of Pennsylvania
Philadelphia, 19104 - 6323, USA

INTRODUCTION

C_{60} is the most spherical of the family of fullerenes, and certainly the most studied since Kratschner et al. reported that it can be easily prepared in macroscopic quantities.[1] The point group symmetry of C_{60} is I_h: it contains 15 two- fold axes C_2, 12 five-fold axes C_5, 20 three-fold axes C_3 and an inversion center at the origin. The simplest way to understand the molecular structure is to start from an icosahedron (12 vertices, 20 triangular faces, Fig.1) and cut off the 12 corners, thereby obtaining the 12 pentagonal faces. The atomic orbital hybridization is such that bonds around hexagons are alternatively short (double) and large (single), maintaining then the C_3 symmetry; the bonds around pentagons are all large. The C-C bond lenghts are, respectively, 1.455 and 1.391Å and the molecular radius is 3.55Å [2-4].

Here we are interested in the condensed phases of pure C_{60}. At room temperature this compound packs in an orientationally disordered phase O_h^5 (Fm3m), with centers of mass located on a fcc lattice, a=14.17Å [2,3]. The heat of sublimation at 707K is 168 kJ/mol [5].

At 249K a weak first order phase transition takes place, with a heat of transition of 2.71(6) kJ/mol [8]. The low temperature phase is cubic: T_h^6 (Pa3) with Z=4. [2-4] NMR measurements indicate nearly continous rotational diffusion in the high temperature phase and large amplitude reorientations even down to about 140K [6].

At 11K the crystal structure is ordered, with a=14.04(1)Å. The relative orientation of neighboring molecules is such that the electron-rich short bond of one molecule faces the electron- poor pentagon center of other one, the electrostatic intermolecular interactions are thus minimized[4]. This molecular orientation can be achieved in the Pa3 structure by a 98 degree anticlockwise rotation of the molecule at the origin, around the [1,1,1] crystallographic C_3 axis of the primitive unit cell; the other molecules are obtained by the corresponding symmetry relationships. The starting configuration for the rotation of the molecule at the origin, Fig.1, corresponds to the T_h^3 (Fm3) structure, with the molecular C_2 axes aligned along the [1,0,0] directions[4]. Neutron powder diffraction measurements were analized using this orientational model, and the refinement factor showed a clear minimum at 98 degrees, implying that the long range order is as described above[4].

Condensed Matter Theories, Vol. 8, Edited by
L. Blum and F.B. Malik, Plenum Press, New York, 1993

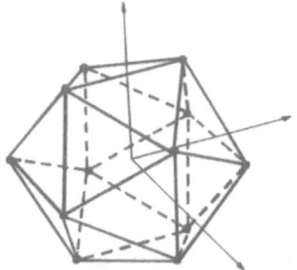

Figure 1. Simplified model
of 12 interaction centers[13].

Several calculations have also been performed on the solid phases of pure C_{60}. Ab initio molecular dynamics (MD) simulations[7], with one molecule per unit cell and periodic boundary conditions, showed that at very high temperatures the molecular vibrations cause large molecular distortions, but the cage structure and its average diameter is preserved.

Cheng and Klein performed classical constant- pressure MD calculations for pure[8,9] and alkali doped[10] solid phases of C_{60}. The molecules were considered as rigid bodies, and their interactions given by an intermolecular potential of the atom- atom type, between the 60 C atoms of neighboring molecules. This model implies the calculation of 3600 terms for each molecule- molecule interaction that is taken into account. Due to this problem, many of the calculations had to be performed in a sample of 32 C_{60} molecules. In this way, they were able to reproduce the molecular dynamics at high and low temperatures, packing energies, crystal structures for doped $K_n C_{60}$, n=1,6, crystals and high temperature unit cell parameters. One drawback of this model is that the low temperature unit cell of pure C_{60} turns out to be tetragonal.[9] This problem is most probably due to the intermolecular potential model used. Figure 2 shows the packing energy calculated in a Pa3 structure, with this model, as a function of the anticlockwise rotation around the C_3 crystallographic axis [1,1,1]. It can be seen that the calculated curve has a minimum around 38 deg, with a secondary minimum at the experimental angle. At 38 degrees, the relative orientation of two neighboring molecules is such that the short bond of one molecule nearly faces a hexagon of the other one; this short range structure leads to a distortion of the crystallographic cell.

Very recently, Sprik et al.[11] proposed a refined intermolecular potential model which reproduced not only all the experimental properties well covered by the first model, but also the low temperature crystal structure. This intermolecular potential model consists of 90 charged interaction centers: to the 60 Lennard- Jones (LJ) atom- atom interactions of Ref. 8, they added 30 LJ interactions centers and 30 charges in the middle of the short bonds (to simulate the dispersion and electrostatic interactions of the electron-rich short bonds). The 60 C are charged, in order to achieve molecular neutrality. No problems related with unit cell deformation were found this time. Figure 3 shows the curve obtained for this potential model, calculated as in Fig.2. It can be seen that the electrostatic contribution is relatively small, but the minimum of the curve is now at 98 degrees, the same as determined by X-ray measurements[4].

In our study we searched for a simplified intermolecular potential model that would allow the calculation of large samples of molecules. This could be very useful for studying more accurately many problems related to the collective behaviour of C_{60}. We are thinking, for example, in an accurate calculation of the lattice vibrational density of states, phase diagram of C_{60}, Gibbs free energies for pure and doped phases, the growth of ordered and disordered phases of C_{60} in multilayers,[12] etc. This simple model is also

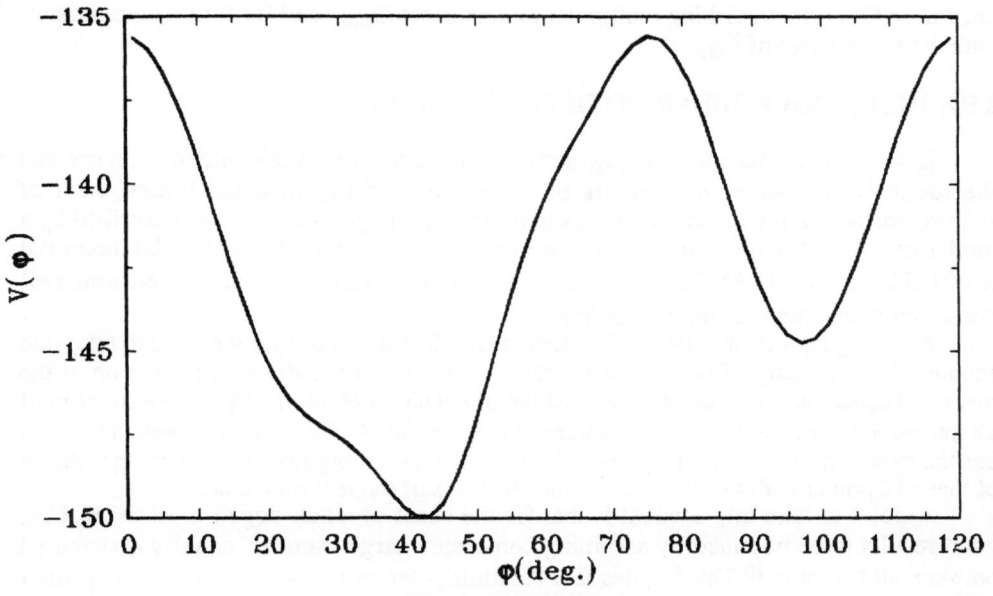

Figure 2. Model of 60 LJ
interaction centers[8,9].

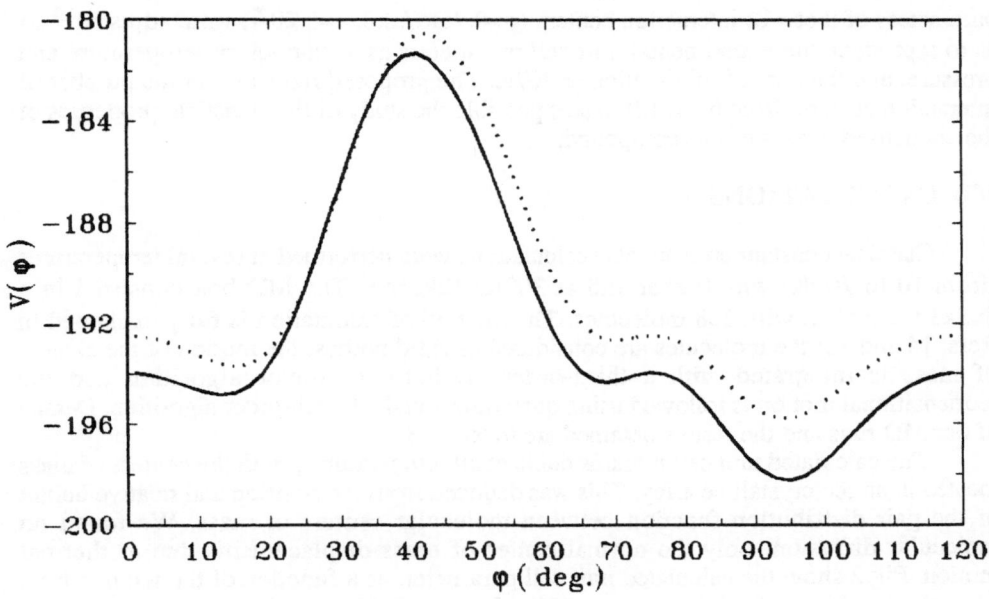

Figure 3. Model of 90 charged sites[11],
___total energy, ... atom-atom term .

of intrinsic interest. For example, a MD study of molecules with symmetry of order 2^6, i.e. whose first non- vanishing multipolar moments are Q_{6m}, and for future comparisons with the real system of C_{60}.

THE INTERMOLECULAR POTENTIAL MODEL

If we take into account that C_{60} is very nearly a spherical molecule, we can see that the rough behaviour of the centers of mass (i.e. packing in a fcc lattice, heat of sublimation and unit cell volume versus temperature and pressure) can be described by a model of spherical molecules of mass m=720 au., interacting with a LJ potential (σ=21kJ/mol, ε=9.15Å). Figure 4 shows the experimental and calculated unit cell parameter using this first approximation[13].

For studying the orientationally ordered and disordered phases we need to take into account the symmetry of the isolated molecule. In a second order approximation to the real intermolecular potential, we replaced the molecule of 60 atoms by an icosahedron of 12 interaction sites, each one in the centre of a pentagon.[13] This simple molecular model has the symmetry of the full molecule. There is also no ambiguity in the correspondence of these 12 points with the positions of the 60 atoms of the real molecule.

A point to take into account is that for the cases of alkali doped crystals, $K_x C_{60}$, good results were obtained by assuming complete charge transfer, equally distributed between all C atoms.[10] This implies that the multipolar moments Q_{6m}, are the first ones different from zero (except for the total charge, the moment Q_0). In our model of an icosahedron, these multipolar moments can be reproduced by locating the 12 sites inside the molecule (~ 30%), each one with a charge 5e/60.

Figure 1 shows the simplified geometry of the molecule, the coordinates are (with normalized radii): (x,0,z), (z,x,0), (0,z,x), (-x,0,z), (z,-x,0), (0,z,-x), and those generated by the inversion center at the origin; x=0.52573 and $x^2+z^2=1$. These interaction centers are located at 2.546Å from the origin, so as to reproduce the electrostatic multipolar moments Q_{6m} of the real molecule, when it is charged. The LJ parameters of these 12 interaction centers (σ=0.75kJ/mol, ε=6.20Å) were adjusted[13] so as to reproduce the experimental unit cell parameters as a function of temperature and pressure, and the heat of sublimation at 707K. The proposed reduction in the number of interaction centers (from 60 to 12) makes possible the study of the statistical properties of the condensed phases of this compound.

MD CALCULATIONS

Classical constant pressure MD calculations were performed at several temperatures (from 10 to 700K) with 0 kbar and at 300K, 12kbar.[13] The MD box consisted in a 3x3x3 fcc lattice, with 108 molecules. The method of calculation is fully explained in Refs. 14 and 15: the molecules are considered as rigid bodies, the motion of the centers of mass is integrated with a third-order predictor corrector algorithm and the reorientational motion is followed using quaternions and a fourth-order algorithm. Details of our MD runs and the results obtained are in Ref. 13.

The calculated unit cell remains cubic at all temperatures, with the centers of mass located in an fcc crystalline array. This was deduced from the position and relative height of the pair distribution function between molecular centers of mass. We found no molecular diffusion, only the normal center of mass displacements due to thermal motion. Fig.2 show the calculated unit cell parameter, as a function of the temperature. The calculated heat of sublimation is 175(1)kJ/mol at 740K, which compares reasonably well with an experimental value of 168(5)kJ/mol at 707K. [5]

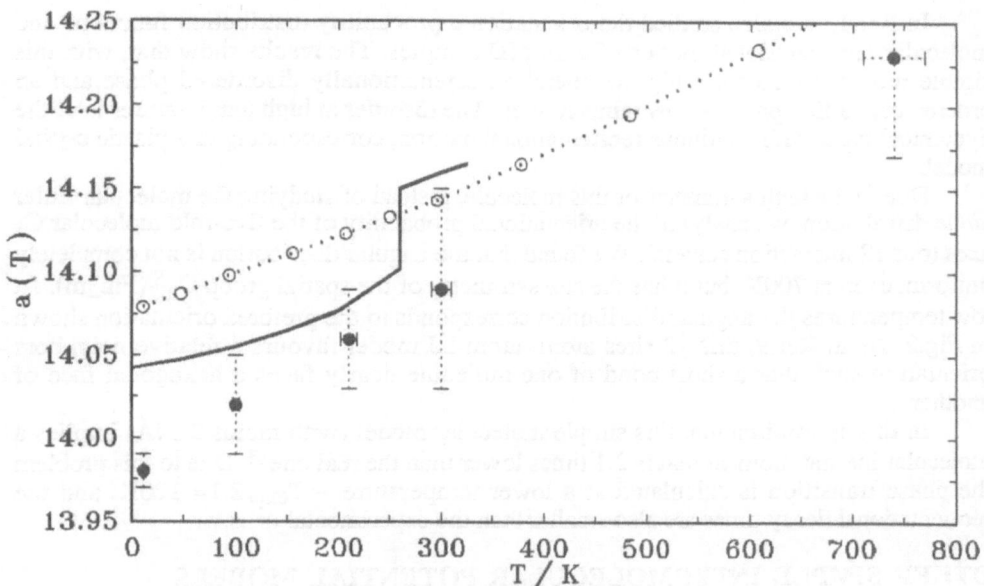

Figure 4. Lattice constant a(T); ___ experimental[16],...one LJ site, * 12 LJ sites.[13]

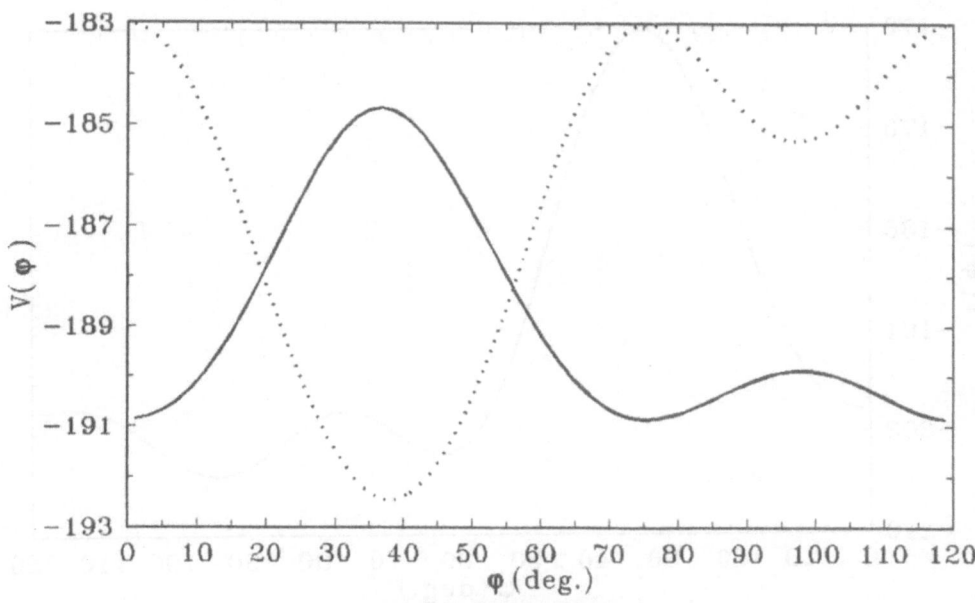

Figure 5. ....Icosahedron model,
___ 20 sites- polyhedra model.

In Ref.13 we also studied the orientational probability distribution functions and molecular reorientational motions for all MD samples. The results show that, with this simple model we obtain a high temperature orientationally disordered phase and an ordered crystalline phase at low temperatures. The disorder at high temperatures is of the dynamical type, with continous reorientational motion, corresponding to a plastic crystal model.

Due to the high symmetry of this molecule, instead of studying the molecular Euler angle distribution, we analyzed the orientational probability of the five-fold molecular C_5 axes (our 12 interaction centers). We found that the angular distribution is not completely uniform, even at 700K, but it has the site symmetry of the spatial group O_h^5 (Fm$\underline{3}$m). At low temperatures the angular distribution corresponds to the prefered orientation shown in Fig.2. As in Ref.9, this 12 sites atom- atom LJ model favours a relative neighbors orientation such that a short bond of one molecule nearly faces a hexagonal face of another.

In this approximation, this simple molecular model (with radius 2.54Å) implies a molecular inertial moment that is 2.1 times lower than the real one[13]. Due to this problem the phase transition is calculated at a lower temperature ~ $T_{exp}/2.1= 120K$, and the reorientational decay times are also smaller than the experimental ones.

OTHER SIMPLE INTERMOLECULAR POTENTIAL MODELS

In this section we present other simple models that we studied in order to reproduce the low temperature experimental crystal structure. By considering that the low temperature phase is ordered, different intermolecular potential models can be analyzed with the following procedure: a) search of the minimum packing energy crystal structure, b) plot of the packing energy as a function of the molecular rotation around the crystallographic C_3 axis. It is the plot shown in Figs.2, 3 and also used in the

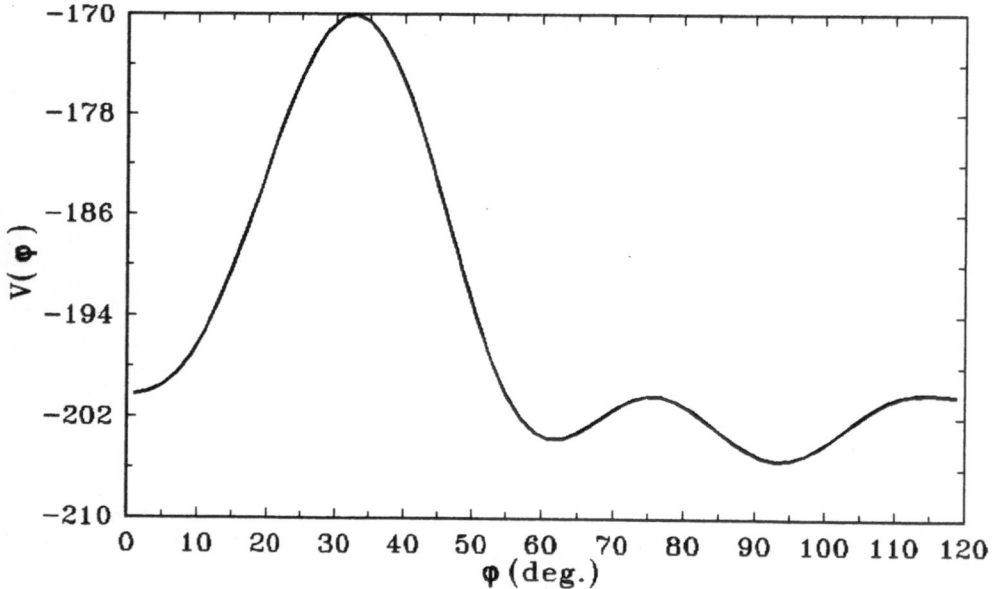

Figure 6. electrostatic model

experimental work of David et al.[16] Using this procedure we studied several cases, finding that, in general, they can be grouped in three typical models:

a) **Icosahedral model:** First we improved the model presented in Ref.13, which has as main drawback the small molecular radius. The new version is a model with one LJ interaction site (σ=19.3kJ/mol , ε=9.15Å) in the origin, which carries the main interaction term, and a perturbation of 12 LJ sites at the molecular radius = 3.55Å (σ=0.02kJ/mol, ε=4.30Å). Figure 5a shows the plot obtained for this model, the main minimum is always at ~38 deg., and the secondary at ~98 degrees. Although the value of these minima can change, we could not find a set of atom - atom LJ parameters that reverses the relative values of the two minima.

The same pattern can be obtained for a model of 60 LJ interaction centers at the C atoms, or with 60 interaction sites in the middle of the large bonds. It should be noticed that the molecular inertial moment has now the correct value. The problem of reproducing the molecular electrostatic multipolar moments when the molecule is charged (in doped samples, for example) can be solved by locating smaller charges at the 12 sites, and balancing the total molecular charge with an extra charge at the center of mass. This procedure seems to be more adequate than reducing the molecular radius in order to simulate any molecular multipole.

b) **20 site-polyhedra:** Following the idea of model a), the next model in complexity is to take into account the molecular symmetry with 20 LJ interaction centers located at the center of the hexagonal faces. The LJ site at origin (σ=16.0kJ/mol, ε=9.15Å) is supplemented with 20 LJ sites of parameters σ=0.01kJ/mol, ε=4.30Å. The obtained pattern is shown in Fig. 5b. This curve has two equivalent minimums at ~70 and 120 deg, their value and exact position depend on the set of LJ parameters, but the general pattern (the two minima at both sides of an angle ~95 deg) do not change. A quite similar pattern can be obtained for a model of 30 interaction centers in the middle of the short bonds.

c) **Electrostatic model:** In this model the atom- atom term is given by a LJ potential with σ=19.3kJ/mol, ε=9.15Å at the origin (a site with mass = 0. au.), and all the anisotropy of the interactions is given by charges of 0.5e at the 12 interaction centers, with a charge -6e at the origin, to attain neutrality. These 12 sites are located at 3.55Å from the origin, each one with a mass of 60au, so the inertial moment has the correct value. These distributed charges were chosen so as to have the same molecular multipolar moment than the 'real ' charge distribution of the neutral molecule; in Ref.11 it is estimated by locating 30 charges of -0.35e at the center of the short bonds and 0.175e at the 60 C atoms. Figure 6 shows the pattern obtained. It can be seen that the main minimum is now at 98 deg, although the calculated curve is very flat. It is possible that taking into account the molecular polarizability, the calculated values of the 'effective' molecular multipolar moments would be larger than the estimated ones, stretching the scale of the plot in Fig.6.

With an adequate combination of these three models it seems possible to simulate, in a more simple way, the 'real' intermolecular potential model of C_{60}.

THE LOW TEMPERATURE CRYSTAL STRUCTURE

The experimental data of Ref.4 determined that at 10 K the crystal structure is Pa3 and that there is a preferential long range orientational order, with the molecule rotated 98 deg. around the crystallographic C_3 axis. But recent experimental data[16,17] imply that there is a certain amount of statical orientational disorder even at 10 K.

David et al.[16] measured the crystal structure of pure C_{60} between 5 and 320K.

They confirmed the first order phase transition at 260K and found a second order one at 90K. They proposed the following dynamical model: Above 260K, the molecules perform continous reorientational motion and below that temperature, the molecules rotate about a single [1,1,1] axis, with large residence times at sites corresponding to 98 and 38 deg. orientations. In the low temperature phase (T < 90 K), the rotational motion is frozen, but only 5/6 of the molecules take the 98 deg. orientation of Fig.3. For the short range disorder, they propose a model in which the rest of the molecules have an orientation at ~38 deg, close to the minimum van der Waals configuration.

On the other hand, Hu et al.[17] have also measured the crystal structure at 10K, finding that the percentage of molecules with a 98 deg. orientation is only 60%.

It is possible that the differences of the measured short range disorder depend on the way the crystals were obtained, but there is no doubt about a short range statical disorder whose nature needs to be clarified. Work on this line is under progress.

Acknowledgments: ZG thanks Ailan Cheng for interesting discussions and for sending to Argentina current literature on this rapidly changing field. This work, and the assistance to this meeting was partially supported by Fundacion Antorchas.

REFERENCES

1) W. Kratschmer, L. D. Lamb, K.Fostiropoulos and D. R. Huffman, Nature 347, 354, 1990).

2) P. A. Heiney, J. E. Fischer, A. R. McGhie, W.J. Romanov, A. M. Dueustein, J. P. McCauley Jr. and A. B. Smith III, Phys. Rev. Lett 66, 2911, (1991).

3) R. Sachidanardam and A. B. Harris, Phys. Rev. Lett. 67, 1467, (1991).

4) W. I. F. David, R. M. Ibberson, J. C. Matthewman, K. Prassides, T. J. S. Dennis, J. P. Hare, H. W. Kroto, R. Taylor and R. M. Walton, Nature 353, 147 (1991)

5) C. Pan, M. P. Sampson, Y. Chai, R. H. Hauge and J. L. Margrave, J. Phys. Chem.95, 2944 (1991).

6) R. Tycko, G. Dabkagh, R. M. Fleming, R. C. Haddon, A. V. Makhija and S. M. Zahurak, Phys. Rev. Lett. 67, 1886, (1991).

7) Q. M. Zhang, J. Y. Yi and J. Bernholc, Phys. Rev. Lett. 66, 2633, (1991).

8) A. Cheng and M. L. Klein, J. Phys. Chem. 95, 6750 (1991).

9) A. Cheng and M. L. Klein, Phys. Rev. B, 45, 1889 (1992).

10) A. Cheng and M. L. Klein, J. Phys. Chem., 95, 9622 (1991).

11) M. Sprik, A. Cheng and M. L. Klein, J. Phys. Chem. 96, 2027 (1992).

12) Y. Z. Li, M. Chander, J. C. Patin, J. H. Weaver, L. P. F. Chibante, R. E. Smalley, Science 253, 429 (1991).

13) Z. Gamba, J. Chem. Phys., in press (1992).

14) S. Nose and M. L. Klein, Mol. Phys. 50, 1055 (1983).

15) R. W. Impey, S. Nose and M. L. Klein, Mol. Phys. 50, 243, (1983).

16) W. I. F. David, R. M. Ibberson, T. J. S. Dennis, J. P. Hare and K. Prassides, Europhys. Lett, 18, 219 (1992).

17) R. Hu, T. Egami, F. Li and S. Lannin, preprint submitted to Phys. Rev. B (1992).

A MODIFIED BGY EQUATION FOR CLASSICAL FLUIDS

J.A.Hernando and Z.Gamba

Departamento de Física
Comisión Nacional de Energía Atómica
Av. Libertador 8250
1429 Buenos Aires, Argentina

INTRODUCTION

When comparing the available theories of the liquid state, the hypernetted chain-like theories (HNC) are, globally speaking, the more satisfactory ones [1]. In particular, the modified HNC (MHNC) is the best available theory. In the HNC equation the bridge function vanishes. This function can be written as a virial-like series expansion [2,3]

$$B(r) = \sum_{k=3}^{\infty} \frac{\rho^{k-1}}{(k-1)!} \int dr_1...dr_{k-1} c_k(r_1, ..., r_{k-1}, r) \Pi_{i=1}^{k-1} h_2(r_i) \tag{1}$$

where $h_2(r) = g_2(r) - 1$ and $c_k(r)$ are the 2-particle total and k-particle direct correlation functions respectively. It is seen that any improvement on the HNC calls for a discussion of k-particle correlations ($k \geq 3$). In the MHNC [1] - $\beta B(r)$ was interpreted as a repulsive effective potential and, based on a comprehensive set of numerical simulations, they concluded that the short range behavior of $B(r)$ is universal. They also modelled $B(r)$ by a hard sphere (HS) fluid in the Percus-Yevick approximation and the HS density was seen as an adjustable parameter. This was a considerable improvement upon the HNC but, at high densities, the fitting HS density begins to be indeterminate [4]. Other important equation is the Born-Green-Yvon equation (BGY) [5] which stems from applying the Kirkwood superposition approximation (KSA)

$$g_3^{KSA}(r_{12}, r_{13}, r_{23}) = \Pi_{i<j=1}^{3} g_2(r_{ij}) \tag{2}$$

to the stationary Bogoliubov-Born-Green-Kirkwood-Yvon hierarchy. The results obtained are reasonable at low densities and they rapidly deteriorate at higher densities. But, on the other hand, the BGY equation has good convergence properties at high densities [6]. It is also clear that to improve on the KSA requires a discussion of three body configurations. Therefore, although we keep in mind that the HNC results are better than the BGY ones, we have seen that both the BGY and HNC equations can be improved by considering three body effects and that there seems to be a complementarity in the sense of their density convergence properties. Another point to keep in mind is that, as the convergence radius of virial-like series expansions is unknown, there is no guarantee that a series expansion will be able to reproduce the behavior found at liquid state densities. The MHNC circumvents this problem by fitting the bridge function to some parametric function. As subjects like freezing theories [7], inhomogeneous liquids near surfaces [8], study of the force between two surfaces very near to each other [3,9] are all very active fields which need accurate liquid state theories at high densities, we can see the interest of improving upon these theories. In a very recent article [10] we suggested a modified KSA (MKSA)

$$g_3^{MKSA}(r_{12}, r_{13}, r_{23}) = \Pi_{i<j=1}^3 \alpha(r_{ij}) g_2[\alpha(r_{ij})r_{ij}]$$
$$\alpha(r) = 1 + \gamma/r. \tag{3}$$

and tested it against three numerical simulations of a dense Lennard-Jones (LJ) fluid. Here in this article we discuss some general conditions to be met by a modified BGY (MBGY) equation.

THE MKSA

The diagrammatic expansion of $g_3(r_{12}, r_{13}, r_{23})$ closer in spirit to equation (1) can be written as

$$g_3(r_{12}, r_{13}, r_{23}) = \Pi_{i<j=1}^3 g_2(r_{ij}) exp[\tau(r_{12}, r_{13}, r_{23})] \tag{4}$$

$$\tau(r, s, t) = \sum_{k=1}^\infty \rho^k \delta_{k+3}(r, s, t) \tag{5}$$

where τ is essentially the irreducible three - body component of the mean force potential, δ_n is defined as a sum over nth order diagrams [11] and an nth order diagram consists of three root points and n field points connected by h_2-bonds subject to the conditions: 1) each pair of field points is connected by at least one h_2-bond which does not passs through the root points; 2) through each field point passes at least one h_2-bond connecting each pair of root points; 3) the diagrams are free of articulation pair of circles. The diagrams up to seven h_2-bonds have been evaluated in Ref. 11. Therefore, the KSA is the first term in the expansion of g_3 and, for large particle separations, it is asymptotically correct. So it is reasonable to look for a modification to the KSA that will be essentially felt in the neighborhood of the first peak of g_2 via an adjustable density-like parameter.

Let us physically discuss the origin of the irreducible three-body contribution [10]. If we have three particles very near to each other, any pair of them partially screens the other particle and so, when we consider the three particles simultaneously, there are more collisions trying to cluster the three particles together than those trying to separate them when one pair at a time is considered. But the partial screening holds true when the separation is small and, therefore, as we are in the repulsive region of the potential, the clustering implies a stronger inwards force (by comparison with the KSA case) that must be balanced by a stronger repulsion. In other words, we can try shifting outwards the pair distribution function while keeping the functional form of the KSA. The shifting was assumed quite small, we wrote it as a power series in $1/r$ and tested a MKSA of the form given in equation (3). A typical result is given in figures 1a,b and c. MD runs were made on three LJ points with the same qualitative results. The improvement on g_3 is quite remarkable and, based on short range universality arguments, we proposed this MKSA for all simple fluids with $\gamma=0.07$.

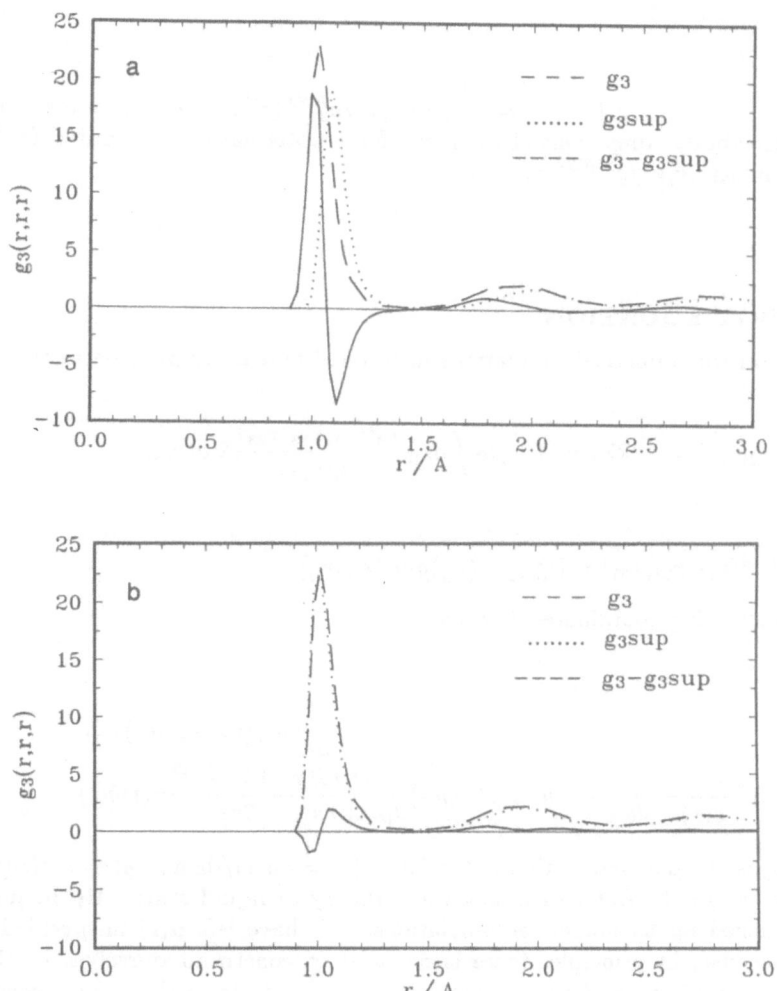

Figure 1. MD results for a LJ fluid with $\rho^*=1.02$, $T^*=1.84$ (high T - high ρ fluid. Distances are measured in units of σ. a) shown are $g_3(r,r,r)$, $g_3sup(=g_3^{MKSA}(r,r,r))$, $g_3 - g_3sup$ with $\gamma=0$; b) same as a) but with $\gamma=0.07$.

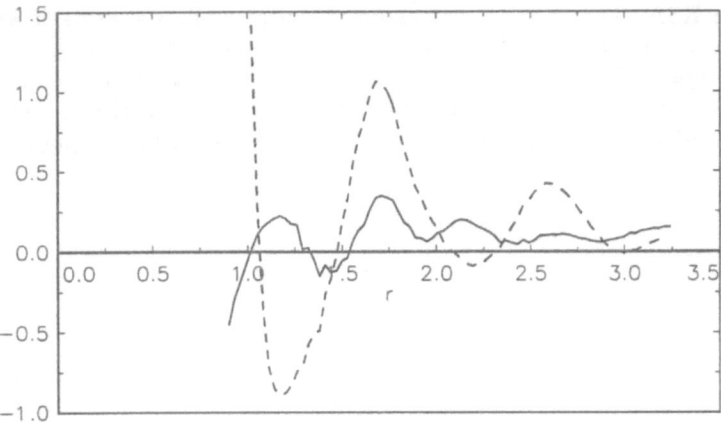

Figure 2. - - - - $\tau\ (r,r,r) = ln[g_3(r,r,r)/g_3^{KSA}(r)^3]$, —— the remaining irreducible three-body component of the mean force potential not accounted for by the
MKSA $= \ln[g_3(r,r,r)/g_2^{MKSA}(r,r,r)]$.

THE MBGY EQUATION

Based on the numerical simulations mentioned above, we propose as the MBGY
equation

$$\nabla ln g_2(r_{12}) = -\beta \nabla u(r_{12}) - \beta \rho \int dr_{13} \frac{g_3(r_{12}, r_{13}, r_{23})}{g_2(r_{12})} \nabla u(r_{13}) \tag{6}$$

with

$$g_3^{MKSA}(r_{12}, r_{13}, r_{23}) = \Pi_{i<j=1}^3 \alpha(r_{ij})g_2[\alpha(r_{ij})r_{ij}] \tag{7}$$

After transforming coordinates it reads

$$\frac{d}{dr}(ln g(r) + \beta u(r)) =$$
$$-2\pi\rho \frac{\alpha(r)\hat{g}(r)}{g(r)} \int_0^\infty ds s^2 u'(s)\alpha(s)\hat{g}(s) \int_{|r-s|}^{r+s} \frac{t dt}{rs} \frac{r^2 + s^2 - t^2}{2rs} \alpha(t)\hat{g}(t) \tag{8}$$

where u(r) is the pair interaction potential, $u'(r) = du(r)/dr$ and $\hat{g}(r) = g(\alpha(r)r)$.

So far, so good. But this is not yet a theory of liquid state. Up to now it is
a guess backed up by numerical simulations. We have left $\alpha(r)$ unspecified in the
MBGY because, in principle, there is no need to constraint ourselves to the $\alpha(r)$
used in the simulations. Among several possible points to analyze, we discuss here,
in a preliminary way, thermodynamical consistency and the role played by the unit
of length used in this approximation. It is now well established that the inaccuracy
of approximate liquid state theories can be gauged by comparing different routes to
thermodynamics [12].

One possible approach could be to assume a specific $\alpha(r)$ and to evaluate the free energy as a function of γ looking for the γ-value that makes it stationary. But, which free energy expression do we use?. A free energy expression related to the KSA is known [13] but it must be translated to the MKSA. The difficulty lies in that terms like

$$\begin{aligned}
S_3 = \ & g_3(1,2,3) - g_2(1,2)g_2(1,3) - g_2(1,2)g_2(2,3) - \\
& g_2(1,3)g_2(2,3) + g_2(1,2) + g_2(1,3) + g_2(2,3) - 1
\end{aligned} \qquad (9)$$

(eq.(39) of Ref. 14) have extensive cancellations when treated in the KSA but such is not the case in the MKSA. And, at least in this stage of the theory, there is no theoretical justification in just replacing the $g_2(r_{1,2}; \gamma = 0)$ obtained by solving the BGY equation by $g_2(r_{1,2}; \gamma \neq 0)$ (MBGY result) in the free energy expression. Also, the HNC equation has been recently interpreted as an optimized superposition approximation in Ref. 13 in the sense that, if the free energy is written in the KSA and leading terms retained, after varying it as a functional of g_2, the HNC equation (plus a term that vanishes everywhere except in a neighborhood of the critical point) is recovered. There also free energy expressions related to the HNC and MHNC equations [12] but, leaving aside the problem of internal consistency that arises when one analyzes an approximate theory via equations related to another approximate theory, an expression is needed for the bridge function (and its temperature derivative) in the MBGY theory if one wants to use these formulae. That can be numerically done after solving the MBGY equation.

Another approach is to try to fulfill the conditions for a thermodynamically consistent truncation of the BBGKY hierarchy. It is known that truncation of the BBGKY hierarchy at the fourth-particle distribution function level gives non-conservative mean force potentials [15]. This does not happen at the three-particle level. So, we think that it would be worthwhile to try achieving thermodynamical consistency of the hierarchy truncation in a more restricted sense than discussed by Jones et al. [16], i.e. to look for some $\alpha(r)$ that optimizes the truncation at the three-particle level. In Ref. 16 they looked for a minimum set of equations which, if the hierarchy truncation is done at the k-particle level, will involve only distribution functions g_n with $n \leq k$ and the goal is to try to satisfy simultaneously all of them. The delicate point is to choose an appropriate set of variables (like ρ, β) and, in our case, to decide on the role played by $\alpha(r)$. In our opinion, $\alpha(r)$ must not be considered an independent variable in that sense and, therefore, the minimum set of equations put forward by them (their eqs. (2.18) - (2.22)) is the set to solve with z (fugacity), β and V as independent variables. These equations are the BGY equation (eq. (7)) plus four thermodynamic relations expressed in terms of p, ρ, g_2, g_3. As far as we know this set of equations has never been numerically solved and even the proposal of solving it for $\alpha(r)$ in the MKSA is still rather complicated.

Another approximate approach, but that can be accurate enough, is to adjust $\alpha(r)$ so as to obtain the same pressure (which depends in a very sensitive way on the first g_2-peak position [11]) independently of the route followed for calculating it. Due to our force balance arguments this proposal is particularly appealing. Moreover, a study of up to the 10th virial coefficientes for rigid disks and rigid spheres [17] showed a great improvement upon fulfilling the pressure consistency requirement. Expressions useful in finding the pressure are

$$\beta \frac{p}{\rho} - 1 = \rho \frac{\partial(\beta F^E / N)}{\partial \rho} |_\beta \tag{10}$$

$$\beta \frac{p}{\rho} - 1 = -\beta \frac{\rho}{6} \int d\vec{r} g_2(r) r \frac{d\phi(r)}{dr} \tag{11}$$

$$\beta \frac{\partial p}{\partial \rho} |_\beta - 1 = \lim_{k \to 0}(-\rho \tilde{c}_2(k)) \tag{12}$$

$$\frac{1}{\beta} \frac{\partial \rho}{\partial p} - 1 = \frac{\beta \rho}{3} \int d\vec{r} g_2(r) \frac{du(r)}{dr} +$$
$$\frac{\rho^2 \beta}{6} \int \int d\vec{r}_2 d\vec{r}_3 [g_3(r_{12}, r_{13}, r_{23}) - g_2(r_{12})] \frac{u(r_{12})}{dr_{12}} \tag{13}$$

and the equality between these expressions is a very severe test. Here F^E is the excess free energy and $\tilde{c}_2(k)$ is the Fourier transform of the pair direct correlation function. Equation (13), not so widely known, is taken from Ref. 18. Another possibility is the use, first suggested by Hiroike [15], of the thermodynamic equality

$$\frac{1}{\rho^2} \frac{\partial U/N}{\partial \rho} |_\beta = \frac{\partial(\beta p)}{\partial \beta} |_\rho \tag{14}$$

The MHNC satisfies this relation [12]. All these expressions require solving the MBGY equation and, as an additional test, their results must be compared to numerical simulations.

A point we also want to discuss is the role played by the unit of length implicit in this treatment. In a simple liquid there is only one unit of length and, if the potential is not very soft, it is irrelevant wether we consider the unit of length given either by r_1 (V(r_1)=0) or r_2 (V(r_2)=minimum), $r_1=\sigma$, $r_2=r_1 2^{1/6}$ in the LJ fluid. But things are not so clear in the case of a binary fluid mixture. In that case we have three interaction potentials each one of them with its typical length. It could be possible that a good compromising length cannot be found. This is equivalent to the discussion of soft vs. hard potentials in the MHNC [12]. Therefore, the universality claim must not be pressed too hard. But if we have a binary mixture of charged particles, then we again have a relevant unit of length given as the typical distance of the total anion-cation interaction potential; the other two potentials are strongly repulsive at short distances. This is in contrast with the short distance behavior in such systems where it is claimed that the bridge function $B_{+-}(r)$ does not belong to the same universality class that $B_{++}(r)$ and $B_{--}(r)$ [20]. Therefore, it is quite possible that the MBGY has to be solved assuming an $\alpha(r)$ different for each possible pair.

In summary, there is an obvious need of solving the MBGY equation for several points in the thermodynamic space in order to assess its possible status as a theory of dense fluids. Work along these lines is in progress and will be reported elesewhere.

REFERENCES

1. Y.Rosenfeld and N.W.Ashcroft, Phys. Rev. A 20, 1208 (1979).
2. H.Iyetomi, Prog. Theor. Phys. 71, 427 (1984).
3. R. Kjellander and S. Sarman, Mol. Phys. 70, 215 (1990).
4. S.M.Foiles, N.W.Aschcroft and L.Reatto, J.Chem.Phys. 81, 6140 (1984).

5. R. Balescu, Equilibrium and Non-Equilibrium Statistical Mechanics, John Wiley and Sons, N. York (1975).

6. A.Jangkamolkulchai, K.A.Green and K.D.Luks, Mol. Phys. 68, 791 (1989).

7. T.V.Ramakrishnan and M.Yussouff, Phys. Rev. B19, 2775 (1979); W.A.Curtin, J. Chem. Phys. 93, 1919 (1990); H.Iyetomi and P.Vashishta, J. Phys.: Condens. Matter 1, 1899 (1989); H.Iyetomi and P.Vashishta, Phys. Rev. A (preprint); Y.Rosenfeld, Phys. Rev. Lett. 63, 980 (1989); Y.Rosenfeld, D.Levesque and J.J.Weis, J.Chem.Phys. 92, 6818 (1990) and references therein.

8. J.K. Percus, in The Liquid Stateof Matter: Fluids Simple and Complex, E.W. Montroll and J.L. Lebowitz (eds.), North Holland, Amsterdam (1982)

9. R. Kjellander and S. Sarman, Mol. Phys. 74, 665 (1991).

10. J.A. Hernando and Z. Gamba, J. Chem.Phys. (in press)

11. A.D.Haymet, S.A.Rice and W.G.Madden, J.Chem.Phys. 74, 3033 (1981); W.J.McNeil, W.G.Madden, A.D.J.Haymet and S.A.Rice, ibid. 78, 388 (1983); A.D.J.Haymet, ibid. 80, 3801 (1984); A.D.Haymet, S.A.Rice and W.G.Madden, J.Chem.Phys. 75, 4696 (1981).

12. F.Lado, S.M.Foiles and N.W.Aschroft, Phys. Rev. A 28, 2374 (1983); Y.Rosenfeld, ibid. 29, 2877 (1984); S.M.Foiles, N.W. Aschcroft and L.Reatto, J.Chem.Phys. 80, 4441 (1984).

13. J.A.Hernando, Mol. Phys. 69, 327 (1990).

14. J.A.Hernando, Mol. Phys. 69, 319 (1990).

15. H.J. Raveche and M.S. Green, J. Chem. Phys. 50, 5334 (1969).

16. G.L. Jones, J.J. Kozak and E.K. Lee, J.Chem.Phys. 80, 2092 (1984).

17. J.A. Devore, J. Chem. Phys. 80, 1304 (1984).

18. T.R. Choy and J.E. Mayer, J. Chem. Phys. 46, 110 (1967).

19. K. Hiroike, J. Phys. Soc. Japan 12, 326 (1957).

20. M. Kinoshita, M. Harada and A. Shioi, Mol. Phys. 70, 1121 (1990).

A MODIFIED POISSON-BOLTZMANN TREATMENT OF AN ISOLATED CYLINDRICAL ELECTRIC DOUBLE LAYER

L.B. Bhuiyan
Laboratory of Theoretical Physics
Department of Physics
University of Puerto Rico
Rio Piedras, Puerto Rico 00931

and

C.W. Outhwaite
Department of Applied and Computational Mathematics
University of Sheffield
Sheffield S10 2TN, UK

ABSTRACT

The modified Poisson-Boltzmann theory is applied to a study of the ionic environment around a uniformly charged infinite cylindrical polyion immersed in a 1:1 restricted primitive model electrolyte. The mean electrostatic potential and the singlet ionic density profiles with respect to the cylinder axis are calculated for a range of values of the polyion radius, polyion surface charge, and electrolyte concentration. The predictions of MPB theory are seen to be in very good agreement overall with those from the HNC/MSA theory, while the classical PB theory is useful only at low electrolyte concentrations.

INTRODUCTION

The characterization of the ionic atmosphere in polyelectrolytes has significance in such diverse areas as colloid chemistry, biological systems, e.g., DNA, and industrial products, e.g., detergents.[1-4] In this article we shall consider polyelectrolyte-simple electrolyte systems where the polyion has cylindrical symmetry and also has a very low relative concentration so that the (cylindrical) electric double layer (CDL) around it is essentially isolated. The usual analysis of such cylindrical polyelectrolyte solutions is based on the classical Poisson-Boltzmann (PB) equation and/or the counterion condensation theory of Manning (see for example, references [5-8]). These theories are generally useful only at low electrolyte concentrations.

Of late, the traditional approach is increasingly being transcended by the methods of formal statistical mechanics and the techniques used in the planar double layer (PDL). A basic model in use has been that of a uniformly charged hard cylinder of infinite extent interacting with equisized charged hard spheres in a dielectric continuum. This primitive model CDL has been studied in the hypernetted chain/mean spherical approximation (HNC/MSA) by Losada-Cassou, [9] and Gonzales-Tovar, Lozada-Cassou, and Henderson.[10] Rossky and coworkers [11,12] have used the HNC/HNC theory but with a soft repulsive potential for the ions. Here we employ the modified Poisson-Boltzmann (MPB) theory for the solution of the model polyelectrolyte. The MPB theory improves upon the PB theory by considering fluctuation potential and exclusion volume effects neglected in the latter, and has earlier proved enormously successful in describing both homogeneous bulk electrolytes, and inhomogeneous systems of planar and spherical double layers (SDL) (see for instance, reviews [13,14] and reference [15]). There is an attraction from a theoretical point of view also, to see how the MPB performs here, since the substantial averaging associated with the spherical and planar symmetries is expected to be considerably less in cylindrical symmetry. It is worth mentioning here that Bratko and Vlachy [16] have considered an approximate version of the MPB theory, which is a truncated form of the planar equation, for cylindrical polyelectrolytes in the cell model.

THEORY

The MPB equation appropriate for the cylindrical geometry above and for no imaging has been formulated by Outhwaite.[17] This equation is more complex than either the planar or the spherical MPB equation. However, Outhwaite has shown that by neglecting certain higher order terms in the fluctuation potential corrections the cylindrical MPB equation becomes structurally the same as that for planar [18] and spherical symmetries[15]. In the present work we have utilized the reduced form of the equation, which would be at similar level of approximation as the planar MPB5 version [18] ——the most successful version in the planar situation. We quote here the final equations of the theory, and refer the reader to reference [17] for details. We take the cylinder to be of radius R with a uniform surface charge density σ. The common diameter of the surrounding ions is taken to be a with ϵ being the relative permittivity. The cylindrical MPB equation then reads

$$\nabla^2 \psi(x) = -\frac{4\pi}{\epsilon} \sum e_s n_s^0 g_s(x), \tag{1}$$

$$
\begin{aligned}
g_s(x) =\ & \xi_s(x) \exp\left[-\frac{\beta e_s^2}{2\epsilon a}(F - F_0) - \frac{\beta e_s F}{2\sqrt{x}} \{u(x+a) + u(x-a)\} \right. \\
& \left. + \frac{e_s \beta (F-1)}{2a\sqrt{x}} \int_{x-a}^{x+a} u(R)dR \right]
\end{aligned}
\tag{2}
$$

Here $\psi(x)$ denotes the mean electrostatic potential at a perpendicular distance x from the cylinder axis and $u(x) = \sqrt{x}\psi(x)$, g_s is the cylinder-ion singlet distribution function for ion species s, $\beta = 1/k_B T$ (k_B is the Boltzmann constant and T the absolute temperature), and e_s, n_s^0 are the charge and the mean number density respectively of ion s. We remark here that there are typographical errors in Eqns. (19),(20), and (22) of ref.[17]. The sign of the integral term in equations (19) and (20) should be negative, and the first term on the right hand side of equation (22) should be absent. In Eq. (2) above, F and F_0 are given by

$$F = \begin{cases} 1/\{(1 + \kappa a) - (\kappa a/\pi)S\} & R + a/2 \le x \le R + 3a/2 \\ 1/(1 + \kappa a) & x \ge R + 3a/2 \end{cases} \tag{3}$$

$$F_0 = 1/(1 + \kappa_0 a) \tag{4}$$

with

$$S = \int_{\theta_0}^{\pi/2} \sin\theta \cos^{-1}\left\{\frac{c - \cos^2\theta}{(2x/a)\sin\theta}\right\} d\theta$$

$$\theta_0 = \sin^{-1}\left[\frac{x - (R + \frac{a}{2})}{a}\right]$$

$$c = 1 - \left(\frac{R}{a} + \frac{1}{2}\right)^2 + \left(\frac{x}{a}\right)^2 \tag{5}$$

κ and κ_0 are the local and bulk Debye-Hückel screening parameters,

$$\kappa^2 = \frac{4\pi\beta}{\epsilon}\sum e_s^2 n_s^0 g_s(x) \tag{6}$$

$$\kappa_0 = \lim_{x\to\infty}\kappa \tag{7}$$

The exclusion volume term $\xi_s(x)$ is approximated by its form in the planar MPB theory, which in turn, is calculated using the Bogoliubov-Born- Green-Yvon heirarchy[19].

$$\xi_s(x) = g_s(x|e_s = 0)$$

$$= H\left(x - \left(R + \frac{a}{2}\right)\right)\exp\left[2\pi\int_x^\infty\sum n_t^0\int_{max(R+\frac{a}{2},x-a)}^{x+a}(X - y)g_t(X)\right.$$

$$\exp\left\{-\beta|e_t| z\phi(y, X)\right\}dXdy] \tag{8}$$

$$\phi(y, X) = \frac{F}{4\pi a}\int_V \nabla^2\psi dV \tag{9}$$

where $\phi(y, X)$ is the fluctuation potential on the exclusion surface of the discharged ion. This approximation will be reasonable for small a/R, becoming exact as $R \to \infty$. By considering the linearized equation it was shown [17] that the asymptotic property of the MPB equation is analogous to that in the planar and spherically symmetric cases, viz., damped oscillatory solutions are predicted at higher concentrations. For symmetric valency salts, this occurs for $1.241 < \kappa_0 a < 7.83$. We further note that as expected, the MPB theory reduces to the corresponding non-linear PB theory upon taking $\xi_s(x) = 1$, $F = F_0$, and $a \to 0$.

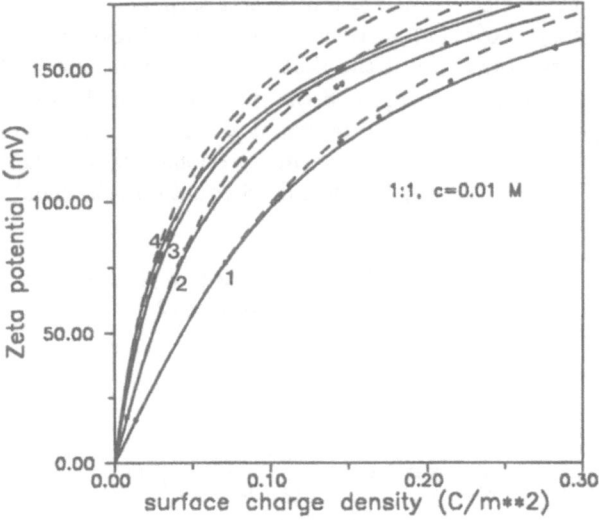

Figure 1. Zeta potential $\psi(R + a/2)$ versus surface charge density σ, for different values of cylinder radius R at electrolyte concentration $c = 0.01M$: Solid lines MPB; dashed lines, PB; stars, HNC/MSA. Curves: 1, $R = 5A^0$; 2, $R = 15A^0$; 3, $R = 80A^0$; 4, planar limit.

RESULTS AND DISCUSSION

The MPB and the non-linear PB equations were solved numerically by a quasilinearization iteration scheme and following procedure used previously in the solutions of the PDL and SDL MPB equations. The algorithm was checked in the large R limit when the PDL solutions were approached.

The physical parameters chosen for the 1:1 salt were $a = 4.25A^0$, $T = 298K$, and $\epsilon = 78.5$. These parameters have been used by Gonzales-Tovar et al in their HNC/MSA calculations. [10] In Figs. (1) and (2), the variation of the MPB Zeta potential $\psi(R+a/2)$ with the surface charge density σ at different R is shown for salt concentrations $c = 0.01M$ and $1M$ respectively. Also shown are the corresponding PB results, and some HNC/MSA results at $R = 5A^0$ and $R = 15A^0$. We note first that the MPB and the HNC/MSA results are very similar except in the region $\sigma > 0.25C/m^2$ for the $1M$ case where the MPB curve lies above the HNC/MSA points. The pattern at $R = 80A^0$ at both the concentrations (not shown here) was found to be the same. These deviations as R increases seem consistent with the known predictions of the theories in the planar limit, shown in Fig. 7 of reference [14]. Also from the Monte Carlo results of the same figure it would seem that the CDL MPB theory is more accurate at the higher surface charges. The PB results are, not unexpectedly, closer to the statistical mechanical theories at the lower concentration. Also, for a given cylinder charge, the PB zeta potential is over estimated. These aspects of the classical theory are, of course, well documented from its application to the PDL. The PB results, in addition, diverge prgressively from those of the MPB, HNC/MSA as R increases at a particular concentration. This point has been noted before by Gonzales-Tovar et al.[10] We also noticed an analogous feature in our MPB

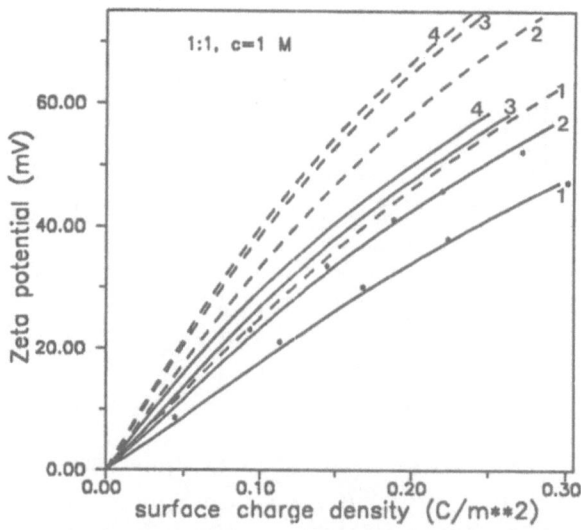

Figure 2. Zeta potential $\psi(R + a/2)$ versus surface charge density σ for different values of cylinder radius R at electrolyte concentration $c = 1.0M$. The curves and the symbols have the same meaning as in Fig. (1).

study of the SDL.[15] Clearly the effects of the relative sizes of the simple ions and the polyion become significant with increasing polyion size.

Turning now to the profiles, once again we find a remarkable correspondence between the MPB and the HNC/MSA results for the mean potential ψ and the singlet densities g_s ($R = 15A^0$, $\sigma = 0.220C/m^2$, $c = 1M$), shown in Figs. (3) and (4) respectively. The PB theory is qualitative and shows a thicker double layer than do the MPB and HNC/MSA theories. The MPB plots at $\sigma = 0.3031C/m^2$ show the effects of a higher charge on the profiles. The effects are substantial near the cylinder surface and lead to raising of the contact values of the potential and counterion density profiles and lowering of the coion density profile. The distributions at $R = 5A^0$, $\sigma = 0.220C/m^2$, $c = 1M$ were found to be similar, with the MPB and HNC/MSA results showing excellent correspondence throughout the diffuse layer, and are therefore not shown. The MPB and the HNC/MSA profiles in Figs. (3) and (4) display faint oscillations at large distances from the polyion surface (outside the range of the graphs). Notice, for instance, the negative values in ψ. Such oscillations are indicative of charge ordering. To press this point further, in Figs. (5) and (6) we have plotted the MPB and PB distributions at $c = 2M$ ($R = 15^0A$, $\sigma = 0.220C/m^2$). The oscillations in the MPB profiles at this high concentration are very pronounced and clearly visible. Note that the PB theory is now not even qualitative. We have also done calculations at $c = 0.01M$. Here both the PB and the MPB mean potentials and the densities show a monotonic decrease and the theories are close for a wide range of R and not too high σ. The corresponding HNC/MSA results, though not published, are again expected to be very close taking hint from Fig. (1).

The present work thus would appear to indicate that the MPB indeed performs very well in a cylindrical symmetric situation. For 1:1 electrolytes the MPB results parallel the HNC/MSA integral equation results to a great degree consistent with

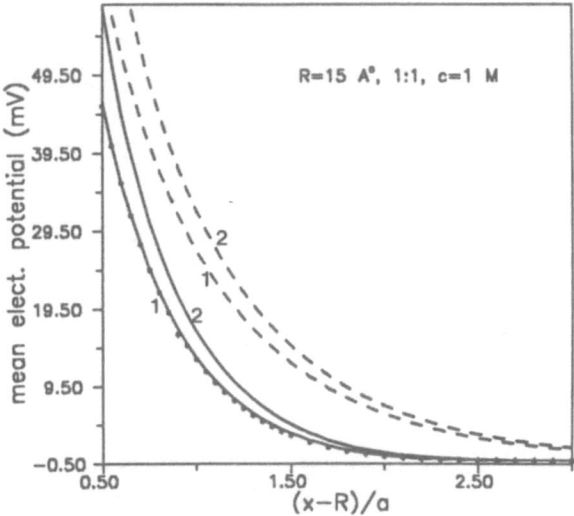

Figure 3. Mean electrostatic potential $\psi(x)$ profiles at electrolyte concentration $c = 1M$, cylinder radius $R = 15A^0$: solid line, MPB; dashed line, PB; stars, HNC/MSA. Curves: 1, $\sigma = 0.220C/m^2$; 2, $\sigma = 0.3031C/m^2$.

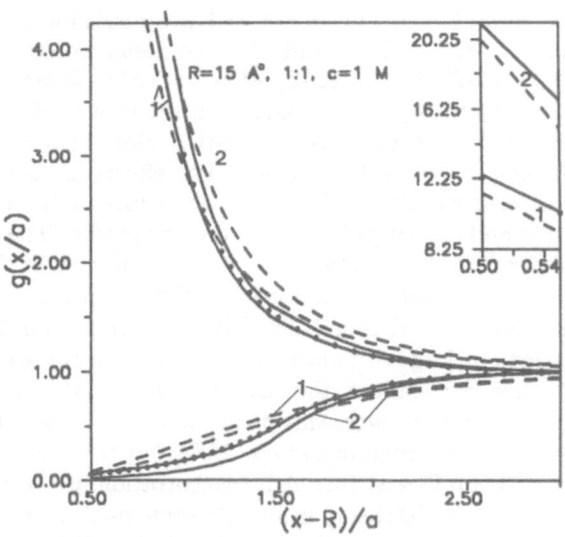

Figure 4. The singlet density $g_s(x)$ profiles at electrolyte concentration $c = 1M$, cylinder radius $R = 15A^0$. The curves and the symbols have the same meaning as in Fig.(3).

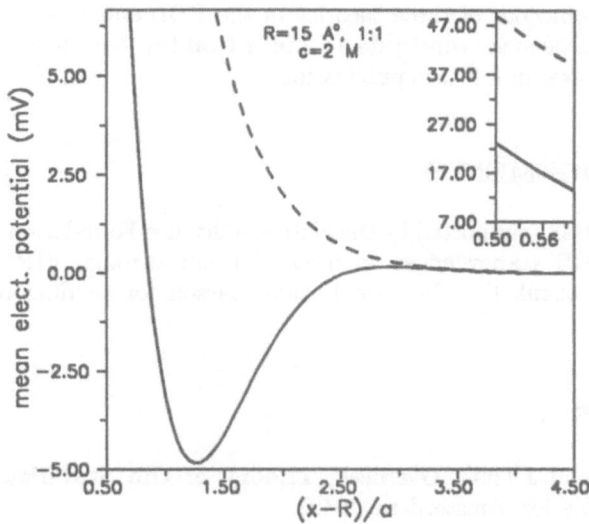

Figure 5. Mean electrostatic potential $\psi(x)$ profiles at electrolyte concentration $c = 2M$, cylinder radius $R = 15A^0$, cylinder surface charge density $\sigma = 0.220C/m^2$: solid line, MPB; dashed line, PB.

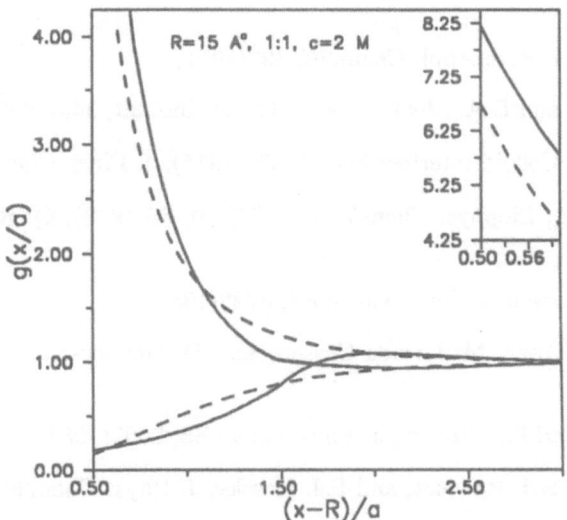

Figure 6. Singlet density $g_s(x)$ profiles at electrolyte concentration $c = 2M$, cylinder radius $R = 15A^0$, cylinder surface charge density $\sigma = 0.220C/m^2$. The curves have the same meaning as in Fig.(5).

the comparative behaviour of these theories in the PDL and SDL. However, higher and/or asymmetric valencies constitute a more critical test of a theory, and we intend to consider these cases in a future publication.

ACKNOWLEDGEMENTS

This work was partially supported by the National Science Foundation through Grant CHE-8907130. LBB acknowledges an internal grant through FIPI, University of Puerto Rico. We thank Dr. Marcelo Lozada-Cassou for sending us some of the HNC/MSA data.

REFERENCES

[1] E.J. Verwey and J.Th.G. Overbeek, **Theory of the Stability of Lyphobic Colloids**, (Elsevier, Amsterdam, 1948).

[2] S.A. Rice and M. Nagasawa, **Polyelectrolyte Solutions**, (Academic, New York, 1961).

[3] C.F.Anderson and M.T.Record, Jr., Ann. Rev. Phys. Chem.**33**, 191(1982).

[4] C. Tanford, **Physical Chemistry of Macromolecules**, (Wiley, New York ,1961).

[5] A. Katchalsky, Pure Appl. Chem.**26**, 327(1971).

[6] S.L. Brenner and D.A. McQuarrie, J. Theor. Biol.**39**, 343(1973).

[7] D. Stigter, J. Colloid Interface Sci.**53**, 296(1975); J. Phys. Chem.**82**, 1603(1978).

[8] G.S. Manning, Biophys. Chem.**7**, 95(1977); **9**, 65(1978); Q. Rev. Biophys.II,**2**, 179(1978).

[9] M. Lozada-Cassou, J. Phys. Chem.**87**, 3729(1983).

[10] E. Gonzales-Tovar, M. Lozada-Cassou, and D. Henderson, J. Chem. Phys.**83**, 361(1985).

[11] R. Bacquet and P.J. Rossky, J. Phys. Chem.**88**, 2660(1984)

[12] C.S. Murthy, R.J. Bacquet, and P.J. Rossky, J. Phys. Chem.**89**, 701(1985).

[13] C.W. Outhwaite, in:**Statistical Mechanics**, (The Chemical Society, London, 1975), Vol.2, p.188.

[14] S.L. Carnie and G.M. Torrie, Adv. Chem. Phys.**56**, 141(1984).

[15] C.W. Outhwaite and L.B. Bhuiyan, Electrochimica Acta **36**,1747(1991); Molec. Phys.**74**, 367(1991).

[16] D. Bratko and V. Vlachy, Chem. Phys. Letters. **90**, 434(1982); **115**, 294(1985).

[17] C.W. Outhwaite, J. Chem. Soc., Faraday Trans.2,**82**, 789(1986).

[18] C.W. Outhwaite and L.B. Bhuiyan, J. Chem. Soc., Faraday Trans.2,**79**, 707(1983).

[19] C.W. Outhwaite and L.B. Bhuiyan, J. Chem. Soc., Faraday Trans.2,**78**, 775(1982).

THEORETICAL CALCULATION OF THE OPTICAL

ABSORPTION OF FRACTAL COLLOIDAL AGGREGATES

USING A MULTIPLE SCATTERING FORMALISM

L. Fonseca[1], L. Cruz[1], W. Vargas[2], and M. Gomez[1]

[1]Dept. of Physics, University of Puerto Rico
Rio Piedras, Puerto Rico 00931
[2]Dept. de Fisica, Universidad de Costa Rica
San Jose, Costa Rica

ABSTRACT

The optical response of fractal aggregated gold colloids is described considering the interaction between particles due to two separated contributions resulting from short and long range effects. The systems will be studied in an effective medium formalism, where the scattering units are clusters of two spheres. The short range interaction will be taken into account via cluster effects, while the long range interaction will be taken into account by considering a multiple scattering effective medium analysis.

I. INTRODUCTION

The optical properties of media containing small metal particles are of great importance in many diverse areas of science and technology.

Among the most studied systems are those where the metal particles are disperse in a dielectric matrix with the particles far apart. These systems in the long wavelength limit behave as homogeneous effective media. Two basic theories are widely used in the case of dilute systems; Maxwell-Garnett[1], and Bruggeman[2] models. Also in colloidal systems composed of near-spherical metal particles in a very dilute limit Rayleigh or Mie theory, depending on the size of the particle, can describe the system.

Recently, more complex systems have been studied, that consist of aggregated structures. The aggregation of the particles are responsible for the new optical characteristics both in visible and infrared regions. Standard mean field theories fail to describe these more complex systems because they neglect the short-range high-order multipolar interactions between the particles that make-up the aggregate. These models fail to describe two experimentally observed resonant peaks in the absorption spectrum in the optical region[3] and the so-called abnormal absorption at far-infrared frequencies[4].

Colloidal aggregates systems have optical properties that are of great importance in diverse phenomena such as the optical effects due to air pollutants. In many cases, these type of aggregates have fractal configuration[5,6] with chain-like and low density distribution of the particles. Gold colloids in water solution is one of this type of systems that has been widely studied. Measurements of the optical absorption for such systems show broadening of the absorption resonance peak due to single-isolated particles and the appearance of a new absorption band that shifts toward lower frequencies as the size of the aggregate increases when the time provided for aggregation increases[3]. Calculations using different approaches have been applied to these structures[7].

Due to the fractal nature of these aggregates, the density of the particles forming the aggregate is low and a large fraction of these particles have on the average only two near neighbors. This fact permits multipolar interactions between the particles to be considered in two parts: the short-range interaction between the nearest neighbors and the long-range interaction with the rest of the system. The multipolar interaction between the nearest neighbors must be considered in detail while the interaction with the other particles can be taken as an effect of the average effective medium[8]. Following this approach, the absorption properties of aggregated gold colloids will be studied by considering the detailed multipolar interaction between nearest neighbors. The effect of the rest of the particles will be taken into account as an effective medium acting on the small clusters of nearest neighbors particles.

We start by extending the multipole scattering theory developed by Varadan et.al[9] to calculate the effective index of refraction of a system composed of small clusters of gold particles with short-range interactions. The theory is developed from a multiple scattering field equation in a T-matrix formalism. Therefore, the effective index of refraction of the structure is obtained in terms of the T-matrix of the scattering unit which are clusters of pairs of particles. From this index of refraction, the absorption properties of the aggregated structure are obtained as a function of the size of the system.

In section II the theoretical formalism is presented. While in Section III the model is applied to obtain the optical absorption of fractal gold aggregates and the results of the calculations are compared with the experimental measurements. Section IV summarizes the results and presents the conclusions.

II. THEORY

To obtain the effective response of the medium, the system is considered to be formed by scattering units constituted by two nearest neighbor spherical particles rather than of isolated ones. Following Varadan et al[9], the electric field $\vec{E}$ at any point in the medium is the sum of the incident field $\vec{E}_o$ plus the fields scattered by these units, $\vec{E}^s$,

$$\vec{E}(\vec{r}) = \vec{E}_o(\vec{r}) + \sum_i \vec{E}_i^s(\vec{r}-\vec{r}_i)$$

where $\vec{r}_i$ is the position of the scattering unit "i". The field exciting unit "i" is:

$$\vec{E}_i^e(\vec{r}) = \vec{E}_o(\vec{r}) + \sum_{j \neq i} \vec{E}_j^s(\vec{r}-\vec{r}_j) \quad , \quad d \leq |\vec{r}-\vec{r}_j| \leq 2d$$

where "d" is the radius of the smallest sphere that inscribes the scattering unit of two particles and superposition of the units is avoided.

Expanding $\vec{E}^s$ and $\vec{E}^e$ in terms of base functions, and relating the exciting field and the scattered field coefficients by T-matrix of the scattering units, a relation between the exciting field expansion coefficients B_n and the incident field expansion coefficients A_n is obtained,

$$B_n^i = \sum_{n''} T_{n'',n}^i \left\{ e^{i\vec{k}_o \cdot \vec{r}_i} A_{n''} + \sum_{j \neq i} \sum_{n'} B_{n'}^j S_{n',n''}(\vec{r}_i-\vec{r}_j) \right\}$$

where $S_{n',n''}$ are the matrix elements describing the translation properties of the base functions. Finally, a configurational average of the above equation is performed assuming that the average with one or two units fixed are approximately the same,

$$< B_n >_{ij} \cong < B_n >_j \quad .$$

Considering no correlation between the scattering units other than the impenetrability condition,

$$<B_n^i>_i = \sum_{n''} T_{n'',n}^i \left\{ e^{i\vec{k}_o \cdot \vec{r}_i} A_{n''} + \frac{1}{V} \int_{V'} \sum_{j \neq i} \sum_{n'} <B_{n'}^j>_j S_{n',n''}(\vec{r}_i-\vec{r}_j) \right\}$$

where V is the volume of the sample and V' is V minus a spherical volume of radius 2d to avoid penetrability. Using an effective medium approach,

$$<B_n^i>_i = X_n e^{i\vec{k}_{eff} \cdot \vec{r}_i} \quad ,$$

where $\vec{k}_{eff}$ is the effective propagation wave vector, a final system of coupled equation is obtained for the unknowns $X_{n'}$. From this system of equation, the dispersion relation is obtained by finding an adequate root of the determinant with the following elements,

$$\sum_{n''} T_{nn''} \; I_{n'n''} \; - \; \frac{(k^2_{eff} - k^2_o)\delta_{nn'}}{\nu} \qquad (1)$$

where $T_{nn''}$ is the T-matrix elements of a cluster of two spherical particles, ν is the number of scattering units per volume and,

$$I_{nn''} = \int_{r=2d} \left\{ S_{nn''}(k_o r) \; \partial_r e^{i\vec{k}_{eff}\cdot\vec{r}} \; - \; e^{i\vec{k}_{eff}\cdot\vec{r}} \; \partial_r S_{nn''}(k_o r) \right\} \; ds$$

In order to obtain the key term of equation (1), that is $T_{nn''}$, we are going to use the formalism developed by P.C. Waterman[10] for isolated particles and extended by Peterson and Ström to clusters of two particles. This T-matrix formalism takes into account multipolar contributions which are essential for any valid calculation of local fields of single non-spherical particles as well as for all clusters even when they are in the long wavelength regime. Also, the formalism takes into account phase retardation effects due to the size of the scatterers which are important for particles and clusters whose sizes are comparable to the wavelength of the incident field.

In this method the scattered, internal and incident fields are expanded in terms of the corresponding elementary fields that are a basis set of solutions for the vector Helmholtz equation,

$$\nabla \times \nabla \times \Psi \; - \; k^2 \Psi \; = \; 0 \; .$$

The incident $\mathcal{E}_o$ and internal $\mathcal{E}_i$ fields are expanded in terms of the regular basis $\text{Re}\Psi$ and the scattered $\mathcal{E}_s$ in terms of the non-regular Ψ one,

$$\mathcal{E}_o = \sum_n A_n \; \text{Re}\Psi_n$$

$$\mathcal{E}_i = \sum_n D_n \; \text{Re}\Psi_n \qquad\qquad |\vec{r}| < r_{min}$$

$$\mathcal{E}_s = \sum_n F_n \; \Psi_n \; . \qquad\qquad |\vec{r}| > r_{max}$$

where A_n are known coefficients, D_n and F_n are unknown, R_{min} is the radius of the maximum sphere inscribed in the scatterer and R_{max} is the minimum sphere inscribing the scatterer.

The elementary wavefunctions are expressed as:

$$\Psi_{\tau\sigma mn}(\vec{r}) \; = \; \gamma^{1/2}_{mn}(k^{-1}\nabla\times)^\tau \; [k \; \vec{r} \; Y_{\sigma mn}(\hat{r}) h_n(kr)]$$

where $\tau=1,2$, $\sigma=\text{even}(e)$ or $\text{odd}(o)$, $n=1,2,\dots$, $m=0,1,\dots,n$,

$$\gamma_{mn} = \varepsilon_m \frac{(2n+1)(n-m)!}{4n(n+1)(n+m)!},$$

and,

$$Y_{\text{emn}}(\hat{r}) = \cos(m\phi)\, P_n^m(\cos\theta)$$

$$Y_{\text{omn}}(\hat{r}) = \sin(m\phi)\, P_n^m(\cos\theta)\ .$$

The index $\tau = 1,2$ describes the type of excitation; magnetic or electric; ε_m is the Neumann symbol defined as $\varepsilon_0=1$, $\varepsilon_m=2$ otherwise, n is the order of the multipole, and σ gives the parity of the elementary functions. The regular form of the basis functions are obtained by substituting the Hankel functions by the Bessel functions.

The surface currents on the scatterers are used to express the expansion coefficients of the internal field with those of the scattered and incident fields respectively by the following relationships,

$$F = -i\ \text{Re}(Q')D \tag{2}$$

$$A = i\ Q'D \tag{3}$$

where Q' represents the transpose of the Q matrix which for a particle with complex dielectric function is given by

$$Q_{nn'} = \frac{k_o}{\pi} \int_s d\vec{s} \cdot \left\{ [\nabla \times \text{Re}\Psi_n(k\vec{r})] \times \Psi_{n'}(k_o\vec{r}) + \text{Re}\Psi_n(k\vec{r}) \times [\nabla \times \Psi_{n'}(k_o\vec{r})] \right\}$$

where $k_o^2 = \varepsilon_{\text{ext}}\omega^2/c^2$ and $k^2 = \varepsilon_{\text{int}}\omega^2/c^2$, s is the surface of the scatterer, and $\Psi_n(k\vec{r})$ is substituted by $\text{Re}\Psi_n(k\vec{r})$ wherever ReQ appears. In our case $\varepsilon_{\text{ext}}=1.77$ (water) and ε_{int} is the corresponding value for gold particles.

Eliminating **D** from equations (2) and (3) a relation between the coefficients of the scattered and incident fields is obtained,

$$F = \underline{T}\ A\ ,$$

where $\underline{T}$ is the T-matrix of the single scatterer defined as,

$$\underline{T} = -Q^{-1}\text{Re}Q\ .$$

The T-matrix formalism has been extended to systems with more than one scatterer by Peterson and Ström[11] using the translation theorems for the vector spherical functions[12]. The translation properties of Ψ_n and $\text{Re}\Psi_n$ are summarized by Ref.13.

$$\text{Re}\Psi_{\tau n}(\vec{r}+\vec{a}) = \sum_{\tau'n'} R_{\tau n, \tau'n'}(\vec{a})\ \text{Re}\Psi_{\tau'n'}(\vec{r})$$

$$\Psi_{\tau n}(\vec{r}+\vec{a}) = \sum_{\tau'n'} \sigma_{\tau n, \tau'n'}(\vec{a})\ \text{Re}\Psi_{\tau'n'}(\vec{r})\quad ;\quad |\vec{a}| > |\vec{r}|$$

$$\Psi_{\tau n}(\vec{r}+\vec{a}) = \sum_{\tau'n'} R_{\tau n, \tau'n'}(\vec{a})\ \Psi_{\tau'n'}(\vec{r})\quad ;\quad |\vec{a}| < |\vec{r}|$$

where σ and R are the elements of the translation matrices as defined in Ref.11.

Peterson and Ström obtained a T-matrix for the cluster of two particles in terms of the T-matrices of each single scatterer,

$$\underline{T}(1,2) = \underline{R}(\vec{a}_1)\Big\{\underline{T}(1)\,[\underline{1}-\underline{\sigma}(-\vec{a}_1+\vec{a}_2)\underline{T}(2)\underline{\sigma}(-\vec{a}_2+\vec{a}_1)\underline{T}(1)]^{-1}$$

$$[\underline{1}+\underline{\sigma}(-\vec{a}_1+\vec{a}_2)\underline{T}(2)\underline{R}(\vec{a}_1-\vec{a}_2)]\Big\}\underline{R}(-\vec{a}_1)\ +$$

$$+\ \underline{R}(\vec{a}_2)\Big\{\underline{T}(2)\,[\underline{1}-\underline{\sigma}(-\vec{a}_2+\vec{a}_1)\underline{T}(1)\underline{\sigma}(-\vec{a}_1+\vec{a}_2)\underline{T}(2)]^{-1}$$

$$[\underline{1}+\underline{\sigma}(-\vec{a}_2+\vec{a}_1)\underline{T}(1)\underline{R}(\vec{a}_2-\vec{a}_1)]\Big\}\underline{R}(-\vec{a}_2)\qquad (4)$$

where $\vec{a}_1$ and $\vec{a}_2$ are the distances from the origin to the center of scatterers 1 and 2 respectively.

The scattered field can be expressed in terms of the incident field using the T-matrix by the following relationship,

$$\mathcal{E}_s = \underline{T}\ \mathcal{E}_o\ .$$

III. CALCULATIONS AND DISCUSSION

It can be shown analytically the presented formalism gives the Maxwell-Garnett result when the scattering units consist of isolated spherical particles[9]. The absorption spectrum obtained in this manner has one resonant frequency corresponding to the isolated scatterer shifted by the average interaction of the surrounding particles, in accordance with most experimental results when no aggregation is evident. For systems whose constituents show aggregation, considerable change in the absorption spectrum is observed due to the multipolar interaction between the particles. Experimental results for gold colloids in fractal aggregates report the appearance of two absorption bands. This number of bands is compatible with the assumption that the multipolar interaction is of short-range nature and occurs mainly between the nearest neighbors. The fractal pattern of these gold aggregates reduces the number of nearest neighbors that a particle could have. We present here calculations of the above model considering as scattering units a cluster of two nearest spherical gold particles.

The choice of the scattering unit as a cluster of two particles instead of three or more is made based on the fact that the absorption spectrum of an isolated cluster of two gold small particles shows two resonant frequencies as is indicated in figure 1, and in the fact that this is the simplest scattering unit, larger that the well known isolated particle case, for which the basic behavior of the experimental system is obtained.

According to the experiments, particles of 15nm in diameter are studied and a small separation of 3A (similar to the gold-gold bond length) between the surfaces of the two nearest particles is used. The dielectric function reported in the literature[14] was corrected to include scattering at the surface. Our convergence tests for the calculation of the T-matrix of such cluster of two gold particles indicate that the calculations converge when multipolar order of n=11 are considered in the calculation.

In figure 1 the comparison between the absorption behavior of an isolated gold particles and the cluster is shown. For a cluster with the axis going through the two particles aligned in the direction of the incident electric field, the multipolar interaction between the particles

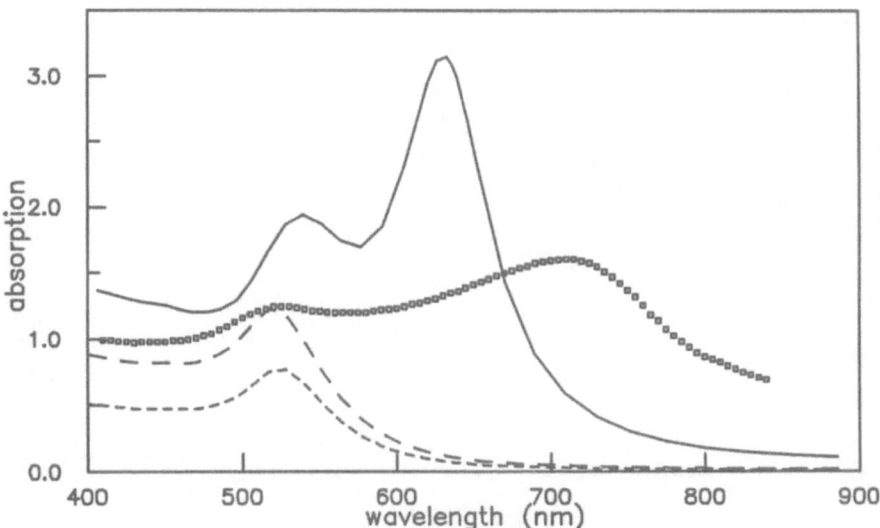

Fig. 1. Comparison between the absorption curves of (a) isolated sphere (short dashes), (b) cluster perpendicularly oriented (large dashes), (c) cluster parallel oriented (solid line), and (d) experimental results from ref.8 (dotted line).

produces two absorption bands. The lower energy one is located at 633nm in wavelength and the higher energy one is near the dipolar band corresponding to the isolated sphere case but broadened and shifted toward lower energies. The case of cluster aligned perpendicularly to the above orientation does not produce two absorption bands and the response is similar to the isolated sphere case. The comparison of the response of the aligned to the field case with the experimental absorption of a system of aggregated colloidal gold shows that this type of cluster is the most important to explain the appearance of two absorption resonances in the experimental measurements.

Using the T-matrix elements calculated with equation (4) for n=11 multipolar order, the root of equation (1) with physical meaning was obtained. This root gives the effective index of refraction of the aggregated structure. The absorption properties of this structure is then obtained considered the structure as an homogeneous region described by the obtained effective index of refraction.

Figure 2 shows the calculated absorption of such aggregated structures for different aggregation sizes. The results show a shifting and broadening of the low energy peak as the size of the structure increases, which corresponds with increasing time of aggregation in the experiment. From this calculation we explain the shifting and broadening of the lower energy absorption peak as the result of the long range interaction and the size of the aggregate. This result is consistent with the experimental measurements that show such a shift with the increasing time of aggregation[3].

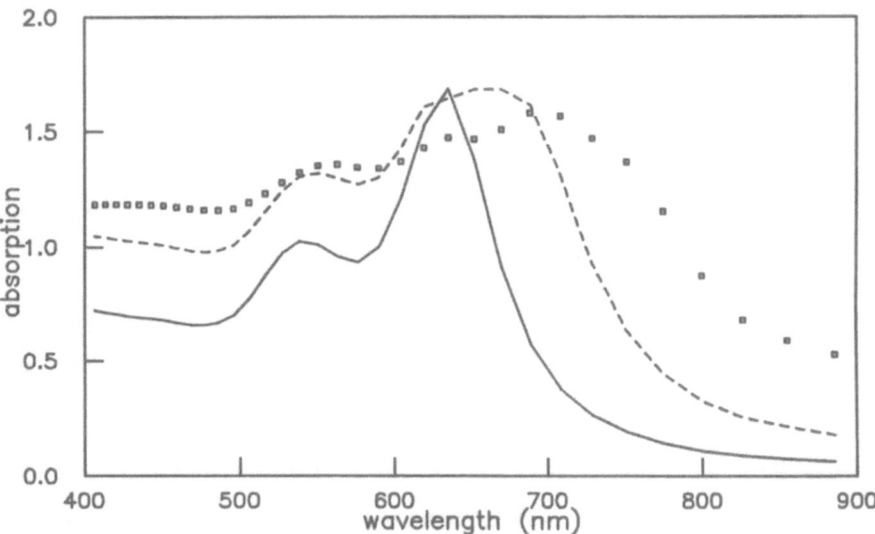

Fig. 2. Position of the absorption bands for three different sizes of the aggregate according to the model calculation. Solid line corresponds to 50nm in size, dashed line corresponds to 100nm and dotted line corresponds to 200nm.

Real systems are composed of a variety of aggregates of different sizes or number of aggregated particles. The distribution of these sizes depends on the time of aggregation of the sample. Figure 3 show the comparison between experimental results of the absorption of colloidal fractal aggregates in water with the model calculation considering the sample to be made up of 80% of aggregates of 200nm in size and 20% of isolated particles. The experimental results correspond to the published data for a sample 25min. after aggregation[8] and the chosen size is consistent with those reported for this time for diffusion-limited aggregated colloids of gold[15].

The calculations have been done considering only clusters aligned with the field but the real system consists of a distribution of orientations. The model is expected to produce good results since, the used orientation in the calculation produces two peaks and the other orientations yield results that are essentially the same as those obtained for isolated spheres. A more detailed calculation should consider an average of orientation but the T-matrix of the clusters would then require considerably more complicated calculations. With this type of average a broadening of the peaks is expected but with no other changes of substance.

IV. CONCLUSIONS

Three factors have been considered in describing the gold aggregated colloidal system : the short-range distance interaction by considering the principal component of the systems to consists of clusters of pair of particles instead

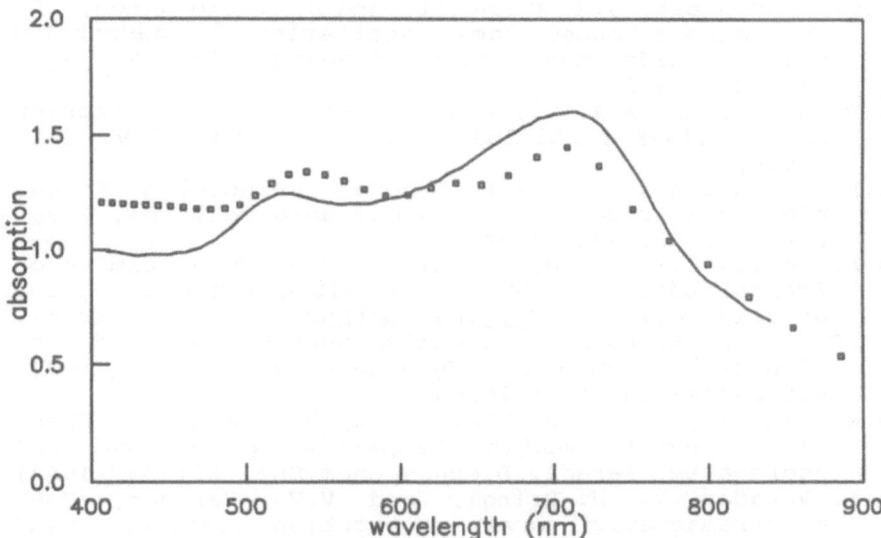

Fig. 3. Comparison between the position of the absorption bands according to experimental result published in ref.8 (solid line), and the predictions of the theoretical model considering 80% of 200nm in size aggregates and 20% of isolated particles (dots).

of isolated ones; the long-range interaction by using the multiple scattering formalism outlined above to obtain the effective mean field index of refraction of a system composed of the above clusters; and the fact that the total system is composed of a variety of these aggregated structures that are large in size with respect to the dimension of the constituent particles and well far apart from each other. The comparison between the calculations and the experimental results show a good agreement giving credence to the assumption that in the aggregated fractal form, the optical interaction between the gold particles can be divided into the short-range multipolar interaction and a long-range average interaction.

The theoretical model presented here is very effective in handling these two types of interactions. It is anticipated that the applicability of the model will yield good results for other types of colloidal systems, and probably for the calculation of the far infrared absorption of aggregated particles[4] also.

ACKNOWLEDGEMENTS

The support of EPSCoR-NSF grant EHR-9108775 is acknowledged.

REFERENCES

1. J.C. Maxwell - Garnett, Colours in metal glasses and in metallic films, Philos.Trans.R.Soc.London 203:385 (1904).
2. See for example, R. Landauer, Electrical conductivity in inhomogeneous media, AIP Conference Proc. 40:2 (1978).
3. C.G. Blatchford, J.R. Campbell, and J.A. Creighton, Plasma resonance-enhanced raman scattering by adsorbates on gold colloids: the effects of aggregation, Surface Sci. 120:435 (1982).
4. R.P. Devaty and A.J. Sievers, Far-infrared absorption by small silver particles in gelatin, Phys.Rev. B41:7421 (1990).
5. D.A. Weitz and M. Oliveria, Fractal structures formed by Kinetic Aggregation of aqueous gold colloids, Phys.Rev. Letters, 52:1433 (1984).
6. D.A. Weitz, J.S. Huang, M. Lin, and J. Sung, Limits of the fractal dimension for irreversible kinetic aggregation of gold colloids, Phys.Rev.Letters B54:1416 (1985).
7. F. Claro and R. Fuchs, Collective surface modes in fractal clusters of spheres, Phys.Rev. B44:4109 (1991) and references there included.
8. H.M. Lindsay, M.Y. Lin, D.A. Weitz, P. Sheng, Z. Chen, R. Klein, and P. Meakin, Properties of fractal colloid aggregates, Faraday Discuss.Chem.Soc. 83:153 (1987).
9. V.K. Varadan, V. N. Bringi, and V.V. Varadan, Coherent electromagnetic wave propagation through randomly distributed dielectric scatterers, Phys.Rev. D19:2480 (1979).
10. P.C. Waterman, Symmetry, unitarity, and geometry in electromagnetic scattering, Phys.Rev. D3:825 (1971).

11. B. Peterson and S. Strom, T matrix for electromagnetic scattering from an arbitrary number of scatterers and representations of E(3), Phys.Rev. D8:3661 (1973).
12. O.R. Cruzan, Translational addition theorems for spherical vector wave equations, Q.Appl.Math 20:33 (1962).
13. V.V. Varadan, Elastic wave scattering, in: "Acoustic Electro- magnetic and Elastic Wave Scattering", V.K.Varadan and V.V.Varadan, ed., Pergamon Press, New York (1980).
14. P.B. Johnson and R.W. Christy, Optical constants of the noble metals, Phys.Rev. B6:4370, (1972).
15. D.A. Weitz, M.Y. Lin, and C.J. Sandroff, Colloidal aggregation revisited: new insights based on fractal structure and surface-enhanced Raman scattering, Surface Sci. 158:147 (1985).

RECENT THEORETICAL DEVELOPMENTS IN THE ANALYSIS OF THE VISCOELASTIC BEHAVIOUR OF MATERIALS

F. Povolo[1,2] and Élida B. Hermida[1,2]

1: Dto. de Física, Fac. de Ciencias Exactas y Naturales, U.B.A.
   Pabellón I, Ciudad Universitaria, 1428 Buenos Aires, Argentina

2: Dto. de Materiales, Comisión Nacional de Energía Atómica
   Av. del Libertador 8250, Buenos Aires, Argentina

Abstract: The viscoelastic behaviour of materials, particularly polymers, is interpreted in terms of relaxation or retardation spectra. To obtain the spectra from the experimental curves, it is necessary to make use of the time-temperature superposition, in order to extend the experimental range of the macroscopical variables. The master curves are, generally, subsequently used to derive the spectra from the experimental data. The spectra obtained give information on the micromechanisms controlling the viscoelastic behaviour and are related to the theoretical models.

It is the purpose of this work to discuss recent theoretical developments in the analysis of the time-temperature superposition, on the validity of the spectra obtained from the master curves, through a comparison with those obtained from the individual curves, on the presence or not of a master curve and on some mathematical properties of distribution functions of relaxation or retardation times.

INTRODUCTION

The mechanical properties of a polymeric material are used to characterize the micromechanisms that control its linear behaviour. Since each mechanism, described by a characteristic time τ, has a different influence on the properties of the whole system, a distribution function or spectrum of the characteristic times determines the qualitative contribution of the processes involved. These distribution functions depend on the mechanical test that is considered: when the amplitude of strain is kept constant, a relaxation distribution function results while a test done at a fixed amplitude of stress leads to a spectrum of retardation times. Though the relaxation and retardation spectra seem to be independent from each other, according to the theory of linear viscoelasticity they can be interconverted[1] and also related to the measured mechanical properties by integral transformations[2]. As these transformations are usually complex, the spectra are calculated, by approximation methods, from the measured moduli or compliances[3-6]. These methods approach the spectra by the first derivative or even derivatives of higher order of the viscoelastic functions in such a way that the interval of characteristic times is essentially the same as the time or frequency interval of the measured property.

Now, in the transition region from the glassy to the rubbery state, the moduli or compliances cover nearly three orders of magnitude while the response of the acting micromechanisms extends over many decades of the characteristic times[7]. This behaviour can be established on considering the mechanical evolution of a linear amorphous polymer tested at different temperatures but cannot be measured at a fixed temperature. In effect, any experimental technique can cover only three to four decades of the log-time (or frequency) scale so the experimental curves, usually represented in a double-log plot, seem to provide scanty information about the complete distribution function. Nevertheless, several investigators[8-12] considered that the viscoelastic functions could be traced out over a much larger effective range by making measurements at different temperatures and using a sort of equivalence between time and temperature. This equivalence, known as the time-temperature superposition principle (TTSP), establishes that a given property measured for short times must be identical with one measured for longer times at a lower temperature, except that the curves are shifted parallel to the log-time (or frequency) axis. Therefore, the matching of the segments of curves measured at different temperatures lead to a master curve extended over many orders of magnitude, at the reference temperature. In this way the TTSP becomes a useful tool to determine the distribution functions of relaxation or retardation times.

In the classical literature it is established that[2]: "Generally the short time processes are revealed in more detail in the relaxation spectrum and the long time processes in the retardation spectrum". However, according to the linear theory of viscoelasticity, both the relaxation and retardation spectra should provide the same information about the micromechanisms involved. Thus, it seems to appear an apparent contradiction between the theoretical interconversion of the spectra and the prevalence of long or short time processes according to the approximated distribution that is considered. Moreover, on one hand, according to their definitions, the spectra provide not only a statistical distribution of the structural processes but include also the contribution of each mechanism to the macroscopic evolution of the viscoelastic property; on the other hand, the interconversion formulae between the spectra depend on the equilibrium modulus or the glassy-state compliance. Therefore, the interconverted spectrum overlaps the information of both the statistical distribution of the micromechanisms and the change of the mechanical response of the sample.

Consequently, it is the purpose of this paper to determine if the master curve can be built, that is, if the measured curves can be extrapolated through horizontal shifts of curves measured at different temperatures. Furthermore, a normalized treatment of the viscoelastic function will be proposed to solve the apparent contradiction between the relaxation and retardation approximated spectra. This treatment will also provide a separation of the variables that describe the micromechanisms and the macroscopic response of the whole system.

THEORETICAL BACKGROUND

In order to facilitate the comprehension of the concepts to be used later on in this paper, this section will provide the main theoretical background needed. The detailed theoretical analysis can be found in the appropriate references.

Scaling Properties and the General Form of a Function with Translation Parallel to the Abscissa

Figure 1 shows two curves in the (x, y) plane, parametrized in z, which are related by scaling along the translation path of slope μ. This means that the points A, B are translated to A', B' by making the increments $(\Delta x, \Delta y)$ and $(\Delta x', \Delta y')$, respectively, in such a way that the following holds:

a) Points of equal derivatives must be superposed, that is,

$$y_x(x + \Delta x, z + \Delta z) = y_x(x, z) \qquad (1)$$
$$y_x(x + \Delta x + \Delta x', z + \Delta z) = y_x(x + \Delta x, z)$$

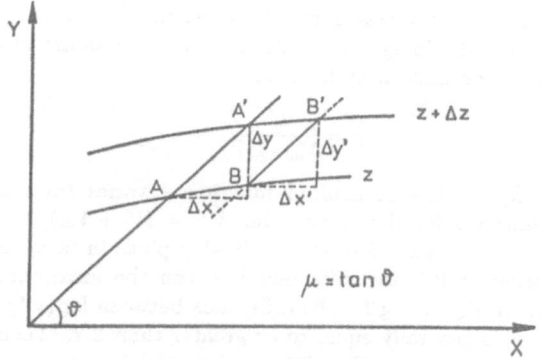

Figure 1. *Two curves in the (x,y) plane, parametrized in z, related by scaling along a translation path of slope* μ. *Points A, A' or B, B' have the same derivatives.*

where the subindex indicates a partial derivative with respect to the corresponding variable.

b) The slope of the translation path must be independent from the variables, that is,

$$\frac{\Delta y}{\Delta x} = \frac{\Delta y'}{\Delta x'} = \mu = \text{constant} \tag{2}$$

c) Since the translation is rigid, the square of the hypotenuse of the triangle defined by the increments of the variables x and y depends only on the increments of the parameter z, that is,

$$[(\Delta x)^2 + (\Delta y)^2]^{1/2} = [(\Delta x')^2 + (\Delta y')^2]^{1/2} = M(\Delta z) \tag{3}$$

where M is a function of Δz.

According to these properties, it has been demonstrated that the most general function with scaling along a certain direction is given by[13]

$$F(y - \mu x, x - \beta h(z)) = 0 \tag{4}$$

where F is a general function, $h(z)$ is an arbitrary function that depends only on z and β is the slope of the translation path in the $(h(z),x)$ plane, that is,

$$\beta = \frac{\Delta x}{\Delta h(z)} \tag{5}$$

In particular, when the translation path in the (x,y) plane is parallel to the abscissa, that is, when $\mu = 0$, Eqn.(4) reduces to

$$F(y, x - \beta h(z)) = 0 \tag{6}$$

Normalized Viscoelastic Functions

Because of the great variation of the modulus or the compliance of an amorphous polymer in the transition region, the mechanical properties are usually represented in a double-log plot. However, the normalization of these functions provide another possible representation. Effectively, for the transient properties or the real component of the dynamic modulus or compliance, this normalized function is given by

$$f = \frac{V - V_l}{V_m - V_l} \tag{7}$$

where the viscoelastic function V varies from a lower value V_l to a maximum V_m. On the other hand, if V represents the imaginary component of a dynamic mechanical property, characterized by a peak, its normalization leads to

$$g = \frac{V}{V_m - V_l} \tag{8}$$

Now, according to Eqn. 7, the normalized function f varies from zero when $V = V_l$ to one for $V = V_m$, being 0.5 for the mean value $V^* = (V_l + V_m)/2$. This proportional distribution of the data is not observed in the double-log plot. In fact, considering that the ratio V_m/V_l is of the order of 10^3, the difference between the maximum ordinate $\log(V_m)$ and $\log(V^*)$ reduces practically to $\log 2$. The difference between $\log(V^*)$ and the minimum ordinate $\log(V_l)$, however, is generally equal to or greater than 2.7. Then in the double-log representation the values of V greater than V^* are clustered in a small region whose width is equal to $\log 2$, while the lower values extend over nearly three orders of magnitude.

To illustrate this distortion, the real part of the dynamic compliance, J', and the lognormal retardation spectrum are considered. The lognormal distribution function is defined as[14]

$$L(\ln(\tau/\tau_m)) = \frac{J_e - J_g}{\beta\sqrt{\pi}} \exp\{-[\ln(\tau/\tau_m)/\beta]^2\} \tag{9}$$

where τ is the retardation time, τ_m the mean retardation time, β the half-width of the distribution, J_g the compliance in the glassy state and J_e the equilibrium compliance for a viscoelastic solid or the steady-state compliance for a viscoelastic liquid. This function is symmetrical and can be expressed as

$$L(y) = L_l(y) + L_r(y) \tag{10}$$

being

$$y = \ln(\tau/\tau_m) \tag{11}$$

$$L_l = \begin{cases} L & \text{for } y \leq 0 \\ 0 & \text{for } y > 0 \end{cases} \tag{12}$$

and

$$L_r = \begin{cases} 0 & \text{for } y \leq 0 \\ L & \text{for } y > 0 \end{cases} \tag{13}$$

This separation into the specular halves of the lognormal spectrum will be used to analyze how the short-time and long-time processes contribute to the mechanical response.

It is known that the relationship between the spectrum L and J' can be written as[2]

$$J'(x) = J_g + \int_{-\infty}^{\infty} \frac{L(y)\,dy}{1 + \exp[2(x+y)]} \tag{14}$$

where $x = \log(\omega\tau_m)$, ω being the angular frequency. Likewise, the contributions of L_l and L_r lead to the partial compliances

$$J'_{(l,r)}(x) = J_g + \int_{-\infty}^{\infty} \frac{L_{(l,r)}(y)\,dy}{1 + \exp[2(x+y)]} \tag{15}$$

The double-log representations of the dynamic compliances J', J'_l and J'_r are shown in Figure 2 (A). From these curves it results that when $x \ll 0$, J' and J'_l are nearly parallel to the horizontal axis with a vertical difference of about $\log 2$. Thus, the approximated spectra associated with both curves will be zero for retardation times longer than τ_m. Consequently, the double-log representation is not sensitive to differenciate between L and L_l because it expands the region where only the short-time mechanisms are involved and compresses the portion of the curve where these two functions are different. The same distortion is found in the double-log representation of the imaginary component of the dynamic modulus or

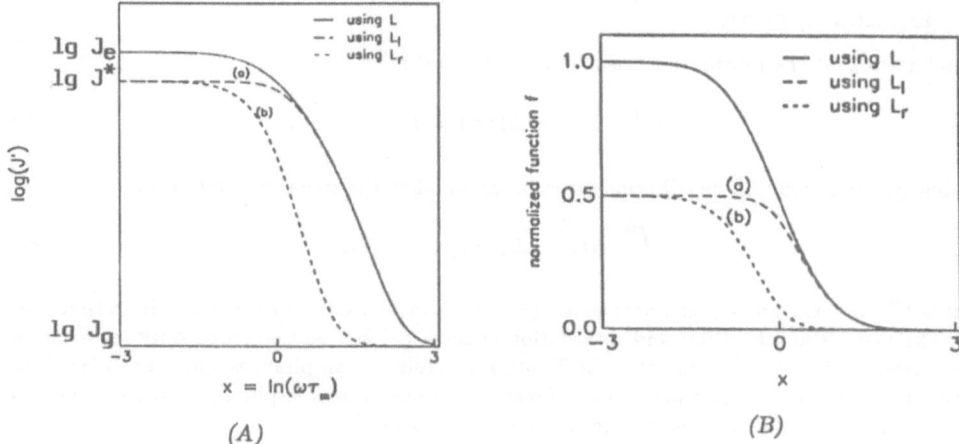

Figure 2. Double-log (A) and normalized (B) representations of the real component of the dynamic compliance: the full line corresponds to the lognormal spectrum L, the dashed curves (a) and (b) refer to the left and right halves of the lognormal spectrum, respectively.

compliance[15]. A different response is found if, for example, the normalized component of the storage compliance is considered. Combining Eqns. 7 and 14, the normalized function results

$$f(x) = \frac{1}{J_e - J_g} \int_{-\infty}^{\infty} \frac{L(y)\,dy}{1 + \exp[2(x + y)]} \tag{16}$$

Furthermore, the partial normalized functions f_l and f_r, defined replacing L by L_l and L_r respectively, verify that

$$f(x) = f_l(x) + f_r(x) \tag{17}$$

for all values of x. These two functions f_l and f_r and also f are represented in Figure 2 (B) showing that when $x \ll 0$,

$$f_l \approx 0.5 \tag{18}$$

and accordingly to Eqn. 17

$$f_r(x) \approx f(x) - 0.5 \tag{19}$$

while for $x \gg 0$

$$f_r(x) \approx 0 \tag{20}$$

so that

$$f_l(x) \approx f(x) \tag{21}$$

Then, on applying the first order approximation of L^3

$$L_1(y) = -(J_e - J_g) \left. \frac{df(x)}{dx} \right|_{x=-y} \tag{22}$$

to Eqns. 18 to 21, the approximated partial spectra L_{1r} and L_{1l} result

$$L_{1l} = \begin{cases} 0 & \text{for } y \gg 0 \\ L_1 & \text{for } y \ll 0 \end{cases} \tag{23}$$

and

$$L_{1r} = \begin{cases} L_1 & \text{for } y \gg 0 \\ 0 & \text{for } y \ll 0 \end{cases} \tag{24}$$

which are in accordance with the definitions of L_l and L_r.

Hence, through the approximation methods, f provides genuine information about the distribution of the times that characterize the viscoelastic mechanisms.

Normalized Spectra

It is known that the retardation spectrum L verifies that[2]

$$\int_{-\infty}^{\infty} L(\ln \tau) \, d(\ln \tau) = J_e - J_g \tag{25}$$

Similarly, the relaxation distribution function H satisfies the following condition[2]

$$\int_{-\infty}^{\infty} H(\ln \tau) \, d(\ln \tau) = G_g - G_e \tag{26}$$

where G_g and G_e are the instantaneous (glassy) and the equilibrium moduli, respectively. Therefore, the relaxation and retardation spectra cannot be described only in terms of the statistical parameters but also the limiting moduli or compliances must be taken into account. These macroscopic parameters, however, do not appear explicitly in the normalized distribution functions. In fact, the normalized retardation spectrum

$$\Psi(\ln \tau) = \frac{L(\ln \tau)}{J_e - J_g} \tag{27}$$

and the normalized relaxation spectrum

$$\Phi(\ln \tau) = \frac{H(\ln \tau)}{G_g - G_e} \tag{28}$$

are defined in such a way that

$$\int_{-\infty}^{\infty} \Phi(\ln \tau) \, d(\ln \tau) = 1 = \int_{-\infty}^{\infty} \Psi(\ln \tau) \, d(\ln \tau) \tag{29}$$

Thus, Φ and Ψ constitute real distribution functions because the integral of Φ (or Ψ) between the characteristic times τ_1 and τ_2 gives directly the probability that the processes with relaxation (or retardation) times between τ_1 and τ_2 are involved in the evolution of any modulus (or compliance).

Now, even when the differences $G_g - G_e$ and $J_e - J_g$ provide a quantitative description of the relaxation or retardation process, it is more suitable to employ the intensity of relaxation Δ defined as[14]

$$\Delta = \frac{G_g - G_e}{G_e} = \frac{J_e - J_g}{J_g} \tag{30}$$

because this parameter gives an idea of the relative change of the respective mechanical property. Then, on considering eqn.(30), the normalized spectra are redefined as

$$\Phi(\ln \tau) = \frac{H(\ln \tau)}{G_e \, \Delta} \tag{31}$$

$$\Psi(\ln \tau) = \frac{L(\ln \tau)}{J_g \, \Delta} \tag{32}$$

Any of these normalized spectra will be characterized by statistical parameters such as its mean value, its variance, etc. but will not depend explicitly on the limits of the viscoelastic function that has been measured. Furthermore, as pointed out in the Introduction, the connection between relaxation and creep processes leads to an interrelation between the relaxation and the retardation spectra[1]. Hence, on substituting Eqns. (31) and (32) in those conversion formulae, the following equations can be easily derived

$$\Phi(\ln \tau) = \frac{(1 + \Delta) \Psi(\ln \tau)}{\left\{ 1 + \Delta \int_{-\infty}^{\infty} \frac{\Psi(\ln u) \, d(\ln u)}{(1 - \frac{u}{\tau})} - \frac{\tau}{\eta_o J_g} \right\}^2 + \pi^2 \Delta^2 \Psi^2(\ln \tau)} \tag{33}$$

$$\Psi(\ln \tau) = \frac{(1 + \Delta)\, \Phi(\ln \tau)}{\left\{ 1 + \Delta \int_{-\infty}^{\infty} \frac{\Phi(\ln u)\, d(\ln u)}{\left(\frac{\tau}{u} - 1 \right)} \right\}^2 + \pi^2 \Delta^2 \Psi^2(\ln \tau)} \tag{34}$$

where η_o is the viscosity of an uncrosslinked polymer (for a cross-linked polymer η_o tends to infinity).

Therefore, for a cross-linked polymer, the conversion of the known spectrum depends neither on G_g or J_e nor on its difference, but on its ratio or, equivalently, on Δ. In this way, the influence that the statistical parameters of the original distribution and Δ has on the conversion of the normalized spectra can be analized separetely.

APPLICATIONS

Time-temperature Superposition

It has been established in the Introduction that the characteristics of the micromechanisms can be derived from the experimental curves of any viscoelastic property only after building the master curve according to the TTSP. In order to analyze the applicability of this principle, the data of Ferry and coworkers, given in the literature[2] as a classical example of the validity of the TTSP, will be considered. These data correspond to the reduced dynamic storage compliance, J_p', measured as a function of frequency in poly(n-octyl methacrylate), in the temperature region between 258.9 K and 402.7 K. In the original reference a master curve was constructed by horizontal shifts of the curves measured at different temperatures with respect to the one measured at 332.8 K taken as the reference, so it will not be repeated here. It should be pointed out, however, that the matching is not good enough to guarantee the validity of the TTSP, as it is shown in Figure 3 where a partial view of the master curve is indicated. This partial matching has already been noticed by Ferry[2] in the application of reduced variables specially for several methacrylate polymers.

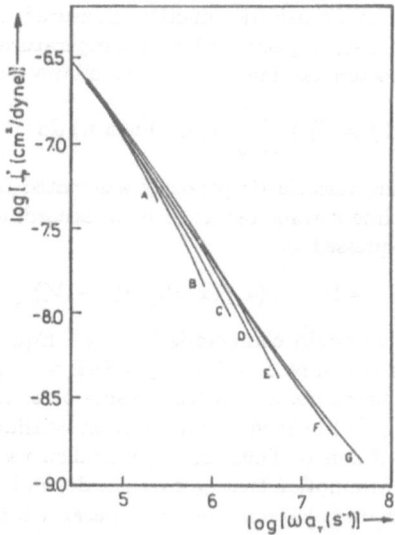

Figure 3. *Expanded view of parts of the master curve constructed by horizontal shifts of some curves of the reduced shear storage compliance of poly(n-octyl methacrylate) characterized by the temperatures: A = 338.6K; B = 332.8K; C = 327.4K; E = 317.4K; F = 307.2K; G = 303K.*

The definitive proof of the existence of the master curve, however, is the validity of the scaling conditions, particularly, the superposition of the derivatives when the respective horizontal shifts are considered. Figure 4 shows that if the translation paths proposed by

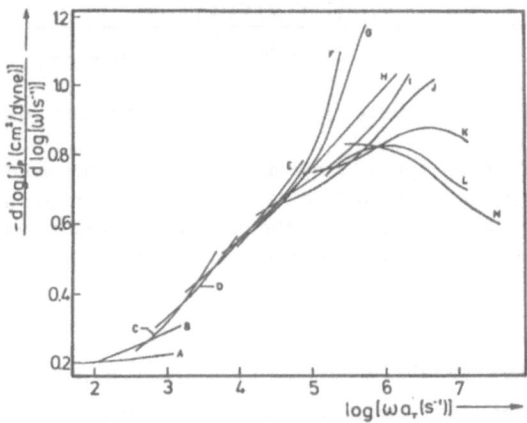

Figure 4. *Derivatives of the reduced dynamic storage compliance of poly(n-octyl methacrylate) measured as a function of frequency at the temperatures: $A = 382.4K$; $B = 372.8K$; $C = 362.4K$; $D = 353.2K$; $E = 343.9K$; $F = 338.6K$; $G = 332.8K$; $H = 327.4K$; $I = 323.2K$; $J = 317.4K$; $K = 311.8K$; $L = 307.2K$; $M = 303K$. Curve B is taken as the reference and the horizontal displacements given by Ferry[2] were considered.*

Ferry et al[2] are considered, the derivatives of each set of curves do not superpose at all, that is, the different curves used to construct the master curve do not belong to the same family.

Nevertheless, the question is: Why is an apparent matching of the different curves observed? A possible answer to this question can be obtained from the following arguments: In general, with the procedure normally used to construct the master curve the limits V_l and V_m are fixed while the temperature dependence appears only in the parameters of the spectrum. Hence, if $V(x,T)$ is the viscoelastic function measured at $x = \log t$ or $-\log \omega$ (for a cuasi-static or a dynamic test, respectively) at a temperature T, and $D(u,T)$ is the distribution function of the characteristic times $\tau = e^u$, it follows that

$$V(x,T) = V_l + \int_{-\infty}^{\infty} D(u,T)\, v(x,u)\, du \tag{35}$$

where $v(x,u)$ is the corresponding viscoelastic property associated to a single anelastic element (SAE)[14] of characteristic time τ evaluated at x. Now, according to the theorem of the mean value[16], Eqn. 35 can be expressed as

$$V(x,T) = V_l + v(x,\xi(x,T))\,(V_m - V_l) \tag{36}$$

being $\xi(x,T)$ the logarithm of a certain characteristic time. Eqn. 36 means that at any value of x, the mechanical property can be described by a SAE with a variable characteristic time given by $\exp[\xi(x,T)]$. Furthermore, around each value of x, the function $\xi(x,T)$ can be approximated by a function $\xi_x(T)$ in such a way that, on considering $h(z) = \xi_x(T)$ and $y = \ln V$, Eqn. 36 takes the form of Eqn. 6. Thus, the individual curves can be superposed only locally because of the restricted assumption that ξ_x does not depend on x. Consequently, the master curve represents a certain kind of average response because it is constructed under the assumption of local superposition of segments of curves measured at different temperatures.

Normalized Representation

Even when the superposition is not rigorously valid, the construction of the master curve is widely used not only to extrapolate the mechanical behaviour but also to analyze molecular models of the polymeric structure. This can lead to severe inconsistencies as it will be shown in the following example.

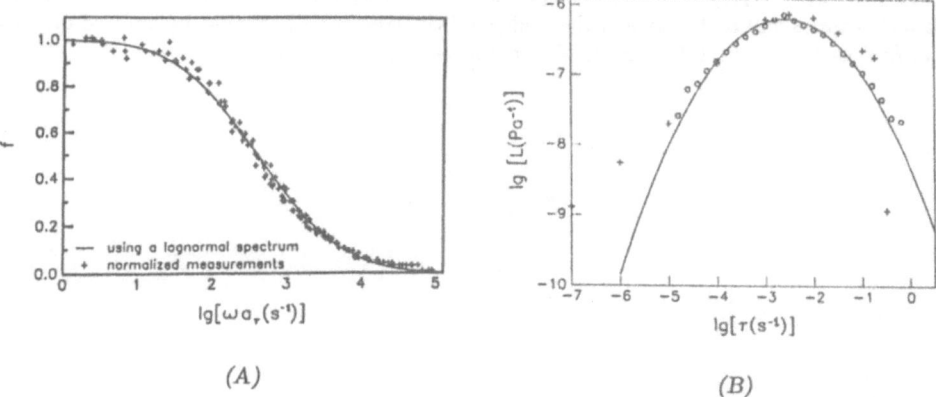

Figure 5. *Master curve of the dynamical storage compliance (A) and retardation spectrum (B) of polyisobutylene. The full lines correspond to a lognormal retardation spectrum characterized by $\log \tau_m = -2.6$ and $\beta = 2.7$. The crosses (+) in Figs. (A) and (B) represent the measured data and the first order approximated spectra derived from the double-log plot, respectively. The open circles (○) represent the first order spectrum calculated from the normalized function.*

The data of the dynamic storage compliance measured by Fitzgerald et al.[17] had been used to verify the time-temperature superposition, to calculate the distribution function using several approximation methods and also to determine the parameters of a molecular theory[18]. In this case the data, usually represented in a double-log plot, were shifted using the horizontal paths proposed in the original reference[19], but in a normalized plot, leading to the matching shown in Figure 5 (A). From this normalized plot, Alfrey's rule leads to the spectrum indicated by the circles in Figure 5 (B). This spectrum is compared with the one derived when Alfrey's rule is applied to the double-log plot[19], which emphasises the influence of the short-time mechanisms, falling sharply to zero after the maximum, as it is shown by the crosses in Figure 5 (B). Then, one spectrum is symmetrical while the other is not, leading to very different molecular descriptions of the polymeric structure. Therefore, in order to define which distribution function is correct, an attempt to fit the curve with a lognormal spectrum will be made. In effect, the parameters τ_m and β of the distribution were selected to provide a rather good fit to the normalized master curve, as it is shown in Figure 5 (A). The full line of Figure 5 (B) represents the lognormal distribution which is centered at the same mean time as the approximated distribution function derived from the normalized representation, pointing out that the true distribution of retardation times is a gaussian which do not emphasize the response of the fast mechanisms.

Interconversion of Spectra

The comparison of the approximated spectra derived from different representations of the same data is a proof of the distortion not only of the viscoelastic functions but also of the distribution functions derived from the double-log plot of any mechanical property and, consequently, of the molecular models involved. For instance, the relaxation spectra, derived through approximation methods from the double-log plot of a modulus, usually cuts off at short times so, it is usually approximated by a wedge distribution function as the one shown in Figure 6.

The wedge spectrum is defined as

$$\Phi(y) = \frac{\alpha}{2 \sinh(\alpha \gamma)} \exp(\alpha y)[\theta(y + \gamma) - \theta(y - \gamma)] \tag{37}$$

where $y = \ln(\tau/\sqrt{\tau_1 \tau_2})$, $\gamma = \ln(\sqrt{\tau_2/\tau_1})$ being τ_1 and τ_2 the minimum and the maximum relaxation times and α the slope of the wedge of $\ln \Phi$ against y. It should be pointed out that Eqn. (37) satisfies Eqn. (29), that is, Φ is a normalized distribution function.

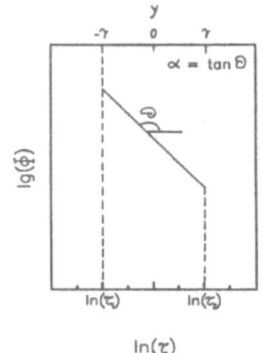

Figure 6. *Scheme of the normalized wedge relaxation spectrum characterized by a slope α.*

Then, according to the interconversion relationship given by eqn. (34), the normalized retardation spectrum associated with Φ results

$$\Psi(y) = \frac{\frac{\alpha\, e^{\alpha y}}{2 \sinh(\alpha\gamma)} (1 + \Delta)\, [\theta(y + \gamma) - \theta(y - \gamma)]}{\pi^2 \Delta^2 \left[\frac{\alpha e^{\alpha y}}{2 \sinh(\alpha\gamma)}\right]^2 [\theta(y+\gamma) - \theta(y-\gamma)]^2 + \left\{1 - \frac{\Delta\alpha\, e^{\alpha y}}{2 \sinh(\alpha\gamma)} \int_{-\gamma}^{\gamma} \frac{e^{-\alpha(y-w)}\, dw}{e^{y-w}-1}\right\}^2} \tag{38}$$

The integral which appears in Eqn. (38) can be expressed in terms of the hypergeometric function[20] $F_1(1,1,-\alpha,2;u,v)$ that depends on two variables, u and v, and on the parameter α. The general expression, however, cannot be used to describe any converted spectrum in terms of known functions, except if a particular value of α is considered.

Several examples have been given in the literature where the wedge spectra are associated to Rouse's molecular theory[18] characterized by $\alpha = -0.5$. In this case, the converted retardation spectrum results

$$\Psi(y) = \frac{\frac{1+\Delta}{4 \sinh(\gamma/2)} e^{-y/2} [\theta(y + \gamma) - \theta(y - \gamma)]}{\frac{\pi^2 \Delta^2}{16 \sinh^2(\gamma/2)} e^{-y} [\theta(y+\gamma) - \theta(y+\gamma)]^2 + \left\{1 - \frac{\Delta e^{-y-\gamma/2}}{4 \sinh(\gamma/2)} \ln \left|\frac{\sinh(\gamma/2)+\sinh(y/2)}{\sinh(\gamma/2)-\sinh(y/2)}\right|\right\}^2}$$
$$\tag{39}$$

When $-\gamma < y < \gamma$, the retardation spectrum is a continuous function whose integral decreases when Δ increases, as illustrated in Figure 7 for $\gamma = 3$. The area under the normalized converted spectrum over the interval $(-\gamma, \gamma)$, that is, the probability of having retardation processes with characteristic times between $\tau_m\, e^{-\gamma}$ and $\tau_m\, e^{\gamma}$, diminishes as Δ increases. According to Eqn. (29), however, the integral of the normalized converted spectrum must be unity so, another contribution must be included.

Even when the numerator is zero outside the interval $(-\gamma, \gamma)$, the retardation distribution function Ψ has a singularity when the denominator is also zero. Effectively, Eqn. (39) can be written as

$$\Psi(y) = \frac{A\, f(y)}{f^2(y) + g^2(y)} \tag{40}$$

being

$$A = \frac{4\,(1 + \Delta)\, \sinh(\gamma/2)}{\pi^2 \Delta^2}$$

$$f(y) = e^{-y/2}\, [\theta(y + \gamma) - \theta(y - \gamma)]$$

$$g(y) = \frac{4 \sinh(\gamma/2)}{\pi \Delta} - \frac{e^{-y/2}}{\pi} \ln \left|\frac{\sinh(\gamma/2) + \sinh(y/2)}{\sinh(\gamma/2) - \sinh(y/2)}\right|$$

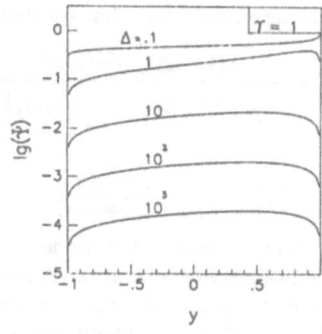

Figure 7. *Continuous portion of the normalized retardation spectrum corresponding to a wedge relaxation distribution function with a half-width $\gamma = 3$, for different values of Δ.*

Eqn. (40) can be written in terms of a Dirac's delta function[21] centered at the zero of g, namely at $y = y^*$. The dependence of y^* on Δ is illustrated in Fig. 8 (A) for different values of γ. Hence, the wedge relaxation spectrum is converted to a retardation distribution function which is divided into two pieces: a continuous spectrum, defined over the same interval of time as the relaxation distribution and a single retardation process with a characteristic time higher than any time of the continuous piece. Furthermore, on using Eqn. (29), the intensity of relaxation of the Dirac's function results

$$I = \frac{A\pi}{|g'(y^*)|} = \frac{2(1+\Delta)}{\Delta} \left\{ 1 + \frac{\Delta e^{-y^*/2} \cosh(y^*/2)}{2[\sinh^2(y^*/2) - \sinh^2(\gamma/2)]} \right\} \tag{41}$$

giving the probability of having a retardation process with a characteristic time $\tau_m e^{y^*}$. This probability increases with Δ as illustrated in Figure 8 (B), being nearly unity for high values of Δ.

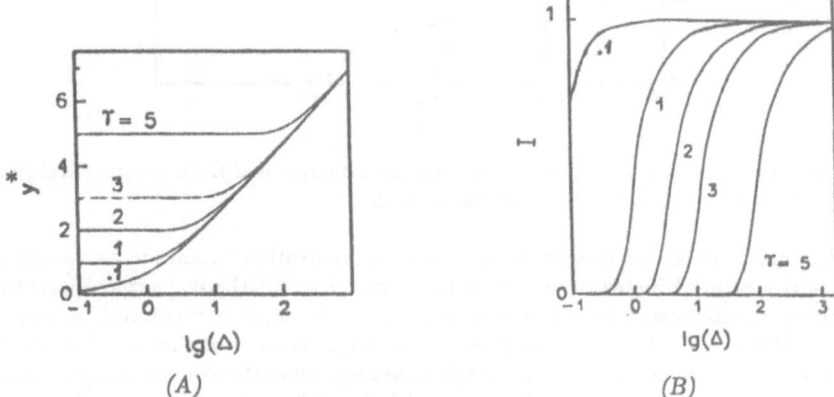

Figure 8. *Curves of y^* and I against $\log \Delta$ in (A) and (B), respectively. The curves correspond to different half-widths γ of the original wedge relaxation spectrum.*

From the interconversion of other bounded spectra treated in previous papers[22], it can be stated that any bounded distribution function of relaxation times leads to a retardation spectrum divided into a continuous piece plus a single retardation process whose probability increases with Δ. Analogously, when a bounded distribution function of retardation times is converted, the relaxation spectrum is divided into a discrete and a continuous piece. In this case, the transformation gives a continuous piece defined over the same temporal interval as the retardation spectrum plus a single process with a relaxation time located below the lower limit of the continuous distribution.

For a lognormal distribution, on substituting Eqn.(9) into Eqn.(33) and introducing a new variable $w = \ln(\tau/\tau_m)/\beta$, the converted relaxation spectrum results

$$\Phi(w) = \frac{[(1+\Delta)/\sqrt{\pi}]\,e^{-w^2}}{\frac{\pi\Delta^2}{\beta}e^{-2w^2} + \left\{1 + \frac{\Delta}{\sqrt{\pi}}I(w)\right\}^2} \tag{42}$$

where

$$I(w) = \int_{-\infty}^{\infty} \frac{e^{-u^2}\,du}{1 - e^{\beta(u-w)}} \tag{43}$$

As $I(w)$ cannot be expressed in terms of known functions, the dependence of $\Phi(w)$ on the statistical parameter β and on the intensity of relaxation Δ must be determined numerically. Particularly, if the statistical parameter β is fixed, the converted retardation spectrum will depend only on Δ. This dependence of Φ on the macroscopical response of the system can be illustrated on considering the dynamical measurements in polyisobutylene (PIB) reported by Fitzgerald et al.[19] as an example. Using the recurrency relationships of the viscoelastic functions associated to a lognormal spectrum[23-24] it can be established that the master curve of the real component of the dynamic compliance of PIB is related to a lognormal retardation spectrum with $\beta = 2.7$ and an intensity of relaxation $\Delta = 2.8 \times 10^3$. Then, on considering $\beta = 2.7$ as a rather good parameter to characterize the statistical distribution of retardation times involved, the dependence of Φ on the intensity of relaxation is represented in Figure 9.

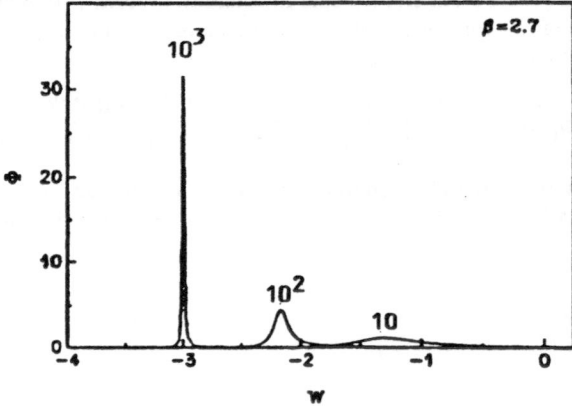

Figure 9. *Normalized relaxation distribution function determined from a lognormal retardation spectrum with $\beta = 2.7$, for different values of Δ.*

The converted relaxation spectrum shifts to lower relaxation times, looses its symmetry and gets narrower as Δ increases. Particularly, when $\Delta = 10^3$ the converted spectrum can be practically approximated by a single relaxation process with a characteristic time $\tau_\varepsilon = \tau_m/(1+\Delta)$. Hence, for a high value of Δ the relaxation behaviour of PIB could be described by a single relaxation time. However, the original description of the system was given in terms of a lognormal distribution of retardation times with $\beta = 2.7$ so, there is a contradiction in the characterization of the structural mechanisms of PIB. Therefore, as no other condition was imposed, this contradiction seems to be a consequence of assuming the validity of the master curve. The principal reason to introduce the master curves is that the individual curves do not provide enough information to determine the distribution functions through approximation methods, in previous papers[25-24], however, it was shown how the parameters β and Δ can be calculated by using the mathematical properties of the viscoelastic functions associated to a lognormal distribution. In effect, under the assumption of a lognormal spectrum, the mechanical properties must fulfill certain recurrency relationships which involve critical points such as the inflection point of a sigmoidal curve (associated with a stress relaxation modulus, creep compliance or the real component of a dynamic property) or, the maximum and half-width of a peak (corresponding to the imaginary component of the dynamic compliance or the loss tangent). That is to say, the determination of β and Δ from an individual curve is possible when a sigmoidal curve includes the inflection point or when a peak is measured at least from the half of its maximum value.

Particularly, on considering the data of the real component of PIB measured by Fitzgerald et al.[17] at 313 K and 323 K (including the critical points mentioned above), the intensity of relaxation and the half-width of the lognormal distribution can be calculated at each temperature. In both cases it results $\beta = 2.08$ while Δ takes the values 9.4 at 313 K and 23.1 at 323 K. Although the value of β is a little bit lower than the one obtained from the master curve, the most important difference is found in the intensity of relaxation which is two orders of magnitude lower for the individual curves. With these new values of Δ, the relaxation of PIB cannot be associated with a single micromechanism but to the relaxation spectra represented in Figure 10.

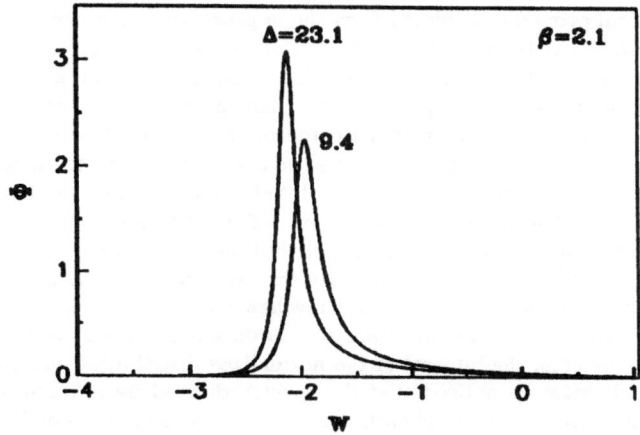

Figure 10. *Normalized relaxation spectra converted from the lognormal retardation spectra with $\beta = 2.08$, corresponding to the dynamic compliance of PIB measured at 313 K ($\Delta = 9.4$) and 323 K ($\Delta = 23.1$).*

Consequently, the apparent contradiction in the description of the retardation spectrum arises from the analysis of the master curve that leads to an extremely high value of Δ. Furthermore, Figure 10 shows that the normalized relaxation spectra depend on T in such a way that they cannot be superposed at all. Moreover, as it has been demonstrated elsewhere[26], this loss of matching to a master spectrum enables the use of a master curve to describe the mechanical behaviour of the system properly. Then, in order to characterize the viscoelastic mechanisms of a polymer, the individual curves measured at different temperatures should be considered. In this sense, it has been shown that some curves provide enough information about the mechanisms acting at a certain temperature. These mechanisms, however, may be absent or, on the contrary, may be not enough to describe the evolution of the system at other temperatures. Consequently, data of any mechanical property measured at different temperatures, interpreted in terms of the same distribution function and the same intensity of relaxation may be of doubtful physical meaning.

1   DISCUSSION AND CONCLUSIONS

The dynamical measurements in poly(n-octyl methacrylate) constitute one of the examples reported in the literature[25] to demonstrate that the time-temperature superposition is not strictly obeyed, that is, the experimental curves measured at different temperatures do not satisfy the scaling conditions. This is particularly demonstrated by the total failure in the superposition of the derivatives of the experimental curves. In these conditions, it is difficult to give a physical meaning to the master curve and any extrapolation should be considered with caution. Furthermore, it was clearly demostrated that the partial matching of the different curves, to form in a rather artificial way the master curve, can be described by a single anelastic element with a variable characteristic time.

Moreover, because of the importance of the distribution functions, the concepts related to the determination and interpretation of the spectra must be critically reviewed, particularly the incomplete information provided by the log-log plots of viscoelastic functions. The distortion in these plots, produced by the logarithm leads to molecular models that do not describe the polymeric structure properly. The normalized functions, however, provide a better approach to the viscoelastic behavior and, consequently, to the spectra determined by approximation methods or by a fitting to a master curve adjusting the parameters of a proposed distribution function. This fitting to the data, however, depends on the translation of the individual curves, that is, on the time-temperature superposition. In effect, the normalized mechanical property calculated from the approximate spectrum does not fit directly the measured values but the values shifted parallel to the horizontal axis, according to certain translation paths. Then, even when the time-temperature superposition is not strictly obeyed, because of the smooth shape of the transient properties and the real component of the dynamical properties, an apparent matching of the individual segments is found while the imaginary components lead to less defined master curves due to their sharper shapes. Consequently, the distribution functions calculated from the normalized representation of the quasi-static properties or the real components of the dynamic functions seem to provide better results. Effectively, these normalized distribution functions are more suitable than the generally used non-normalized spectra because they give directly the probabilistic information on the mechanisms that evolve during a mechanical test. Furthermore, the normalizing condition guarantees that all the relaxation or retardation processes are considered because, if that were not the case, the integral of the normalized distribution function would not be unity. However, it must be noticed that the spectra derived by a fitting to master curves establish implicitly that the limits of each individual curve do not depend on temperature.

The interconversion of the normalized spectra depends not only on the statistical parameters associated with the geomety of the known distribution but also on a macroscopic factor, the intensity of relaxation Δ. This dependence on Δ, illustrated in this paper for bounded and unbounded distribution functions, leads to a discrepancy in the statistical characterization if the intensity of relaxation is determined from a master curve. In effect, the original spectrum re-calculated from the converted distribution function practically represents a single mechanism though the known distribution may be a rather more complex function. This inconsistency does not appear, however, if the curves of the viscoelastic properties measured at different temperatures are considered, because their intensities of relaxation are much lower. Moreover, as the value of Δ depends on the temperature, the converted spectra are different and cannot be superposed. Consequently, any mechanical property measured at several temperatures cannot be interpreted in terms of the same set of micromechanisms. That is to say, the master curve might lead to an extrapolation that is suitable for the technical analysis of the mechanical behaviour. However, to describe the physical micromechanisms, the curves measured at different temperatures must be treated individually. This analysis of the data is not always possible and, though it has been the aim of previous work[26-24], further research must be developed to characterize the processes that govern the viscoelastic behaviour.

Acknowledgements: This work was supported in part by the Consejo Nacional de Investigaciones Científicas y Técnicas (CONICET), the Proyecto Multinacional de Materiales (OAS-CNEA) and the Fundación Antorchas.

REFERENCES

[1] B. Gross, *Mathematical Structure of the Theories of Viscoelasticity*, Herman et Cie., Paris (1953), p. 41.

586

[2] J. D. Ferry, *Viscoelastic Properties of Polymers*, John Wiley & Sons, New York (1980), Chs. 3 and 11.

[3] T. Alfrey and P. Doty, J. Appl. Phys. 16 (1945) 700.

[4] F. Schwarzl and A. J. Staverman, Appl. Sci. Res. A4 (1953) 127.

[5] M. L. Williams and J. D. Ferry, J. Polym. Sci. 11 (1953) 169.

[6] N. W. Tschoegl, *The Theory of Linear Viscoelastic Behaviour*, Academic Press, New York, (1981).

[7] G. M. Bartenev and Yu. V. Zelenev, *Relaxation Phenomena in Polymers*, John Wiley & Sons, New York (1974).

[8] K. W. Wagner, Elektrotech. Z. 36 (1915) 135, 163.

[9] H. Leaderman, *Elastic and Creep Properties of Filamentous Materials and Other High Polymers*, The Textile Foundation, Washington, (1943).

[10] F. H. Müller, Kolloid-Z. 114 (1949) 2.

[11] A. V. Tobolsky and R. D. Andrews, J. Chem. Phys. 11 (1943) 125.

[12] J. D. Ferry, J. Am. Chem. Soc. 72 (1950) 3746.

[13] F. Povolo and M. Fontelos, Il Nuovo Cimento 13D (1991) 1513.

[14] A. S. Nowick and B. S. Berry, *Anelastic Relaxation of Crystalline Solids*, Pergamon Press, New York (1972), Ch. 1.

[15] F. Povolo and Élida B. Hermida, Polymer Journal 24 (1992) 1.

[16] I. N. Bronshtein, K. A. Semendyayev *Handbook of Mathematics*, Leipzig Ed., Leipzig (1985).

[17] E. R. Fitzgerald, L. D. Grandine and J. D. Ferry, J. Appl. Phys. 24 (1953) 650.

[18] P. E. Rouse, J. Chem. Phys. 21 (1953) 1272.

[19] J. D. Ferry, L. D. Grandine, Jr. and E. R. Fitzgerald, J. Appl. Phys. 24 (1953) 911.

[20] J. S. Gradshtejn and J. M. Ryzhik, *Table of Integrals, Series and Products* (Academic Press, New York, 1965), p.1053.

[21] E. Roubine, *Distribution Signals*, Eurolles, Paris (1982), p. 65.

[22] F. Povolo and Élida B. Hermida, Polymer Journal 24 (1992) 11.

[23] F. Povolo and C. L. Matteo, in *Internal Friction and Ultrasonic Attenuation in Solids*, T. S. Ke, Ed., International Academic Pub., Beijing (1990), p. 579.

[24] F. Povolo and C. L. Matteo, Il Nuovo Cimento 13 (1991) 1491.

[25] F. Povolo and Élida B. Hermida, Phys. Stat. Sol. (b) 171 (1989) 71.

[26] F. Povolo and Élida B. Hermida, J. Mater. Sci. 25 (1990) 4036.

[27] F. Povolo and Élida B. Hermida, Mech. Mater. 12 (1991) 35.

LINE TENSION AT WETTING AND PREWETTING TRANSITIONS

B. Widom

Department of Chemistry
Baker Laboratory
Cornell University
Ithaca, New York 14853

Three coexisting phases α, β, and γ meet along a line of common contact with definite contact angles whenever the tensions $\sigma_{\alpha\beta}$, etc., of the three interfaces satisfy

$$\sigma_{\alpha\gamma} < \sigma_{\alpha\beta} + \sigma_{\beta\gamma} \qquad \text{(all permutations of } \alpha, \beta, \gamma \text{).} \tag{1}$$

If, instead,

$$\sigma_{\alpha\gamma} = \sigma_{\alpha\beta} + \sigma_{\beta\gamma} , \tag{2}$$

then the phases do not meet at a contact line, but rather the phase β, in this case, is spread as a film at the $\alpha\gamma$ interface. The transition between these two circumstances is termed the wetting transition.

When (1) holds, there is a (positive or negative) excess free energy due to inhomogeneities and distortions of the compositions and structures of the three interfaces where they meet at the contact line. The excess free energy per unit length of that line is termed the line tension τ. We shall be interested in what happens to τ as the wetting transition is approached, where the three-phase line itself disappears.

We shall also consider a related question with respect to a so-called prewetting (or premonitory wetting) transition. In the accompanying Fig. 1 we show a plane of two thermodynamic field variables (labeled μ and T to suggest a chemical potential and the temperature but they may be any two such fields) for a system consisting of two independent chemical components and two coexisting phases, α and γ or α and β. Such a system has two degrees of freedom, so a two-dimensional thermodynamic space, as shown. The curve (solid and hatched) labeled $\alpha\beta\gamma$ is the triple-point curve, along

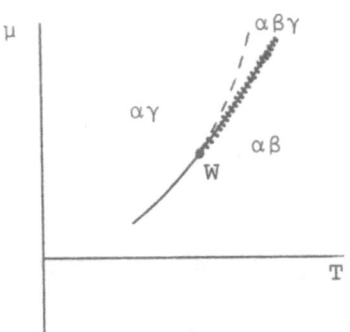

μ

αβγ

αγ

αβ

W

T

Fig. 1

which α, β, and γ coexist. On one side of the curve only phases α and γ
are present at equilibrium, on the other side only α and β. The point W
is the point of wetting transition; on the αβγ triple-point curve below W
(solid curve) the three interfacial tensions satisfy (1) while above W
(hatched curve) they satisfy (2).

The dashed curve in the αγ region of the figure, tangent to the
triple-point curve at W,[1] is the locus of prewetting transitions.[2,3] In
states represented by points between this curve and the triple-point curve
the αγ interface incorporates a film of finite thickness that resembles
the β phase, although β would not be stable as a bulk phase in that region
of the thermodynamic plane. The closer the thermodynamic state is to the
triple-point line the thicker is that β-like layer in the αγ interface,
and as the triple-point line is approached that layer becomes a macroscopically (in principle, infinitely) thick layer of bulk β.

On the other side of the line of prewetting transitions, away from
the triple-point line, the αγ interface has a different structure, not,
in general, resembling bulk β. At the prewetting-transition line itself,
those two alternative structures of the αγ interface are of equal free
energy and may coexist. When they do, there is a one-dimensional boundary
separating them, with a boundary tension, τ_b, which, like the line tension
τ defined above, has the dimensions of energy per unit length. Besides
being interested in how τ behaves as W is approached along the lower part
of the triple-point curve (Fig. 1), we are also interested in the behavior
of τ_b as W is approached along the prewetting-transition locus and in the
connection between τ and τ_b at W.

The two line tensions τ and τ_b must be equal at W, as the following
argument shows.[4]

In Fig. 2a, appropriate to the inequality (1), we are looking along
the contact line, which is perpendicular to the plane of the figure.

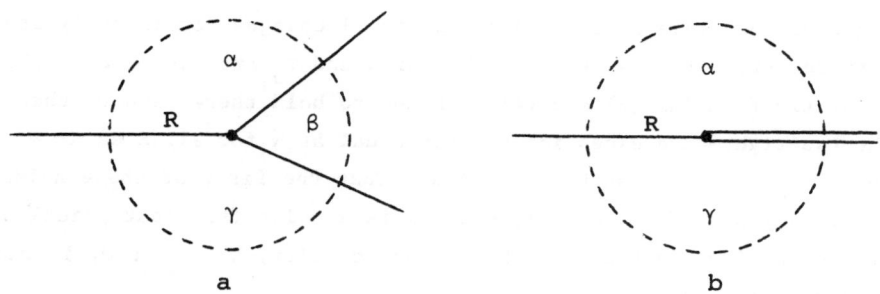

Fig. 2

The three two-phase interfaces intersect the figure plane in the three
lines shown. The bulk phases occupy the dihedral angles between interfaces. We contemplate a circular cylindrical sample of radius R (the
radius of the dashed circle in the plane of the figure), with the three-
phase line as the cylinder axis. Figure 2(b) represents the coexistence
of the two alternative structures of an $\alpha\gamma$ interface at a point of the
prewetting line of Fig. 1. Here we are looking along the one-dimensional
boundary that separates the two coexisting structures. Each of those
structures is a two-dimensional phase confined to the $\alpha\gamma$ interface. The
single line extending to the left in Fig. 2(b) and the double line extending to the right are those two surface phases as they appear in the plane
of the figure: a "thin" phase that does not resemble bulk β and a "thick"
one that does. We again contemplate a circular cylindrical sample, of
radius R, with the boundary between the two surface phases as the axis of
the cylinder, perpendicular to the plane of the figure.

Corresponding to Fig. 2(a) is a free energy F_R, which is the total
free energy of inhomogeneity of the cylindrical sample per unit length of
the three-phase line. By definition of the line tension τ this inhomogeneity free energy F_R is

$$F_R \sim (\sigma_{\alpha\beta} + \sigma_{\beta\gamma} + \sigma_{\alpha\gamma}) R + \tau \qquad (R \to \infty), \qquad (3)$$

with correction terms that vanish as $R \to \infty$. Corresponding to Fig. 2(b) is
an analogous free energy, F_R', which by definition of the boundary tension
τ_b is

$$F_R' \sim 2\sigma_{\alpha\gamma} R + \tau_b \qquad (R \to \infty). \qquad (4)$$

(The two $\alpha\gamma$ interfacial structures coexist and so have a common free energy
per unit area, $\sigma_{\alpha\gamma}$.) Figure 2(a) and Eq. (3) apply along the lower (solid)
part of the $\alpha\beta\gamma$ triple-point curve in Fig. 1, while Fig. 2(b) and Eq. (4)
apply along the locus of prewetting transitions (dashed curve). The left-
hand sides of Eqs. (3) and (4) remain finite and pass continuously into

each other as the state of the system in Fig. 1 changes continuously from one of those curves to the other via W. If τ and τ_b are finite at W then the asymptotic formulas (3) and (4) continue to hold there, and we then have the two right-hand sides identically equal at W for all R as $R \to \infty$. Therefore $\sigma_{\alpha\gamma} = \sigma_{\alpha\beta} + \sigma_{\beta\gamma}$ and $\tau = \tau_b$ at W. That the first of these holds at W we already know from Eq. (2), since W is a point (the last point) on the hatched $\alpha\beta\gamma$ line in Fig. 1. The second equality, $\tau = \tau_b$ at W, is what we wished to show.

It may be (see below) that τ and τ_b are infinite at W. In that case the asymptotic ($R \to \infty$) formulas (3) and (4), while correct at all other points along the solid and dashed curves, respectively, are of the wrong form at W; the corrections to the leading O(R) terms at W would be greater than O(1), perhaps O($\sqrt{R}$) or O(log R). Then the foregoing continuity argument shows that τ and τ_b both diverge at W if either does.

The common value (finite or infinite) of τ and τ_b at W must be positive. That is because the boundary tension τ_b must be positive if the equilibrium of the two surface phases is to be stable. Thus, if the tension τ of the three-phase line is negative anywhere along the solid curve in Fig. 1, which it may be, it must change sign before W is reached.

Szleifer and I evaluated numerically the line tension τ from an assumed free-energy functional[5] and found the results in the accompanying Fig. 3. There is an arbitrary scale factor in the ordinate; the abscissa is the contact angle (dihedral angle in Fig. 2(a)) of the β phase, which vanishes at the wetting transition. We see that τ, which is negative at large β, changes sign at around 55° and becomes large and positive, or diverges to $+\infty$, at the wetting transition. With S. Perković we are also calculating the boundary tension τ_b in a related model with a prewetting transition, with the object of determining the behavior of τ_b on approach to the wetting transition at W.

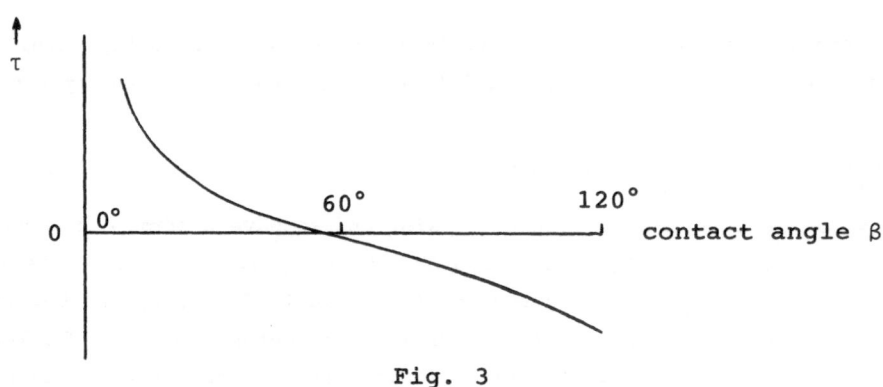

Fig. 3

Varea and Robledo[6] have evaluated τ and τ_b numerically for a spin $-\frac{1}{2}$ Ising model in mean-field approximation, and find both to become large and positive, and perhaps to diverge, at the wetting transition. Joanny and de Gennes[7] and Indekeu[8] have evaluated τ in interface-displacement models; Indekeu[8] also evaluated τ_b. They find that with short-range forces τ increases to a finite, positive limit as W is approached, but that τ diverges at W if the forces are of sufficiently long range. Indekeu finds the boundary tension at the prewetting transition, τ_b, to be continuous with τ at W when both are finite and to diverge when τ diverges, in agreement with our present expectation.

Acknowledgments

I am grateful to J.O. Indekeu for helpful conversations and correspondence. This work was supported by the U.S. National Science Foundation and the Cornell University Materials Science Center.

References

1. E.H. Hauge and M. Schick, Continuous and first-order wetting transition from the van der Waals theory of fluids, Phys. Rev. B 27:4288 (1983).
2. J.W. Cahn, Critical point wetting, J. Chem. Phys. 66:3667 (1977).
3. C. Ebner and W.F. Saam, New phase-transition phenomena in thin argon films, Phys. Rev. Lett. 38:1486 (1977).
4. B. Widom and A.S. Clarke, Line tension at the wetting transition, Physica A 168:149 (1990).
5. I. Szleifer and B. Widom, Surface tension, line tension, and wetting, Mol. Phys. 75:925 (1992).
6. C. Varea and A. Robledo, Evidence for the divergence of the line tension at the wetting transition, Phys. Rev. A 45:2645 (1992).
7. J.F. Joanny and P.G. de Gennes, Role of long-range forces in heterogeneous nucleation, J. Colloid Interface Sci. 111:94 (1986); P.G. de Gennes, private communication (1990).
8. J.O. Indekeu, Line tension near the wetting transition: results from an interface displacement model, Physica A, in press (1992).

INTERFACIAL PHASE TRANSITIONS UNDERLYING

AMPHIPHILE MICELLAR SELF-ASSEMBLY

Alberto Robledo and Carmen Varea

Instituto de Física, Universidad Nacional Autónoma de México
Apartado Postal 20-364, México 01000 D.F., México

INTRODUCTION

We describe two types of curvature-related interfacial transitions occurring in soluble amphiphile monolayers. The first type corresponds to the buckling of areas with sizes up to the monolayer's persistence length squared, and starts to occur when the interfacial tension is about one-half its bare value. A relationship is hinted at between the properties of this transition and the stability of foam films. The second, topologically-driven, transition separates a simply-connected surface state and a multiply-disconnected (or multiply connected) volume-filling state. Its order parameter is given by the surface's genus (or number of micelles). The three-dimensional gain in configurational entropy of surface fragmentation masks the singularity and provides the known continuous properties of amphiphile solutions at the critical micellar concentration. The resulting structure, droplet dispersion or interconnected bicontinuous, is determined by the sign of the monolayer's saddle-splay bending constant.

Descriptions of the onset of micelle assemblage in amphiphile solutions, at the so-called critical micellar concentration (cmc), are often preceded by that of the development of compact interfacial monolayers. As it is well-documented,[1] within this region various bulk and interfacial physical properties of the solution exhibit remarkable alterations, which have been seen and understood to correspond to the emergence of micellar aggregates in the solution.[1] Within this range of amphiphile concentration it is often observed a marked enhancement in the stability of foam films, which may result from a modification in the elastic properties of the monolayers involved. Here we describe a model amphiphile monolayer in terms of a free energy that considers both interfacial tension and curvature elastic energy contributions. An important feature of this model is the consideration that the amphiphiles are soluble in the monolayer's supporting solvent or solvents, and that the entire interface between, say, water and air, or water and oil, is constituted by the monolayer. Two types of curvature-related transitions have been

Condensed Matter Theories, Vol. 8, Edited by
L. Blum and F.B. Malik, Plenum Press, New York, 1993

predicted to occur[2] under reduction of the monolayer area per amphiphile. Here we extend the characterization of these interfacial transitions.

THE BUCKLING TRANSITION

In a system of soluble amphiphiles, for which the entire interface is constituted by the monolayer, increments in the amphiphile bulk concentration result in increments in interfacial adsorption and in a reduction of the interfacial area per amphiphile molecule. This process is similar in some respects to the compression of an insoluble amphiphile monolayer in a Langmuir trough, and a reduction of the interfacial tension γ takes place, this can be decomposed as $\gamma = \gamma_0 - \Pi$, where γ_0 is the bare (in the absence of amphiphile) interfacial tension and Π is the Langmuir surface pressure of the film. A simple (ideal lattice gas) interfacial free energy model that reproduces this development can be written as

$$f_0 = \left\{ \gamma_0 + kT \cdot \left[\eta \ln \eta + (1 - \eta) \ln(1 - \eta) \right] \right\},$$ (1)

where f_0 is the (flat) monolayer's free energy per unit area, k is Boltzmann's constant, T the temperature, and η the monolayer amphiphile concentration. Minimization of the grand potential $\omega = f_0 - \mu \eta$, where μ is the monolayer amphiphile chemical potential, yields

$$\omega_{min} = \gamma = \gamma_0 - \Pi, \text{ where } \Pi = -kT \ln(1 - \eta).$$ (2)

To study the shape response of the monolayer to increments in the amphiphile concentration we need to consider the difference in free energy between different shape states. The free energy term (per unit area) that quantifies the curvature elastic energy of the system is[3]

$$f_{curv} = \kappa(c - c_0)^2 + \bar{\kappa} \cdot C,$$ (3)

where κ and $\bar{\kappa}$ are, respectively, the splay and saddle-splay bending constants and c, c_0 and C are, respectively, the mean, spontaneous and Gaussian curvatures of the monolayer. The last term in Eq. (3) does not intervene when variations in shape that do not change the topology of the film are considered, since, according to the Gauss-Bonnet theorem, the integral of C over the whole monolayer area is then a topological invariant. Whenever a portion of the monolayer undergoes a shape departure from its flat state two additional free energy terms need to be considered: (i) The work of compression exerted by the surrounding edges of the remainder of the monolayer as the planar projection area of the given portion is decreased. And (ii) the change in surface tension energy due to the increment in the total monolayer area necessary to keep its total planar projection area constant. These terms are, respectively, $f_{comp} dS = -\Pi (dS - dS_0)$ and $f_{inflow} dS = \gamma (dS - dS_0)$, where dS and dS_0 are, respectively, area elements of the monolayer and its planar projection. Local bending, since no ruptures or pores are allowed, necessarily results in monolayer enlargement via an inflow of amphiphile from the sustaining solvent or solvents.

Because the monolayer undergoes thermal shape fluctuations,[4] it actually consists of a collection of independent pieces with linear sizes, on average, of the order of the de

Gennes-Taupin persistence length ξ.[5] Also, interfacial portions with areas increasingly smaller than the maximum size of $O(\xi^2)$ appear to be more tight or rigid, since thermal undulations produce a reduction of κ with increasing area size. It has been found[4] that the (effective) rigidity κ vanishes when the area size considered is of $O(\xi^2)$. Therefore, in order to examine the shape of coherent pieces of monolayer as a function of amphiphile concentration η it is necessary to minimize the free energy in excess per unit area f between bent and flat states only over areas of $O(\,l^2 < \xi^2\,)$, i.e.

$$\min\left\{\int f \cdot dS \right\} = \min\left\{\int\int \left[f_{comp} + f_{inflow} + f_{curv}\right] \cdot dS \right\}, \tag{4}$$

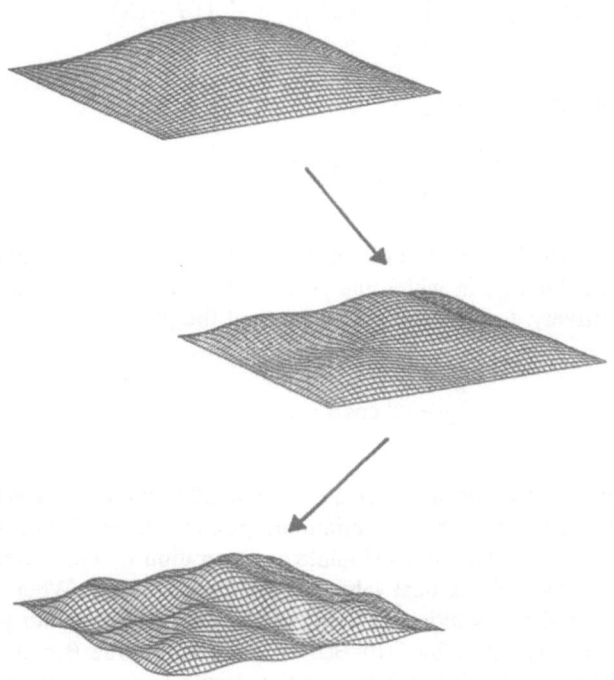

Fig. 1   When the interface becomes progressively compressed via amphiphile additions it buckles first at a critical value of the interfacial tension with sizes of the order of its persistence length. Further decrements in the area per amphiphile produces buckling at smaller scales.

This minimization procedure indicates that when the monolayer has a vanishing spontaneous curvature c_0 it is composed of flat pieces each with area of $O(\xi^2)$ if $\gamma > \gamma_c = \gamma_0 / 2$, but these pieces buckle progressively if $\gamma < \gamma_c = \gamma_0 / 2$. There is also buckling associated with smaller scale lengths $l < \xi$, and one obtains a line of self-similar buckling transitions defined by $\gamma_c = \gamma_0 / 2 - \delta \, \kappa(l)/\, l^2$, where δ is a constant of $O(1)$ and $\kappa(l)$ is the effective rigidity for size l. When $c_0 = 0$ the monolayer is never flat and it bends progressively in a continuous manner as γ decreases with increments in η.[2]

Detailed descriptions of the buckling transition can be obtained for specific choices of the surface for the piece of monolayer over which the minimization in Eq. (4) is performed. For regular shapes the solutions can be found in the literature.[6] Thus, for a circle, the problem is equivalent to that of the buckling of a circular elastic plate under compression, and the critical value for γ_e and the surface shape close to the transition can be found in terms of the Bessel function of first order.[2,6] An even more detailed description can be given for the particular case of a one-dimensional monolayer, the buckling of an elastic beam, which has as a mechanical analog the classical simple pendulum. Since the one- and two-dimensional problems share many relevant properties we give here a brief description of the former.

The sum of free energy terms in Eq. (4) for a segment of length L is

$$F = \gamma L + (\gamma - \Pi)\int_0^L (1 - \cos\theta)\,dl + \frac{\kappa}{2}\int_0^L (\dot\theta - C_0)^2\,dl,$$

(5)

which, when minimized leads to

$$\ddot\theta = \frac{\gamma - \Pi}{\kappa}\sin\theta = \alpha^2 \sin\theta,$$

(6)

Since Eq. (6) describes the motion of a simple pendulum, there is a mechanical analogy for the buckling of the line, in which the position l, and the parameter $\alpha=[(\Pi-\gamma)/\kappa]^{1/2}$, correspond, respectively, to time and frequency, and the phase portrait obtained from the energy equation

$$\frac{\dot\theta^2}{2} - \alpha^2 \cos\theta = \alpha^2 \cos\theta_0$$

(7)

determines the pendulum orbits or line shapes. For small θ one has: (i) When $c_0=0$ the line is flat when its length L is less than the minimum period π/α and it buckles progressively when L exceeds π/α, e.g. when the amphiphile concentration η is sufficiently large for the interfacial tension to reach the critical value $\gamma_c = \gamma_0 / 2 - \pi L^{-2}\kappa$. (ii) When $c_0 \neq 0$ there is no buckling transition and the solution corresponds to that in which the pendulum has an initial and final velocity given by $\theta(0)=\theta(L)=c_0$, and amplitude $\theta_0=c_0 L/2$. The buckling transition bears some similarities with that of the simple Ising model in that, when $c_0=0$, there is a zero curvature $c=0$ phase for low Π or η (the high-temperature phase with zero magnetization at vanishing magnetic field), and there are two $c \neq 0$ phases with equal magnitude of the order parameter but with opposite signs, i.e. $c=-c'$, for large Π or η (the low-temperature phase with spontaneous magnetization). the ordering field is proportional to the spontaneous curvature c_0 and when this is different from zero there is no buckling transition. The Gibbs elasticity E is given by

$$E = -\eta\left(\frac{\partial\gamma}{\partial\eta}\right) = \delta kT \frac{\eta}{1-\eta},$$

(8)

where δ is unity when $\gamma > \gamma_c$ and one half when $\gamma < \gamma_c$, and therefore there is a discontinuity in E at the buckling transition, registered by our mean-field description as a finite jump, but which may be replaced by a divergence when the effect of thermal fluctuations is considered. As mentioned, these fluctuations produce a reduction of the (effective) rigidity $\kappa(l)$ with increasing monolayer size l, so that buckling occurs first at the maximum length ξ, when $\gamma_c = \gamma_0 / 2$, followed, as η increases, by the buckling associated to smaller sizes.

Thus, thermodynamically, the equilibrium shape of the monolayer is analogous to an Ising model where the role of the inverse temperature is taken by Π, or equivalently, by η. The roles of the magnetization, the magnetic coupling, the external field and the specific heat in the magnet, are taken, respectively, by the mean curvature, the rigidity, the spontaneous curvature, and the Gibbs elasticity of the membrane-like interface. Symmetrical interfacial objects with vanishing spontaneous curvature, like foam films, may show enhanced stability for sufficiently low interfacial tensions, or large amphiphile interfacial concentration. This would be because along the line of buckling transitions the elasticity of Gibbs may take large (divergent) values. The modification of their condition $c_0 = 0$, would lead to non critical interfacial states with restricted tolerance for shape fluctuations, and their film stability may become greatly reduced.

TOPOLOGICAL TRANSITION

We now discuss the implications of the Gaussian term $\overline{\kappa}C$ in Eq. (3). For a system of surfaces that consists of a simply connected piece to which N_h handles have been attached and of N_s boundaryless disjoint pieces, the Gauss-Bonnet theorem states that

$$\int CdS = 2\pi(1 + 2N_s - 2N_h) \tag{9}$$

Thus, according to the sign of $\overline{\kappa}$ in Eq. (3) the surface can lower its free energy by fragmentation or proliferation of handles. Here, to be more specific, we chose the sign of $\overline{\kappa}$ to be negative and only the number of spheres to be relevant. We write a model grand potential Ω for the system as follows

$$\Omega = kT \cdot V\left\{[n_s \ln n_s + (1 - n_s)\ln(1 - n_s)\right\}$$

$$+[n_p a_p A + n_s a_s V]\left\{\gamma_0 + kT \cdot [\eta \ln \eta + (1 - \eta)\ln(1 - \eta)]\right\}$$

$$+A \cdot \kappa \cdot c_0^2 \cdot a_p n_p + 4\pi \cdot V \cdot n_s \cdot [\kappa \cdot (1 - c_0 / c)^2 + \overline{\kappa}]$$

$$-A \cdot \Pi \cdot a_p n_p^2 + V \cdot e \cdot a_s n_s^2 - (n_p \cdot a_p \cdot A + n_s \cdot a_s \cdot V) \cdot \eta \cdot \mu \tag{10}$$

where V is the volume of the system, $a_s = 4\pi c^{-2}$ is the area of the spheres considered to be monodisperse with curvature c and with volume concentration n_S. The spheres experience a repulsion e due to confinement of the solvent. A is the area of the simply-connected monolayer, which is considered to be made of N_p patches of area a_p and surface concentration $n_p = N_p / A$, these patches are held together via the surface pressure Π. The first line in the equation above quantifies the three-dimensional configurational entropy of the spheres or micelles. The second line contains the bare interfacial tension of the total

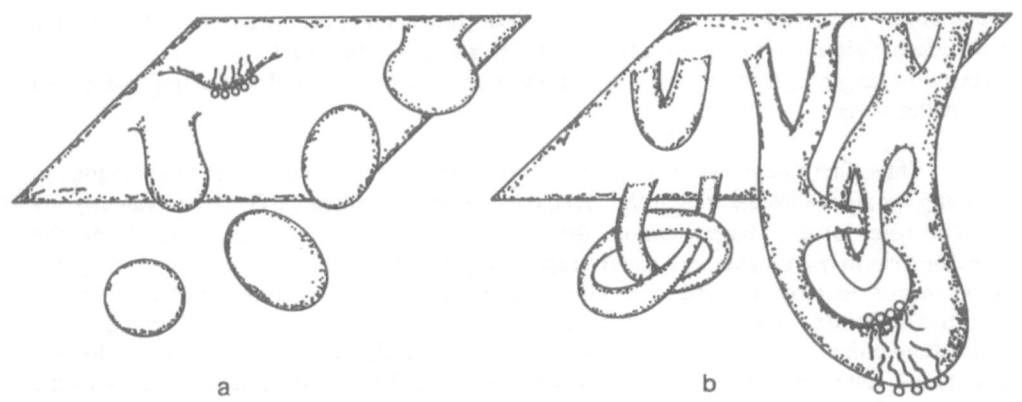

Fig. 2 Topologically driven interfacial transitions. a) A dispersion of droplets is obtained when the saddle-splay constant is negative. b) Proliferation of handles produces a bicontinuous structure when this constant is positive.

surface and the configurational entropy of the individual amphiphiles that constitute the monolayer and the spheres. The third line contains the curvature energy of the system of surfaces, and the fourth line the interaction terms and the chemical potential contribution of the amphiphiles.

Generally, the volume terms in Eq. (10) predominate over the monolayer area terms, and minimization of this equation with respect to both η and n_s when the area terms are neglected leads to

$$kT \ln \frac{n_s}{1 - n_s} + 2a_s e n_s = f_{top}$$

(11)

where

$$-f_{top} = a_s \gamma + 4\pi \left[\kappa(1 - \frac{c_0}{c})^2 + \bar{\kappa} \right].$$

(12)

Therefore, the system behaves in a manner equivalent to a fluid of repulsive particles at chemical potential f_{top}, which can be seen to be the free energy cost for the creation of a sphere, and we note that $n_s \cong 0$ when $f_{top} < 0$ and $n_s \cong 1$ when $f_{top} > 0$. Thus, if the amphiphile concentration η is increased from a sufficiently low value, the system suffers a gradual but sharp transformation from a state with very low concentration of micelles to one in which there is a large number of them. This occurs when f_{top} changes sign. Our description leads to results similar to those of more traditional treatments of the cmc phenomenon based on the law of mass action.[7]

On the other hand, in the opposite limiting situation in which the volume V available for the spheres is of the order of the area A, Eq. (10) can be minimized under the condition $n_p = 1 - n_s$, where now n_s is defined as the concentration of micelles per unit area. This procedure leads to

$$kT \ln \frac{n_s}{1 - n_s} - 2a_p \Pi n_s = g_{top} \tag{13}$$

where

$$-g_{top} = 2a_p \Pi + 4\pi \left[\kappa(1 - \frac{c_0}{c})^2 + \bar{\kappa} \right]. \tag{14}$$

In magnetic language, $m = 2n_s - 1$, the above equations can be written as

$$kT \ln \frac{1 + m}{1 - m} - a_p \Pi m = h_{top} \tag{15}$$

where h_{top} is given by

$$-h_{top} = a_p \Pi + 4\pi \left[\kappa(1 - \frac{c_0}{c})^2 + \bar{\kappa} \right]. \tag{16}$$

The field h_{top} can be seen to be the free energy cost for pulling out a patch away from the monolayer and forming a sphere with it. Therefore, for sufficiently low temperature, a phase transition takes place, when $h_{top}=0$, from a state composed mainly of a simply connected surface with very small genus to another state which is highly fragmented and has a large genus. The locus $h_{top}=0$ coincides with that of $f_{top}=0$ provided $a_p \Pi = a_s \gamma$.

Our aim has been to offer a minimal model for the characterization of the curvature interfacial transitions. Within a phenomenological, mean-field, description, these transitions appear to take place when the area extension and bending free energy costs for interfacial deformations happen to be of the same order of magnitude, and a competition between different equilibrium interfacial shapes can be established. Generally the surface - extension free-energy term is much larger than the bending terms, however, amphiphiles efficiently suppress the interfacial tension, and this competition may become observable in systems that contain them.

ACKNOWLEDGEMENTS

This work was supported partially by DGAPA-UNAM under contracts Nos. IN-104189 and IN-102291 and by CONACyT under contract No.0594-E9109 Convenio 2146.

REFERENCES

1. A.W. Adamson, *Physical Chemistry of Surfaces* (Wiley, New York, 1976).

2. A. Robledo, C. Varea and V. Talanquer, Phys. Rev. **A43**, 5736 (1991).

3. P.B. Canham, J. Theor. Biol. **26**, 61 (1970); W. Helfrich, Naturforsch. **28a**,693 (1973).

4. W. Helfrich. J. Phys. (Paris) **46**, 1263 (1985); L. Peliti and S. Leibler, Phys. Rev. Lett. **54**, 1960 (1985).

5. P.G. de Gennes and C. Taupin, J. Phys. Chem. **86**, 2294 (1982).

6. S. Timoshenko and S. Woinowsky-Krieger, *Theory of Plates and Shells* (Mc. Graw Hill, New York, 1959).

7. M. Borkovec, J. Chem. Phys. **91**, 6268 (1989), and references therein.

HOT SOLID PROPERTIES FROM LIQUID STRUCTURE
WITHIN DENSITY FUNCTIONAL THEORY

M. P. Tosi

Scuola Normale Superiore
I-56100 Pisa, Italy

INTRODUCTION

The idea that structural correlations in a liquid near freezing carry useful information on properties of its solid near melting is an old one. Early examples are the Kirkwood-Monroe theory of the liquid-solid transition[1] and Faber's treatment of the vacancy formation energy in hot close-packed metals,[2] which were framed in terms of the liquid pair distribution function g(r) (or equivalently the liquid structure factor S(k)) and a pairwise potential of interaction between the particles. The underlying assumption of these theories is that the character of the interatomic forces should not be altered across the phase transition. The approximate notion of pair potentials is transcended in the functional cluster expansion of Lebowitz and Percus.[3] This leads to a formal expression for the free energy of a classical system in an inhomogeneous state as a function of its density profile n(r), involving the many-particle direct correlation functions of its homogeneous liquid. Their work preluded to the development of the density functional theoretical approach[4,5] (DFT), which focusses on the free energy functional F[n(r)] of the inhomogeneous system and aims at approximately evaluating it from a knowledge of thermodynamic and microscopic correlation-response functions of a corresponding homogeneous system.

The relevance of DFT to the problem of the liquid-solid transition was first realized by Ramakrishnan and Yussouff[6] within the context of the cluster expansion truncated at lowest order (see also Haymet and Oxtoby[7] and March and Tosi[8]). When one assigns a free energy to both the liquid and the solid in each thermodynamic state, the crucial thermodynamic quantity in determining the coexistence of the two phases is the difference $\Delta\Omega$ in their grand potentials as a function of the appropriate thermodynamic state variables. The phase transition is signalled by the spontaneous appearance of finite values for its order parameters, which are the fractional density change $(n_s - n_l)/n_l$ between solid and liquid and the microscopic Fourier components n_G of the periodic density profile in the crystal,

$$n(\mathbf{r}) = n_s + \sum_{G \neq 0} n_G \exp(i\mathbf{G}.\mathbf{r}) \quad , \tag{1}$$

the G's being the vectors of the reciprocal lattice (RLV). The quantities $|n_G|^2$ give the Debye - Waller factors of the crystal and are a measure of the localization of the particles on lattice sites. In the simplest representation of the crystalline density profile as a superposition of Gaussian distributions centered on the lattice sites, and introducing the Lindemann parameter L as the ratio between the root-mean-square displacement of the particles from the lattice sites and the first-neighbour distance d, one has for a cubic crystal

$$n_G = \exp[-\frac{1}{6}L^2G^2d^2] \quad . \tag{2}$$

Thus, an expression of $\Delta\Omega$ in terms of the density profiles (or, more simply, of the average densities and the Lindemann parameter, for a given crystal structure) allows a variational determination of the order parameters of the phase transition and of the coexistence curve between the two phases.

The DFT approach to freezing implements this programme by constructing an approximate expression for $\Delta\Omega$ which involves only liquid-state properties. This treatment of the phase transition can test the Gaussian approximation and provide a microscopic justification of the Lindemann criterion for melting - L is approximately 0.15 for many crystals at melting in the classical regime. At the same time, since an essential input of the theory is the liquid structure factor S(k), the DFT approach can justify the Hansen-Verlet criterion for freezing, stating that a classical monatomic liquid crystallizes when the height of the main peak in S(k) reaches a value of about 3. Evidently, a minimal amount of localization in the crystal near melting requires a minimal amount of short range order in the liquid near freezing.

The extension of the DFT approach to freezing of quantal fluids has been one of the most interesting recent developments in this field.[9] In the quantal regime the critical value of the Lindemann parameter is about 0.3 and the role of S(k) in the theory is taken up by the non-ideal part of the density-density response function. In particular, Wigner crystallization for degenerate electrons is driven by the "local field" factor accounting for exchange and correlation in the static dielectric function of the homogeneous electron gas. With increasing coupling strength, exchange and correlations lead to a mean-field attractive potential on electrons in the region of wavenumber overlapping the first RLV stars of the bcc or fcc structure.

There exist already several reviews of results achieved by the DFT approach to liquid-solid coexistence.[10-14] Therefore, after a brief summary on thermodynamic potentials for an inhomogeneous system, I confine myself below to a brief review of a new line of development, dealing with deformed classical crystals at high temperature in relation to their phonon dispersion curves and elastic constants.

THERMODYNAMIC POTENTIALS

The relevant functionals for a system in a given thermodynamic state and subject to an applied one-body potential $U(r)$ are the grand potential $\Omega[n(r)]$ and the free energy functional $F[n(r)]$, which are related by

$$F[n(r)] = \Omega[n(r)] - \int dr \; n(r) \; u(r) \; , \tag{3}$$

where $u(r) = U(r) - \mu$ and μ is the chemical potential. F is the Helmholtz free energy of the system aside from its interaction energy with the external potential. By taking functional derivatives Lebowitz and Percus found the basic relations

$$\frac{\delta\Omega[n(r)]}{\delta u(r)} = n(r) \tag{4}$$

and

$$\frac{\delta F[n(r)]}{\delta n(r)} = - u(r). \tag{5}$$

One proceeds from these general statements of principle towards a workable theory by separating out from $F[n(r)]$ the ideal-gas contribution $F_{id}[n(r)]$ for non-interacting particles at density $n(r)$. In the classical regime,

$$F_{id}[n(r)] = k_B T \int dr \; n(r) \; \{\ln[\Lambda n(r)] - 1\} \tag{6}$$

where $\Lambda = (h^2/2\pi m k_B T)^{3/2}$. Defining $\phi[n(r)] = F_{id} - F$ and

$$C(r) = (k_B T)^{-1} \frac{\delta\phi[n(r)]}{\delta n(r)} \; , \tag{7}$$

one can write for the real interacting system[15,7]

$$n(r) = \Lambda^{-1} \exp[C(r) - u(r)/k_B T] \tag{8}$$

and

$$\Omega = -\phi + k_B T \int dr \, n(r) \, [C(r) - 1] \quad . \tag{9}$$

Comparison of (8) with the corresponding expression for the ideal gas shows that $C(r)$ acts as a self-consistent potential arising from the interactions between the particles.

Finally, a functional expansion of the inhomogeneous system around a homogeneous state expresses $C(r)$ and $\Delta\Omega[n(r)]$ through correlation functions of the latter. Indeed, the higher functional derivatives of the non-ideal free energy ϕ define a hierarchy of correlation functions,

$$c(r_1, r_2) = \frac{\delta C(r_1)}{\delta n(r_2)} = (k_B T)^{-1} \frac{\delta^2 \phi[n(r)]}{\delta n(r_1) \delta n(r_2)} \tag{10}$$

etcetera. The expansion of $C(r)$ around its value C_1 in the homogeneous fluid thus involves these correlation functions evaluated on the fluid,

$$C(r) = C_1 + \int dr' \, c(|r - r'|) \, [n(r') - n_1] + \dots \quad . \tag{11}$$

A similar expansion for ϕ leads to

$$\Omega[n(r)] = \Omega_1 - k_B T \int dr \, [n(r) - n_1] + \frac{1}{2} k_B T \int\int dr dr' c(|r-r'|)[n(r)+n_1][n(r')-n_1] + \dots \tag{12}$$

In these equations $c(r)$ is the Ornstein-Zernike direct correlation function of the fluid. Its Fourier transform $c(k)$ is directly related to the liquid structure factor by $c(k) = 1 - 1/S(k)$. Clearly, truncation of the expansion at second-order terms as shown above becomes strictly valid only for a weakly inhomogeneous system.

The Ramakrishnan-Yussouff theory of liquid-solid coexistence follows from (11) and (12). The two phases have been treated in the expansion as being at the same temperature and chemical potential: therefore, the coexistence line is determined by the equality of their grand potentials, the order parameters being determined by inserting (11) into (8) at $U(r) = 0$, when the density profile $n(r)$ takes the form of (1). There has been great concern in the literature with the role of the higher order terms in the expansion, which have been omitted in writing (11) and (12). The three-body correlation terms account for (i) the difference in compressibility between liquid and solid, (ii) the density dependence of the direct correlation function, and (iii) couplings between the microscopic order parameters. Third and higher order correlations are taken approximately into account in various extensions of the low-order theory, such as the weighted density approximation[16] or its recent extension to isochoric freezing in the case of the classical one-component plasma.[17] The practical relevance of such corrections is system-dependent, tending in particular to increase in simple pair-potential models as the core repulsions are softened from the hard-sphere system to the classical plasma.

PHONON DISPERSION CURVES

The central role in the standard theory of lattice dynamics is played by the potential energy of the crystal as a function of the nuclear positions, which is expanded in powers of the atomic displacements from the equilibrium lattice sites.[18] Anharmonic effects beyond thermal expansion are treated either by perturbation theory[19] or by a self-consistent phonon approach.[20] Anharmonicity results in a renormalization of the phonon frequencies and a broadening of the phonons, which become increasingly important with increasing temperature. The DFT approach as outlined above provides an entirely different method to evaluate renormalized phonon frequencies in high-temperature crystals, by relating them to the correlation functions in the liquid and thus leaving the role of the interatomic forces implicit.[21-23] A similar approach to phonon broadening could be based on time-dependent DFT.

Consider a monatomic crystal having a simple Bravais lattice structure described by a set of lattice sites $\mathbf{R_i}$, which is deformed by giving to each site a displacement $\mathbf{d_i}$ at constant temperature, volume and chemical potential. The set of displacements is chosen in the form

$$\mathbf{d_i} = \alpha \, N^{-1/2} \, \hat{\varepsilon}_{qs} \cos(\mathbf{q} \cdot \mathbf{R_i}) \tag{13}$$

where α is an arbitrarily small constant, N is the number of lattice sites and $\hat{\varepsilon}_{qs}$ is the normalized eigenvector of a lattice vibration having wave vector $\mathbf{q}$ and branch index s ($\mathbf{q}$ lies in the first Brillouin zone and the eigenvector is invariant under the transformation $\mathbf{q} \rightarrow \mathbf{q} + \mathbf{G}$ for any RLV). The density profile of the deformed crystal is

$$n(\mathbf{r}) = \Big\langle \sum_i \delta(\mathbf{r} - \mathbf{R_i} - \mathbf{d_i} - \mathbf{u_i}(t)) \Big\rangle \tag{14}$$

where $\mathbf{u_i}(t)$ are the atomic displacements due to thermal fluctuations and the brackets denote the statistical average. To quadratic terms in α, the work $\Delta\Omega$ done in deforming the crystal is

$$\Delta\Omega = - \pi^2 \, M \, v_{qs}^2 \, \alpha^2 \tag{15}$$

where M is the atomic mass and v_{qs} is the eigenfrequency of the lattice vibration indexed by $\mathbf{q}$ and s. Equation (15) follows from the fact that, according to (3), $\Delta\Omega$ contains the intrinsic free energy change stored in the deformed crystal as well as the interaction of the deformation with the external potential causing it. On the other hand, by expanding both the deformed and the undeformed crystal at high temperature around the homogeneous liquid by the method outlined in the preceding section, one may relate $\Delta\Omega$ and hence v_{qs} to the direct correlation functions of the liquid.

The simplest approximations that may be invoked for the realization of this scheme are (i) the truncation of the functional expansion at two-body correlation terms, and (ii) the representation of $n(\mathbf{r})$ in both the deformed and the undeformed crystal as a superposition of Gaussian clouds centered on the lattice sites, thus neglecting the deformation of these clouds upon deformation of the crystal. The latter approximation reduces (14) for a cubic crystal to

$$n(\mathbf{r}) = \frac{1}{V} \sum_{\mathbf{k}} f(\mathbf{k}) \sum_{i} \exp[i\mathbf{k}.(\mathbf{r} - \mathbf{R}_i - \mathbf{d}_i)] \tag{16}$$

where

$$f(\mathbf{k}) = \langle \exp[-i\mathbf{k}.\mathbf{u}_i(t)] \rangle = \exp[-\tfrac{1}{6}L^2 k^2 d^2] \ . \tag{17}$$

The final result for the phonon frequencies then is

$$v_{\mathbf{qs}}^2 = -\frac{n_s k_B T}{4\pi^2 n_l M} \left\{ \sum_{\mathbf{G}} f^2(\mathbf{q}+\mathbf{G}) c(|\mathbf{q}+\mathbf{G}|)[(\mathbf{q}+\mathbf{G}).\hat{\varepsilon}_{\mathbf{qs}}]^2 - \sum_{\mathbf{G}\neq 0} f^2(\mathbf{G}) c(\mathbf{G}) (\mathbf{G}.\hat{\varepsilon}_{\mathbf{qs}})^2 \right\} \tag{18}$$

It is easily shown that the expression (18) for the dispersion relations of the vibrational modes is invariant under the transformation $\mathbf{q} \to \mathbf{q} + \mathbf{G}$ for any RLV and has the rotation and reflection symmetry properties associated with the Brillouin zone. Indeed, Mahato et al.[23] have explicitly derived the dynamical matrix of the crystal by the same DFT approach. Both longitudinal modes (eigenvector parallel to $\mathbf{q}$) and transverse modes (eigenvector perpendicular to $\mathbf{q}$) are contained in (18), the latter arising from the "Umklapp" scalar products of the eigenvector with the RLV. It is also easily shown that for a monatomic system in the limit $q \to 0$, with $c(q \to 0)$ tending to a constant value, (18) yields an acoustic dispersion relation. On the other hand, in the classical Wigner crystal the $\mathbf{G} = 0$ term in the first sum in (18), on account of the relation $c(q \to 0) \to -4\pi n_l e^2/(q^2 k_B T)$ for Coulombic interactions, yields in the limit $q \to 0$ the longitudinal optic mode at the plasma frequency.

The usefulness of (18) in describing the phonon dispersion curves in crystals at high temperature has been examined for a number of systems by Ferconi and Tosi.[22] The systems considered are (i) the bcc alkali metals Na and K, with liquid structure input from both X-ray and neutron diffraction data; (ii) fcc Ar, with liquid structure input from a Lennard-Jones pair potential model; and (iii) the classical Wigner crystal formed by the one-component classical plasma on a uniform neutralizing background, in both the bcc and fcc structure at their respective melting points as determined from computer simulation. For both the Lennard-Jones model and the classical plasma model the liquid structure factor was evaluated by accurate integral-equations approaches, embodying in particular the requirement of self-consistence between structural and thermodynamic properties. A sensible choice of the Lindemann parameter in (17) is crucial to ensure rapid convergence of the sums over RLV in (18), thus avoiding the need to know the liquid structure at very large wave number. Measurements of the liquid structure factor usually extend in wave number to cover 10÷20 stars of RLV, ensuring convergence of the sums in (18) to at least a few parts in 10^3 when L ≈ 0.15. A model calculation of $S(k)$ is free from such truncation problems, but is affected by uncertainties at large k from imprecise knowledge of the interatomic potential. Of course, the magnitude of L is also quantitatively important for the results: in particular, the increase in L with increasing temperature, for given liquid structure input, yields a softening of phonon frequencies as temperature increases.

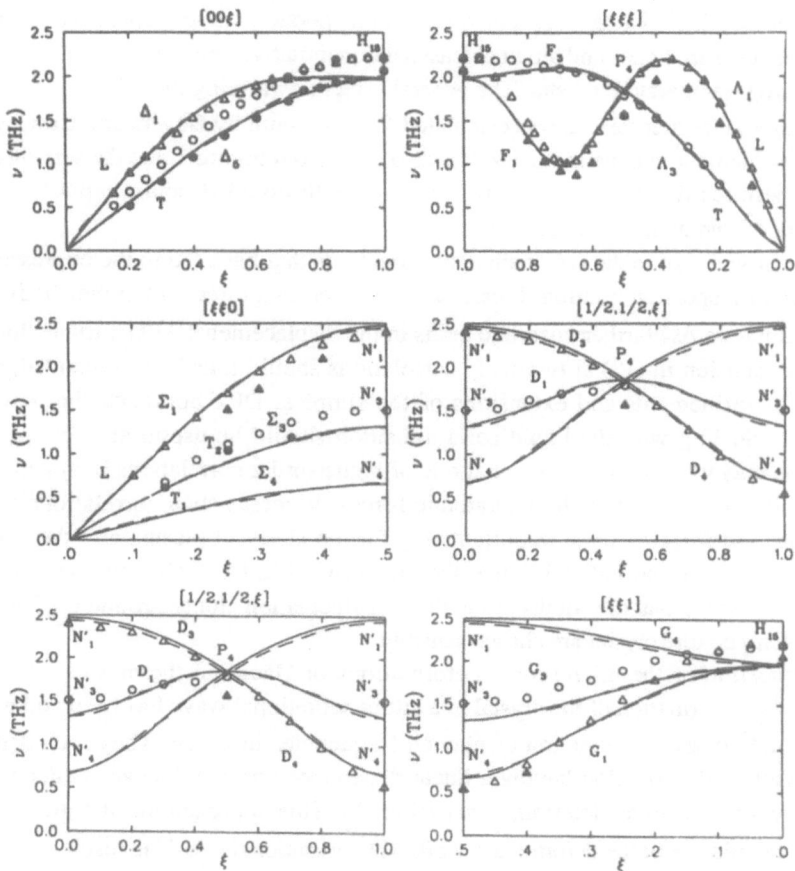

Figure 1. Calculated dispersion curves for phonons in potassium at melting (T = 336 K), from liquid structure data obtained by X-ray diffraction (full curves) and neutron diffraction (broken curves), compared with those measured in inelastic scattering experiments at 9 K (open circles and triangles) and at 299 K (full circles and triangles). The direction of the wave vector q, with components in units of $\pi\sqrt{3}/d$, is indicated at the top of each graph. The dispersion curves along the [1/2,1/2,ξ] direction are given twice to show their degeneracies at the Brillouin zone boundaries with those along the [$\xi,\xi,0$] and [$\xi,\xi,1$] directions. From Ferconi and Tosi.[21]

An illustration of dispersion relations obtained by this approach is given in Figure 1 for potassium at melting (T = 336 K) in comparison with data from neutron inelastic scattering at 9 K by Cowley, Woods and Dolling[24] and at 299 K by Buyers and Cowley.[25] The Lindemann parameter for potassium at melting was taken to be L = 0.15, as obtained from phonon frequency spectra constructed from neutron inelastic scattering data. A calculation of liquid-solid coexistence in the alkali metals by the simplest DFT approach shows that the Gaussian approximation is quite good (although the theory has difficulties in dealing with the order parameter at the [200] star) and estimates L ≈ 0.15 for these metals at melting.[26]

It is evident from Figure 1, as well as from the results for the other systems evaluated in the original work,[22] that the calculated phonon dispersion curves reproduce the general shapes of the measured ones and are in at least semiquantitative agreement with the available data on the high-temperature crystal. The general effect of warming the crystal is a softening of the vibrational frequencies, as was calculated for potassium by Buyers and Cowley[25] in a perturbative treatment of anharmonicity. It is also clear from Figure 1 that there is substantial agreement between the theoretical results obtained with liquid structure input from X-ray diffraction and from neutron diffraction.

Let us now return to the two main approximations that have led to the expression (18) for the phonon dispersion relation. It can easily be seen that three and higher-body correlations contribute to $\Delta\Omega$ further quadratic terms in the displacements.[22] The truncation of the functional expansion in (12) at two-body correlations should thus be transcended, and the same type of refinements and extensions of the simplest DFT approach that have been developed in dealing with the liquid-solid transition should be useful also in the present context. One may well expect an essential role of higher-order correlations in systems with a strong angular dependence of the interatomic forces. Whereas such correlations enter the DFT approach to freezing in an important way for the classical plasma and the bcc alkali metals, there is as yet no indication that they are crucial in the DFT approach to phonon frequencies, provided that the widths of the thermal-fluctuation clouds around the lattice sites in the high-temperature crystal are chosen sensibly.

The question of the microscopic deformations of Gaussian thermal clouds that are induced by the deformation of the crystal in a lattice vibrational wave has been addressed by Mahato et al.[23] in their calculation of phonon frequencies in argon. They included in the Gaussian width a deformation having a linear dependence on the displacements, which is described by a tensor to be determined variationally. This approach has the great merit of making the calculation of the deformation work $\Delta\Omega$ a variational one. The result is a lowering of the calculated phonon frequencies, although not by a major quantitative amount in this case. Mahato et al. used as input experimental data on the liquid structure factor and evaluated first the crystalline density and the Lindemann parameter at melting from a DFT calculation on liquid-solid coexistence. The somewhat high value of the density and the exceedingly low value of the Lindemann parameter that they obtained from this calculation affect the quantitative disagreement of their theoretical dispersion curves with experiment, which amounts to an overestimate of the phonon frequencies by roughly 50% near the Brillouin zone boundaries.

ELASTIC CONSTANTS

There currently is an active theoretical interest in the elastic constants of crystals on the approach to thermodynamic melting and in expanded states, in relation to disordering of the crystal by a mechanical instability and its parallels with solid-state amorphization.[27] The possibility of mechanical melting is envisaged to arise at temperatures both above the thermodynamic melting temperature and below the triple-point temperature, under conditions in which the kinetic processes which accompany thermodynamic melting or sublimation are suppressed. This field goes back to the early discussion given by Born[18] of the criteria for mechanical stability. The relevant elastic parameter for a cubic crystal is the $(c_{11} - c_{12})/2$ shear elastic constant, whose vanishing marks the absolute limit for mechanical stability.

The evaluation of elastic constants may be pursued by two alternative methods.[18] These actually correspond to alternative experimental techniques for their determination, i.e. by measuring the deformation of the crystal under homogeneous static stresses or by ultrasonic measurements of sound velocities. In the so-called homogeneous deformation (HD) method, one has to evaluate the strain free energy arising from the application of constant stresses and compare the results with the well-known general expression given by elasticity theory. In the so-called method of long waves (LW), instead, the slopes of appropriate phonon acoustic branches at long wavelengths yield the velocities of sound waves propagating along chosen crystalline directions with longitudinal or transverse polarization. The elastic constants are related to the sound velocities by well-known formulae. While these two theoretical methods should in principle lead to the same result, appreciable discrepancies between HD and LW results often arise as a consequence of approximations in the theory - especially when the theory takes the form of a truncated expansion. It seems to be an empirical fact that the LW method usually gives in such cases a closer and often quite accurate representation of the true result.

Within the DFT approach it was first pointed out by Ramakrishnan[10] that the elastic constants of a hot crystal could be related to its liquid structure by an evaluation of the strain free energy associated with homogeneous deformations of the crystal. Such an HD route to the elastic constants has been further discussed and evaluated by a number of authors.[28-32] In the treatment proposed by Lipkin et al.,[28] a homogeneous strain is applied to the crystal at constant temperature, volume and chemical potential, and the strained density profile is simply obtained by allowing the RLV to change from G to $G \cdot (1 + \varepsilon)^{-1}$, where ε is the strain tensor. The work of deformation is then evaluated by the cluster expansion truncated at two-body correlations. An exhaustive discussion of the thermodynamics of strained crystals has been given by Jaric and Mohanty.[32] Allowing for microscopic deformations of the Gaussian clouds in the strained crystal leads in the case of the hard-sphere system to major quantitative changes in the calculated elastic constants.[29,30,32] Indeed, the c_{12} elastic constant and hence the Poisson ratio are brought down to negative values. Velasco and Tarazona[31] have consequently reexamined this question using a weighted-density approximation to account for higher-order correlations. They find that a positive value is thereby recovered for the Poisson ratio of the hard-sphere crystal. Their calculated value for this elastic parameter is about 0.3, which is typical for fcc crystals. Relaxation of the Gaussian clouds seems to entail much less drastic consequences in a Lennard-Jones model for argon.[32]

The LW route to the elastic constants of high-temperature crystals has been developed very recently[21-23] as a subproduct of the newly discovered DFT approach to phonon dispersion relations. As already remarked in the preceding section, the theoretical phonon frequencies at long wavelength are linear in the wave vector q in a monatomic crystal and the limit can be taken analytically. Ferconi and Tosi[22] have also comparatively discussed the results of the LW and HD methods and their merits in comparison with experimental and computer simulation data on the elastic constants of crystals near melting, using the simple DFT approach leading to (18) in the preceding section and the HD treatment of Lipkin et al.[28] The main discrepancy between these particular realizations of the LW and HD methods is that the latter misses the contribution coming from the $G = 0$ term in the first sum over RLV on the rhs of (18). In the LW method this term gives a contribution $n_s^2/(n_l^2 K_T)$ to the

Table 1. Elastic constants of Lennard-Jones models of argon near melting (in GPa) from DFT approaches including only two-body correlations, compared with data on argon at 82.3 K from Brillouin scattering experiments.[33] The first and second column give results obtained by the LW method and the HD method, respectively, using unrelaxed Gaussian clouds.[22] The third and fourth column give results obtained by the HD method using unrelaxed and relaxed Gaussian clouds, respectively.[32]

	LW-unrel.	HD-unrel.	HD-unrel.	HD-rel.	Experiment
c_{11}	2.4	1.8	4.8	4.3	2.38 ± 0.04
c_{12}	1.9	1.2	4.0	3.6	1.56 ± 0.03
c_{44}	1.3	1.3	3.5	3.1	1.12 ± 0.03

isothermal bulk modulus of the crystal from the compressibility K_T of the liquid. The same term is also responsible for yielding the long-wavelength longitudinal mode at the standard value of the plasma frequency in the classical Wigner crystal, as already remarked in the preceding section.

Table 1 gives an illustration of available theoretical results for the elastic constants of Lennard-Jones models of argon near melting, in comparison with the data of Gewurtz and Stoicheff[33] for argon at 82.3 K from Brillouin scattering experiments. The first two columns in the Table serve to contrast LW and HD results obtained at comparable levels of approximation (truncation at two-body correlations and unrelaxed Gaussian clouds), while the third and fourth column illustrate the effect of relaxing the Gaussian clouds within truncation at two-body correlations. Other comparisons between the various columns should not be taken strictly in a quantitative way, since the theoretical results refer to two sets of calculations involving different inputs and no attempt has been made to correct for the difference between isothermal and adiabatic elastic constants. The following remarks may nevertheless be made: (i) there are serious discrepancies between LW and HD results in the simplest theory; (ii) relaxation of the Gaussian clouds leads to a softening of the calculated elastic constants by roughly 10% in this case; (iii) the magnitude of the Lindemann parameter is again quite important in determining the quantitative results in this type of calculation; and (iv) the LW method in its simplest realization performs reasonably well in comparison with the data. The latter remark is comforted by results of similar quality that have been obtained for alkali metals and for the classical Wigner crystal.[22]

It is also remarkable from Table 1 that, even though a pair potential has been used in the evaluation of liquid structure, an appreciable deviation from the Cauchy relation $c_{12} = c_{44}$ is found, which is broadly in line with experiment. The deviation is arising in the simplest DFT approach from the fact that the positions of peaks and valleys in the direct correlation function of the liquid do not coincide with the RLV, i. e. the derivatives $c'(k)$ calculated at $k = G$ are usually different from zero. The consequences are equivalent to inclusion of non-central forces in a standard lattice-theory calculation of elastic constants.

CONCLUDING REMARK

Directions of further development in this area may be expected to concern refinements of the theoretical approach, taking advantage from progress in the theory of the liquid-solid transition, as well as applications to other physical systems of special interest and extensions to other physical properties of solids at high temperature. Application to classical ionic systems seems useful[34] in relation to (i) a general discussion of the symmetry properties of phonon dispersion relations that have been noticed in the family of the alkali halides,[35] and (ii) the strong temperature effects on the phonon dispersion curves in fluorite-type fast ion conductors.[36] The DFT approach to phonon frequencies of crystals near melting in the classical regime can also be extended to quantal crystals. Ferconi and Vignale[37] have already used this approach to treat magnetophonons in the quantal Wigner crystal formed by 2D electron systems in strong magnetic fields.

ACKNOWLEDGMENTS

This work was sponsored by the Ministero dell'Università e della Ricerca Scientifica e Tecnologica of Italy through the Consorzio Interuniversitario Nazionale di Fisica della Materia.

REFERENCES

1. J.G. Kirkwood and E. Monroe, Statistical mechanics of fusion, *J. Chem. Phys.* **9**:514 (1941).
2. T.C. Faber, "An Introduction to the Theory of Liquid Metals", University Press, Cambridge (1972).
3. J.L. Lebowitz and J.K. Percus, Statistical thermodynamics of nonuniform fluids, *J. Math. Phys.* **4**:116 (1963).
4. R. Evans, The nature of the liquid-vapour interface and other topics in the statistical mechanics of nonuniform classical fluids, *Adv. Phys.* **28**:143 (1979).
5. S. Lundqvist and N.H. March, "Theory of the Inhomogeneous Electron Gas", Plenum, New York (1983).
6. T.V. Ramakrishnan and M. Yussouff, First-principles order-parameter theory of freezing, *Phys. Rev.* B **19**:2775 (1979).
7. A.D.J. Haymet and D.W. Oxtoby, A molecular theory for the solid-liquid interface, *J. Chem. Phys.* **74**:2559 (1981).
8. N.H. March and M.P. Tosi, Liquid direct correlation function, singlet densities and the theory of freezing, *Phys. Chem. Liquids* **11**:129 (1981).
9. G. Senatore and G. Pastore, Density-functional theory of freezing for quantum systems: the Wigner crystallization, *Phys. Rev. Lett.* **64**:303 (1990).
10. T.V. Ramakrishnan, Density wave theory of freezing and the solid, *Pramana* **22**:365 (1984).
11. M. Rovere, G. Senatore and M.P. Tosi, Ordering transitions induced by Coulomb interactions, *in*: "Progress on Electron Properties of Solids", E. Doni, R. Girlanda, G. Pastori Parravicini and A. Quattropani, eds., Kluwer, Dordrecht (1989).

12. M. Baus, The present status of the density-functional theory of the liquid-solid transition, *J. Phys.: Condens. Matter* **2**:2111 (1990).

13. M.P. Tosi, Freezing of Coulomb liquids, *in*: "Strongly Coupled Plasma Physics", S. Ichimaru, ed., Yamada Science Foundation, Tokyo (1990).

14. Y. Singh, Density-functional theory of freezing and properties of the ordered phase, *Phys. Rept.* **207**:351 (1991).

15. A.J.M. Yang, P.D. Fleming and J.H. Gibbs, Molecular theory of surface tension, *J. Chem. Phys.* **64**:3732 (1976).

16. A.R. Denton and N.W. Ashcroft, Modified weighted-density-functional theory of nonuniform classical liquids, *Phys. Rev.* A **39**:4701 (1989).

17. C.N. Likos and N.W. Ashcroft, Self consistent theory of freezing of the classical One Component Plasma, in press.

18. M. Born and K. Huang, "Dynamical Theory of Crystal Lattices", University Press, Oxford (1954).

19. R. A. Cowley, Anharmonic crystals, *Rept. Progr. Phys.* **31**:123 (1968).

20. P.F. Choquard, "The Anharmonic Crystal", Benjamin, New York (1967).

21. M. Ferconi and M.P. Tosi, Phonon dispersion curves in high-temperature solids from liquid structure factors, *Europhys. Lett.* **14**:797 (1991).

22. M. Ferconi and M.P. Tosi, Density functional approach to phonon dispersion relations and elastic constants of high-temperature crystals, *J. Phys.: Condens. Matter* **3**:9943 (1991).

23. M.C. Mahato, H.R. Krishnamurthy and T.V. Ramakrishnan, Phonon dispersion of crystalline solids from the density-functional theory of freezing, *Phys. Rev.* B **44**:9944 (1991).

24. R.A. Cowley, A.D.B. Woods and G. Dolling, Crystal dynamics of potassium I: pseudopotential analysis of phonon dispersion curves at 9 K, *Phys. Rev.* **150**:487 (1966).

25. W.J.L. Buyers and R.A. Cowley, Crystal dynamics of potassium II: the anharmonic effects, *Phys. Rev.* **180**:755 (1969).

26. Z. Badirkhan, M. Rovere and M.P. Tosi, Freezing of liquid alkali metals as screened ionic plasmas, *J. Phys.: Condens. Matter* **3**:1627 (1991).

27. D. Wolf, P.R. Okamoto, S. Yip, J.F. Lutsko and M. Kluge, Thermodynamic parallels between solid-state amorphization and melting, *J. Mater. Res.* **5**:286 (1990).

28. M.D. Lipkin, S.A. Rice and U. Mohanty, The elastic constants of condensed matter: a direct-correlation function approach, *J. Chem. Phys.* **82**:472 (1985).

29. G.L. Jones, Elastic constants in density-functional theory, *Molec. Phys.* **61**:455 (1987).

30. M.V. Jaric and U. Mohanty, "Martensitic" instability of an icosahedral quasicrystal, *Phys. Rev. Lett.* **58**:230 (1987).

31. E. Velasco and P. Tarazona, Elastic properties of a hard-sphere crystal, *Phys. Rev.* A **36**:979 (1987).

32. M.V. Jaric and U. Mohanty, Density-functional theory of elastic moduli: hard-sphere and Lennard-Jones crystals, *Phys. Rev.* B **37**:4441 (1988).

33. S. Gewurtz and B.P. Stoicheff, Elastic constants of argon single crystals determined by Brillouin scattering, *Phys. Rev.* B **10**:3487 (1974).

34. M. P. Tosi and V. Tozzini, to be published.

35. L.L. Foldy and B. Segall, Anion-cation mirror symmetry in alkali halide ion dynamics, *Phys. Rev.* B **25**:1260 (1982).

36. M.H. Dickens, M.T. Hutchings and J.B. Suck, Temperature variation of phonon frequency distribution in the fast ion conductor Lead Fluoride, *Solid State Commun.* **34**:559 (1980).

37. M. Ferconi and G. Vignale, in the course of publication.

HIGHLY ASYMMETRIC ELECTROLYTE SUSPENSIONS
IN THE PRIMITIVE MODEL

Gaetano Senatore and Giorgio Pastore

Dipartimento di Fisica Teorica
Università di Trieste
Strada Costiera 11
I-34014 Trieste, Italy

ABSTRACT. We study monodisperse polyelectrolyte suspensions within the primitive model (PM). We employ a rescaled mean spherical approximation (RMSA), which involves the MSA treatment of a mixture of charged hard sphere with non-additive radii (NAR-MSA). For point counterions and coions the NAR-MSA has an appealing analytic solution, in terms of one separation parameter, thus allowing for a very handy estimate of both structural and thermodynamic properties. We compare our structural results with those of other more demanding approximations, as well as with those of new Monte Carlo Simulations. We also emphasize the importance of a multi-component description of the suspension to predict thermodynamic properties, and in particular fluid-fluid equilibrium.

1. Introduction

Charge-stabilized colloidal suspensions are very asymmetric ionic systems made up of highly charged polyions (or macroions) dissolved in water or some other polar solvent. The polyions are mesoscopic particles, in the size range $\sigma_p \approx 10^2 - 10^4 Å$, if σ_p denotes some characteristic diameter. Microscopic ions are also present, originating from the ionization of the polar groups on the surface of the polyelectrolite (counterions) and the dissociation of added salt (coions and counterions). The typical polyion charge $z_p e$ may range from few tens to thousands of elementary charges. The great complexity of colloidal suspensions makes the *ab initio* study of such systems a formidable task, involving the microscopic modelling of water as well as the simultaneous treatment of objects with well separated scales of length, charge, and mass. Recourse to simplified models, retaining the characteristic features of real suspensions, is thus necessary. Restriction will be made in the following to rigid spherical colloidal particles.

In the primitive model (PM) of electrolytes, the molecular solvent is crudely replaced by a dielectric continuum of dielectric constant ϵ, while the various ionic species are described as charged hard spheres, with charges $z_i e$ and diameters σ_i. Thus, particles of the species i and j can only approach to a distance $(\sigma_i + \sigma_j)/2$, while interacting Coulombically at larger distances, $v_{ij}(r) = z_i z_j e^2/\epsilon r$ for $r \geq (\sigma_i + \sigma_j)/2$. Apart from the excluded volume and statically screened electrostatic interactions, other effects present in a real suspension are neglected. Further simplification in the statistical description of charge-stabilized suspension is usually achieved by invoking the adiabatic approximation. This allows for the averaging over the degree of freedom of the small co- and counterions, yielding effective interaction potentials between the polyions. In particular, at low polyion density and using linear screening by point-like ions, one obtains a Yukawa pair repulsion. When the excluded volume constraint is also enforced, one recovers the electrostatic part of the famous Derjaguin-Landau-Verwey-Overbeek (DLVO) potential[1], which for years has been the standard in colloidal physics[2].

While allowing the calculation of particle-particle correlations for the species of interest, effective potentials do not fully determine the thermodynamic properties of the original many-component system. This precludes the study of phases equilibrium, which requires knowledge of the free energies of the full system. Thus, the form of the *effective pair interaction* has no direct bearing to the possibility of liquid-liquid coexistence, as it has been erroneously implied by a number of authors in recent years[3,4,5]. Moreover, originating from a partial averaging, effective potentials generally depend on thermodynamic variables such as density and temperature, which spoils the equivalence of different statistical ensembles in treating the effective one-component system. Even for the calculation of structural properties, the use of the DLVO potential has been frequently questioned, on both theoretical[6] and experimental[7] grounds. Apart for linear screening, the DLVO potential neglects fluctuations of the point-like ions, as well as the interference between screening clouds of different polyions, which would yield triplet and higher order forces. In fact, the DLVO potential has frequently been used with an adjustable macroion charge to interpret experimental data for the static structure factor $S_{pp}(q)$.

The deficiencies of the DLVO approach can be circumvented by going back to the original multi-component primitive model, and treating both polyions and ions on the same footing. Unfortunately, direct computer simulations become rapidly intractable for large charge and size asymmetry[8,9], and the same happens with non-linear integral equations for the liquid structure[10]. One of us has recently shown[11] that a rescaled mean spherical approximation (RMSA) may be capable of giving results of quality comparable to non-linear schemes, while retaining the simplicity of analytic solution. Here we compare the structural prediction of such RMSA with those of a recent developed classical version[12] of the famous Car-Parrinello[13] method, and with results of a new Monte Carlo study[9], performed in a regime of asymmetry where direct numerical simulations are still feasible. We also examine briefly the predictions of RMSA on liquid-liquid coexistence.

2. Primitive Model and RMSA

The aim of a theory of liquid structure[14] is to predict correlation functions and thermodynamics, starting from a know interaction low. In the primitive model, the interaction potential can be written as the sum of pair terms of the form

$$\begin{aligned}
v_{ij}(r) &= \infty, & r < \sigma_{ij}, \\
v_{ij}(r) &= z_i z_j e^2/r, & r > \sigma_{ij},
\end{aligned} \tag{1}$$

where

$$\sigma_{ij} = (\sigma_i + \sigma_j)/2, \tag{2}$$

and σ_i and z_i are hard core diameter and valence of the species i. Also, if ρ_i is the density of the species i

$$\sum_i \rho_i z_i = 0, \tag{3}$$

to ensure charge neutrality. Central quantities in a liquid state theory are the pair correlation functions[14] $g_{ij}(r)$, which give the probability of finding particles of type i and j at distance r, and are intimately related to the intensity scattered in a diffraction experiment. In fact, for pair interactions their knowledge—together with that of the interparticle potentials—is sufficient to calculate thermodynamics. Rather than seeking approximate relations directly between the g_{ij}'s and the v_{ij}'s, it is convenient to define intermediate functions c_{ij}'s, known as *direct* correlation functions, satisfying the Ornstein-Zernike (OZ) equations[14]

$$g_{ij}(r) - 1 = c_{ij}(r) + \sum_l \rho_l \int d\mathbf{r}'[g_{il}(r') - 1]c_{lj}(|\mathbf{r} - \mathbf{r}'|), \tag{4}$$

and look for approximations relating the g_{ij}'s and the c_{ij}'s with the v_{ij}'s. Once such *closure* relations are given, the OZ equations can be solved for the pair correlation functions.

In the mean spherical approximation[15] (MSA) to the PM, the closure is taken to be

$$g_{ij}(r) = \quad 0, \qquad\qquad r < \sigma_{ij}, \tag{5}$$

$$c_{ij}(r) = \quad -L_B z_i z_j/r, \qquad r > \sigma_{ij}, \tag{6}$$

with L_B the so-called Bjerrum length ($L_B \equiv \beta e^2/\epsilon \simeq 7.16\text{Å}$ in water at $300°K$), $\beta \equiv 1/K_B T$ and σ_{ij} given by Eq.(2). Thus, in the MSA one enforces exactly the hard core condition, while approximating the direct correlation functions outside the cores with the RPA form [14], which is believed to be correct at large separation. The MSA for the primitive model has a particularly simple analytic solution[16]. However it generally breaks down at large Coulomb coupling and low density, by yielding negative values of the $g_{ij}(r)$ at distances $r \gtrsim \sigma_{ij}$, for particles which repel each other, i.e., $z_i z_j > 0$. This is due to a poor description of Coulomb correlations. The exact $g_{ij}(r)$, in fact, is essentially vanishing over such distances, by virtue of correlations which are neglected in the MSA.

In a colloidal suspension the emphasis is on the polyion-polyion pair function, $g_{pp}(r)$, which is experimentally accessible through light scattering and small-angle neutron or X ray scattering[17]. Accordingly, we shall concentrate our attention on this quantity. To further simplify the discussion, we shall assume for the moment a colloidal suspensions with no salt added, so that only counterions are present. At the regime of interest, Coulomb correlations have the effect of keeping the polyions apart, well beyond the contact distance. In other words the polyion-polyion correlation hole is larger than the polyion physical diameter. A simple manner to correct the deficiencies of the MSA is to take approximately into account the important Coulomb correlations by resorting to an effective system in which the polyion-polyion approach distance is not given by Eq.(2) anymore, but[18,19]

$$\sigma_{pp} > \sigma_p. \tag{7}$$

At the same time Eq.(2) is retained for counterion-polyion

$$\sigma_{pc} = (\sigma_p + \sigma_c)/2. \tag{8}$$

on the ground that unlike particle attract each other[6], and for counterion-counterion

$$\sigma_{cc} = \sigma_c. \tag{9}$$

in that one renounces to possibly correct for their correlation hole. In practice, if $\sigma_{pp} = \sigma_p$ yields a negative $g_{pp}(r)$ for $r \gtrsim \sigma_{pp}$, one starts rescaling σ_{pp} to larger values, until $g_{pp}(\sigma_{pp}^+) = 0$. The rescaled MSA (RMSA) thus introduced involves the MSA treatment of an *effective primitive model* with non-additive radii (NAR), since Eq.(2) will not be satisfied for the pair polyion-polyion (cfr. Eq.(7)), or equivalently for the pair polyion-counterion in that according to Eqs.(7)-(9)

$$\sigma_{pc} \neq (\sigma_{pp} + \sigma_{cc})/2, \tag{10}$$

in contrast with Eq.(2), which implies the equality. It should be stressed, however, that the underlying physical model remains unchanged, being in particular one with additive radii. Thus the RMSA is just expedient to correct the deficiencies of MSA, which is a linear scheme[14], introducing non-linear effects through the correlation diameter σ_{pp}.

In general, the MSA solution of the primitive model with non-additive radii must be sought by numerically integrating the OZ equations[6]. However, there is one notable case in which an analytic solution exists, as we shall briefly review below. This is when one has only one kind of polyion and all small ions are taken point-like[20,21,11] .

3. NAR-MSA for polyions and point-like ions

Let us consider a monodisperse suspension of polyions, with charge $z_p e$, diameter $\sigma \equiv \sigma_{pp}$, and density ρ_p, in which m other charged species are present, all point-like, with charges $z_i e$, and densities ρ_i, $i = 1, 2, \ldots, m$. The implementation of the RMSA sketched above (see, in particular Eqs.(7)-(9)) requires the mean spherical approximation for a PM with non-additive radii satisfying

$$\sigma_{pi} = R \leq \sigma/2, \quad i = 1, 2, \ldots m, \tag{11}$$

$$\sigma_{ii} = 0, \quad i = 1, 2, \ldots m. \tag{12}$$

It easy to show that such a problem can be always transformed into an equivalent three component MSA problem[20], with the polyions (p), a charged point like species (c) and a neutral point like species (n). One finds in particular

$$\rho_c = (\rho_c z_c)^2 / \sum_{i=1}^{m} \rho_i z_i^2, \tag{13}$$

$$z_c = -\sum_{i=1}^{m} \rho_i z_i^2 / \rho_c, \tag{14}$$

$$\rho_n = \sum_{i=1}^{m} \rho_i - \rho_c, \tag{15}$$

$$\sigma_{pc} = \sigma_{nc} = R, \tag{16}$$

and the parameter of the polyions remain unchanged.

This three-component problem is further reduced to a two-component problem, and finally to a one-component MSA with Yukawa closure[21],

$$g(r) = \quad 0, \qquad r < \sigma; \tag{17}$$
$$c(r) = \quad Ke^{-\kappa(r-\sigma)}/r \quad r > \sigma, \tag{18}$$

with $g(r) \equiv g_{pp}(r)$, and $\kappa = [4\pi L_B \rho_c z_c^2]^{1/2}$ the inverse Debye length. The constant K appearing in Eq.(18) needs to be determined selfconsistently. In fact,

$$K = -L_B z_p^2 e^{-\kappa(\sigma-2R)}/[2\gamma e^{\kappa R}(sinh(\kappa R) - \kappa R cosh(\kappa R)) + 1 + \kappa R]^2, \tag{19}$$

and

$$\gamma = 2\pi \rho_p \sigma \int drr g(r) e^{-\kappa r}. \tag{20}$$

Fortunately enough, γ coincides with one of the possible choices[22] for the separation parameter entering the analytic solution of the OZ equation with a Yukawa closure. In fact following the analysis of Ref.[22], one has to solve a quartic in γ which contains K linearly. By substituting the expression (19) for K in such a quartic equation, one obtains a new quartic, with new coefficients. Once such an equation has been solved, by selecting among the real roots the physical one[23], analytic expression in terms of γ are available for various quantities[22], including the structure factor

$$S_{pp}(q) = S(q) = 1 + \rho_p \int dr \, [g(r) - 1] e^{i\mathbf{q}\cdot\mathbf{r}}. \tag{21}$$

Exploiting the correspondence between the one and the three component system, it is a simple matter to further obtain analytic expressions for the structure factors of the three component system[20,11], and in fact for all those of the original system with $m + 1$ components, if wanted. Similarly, one may easily calculate thermodynamic properties. The formulae for structure and thermodynamics may be found in[11,24].

4. Static Structure

The RMSA that we have outlined above, can be applied to the study of monodisperse polyion suspensions containing any number of small ion species, provided that taking the point ion limit for such species can be regarded as a reasonable approximation. This should be true, on account of the much larger size of the polyions; in fact it seems to be already so for micellar sizes $\sigma_p \approx 50\text{Å}[6]$. Once the point ion limit has been taken for the small species, one can study with the same easy both salt-free monodisperse polyion suspensions and suspensions with salt added. Here, it may be noted[11] that the structural results of our scheme are essentially coincident with those of Kanh et.al.[20], a fact that was not unexpected[6]. This in spite of the fact the present scheme is much simpler, in that it avoids the use of the HNC[14] to treat the effective one component system, and in fact it remains analytic.

As we have observed before, non-linear integral equations such as the HNC also break down in the treatment of highly asymmetric electrolyte solutions[10]. Thus, trying to correct upon the HNC, recourse has been made[10] to so-called mixed integral equations of the RY type[25,26] . One of us[11] has shown that the present RMSA scheme is able to yield results of quality equivalent to those of mixed integral

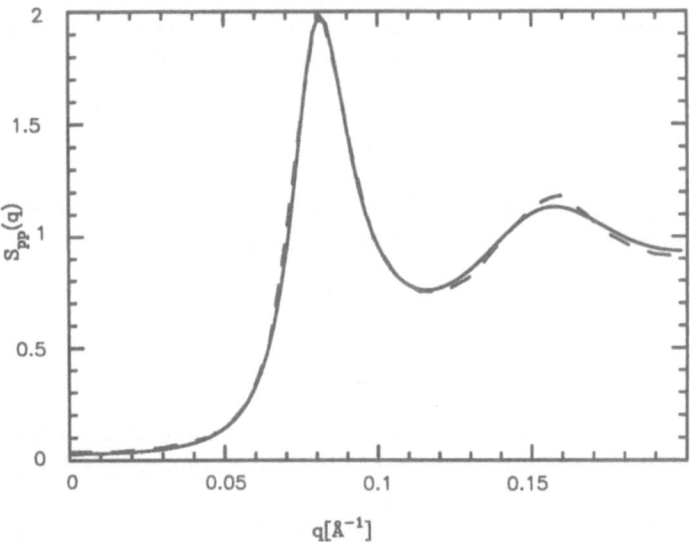

Figure 1. Polyion-polyion structure factor for the binary system: $\sigma_i = 50/5\text{Å}$, $z_i = -70/1$, $\eta = 0.1$, $T = 300°K$. The full curve gives the results of the present RMSA with point counterions (see text), whereas the dashed curve gives the results of an RY-type calculation[10].

equations, while retaining the merit of being analytic. For the easy of the reader, we report in Fig.1 the comparison between RMSA and RY polyion-polyion structure factors, for the largest charge asymmetry studied with the RY[10], i.e., $z_p = -70$, $z_c = 1$, $\sigma_p = 50\text{Å}$, $\sigma_c = 5\text{Å}$, $\eta = (\pi/6)\rho_p\sigma_p^3 = 0.1$. Here, η is the polyion packing fraction or equivalently the volume fraction occupied by the polyions. Also note that ρ_c remains fixed by charge neutrality, i.e., by Eq.(3). It should be mentioned that, in adapting our scheme with point counterions to the description of the above suspension, we have chosen $R = (\sigma_p + \sigma_c)/2 = 27.5\text{Å}$, to preserve the correct polyion-counterion approach distance, and found it necessary to rescale σ_{pp} by a factor 1.47 with respect to its initial value $\sigma_p = 50\text{Å}$. It is evident that the results of RMSA and RY are in close agreement. The same was found[11] for the other choice of parameters studied with the RY[10].

Recently a classical version of the Car-Parrinello method[13] has been applied[12] to perform *ab initio* molecular dynamics (AIMD) of colloidal suspensions. The advantage of such a scheme is that the motion of the polyions is dictated by forces which transcend the pair approximation and are in fact state dependent. In particular, model suspensions were considered, with one kind of polyions and only counterions. Differences were found with the predictions of effective one-component descriptions such as DLVO. We have thus applied the RMSA scheme to the two suspensions studied in [12].

Before we can compare our results with those of the AIMD we should briefly comment on the choice of parameters which characterize the suspensions. Two cases were studied in [12]: (a) $z_p = -200$, $z_c = 1$, $\sigma_p = 530\text{Å}$, $\sigma_c = 0$, $\eta = 0.1$; and (b) $z_p = -100$, $z_c = 1$, $\sigma_p = 530\text{Å}$, $\sigma_c = 0$, $\eta = 0.3$; both cases at

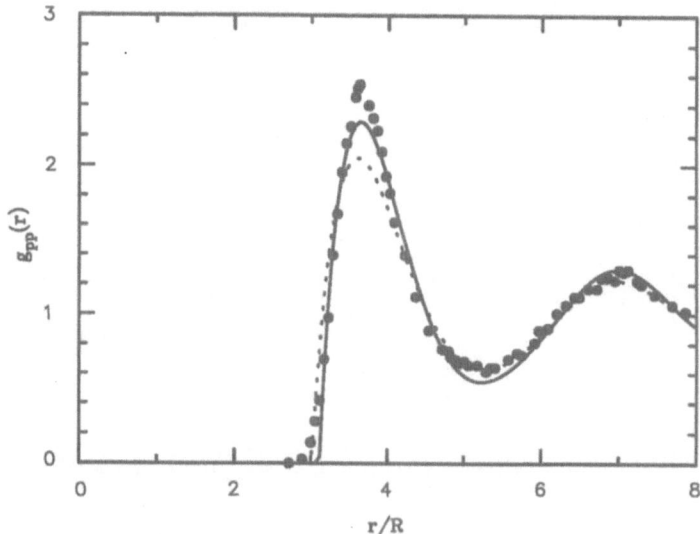

Figure 2. Polyion-polyion pair correlation function for the binary system (a): $\sigma_i = 200/0\text{Å}$, $z_i = -200/1$, $\eta = 0.1$, $T = 300°K$. The full circles give the results of *ab initio* MD[12]. Dotted and full curves give respectively the results of RMSA with reference and effective parameters (see text). Note that $R = \sigma_p/2$.

room temperature $(T = 300°K)$. However, the classical AIMD—as it is true also for the original Car-Parrinello—to be practical requires the use of soft counterion-polyion pseudopotentials. This implies the use of *effective* polyion charges[12], (a) $z_p^* = -266.6$, and (b) $z_p^* = -167.3$, which in turn implies the change of the counterions density. Clearly, in the AIMD, the effective parameters are used with a soft counterion-polyion pseudopotential.

We have performed RMSA calculations both with the *reference* parameters and with the *effective* ones. Our results are compared with those of AIMD in Fig.2 and Fig.3. We find that the RMSA pair correlation functions obtained using the reference parameters underestimate the ordering predicted by the AIMD, in particular at the first peak. However, using the effective parameters, we find that the RMSA pair correlations get much closer to the AIMD prediction. This is somewhat intriguing and we have no simple explanation for it at present. It is evident that in case (b), for which the packing is appreciable, the improvement is really impressive, bringing the RMSA prediction in excellent agreement with AIMD. On the other hand, at smaller packing, though the improvement is still sizeable, appreciable discrepancies remain around the first peak, which is still underestimated. In both cases, the rigidity of the hard core correlation diameter—in the RMSA—allows less penetration of the polyions in the correlation hole of their partners than found in the AIMD. We note than in our RMSA calculations we had to rescale the polyion diameter by a factor 1.51 for case (a) and by a factor 1.02 and 1.09 for case (b), respectively in the calculations with reference and effective parameters.

It is evident that the RMSA for highly asymmetric electrolytes is of equivalent quality to the RY[26] in the cases in which this scheme has been applied. On the

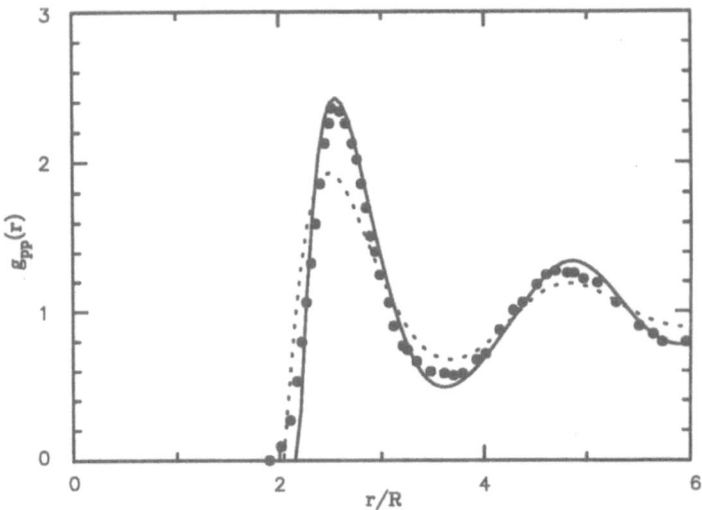

Figure 3. Polyion-polyion pair correlation function for the binary system (a): $\sigma_i = 200/0\text{Å}$, $z_i = -100/1$, $\eta = 0.3$, $T = 300°K$. The full circles give the results of *ab initio* MD[12]. Dotted and full curves give respectively the results of RMSA with reference and effective parameters (see text). Note that $R = \sigma_p/2$.

other hand its agreement with the novel *ab initio* molecular dynamics for classical asymmetric systems depends on the choice of parameters in a sensitive manners, as it has been illustrated. In this respect, it may be worth mentioning that the AIMD exploits the adiabatic principle for the small ions to combine molecular dynamics with the classical density functional theory in its local density approximation (LDA). However, at variance with the quantum case, the quality of the LDA for the correlation free energy of a classical system is not as well established—to our knowledge. Thus, we cannot consider the comparison of our RMSA results with those of the AIMD conclusive. On this ground, to have an independent test, we have performed standard Monte Carlo simulation for a binary asymmetric primitive model at intermediate packing, $\eta = 0.042$. The charge asymmetry was chosen as large as we could ($z_p = -20$, $z_c = 1$), to get still acceptable statistical noise, whereas the size asymmetry was moderate ($\sigma_p = 30\text{Å}$, $\sigma_c = 4\text{Å}$). The detail of the MC calculations will be given elsewhere[9]. Here we shall just mention that (i) we used 128 polyions + 1536 counterions in the cubic simulation box, and (ii) that Ewald summation of the interactions on the periodic copies of the box was performed, as it is appropriate for particles interacting Coulombically.

In Fig.4 we show the results of our Monte Carlo simulation together with the prediction of the RMSA. Some smoothing was performed to obtain the MC data shown in the figure. Also, in the RMSA, the finite size of the counterions was accounted for by taking $R = (\sigma_p + \sigma_c)/2 = 17\text{Å}$, whereas rescaling of the polyion-polyion diameter by a factor 1.64 was necessary to ensure positivity of $g_{pp}(r)$. It is clear that in this case, in contrast to the comparison with the AIMD, the RMSA is overestimating the structure in the pair correlations. This is in agreement with what was found by Linse who

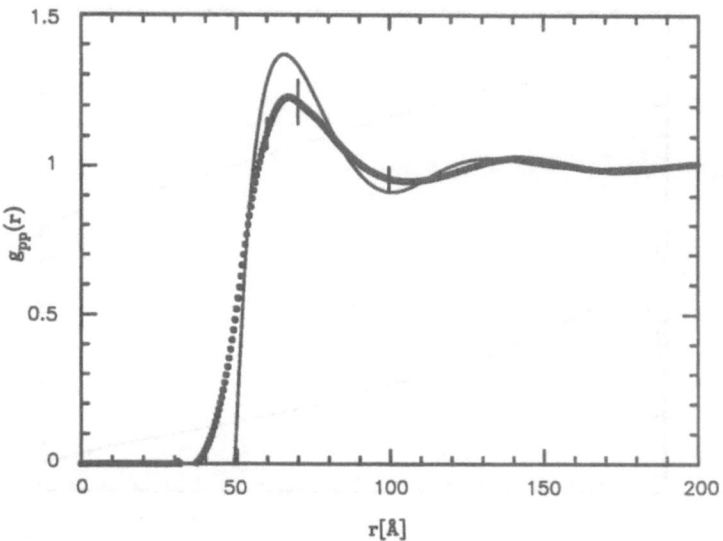

Figure 4. Polyion-polyion pair correlation function for the binary system: $\sigma_i = 20/4\text{Å}$, $z_i = -20/1$, $\eta = 0.042$, $T = 300°K$. The full circles are the results of Monte Carlo[9], with the bars showing typical uncertainties. The curve is the RMSA prediction.

performed mixed calculations[8]—combining simulations with analytic theories—for a similar system with a softer potential and fully numerical RMSA calculations[27] for the corresponding primitive model. We believe this overestimation of pair correlations to be a definite shortcoming of the RMSA at these regimes.

5. Thermodynamics

As we have already mentioned in the foregoing, one advantage of the present multi-component RMSA is the possibility to study the thermodynamics of the full solution, i.e., small ions included. Of course, as it is happens with all approximate theories[14], also with the RMSA there are several routes to thermodynamics[11]. We have already examined in some detail the predictions for a number of thermodynamic quantities within the RMSA for highly asymmetric electrolytes, comparing them with the predictions of HNC and RY[11]. Here we shall just reexamine the possibility of coexistence between liquids with different densities, as predicted by the RMSA.

To date there have been both experimental claims of a liquid-vapor separation in colloidal suspensions[3,5], and theoretical predictions of such a phenomenon[28,29,11]. As we have argued before, in a many-component system, there is no need to invoke[4,5] effective pair interactions with attractive character to explain such a phenomenon. Rather, it follows naturally from consideration of the full free energy of the many-component system[29,11]. From the theoretical point of view, due to the approximate nature of calculations[28,29,11], the detail of such phase coexistence are not fully established. In the past we only considered the energy route, in study phases coexistence[11], while from our MC simulations[9] we have now indications that for the RMSA the energy and virial routes to the equation of state might bracket the

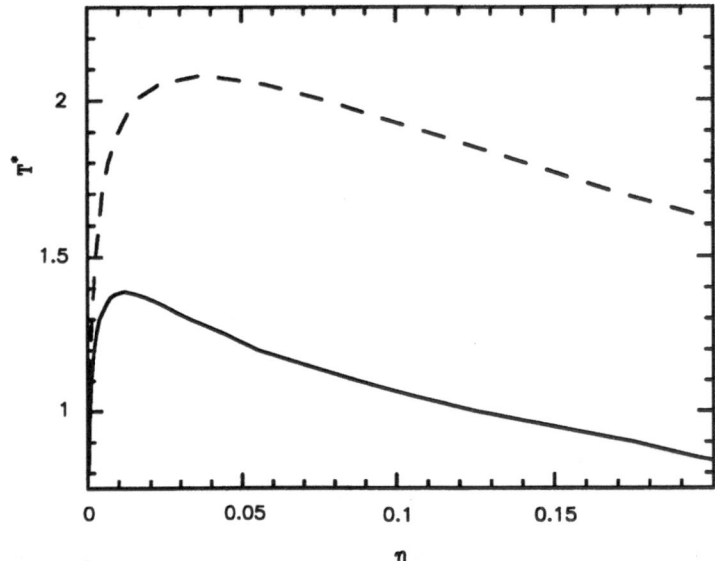

Figure 5. RMSA spinodal line for a binary suspension of polyions and point counterions with $\xi = |z_p/z_c| = 20$. T^* is a reduced temperature defined in the text and η the packing fraction. The full and dashed curves give the predictions obtained respectively from the energy and virial equation of state.

exact result. Therefore we have constructed the liquid-vapor spinodal as predicted by the virial route, within the RMSA. This was done for the same case studied by other authors[28,29,11], i.e., $|z_p/z_c| = 20$, in terms of the natural variables η and $T^* = \sigma/2L_B z_c^2$.

In Fig.5 we show the the spinodal lines obtained from the energy and virial equation of states, in the RMSA, for the binary asymmetric primitive model specified above. There are quantitative differences between the two curves, but both equation of state predict a liquid-vapor coexistence. The critical parameter from the virial route are $T_c^* = 2.08$ and $\eta_c = 0.037$. These values are to be compared with those obtained previously from the energy route, $T_c^* = 1.39$ and $\eta_c = 0.012$. Both, critical temperature and packing are predicted to be larger by the virial equation of state, the first by a 50% and the second by a factor 3. Yet, such critical temperatures are at most a half of the one obtained from the HNC[28], $T_c^* \simeq 4.25$. On the contrary, the HNC packing fractions was found to be smaller, $\eta_c \simeq 0.006$. The smallest critical temperature, on the other hand, has been obtained in an MSA study[29], with $T_c^* = 1.054$. The same study gave a critical packing $\eta_c \simeq 0.03$, in substantial agreement with the RMSA prediction from the virial route.

6. Conclusions

We have applied an analytic multi-component rescaled MSA to the study colloidal suspensions, within the description afforded by the primitive model. The present RMSA crucially relies on the existence of an analytic MSA solution for the mixture of charged hard spheres and point-like ions with non-additive radii[21]. We should stress that, to our knowledge, this solution is the only one in closed form present in the literature for mixtures with non-additive diameters.

The applicability of this RMSA requires the approximation of point-like size for the small species, which seems a reasonable one. In any case, a partial account of the small species size is possible, by keeping such size in the minimum approach distance between the small ions and the polyion. When this is done, the predictions of the RMSA for the polyion-polyion structure factor appear to be of quality comparable to that of mixed integral equations, though being essentially analytic. Comparison with the results obtained with a recently developed classical Car-Parrinello method[12], on the other hand, give for the moment contradictory indications. However, it seem established beyond doubt that at moderate charge asymmetry, as found in micellar solutions, the RMSA tend to overestimate the ordering in the suspension. In spite of this, in consideration of (i) its analytic nature[11], and (ii) its equivalence with available mixed-integral equations[26], the RMSA scheme remains a natural candidate for the fitting of experimental data.

The multi-component RMSA, allowing a prediction of all pair correlation functions, also permits the study of the thermodynamic properties of the suspensions, with respect to an inert dielectric continuum. Thus, the region of instability (two-phase region) predicted by the HNC[28] is confirmed by RMSA, though substantially displaced toward lower temperatures. On the contrary, we find a critical temperature somewhat larger than in MSA[29]. While we believe that a fluid-fluid phase separation at low temperature and density is a true property of the primitive model, it is not yet completely clear what is the relation of such a phenomenon with observations in real polyion suspensions[3,5].

Acknowledgements

GS dedicates this paper to Professor L. Blum. To him and to Prof. F.B.Malik, GS is also grateful for the partial support to the participation in the *XVI International Workshop on Condensed Matter Theories*. We acknowledge computer time allocation by CNR. This work has been supported in part by the Ministero della Ricerca Scientifica e Tecnologica.

References

[1] E.J.W.Verwey and J.T. Overbeek, *Theory of the stability of Lyophobic Colloids* (Elsevier, New York, 1948).

[2] See, for instance, P.N.Pusey, in *Liquids, Freezing and the Glass Transition*, edited by J.P.Hansen et al (North-Holland, Amsterdam, 1991).

[3] N.Ise, T.Okubo, M.Sugimura, K.Ito, and H.J.Nolte, J. Chem. Phys **78**, 536 (1983).

[4] I.Sogami, Phys. Lett. **96A**, 199 (1983); I. Sogami and N. Ise, J. Chem. Phys. **81**, 6320 (1984).

[5] A.K.Arora, B.V.R.Tata, A.K.Sood, and R. Kesavamoorthy, Phys. Rev. Lett. **60**, 2438 (1988).

[6] L.Belloni, J. Chem Phys. **85**, 519 (1986).

[7] E.B.Sirota, H.D.Ou-Yang, S.K.Shina, P.M.Chaikin, J.D.Axe, and Y.Fuji, Phys. Rev. Lett. **62**, 1524 (1989).

[8] P. Linse, J. Chem. Phys. **93**, 1376 (1990).

[9] G. Pastore and G. Senatore, to be published.

[10] L.Belloni, J. Chem. Phys. **88**, 5143 (1988).

[11] G. Senatore, in *Structure and dynamics of Strongly Interacting Colloids and Supramolecular Aggregates in Solution*, edited by S.H. Chen et al (Kluwer, Dordrecht, 1992).

[12] H.Löwen, P.A.Madden, and J.P.Hansen, Phys. Rev. Lett. (1992).

[13] R. Car and M. Parrinello, Phys. Rev. Lett. **55**, 2471 (1985).

[14] See, e.g., J.P.Hansen and I.R.McDonald, *Theory of Simple Liquids*, Academic Press (London, 1986).

[15] E. Waisman and J.L.Lebowitz, J. Chem. Phys. **56**, 3086 (1972).

[16] L.Blum and J.S.Hoye, J. Phys. Chem. **81**, 1311 (1977).

[17] See, for instance, *Structure and dynamics of Strongly Interacting Colloids and Supramolecular Aggregates in Solution*, edited by S.H. Chen et al (Kluwer, Dordrecht, 1992).

[18] M.Gillan, J. Phys. **C7**, L1 (1974).

[19] G. Senatore and L. Blum, J. Phys. Chem. **89**, 2676 (1985).

[20] S.Kahn, T.L.Morton and D. Ronis, Phys. Rev. **A35**, 4295 (1987).

[21] S.Kahn and D. Ronis, Mol. Phys. **60**, 637 (1987); .

[22] J.S.Hoye and L.Blum, J. Stat. Phys. **16**, 399 (1977).

[23] G. Pastore, Mol. Phys. **63**, 731 (1988).

[24] Note that equation (22) in the appendix of the Reference[11] should be corrected by multiplying the expression on the right hand side by $c_{pc}(q)$.

[25] G.Zerah and J.P.Hansen, J. Chem. Phys. **84**, 2336 (1986).

[26] F.J.Rogers and D.A.Young, Phys. Rev. **A30**, 999 (1984).

[27] P. Linse, J. Chem. Phys. **94**, 3817 (1991).

[28] L.Belloni, Phys. Rev. Lett. **57**, 2026 (1986).

[29] Z.Badirkhan and M.P.Tosi, Phys. Chem. Liq. **21**, 177 (1990).

COMPUTATIONAL MODELING OF OXYGEN ORDERING

AND STRUCTURAL PHASE TRANSITIONS IN YBa$_2$Cu$_3$O$_{6+x}$

Per Arne Rikvold[1,2] and Mark A. Novotny[2]

[1] Physics Department and Center for Materials Research and Technology
[2] Supercomputer Computations Research Institute
Florida State University, Tallahassee, Florida 32306, USA

INTRODUCTION

The superconducting critical temperature of the high-temperature superconductor YBa$_2$Cu$_3$O$_{6+x}$ depends crucially on the concentration and ordering of oxygen ions in the CuO$_x$ basal planes.[1] If x is less than approximately $\frac{1}{2}$, the material is in a nonsuperconducting tetragonal phase (T). Orthorhombic phases with $x \approx \frac{1}{2}$ (OII) and $x \approx 1$ (OI) become superconducting at approximately 60 K and 90 K, respectively. The existence of both OI and OII as equilibrium ordered phases has been confirmed by a number of experimental methods.[2-9] Other ordered phases, including $(2\sqrt{2} \times 2\sqrt{2})$,[10] have also been observed, but their thermodynamic stability has not yet been firmly established.[11]

A simple description of the oxygen ordering in the CuO$_x$ basal planes in terms of a two-dimensional lattice-gas model with locally anisotropic next-nearest-neighbor pair interactions (often referred to as the Asymmetric Next-Nearest-Neighbor Ising, or ASYNNNI, model) was introduced by de Fontaine, Wille, and coworkers.[12,13] The rationale behind this simple model is the assumption that the formation of oxygen chains in the CuO$_x$ planes, which breaks the symmetry between the in-plane axes, drives the structural phase transitions in the three-dimensional material. This requires weak interlayer interactions to ensure that the resulting anisotropy remains correlated from plane to plane. The view that the effects of these couplings on the in-plane oxygen ordering can be ignored, as far as the positions of the transition lines in the phase diagram are concerned, clearly amounts to an approximation[14] which must be tested by comparison with experiments. We are of the opinion that the close agreement with a large body of experimental data concerning both critical and non-critical properties of YBa$_2$Cu$_3$O$_{6+x}$, which we present here, continues to justify this view, at least for the relatively high temperatures at which experimental results are available. Other approximations inherent in the model are the restriction to pair interactions[15] and the absence of longer-ranged interactions. The former approximation leads to a phase diagram which is symmetric about $x=1$. However, since experimental data for $x>1$ are not available, and such high oxygen concentrations are most likely unphysical in the real material, there is little need to include this complication. The latter approximation presupposes strong screening of the electrostatic interactions between oxygen ions and is expected to be most applicable for intermediate and large values of x. A lattice-gas model with screened Coulomb interactions has been introduced by Aligia and coworkers,[11,16] and has also been used by Khachaturyan et al.[14]

The ASYNNNI lattice-gas model is defined by the Hamiltonian

$$\mathcal{H} - u\frac{x}{2}N = -\Phi_{NN}\sum_{\langle NN\rangle} c_i c_j - \Phi_{Cu}\sum_{\langle NNN_{Cu}\rangle} c_i c_j - \Phi_V\sum_{\langle NNN_V\rangle} c_i c_j - u\sum_i c_i , \qquad (1)$$

where $c_i \in \{0, 1\}$ are the O site-occupation variables, and $x = \frac{2}{N}\sum_i c_i$. The effective O–O nearest-neighbor interaction is Φ_{NN}, and Φ_{Cu} and Φ_V are locally anisotropic next-nearest-neighbor effective interactions, with and without a Cu atom between the O sites, respectively. The oxygen chemical potential is u. With $\Phi_{NN} < 0$ (repulsive), $\Phi_{Cu} > 0$ (attractive), and $-1 < \Phi_V/|\Phi_{NN}| < 0$ (weakly or moderately repulsive), this model reproduces the phases T, OI, and OII. The ground-state configurations in the CuO_x planes, corresponding to these phases, are shown in Fig. 1. The OI phase consists of densely

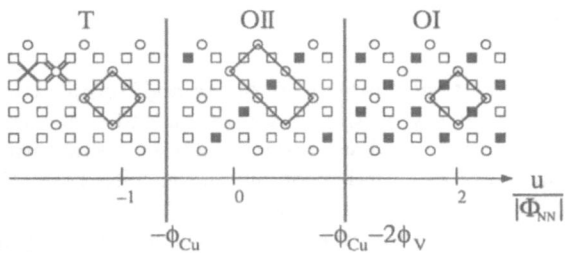

Figure 1. The ground-state configurations in the CuO_x planes, corresponding to the phases T, OII, and OI, are shown together with the ground-state diagram for the ASYNNNI model, which indicates the intervals in chemical potential u, within which each is stable. Here $\phi_{Cu} = \Phi_{Cu}/|\Phi_{NN}|$ and $\phi_V = \Phi_V/|\Phi_{NN}|$. Cu atoms are open circles, O atoms filled squares, and vacant O sites open squares. In the figure which shows the phase T are also indicated the next-nearest-neighbor interactions Φ_{Cu} (double lines) and Φ_V (thick lines).

packed parallel chains of O atoms connected by Cu atoms, whereas in the OII phase every other such chain is empty. The T phase corresponds to a disordered, low-density gas of O atoms although, as will be discussed below, considerable short-range order may be present.

In this paper we review results of nonperturbative transfer-matrix (TM) and Monte Carlo (MC) finite-size-scaling (FSS) studies of the ASYNNNI lattice-gas model of oxygen ordering in the CuO_x basal planes of $YBa_2Cu_3O_{6+x}$ and provide comparisons with a number of experimental results, as well as with theoretical studies by perturbative methods.

EARLY RESULTS

Early studies[12,13,17–21] of the ASYNNNI model used interaction constants $\Phi_{NN}<0$, $\Phi_{Cu}/|\Phi_{NN}|=+0.5$ and $\Phi_V/|\Phi_{NN}|=-0.5$. Phase diagrams in reasonable agreement with early experiments[3] were obtained,[18,19] and the transitions were shown to be continuous for all nonzero temperatures (in contrast to some cluster-variation results), and to belong to the universality class of the two-dimensional Ising model for the T/OI and OI/OII transitions, and to that of the XY model with cubic anisotropy for the T/OII transition. The three transition lines were found to meet at a multicritical point, giving

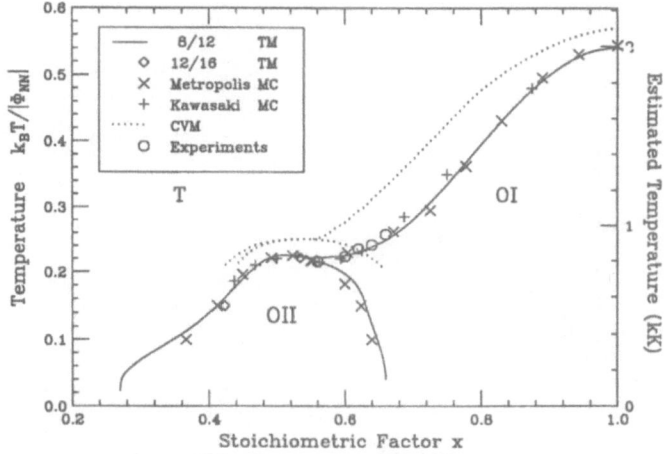

Figure 2. The phase diagram in the (x,T) plane for effective interactions $\Phi_{NN}<0$, $\phi_{Cu}=\Phi_{Cu}/|\Phi_{NN}|=+0.5$ and $\phi_V=\Phi_V/|\Phi_{NN}|=-0.5$, after Aukrust et al.[19] Also shown are experimental points due to Specht et al.[3] (circles), which were used to establish the estimated temperature scale along the right-hand vertical axis. The MC FSS results were obtained with both Metropolis ($\times$) and Kawasaki (+) dynamics. The solid line gives TM FSS results for strip widths $N/N'=8/12$. Two data points for 12/16 TM FSS are also shown ($\diamond$). The dotted lines are CVM results by Wille et al.[13] for exactly the same interactions.

a phase diagram resembling that of the Ashkin-Teller model.[18] The phase diagram for this set of interactions, obtained by the nonperturbative TM and MC FSS methods,[19] is shown in Fig. 2. For reference we also include results obtained by the perturbative Cluster Variation Method (CVM) by Wille et al.[13] for exactly the same interactions. Note that both the non-perturbative results (TM and MC) agree with each other, but are quantitatively different from the CVM results, which overestimate the transition temperatures by 6–35%.

NON-CRITICAL PHENOMENA: DISORDER LINES

In recent thermodynamic studies[4,9] low, broad maxima in the response function $k_B T \left(\frac{\partial x}{\partial u} \right)_T$ were observed for temperatures in the range 723–923 K. These maxima were located within the disordered T phase region of the phase diagram, and were clearly distinguishable from the peak at the T/OI phase transition. These observations were explained by Rikvold et al.[21] in terms of short-range order and anisotropy in the distribution of oxygen atoms in the CuO_x basal planes. Such local anisotropy appears when the oxygen concentration is sufficiently high to allow formation of short

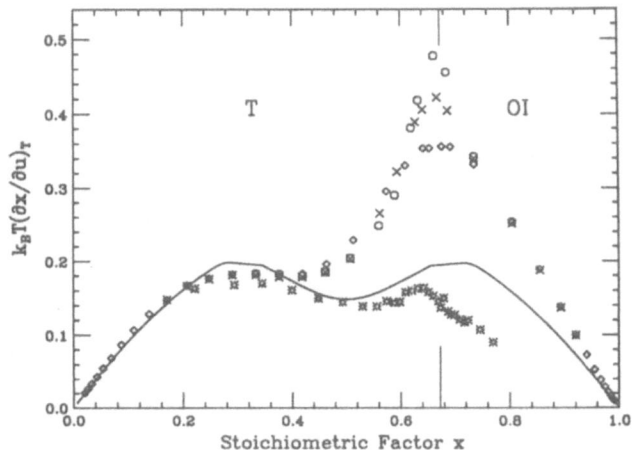

Figure 3. The response function $k_B T \left(\frac{\partial x}{\partial u} \right)_T$ vs x at T=923 K. After Rikvold et al.[21] Experimental data from McKinnon et al.[4] are represented by asterisks. Numerical results are given for the ASYNNNI lattice-gas model with the effective interactions obtained from LMTO calculations by Sterne and Wille[23] (see Table I). The solid line represents the contributions from local fluctuations, obtained with our special mean-field method, whereas the data points are Monte Carlo results for $2(L \times L)$ systems with $L = 8$ (◇), 16 (×), and 32 (○). The lack of system-size dependence of the MC results below $x \approx 0.5$ indicates the local nature of the corresponding fluctuations. The vertical lines mark the phase transition between the T and OI phases, obtained from TM FSS calculations with strip widths N/N'=8/12,[24] using the same interactions. The modest height of the experimental peak near the phase transition is probably due to imperfections in the sample. The corresponding maxima in the more recent data by Schleger et al.[9] are considerably sharper.

fragments of oxygen chains due to the attractive interaction Φ_{Cu}, but too low for the formation of an ordered orthorhombic phase. Although no thermodynamic singularities (phase transitions) are associated with these fluctuations, their effects are nevertheless macroscopically observable. As discussed in detail in Ref. 21, these local, essentially one-dimensional fluctuations can be identified with disorder lines[22] in the phase diagram of the ASYNNNI model. In Fig. 3 we present the experimental results of McKinnon et al.,[4] together with MC results for lattices of different sizes, as well as results obtained by a special mean-field technique for the study of local, noncritical fluctuations.[21] The contrast between the physics causing the disorder peak and the peak representing the T/OI phase transition is best documented by the absence of finite-size effects in the MC results for the former, which indicates the local, short-range nature of the corresponding fluctuations. The numerical results presented in Fig. 3 were calculated with a set of effective lattice-gas interactions obtained by Sterne and Wille, using a Linear Muffin-Tin Orbital (LMTO) method.[23] (A listing of the different sets of effective interaction constants referred to in this paper is given in Table I.)

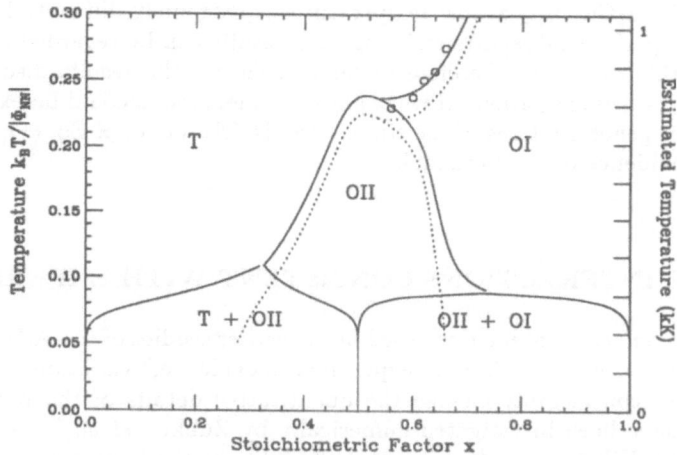

Figure 4. The phase diagram in the (x, T) plane for the same effective interactions as in Fig. 2, but with an additional, weakly attractive chain-chain interaction $\Phi_{CC} = +0.02|\Phi_{NN}|$, which causes first-order phase transitions and phase coexistence at low temperatures. Also shown are experimental points due to Specht et al.[3] (circles), which were used to establish the estimated temperature scale along the right-hand vertical axis. The solid line gives TM FSS results for strip widths $N/N' = 8/12$. The pointed shape of the T/OII coexistence loop is a finite-size artefact. The dotted lines are the corresponding results with $\Phi_{CC} = 0$. After Günther et al.[20]

EFFECTS OF LONGER-RANGE INTERACTIONS

The ASYNNNI model as defined in Eq. (1) cannot model ordered phases with a periodicity larger than two in the direction along the next-nearest-neighbor bonds,[10,25] neither can it account for the possibility of first-order phase transitions and phase coexistence at low temperatures.[26] A simple modification of the model is given by the

Hamiltonian

$$\mathcal{H} - u\frac{x}{2}N = -\Phi_{NN}\sum_{(NN)} c_i c_j - \Phi_{Cu}\sum_{(NNN_{Cu})} c_i c_j - \Phi_V\sum_{(NNN_V)} c_i c_j - \Phi_{CC}\sum_{(CC)} c_i c_j - u\sum_i c_i ,$$

$$(2)$$

where the additional interaction Φ_{CC} connects next-nearest-neighbor oxygen chains. Zubkus et al.[25] have demonstrated that a very weak, repulsive Φ_{CC} may lead to intermediate phases between the OI and OII phases at low temperature. However, their finite-temperature calculations were only performed by the perturbative CVM method. Similarly, in a TM FSS calculation, Günther et al.[20] showed that a very weak, attractive Φ_{CC} leads to island formation, thus changing the order of the OI/OII and T/OII transitions from second to first at low temperatures. The resulting phase diagram, which contains regions of coexistence between OI and OII, and between T and OII, is shown in Fig. 4. However, it should be noted that due to slow equilibration, experimental identification of ordered phases, and clear distinction between second- and first-order phase transitions in $YBa_2Cu_3O_{6+x}$ at low temperatures is extremely difficult. It is therefore unlikely that experimental reports claiming such results can be regarded as completely conclusive at this time. We therefore prefer to consider the results discussed in this section merely as demonstrations of what physical phenomena could be explained with relatively minor generalizations of the simple ASYNNNI model of Eq. (1), should clear experimental evidence become available.

EFFECTIVE INTERACTIONS CONSISTENT WITH EXPERIMENTS

Whereas the effective interactions used in the earlier studies of the ASYNNNI model produce results consistent with early experimental evidence,[3] variation of the effective interactions can considerably change the quantitative details of the phase diagram. These effects have been investigated numerically by Zubkus et al.,[27] using the CVM method, and by Hilton et al.,[24] using the TM FSS method combined with detailed ground-state calculations. (Ref. 24 also contains prescriptions for producing approximate sketches of the finite-temperature phase diagrams.) The phase diagrams obtained by the two groups are in good qualitative agreement with one another. It is, however, most likely that the regions of phase coexistence featured in the phase diagrams of Zubkus et al. are artefacts produced by their perturbative numerical method. An important result, confirmed by both these investigations, is that the detailed shape of the phase diagram is quite sensitive to the values of the effective interactions. Thus it is feasible to obtain a set of interactions by a numerical fit of the ASYNNNI phase diagram to the now considerable available body of experimental information obtained by a number of different techniques.

A calculation of the type discussed at the end of the preceding paragraph was successfully undertaken by Hilton et al.,[24] who estimated the effective interactions by a nonlinear least-squares fit to experimental data[4-9] of the TM FSS phase dia-

Table 1. Effective interactions as originally proposed in Ref. 12, and as calculated by the LMTO method,[23] and by χ^2 fits to experimental data using 8/12 TM FSS.[24]

Source	Method	Φ_{NN} (kK)	Φ_{Cu} (kK)	Φ_V (kK)
de Fontaine et al.[12]		adjustable < 0	$+0.50\|\Phi_{NN}\|$	$-0.50\|\Phi_{NN}\|$
Sterne and Wille[23]	LMTO	-4.36	$+1.52$	-0.70
Hilton et al.[24]	8/12 TM FSS	-2.8 ± 0.4	$+2.38 \pm 0.06$	-0.27 ± 0.04

gram calculated with strip widths $N/N'=8/12$ (8/12 TM FSS). The resulting phase diagram, which agrees well with the experiments over the whole temperature range covered by the data, is shown in Fig. 5. The corresponding effective interactions are $\Phi_{NN} = (-2.8\pm0.4)$ kK, $\Phi_{Cu} = (+2.38\pm0.06)$ kK, and $\Phi_V = (-0.27\pm0.04)$ kK. These results should be compared with the interactions obtained by Sterne and Wille[23] $\Phi_{NN} = -4.36$ kK, $\Phi_{Cu} = +1.52$ kK, and $\Phi_V = -0.70$ kK. The differences are statistically significant, but considering that the latter set of interactions was calculated with no adjustable parameters, we find the agreement encouraging. The fact that the interaction constants obtained by 8/12 TM FSS agree with the LMTO calculations to within 35% for Φ_{NN}, and to within 60% for the weak Φ_V and Φ_{Cu} interactions, provides mutual support for these two fundamentally different approaches to the problem of de-

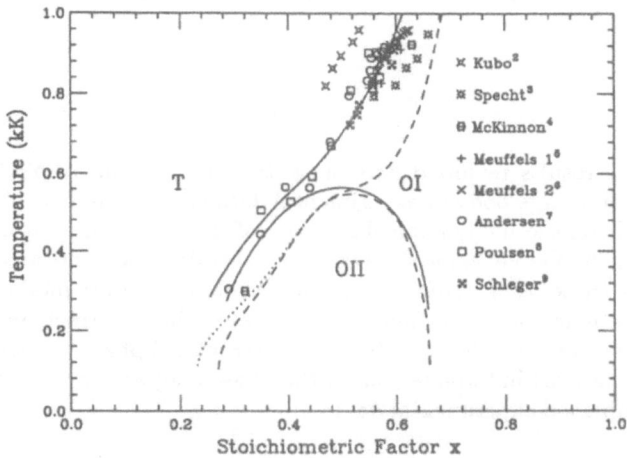

Figure 5. The (x,T) phase diagram, showing second-order phase boundaries (solid lines) calculated by 8/12 TM FSS with $\Phi_{NN}=(-2.8\pm0.4)$kK, $\Phi_{Cu}=(+2.38\pm0.06)$kK, and $\Phi_V=(-0.27\pm0.04)$kK. These effective interactions were obtained by a nonlinear least-squares fit[24] to the experimental data of Refs. 4–9. The dashed lines are phase boundaries obtained by the same method with the LMTO interaction constants of Sterne and Wille[23] (see Table I). See the text for a discussion of the significance of the dotted line. After Hilton et al.[24]

termining the effective lattice-gas interaction constants. The different sets of effective interactions considered in this review are summarized in Table I. For reference, the phase diagram for the LMTO interactions, calculated by 8/12 TM FSS, is also included in Fig. 5. The considerable difference between the phase diagrams obtained with these two rather similar sets of interaction constants reflects the fact that both sets lie in a region of the (ϕ_{Cu},ϕ_V) plane where the critical temperature for the OII phase is quite sensitive to changes in the interaction strengths.

When one compares the fitted phase diagram in Fig. 5 with that obtained with interactions from LMTO calculations (dashed lines in the same figure), or with that shown in Fig. 2, a striking feature is the narrow "tail" of the OI phase which extends between the T and OII phase regions at low x. If, as we believe, this is not a result of finite-size effects in the TM calculation, the OI phase in this part of the phase diagram should be structurally similar to the low-x orthorhombic phase previously proposed by other authors.[17,18,28] The dotted line in Fig. 5 indicates a region where such a phase conceivably (although, in our opinion, not entirely convincingly) might be inferred from our TM FSS calculations for the LMTO interaction set. For detailed discussions of our views on the possible existence of a separate, low-x orthorhombic phase in the ASYNNNI model, we refer the reader to Refs. 21 and 24. A simpler lattice-gas model, in which a similar ordered low-density phase has been quite clearly established, has been described by Bartelt et al.[29] However interesting this open question may be in the context of computational statistical mechanics, for the purpose of understanding the structural phase behavior of $YBa_2Cu_3O_{6+x}$ it is probably of very limited interest. This is because the features under discussion occur at temperatures sufficiently low to render thermodynamic equilibrium virtually unachievable in experiments, and at values of x for which the applicability of the ASYNNNI model to the real material is quite doubtful, due to the lack of electrostatic screening, as pointed out in the introduction.

CONCLUSION

The numerical results reviewed here show that the simple ASYNNNI lattice-gas model accounts for a large body of experimental data concerning both critical and non-critical structural and thermodynamical behavior of the high-temperature superconductor $YBa_2Cu_3O_{6+x}$ for a large region of oxygen concentrations and elevated temperatures. We consider that these results support the view that this simple model, despite its inherent approximations and limitations, gives an essentially correct description of the mechanism causing the orthorhombic–tetragonal structural phase transitions in this important material, at least in those regions of the phase diagram covered by the presently available and confirmed experimental data.

ACKNOWLEDGEMENTS

We thank our collaborators in the studies reviewed here: T. Aukrust, B. M. Gorman, C. C. A. Günther, D. K. Hilton, and D. P. Landau. We appreciate discussions with N. H. Andersen, who also has made experimental data available to us before publication, T. L. Einstein, D. de Fontaine, W. R. McKinnon, O. G. Mouritsen, M. Schick, P. Schleger, P. A. Sterne, and L. T. Wille.

Supported in part by Florida State University through supercomputer time and through its Supercomputer Computations Research Institute (U.S. Department of Energy Contract DE-FC05-85ER25000) and Center for Materials Research and Technology, and by the U.S. National Science Foundation Grants DMR-9013107 and DMR-9146922, The Petroleum Research Fund, administered by the American Chemical Society, and the Florida Graduate Scholars' Fund.

REFERENCES

1. R. J. Cava, B. Batlogg, C. H. Chen, E. A. Rietman, S. M. Zahurak, and D. Werder, *Phys. Rev. B* 36:5719 (1987); H. Claus, S. Yang, A. P. Paulikas, and B. W. Veal, *Physica C* 171:205 (1990).

2. Y. Kubo, Y. Nakabayashi, J. Tabuchi, T. Yoshitake, A. Ochi, K. Utsumi, H. Igarashi, and M. Yonezawa, *Jpn. J. Appl. Phys.* 26:L1888 (1987); Y. Kubo, T. Ichihashi, T. Manako, K. Baba, J. Tabuchi, and H. Igarashi, *Phys. Rev. B* 37:7858 (1988).

3. E. D. Specht, C. J. Sparks, A. G. Dhere, J. Brynestad, O. B. Cavin, D. M Kroeger, and H. A. Oye, *Phys. Rev. B* 37:7426 (1988).

4. W. R. McKinnon, M. L. Post, L. S. Selwyn, G. Pleizier, J. M. Tarascon, P. Barboux, L. H. Greene, and G. W. Hull, *Phys. Rev. B* 38:6543 (1988).

5. P. Meuffels, B. Rupp, and E. Pörschke, *Physica C* 156:441 (1988).

6. P. Meuffels, R. Naeven, and H. Wenzl, *Physica C* 161:539 (1989).

7. N. H. Andersen, B. Lebech, and H. F. Poulsen, *J. Less-Common Met.* 164/165:124 (1990); *Physica C* 172:31 (1990).

8. H. F. Poulsen, N. H. Andersen, and B. Lebech, *Physica C* 173:387 (1991).

9. P. Schleger, W. N. Hardy, and B. X. Yang, *Physica C* 176:261 (1991).

10. R. Sonntag, D. Hohlwein, T. Brückel, and G. Collin, *Phys. Rev. Lett.* 66:1497 (1991).

11. A. A. Aligia, J. Garcés, and H. Bonadeo, *Physica C,* 190:234 (1992), and references cited therein.

12. D. de Fontaine, L. T. Wille, and S. C. Moss, *Phys. Rev. B* 36:5709 (1987).

13. L. T. Wille, A. Berera, and D. de Fontaine, *Phys. Rev. Lett.* 60:1065 (1988).

14. A. G. Khachaturyan and J. W. Morris, Jr., *Phys. Rev. Lett.* 59:2776 (1987); *Phys. Rev. Lett.* 61:215 (1988); A. G. Khachaturyan, S. V. Semenovskaya, and J. W. Morris, Jr., *Phys. Rev. B* 37:2243 (1988).

15. M. Asta, C. Wolverton, D. de Fontaine, and H. Dreyssé, *Phys. Rev. B* 44:4907 (1991); C. Wolverton, M. Asta, H. Dreyssé, and D. de Fontaine, *Phys. Rev. B* 44:4914 (1991).

16. A. A. Aligia, A. G. Rojo, and B. R. Alascio, *Phys. Rev. B* 38:6604 (1988); A. A. Aligia, J. Garcés, and H. Bonadeo, *Phys. Rev. B* 42:10226 (1990); A. A. Aligia, H. Bonadeo, and J. Garcés, *Phys. Rev. B* 43:542 (1991); A. A. Aligia and J. Garcés, Phys. Rev. B **44**, 7102 (1991); A. A. Aligia, *Europhys. Lett.* in press (1992).

17. R. Kikuchi and J.-S. Choi, *Physica C* 160:347 (1989); V. E. Zubkus, S. Lapinskas, and E. E. Tornau, *Physica C* 159:501 (1989).

18. N. C. Bartelt, T. L. Einstein, and L. T. Wille, *Phys. Rev. B* 40:10759 (1989).

19. T. Aukrust, M. A. Novotny, P. A. Rikvold, and D. P. Landau, *Phys. Rev. B* 41:8772 (1990).

20. C. C. A. Günther, P. A. Rikvold, and M. A. Novotny, *Phys. Rev. B* 42:10738 (1990); P. A. Rikvold, *in:* "Computer Simulation Studies in Condensed Matter Physics IV", D. P. Landau, K. K. Mon, and H. B. Schüttler, ed., Springer, Heidelberg (1993, in press).

21. P. A. Rikvold, M. A. Novotny, and T. Aukrust, *Phys. Rev. B* 43:202 (1991).

22. J. Stephenson, *J. Math. Phys.* 11:420 (1970); *Phys. Rev. B* 1:4405 (1970).

23. P. A. Sterne and L. T. Wille, *Physica C* 162–164:223 (1989).

24. D. K. Hilton, B. M. Gorman, P. A. Rikvold, and M. A. Novotny, *Phys. Rev. B* 46: 381 (1992).

25. V. E. Zubkus, S. Lapinskas, and E. E. Tornau, *Physica C* 166:472 (1990).

26. H. You, J. D. Axe, X. B. Kan, H. Hashimoto, S. C. Moss, J. Z. Liu, G. W. Crabtree, and D. J. Lam, *Phys. Rev. B* 38:9213 (1988); A. K. Sood, K. Sankaran, V. S. Sastry, M. P. Janawadkar, C. S. Sundar, J. Janaki, S. Vijayalakshmi, and Y. Hariharan, *Physica C* 156:720 (1988).

27. V. E. Zubkus, E. E. Tornau, S. Lapinskas, and and P. J. Kundrotas, *Phys. Rev. B* 43:13112 (1991).

28. L. T. Wille and D. de Fontaine, *Phys. Rev. B* 37:2227 (1988); A. Berera, L. T. Wille, and D. de Fontaine, *J. Stat. Phys.* 50:1245 (1988); D. de Fontaine, M. E. Mann, and G. Ceder, *Phys. Rev. Lett.* 63:1300 (1989); D. de Fontaine, G. Ceder, and M. Asta, *Nature* 343:544 (1990); G. Ceder, M. Asta, W. C. Carter, M. Kraitchman, D. de Fontaine, M. E. Mann, and M. Sluiter, *Phys. Rev. B* 41:8698 (1990); D. de Fontaine, G. Ceder, and M. Asta, *J. Less-Common Met.* 164/165:108 (1990).

29. N. C. Bartelt, L. D. Roelofs, and T. L. Einstein, *Surf. Science* 221:L750 (1989).

PHASE TRANSITIONS AT THE FLUID-SOLID INTERFACE IN A MODEL

FOR THE ADSORPTION OF HARD SPHERES

Dale A. Huckaby[1] and Lesser Blum[2]

[1]Department of Chemistry, Texas Christian University
Fort Worth, Texas 76129
[2]Department of Physics, University of Puerto Rico
Rio Piedras, Puerto Rico 00931-3343

Phase transitions at the fluid-solid interface are studied using an adsorption model consisting of a fluid of hard spheres in contact with a planar wall which contains a lattice of sticky adsorption sites. The model is equivalent to a lattice gas with n-body interactions that are related to the n-body correlation functions of the fluid.

I. INTRODUCTION

The present paper is an overview of some of the results we have obtained in the past three years using a statistical mechanical model to study phase transitions which occur at the fluid-solid interface.[1-3] The model,[4,5] as discussed in Sec. II, consists of a dense fluid of hard spheres of diameter σ near a planar wall that contains a triangular lattice of sticky sites. This three-dimensional model is equivalent to a two-dimensional lattice gas with many-body interactions that are related to the many-body contact correlation functions of the fluid.

The nature of the phases which occur at the fluid-solid interface in this model depends on the fluid density, the strength of the sticky attraction at the lattice sites, and also on the ratio of the hard sphere diameter to the lattice spacing d. If σ slightly exceeds d, if the pair correlation functions are assumed to be unity at distances exceeding d, and if many-body correlation functions are approximated using the Kirkwood superposition approximation,[6] the adsorption model is then equivalent to the hard-hexagon lattice gas solved by Baxter.[7] The isotherms for this case of the adsorption model, which undergoes an order-disorder phase transition, can be calculated using some exact expressions obtained by Joyce[8] for the hard-hexagon model. These results are presented in Sec. III along with some simplified expressions for the isotherms which we obtained by exploiting a symmetry present in Joyce's original expressions.

If the sphere diameter σ is much smaller than the lattice spacing d, so that the correlation functions are approximately unity at distances as large as d, then the model has no phase transition and the Langmuir adsorption isotherm results.[2]

We have treated in some detail the special case for which $\sigma = d$.[1-3] Simple, but accurate analytical expressions are known for the contact pair correlation function as a function of the fluid density.[9-11] Assuming the pair correlation functions decay to approximately unity at distances approaching the second neighbor lattice spacing, and using the Kirkwood superposition approximation, the adsorption model is equivalent to a lattice gas with first-neighbor pairwise interactions.[1,5] The coexistence surface for the first-order phase separation which occurs in this lattice gas is known exactly.[12] Using several exactly known coefficients in the series expansion of the properties of the lattice gas,[13] we obtained accurate adsorption isotherms which are a generalization of Langmuir's isotherm.[2] These results are presented in Sec. IV.

Recent calculations by Attard and Stell indicate that the Kirkwood superposition approximation is not accurate for a triangle of hard spheres all in mutual contact.[14,15] Since such a configuration is present in the adsorption model if $\sigma = d$, we have included the effects of three-body interactions in this case of the model. The critical point of the equivalent lattice gas, which contains pair and triplet interactions, has been approximated[16] using the interface method of Müller-Hartmann and Zittarz.[17] In addition, we discovered a simple but accurate analytical approximation to the three-body correlation function for three spheres in mutual contact.[3] Together these results yielded an estimate of the fluid density at the critical point of the two-phase coexistence surface for the adsorption model[3] which is significantly higher than that predicted using only the contact pair correlation function and the Kirkwood superposition approximation.[1] These results are presented in Sec. V.

II. THE MODEL

We consider a model for adsorption in which a fluid of N hard spheres of diameter σ in a volume V interacts with a hard wall, located at $z = -\sigma/2$, containing a lattice Λ of sticky adsorption sites.[1,4,5] The partition function for the system is

$$Z = \frac{1}{N!} \int e^{-\beta H} dr^N , \tag{1}$$

where $\beta = (kT)^{-1}$. The Hamiltonian can be written as

$$H = H_0 + \sum_{i=1}^{N} U^s(r_i) , \tag{2}$$

where H_0 is the Hamiltonian for the system in the absence of the sticky sites (the smooth wall problem), and $U^s(r_i)$ is the potential for the interaction of a hard sphere i at r_i with the lattice of sticky sites $\{R_s\}$. This sticky potential can be written as

$$e^{-\beta U^s(r_i)} = 1 + \lambda \sum_{R_s \in \Lambda} \delta(r_i - R_s) , \tag{3}$$

where δ is the Dirac delta function. The stickiness parameter λ has units of volume and, except for a constant factor, is the fugacity of adsorption of a hard sphere onto a sticky site.

Performing the integrations in Eq. (1) to remove the delta functions and rearranging terms yields

$$Z/Z_0 = \sum_{n=0}^{N} \frac{\lambda^n}{n!} \sum_{\{R_i\} \subset \Lambda} \rho_n^0(R_1,...,R_n) , \tag{4}$$

where

$$\rho_n^0(r_1,...,r_n) = [Z_0(N-n)!]^{-1} \int e^{-\beta H_0} dr_{n+1}...dr_N \tag{5}$$

$$= g_n^0(r_1,...,r_n) \prod_{i=1}^{n} \rho_1^0(r_i) .$$

Here, Z_0, $g_n^0(r_1,...,r_n)$, and $\rho_1^0(r_i)$ are respectively the partition function, an n-body correlation function, and the single particle density for the smooth wall problem.

Defining the potential of mean force $U(R_1,...,R_n)$ as

$$g_n^0(R_1,...,R_n) = e^{-\beta U(R_1,...,R_n)} \tag{6}$$

yields

$$Z/Z_0 = \sum_{n=0}^{N} \frac{[\lambda \rho_1^0(0)]^n}{n!} \sum_{\{R_i\} \subset \Lambda} e^{-\beta U(R_1,...,R_n)} , \tag{7}$$

639

where $\rho_1^0(0)$ is the single particle density at the contact plane ($z = 0$). Changing from a sum over the positions of labelled hard spheres on Λ to a sum over lattice sites of Λ, Eq. (7) yields

$$\Xi = Z/Z_0 = \sum_{\{t_i\}} [\lambda \rho_1^0(0)]^{\Sigma t_i} \, e^{-\beta U(\{t_i\})} \, , \tag{8}$$

where t_i is the occupation number of site i in a given configuration $\{t_i\}$.

The adsorption model is thus equivalent to a two-dimensional lattice gas with a grand canonical partition function Ξ, a many-body interaction energy $U(\{t_i\})$, and a chemical potential μ given as

$$e^{\beta \mu} = \lambda \rho_1^0(0) \, . \tag{9}$$

The fraction of sites of Λ which are occupied by spheres is given by[1,5]

$$\theta = \frac{\lambda}{|\Lambda|} \frac{\partial}{\partial \lambda} \ln \Xi \, . \tag{10}$$

III. ADSORPTION OF LARGE SPHERES

If the hard sphere diameter σ slightly exceeds the lattice spacing d, then occupancy of two first-neighbor sites is excluded, and hence $U(\{t_i\})$ is infinite for all such excluded configurations. If $U(\{t_i\})$ is assumed to be zero for all allowed configurations, this is equivalent to assuming the pair correlation function is unity at distances greater than or equal to the second-neighbor separation, $\sqrt{3}\, d$, and that the n-body correlation functions are given by the Kirkwood superposition approximation[6]

$$g_n^0(R_1,...,R_n) = \prod_{\langle ij \rangle} g_2^0(R_i,R_j) \, . \tag{11}$$

Within the above approximations, the adsorption model is equivalent to the hard hexagon lattice gas, which has been solved exactly by Baxter.[7] The isotherms for the adsorption model can then be computed using exact expressions for $\lambda \rho_1^0(0)$ as a function of θ which were obtained by Joyce[8] for the equivalent hard hexagon lattice gas. We noticed that these expressions have a more compact form when $\lambda \rho_1^0(0)$ is written as a function of the variable

$$\phi = \theta(1 - \theta) \, . \tag{12}$$

The adsorption model undergoes an order-disorder transition at the fluid-solid interface at a critical coverage $\theta_c = (5 - \sqrt{5})/10 = 0.2764$, which corresponds to the value $\phi_c = 1/5$.

The exact expression for $\lambda\rho_1^0(0)$ as a function of ϕ in the disordered region, $\phi \leq 1/5$, can be written as

$$\lambda\rho_1^0(0) = Q[Q_0{}^2 Q_1{}^{1/2} + Q_2 - Q_0(2Q_3 + 2Q_2Q_1{}^{1/2})^{1/2}] , \tag{13}$$

where

$$Q = (8\phi^6)^{-1}[1 - 5\phi + 5\phi^2 + (1 - 4\phi)^{1/2}(1 - 3\phi + \phi^2)]$$

$$Q_0 = 1 - 5\phi$$

$$Q_1 = (1 - \phi)(1 - 5\phi) \tag{14}$$

$$Q_2 = (1 - 4\phi)^{1/2}(1 - 11\phi + 33\phi^2 - 11\phi^3)$$

$$Q_3 = 1 - 16\phi + 90\phi^2 - 198\phi^3 + 119\phi^4 - 10\phi^5 .$$

The expression in the ordered region, $\phi \geq 1/5$, is given as

$$\lambda\rho_1^0(0) = \frac{-2 + 9\phi - 6\phi^2 - (2 - 5\phi)(1 - 4\phi)^{1/2}}{1 - 12\phi + 33\phi^2 + (5\phi - 1)^{3/2}(9\phi - 1)^{1/2}} . \tag{15}$$

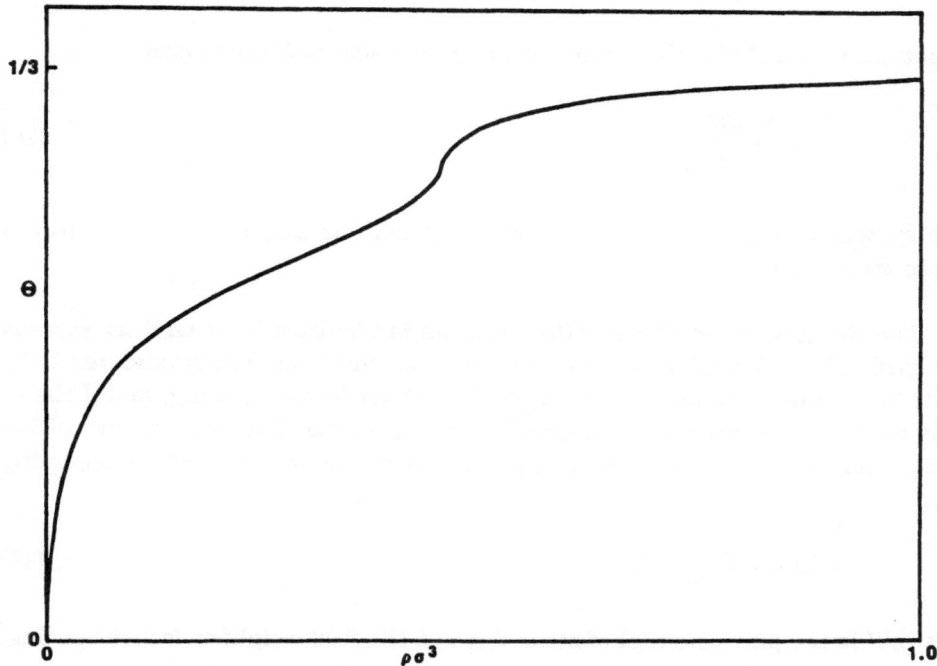

Figure 1. An isotherm with $\lambda/\sigma^3 = 10$ for the case $d < \sigma < \sqrt{3}\, d$.

At the transition, $[\lambda\rho_1^0(0)]_c = (11 + 5\sqrt{5})/2 = 11.09... $.

The contact single particle density as approximated by the Percus-Yevick (PY) theory is [9,10]

$$\rho_1^0(0)\sigma^3 = \frac{6\eta(1 + 2\eta)}{\pi(1 - \eta)^2} , \tag{16}$$

where $\eta = (\pi/6)\rho\sigma^3$ is the packing fraction. The maximum density possible for hard spheres is $\rho\sigma^3 = \sqrt{2}$, which occurs at closest-packing.

Isotherms in the θ versus $\rho\sigma^3$ plane can be easily calculated using Eqs. (13) - (16). An isotherm with $\lambda/\sigma^3 = 10$ is illustrated in Fig. 1.[1]

IV. ADSORPTION OF SMALL SPHERES

If the lattice spacing d greatly exceeds the hard sphere diameter σ, then the correlation functions can all be assumed to be unity for distances as large as d. This is equivalent to assuming $U(\{t_i\})$ in Eq. (6) is zero for all allowed configurations, and Eq. (8) becomes[2]

$$\Xi = [1 + \lambda\rho_1^0(0)]^{|\Lambda|} . \tag{17}$$

Equations (10) and (17) then yield the Langmuir adsorption isotherm

$$\theta = \frac{\lambda\rho_1^0(0)}{1 + \lambda\rho_1^0(0)} . \tag{18}$$

The system in this case has no lateral interactions and does not undergo a phase transition.

For the case in which the lattice spacing is identical to or slightly exceeds the hard sphere diameter, if the pair correlation function is approximated to be unity for distances as large as the second-neighbor lattice spacing, and if the n-body correlation functions are approximated using the Kirkwood superposition approximation of Eq. (11), these approximations are equivalent to assuming that

$$U(\{t_i\}) = W \sum_{nn} t_i t_j , \tag{19}$$

where W is the pair potential of mean force at the first-neighbor lattice spacing, i.e.,

$$e^{-\beta W} = g_2^0(d) .$$ (20)

The equivalent lattice gas thus has a partition function given by Eq. (8) and Eq. (19). A first-order phase transition occurs in this lattice gas on the triangular lattice if [1]

$$\lambda \rho_1^0(0) = [g_2^0(d)]^{-3} .$$ (21)

The two-phase coexistence surface for this transition has been calculated exactly and is given by [12]

$$\theta = \tfrac{1}{2}(1 \pm \{1 - 16g_2^0(d)[g_2^0(d) - 1]^{-3}[g_2^0(d) + 3]^{-1}\}^{1/8}) .$$ (22)

The parameters at the critical point of this transition, which occurs at $\theta = 1/2$, are given from Eq. (21) and Eq. (22) as [1]

$$[g_2^0(d)]_c = 3$$

$$[\lambda \rho_1^0(0)]_c = 1/27 .$$ (23)

For the special case $\sigma = d$, we let $g_2 = g_2^0(\sigma)$, and the PY approximation to the contact pair correlation function [9,10]

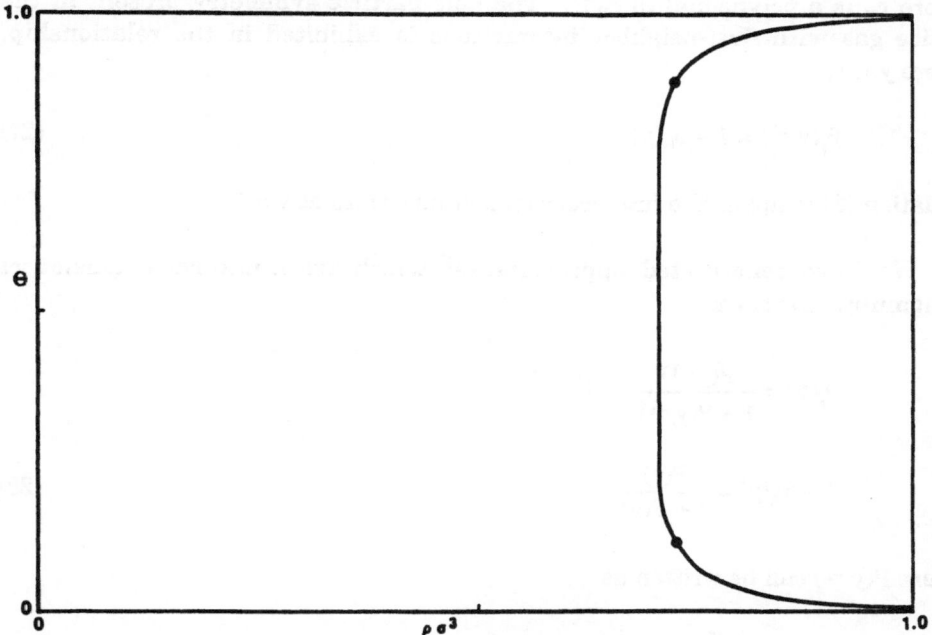

Figure 2. The coexistence curve for the case $\sigma = d$. The two coexisting phases on the isotherm with $\lambda/\sigma^3 = 0.01$ are marked with dots.

$$g_2 = \frac{1 + \eta/2}{(1 - \eta)^2} \tag{24}$$

can be used with Eq. (22) to plot the two-phase coexistence surface in the $(\theta, \rho\sigma^3)$ plane. For an appropriate fixed value of λ/σ^3, the density at which a phase transition occurs on this isotherm can be calculated using Eqs. (16), (21), and (24).[1] For example, on the isotherm for which $\lambda/\sigma^3 = 0.01$, the two-phase coexistence occurs at $\rho\sigma^3 = 0.727$ with $\theta = 0.886$ and $\theta = 0.114$. These two transition points are pictured on the two-phase coexistence curve in Fig. 2.

Although the isotherms for this case have not been calculated exactly (this would be equivalent to solving the Ising model in non-zero field), many exact coefficients in series approximations to θ have been obtained. Letting $y = [\lambda\rho_1^0(0)g_2^3]^{-1}$, at low densities $(y > 1)$[13]

$$\theta_l(y) = \sum_{r=1}^{\infty} r\, y^{-r}\, c_r(g_2^{-1}) \ , \tag{25}$$

and at high densities $(y < 1)$

$$1 - \theta_h(y) = \sum_{r=1}^{\infty} r\, y^r\, c_r(g_2^{-1}) \ , \tag{26}$$

where c_r is a polynomial in g_2^{-1}. The hole-particle symmetry present in the lattice gas with first-neighbor interactions is exhibited in the relationship, where $y < 1$,

$$\theta_l(y^{-1}) = 1 - \theta_h(y) \ . \tag{27}$$

Equation (21) implies the first order transition occurs at $y = 1$.

We have constructed approximants[2] which are a natural extension of Langmuir's isotherm

$$\theta_l(y) = \frac{P(y^{-1})}{1 + P(y^{-1})}$$

$$1 - \theta_h(y) = \frac{P(y)}{1 + P(y)} \ , \tag{28}$$

where $P(y^{-1})$ can be written as

$$P(y^{-1}) = \sum_{r=1}^{m} p_r(g_2)\, [\lambda\rho_1^0(0)]^r \ . \tag{29}$$

The coefficients $p_r(g_2)$ are polynomials in g_2 which are determined by requiring that the coefficients in the series expansions of Eq. (28) match the first m coefficients of Eq. (25) and Eq. (26). Written in terms of $f = g_2 - 1$, the polynomials $p_r(g_2)$, $r \leq 8$, for the triangular lattice were calculated to be

$$p_1 = 1$$

$$p_2 = 6f$$

$$p_3 = -6f + 45f^2 + 6f^3$$

$$p_4 = 6f - 120f^2 + 344f^3 + 108f^4 + 12f^5$$

$$p_5 = -6f + 225f^2 - 1680f^3 + 2478f^4 + 1374f^5 + 315f^6 + 30f^7$$

$$p_6 = 6f - 360f^2 + 4920f^3 - 19788f^4 + 15474f^5 + 14640f^6 + 5298f^7 + 1008f^8 + 84f^9$$

$$p_7 = -6f + 525f^2 - 11270f^3 + 82803f^4 - 205830f^5 + 66926f^6 + 135396f^7 + 71274f^8 + 20776f^9 + 3507f^{10} + 294f^{11} + 7f^{12}$$

$$p_8 = 6f - 720f^2 + 22224f^3 - 254568f^4 + 1179828f^5 - 1905384f^6 - 101754f^7 + 1068366f^8 + 817260f^9 + 330282f^{10} + 83868f^{11} + 13374f^{12} + 1224f^{13} + 48f^{14} . \tag{30}$$

Using $\theta_l(y)$ and $\theta_h(y)$ of Eq. (28), together with a switching function $\eta(y)$ which vanishes at $y = 0$ and becomes unity as $y \to \infty$, a continuous approximation to θ which is accurate both at high and low fluid densities can be constructed as[2]

$$\theta(y) = \theta_l(y)\eta(y) + \theta_h(y)[1 - \eta(y)] . \tag{31}$$

Since $\theta_l(y)$ and $\theta_h(y)$ have the symmetry of Eq. (27), then if $\eta(y)$ satisfies $\eta(y^{-1}) = 1 - \eta(y)$, the approximation to θ given by Eq. (31) also satisfies $\theta(y^{-1}) = 1 - \theta(y)$. A possible choice for the switching function is

$$\eta(y) = \tfrac{1}{2}\{1 + \mathrm{erf}\,[s(y - y^{-1})]\} , \tag{32}$$

where s is a measure of the sharpness of the change between the two limiting values of $\eta(y)$.

V. EFFECTS OF THREE-BODY CORRELATIONS

Using pair correlations only, we can estimate the fluid density at the critical point of the first-order transition for the case $\sigma = d$ by combining the condition $(g_2)_c = 3$ with the PY contact pair correlation function of Eq. (24). This yields the estimate[1]

$$\rho_c \sigma^3 = \frac{13 - \sqrt{73}}{2\pi} = 0.7092 . \tag{33}$$

If the more accurate Carnahan-Starling (CS) pair correlation function[10,11]

$$g_2 = \frac{1 - \eta/2}{(1 - \eta)^3} \tag{34}$$

is used for the calculation, the resulting estimate of this fluid density, $\rho_c \sigma^3 = 0.6678$, is slightly lower than that given by the PY correlation function.

A recent calculation by Attard and Stell[15] using the Percus-Yevick 3 (PY3) theory,[18] which includes three-body correlations, indicates that the Kirkwood superposition approximation of Eq. (11) is not accurate for the triplet correlation function of three spheres in mutual contact, but it is accurate for other possible configurations of three spheres on the triangle lattice.[3]

An improved estimate of $U(\{t_i\})$ is then given as[3]

$$U(\{t_i\}) = W \sum_{nn} t_i t_j + W_3 \sum_{\Delta} t_i t_j t_k , \tag{35}$$

where the second sum is over all triangles of nearest neighbor sites on the lattice. From Eq. (6) and Eq. (35), we can identify W as the pair potential of mean force and $W_3 + 3W$ as the potential of mean force for three spheres in mutual contact.

This is equivalent to the superposition approximation[3]

$$g_n^0(\mathbf{R}_1, ..., \mathbf{R}_n) = \prod_{nn} g_2 \prod_{\Delta} g_3 / g_2^3 , \tag{36}$$

where g_3 is the triplet correlation function for three spheres in mutual contact, $g_2 = e^{-\beta W}$, and $g_3 / g_2^3 = e^{-\beta W_3}$.

Within this triplet correlation approximation, the model is equivalent to a lattice gas with pairwise interactions, W, and three-body interactions, W_3.[3] Using the interface method of Müller-Hartmann and Zittarz,[17] the critical point of the coexistence surface in the lattice gas is predicted to satisfy,[16] where $g_0 \equiv g_3 / g_2^2$,

$$(g_0)_c = 3 . \tag{37}$$

Using the PY3 theory,[18] Attard and Stell[15] calculated g_0 numerically over a wide range of fluid densities. We recently discovered[3] that their numerical results are accurately approximated by the simple analytical expression

$$g_0 = \frac{4 - 7\eta + 7\eta^2 - 2\eta^3}{4(1 - \eta)^3} . \tag{38}$$

Using Eq. (37) and Eq. (38), the fluid density at the critical point of the transition is calculated to be $\rho_c \sigma^3 = 0.8409$.[3] This estimate of the minimum fluid density necessary for a phase transition to occur at the fluid-solid interface is much larger than that predicted using only the pair correlation function and the Kirkwood superposition approximation.[1] The inclusion of triplet correlations is thus important for studying adsorption in this case of the model.

ACKNOWLEDGMENTS

D.H. was supported by the Robert A. Welch Foundation, grant P-0446. L.B. was supported by the Office of Naval Research and by EPSCoR EHR-910-8775.

REFERENCES

1. D. A. Huckaby and L. Blum, J. Chem. Phys. **92**, 2646 (1990).
2. L. Blum and D. A. Huckaby, J. Chem. Phys. **94**, 6887 (1991).
3. D. A. Huckaby and L. Blum, J. Chem. Phys., submitted for publication.
4. M. L. Rosinberg, J. L. Lebowitz, and L. Blum, J. Stat. Phys. **44**, 153 (1986).
5. J. P. Badiali, L. Blum, and M. L. Rosinberg, Chem. Phys. Lett. **129**, 149 (1986).
6. J. G. Kirkwood, J. Chem. Phys. **3**, 300 (1935).
7. R. J. Baxter, J. Phys. A **13**, L61 (1980).
8. G. S. Joyce, J. Phys. A **21**, L983 (1988).
9. M. Wertheim, J. Math. Phys. **5**, 643 (1964).
10. J. A. Barker and D. Henderson, Rev. Mod. Phys. **48**, 587 (1976).
11. N. F. Carnahan and K. E. Starling, J. Chem. Phys. **51**, 635 (1969).
12. R. B. Potts, Phys. Rev. **88**, 352 (1952).
13. C. Domb, in "Phase Transitions and Critical Phenomena," edited by C. Domb and M. S. Green (Academic, New York, 1974) Vol. 3, pp. 1 and 375.
14. P. Attard, Mol. Phys. **74**, 547 (1991).
15. P. Attard and G. Stell, Chem. Phys. Lett .**189**, 128 (1992).
16. J. Dóczi-Réger and P. C. Hemmer, Physica **109A**, 541 (1981).
17. E. Müller-Hartmann and J. Zittarz, Z. Physik **B27**, 261 (1977).
18. P. Attard, J. Chem. Phys. **91**, 3072 (1989).

FERROMAGNETIC-PARAMAGNETIC PHASE TRANSITION IN A RANDOM-BOND POTTS MODEL

Manuel A. Bautista[1], Rafael E. Rangel[2] and Phylip Taylor[3]

[1] Coordinación de Informática, CENAMEC, Apdo. 75055, Caracas 107, Venezuela
[2] Centro de Física, IVIC, Apdo. 21827, Carcas 1020-A, Venezuela
[3] Department of Physics, Case Western University, Cleveland, Ohio, USA

I. INTRODUCTION

Recently there have baeen interesting studying in one-dimensional disorder Ising spin system[1-4]. Transfer matrix has been powerful technique in solving some one-dimensinal model exactly[1-3]. Our goal in this paper is to study ferromagnetic phase transitions of a three-dimensional anisotropic three-state random bond Potts model by applying the mean-field approximation to reduce the model to one-dimensional model and utilizing the transfer matrix technique. In section II, a three-dimensional anisotropic three-state random bond Potts model is presented. Mean-field theory for the Potts model is introduced in section III. In section IV, the resulting effectiv one-dimensional random-bond Potts model is studied via the transfer matrix approach. In section V, we calculate numerically the free energy as a function of the ferromagnetic order for various temperature and different compositions of randomness. According to the numerical results, a first order ferroelectric-paraelectric phase transition is found over a wide range of the compositions. We summarize and discuss our results in section VI. Finally, some perspectives to the future are giving in the last section.

II. THE MODEL

Let us start from the following Hamiltonian of a three-dimensional anisotropic three-state random-bond Potts model

$$H = - \sum_{intra(i,j)} J_{ij} S_i . S_j - \sum_{inter(i,j)} J'_{ij} S_i . S_j, \tag{1}$$

Condensed Matter Theories, Vol. 8, Edited by
L. Blum and F.B. Malik, Plenum Press, New York, 1993

where i and j represent coordinates of three-dimensional cubic lattice sites. S_i is the Potts spin which is a unit vector allowed to have three orientations $(0, 2\pi/3, 4\pi/3)$ on the plane perpendicular to the Potts spin chains. $S_i.S_j$ is the vector product of the two unit spins. J_{ij} represent the nearest neighbor intrachain coupling and are random variables satisfying the same independent distribution $\rho(J_{ij}) = p\delta(J_{ij} - J) - (1-p)\delta(J_{ij} + J)$. So each intrachain coupling has probability p to be ferromagnetic coupling and probability $1 - p$ to be anti-ferromagnetic coupling. J'_{ij} represent the nearest neighbor interchain coupling and are assumed to be uniformly and of ferromagnetic type, that is $J'_{ij} = J' > 0$. Due to weak interchain coupling, we have $J' << J_,$. The sum $\sum_{intra(i,j)}(\sum_{inter(i,j)})$ is the sum over all nearest intrachain (interchain) neighbours.

The ferromagnetic order is then given by the thermal and disorder average of the Potts spin

$$m = \overline{< S_i >}, \tag{2}$$

where $<>$ is the thermal average and the bar represents the average over the random variables J_{ij}.

The model defined by the Hamiltonian (eq.(1)) is a model for a quasi-one-dimensional disorder system. This model has potentials in the studying of a first order ferroelectric-parroelectric phase transition in random copollymers of vinylidene flouride and tetrafluoroethylene[5-7]. The model is still too difficult to be solved, and we will use the mean-field theory to study it in the following section.

III. MEAN-FIELD APPROXIMATION

Since the interchange coupling constant J' is much smaller than the intrachain coupling constant, it is reasonable to make the approximation for the terms involving all the interchain interactions in the Hamiltonian given by eq.(1). Suppose that the system has a simultaneous broken symmetry magnetization along z direction

$$m = mz, \tag{3}$$

where z is a unit vector along the '0' direction in the plane perpendicular to the Potts. Due to this simultaneous broken symmetry, we can think that the effect of all the interchain couplings is to produce a uniform field, $h = h(m, K, p)z$, acting on each individual Potts spin S_i. The mean -field $h(m, K, p)$ is a function of the ferromagnetic order parameter m, temperature parameter K ($K = J/T$ with temperature T and unit Boltzman constant), and the disorder parameter p. This mean field approximation decouples the interchain interactions and reduces the original quasi-one-dimensional disorder system to a one-dimensional disorder system. The resulting effective one-dimensional Hamiltonian is then given by

$$H_{eff} = - \sum_{intra(i,j)} J_{ij}S_i.S_j - h(m, K, p)\sum_i S_i.z, \tag{4}$$

where the sum $\sum_i$ is over all lattice sites.

In mean field theory[8] the energy is given by the disorder average of the Hamiltonian H with respect to the random variable J_{ij} and the effective Hamiltonian H_{eff}. The energy per unit site is

$$e = -\overline{< J_{ij}S_i.S_j >}_{Heff} - 2J'm^2 \tag{5}$$

where S_i and S_j are nearest Potts spins along the Potts chain, $<>_{Heff}$ is the thermal average with respect to the effective Hamiltonian $Heff$, and m is the simultaneous broken symmetry ferromagnetic order given by

$$m = \overline{< S_i.z >}_{Heff}. \tag{6}$$

Next consider the mean entropy of the system. Let N be the total number of the Potts spins, and $N_0, N_{2\pi/3}, N_{4\pi/3}$ be the total number of Potts spins which orient at angles $0, 2\pi/3 and 4\pi/3$ respectively. Mean-field constrains within $N_0, N_{2\pi/3}, N_{4\pi/3}$ in the broken symmetry state are

$$N_{2\pi/3} = N_{4\pi/3},$$

$$\frac{1}{N}[N_0 - \frac{1}{2}(N_{2\pi/3} + N_{4\pi/3})] = m. \tag{7}$$

We have the following expression for the mean-field entropy per unit site

$$s = -2\frac{1-m}{3}\ln\frac{1-m}{3} + \frac{1+2m}{3}\ln\frac{1+2m}{3}, eqno(8)$$

The free energy per unite site is now given by

$$f = -\overline{< J_{ij}S_i.S_j >}_{Heff} - 2J'm^2 + T(-2\frac{1-m}{3}\ln\frac{1-m}{3} + \frac{1+2m}{3}\ln\frac{1+2m}{3}), \tag{8}$$

This expression indicates that the mean-field free energy as a function of the order parameter m is known, once we know the correlation function, we need to know the meanfield h as a function of the parameters m, K and p, indicating that we need to find the magnetization m as a function of the mean-field via eq.(6) as well. Our next two sections are going to find the correlation function and the magnetization.

IV. TRANSFER MATRIX ANALYSIS

Given the one-dimensional Hamiltonian H_{eff} the corresponding 3×3 transfer matrix is given by

$$T_{\pm} = \begin{pmatrix} e^{-\frac{3}{2}\bar{h}} & e^{-\frac{3}{2}k_{\pm}}e^{-\frac{3}{2}\bar{h}} & e^{-\frac{3}{2}k_{\pm}}e^{-\frac{3}{2}\bar{h}} \\ e^{-\frac{3}{2}k_{\pm}} & 1 & e^{-\frac{3}{2}k_{\pm}} \\ e^{-\frac{3}{2}k_{\pm}}e^{-\frac{3}{2}\bar{h}} & e^{-\frac{3}{2}k_{\pm}}e^{-\frac{3}{2}\bar{h}} & e^{-\frac{3}{2}\bar{h}} \end{pmatrix}, \tag{10}$$

where $T_{\pm}$ are the two possible transfer matrix with + representing for the ferromagnetic bond and - for the anti-ferromagnetic bond, $K_{\pm} \equiv \pm K$ is the coupling strength J divided by temperature T, and $\bar{h}$ is the effective mean field $h(m, K, p)$ divided by the temperature. In writing out the above transfer matrix we have pulled out a common factor $e^{K+\bar{h}}$ and neglected it as it does not affect our calculation of the magnetization m and the correlation function. The transfer matrix is arranged in a way such that the first, second, and third column (or line) corresponds to the $0, 2\pi/3, 4\pi/3$ orientation of the Potts spins.

Define a sequence of partition functions

$$\begin{pmatrix} Z_{\searrow n+1} \\ Z_{\uparrow n+1} \\ Z_{\nearrow n+1} \end{pmatrix} = T_n \begin{pmatrix} Z_{\searrow n} \\ Z_{\uparrow n} \\ Z_{\nearrow n} \end{pmatrix}, \tag{11}$$

651

where T_n is the random transfer matrix associated with the nth bond. For a free boundary system the initial partition functions are defined by

$$\begin{pmatrix} Z_{\searrow 0} \\ Z_{\uparrow 0} \\ Z_{\swarrow 0} \end{pmatrix} = \begin{pmatrix} 1 \\ 1 \\ 1 \end{pmatrix}. \tag{12}$$

Since T_n is a random matrix, eq.(11) introduces a sequence of random partition functions, which is a sequence of random three-dimensional vector. To study this sequence of random partition functions, let us introduce the spheric coordinates (r_n, θ_n, ψ_n) for the partition function at a given step n

$$r_n \begin{pmatrix} sin\theta_n cos\psi_n \\ cos\theta_n \\ sin\theta_n sin\psi_n \end{pmatrix} \equiv \begin{pmatrix} Z_{\searrow n} \\ Z_{\uparrow n} \\ Z_{\swarrow n} \end{pmatrix}. \tag{13}$$

The advantages of introducing the spherical coordinates are two folds. First, it pulls out the unimportant common factor r_n reducing the dimensions of the random partition functions from three to two. Second, it utilizes a symmetry preserved by the random matrix T_n, reducing further the dimensions of the random vector from two to one. The symmetry that the random matrix have is that the matrix have an invariant vector subspace which is composed of all the vectors having a form of $\begin{pmatrix} x \\ z \\ x \end{pmatrix}$. By definition, this then says that ψ_n is fixed and is equal to $\pi/4$. After these considerations, we find the following iteration relation between θ_{n+1} and θ_n

$$\theta_{n+1} = arctan\frac{e^{-\frac{3}{2}h}[\sqrt{2}e^{-\frac{3}{2}K_{pm}} + (1 + e^{-\frac{3}{2}K_\pm})tan\theta_n]}{1 + \sqrt{2}e^{-\frac{3}{2}K_\pm}tan\theta_n}, \tag{14}$$

Since all partition functions are positive, all the angles $\{\theta\}$ are defined in $(0, \pi/2)$. When $n \to \infty$, they will satisfy a well defined angle distribution function $rho(\theta)^{1,3}$ on $[0, \pi/2]$. Aa analytic solution for the angle distribution function $\rho(\theta)$ is, in general, difficult to be obtained. Instead, one have to find it via numerical method.

By definition, one can verify that the ferromagnetic order satisfies

$$m = \frac{\overline{1 - \frac{1}{2}tan\theta_i tan\theta_j}}{1 + tan\theta_i tan\theta_j} \tag{15}$$

where θ_1 and θ_2 are angles, each satisfying independently the angle distribution function $rho(\theta)$. The bar indicating the average over the random variable J_{ij} is now replaced by the average over the angle distribution. Denoting the correlation function $<J_{ij}S_i.S_j>_{Heff}$ by G, we find, after some simple calculations,

$$G = \frac{\overline{1 - \frac{1}{2}btan\theta_i - \frac{1}{2}dtan\theta_j + utan\theta_i tan\theta_j}}{1 + btan\theta_i + dtan\theta_j + vtan\theta_i tan\theta_j} \tag{16}$$

$$b \simeq \sqrt{2}e^{-\frac{3}{2}\bar{h}}e^{-\frac{3}{2}\frac{J_{ij}}{T}}, \quad d \simeq \sqrt{2}e^{-\frac{3}{2}\frac{J_{ij}}{T}},$$

$$u \simeq e^{-\frac{3}{2}\bar{h}}(1 - \frac{1}{2}e^{-\frac{3}{2}\frac{J_{ij}}{T}}), \quad v \simeq e^{-\frac{3}{2}\bar{h}}(1 + e^{-\frac{3}{2}\frac{J_{ij}}{T}})$$

V. NUMERICAL RESULTS

In this section, we present the numerical results for calculating the free energy. Starting from the iteration relation (eq.(14)), one can find the angle distribution function $\rho(\theta)$ by counting the histogram of $\{\theta_n\}$. Once the angle distribution function is known, one can calculate the magnetization m and the correlation function G through eqs. (15) and (17). Finally, free energy can be found via eq.(9).

For the numerical simulation, we have set $J = 5, J' = 1$. Figure 1 shows the free

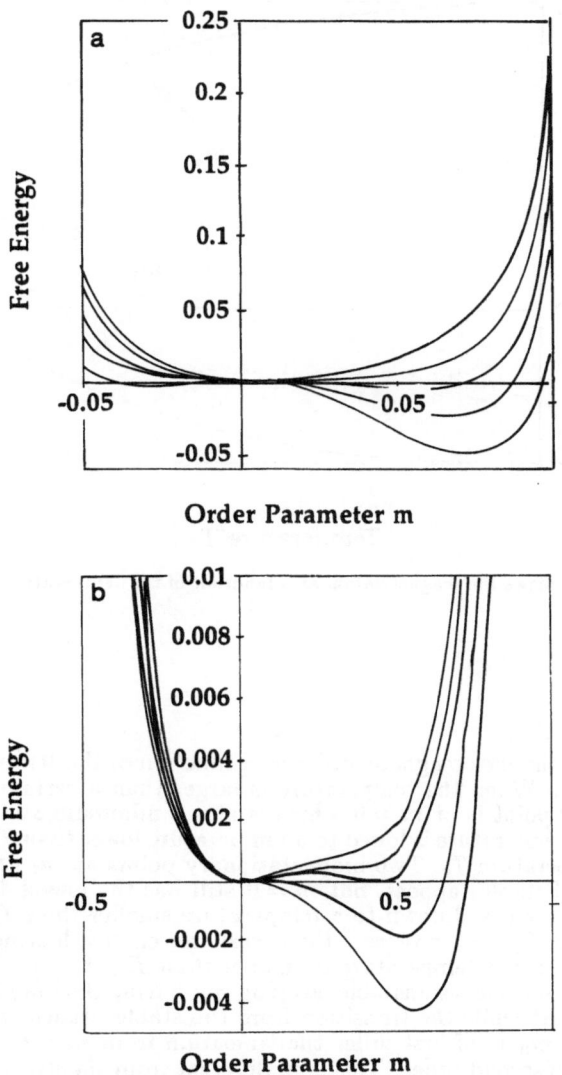

Figure 1. This figure shows free energy as a function of the order parameter m at various temperatures. The composition p is equal to 0.9. The curves correspond to, from to to bottom, various temperatures 0.7 0.6 0.5 0.4 and 0.3 for (a) and 0.5 0.54 0.562 0.57 and 0.58 for (b).

energy as a function of the ferroelectric order parameter m for various temperature at a composition $p = 0.9$. Figure 1a shows at a high temperature, the only minimum occurs at $m = 0$, indicating that there is no broken symmetry ferroelectric order. However, at low temperature, $m = 0$ becomes unstable and two minimums at non zero m (one positive and one negative) appear, implying the existence of a ferroelectric phase. The one has the lowest free energy is located at the positive branch. This says that there is a paramagnetic-ferromagnetic phase transition when temperature is changed from a higher temperature to a lower temperature below a certain transition temperature. A careful exam of the free energy near the transition temperature (figure 1b) shows that the transition is of first order.

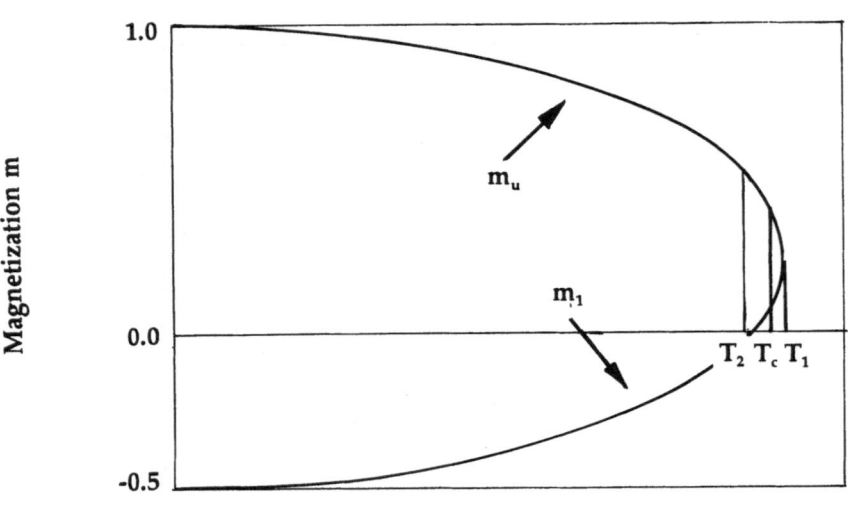

Figure 2. This figure shows the magnetization as a function of the temperature. The composition is chosen at p=0.9.

Figure 2 shows the ferromagnetic order as a function of the temperature at a composition of $p = 0.9$. When the temperature is larger than a certain temperature T_1, the only stationary point is at $m = 0$ which is also a minimum and has a lowest free energy. When the temperature is lowed to a temperature lower than T_1 and higher than the transition temperature T_c, Two more stationary points at m_l (> 0 and unstable) and m_u ($> m_l$ and stable) appear, but $m = 0$ still has the lowest free energy. If the temperature is further lowed down to a temperature smaller than T_c and larger than another temperature T_2, m_u possesses the lowest free energy, leaving other properties unchanged. Finally, if the temperature is smaller than T_2, then m_l becomes negative and stable and $m = 0$ becomes unstable, keeping m_u having the lowest free energy. The figure also shows that while the transition from the stable branch $m = 0$ to the most stable branch $m = m_u$ is of first order the transition from $m = 0$ to the metastable branch $m = m_l$ is of second order. This is in different from the usual Ising spin system where the upper branch $m = m_u$ and the lower branch $m = m_l$ are symmetric about the branch $m = 0$ yielding all second order transitions.

VI. CONCLUSIONS AND DISCUSSIONS

We have studied the ferromagnetic phase transition of a three-dimensional anisotropic three-state random-bond Potts model. Mean-field theory has been employed to decouple the comparatively weak ferroelectric couplings between the random Potts spin chains, reducing the model to an effectively one-dimensional random-bond three-state Potts model in the existence of the mean field. A first order ferromagnetic phase transition is observed over a wide range of compositions via a numerical calculation of the free energy. While the transition to the ferromagnetic phase with positive magnetization (corresponding to the most stable branch) is of first order, the transition to the ferromagnetic phase with negative magnetization (corresponding to the metastable branch) is of second order.

VII. PERSPECTIVES

The results we have obtained rely on the existence of the probability distribution $\rho(\theta)$ obtained from the random map $g(\theta)$[8].

Nothing was said on the transition to chaos for the random map $g(\theta)$, nor on the fractality of the snapshot attractor and the Lyapunov exponent[9].

Also, important is the study of the multifractal properties of the measure $\rho(\theta)$[10]. Another point of interest would be the study of the move general case of a two dimensional random map, which is the case when the symmetry used in reducing the dimension of T_n to one is broken. This type of quenched disorder can be realized, for example, when a Fluor atom in PTFE (tetrafolueroethylene) is replaced by a Hydrogen atom in random way. In that case theory on infinite product of random matrices has to be considered[11,12].

Acknowledgements

This work was supported by CONICIT and NSF Grants DMR 89-022220. The numerical calculation was done at IBM Scientific Center of Venezuela.

VIII. REFERENCES

1. R. Bruinsma and G. Aeppli, Phys. Rev. Lett. **50**, 1494 (1983).
2. T. Tanaka, H. Fujisaka, and M. Inoue, Phys Rev. **A39**, 3170 (1989).
3. G. Gyoryi and P. Rujan, J. Phys. **C17**, 4207 (1984).
4. J. Bene and P. Szepfalusy, Phys. Rev. **A37**, 1703 (1988).
5. H. Kawai, Japan J. Appl. Phys **8**, 975 (1969).
6. J. G. Bergman, J. H. Mcfee, and G. R. Grane, Appl. Phys. Lett. **18**, 203 (1971).
7. R. G. Kepler and R. A. Anderson, *Ferroelectric Polymers*, preprint, (1991).
8. S. K. Ma, *Statistical Mechanics*, World Scientific, Philadelphia (1985).
9. F. Ledrappier and L. S. Young, Common. Math. Phys, **117**, 529 (1988).
10. F. J. Romeivas, C. Grebogi and E. Oh, Phys. Rev. **A41**, 784 (1990).
11. J. Bene and P. Szépfalusy, PR **A37**, 703 (1988).
12. H. Yu, E. Oh and Q. Chen, Physica **D53**, 102 (1991).
13. G. Paladin and A. Volpiani, J. Phys. **A21**, L363 (1988).

CELLULAR AUTOMATA AND SPREAD OF DAMAGE: GENERAL CONSIDERATIONS AND A

RECENT ILLUSTRATION

Constantino Tsallis

Centro Brasileiro de Pesquisas Físicas
Rua Xavier Sigaud, 150
22290 – Rio de Janeiro – RJ, Brazil

ABSTRACT

We characterize Cellular Automata (CA) within a general background
of dynamical systems. We then illustrate CA by presenting various
relevant properties of a stochastic one, namely an extended version of
the   Domany-Kinzel   CA.   We   finally   propose   a   quite   general
classification   of   the   various   types   of   sensitivity   to   initial
conditions that dynamical systems might exhibit; this classification
recovers,   as   particular   cases,   standard   discussions   related   to   the
Hamming distance and the Lyapunov exponent.

1   INTRODUCTION

Phenomena   in   Nature   occur   somewhere   and   at   a   given   time.
Consequently,   the   Theoretical   Physics   basic   mathematical   object   for
studying dynamical systems is a "*field*" $\phi(x,t)$ defined in a *space-time*
(x,t). The field can be *continuous* ($\phi \in \mathbb{R}^{\bar{n}}$, where $\bar{n}$ is the *field
dimension*)   or   *discrete* ($\phi \in \mathbb{N}$ or, equivalently, ϕ is isomorphic to $\mathbb{N}$
or to a part of $\mathbb{N}$);   the space can be *continuous* ($x \in \mathbb{R}^{d}$, where d is
the *space dimension*) or *discrete* ($x \in \mathbb{N}$);   finally,   the   time   can   be
continuous ($t \in \mathbb{R}$)   or   *discrete* ($t \in \mathbb{N}$). As   a   whole,   we   have 2^3
different   cases,   which   are   illustrated   in   Table   I.   It   is   worthy   to
emphasize   the   fact   that   the   (discrete   field) – (discrete   space) –
(discrete time) case is the only one which is strictly   tractable   in a
real   computer.   Cellular   automata   (CA)   belong   to   this
category.   Furthermore,   the   denomination   CA   is   normally   reserved   to
those   dynamical   systems   which   present   a   *time-layered* architecture   in
the   sense   that   we   can   *simultaneously* update   all   the   elements   of   the
system   (thus   being   susceptible   of   *parallel* processing   in
computers).   A   very   general   definition   for   CA   is   outlined   in   what
follows.
Consider   a   *denumerable* set   of   positions $\{x_i\}$ ($i = 1,2,...,N$);
they   could   be   the   sites   of   a   Bravais   lattice   (e.g.,   a   linear   chain)   or
an   hierarchical   lattice   (e.g.,   a   Sierpinski   gasket)   or   any   other
spatial   array.   With   each   position   we   associate   a   variable ϕ_i which   can

Table I   Dynamical physical systems: (a) continuous time (t ∈ ℝ); (b) discrete time (t ∈ ℕ). *In this case trajectories are defined in space-time, consequently a *binary* field can be defined (nonvanishing on the trajectory, and vanishing out of it); **for example, a localized random binary variable which, at arbitrary continuous times, can be zero or one; ***each consecutive iteration can be considered as a discrete "time", ϕ being one (zero) for all values of x ∈ ℝ corresponding to present (absent) points.

(a) Continuous time (t ∈ ℝ)

ϕ \ x	continuous ($x \in \mathbb{R}^d$)	discrete ($x \in \mathbb{N}$)
continuous ($\phi \in \mathbb{R}^n$)	Maxwell equations Schroedinger equation Navier-Stokes equation Field theory	Classical phonons in a Bravais lattice Lattice field theory
discrete ($\phi \in \mathbb{N}$)	Newton equation[*] Special relativity equation[*]	Random binary noise[**]

(b) Discrete time (t ∈ ℕ)

ϕ \ x	continuous ($x \in \mathbb{R}^d$)	discrete ($x \in \mathbb{N}$)
continuous ($\phi \in \mathbb{R}^n$)	Discrete-time field theory	Real and complex logistic map Maps coupled on a lattice Network of continuous neurons
discrete ($\phi \in \mathbb{N}$)	Cantor fractal dust[***]	Hopfield neuronal model Cellular automata

take q_i different values (i.e., $\phi_i = 1, 2, \ldots, q_i$), the most frequent case being $q = q$, $\forall i$, and typically $q = 2$. At time $t = 0$ we must provide the set $\left\{\phi_i^{(0)}\right\}$; the set $\left\{\phi_i^{(1)}\right\}$ at time $t = 1$ is either directly given as part of the initial conditions, or is determined by giving the set of (deterministic or stochastic) rules $\left\{\rho_i^{(1)}\right\}$, i.e., $\phi_i^{(1)} = \rho_i^{(1)}\left(\left\{\phi_i^{(0)}\right\}\right)$. Generally speaking, the initial conditions are given through the set $\left(\left\{\phi_i^{(0)}\right\}, \left\{\phi_i^{(1)}\right\}, \ldots, \left\{\phi_i^{(\tau)}\right\}\right) \equiv \left\{\phi_i^{(0\to\tau)}\right\}$, the most frequent case being $\tau = 0$. The set of rules $\left\{\rho_i^{(t)}\right\}$ at time t might themselves evolve, and can in general be expressed as follows:

$$\phi_i^{(\tau+1)} = \rho_i^{(\tau+1)}\left(\left\{\phi_i^{(0\to\tau)}\right\}\right)$$

$$\phi_i^{(\tau+2)} = \rho_i^{(\tau+2)}\left(\left\{\phi_i^{(0\to\tau+1)}\right\}\right) \tag{1}$$

$$\vdots$$

$$\phi_i^{(t)} = \rho_i^{(t)}\left(\left\{\phi_i^{(0\to t-1)}\right\}\right) \qquad (t = \tau + 1, \tau + 2, \ldots)$$

The most frequent case is

$$\rho_i^{(t)} = \rho\left(\left\{\phi_i^{((t-1-\tau)\to(t-1))}\right\}\right), \quad \forall(i, t).$$

Moreover, the most commonly used rules are the homogeneous *local* ones, in which $\phi_i^{(t)}$ is determined, assuming that $\tau = 0$, by the values $\left\{\phi_i^{(t-1)}\right\}$ where i runs over λ neighbors of site i (including the site i itself). For such CA, the rules are established by giving, for *deterministic* CA, a correspondence of q^λ states into q states (q^{q^λ} possibilities), or, more generally for *stochastic* (also referred to as *probabilistic or random*) CA, by giving the probability set

$\{p(\phi_1, \phi_2, \ldots, \phi_\lambda/1), p(\phi_1, \phi_2, \ldots, \phi_\lambda/2), \ldots, p(\phi_1, \phi_2, \ldots, \phi_\lambda/q)\}$ satisfying

$$\sum_{\phi=1}^{q} p(\phi_1, \phi_2, \ldots, \phi_\lambda/\phi) = 1 \tag{2}$$

with $\phi_j = 1, 2, \ldots, q$ ($j = 1, 2, \ldots, \lambda$), where $p(\phi_1, \phi_2, \ldots, \phi_\lambda/\phi)$ is the probability of having, at a given site, state ϕ at time t if the states, at time $(t-1)$, of its λ neighbors are $(\phi_1, \phi_2, \ldots, \phi_\lambda)$. This type of model is characterized by giving $(q-1)q^\lambda$ independent probabilities, say $\{p(\phi_1 \cdot \phi_2, \ldots, \phi_\lambda/1), \ldots, p(\phi_1, \phi_2, \ldots, \phi_\lambda/(q-1)\}$ for each one of the q^λ states of $(\phi_1, \phi_2, \ldots, \phi_\lambda)$. Its physical space is, consequently, a hyperpolyhedron in $(q-1)q^\lambda$ dimensions which recovers, at each one of its q^{q^λ} corners, deterministic CA. The extended Domany-Kinzel CA discussed in Section 2 is an example of the $\lambda = q = 2$ class and, as we shall see, it is characterized by giving, for instance, $p(00/1)$, $p(01/1)$, $p(10/1)$ and $p(11/1)$; it recovers, as particular cases, 16 deterministic CA.

What are the most relevant properties that can be studied for a particular CA? First of all the attractors ($t \to \infty$) at the thermodynamic limit ($N \to \infty$): they can present spatial and/or temporal modulations of various kinds as well as spatial and/or temporal chaos. Also, the influence of the initial conditions on the attractors often is interesting. Order parameters characterizing the various possible attractors ("phases") can be studied as well, thus enabling the establishment of the CA phase diagram with all sorts of critical phenomena, critical exponents and universality classes. Various types of susceptibilities and relaxation times can also be studied. Finally, the spread of damage (sensitivity to initial conditions) often exhibits interesting peculiarities.

The study of CA is a very interesting one. On one hand they provide simple models for a great variety of systems, including chemical reactions, crystal growth models, artificial intelligence, turbulence, computers, cybernetics, biological systems, various other non-linear processes far from equilibrium, phase transitions (see, for instance, Refs. [1] for reviews, [2] for chemical reactions, [3] for the Q2R CA, [4] for spin glasses and [5] for various other spin systems). On the other hand, CA act as prototypes for a "finite difference equations" Physics in opposition to the traditional "differential equations" Physics. Indeed, if the deep nature of space-time turns out to be discrete (which we believe to be the case, essentialy due to quantum-like fluctuations effects), differential necessarily become limiting cases of finite difference equations to be found.

In the next Section we present results of a phase diagram study[6] of an extended version of the Domany-Kinzel CA[7].

2 EXTENDED DOMANY-KINZEL CA

We consider a one-dimensional chain of N lattice sites ($i = 1, 2, \ldots, N$) with periodic boundary conditions. Each site has two possible states $\phi_i \equiv \sigma_i = 0, 1$ (hence $q = 2$). The state of the system at time t is given by the set $\left\{ \sigma_i^{(t)} \right\}$. At the next time step, the state $\sigma_i^{(t+1)}$ of a given site equals 0 or 1 according to the conditional probabilities $\left\{p\left(\sigma_{i-1}^{(t)}, \sigma_i^{(t)} / \sigma_i^{(t+1)}\right)\right\}$,

namely p(00/1) = 1 - p(00/0), p(01/1) = 1 - p(01/0), p(10/1) =1-p(10/0) and p(11/1) = 1 - p(11/0) (hence. λ = 2). This CA is closely related to directed percolation[7] as well as to directed compact percolation[8]. It is possible to define at least two relevant time-dependent order-like parameters, namely M $\equiv$ (fraction of sites with value 1) and ψ $\equiv$ Hamming distance (i.e., fraction of sites which exhibit different values on two replicas of the system while using the same sequence of random numbers). The equilibrium values (i.e., in the t $\rightarrow$ ∞ limit) of M and ψ in the (p(00/1), p(01/1), p(10/1), p(11/1)) space enable the characterization of the "phase diagram" of the system. No analytical results are available, excepting for the (p(00/1), p(11/1)) = (0,1) critical line which, due to duality arguments, is given by[8,9]

$$p(01/1) + p(10/1) = 1 \qquad (3)$$

In the present study, a Monte Carlo technique has been used by always starting, at t = 0, with half of the sites with value 1, randomly chosen. After arrival to equiblibrium (where M is conveniently determined), a damageis produced and the two replicas (damaged and undamaged) are followed in time until stationnarity is achieved (and ψ is then determined). Let us first present the p(00/1) = 0 (*legal* rules) phase diagram: see Fig. 1 (from [6]), where three phases are present, namely the *frozen*, *active* and *chaotic* ones. M equals unity on half of the p(11/1) = 1 square (in particular at the p(01/1) = p(10/1) = p(11/1) = 1 corner) and vanishes on the frozen-active critical surface; ψ equals 1/2 at the (p(01/1), p(10/1), (11/1)) = (1,1,0) corner and vanishes on the active-chaotic critical surface.

If we now consider p(00/1) $\neq$ 0, the frozen-active critical surface disappears (since p(00/1) acts, on M, as an external conjugated field) but the active-chaotic one remains: see Fig. 2 (from [10]).

The results we have presented up to now have been obtained by using *independent* random numbers for updating each one of the N sites at time t. Let us now generalize this in the sense that the *same* random number will be used to update n (1 $\leq$ n $\leq$ N) neighboring sites (the same set of groups of n sites each for all times). The n = 1 model recovers the previous one; the n = N model is an extreme case for which a single random number is used for updating the entire generation. The n-evolution of the phase diagram is indicated in Fig. 3 (from [10]). In the n = N $\rightarrow$ ∞ limit, the p(00/1) = 0 phase diagram exhibits a frozen phase almost everywhere since the frozen-active and the active-chaotic critical surfaces have collapsed onto the p(11/1) = 1 plane and/or onto the p(01/1) = 1 and the p(10/1) = 1 planes. This fact cannot be considered as surprising since, in the n = N $\rightarrow$ ∞ limit, the system becomes one-dimensional-like in space-time(whereas it is two-dimensional for finite n and N $\rightarrow$ ∞). It is worth stressing that, for the (p(00/1) = 0, p(01/1) = p(10/1) phase diagram (Fig. 3(a)), the frozen area A_f tends to unity whereas the active area A_a as well as the chaotic area A_c tend to zero when n increases from 1 to infinity; in addition to that, it can be shown that the ratio A_a/A_c decreases with increasing n. Hence tendency towards a

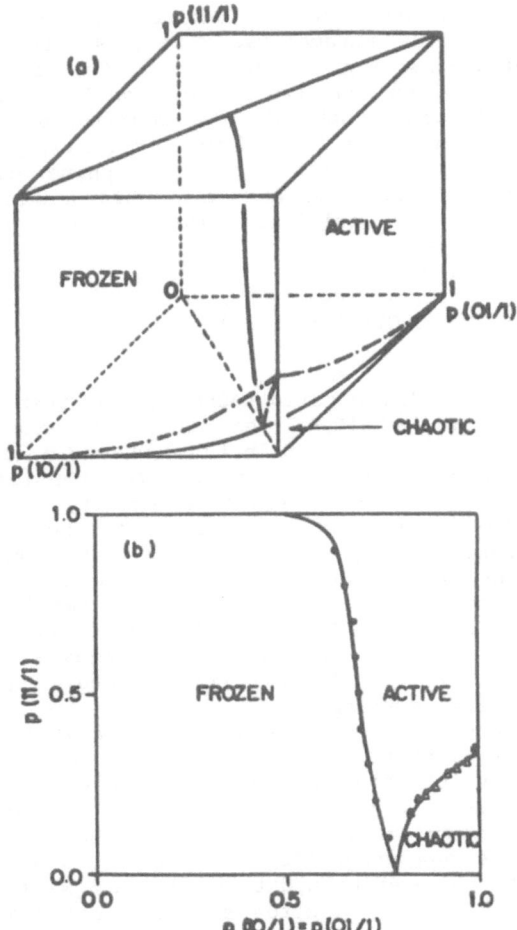

Fig. 1   (a) p(00/1) = 0 phase diagram of the extended Domany-Kinzel
          CA; the solid line belongs to the critical surface
          separating the *frozen* (M = 0 and ψ = 0) and *active* (M ≠ 0
          and ψ = 0) phases; the dashed lines belong to the boundary
          between the active and *chaotic* (M ≠ 0 and ψ ≠ 0) phases.
          (b) p(00/1) = 0 and p(01/1) = p(10/1) phase diagram. The
          data correspond to simulations with 3200 sites; transients of
          10000 (3000) time steps were used for the frozen-active
          (active-chaotic) phase transitions. The damage was averaged
          over another 3000 time steps.

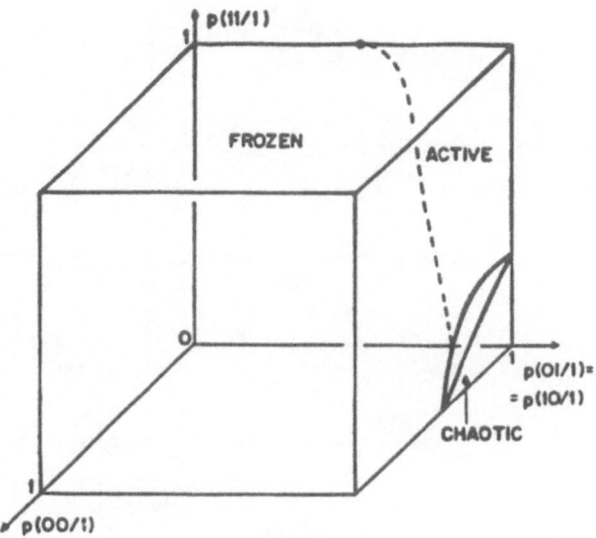

Fig. 2   Phase   diagram,   for   p(01/1) = p(10/1),   of   the   extended
Domany-Kinzel CA.

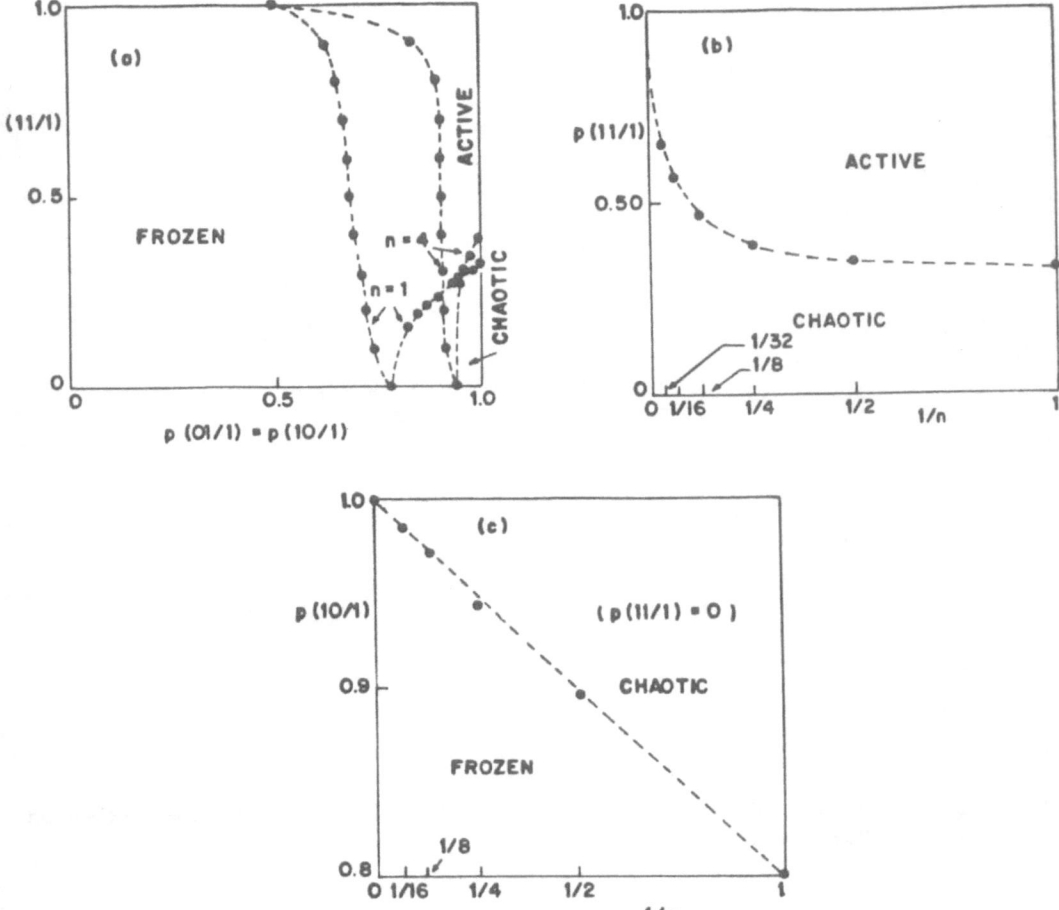

Fig. 3   p(00/1) = 0   and   p(01/1) = p(10/1) phase diagram of the
          CA for n ≥ 1 (n = 1 recovers the extended Domany-Kinzel CA).
          (a) Full   p(01/1) = (10/1)   space;   (b) n-evolution of the
          p(01/1) = p(10/1) = 1 critical point; (c) n-evolution   of the
          (p(01/1) = p(10/1); (p/11/1) = 0) critical point. The dashed
          lines are guides to the eye.

"totalitarian" limit (same random number for *all* the elements of a given generation) decreases chaos, but decreases even more (certain type of) activity!

3 SPREAD OF DAMAGE: A NEW CLASSIFICATION

There are dynamical systems (deterministic or stochastic) which can be very sensitive to small numerical departures of the quantities involved in the determination of the actual trajectory. These quantities include the initial conditions, the roundings of the real numbers (say 6,8,16 algarisms) at every calculational step, the particular sequence of random numbers that might be used, the parameters which are fixed along the evolution, etc. Whenever a system is sensitive to the initial conditions, it is necessarily sensitive to the numerical roundings adopted for calculating its state at a given time from the previous time(s) by using (among other possible ingredients) a finite-difference or differential equation: this is the globally so called *sensitivity to initial conditions*, and constitutes one of the two essential properties that qualify the use of the word "chaos" (the other property being the existence of a large attractor). This sensitivity is checked on an actual system by introducing a "damage" in it and following its spreading. It is now known[11] that the spread of damage in various spin systems has a deep relation with thermodynamical properties. For these systems, the quantity whose time evolution is followed (in order to characterize the damage) typically is the Hamming distance (ψ introduced in Section 2). By comparing the *initial* Hamming distance between two replicas A and B of the system and the *final* (after a long time) Hamming distance, Herrmann presented[12] various typical situations to which he referred to as *Chaotic* I (e.g., in Barber and Derrida 1988 in [5]), *Chaotic* II (e.g., in Derrida and Weisbuch 1987 in [3]), *Frozen* I and *Frozen* II (e.g., in Boissin and Herrmann 1991 in [5]). By following, in this Section, along this line we define a generalized Hamming distance in order to cover both discrete and continuous systems (unifying, in particular, the Hamming distance and the Lyapunov exponent), and then propose a quite general classification (based on the sensitivity to initial conditions) of dynamical systems.

Consider a discrete or continuous "field" $\phi_i(t)$ defined on a discrete space ($i = 1,2,...,N$) and a discrete or continuous time t (everything that follows can be trivially adapted to a continuous space, but we speak here of a discrete space in order to be adapted to the most frequent systems on which spread of damage is studied). We construct, at $t = t_\infty$ (typically $t_\infty \gg 1$ and corresponds to the time necessary for the system to practically arrive to its attractor or equilibrium), two replicas A and B of the system (typically, one or both of the replicas are damaged versions of the original system at time t_∞). And we define the following normalized *generalized Hamming distance*:

$$D(\bar{t}) \equiv \frac{\left\langle\left\langle \sum_{i=1}^{N} |\phi_i^A(t_\infty+\bar{t}) - \phi_i^B(t_\infty+\bar{t})| \right\rangle_{\substack{\text{time} \\ \text{sequences}}} \right\rangle_{\substack{\text{initial} \\ \text{configurations}}}}{\sup\left(\sum_{i=1}^{N} |\phi_i^A - \phi_i^B|\right)} \qquad (4)$$

$\langle ... \rangle_{\text{time sequences}}$ refers to only one trajectory if the system is deterministic, and refers to an average using a sufficiently large set of random number sequences (the *same* for both replicas) if the system is stochastic; $\langle ... \rangle_{\text{initial configurations}}$ refers to an average using a sufficiently large set of initial configurations (at $t = 0$) satisfying the external parameters that are fixed; sup refers to the maximal value its argument can achieve at any conditions (sup $(...) = N$ for binary variables taking values 0 or 1, or $\pm 1/2$: $D = \psi$ for the case discussed in Section 2), hence $D(\bar{t}) \in [0,1]$. We have defined "distance" in the traditional way, i.e., by using the modulus, but, clearly, other definitions (e.g., $|...|^k$ with $k > 0$) could be as well used; if $\phi_i(t)$ is a cyclic or angular-like variable, the *smallest* angle can be conveniently used. It follows, from definition (4), that

$$D(0) = 0 \Rightarrow D(\bar{t}) = 0, \ \forall \bar{t} \geq 0 \tag{5}$$

We assume the quite frequent case satisfying:

(i) $D(\infty) \equiv \lim_{\bar{t} \to \infty} D(\bar{t})$ exists and depends on $\{\phi_i^A(t_\infty), \phi_i^B(t_\infty)\}$ *only*

through $D(0)$; $\tag{6.a}$

(ii) The only finite-cycle attractors are fixed points. $\tag{6.b}$

We follow, on a $D(\bar{t})$ vs. $D(0)$ representation, the time evolution of the generalized Hamming distance. By assuming the frequent case in which

$$\Delta \equiv \lim_{D(0) \to 0} \frac{D(\infty)}{D(0)} \tag{7}$$

is well defined, we have the following possibilities (see Fig. 4):

i) *Strongly sensitive*: $\Delta = \infty$ and $D(\infty)$ is a discontinuous function of $D(0)$ at $D(0) = 0$;

ii) *Sensitive*: $\Delta = \infty$ and $D(\infty)$ is a continuous function of $D(0)$ at $D(0) = 0$;

iii) *Marginal*: Δ is finite ($\infty > \Delta > 0$);

iv) *Nonsensitive*: $\Delta = 0$

The generic strongly sensitive case (Fig. 4(a)) corresponds to Herrmann's Chaotic II situation, and its particular case for which $D(\infty)$ is a nonvanishing constant for all $D(0) \in (0,1]$ corresponds to Herrmann's Chaotic I situation (the chaotic phase of the CA discussed in Section 2 as well as the *positive* Lyapunov exponent cases of say the logistic equation are typical illustrations). Although we are not aware of an illustration for the sensitive case (Fig. 4(b)), there is no reason for its non existence. The marginal case (Fig. 4(c)) corresponds to Herrmann's Frozen II situation (the *zero* Lyapunov exponent case can belong to this class). The generic nonsensitive case could in principle exist (as for the sensitive case, we are not aware of any example at the present moment); its particular case for which $D(\infty) = 0$ for any $D(0) \in [0,1]$ corresponds to Herrmann's Frozen I situation (the *negative* Lyapunov exponent case of say the logistic case is a typical illustration).

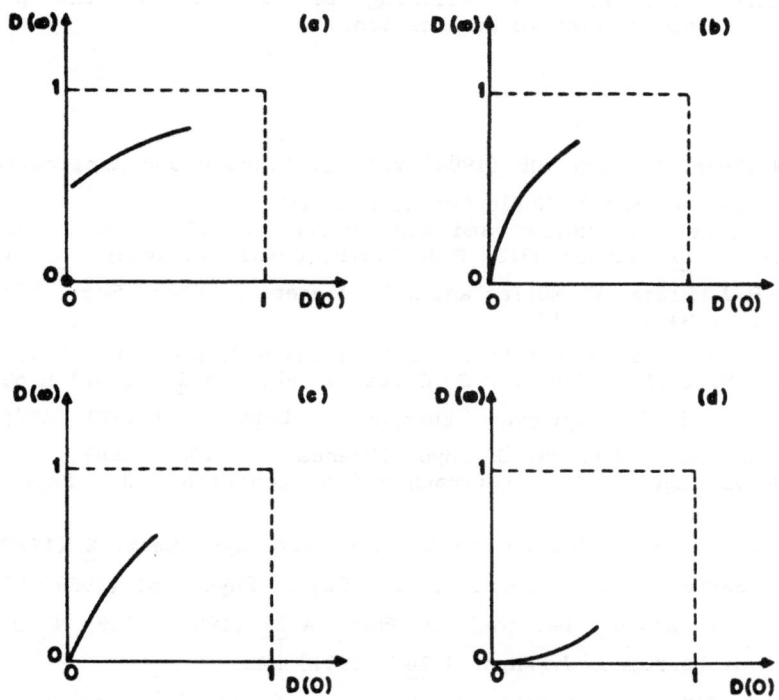

Fig. 4   Final ($D(\infty)$) vs. initial ($D(0)$) possible behaviors for the spread of damage. (a) strongly sensitive; (b) sensitive; (c) marginal; (d) nonsensitive.

4   CONCLUSION

We have defined CA within an unified picture for dynamical physical systems, and discussed the phase diagram of a recent extension of the Domany Kinzel stochastic CA. One of the relevant phenomena is the spread of damage characterizing its chaotic phase. We have introduced a generalized Hamming distance which enables, along Herrmann's lines, a convenient classification of dynamical systems. This classification is based on the type of sensitivity to the initial conditions and recovers the Lyapunov exponent concept as a particular case. Specific illustrations of the *sensitive* and the generic *nonsensitive* classes would be very welcome.

ACKNOWLEDGEMENTS

I am very indebted to H.J. Herrmann for fruitful remarks concerning the present proposal for classification of the spread of damage behavior, as well as to E.M.F. Curado and J.S. Helman for useful remarks on Table I. Finally, I am thankful to my collaborators M.L. Martins and T.J.P. Penna for allowing me to use, in the present talk, Figs. 2 and 3 prior to publication.

REFERENCES

1 - S. Wolfram, Physica 10D (1984) Vol. 1; "Theory and Application of Cellular Automata", World Scientific (1986).

2 - N.I. Jaeger, K. Möller and P.J. Plath, J. Chem. Soc., Faraday Trans. 1, 82 (1986) 3315; P.J. Plath, Catalysis Today 3 (1988) 475; P.J. Plath, K. Möller and N.I. Jaeger, J. Chem. Soc., Faraday Trans. 1, 84 (1988) 1751.

3 - H.E, Stanley, D. Stauffer, J. Kertesz and H.J. Herrmann, Phys. Rev. Lett. 59 (1987) 2326; U.M.S. Costa, J. Phys. A 20 (1987) L 583; B. Derrida and G. Weisbuch, Europhys. Lett. 4 (1987) 657; A.U. Newmann and B. Derrida, J. Phys. (France) 49 (1987) 1647.

4 - L. de Arcangelis, H.J. Herrmann and A. Coniglio, J. Phys. A 22 (1989) 4659.

5 - L.R. da Silva, A. Hansen and S. Rony, Europhys. Lett. 8 (1989) 47; M.N. Barber and B. Derrida, J. Stat. Phys. 51 (1988) 877; D. Golinelli and B. Derrida, J. Phys. A 22 (1989) L939; N. Boissin and H.J. Herrmann, J. Phys. A 24 (1991) L43.

6 - M.L. Martins, H.F. Verona de Resende, C. Tsallis and A.C.N. de Magalhães, Phys. Rev. Lett. 66 (1991) 2045; H.F. Verona de Resende, M.L. Martins, A.C.N. de Magalhães and C. Tsallis, in "Non linear Phenomena in Fluids, Solids and Other Complex Systems", ed. P. Cordero, B. Nachtergaele (Elsevier, 1991).

7 - E. Domany and W. Kinzel, Phys. Rev. Lett. 53 (1984) 447; W. Kinzel, Z. Phys. B58 (1985) 229.

8 - J.W. Essam, J. Phys. A 22 (1989) 4927; J.W. Essam and W. Tanlakishani, "Disorder in Physical Systems", ed. G.R. Grimmett and D.J.A. Welsh (Oxford University Press, 1990).

9 - M.L. Martins, H.F.V. de Resende, C. Tsallis and A.C.N. de Magalhães, communicated at CECAM Workshop on Computer Simulation of Cellular Automata (26 September to 7 October 1988, Paris); Abstract published in J. Stat. Phys. 55 (1989) 1358.

10 - M.L. Martins, T.J.P. Penna and C. Tsallis, to be published (communicated at the XVth Brazilian Meeting of Condensed Matter Physics, Caxambu, 5-9 May 1992).

11 - A. Coniglio, L. de Arcangelis, H.J. Herrmann and N. Jan, Europhys. Lett. 8 (1989) 315; A.M. Mariz, J. Phys. A 23 (1990) 979.

12 - H.J. Herrmann, communicated at the Workshop on Physics of Inhomogeneous Materi/als (11 to 14 June 1991, Trieste).

HANNAY ANGLE IN CLASSICAL MECHANICS

AND ITS GAUGE-INVARIANT GENERALIZATION

Donald H. Kobe

Department of Physics
University of North Texas
Denton, Texas 76203-5368
U.S.A.

ABSTRACT

The geometrical angle in classical mechanics discovered by Hannay is gauge dependent. It is generalized here to be invariant under gauge transformations, and can be applied to nonadiabatic and noncyclic cases. As an example, the time-dependent generalized harmonic oscillator is considered. When the distinction between the Hamiltonian and the energy is made, the generalized Hannay angle is equal to zero.

1. INTRODUCTION

After the discovery of the geometric phase in quantum mechanics by Berry,[1] Hannay [2] showed that there is a classical analogue of the geometrical phase, now called the Hannay angle. Hannay[2] showed that when Hamilton's equations are expressed in terms of angle and action variables, the angle in the adiabatic limit can be written as the sum of a dynamical angle and a geometrical (Hannay) angle. The relationship between the Berry phase in quantum mechanics and the Hannay angle in classical mechanics was shown by Berry.[3] The Hannay angle is equal to the negative of the derivative of the Berry phase with respect to the quantum number.

It has been pointed out[4-6] that the Berry phase is not invariant under unitary transformations, so its value is ambiguous. Nevertheless, in the adiabatic limit the total phase, which is the sum of the dynamical phase and the Berry phase, is invariant.[7] This problem was recently resolved by generalizing the Berry phase so that it is invariant under unitary transformations.[8] The resolution involves making the distinction in time-dependent problems between the Hamiltonian and the energy operator. When eigenstates of the energy operator are used, both the dynamical phase and the Berry phase are separately invariant under unitary transformations.

A similar difficulty has been noticed for the Hannay angle.[9] This problem is resolved here by showing that a suitably generalized Hannay angle is invariant under gauge transformations. For time-dependent classical systems a distinction between the Hamiltonian and the energy must also be made.[10] The Hamiltonian describes the time development of the system through Hamilton's equations. The energy is that quantity whose time rate of change gives the power transferred between the system and its environment. The Hamiltonian is not invariant under gauge transformations, whereas the energy is gauge invariant. The Hannay angle is generalized in a way similar to the way in which the Berry phase[8,11] is generalized to make it invariant under unitary transformations. In the adiabatic case, the total angle is the sum of the dynamical angle and the Hannay angle.

Condensed Matter Theories, Vol. 8, Edited by
L. Blum and F.B. Malik, Plenum Press, New York, 1993

The general formalism is applied to the time-dependent generalized harmonic oscillator.[2,3] The "generalized" harmonic oscillator is one that has a cross term between the coordinate and the canonical momentum, in addition to the usual quadratic terms. Each term in the Hamiltonian is multiplied by a time-dependent coefficient. The equation of motion for the generalized harmonic oscillator is obtained, and it describes a harmonic oscillator with a time-dependent mass and spring "constant." The energy of the generalized harmonic oscillator is taken to be the sum of the kinetic energy and a potential energy which has the same spring constant as in the equation of motion. The time derivative of this energy is the power transferred between the particle and its environment. When the generalized Hannay angle is calculated, the result is zero. This result is consistent with the result for the generalized Berry phase.[8]

In Sec. 2 some background of the Hannay angle is given to show that, as usually formulated, it is gauge dependent. In Sec. 3 it is generalized to be invariant under gauge transformations. The new formalism is applied in Sec. 4 to the generalized harmonic oscillator. Finally, the conclusion is given in Sec. 5.

2. BACKGROUND

In order to show that the usual formulation of action-angle variables[12] in classical mechanics requires some modification, we shall consider a time-independent system with one degree of freedom.

2.1. Action-angle variables

If the Hamiltonian is $H = H(q,p)$, then Hamilton's equations are

$$\dot{q} = \partial H/\partial p , \quad \dot{p} = -\partial H/\partial q , \qquad (2.1)$$

where p is the canonical momentum conjugate to the generalized coordinate q. We shall now make a canonical transformation from the old canonical variables (q,p) to the new canonical variables $(\bar{q},\bar{p}) = (\theta,J)$, where θ is the angle variable and J is the action. The generating function is $F_2(q,\bar{p}) = S(q,J)$, so that

$$p = \partial S/\partial q , \qquad (2.2)$$

$$\theta = \partial S/\partial J , \qquad (2.3)$$

and the new Hamiltonian $\bar{H} = H$ since $\partial S/\partial t = 0$. The generating function $S(q,J)$ can be obtained by integrating Eq. (2.3), which gives

$$S(q,J) = \int_0^q p(q,J)dq , \qquad (2.4)$$

where we choose $S(0,J) = 0$ for convenience. In order to evaluate the integral it is necessary to have the canonical momentum p as a function of q and J. This function is obtained by setting

$$H(q,p) = \omega J , \qquad (2.5)$$

where ω is the angular frequency, and solving for $p = p(q,J)$.

From Eq. (2.3) it can be seen that for $\theta = 2\pi$ we have

$$2\pi = \partial (\oint pdq)/\partial J \qquad (2.6)$$

where the integral in Eq. (2.6) is the integral in Eq. (2.4) performed over one complete cycle. Equation (2.6) implies that

$$\oint pdq = 2\pi J , \qquad (2.7)$$

which is often taken as the definition of J. The angular frequency can then be obtained from Eq. (2.5) as $\omega = H/J$.

Hamilton's equations for the new canonical variables (θ,J) are

$$\dot{\theta} = \partial H/\partial J , \quad \dot{J} = -\partial H/\partial \theta , \qquad (2.8)$$

since $\partial S/\partial t = 0$. From Eq. (2.5) we have

$$\dot{\theta} = \omega, \quad \dot{J} = 0 , \tag{2.9}$$

as Hamilton's equations. The action J is an invariant, and $\theta = \omega t + \alpha$, where α is a constant. The solution to the original problem can be obtained by solving Eq. (2.3) for $q = q(\theta,J)$.

2.2. Gauge transformation

A gauge transformation in classical mechanics is a special type of canonical transformation, which leaves the coordinate unchanged, but shifts the canonical momentum by a function of the coordinate and time.[13] A gauge transformation can be made on the original canonical variables (q,p) to a new set (q',p') by using a generating function of the second type

$$F_2(q,p',t) = qp' - \Lambda(q,t) , \tag{2.10}$$

where $\Lambda(q,t)$ is an arbitrary function called a gauge function. The old momentum p is

$$p = \partial F_2/\partial q = p' - \partial \Lambda/\partial q , \tag{2.11}$$

so that the new canonical momentum p' is shifted from p by a function of coordinate and time $\partial \Lambda(q,t)/\partial q$. The new coordinate is

$$q' = \partial F_2/\partial p' = q , \tag{2.12}$$

so it is unchanged under a gauge transformation. The new Hamiltonian is

$$H' = H - \partial \Lambda/\partial t , \tag{2.13}$$

which is shifted from the old Hamiltonian by $-\partial \Lambda(q,t)/\partial t$.

If we follow the procedure in Sec. 2.1 for the new Hamiltonian H', Eq. (2.5) would become

$$H' = \omega' J' , \tag{2.14}$$

where J' is the new action and ω' is the "new" angular frequency. From Eq. (2.7) the new action J' is

$$2\pi J' = \oint p'dq' = \oint pdq + \oint dq\, \partial \Lambda/\partial q = 2\pi J \tag{2.15}$$

from Eqs. (2.11) and (2.12). The integral of $\partial \Lambda/\partial q$ around a closed path is zero because the gauge function $\Lambda(q,t)$ must be single valued. Equation (2.15) shows that the action J' = J is unchanged under a gauge transformation. The angular frequency in Eq. (2.14) must also be unchanged under a gauge transformation $\omega' = \omega$, because a physical observable should not depend on the gauge. For a time-dependent gauge function in Eq. (2.10) the new Hamiltonian H' in Eq. (2.13) is different from the old Hamiltonian H. Therefore Eq. (2.14) contradicts Eq. (2.5). The contradiction arises because the Hamiltonian is gauge dependent, but the right-hand side of Eq. (2.5) is gauge invariant. In the next section this contradiction is resolved.

3. GAUGE-INVARIANT GENERALIZED HANNAY ANGLE

In this section we present the resolution of the contradiction given in Sec. 2 for the conventional treatment of action-angle variables. A time-dependent system with one degree of freedom is considered.

3.1. Action-angle variables

The Hamiltonian for a time-dependent system with one degree of freedom is $H = H(q,p,t)$, where p is the canonical momentum conjugate to the generalized coordinate q and t is time. Hamilton's equations are

$$\dot{q} = \partial H/\partial p \tag{3.1}$$

and

$$\dot{p} = -\partial H/\partial q \ , \tag{3.2}$$

which describe the time development of the system.

A canonical transformation can be made from (q,p) to a new set of canonical variables $(\bar{q},\bar{p}) = (\theta,J)$, the angle θ and action J variables. The generating function $F_2(q,\bar{p},t) = S(q,J,t)$ which gives the canonical transformation is of the second type, so that it satisfies

$$p = \partial S/\partial q \ , \tag{3.3}$$

$$\theta = \partial S/\partial J \ , \tag{3.4}$$

and the new Hamiltonian is

$$H' = H + \partial S/\partial t \ . \tag{3.5}$$

In order to obtain the generating function it is necessary to integrate Eq. (3.3),

$$S(q,J,t) = \int_0^q pdq + S(0,J,t) \ . \tag{3.6}$$

The generating function at $q = 0$ can often be chosen zero. To evaluate Eq. (3.6) it is necessary to obtain the canonical momentum $p = p(q,J,t)$.

3.2. *Energy*

The Hamiltonian in a time-dependent problem is not necessarily the energy. The energy $E = E(q,\dot{q},t)$ is a function such that its total time-derivative is the power P,

$$dE/dt = P \ . \tag{3.7}$$

The energy is taken to be the sum of the kinetic energy $\tfrac{1}{2}m\dot{q}^2$ plus (conservative) potential energy V,

$$E = \tfrac{1}{2}m\dot{q}^2 + V(q) \ . \tag{3.8}$$

The total time derivative of the energy is

$$dE/dt = \tfrac{1}{2}\dot{m}\dot{q}^2 + (\partial V/\partial t)_q + m\ddot{q}\dot{q} + (\partial V/\partial q)\dot{q} \ . \tag{3.9}$$

The equation of motion for the particle is

$$d(m\dot{q})/dt = F_{nc} - \partial V/\partial q \ , \tag{3.10}$$

where F_{nc} is the nonconservative force (if present) and m is the mass (possibly time-dependent). If Eq. (3.10) is used in Eq. (3.9), it becomes

$$dE/dt = \tfrac{1}{2}\dot{m}\dot{q}^2 + (\partial V/\partial t)_q + F_{nc}\dot{q} + (-\dot{m}\dot{q})\dot{q} = P \ . \tag{3.11}$$

where P is the total power. Each of the four terms on the right-hand side of Eq. (3.11) can be interpreted as a contribution to the power. The first term is the power due to the time rate of change of the mass at constant velocity. The second term is the power due to the time rate of change of parameters in the (quasi) conservative potential energy. The third term is the power due to the nonconservative forces. The fourth term is the power due to the fictitious force $-\dot{m}\dot{q}$ from the left-hand side of Eq. (3.10) due to the time rate of change of the mass.

We shall define the action variable J to be related to the energy E by

$$E = \omega J \ , \tag{3.12}$$

where ω is the angular frequency of the system. The energy $E = E(q,\dot{q},t)$ depends on the velocity $\dot{q}$, but by Hamilton's equation in Eq. (3.1) the velocity $\dot{q} = \dot{q}(q,p,t)$. Therefore the energy can be written as a function of canonical momentum $E = E(q,p,t)$. When the energy expressed in terms of p is used in Eq. (3.12), the equation can be solved for $p = p(q,J,t)$, which is needed in Eq. (3.6) to obtain the generating function $S(q,J,t)$.

We shall assume that for many problems the relationship between the Hamiltonian H and the energy E (expressed in terms of q, p and t) is

$$H = E + \Phi(q,t) \; , \tag{3.13}$$

where $\Phi(q,t)$ is a "scalar potential" which should not be confused with the (conservative) potential energy V.

3.3. Hamilton's equations

Because the transformation to the angle and action variables is canonical, Hamilton's equations in Eqs. (3.1) and (3.2) for the new variables are

$$\dot{\theta} = \partial H'/\partial J \tag{3.14}$$

and

$$\dot{J} = -\partial H'/\partial \theta \; . \tag{3.15}$$

The new Hamiltonian H' is given in Eq. (3.5). The relationship between the energy and the Hamiltonian is given in Eq. (3.13), and the relationship between the energy and the action is given in Eq. (3.12). When these equations are used in Eqs. (3.14) and (3.15), Hamilton's equations become

$$\dot{\theta} = \omega + \partial(\Phi + \partial S/\partial t)/\partial J \tag{3.16}$$

and

$$\dot{J} = -\partial(\Phi + \partial S/\partial t)/\partial \theta \; , \tag{3.17}$$

since $\partial E/\partial J = \omega$ and $\partial E/\partial \theta = 0$. Equations (3.16) and (3.17) can be shown to be invariant under gauge transformations, since $\Phi + \partial S/\partial t$ is a gauge invariant quantity.

3.4. Hannay angle

The time derivative of the Hannay angle $\Delta\theta_{\parallel}$ is defined as

$$d\Delta\theta_{\parallel}/dt = <\partial(\Phi + \partial S/\partial t)/\partial J> \; , \tag{3.18}$$

where the angular average is defined as

$$< \cdots > = (2\pi)^{-1} \int_0^{2\pi} d\theta \; \cdots \; .$$

Equation (3.16) can be written as

$$\dot{\theta} = \omega + d\Delta\theta_{\parallel}/dt$$
$$+ \{\partial(\Phi + \partial S/\partial t)/\partial J - d\Delta\theta_{\parallel}/dt\} \tag{3.19}$$

and Eq. (3.17) can be written as

$$\dot{J} = - <\partial(\Phi + \partial S/\partial t)/\partial \theta>$$
$$+ \{-\partial(\Phi + \partial S/\partial t)/\partial \theta + <\partial(\Phi + \partial S/\partial t)/\partial \theta>\} \; . \tag{3.20}$$

3.5. Adiabatic approximation

In the adiabatic approximation the terms in the braces in Eqs. (3.19) and (3.20) may be neglected. Equation (3.19) may then be integrated to give

$$\theta(t) = \int_0^t \omega dt' + \Delta\theta_{\parallel}(t) \; , \tag{3.21}$$

where the first term is the dynamical angle and the second term is the Hannay angle.

When the term in the braces in Eq. (3.20) is neglected, $\dot{J} = 0$ since

$$<\partial(\Phi + \partial S/\partial t)/\partial\theta> = 0 \ . \tag{3.22}$$

Therefore J is an adiabatic invariant.

A detailed calculation shows that Eqs. (3.16), (3.17), and (3.18) are all manifestly gauge invariant.

4. TIME-DEPENDENT GENERALIZED HARMONIC OSCILLATOR

The theory developed in Sec. 3 is applied here to a time-dependent generalized harmonic oscillator.

4.1. *Hamiltonian and equation of motion*

The classical Hamiltonian for a generalized harmonic oscillator is

$$H = \tfrac{1}{2}\{a(t)p^2 + 2b(t)pq + c(t)q^2\} \ , \tag{4.1}$$

where p is the canonical momentum conjugate to the generalized coordinate q and a, b, and c are arbitrary time-dependent parameters. Hamilton's equations with Eq. (4.1) give

$$\dot{q} = \partial H/\partial p = ap + bq \tag{4.2}$$

and

$$\dot{p} = -\partial H/\partial q = -bp - cq \ . \tag{4.3}$$

If Eq. (4.2) is solved for the canonical momentum p in terms of the velocity $\dot{q}$ and substituted into Eq. (4.3), we obtain the equation of motion

$$d(m\dot{q})/dt = -kq \ , \tag{4.4}$$

where the time-dependent mass $m = a^{-1}$ and the spring "constant" is

$$k = c - mb^2 - d(mb)/dt \sim 0$$

Equation (4.4) describes a harmonic oscillator with a time-dependent mass and spring constant, with a time-dependent angular frequency

$$\omega(t) = [k(t)/m(t)]^{1/2} \ . \tag{4.6}$$

4.2. *Energy*

The energy E is defined so that its time rate of change is the power transferred between the particle and its environment. We take the energy E to be the sum of the kinetic energy and the potential energy for a spring with spring "constant" k,

$$E = \tfrac{1}{2}m\dot{q}^2 + \tfrac{1}{2}kq^2 \ . \tag{4.7}$$

The time rate of change of this energy is a special case of Eq. (3.11),

$$dE/dt = \tfrac{1}{2}\dot{m}\dot{q}^2 + \tfrac{1}{2}\dot{k}q^2 + (-\dot{m}\dot{q})\dot{q} = P \ , \tag{4.8}$$

where P is the power. The second term on the right-hand side of Eq. (4.8) is the power required to change the spring "constant" at constant displacement. There is no true nonconservative force, and only the fictitious force $-\dot{m}\dot{q}$ contributes to the power in Eq. (4.8).

If Eq. (4.2) for the velocity is substituted into Eq. (4.7) for the energy, we obtain

$$E = (2m)^{-1}(p + mbq)^2 + \tfrac{1}{2}kq^2 \ , \tag{4.9}$$

676

which gives the energy as a function of the canonical momentum p, generalized coordinate q, and time t.

4.3. Generating function

If Eq. (4.9) is substituted into Eq. (3.12) and solved for p, the result is

$$p = \rho^{-1}[2J - (q/\rho)^2]^{1/2} - mbq , \qquad (4.10)$$

where $\rho = (m\omega)^{-1/2}$ is time dependent. Equation (4.10) can be substituted into Eq. (3.6) for the generating function to obtain

$$S(q,J,t) = -mbq^2/2 + (q/2\rho)[2J - (q/\rho)^2]^{1/2}$$
$$+ J \sin^{-1}[q/\rho(2J)^{1/2}] . \qquad (4.11)$$

The angle variable θ in Eq. (3.4) can be obtained by differentiating Eq. (4.11) with respect to the action J, which gives

$$\theta = \sin^{-1}[q/\rho(2J)^{1/2}] . \qquad (4.12)$$

The generalized coordinate q is therefore

$$q = \rho(2J)^{1/2} \sin \theta \qquad (4.13)$$

in terms of θ and J.

4.4. Hamilton's equations for angle and action variables

Hamilton's equations for the angle and action variables are given in Eqs. (3.16) and (3.17). The "scalar potential" $\Phi(q,t)$ which occurs in the equations is defined in Eq. (3.13). When Eqs. (4.1), (4.2), and (4.7) are used, the scalar potential is

$$\Phi = H - E = \tfrac{1}{2}[d(mb)/dt] \, q^2 . \qquad (4.14)$$

When the generating function S in Eq. (4.11) is differentiated partially with respect to time, keeping q and J fixed and added to Φ, we obtain

$$\Phi + \partial S/\partial t = -(q/\rho)(\dot{\rho}/\rho)[2J - (q/\rho)^2]^{1/2} . \qquad (4.15)$$

Before we differentiate Eq. (4.15) with respect to J keeping θ fixed, it is necessary to substitute Eq. (4.13) for q into Eq. (4.15) in order to obtain a function of θ, J and t. When this is done, Eq. (4.15) becomes

$$\Phi + \partial S/\partial t = -J(\dot{\rho}/\rho) \sin 2\theta . \qquad (4.16)$$

Differentiating Eq. (4.16) with respect to J and substituting it into Eq. (3.16), we obtain the equation of motion

$$\dot{\theta} = \omega - (\dot{\rho}/\rho) \sin 2\theta , \qquad (4.17)$$

which is a nonlinear equation for θ. When Eq. (4.16) is differentiated with respect to θ, and substituted into eq. (3.17), we obtain

$$\dot{J} = 2J(\dot{\rho}/\rho) \cos 2\theta . \qquad (4.18)$$

Equations (4.17) and (4.18) are Hamilton's equations for this model and are exact.

4.5. Hannay angle

The Hannay angle $\Delta\theta_{11}(t)$ is obtained by integrating Eq. (3.18), which gives

$$\Delta\theta_{11}(t) = \int_0^t dt' \, \langle \partial(\Phi + \partial S/\partial t')/\partial J \rangle . \qquad (4.19)$$

Differentiating Eq. (4.16) partially with respect to J at constant θ and t, we obtain the Hannay angle

$$\Delta\theta_H(t) = -\int_0^t dt'(2\pi)^{-1} \int_0^{2\pi} d\theta(\dot{p}/p) \sin 2\theta = 0 \ . \tag{4.20}$$

This result of zero for the Hannay angle is in contrast to the value obtained by Hannay[2] and Berry[3] for the same problem. They obtain a nonzero Hannay angle and, in the quantum mechanical case, Berry phase because they assume that the Hamiltonian in Eq. (4.1) is the energy of the system. As we have seen in Sec. 4.2 the energy is not the Hamiltonian. The Hamiltonian depends on the gauge, whereas the energy is gauge invariant. The Hamiltonian for the generalized harmonic oscillator in Eq. (4.1) can be gauge transformed to the Hamiltonian for a *simple* harmonic oscillator with an angular frequency ω given by Eq. (4.6). In this case the Hamiltonian is equal to the energy. The Hannay angle for a simple harmonic oscillator is zero. Therefore, by gauge invariance, the generalized Hannay angle for the Hamiltonian in Eq. (4.1) is also zero.

5. CONCLUSION

The Hannay angle[2] in classical mechanics is not invariant under gauge transformations. It is generalized here so that it is gauge invariant. A distinction must be made between the Hamiltonian and the energy in time-dependent problems. The Hamiltonian describes the time-development of the system through Hamilton's equations. On the other hand, the energy is a function whose time derivative is equal to the power transferred between the particle and its environment. The Hamiltonian depends on the gauge, whereas the energy is gauge invariant. It is possible to find a gauge in which the Hamiltonian is equal to the energy, when the canonical momentum is expressed in terms of the velocity.

As an example of the method, the time-dependent generalized harmonic oscillator is considered. When the energy is chosen to be the sum of the kinetic energy and the potential energy with the same spring "constant" as in the equation of motion, the generalized Hannay angle is zero. On the other hand, if the original Hamiltonian in Eq. (4.1) is chosen to be the energy a nonzero generalized Hannay angle is obtained. This choice of the energy does not result in a proper expression for the power transferred between the particle and its environment.

ACKNOWLEDGMENTS

I Would like to express my glatitude to the U.S. Army Researcn Office for partial support of my travel.

REFERENCES

1.   M. V. Berry, *Proc. R. Soc. London, Ser.* A **392**, 45 (1984).
2.   J. H. Hannay, *J. Phys. A: Math. Gen.* **18**, 221 (1985).
3.   M. V. Berry, *J. Phys. A: Math. Gen.* **18**, 15 (1985).
4.   M. V. Berry, *Proc. R. Soc. London, Ser.* A **414**, 31 (1987).
5.   R. Jackiw, *Int. J. Mod. Phys.* A **3**, 285 (1988).
6.   P. de S. Gerbert, *Ann. Phys. (N.Y.)* **189** 155 (1989).
7.   G. Giavarini, E. Gozzi, D. Rohrlich, and W. D. Thacker, *Phys. Lett.* A **137**, 235 (1989); *J. Phys. A: Math. Gen.* **22**, 3513 (1989).
8.   D. H. Kobe, *J. Phys. A: Math. Gen.* **23**, 4249 (1990).
9.   A. Bhattacharjee and T. Sen, *Phys. Rev.* A **38**, 4389 (1988).
10.   D. H. Kobe and K.-H. Yang, *Eur. J. Phys.* **8**, 236 (1988).
11.   D. H. Kobe, *J. Phys. A: Math. Gen.* **24**, 2763 (1991).
12.   H. Goldstein, "Classical Mechanics," Addison-Wesley, Reading, MA (1980), 2nd ed., pp. 457-462.
13.   D. H. Kobe, *Am. J. Phys.* **56**, 252 (1988).

INDEX